Zoological Monographs

Volume 7

The book series *Zoological Monographs* is devoted to the study of the animal kingdom and covers all aspects of animals, both living and extinct, including taxonomy, systematics and evolution, biodiversity and biogeography, ecology, morphology and anatomy, physiology and behavior. The volumes in the series either discuss the biology of a single animal group, approach a topic from an interspecies level, or present recent methodologies. The series is intended for graduate students, university teachers, scientists and professionals.

Maxim V. Vinarski • Antonio A. Vázquez
Editors

The Lymnaeidae

A Handbook on Their Natural History and Parasitological Significance

Springer

Editors
Maxim V. Vinarski
Saint-Petersburg State University and
S.I. Vavilov Institute for the History
of Science and Technology
Saint-Petersburg, Russia

Antonio A. Vázquez
Institute of Tropical Medicine "Pedro Kourí"
La Habana, Cuba

ISSN 2523-3904 ISSN 2523-3912 (electronic)
Zoological Monographs
ISBN 978-3-031-30291-6 ISBN 978-3-031-30292-3 (eBook)
https://doi.org/10.1007/978-3-031-30292-3

This Springer imprint is published by the registered company Springer Nature Switzerland AG
The registered company address is: Gewerbestrasse 11, 6330 Cham, Switzerland

Preface

Seventy years ago, the prominent Swedish malacologist Bengt Hubendick (1916–2012) published a monograph titled "Recent Lymnaeidae: their variation, morphology, taxonomy, nomenclature, and distribution" (Fig. 1), which would quickly become a classic of systematic malacology. This seminal book was the first attempt to synthesize all available knowledge on the taxonomy, phylogeny, variability, and biogeography of the living pond snails, the family Lymnaeidae Rafinesque, 1815. This group of freshwater Mollusca is characterized by almost cosmopolitan distribution, relatively high species richness, and profound economic and medical importance. Hubendick's *chef-d'oeuvre* became the milestone in the lymnaeid studies setting a high standard of systematic publications in this field. Two subsequent taxonomic monographs on the subject, those of Maria Jackiewicz in Poland (published in 1998) and Nikolay D. Kruglov in Russia (published in 2005), are also very detailed and helpful, but their geographic scope is much narrower. Jackiewicz dealt with the Lymnaeidae of Europe only, whereas Kruglov's monograph covered the lymnaeid fauna of Europe and Northern and Central Asia. Thus, Hubendick's treatise remains the only comprehensive sourcebook on the extant Lymnaeidae available today. Seventy years is a substantial period. More than a century ago, a prominent plant taxonomist (Lyberty Hyde Bailey) stated that "every large or variable group needs to be reworked at least every twenty-five years... Frequent review and sifting of evidence are as necessary in systematology and taxonomy as in morphology".[1]

The classical works of the leading lymnaeid taxonomists of the last century were based almost exclusively on the use of morphological (sensu lato) characters, with a strong emphasis on the reproductive anatomy of the pond snails which, as it seemed in these times, provides a firm basis for their objective and unequivocal classification. Such prominent workers as Frank C. Baker in the USA, Hubendick, and Jackiewicz adhered to this approach to lymnaeid classification.

[1] See *Science (New Series)*, 46(1200): 623–629. The quotation is on page 626.

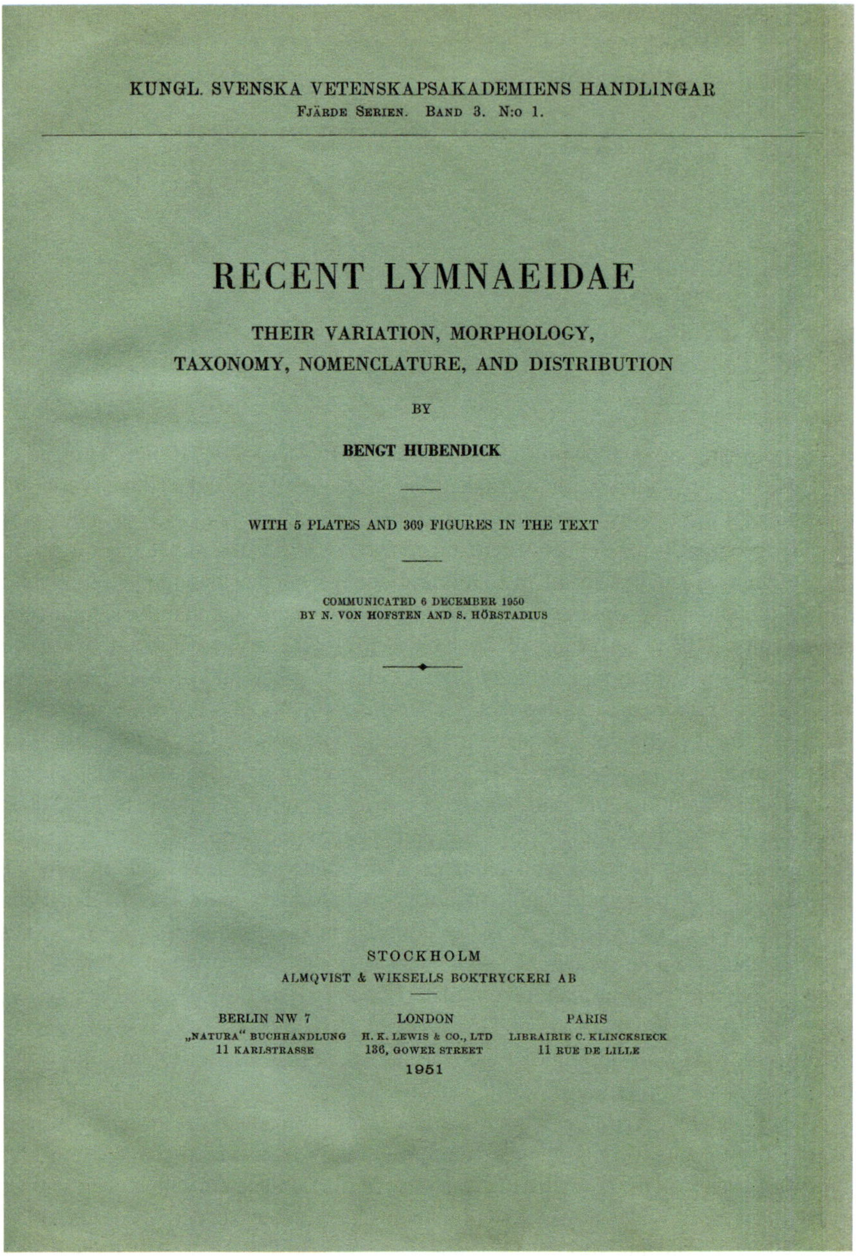

KUNGL. SVENSKA VETENSKAPSAKADEMIENS HANDLINGAR
FJÄRDE SERIEN. BAND 3. N:O 1.

RECENT LYMNAEIDAE

THEIR VARIATION, MORPHOLOGY, TAXONOMY, NOMENCLATURE, AND DISTRIBUTION

BY

BENGT HUBENDICK

WITH 5 PLATES AND 369 FIGURES IN THE TEXT

COMMUNICATED 6 DECEMBER 1950
BY N. VON HOFSTEN AND S. HÖRSTADIUS

STOCKHOLM
ALMQVIST & WIKSELLS BOKTRYCKERI AB

BERLIN NW 7	LONDON	PARIS
„NATURA" BUCHHANDLUNG	H. K. LEWIS & CO., LTD	LIBRAIRIE C. KLINCKSIECK
11 KARLSTRASSE	136, GOWER STREET	11 RUE DE LILLE

1951

Fig. 1 The title page of Hubendick's 1951 monograph

The last seventy years have witnessed great advances in lymnaeid studies that owed chiefly to the invention of new technologies. The taxonomist' toolkit has drastically been expanded by adding different advanced methods of molecular and statistical analysis to it. The face of animal systematics—arguably the oldest biological discipline—is rapidly changing: recent years have witnessed the appearance of the so-called turbo-taxonomy and cyber-taxonomy, and artificial intelligence has entered the field of animal systematics, including systematic malacology. These revolutionary shifts in methodology have revolutionized our knowledge of the diversity, classification, and phylogeny of the family Lymnaeidae. The significance of conchological and anatomical characters as sources of the phylogenetic and taxonomic signal has been reassessed. Notwithstanding that many important questions remain open, we believe the time is ripe to undertake a new synthesis of the available knowledge.

The original intention of this edition has been rather ambitious. The editors have aimed to produce a "new generation" handbook of the lymnaeid family which would replace Hubendick's book as the general source of information on the pond snails. We tried to collect and synthesize all advancements in the field achieved since 1951 and to present the twenty-first century look on taxonomy, phylogeny, biogeography, and ecology of the family. The term "ecology" covers, naturally, many aspects of the relationships between the pond snails and the trematode larvae which use the snails as their primary (or, sometimes, secondary) hosts. The latter circumstance determines the high interest in this family from practitioners working in the fields of parasitology, veterinary, and public health. However, the content of this book does not cover some other topics, including the biochemistry, embryology, immunology, and physiology of the lymnaeids. These experimental fields are developing increasingly, and the reviews on the state of the art in these topics must be found elsewhere.

In other words, the primary focus of this edition is placed on the **diversity** of the Lymnaeidae, which is discussed below from different points of view—taxonomical, morphological, paleontological, geographic, phylogenetic, and ecological. Though the reader will find some articles devoted to the lymnaeid fossil record (Neubauer) and evolution (Albrecht and Clewing), the general scope of our handbook is restricted to the living representatives of the family. The Lymnaeidae proved to constitute an excellent model taxon for current studies in evolutionary biology, speciation, adaptive radiations, and historical biogeography. One of the contributions to this volume (Aksenova et al.) aims to present an updated version of the system of the living Lymnaeidae, with a new taxonomic structure of this family. As any particular system, that of Aksenova et al. is based on the currently available information, and the future will inevitably bring some changes to it, more or less substantial.

Lymnaeidae do not constitute a very species-rich family; probably, there are around 175 lymnaeid species in the recent malacofauna. Nonetheless, the practical significance of the Lymnaeidae snails is enormous and multi-faceted. These mollusks are notoriously known as vectors of different trematode species causing health problems in humans, wild and domestic animals. Fossil shells of the lymnaeids are useful in biostratigraphic research, whereas the recent representatives of this family can be employed as biological indicators of the environmental conditions. Some species of the family, especially the great pond snail, *Lymnaea stagnalis* (Linnaeus, 1758), have become important model objects for experimental studies in many fields of biology, from molecular genetics and gerontology to neuroscience and behavioral research. The ecosystem role played by the lymnaeids is determined by their use as food items by a variety of aquatic and amphibious organisms—freshwater and ubiquitous (e.g., synanthropic rats) mammals, waterfowl, fish, large carnivore insects. Thus, the pond snails are actively engaged in the food chains and energy flows in the natural ecosystems. Some lymnaeid species are known as effective and successful invaders, becoming alien species in different parts of the world.

All these facts explain the ever-increasing interest in the lymnaeid studies observed worldwide. A special emphasis in this book has been put on the issue of lymnaeid identification, based on either morphological or molecular methods. Surely, these topics will be of great interest to practitioners. The second part of this book (chapters 10 to 16) is essentially devoted to various applied aspects of lymnaeid research, chiefly to the studies of their parasitological significance.

We are proud to present to the readership this book containing articles on various aspects of the Lymnaeidae studies authored by the most experienced and actively working students of this group. The international authors' team includes experts from Russia, Cuba, France, Argentina, the Netherlands, and Germany, who have long been engaged in studies of taxonomy, phylogeny, evolution, ecology, and parasitological significance of the family. We hope that this edition will be both interesting and practically useful for the international audience of scientists working in the fields of malacology, aquatic biodiversity and ecology, limnology, parasitology, and other disciplines, both basic and applied.

Saint-Petersburg, Russia — Maxim V. Vinarski
Havana, Cuba — Antonio A. Vázquez
Montpellier, France

Contents

Editors and Contributors

Editors

Antonio A. Vázquez Laboratory of Malacology, Institute of Tropical Medicine "Pedro Kourí", Havana, Cuba
MIVEGEC, University of Montpellier, CNRS, IRD, Montpellier, France

Maxim V. Vinarski Laboratory of Macroecology & Biogeography of Invertebrates, Saint-Petersburg State University, Saint-Petersburg, Russia
S.I. Vavilov Institute for the History of Science and Technology, St. Petersburg Branch, Russian Academy of Sciences, Saint-Petersburg, Russia

Contributors

Olga V. Aksenova N. Laverov Federal Center for Integrated Arctic Research, The Ural Branch of the Russian Academy of Sciences, Arkhangelsk, Russia

Annia Alba Laboratory of Malacology, Institute of Tropical Medicine "Pedro Kourí", Havana, Cuba

Christian Albrecht Department of Animal Ecology and Systematics, Systematics and Biodiversity Group, Justus Liebig University Giessen, Giessen, Germany

Pilar Alda Genética y Ecología Evolutiva (CERZOS–CCT–CONICET Bahía Blanca), Bahía Blanca, Argentina
MIVEGEC, University of Montpellier, CNRS, IRD, Montpellier, France

Ivan N. Bolotov N. Laverov Federal Center for Integrated Arctic Research, The Ural Branch of the Russian Academy of Sciences, Arkhangelsk, Russia

Nicolás Bonel Centro de Recursos Naturales Renovables de la Zona Semiárida (CERZOS, CCT, UNS-CONICET Bahía Blanca), Bahía Blanca, Argentina
Consejo Nacional de Investigaciones Científicas y Técnicas (CONICET), Buenos Aires, Argentina

Elodie Chapuis MIVEGEC, University of Montpellier, CNRS, IRD, Montpellier, France

Catarina Clewing Department of Animal Ecology and Systematics, Systematics and Biodiversity Group, Justus Liebig University Giessen, Giessen, Germany

Patrice David CEFE, University of Montpellier, CNRS, EPHE, IRD, Montpellier, France

Giles Dreyfuss Laboratory of Parasitology, University of Limoges, Limoges, France

Dominique Faugère MIVEGEC, University of Montpellier, CNRS, IRD, Montpellier, France

Benjamin Gourbal Interactions Hosts Pathogens Environments, University of Montpellier, CNRS, Ifremer, University of Perpignan Via Domitia, Perpignan, France

Sylvie Hurtrez-Boussès MIVEGEC, University of Montpellier, CNRS, IRD, Montpellier, France
Département de Biologie-Ècologie, Faculté des Sciences, Université de Montpellier, Montpellier, France

Philippe Jarne Centre d'Ecologie Fonctionnelle et Evolutive, CNRS, Université de Montpellier, Montpellier, France

Irina S. Khrebtova N. Laverov Federal Center for Integrated Arctic Research, The Ural Branch of the Russian Academy of Sciences, Arkhangelsk, Russia

Joris M. Koene A-LIFE, Faculty of Science, Vrije Universiteit, Amsterdam, The Netherlands

Aleksander V. Kondakov N. Laverov Federal Center for Integrated Arctic Research, The Ural Branch of the Russian Academy of Sciences, Arkhangelsk, Russia

Manon Lounnas Unité mixte de recherche Maladies Infectieuses et Vecteurs: Écologie, Génétique, Évolution et Contrôle (MIVEGEC), Université de Montpellier, Montpellier, France

Yumi Nakadera A-LIFE, Faculty of Science, Vrije Universiteit, Amsterdam, The Netherlands

Thomas A. Neubauer Department of Animal Ecology and Systematics, Systematics and Biodiversity Group, Justus Liebig University, Giessen, Germany
Naturalis Biodiversity Center, Leiden, The Netherlands
SNSB - Bavarian State Collection for Paleontology and Geology, Munich, Germany

Julia Pizá Centro de Recursos Naturales Renovables de la Zona Semiárida (CERZOS, CCT, UNS-CONICET Bahía Blanca), Bahía Blanca, Argentina
Consejo Nacional de Investigaciones Científicas y Técnicas (CONICET), Buenos Aires, Argentina

Jean-Pierre Pointier PSL Research University, USR 3278 CNRS–EPHE, CRIOBE Université de Perpignan, Perpignan, France

Daniel Rondelaud Laboratory of Parasitology, University of Limoges, Limoges, France

Jorge Sánchez Laboratory of Malacology, Institute of Tropical Medicine "Pedro Kourí", Havana, Cuba

Bjorn Stelbrink Department of Animal Ecology and Systematics, Systematics and Biodiversity Group, Justus Liebig University Giessen, Giessen, Germany
Museum für Naturkunde – Leibniz Institute for Evolution and Biodiversity Science, Berlin, Germany

Philippe Vignoles Laboratory of Parasitology, University of Limoges, Limoges, France

Marion Vittecoq Institut de Recherche de la Tour du Valat, Arles, France

Chapter 1
A Brief History of the Lymnaeid Research

Maxim V. Vinarski, Jean-Pierre Pointier, and Daniel Rondelaud

Abstract The history of the lymnaeid research, from the mid-sixteenth century to the present, is briefly outlined, with a special emphasis on the development of the systematics of this family and the studies of parasitological significance of the pond snails. An index to the lymnaeid literature published during the last 120 years, which includes references to the most important publications, is also provided.

It appears that the lymnaeid gastropods (the pond snails is the English vernacular name for these animals) have been totally overlooked by the ancient and medieval naturalists. Aristotle, "the father of zoology," did not mention them, as well as many of the natural historians who lived prior to the mid-sixteenth century (Vinarski 2015). Ulisse Aldrovandi (1522–1605), a scientist of the Late Renaissance Italy, was, most probably, the first naturalist to mention a lymnaeid snail in his book, and to illustrate its shell. Aldrovandi (1606) published an image of the shell of *Lymnaea stagnalis* (Linnaeus, 1758), the great pond snail, and named it "Turbo levis item in stagnis degens" (i.e., "Turbo with smooth shell, living in stagnant waters"). Though one cannot find a scientific description of this mollusc in Aldrovandi's book, but in that epoch the very Latin name (non-binomial, i.e., consisting of more than two words) served as a brief account of an animal and contained some information helping to identify it. It is not surprising that the shell in Aldrovandi's picture

M. V. Vinarski (✉)
Laboratory of Macroecology & Biogeography of Invertebrates, Saint-Petersburg State University, Saint-Petersburg, Russia

S.I. Vavilov Institute for the History of Science and Technology, St. Petersburg Branch, Russian Academy of Sciences, Saint-Petersburg, Russia

J.-P. Pointier
PSL Research University, USR 3278 CNRS–EPHE, CRIOBE Université de Perpignan, Perpignan, France
e-mail: pointier@univ-perp.fr

D. Rondelaud
Laboratory of Parasitology, University of Limoges, Limoges, France
e-mail: daniel.rondelaud@unilim.fr

M. V. Vinarski, A. A. Vázquez (eds.), *The Lymnaeidae*, Zoological Monographs 7,
https://doi.org/10.1007/978-3-031-30292-3_1

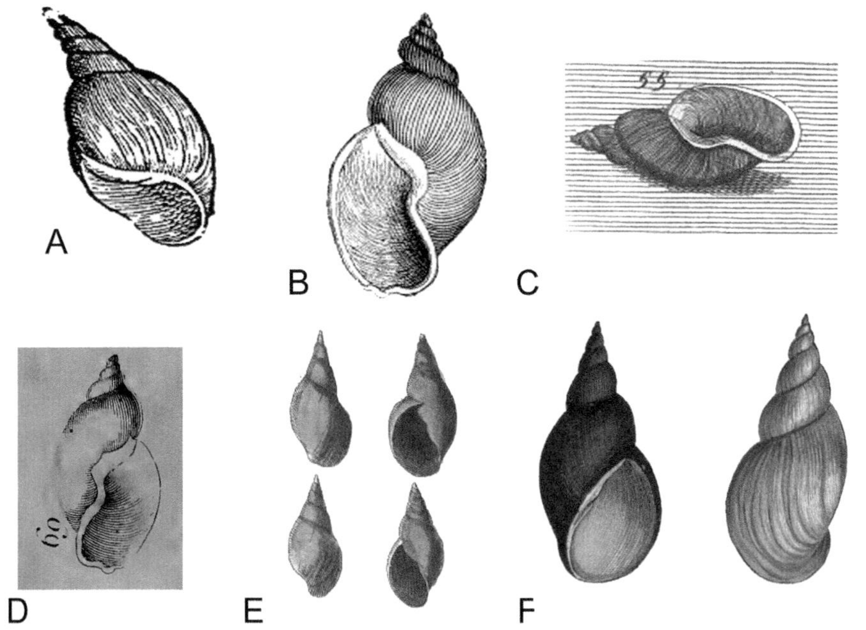

Fig. 1.1 Evolution of accuracy in illustrations of *Lymnaea stagnalis* shells made by naturalists of the seventeenth–eighteenth centuries. (**a**) Ulisse Aldrovandi (1606). (**b**) Martin Lister (1678). (**c**) Filippo Bonanni (1709). (**d**) Jacob Theodor Klein (1753). (**e**) Albert Seba (1758). (**f**) Johann Samuel Schröter (1779)

(Fig. 1.1) is sinistral (pond snails have normally dextral, or right-coiled, shells). The usual technique of engraving in the sixteenth and seventeenth centuries demanded that the plate must be a mirror image of the object to be illustrated. The printers usually were "not preparing a reversed engraving (on wood or copper), but carving the image [of a shell] as it appeared, which would produce a reversed image when printed" (Allmon 2007, p. 175).

Only 70 years later, a book authored by Martin Lister (1639–1712) was published in England, which contained much more detailed accounts on *L. stagnalis* and some other species of freshwater snails. Martin Lister was an English physician and naturalist, vice-president of the Royal Society from 1685 to 1686. Being a devoted conchologist, he wrote the first European treatises on molluscs (*Historiae Conchyliorum*, 1685; *Conchyliorum Bivalvium*, 1696). O.F. Müller (1774, p. xiii) called him "Conchyliologorum princeps» (head of conchologists), which is a clear analogy with Linnaeus' informal title "Princeps botanicorum."

In Lister's *Historiae animalium Angliae tres tractatus* (Lister 1678), one may find a detailed account of *L. stagnalis* that follows much higher standards of zoological descriptions compared with Aldrovandi's. This text contains not only the polynomial name (= short diagnosis) for this species, but also a relatively long two-page sketch of the great pond snail's bionomics. Lister provides a lengthy general description of its external morphology (including the pattern of mantle

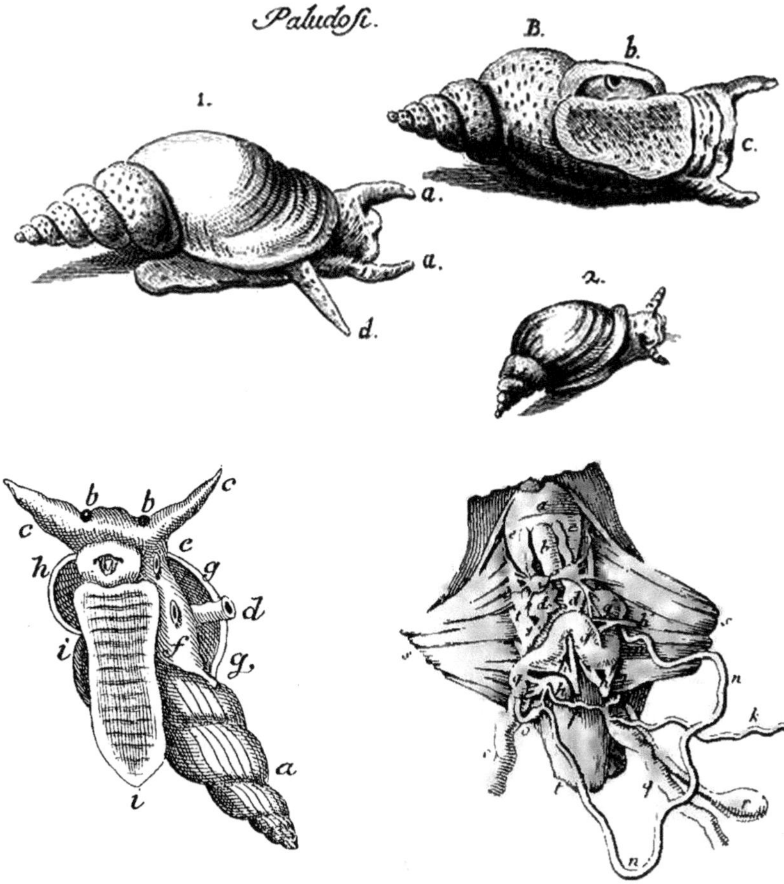

Fig. 1.2 Early illustrations demonstrating some details of the external and internal anatomy of *L. stagnalis*. Upper row – from Ginanni (1757), lower row left – after Swammerdam (1738), lower row right – after Cuvier (1817)

pigmentation), the shape of its excrements, the mode of copulation, the structure of egg-masses alongside a list of aquatic plants being its food. Some localities of *L. stagnalis* in England were also mentioned. Lister's species' account was almost 100 years ahead of its time. This high standard of publication of malacological data was not established until the end of the eighteenth/early nineteenth centuries, when comparable works of naturalists appeared (Müller 1774; Draparnaud 1805). Lister's account looks more detailed and inclusive than the account of the same species published a century later by Da Costa (1778)!

Later on, Lister (1695) published a very detailed report of the *L. stagnalis* internal structure, accompanied by engravings. Trained as a *medicus*, Lister was a brilliant anatomist aiming to dissect molluscs belonging to different taxa, both terrestrial and aquatic. Another perfect anatomist of the age, the Dutch Jan Swammerdam

(1637–1680), was also interested in freshwater molluscs (Fig. 1.2), and his study of the *L. stagnalis* anatomy was published posthumously in the author's prominent book *Bybel der natuure* (Swammerdam 1738). However, in both cases, the advance of anatomical research did not enhance the progress in taxonomy. Systematists of the seventeenth and eighteenth centuries typically did not use anatomical information in their works, and the classification of molluscs long remained purely conchological (Vinarski 2014).

In addition to *L. stagnalis*, Lister (1678) described two other lymnaeids, which may be tentatively identified with *Ampullaceana balthica* (Linnaeus, 1758) and *Radix auricularia* (Linnaeus, 1758). However, their descriptions are less detailed as compared with that of *L. stagnalis*.

The story of taxonomic investigation of the Lymnaeidae starts in 1753, when Jacob Theodor Klein separated lymnaeid snails into a taxon of their own—the genus *Auricula* with three species included (Klein 1753). Before Linnaeus' seminal work (Linnaeus 1758), Klein already explicitly used binomial nomenclature and introduced the first two-part name for the great pond snail—*Auricula stagnorum*. This name has a formal priority before the Linnaeus' *Helix stagnalis* but, being published before 1758, it is not available for taxonomic and nomenclatorial purposes.

Two other members of Klein's genus *Auricula* are more difficult to identify. His *Auricula pellucida*, is, perhaps, identical to *A. balthica*, whereas the third species, *Auricula exigua*, is of exotic origin ("Javana") and one is unable to identify it.

The basis for the subsequent taxonomic work on the family was laid within the 50 years after publication of the tenth edition of Linnaeus' *Systema Naturae*. During these years, several European naturalists and conchologists published descriptions of around 20 lymnaeid species, ten of which are accepted valid today (Box 1.1). It covered nearly one half of the overall diversity of this group in Western and Central Europe. The scope and quality of these descriptions ranged from short (1–2 phrases) and not illustrated diagnoses (Linnaeus 1758; Gmelin 1791) to more or less comprehensive accounts which included various data on species' morphology and ecology (Müller 1774; Draparnaud 1805). Toward the end of the eighteenth century, the minuteness of illustrations of the lymnaeid shells increased greatly and almost reached the modern standards (see Fig. 1.1).

All lymnaeid species described during the first 50 years of the post-Linnean taxonomy (i.e., since 1758) were placed within large and heterogeneous genera *Helix* and *Buccinum*, which included marine, terrestrial as well as true freshwater forms. Only in 1799 a new genus *Lymnaea* was erected by Lamarck, followed by other lymnaeid genera like *Galba* Schrank, 1803, *Radix* Montfort, 1810, and *Omphiscola* Rafinesque, 1819.

In the first half of the nineteenth century, the lymnaeid snails of the non-European (including exotic) countries were discovered. In North America, Thomas Say, "the father of American Conchology" (F.C. Baker 1911, p. 117), explored the Nearctic fauna of continental molluscs (Say 1817, 1821). He is the discoverer of many lymnaeid species accepted today as valid (*Bulimnea megasoma*, *Ladislavella catascopium*, *Pseudosuccinea columella*, and others). Travelers brought to Europe collections of freshwater molluscs of tropical countries. Lamarck (1822), Michelin

(1831), Krauss (1848), and some other workers contributed to the taxonomy of lymnaeids of India, East Asia, Africa. In Russia, Middendorff (1851) published a description of the pond snails of Siberia, a country whose invertebrate fauna was then almost unknown to zoologists.

This descriptive activity culminated in the mid-nineteenth century, when the first catalog of all known lymnaeid species appeared (Küster 1862). Küster's book was an annotated enumeration of almost 90 species from all continents, with descriptions and illustrations of their shells. This tradition was continued by illustrated catalogs authored by G.B. Sowerby II (1872) and Clessin (1882–1886). Westerlund (1885) compiled an annotated (but not illustrated) list of all species and varieties of lymnaeids occurring in the Palearctic region.

The volume of the family Lymnaeidae and its type genus, *Lymnaea*, had changed drastically during the last 100–150 years. In the nineteenth century, most workers treated Lymnaeidae as a family embracing almost all freshwater pulmonate snails, i.e. not only lymnaeids but also physids, planorbids, and some other groups (see, for example, Forbes and Hanley 1852–1853; Clessin 1882–1886; Locard 1893). Since the early twentieth century, the volume of the family became narrower. F.C. Baker (1911), Thiele (1931), Hubendick (1951) accepted it in an almost modern sense. Władysław Dybowski (1838–1910) who worked in Poland (then—a part of the Russian Empire) proposed the most radical system. This zoologist split the Lymnaeidae into three families of smaller volume: Limnaeidae, Lymnophysidae, and Amphipeplidae (W. Dybowski 1903). All his contemporaries, except Władysław's own brother Benedykt (B. Dybowski 1913), ignored these novelties.

The genus *Lymnaea* (alternatively spelled as *Limnaea*, *Limnaeus*, *Lymnaeus, Limnea*) had long served as a mega-genus to embrace almost all diversity of the pond snails, except of some conchologically peculiar forms like *Lanx* and *Myxas*. Typically, the genus *Lymnaea* s. lato had been divided into series of subdivisions (*Galba*, *Radix*, *Stagnicola*, and some others), which are ranked as separate genera in the present-day taxonomy. This "broad" concept of this genus dominated in the twentieth century (see review by Vinarski 2013), however most of the modern systematists arrange the family into numerous genera (Burch 1989; Ponder and Waterhouse 1997; Glöer 2002, 2019; Vinarski 2013; Aksenova et al. 2018; Vinarski et al. 2019, 2020).

Typically, the nineteenth century European malacologists accepted a relatively small number of lymnaeid species, albeit they used ad libitum the category of "varietas" (variety) in order to arrange the enormous intraspecific variation in lymnaeid shell size and proportions. The number of intraspecific entities delineated within some widespread and variable species might have been gargantuan; more than 80 "varieties" of *L. stagnalis* were proposed by taxonomists of the nineteenth and the first half of the twentieth century (Vinarski 2015). In most cases such varieties were based on relatively slight morphological deviations from the "type," i.e. in size, shell shape and proportions, shell surface coloration, etc. Not rarely, juvenile and aberrant individuals were used to establish a new "variety" or a new "morph." The extreme of such a splitting approach to taxonomy was observed in the nineteenth century France, where some malacologists created, on the basis of slight shell modifications, not "varieties" but "full" species. For example, Arnould Locard (1841–1904) delineated not less than 130 (sic!) species of Lymnaeidae in France alone (Locard 1893). One reason for it was that he considered the species category as arbitrary, non-objective. Locard wrote "L'espece malacologique est une notion purement arbitraire, indispensable aux naturalistes pour le besoin de la connaisance et de la classification des êtres" ("the malacological species is a purely arbitrary notion demanded by naturalists for the sake of knowledge and classification of [living] beings") (Locard 1893, p. 136).

It should be noted, however, that all abovementioned works were purely conchological and did not contain descriptions of anatomical characters. Until the early twentieth century, the shell characters remained virtually the only basis for genera and species delineation within the family.

At the same time, the anatomical studies on the Lymnaeidae, initiated in the late seventeenth century by Lister and Swammerdam, were continued by the great French zoologist and anatomist Georges Cuvier (1769–1832). In 1817, his "memoir" on morphology of *L. stagnalis* appeared (Cuvier 1817), in which the author described the external structure of the animal as well as the structure of its respiratory, digestive, nervous, and reproductive systems (see Fig. 1.2c). Some data on morphology of two other lymnaeid species were provided in the same paper. However, the anatomical studies had long remained detached from the taxonomic ones. Though the attempts to use the anatomical data for the purpose of lymnaeid classification were undertaken several times during the nineteenth century (van Beneden 1838; Troschel 1839; Klotz 1889), their impact on the practical taxonomy was almost negligible.

Let us, for example, consider a system of the Lymnaeidae proposed in 1905 by William H. Dall (1845–1927), a prominent American malacologist, in his monograph on Alaskan freshwater Mollusca (Dall 1905). This system was praised by F.C. Baker (1911, p. 118) as "the first attempt to place the classification of this group ... on a modern basis." A closer look at Dall's system reveals that the diagnoses of genera and subgenera published by this author include only conchological characters, and Dall apparently did not dissect studied snails.

The system created by Frank C. Baker (1867–1942) himself constituted an important step toward the use of both conchological and anatomical data as the basis for classification of the pond snails. It was built by combining the traits of shell, radula, and genital anatomy (F.C. Baker 1911, 1915). In Europe, malacologists also started to describe the anatomical differences between lymnaeid species, with the strongest emphasis on reproductive organs, considered a very valuable source of taxonomic signal (Roszkowski 1914a, 1914b, 1925, 1929; J. Wagner 1927; de Larambergue 1928). Roszkowski (1914b) was among the first authors to apply anatomical data to clarification of the true taxonomic identity of some peculiar varieties and local races of lymnaeids described on the basis of shell characters. The jaw and radula structure were decided to be of rather low taxonomic value due mainly to its great intraspecific variation (Roszkowski 1929).

A new standard in lymnaeid taxonomy was set in the works of the Swedish malacologist Bengt Hubendick (1916–2012), perhaps the most influential researcher of the Lymnaeidae of the last century. Contrary to F.C. Baker's (1911, 1915) opinion, he believed that conchological differences are of lesser importance for lymnaeid taxonomy, and only qualitative structural distinctions in the reproductive anatomy may constitute a good basis for classification. During his extensive anatomical work, Hubendick (1951) could recognize only two *Bauplans* of the reproductive system within Lymnaeidae, and therefore sorted all pond snails into two genera: *Lymnaea* Lamarck, 1799 and *Lanx* Clessin, 1882. The latter genus is of North American distribution and includes lymnaeids with limpet-like shells and very peculiar anatomy (H.B. Baker 1925). The internal structure of *Lanx* is so unusual as compared to the rest of Lymnaeidae that this genus has often been placed in a separate family Lancidae Hannibal, 1914 (H.B. Baker 1925; Taylor and Sohl 1962; Starobogatov 1967; but see Walter 1969). All other species of lymnaeids from all continents proved to be too uniform in their anatomical structure even to delimit subgenera within the genus *Lymnaea* (Hubendick 1951). Hubendick demonstrated this amazing uniformity of the internal structures of the pond snails does not correspond to morphological diversity of their shells. Hubendick believed that conchological differences among taxa should be neglected if these are not accompanied by qualitative and stable differences in anatomical structures. He adhered to a very broad species concept and, as a result, he reduced a huge amount of 1150 species and varieties of lymnaeids, proposed prior to 1951, to a totality of nearly 40 species. Some of these species were conchologically polymorphic and characterized by very wide ranges. The quantitative anatomical differences between forms were almost ignored by Hubendick (1951).

However, as soon as 8 years later, Maria Jackiewicz (1920–2018) in Poland has shown that there are really more "good" species of pond snails even in the European waterbodies than it was accepted by Hubendick (Jackiewicz 1959). She reported clear anatomical differences among Eastern European species of the (sub-)genus *Stagnicola* Leach in Jeffreys, 1830 and described a new species, *Galba occulta* Jackiewicz, 1959. Since 1959, a new stage of description of novel species of pond snails has started. The classical approach to classification of the pond snails, based primarily on study of their shell characters and internal morphology, was accepted in the former USSR by Yaroslav I. Starobogatov (1932–2004) and Nikolay D. Kruglov (1939–2010). Working together, these authors developed an original system of the family and described several tens of new species as well as new supraspecific taxa (summarized in Kruglov 2005). In total, Kruglov and Starobogatov (1993a, 1993b) distinguished more than 140 species of Lymnaeidae in the fauna of Europe and Northern Asia, which far exceeded the species richness estimates made by the Western European malacologists. Their approach, however, met strong criticism and no malacologist working beyond the former USSR agreed with the lymnaeid classification proposed by Kruglov and Starobogatov (1993a, 1993b), though usually no serious analysis of their arguments was provided. For instance, Jackiewicz (1998, p. 3) stated that "opinions of Russian malacologists on the lymnaeid taxonomy <. . .> raised great doubts and <. . .> have not been taken into consideration." The reasons of these "great doubts" were, however, not explained by Jackiewicz. It is worth to note, however, that some species and genera delineated by Kruglov and

Starobogatov have recently received a molecular support (Vinarski et al. 2011, 2012, 2016; Aksenova et al. 2018).

An alternative morphology-based version of the Lymnaeidae system was published in 1997 by Ponder and Waterhouse (1997). Essentially, it was the last example of a "traditional" classification, since from 1997 onwards the DNA taxonomy starts to gain a foothold in the field (Bargues and Mas-Coma 1997; Remigio and Blair 1997a, 1997b; Bargues et al. 2001, 2003; Remigio 2002).

The limits of the strictly morphological approach to lymnaeid classification were obvious to everyone who is familiar with the high degree of shell and anatomical variability of these snails, both at the intra- and interspecific level. In the second half of the twentieth century, there were numerous attempts to expand the set of taxonomic methods and to apply new sources of information to classification of the family. None else than Kruglov and Starobogatov (1985) tried to use the method of crossing experiment for delineation of lymnaeid species. In doing so, they followed the "biological species concept" that proposes to delimit species on the basis of their reproductive isolation, which may be established experimentally. The authors attempted to demonstrate that *Lymnaea stagnalis* in Europe is a complex of several species, at least two of which, *L. stagnalis* s. str., and *L. fragilis* (Linnaeus, 1758), are reproductively isolated (Kruglov and Starobogatov 1985). The authors, however, acknowledged that this method has serious limitations when applied to hermaphroditic animals like aquatic pulmonate snails. The reason is that "their capacity for self-fertilization leads to difficulty in deciding whether the progeny is the result of self-fertilization of cross-fertilization between different forms" (Kruglov and Starobogatov 1985, p. 22). It can explain why the method has never gained much popularity among students of the lymnaeid snails.

Another approach, advocated by Kruglov (1986, 2005), was grounded on the so-called parasitological criterion. It was assumed that different species of parasitic trematodes are species-specific in their host choice, and each species of pond snails has its own specific circle of trematode larvae. If a larva infests a "wrong" host, it dies. Thus, observed differences in infestation by trematodes may be viewed as indirect evidence for taxonomic distinctness of two or more forms of lymnaeids (Kruglov 1986). However, recent parasitological and molecular studies showed that lymnaeid species serving as intermediate hosts of *Fasciola hepatica* are widely distributed across their phylogeny and basically, all clades contain species that have proven to be naturally or experimentally infected with the parasite (Correa et al. 2010).

A quite another approach to species delimitation used in the twentieth century zoological systematics was the biochemical one (Throckmorton 1968). In the 1960s–1980s, three basic types of experimental biochemical taxonomy techniques had been applied to freshwater Mollusca: chromatography, electrophoresis, and immunology (serology). Davis (1978) and Meier-Brook (1993) published reviews of these techniques in application to aquatic gastropods, with many examples of their practical usage for recognizing species. All these methods were directed toward identification of genotypic characters, including amino acid analysis of proteins, allowing thereby to characterize populations, species, or higher taxa of molluscs, and to assess relationships among them (Davis 1978). Allozyme electrophoresis was the most popular technique. Davis (1994, p. 3) recommended it as "an ideal tool for population genetics as applied to delineating species." A good illustration of this recommendation may be found in a population genetic study that was performed on populations of *Galba* species from several Neotropical countries. This study using starch gel electrophoresis analyzed populations from Venezuela, Cuba, Guadeloupe, Dominican Republic, and Bolivia as well as several *G. truncatula* samples collected from France, Portugal, and Morocco for comparison (Jabbour-Zahab et al. 1997). Multilocus enzyme electrophoresis was determined for 282 snails at 18 loci. Two genotypic groups could be differentiated by their multilocus genotypes (i) a Western genotypic group associating samples from Venezuela, Guadeloupe, Cuba, and Dominican Republic (*G. cubensis*) and (ii) an Eastern genotypic group associating samples from France, Portugal, and Morocco (*G. truncatula*). Surprisingly, the Northern Bolivian Altiplano populations formerly identified as *G. viator* did not present any divergence with the Portuguese sample and belong entirely to the Eastern genetic group (*G. truncatula*). These results showed that the lymnaeids coming from the Northern Bolivian Altiplano undoubtedly belong to the *G. truncatula* species (Jabbour-Zahab et al. 1997). In some countries, allozyme electrophoresis was exploited in the lymnaeid studies until quite recently (Mezhzherin et al. 2008) but today it is completely replaced by more advanced methods based on DNA sequencing.

Cytotaxonomy also had attracted many practitioners in the lymnaeid systematics of the last century (Perrot and Perrot 1938; Inaba 1969; Patterson and Burch 1978; Meier-Brook 1993) and was still in some use at the dawn of the new millennium (Garbar and Korniushin 2002; Garbar et al. 2004). Inaba (1969) attempted to reconstruct the phylogeny of the family based on cytotaxonomic and biogeographic evidence.

As it was mentioned above, since the last decade of the twentieth century, the molecular methods started to be actively used in the reviewed field. The current taxonomic work on the Lymnaeidae is based either on a strictly molecular approach (Bargues et al. 2006; Pfenninger et al. 2006; Puslednik et al. 2009; Correa et al., 2010) or on the "integrated" taxonomic scrutiny combining analysis of genetic, morphological, and zoogeographical information (Bargues et al. 2011; Correa et al. 2011; Schniebs et al. 2011, 2013; Vinarski et al. 2016; Campbell et al. 2017; Aksenova et al. 2018). The molecular methods also dominate in the studies of speciation and biogeography of lymnaeid snails (e.g., Albrecht et al. 2008; Cordellier and Pfenninger 2009; von Oheimb et al. 2011; Aksenova et al. 2018; Mahulu et al. 2019). The most recent advances in the fields of systematics and phylogeny of the family will be reviewed in subsequent chapters. Box 1.2 summarizes the most important works on the Lymnaeidae published over the last 120 years.

Of course, the lymnaeid studies of the last 50–70 years were not focused solely on taxonomy and phylogeny. The great pond snail, *L. stagnalis*, appeared to be a laboratory animal very suitable as a model object for studies in genetics, physiology, embryology, and some other experimental branches of biology (Freeman and Lundelius 1982; Meshcheryakov 1990; Fodor et al. 2020). Though this snail never reached a popularity in the laboratories comparable to that of the fruit fly or *Caenorhabditis elegans*, some important discoveries resulted from experiments with the great pond snail. A classical example is Alfred Sturtevant's (1891–1970) work, in which the phenomenon of "maternal inheritance" was discovered (Sturtevant 1923). Experimental works on the inheritance of sinistrality in lymnaeid snails, performed in England by Arthur Edwin Boycott (1877–1938) and his co-workers (Boycott and Diver 1923, 1927; Boycott et al. 1930), became a further substantial contribution to understanding of this phenomenon (reviewed briefly in Gurdon 2005).

Around a century ago, some lymnaeid species, whose shells were easy to obtain in large quantities, constituted an important object for studies in variability of animal populations. These works were significant for the advance of biometric methods and the progress of the "population thinking" in zoology. The students tried to reveal both the patterns of conchological variation and the factors that may govern them (Zhadin 1923; H. Wagner 1929; Boettger 1930; Peters 1938a; see review in Arthur 1982).

But the most important field of the applied lymnaeid studies is, surely, that aimed at revealing the complex relationships between trematode larvae and their mollusc hosts. The parasitological significance of the pond snails and their impact on public health in various countries are enormous and will be reviewed in detail in the second part of this book. Only a short historical introduction to these studies is worthy to provide here.

The history of studies of Trematoda parasitizing freshwater snails can be traced back to the late seventeenth century. Swammerdam (1738) discovered, described, and depicted some trematode larvae found inside bodies of aquatic Gastropoda. (The adult worms living inside organisms of humans and warm-blooded animals were known much earlier; see Grove 1990). The work on description of their diversity had continued during the next two centuries, but the life cycle of the trematodes and the role of snails as their reservoir hosts and vectors remained obscure until the end of the nineteenth century. For instance, Govert Bidloo (1649–1713), the Dutch physician, in 1688 expressed his opinion that the worms bred in moist earth then were swallowed together with their eggs in water by herbivorous animals such a sheep, stags, calves, and wild boars (Grove 1990). Fischer (1880, pp. 111–112) enlisted several trematode species whose larvae occur in molluscs, however all species included into his list use waterfowl and frogs, not humans or domestic animals, as their definitive hosts.

The liver fluke, *Fasciola hepatica*, became the first trematode in which the life cycle was understood. After a long series of investigations and fruitless attempts of various authors to work out the extremely complex development of this parasite, Rudolf Leuckart (1822–1898) in Germany and Algernon Thomas (1857–1937) in England, who had been working quite independently of each other, published papers (both appeared in October 1882; see Leuckart, 1882; Thomas, 1882) on this subject. Due to the efforts of these two students, the key role of the lymnaeid *Galba truncatula* (O.F. Müller, 1774) in the transmission of the parasite was firmly established. In the last century, this mollusc, also known as the dwarf pond snail (or the mud snail), became a focus of very intensive research in different countries of Europe (i.e., Patzer 1927; Zhadin and Pankratova 1931; Peters 1938b). Similar studies, involving other representatives of the genus *Galba*, were made in non-European countries, for example in Colombia (Brumpt et al. 1940).

Shortly after this discovery, many species of the Lymnaeidae were found to transmit other species of Trematoda of medical and veterinary importance (such as *Fasciola gigantica*, a cause of tropical fascioliasis, and *Fascioloides magna*, or the giant liver fluke). The family of pond snails is considered as one of the most significant from the point of view of public health, second only to planorbids (= Planorbidae s. lato, including bulinid snails) [Brown 1978; Bargues and Mas-Coma 2005; Vázquez et al. 2018].

Research on the role of molluscs as intermediate hosts of *Fasciola hepatica* has subsequently evolved as a result of successive outbreaks of animal or human distomatosis and the development of new analytical techniques. Until the 1980s, many species of Lymnaeidae were tested to determine their potential role as host snails in the parasite cycle (Kendall 1950; Berghen 1964; Boray, 1969, 1978; Busson et al. 1982; Kruglov 2005). Studies on several species of lymnaeids have also been carried out on various continents in order to know the characteristics of the environment in which they live, namely the life cycle of snails, the vegetation of their habitats, the associated fauna and the factors which act on snail ecology (Mehl 1932; Zhadin 1923; Roberts 1950; Chowaniec and Drozdz 1958; Stefanski 1959; Bednarz 1960; Over 1962, 1967; Erhardová-Kotrlá 1971; Pécheur 1974; de Kock et al. 1974; Smith 1981 for *G. truncatula*). The occurrence of an outbreak of human and animal distomatosis in 1968–1969 in central France (Dreyfuss et al. 2015b) prompted studies on the control of molluscs in order to determine whether methods of biological control (predators, competitors) had to be used alone or in combination with molluscicides which were already known at that time (Mage et al. 1989; Rondelaud et al. 2006). Additional research was carried out on histological sections of lymnaeids in order to determine the kinetics of the various larval stages which follow one another in the body of the snail (Rondelaud and Barthe 1982; Rondelaud et al. 2009) and the pathology that the parasite induces in the latter (Sindou et al. 1991). The development of modern analytical techniques (electrophoresis, molecular biology) since the 1990s has made it possible to determine whether the DNA or RNA of the parasite is present in the bodies of the snails studied (Caron et al. 2008; Kim et al. 2014; Alba et al. 2015, for example). The current development of PCR (polymerase-chain reaction) has changed the conventional methods (dissection, cercarial shedding) used by the authors to test for the presence of the parasite in molluscs (Caron et al. 2008). The PCR indicates whether miracidia have penetrated the lymnaeid, even if the larval forms have degenerated later in the body of the snail. However, its use alone does not permit the determination of whether a lymnaeid species in a given region is a potential intermediate host capable of transmitting the parasite or a non-target mollusc into which the miracidia have penetrated through a decoy effect.

Box 1.1 Enumeration of Lymnaeid Species Described During 1758–1803 (Only Species Accepted Today as Valid Are Included)

Original name (in brackets – combination accepted in current taxonomy)	Type locality	Authority
Buccinum glabrum (*Omphiscola glabra*)	Denmark	O.F. Müller (1774)
Buccinum glutinosum (*Myxas glutinosa*)	Denmark	O.F. Müller (1774)
Buccinum lagotis (*Ampullaceana lagotis*)	Germany	Schrank (1803)
Buccinum palustre (*Stagnicola palustris*)	Denmark	O.F. Müller (1774)
Buccinum peregrum (*Peregriana peregra*)	Denmark	O.F. Müller (1774)
Buccinum truncatulum (*Galba truncatula*)	Denmark	O.F. Müller (1774)
Helix auricularia (*Radix auricularia*)	Europe	Linnaeus (1758)
Helix balthica (*Ampullaceana balthica*)	Sweden	Linnaeus (1758)
Helix corvus (*Stagnicola corvus*)	Germany	Gmelin (1791)
Helix stagnalis (*Lymnaea stagnalis*)	Europe	Linnaeus (1758)

Box 1.2 A Short Index to the Lymnaeid Literature of the Last 120 Years

Field of study	References
Morphology	Boettger 1944; Jackiewicz 1954, 1993, 1998; Hubendick 1951, 1978; Walter 1969; Falniowski 1980; Meshcheryakov 1990; Kruglov 2005
Phylogeny and taxonomy	Dall 1905; FC Baker 1911; Thiele 1931; Hubendick 1951, 1978; Starobogatov 1967; Inaba 1969; Kruglov and Starobogatov 1993a, 1993b; Bargues and Mas-Coma 1997, 2005; Ponder and Waterhouse 1997; Jackiewicz 1993, 1998; Remigio 2002; Bargues et al. 2003; Garbar et al. 2004; Kruglov 2005; Puslednik et al. 2009; Correa et al. 2010; Dayrat et al. 2011; Vinarski 2013; Bouchet et al. 2017; Campbell et al. 2017; Aksenova et al. 2018; Saadi et al. 2020; Alda et al. 2021; Saito et al. 2021
Nomenclature and type series	Hubendick 1951; Sitnikova et al. 2012, 2014; Vinarski et al. 2013; Vinarski 2016; Eschner et al. 2020
Biogeography and ecology	Roszkowski 1928; Boycott 1936; Frömming 1956; Over 1967; Starobogatov 1970; Russell-Hunter 1964, 1978; McMahon 1983; Beriozkina and Starobogatov 1988; Banarescu 1990; Økland 1990; Dillon 2000; Stadnichenko 2006; Strong et al. 2008; Aksenova et al. 2018; Lounnas et al. 2017, 2018; Vinarski et al. 2019; Alda et al. 2021

(continued)

Fossil record	FC Baker 1911, Wenz 1923; Korobkov 1955; Wenz and Zilch 1959–1960; Pchelintsev and Korobkov 1960; Taktakishvili 1967; Gray 1988; Taylor 1988
Most important regional surveys	Yen 1939; Boettger 1944; Hubendick 1951; Brandt 1974; Burch 1989; Subba Rao 1989; Økland 1990; Kruglov and Starobogatov 1993a, 1993b; Brown 1994; Jackiewicz 1998; Glöer 2002, 2019; Gittenberger et al. 2004; Stadnichenko 2004; Kruglov 2005; Khokhutkin et al. 2009; Andreeva et al. 2010; Thompson 2011; Welter-Schultes 2012; Johnson et al. 2013; Pointier 2015; Piechocki and Wawrzyniak-Wydrowska 2016; Vinarski and Kantor 2016; Vinarski 2019; Pointier and Vàzquez 2020; Vinarski et al. 2020
Parasitological significance	Mozley 1957; Wright 1971; Brown 1978, 1994; Malek 1980; Mas-Coma and Bargues 1997; Kruglov 2005; Dreyfuss et al. 2015a; Pointier 2015; Caron et al. 2017; Vázquez et al. 2018; Celi-Erazo et al. 2020; Alba et al. 2019; Vázquez et al. 2019; Pereira et al. 2020

References

Aksenova OV, Bolotov IN, Gofarov MY et al (2018) Species richness, molecular taxonomy and biogeography of the radicine pond snails (Gastropoda: Lymnaeidae) in the Old World. Sci Rep-UK 8:11199. https://doi.org/10.1038/s41598-018-29451-1

Alba A, Vázquez AA, Hernández H et al (2015) A multiplex PCR for the detection of *Fasciola hepatica* in the intermediate snail host *Galba cubensis*. Vet Parasitol 211:195–200. https://doi.org/10.1016/j.vetpar.2015.05.012

Alba A, Vázquez AA, Sánchez J et al (2019) Patterns of distribution, populations genetics and ecological requirements of field-occurring resistant and susceptible *Pseudosuccinea columella* snails to *Fasciola hepatica* in Cuba. Sci Rep-UK 9:14359. https://doi.org/10.1038/s41598-019-50894-7

Albrecht C, Wolff C, Glöer P et al (2008) Concurrent evolution of ancient sister lakes and sister species: the freshwater gastropod genus *Radix* in lakes Ohrid and Prespa. Hydrobiologia 615: 157–167. https://doi.org/10.1007/978-1-4020-9582-5_11

Alda P, Lounnas M, Vázquez AA et al (2021) Systematics and geographical distribution of *Galba* species, a group of cryptic and worldwide freshwater snails. Molec Phyl Evol 157:107035. https://doi.org/10.1016/j.ympev.2020.107035

Aldrovandi U (1606) De reliquis animalibus exanguibus libri quatuor post mortem eius editi nempe de mollibvs, crvstaceis, testaceis et zoophytis. Bellagamba, Bologna. https://doi.org/10.5962/bhl.title.118781

Allmon WD (2007) The evolution of accuracy in natural history illustration: reversal of printed illustrations of snails and crabs in pre-Linnaean works suggests indifference to morphological details. Arch Nat Hist 34:174–191

Andreeva SI, Andreev NI, Vinarski MV (2010) Key to freshwater gastropods of Western Siberia (Mollusca: Gastropoda). V. 1. Gastropoda: Pulmonata. Fasc. 1. Families Acroloxidae and Lymnaeidae. The authors, Omsk. [in Russian]

Arthur W (1982) Control of shell shape in *Lymnaea stagnalis*. Heredity 49(2):153–161. https://doi.org/10.1038/hdy.1982.81

Baker FC (1911) The Lymnaeidae of North and Middle America. Special Publication of the Chicago Academy of Sciences 3:1–539. https://doi.org/10.5962/bhl.title.20500

Baker FC (1915) On the classification of the Lymnaeids. The Nautilus 29(2):20–24

Baker HB (1925) Anatomy of *Lanx*, a limpet-like Lymnaeid Mollusc. Proceedings of the Californian Academy of Sciences. Fourth Series 14(8):143–169

Banarescu P (1990) Zoogeography of fresh waters. Vol. 1. General distribution and dispersal of freshwater animals. AULA-Verlag, Wiesbaden

Bargues MD, Mas-Coma S (1997) Phylogenetic analysis of lymnaeid snails based on 18S rDNA sequences. Mol Biol Evol 14(5):569–577. https://doi.org/10.1093/oxfordjournals.molbev.a025794

Bargues MD, Mas-Coma S (2005) Reviewing lymnaeid vectors of fascioliasis by ribosomal DNA sequence analyses. J Helminthol 79:257–267. https://doi.org/10.1079/JOH2005297

Bargues MD, Vigo M, Horák P et al (2001) European Lymnaeidae (Mollusca: Gastropoda), intermediate hosts of trematodiases, based on nuclear ribosomal DNA ITS-2 sequences. Infect Genet Evol 1:87–107. https://doi.org/10.1016/S1567-1348(01)00019-3

Bargues MD, Horák P, Patzner RA et al (2003) Insights into relationships of Palearctic and Nearctic lymnaeids (Mollusca: Gastropoda) by rDNA ITS-2 sequencing and phylogeny of stagnicoline intermediate host species of *Fasciola hepatica*. Parasite 10:243–255. https://doi.org/10.1051/parasite/2003103243

Bargues MD, Artigas P, Jackiewicz M et al (2006) Ribosomal DNA ITS-1 sequence analysis of European stagnicoline Lymnaeidae (Gastropoda). Heldia 6(1–2):57–68

Bargues MD, Artigas P, Khoubbane M et al (2011) *Lymnaea schirazensis*, an overlooked snail distorting fascioliasis data: genotype, phenotype, ecology, worldwide spread, susceptibility, applicability. PLoS One 6:e24567. https://doi.org/10.1371/journal.pone.0024567

Bednarz S (1960) On the biology and ecology of *Galba truncatula* Müll. and cercariae of *Fasciola hepatica* L. in basin of the river Barycz. Acta Parasitologica Polonica 8:279–288

Berghen P (1964) Some Lymnaeidae as intermediate hosts of *Fasciola hepatica* in Belgium. Exp Parasitol 15:118–124

Beriozkina GV, Starobogatov YI (1988) Reproductive ecology and egg clusters of freshwater Pulmonata. Trudy Zoologicheskogo Instituta AN SSSR 174:1–306. [in Russian]

Boettger CR (1930) Die Standortsmodifikationen der Wasserschnecke *Radix auricularia* L. Helios 30:49–64

Boettger CR (1944) Basommatophora. Die Tierwelt der Nord- und Ostsee. Akademische Verlagsgesellschaft Becker and Erler Kom.-Ges., Leipzig 9(b2):241–478

Bonanni P (1709) Musaeum Kircherianum, sive Musæum a P. AthanasioKirchero in Collegio Romano Societatis Jesu jam pridem incoeptum, nuper restitutum, auctum, descriptum, & iconibus illustratum.Plachi, Roma

Boray JC (1969) Experimental fascioliasis in Australia. Adv Parasit 7:95–210

Boray JC (1978) The potential impact of exotic *Lymnaea spp.* on fascioliasis in Australasia. Vet Parasitol 4:127–141. https://doi.org/10.1016/0304-4017(78)90004-3

Bouchet P, Rocroi JP, Hausdorf B et al (2017) Revised classification, nomenclator and typification of gastropod and monoplacophoran families. Malacologia 61(1–2):1–526. https://doi.org/10.4002/040.061.0201

Boycott AE (1936) The habitats of fresh-water Mollusca in Britain. J Anim Ecol 5(1):116–186. https://doi.org/10.2307/1096

Boycott AE, Diver C (1923) On the inheritance of sinistrality in *Limnaea peregra*. P Roy Soc Lond B Bio 95:207–213. https://doi.org/10.1098/rspb.1923.0033

Boycott AE, Diver C (1927) The origin of an albino mutation in *Limnaea peregra*. Nature 119:9. https://doi.org/10.1038/119009a0

Boycott AE, Diver C, Garstang SL et al (1930) The inheritance of sinistrality in *Limnaea peregra* (Mollusca, Pulmonata). Philos T Roy Soc B 219:51–131

Brandt RAM (1974) The non-marine aquatic Mollusca of Thailand. Arch Molluskenkd 105:1–423

Brown DS (1978) Pulmonate molluscs as intermediate hosts for digenetic trematodes. In: Fretter V, Peake J (eds) Pulmonates, vol 2A. Academic Press, London, pp 287–333
Brown DS (1994) Freshwater snails of Africa and their medical importance. Taylor and Francis, London. https://doi.org/10.1201/9781482295184
Brumpt E, Velásquez J, Ucross H et al (1940) Découverte de l'hôte intermédiaire *Limnaea bogotensis* Pilsbry, de la grande douve *Fasciola hepatica*, en Colombie. Ann Parasit Hum Comp 17:563–579
Burch JB (1989) North American freshwater snails. Malacological Publications, Hamburg, MI
Busson P, Busson D, Rondelaud D et al (1982) Données expérimentales sur l'infestation des jeunes de cinq espèces de limnées par *Fasciola hepatica* L. Ann Parasit Hum Comp 57:555–563
Campbell DC, Clark SA, Lydeard C (2017) Phylogenetic analysis of the Lancinae (Gastropoda, Lymnaeidae) with a description of the U.S. federally endangered Banbury Springs lanx. Zookeys 663:107–132. https://doi.org/10.3897/zookeys.663.11320
Caron Y, Rondelaud D, Losson B (2008) The detection and quantification of a digenean infection in the snail host with special emphasis on *Fasciola sp*. Parasitol Res 103:735–744. https://doi.org/10.1007/s00436-008-1086-1
Caron Y, Celi-Erazo M, Hurtrez-Boussès S et al (2017) Is *Galba schirazensis* (Mollusca, Gastropoda) an intermediate host of *Fasciola hepatica* (Trematoda, Digenea) in Ecuador? Parasite 24:24. https://doi.org/10.1051/parasite/2017026
Celi-Erazo M, Alda P, Montenegro-Franco M et al (2020) Prevalence of *Fasciola hepatica* infection in *Galba cousini* and *Galba schirazensis* from an Andean region of Ecuador. Vet Parasitol Reg Stud Rep 20:100390. https://doi.org/10.1016/j.vprsr.2020.100390
Chowaniec W, Drozdz J (1958) Studies on biology and ecology of *Galba truncatula* and on larval forms of *Fasciola hepatica*. Wiad Parazytol 4:433–434. [in Polish]
Clessin S (1882–1886) Die Familie der Limnaeiden enthaltend die Genera *Planorbis*, *Limnaeus*, *Physa* und *Amphipeplea*. Systematisches Conchylien-Cabinet von Martini und Chemnitz, vol 1(17), 2nd edn. Bauer and Raspe, Nürnberg, pp 63–430
Cordellier M, Pfenninger M (2009) Inferring the past to predict the future: climate modelling predictions and phylogeography for the freshwater gastropod *Radix balthica* (Pulmonata, Basommatophora). Mol Ecol 18(3):534–544. https://doi.org/10.1111/j.1365-294X.2008.04042.x
Correa AC, Escobar JC, Durand P et al (2010) Bridging gaps in the molecular phylogeny of the Lymnaeidae (Gastropoda: Pulmonata), vectors of fascioliasis. BMC Evol Biol 10:381. https://doi.org/10.1186/1471-2148-10-381
Correa AC, Escobar JS, Noya O (2011) Morphological and molecular characterization of Neotropic Lymnaeidae (Gastropoda: Lymnaeoidea), vectors of fasciolosis. Infect Genet Evol 11:1978–1988. https://doi.org/10.1016/j.meegid.2011.09.003
Cuvier G (1817) Mémoire sur le Limnée (*helix stagnalis*, Lin.) et le Planorbe (*helix cornea*, ejusd.). In: Cuvier G Mémoires pour servir a l'histoire et a l'anatomie des mollusques. Lebland, Paris, pp 1–14. (separate pagination for each 'memoir')
Da Costa EM (1778) Historia naturalis testaceorum Britanniae, or, the British conchology; containing the descriptions and other particulars of natural history of the shells of Great Britain and Ireland: illustrated with figures. Millan, White, Elmsley and Robson, printed for the author, London. https://doi.org/10.5962/bhl.title.13179
Dall WH (1905) Land and fresh water molluscs of Alaska and adjoining regions. Harriman Alaska Expedition with cooperation of Washington Academy of Sciences 13(i–vii):1–171
Davis GM (1978) Experimental methods in molluscan systematics. In: Fretter V, Peake J (eds) Pulmonates, vol 2A. Academic Press, London, pp 99–169
Davis GM (1994) Molecular genetics and taxonomic discrimination. The Nautilus Supplement 2:3–23
Dayrat B, Conrad M, Balayan S et al (2011) Phylogenetic relationships and evolution of pulmonate gastropods (Mollusca): new insights from increased taxon sampling. Mol Phylogenet Evol 59:425–437. https://doi.org/10.1016/j.ympev.2011.02.014

De Kock KN, Pretorius SJ, Van Eeden JA (1974) Voorlopige kommentar aangaande die voorkomens van die varswaterslakke in den Oranjerivier. 1. In: Van Zinderen Bakker EM (ed) The Orange River. University of The Orange Free State, Bloemfontein, pp 187–212

de Larambergue M (1928) Etude de l'appareil génital de quelques limnées, ses rapports avec la systématique. B Soc Zool Fr 53:491–509

Dillon RT (2000) The ecology of freshwater molluscs. Cambridge University Press, Cambridge etc. https://doi.org/10.1017/CBO9780511542008

Draparnaud JPR (1805) Histoire naturelle des mollusques terrestres et fluviatiles de la France. Plassan; Renaud, Paris et Monpellier. https://doi.org/10.5962/bhl.title.12856

Dreyfuss G, Correa AC, Djuikwo-Teukeng FF (2015a) Differences in compatibility of infection between the liver flukes *Fascioloides magna* and *Fasciola hepatica* in a Colombian population of *Galba* sp. J Helminthol 89:720–726. https://doi.org/10.1017/S0022149X14000509

Dreyfuss G, Vignoles P, Rondelaud D et al (2015b) The mud snail (*Galba truncatula*). Ecology, parasitism and control. Lambert Academic Publishing, Saarbrücken

Dybowski W (1903) Bemerkungen über die gegenwärtige Systematik der Süsswasserschnecken. Nachrichtsblatt der Deutschen Malakozoologischen Gesellschaft 35:130–145

Dybowski B (1913) Bemerkungen und Zusatze zu der Arbeit von Dr. W. Dybowski "Molluscen aus der Uferregion des Baicalsees". Ezhegodnik Zoologicheskogo muzeya Imperatorskoy Akademii nauk 17:165–218

Erhardová-Kotrlá B (1971) The occurrence of *Fascioloides magna* (Bassi, 1875) in Czechoslovakia. Academia, Praha

Eschner A, Vinarski MV, Schnedl SM (2020) Addendum to the examination of the type material of freshwater mollusc species described by J.P.R. Draparnaud. Annalen des Naturhistorischen Museums in Wien, Serie B 122:183–191

Falniowski A (1980) Podrodzaj *Radix* s.str. (Gastropoda, Basommatophora) w Polsce. 1. Pigmentacja i anatomia. Opis *Lymnaea peregra roszkowskiana* subsp. nov. Zeszyty Naukowe Uniwersytetu Jagiellonskiego. Prace Zoologiczne 26:67–108

Fischer P (1880) Manuel de Conchyliologie et de Paléontologie Conchyliologique. Savy, Paris 1:1–112. https://doi.org/10.5962/bhl.title.13213

Fodor I, Hussein AAA, Benjamin PR et al (2020) The unlimited potential of the great pond snail, *Lymnaea stagnalis*. eLife 9:e56962. https://doi.org/10.7554/eLife.56962.sa2

Forbes E, Hanley S (1852–1853) A history of British molluscs and their shells. van Voorst, London. 4:vi+1–616

Freeman G, Lundelius JW (1982) The developmental genetics of dextrality and sinistrality in the gastropod *Lymnaea peregra*. W Roux Arch Dev Biol 191:69–83. https://doi.org/10.1007/BF00848443

Frömming E (1956) Biologie der mitteleuropäischen Süßwasserschnecken. Dunker-Humblot, Berlin

Garbar AV, Korniushin AV (2002) Karyotypes of two European species of the genus *Lymnaea* with disputable taxonomic status (Gastropoda, Pulmonata, Lymnaeidae). Malakologische Abhandlungen / Museum für Tierkunde Dresden 20(2):235–246

Garbar AV, Manilo VV, Korniushin AV (2004) Probable directions of evolution of karyotypes of the European Lymnaeidae (Mollusca, Gastropoda, Pulmonata) in the light of modern concepts of phylogeny of the family. Vestnik zoologii 38(2):29–37. [in Russian]

Ginanni G (1757) Opere postume del conte Giuseppe Ginanni ravennati. Tomo secondo, nel quae si contegnano Testacei marittimi, paludosi e terrestri dell' Adriatico e del territorio di Ravenna, da lui osservati e descritti. G. Zerletti, Venezia

Gittenberger E, Janssen AW, Kuiper JGJ et al (2004) De Nederlandse Zoetwatermolluscen. Recente en fossiele Weekdieren uit Zoet en Brak Water. Nederlandse Fauna. Nationaal Natuurhistorisch Museum Naturalis, KNNV Uitgeverij & EIS-Nederland, Leiden 2:1–288

Glöer P (2002) Die Sußwassergastropoden Nord- und Mitteleuropas: Bestimmungschlussel, Lebenweise, Verbreitung. Die Tierwelt Deutschlands. Conchbooks, Hackenheim 73:1–327

Glöer P (2019) The freshwater gastropods of the west-Palaearctis. Volume 1. Fresh- and brackish waters except spring and subterranean snails. Identification key, anatomy, ecology, distribution. The author, Hetlingen

Gmelin JF (1791) Caroli a Linné Systema Naturae per regna tria naturae, secundum classes, ordines, genera, species, cum characteribus, differentiis, synonymis, locis. Editio decima tertia, aucta, reformata, vol 1(6). G.E. Beer, Leipzig, pp 3021–3910

Gray J (1988) Evolution of the freshwater ecosystem: the fossil record. Palaeogeogr Palaeocl 62:1–214. https://doi.org/10.1016/0031-0182(88)90054-5

Grove DI (1990) A history of human helminthology. C.A.B International, Wallingford

Gurdon JB (2005) Sinistral snails and gentlemen scientists. Cell 123(5):751–753. https://doi.org/10.1016/j.cell.2005.11.015

Hubendick B (1951) Recent Lymnaeidae. Their variation, morphology, taxonomy, nomenclature and distribution. Kungliga Svenska Vetenskapsakademiens Handlingar. Fjärde Serien 3(1): 1–223

Hubendick B (1978) Systematics and comparative morphology of the Basommatophora. In: Fretter V, Peake J (eds) Pulmonates, vol 2A. Academic Press, London, pp 1–47

Inaba A (1969) Cytotaxonomic studies of lymnaeid snails. Malacologia 7(2/3):143–168

Jabbour-Zahab R, Pointier JP, Jourdane J et al (1997) Phylogeography and genetic divergence of some lymnaeid snails, intermediate hosts of human and animal fascioliasis with special reference to lymnaeids from the Bolivian Altiplano. Acta Trop 64:191–203. https://doi.org/10.1016/S0001-706X(96)00631-6

Jackiewicz M (1954) Z badań anatomiczno-porównawczych nad niektórymi gatunkami z rodzaju *Radix* Montfort na terenie Wielkopolski. Prace Komisji Biologicznej, Poznańskie Towarzystwo Przyjaciół Nauk 15(3):1–20

Jackiewicz M (1959) Badania nad zmiennością i stanowiskiem systematycznym *Galba palustris* O.F. Müll. Prace Komisji Biologicznej, Poznańskie Towarzystwo Przyjaciół Nauk 19(3):1–86

Jackiewicz M (1993) Phylogeny and relationships within the European species of the family Lymnaeidae. Folia Malacologica 5:61–95. https://doi.org/10.12657/folmal.005.003

Jackiewicz M (1998) European species of the family Lymnaeidae (Gastropoda, Pulmonata, Basommatophora). Genus 9(1):1–93

Johnson PD, Bogan AE, Brown KM et al (2013) Conservation status of freshwater gastropods of Canada and the United States. Fisheries 38(6):247–282. https://doi.org/10.1080/03632415.2013.785396

Kendall SB (1950) Snail hosts of *Fasciola hepatica* in Britain. J Helminthol 24:63–74. https://doi.org/10.1017/S0022149X00019131

Khokhutkin IM, Vinarski MV, Grebennikov ME (2009) Molluscs of the Urals and the adjacent areas. The family Lymnaeidae (Gastropoda, Pulmonata, Lymnaeiformes). Fasc. 1. Goshchitskiy Publishers, Yekaterinburg. [in Russian]

Kim HY, Choi IW, Kim YR et al (2014) *Fasciola hepatica* in snails collected from water-dropwort fields using PCR. Korean J Parasitol 52:645–652

Klein JT (1753) Tentamen methodi ostracologicæ sive dispositio naturalis cochlidum et concharum in suas classes, genera et species, iconibus singulorum generum aeri incisis illustrata etc. Wishoff, Leiden. https://doi.org/10.5962/bhl.title.117134

Klotz J (1889) Beitrag zur Entwicklungsgeschichte und Anatomie des Geschlechtsapparates von *Lymnaeus*. Jenaische Zeitschrift für Naturwissenschaft 23:1–40

Korobkov IA (1955) A handbook and methodical guide to tertiary molluscs. Gastropoda. Gostoptekhizdat, Leningrad. [in Russian]

Krauss F (1848) Die Südafrikanischen Molluscen. Ein Beitrag zur Kenntniss der Molluscen des Kap- und Natallandes und zur Geographischen Verbreitung derselben mit Beschreibung und Abbildung der neuen Arten. Ebner and Seubert, Stuttgart. https://doi.org/10.5962/bhl.title.13936

Kruglov ND (1986) On the specificity of pond snails (Pulmonata, Lymnaeidae) to infestation by the trematode parthenitae and its analysis for the purposes of systematics and parasitology. Zool Zh 65(12):1799–1807. [in Russian]

Kruglov ND (2005) Lymnaeid snails of Europe and Northern Asia. Smolensk State Pedagogical University Press, Smolensk. [in Russian]

Kruglov ND, Starobogatov YI (1985) Methods of experimental hybridization and some results of its applications in the taxonomy of Lymnaeidae. Malacol Rev 18:21–35

Kruglov ND, Starobogatov YI (1993a) Annotated and illustrated catalogue of species of the family Lymnaeidae (Gastropoda Pulmonata Lymnaeiformes) of Palaearctic and adjacent river drainage areas Part I. Ruthenica Russian Malacol J 3(1):65–92

Kruglov ND, Starobogatov YI (1993b) Annotated and illustrated catalogue of species of the family Lymnaeidae (Gastropoda Pulmonata Lymnaeiformes) of Palaearctic and adjacent river drainage areas Part II. Ruthenica Russian Malacol J 3(1):161–180

Küster HC (1862) Die Gattungen *Limnaeus*, *Amphipeplea*, *Chilina*, *Isidora* und *Physiopsis*. Martini und Chemnitz Systematisches Conchylien-cabinet. Bauer und Raspe, Nürnberg 1(17):1–77

Lamarck JBPA d M (1822) Histoire naturelle des animaux sans vertèbres. chez l'Auteur, Paris 6(2): 1–232. https://doi.org/10.5962/bhl.title.12712

Leuckart R (1882) Zur Entwickelungsgeschichte des Leberegels Zweite Mittheilung. Zool Anz 5: 524–528

Linnaeus C (1758) Systema naturæ per regna tria naturæ, secundum classes, ordines, genera, species, cum characteribus, differentiis, synonymis, locis. Editio decima, reformata. Laurentius Salvius, Stockholm. https://doi.org/10.5962/bhl.title.542

Lister M (1678) Historiae animalium Angliae tres tractatus. Unus de Araneis. Alter de Cochleis tum terrestribus tum fluviatilibus. Tertius de Cochleis marinis. Quibus adjectus est quartus de Lapidibus ejusdem insulae ad cochlearum quandam imaginem figuratis J. Martyn, London. https://doi.org/10.5962/bhl.title.39518

Lister M (1695) Exercitatio anatomica altera, in qua maximè agitur de buccinis fluviatilibus and marinis etc. In: Smith S, Walford B (eds) London. https://doi.org/10.5962/bhl.title.39640

Locard A (1893) Coquilles des eaux douces et saumâtres de France. Description des familles, genres et espèces. Alexandre Rey, Lyon. https://doi.org/10.5962/bhl.title.12982

Lounnas M, Correa AC, Vázquez AA et al (2017) Self-fertilization, long-distance flash invasion and biogeography shape the population structure of *Pseudosuccinea columella* at the worldwide scale. Mol Ecol 26:887–903. https://doi.org/10.1111/mec.13984

Lounnas M, Correa AC, Alda P et al (2018) Population structure and genetic diversity in the invasive freshwater snail *Galba schirazensis*. Can J Zool 96(5):425–435. https://doi.org/10.1139/cjz-2016-0319

Mage C, Reynal P, Rondelaud D et al (1989) Mise en pratique du contrôle de l'infestation par *Fasciola hepatica* chez des bovins limousins. Bulletin des Groupements Techniques Vétérinaires 347:5–10

Mahulu A, Clewing C, Stelbrink B et al (2019) Cryptic intermediate snail host of the liver fluke *Fasciola hepatica* in Africa. Parasite Vectors 12:573. https://doi.org/10.1186/s13071-019-3825-9

Malek EA (1980) Snail-transmitted parasitic diseases, in two volumes. CRC Press, Boca Raton, FL

Mas-Coma S, Bargues MD (1997) Human liver flukes: a review. Res Rev Parasitol 57:145–218

McMahon RF (1983) Physiological ecology of freshwater pulmonates. In: Russell-Hunter WD (ed) The Mollusca, vol 6. Academic Press, Orlando etc, pp 360–430

Mehl S (1932) Die Lebensbedingungen der Leberegelschnecke (*Galba truncatula* Müller). Untersuchungen über Schale, Verbreitung, Lebensgeschichte, natürliche Feinde und Bekämpfungsmöglichkeiten. Arbeiten aus der Bayerischen Landesanstalt für Pflanzenbau und Pflanzenschutz 2:1–177

Meier-Brook C (1993) Artaufassungen in Bereich der limnischen Molluscen und ihr Wandel im 20 Jahrhundert. Arch Molluskenkd 122:133–147

Meshcheryakov VN (1990) The common pond snail, *Lymnaea stagnalis*. In: Dettlaff TA, Wassetsky SG (eds) Animal species for developmental studies: invertebrates. Plenum Publishing Corp, New York, pp 69–132. https://doi.org/10.1007/978-1-4613-0503-3_5

Mezhzherin SV, Garbar AV, Korshunova ED et al (2008) The analysis of morphological and genetic variation of the snail, *Lymnaea stagnalis* s. 1. (Gastropoda, Lymnaeidae), in Ukraine. Vestnik Zoologii 42:339–345. [in Russian]

Michelin H (1831) *Lymnaeus rubiginosus*. Magasin de Zoologie, journal destiné à Établir une correspondance entre les Zoologists de tous les pays, et à leur faciliter les moyens de publier les espèces nouvelles ou peu connues qu'ils possedent. 2me sect. Mollusques et Zoophytes 1: No. 22

Middendorff AT (1851) Mollusken. In: Middendorff AT (ed) Reise in den äussersten Norden und Osten Sibiriens. Band II. Zoologie. Theil 1. Wirbellose Thiere. Kaiserlische Akademie der Wissenschaften, Saint-Pétersburg, pp 163–465. https://doi.org/10.5962/bhl.title.52029

Mozley A (1957) Liver fluke snails in Britain. H. K. Lewis and Co., London

Müller OF (1774) Vermium terrestrium et fluviatilium seu animalium infusorium, helminthicorum et testaceorum non marinorum succincta historia. Heineck et Faber, Havniae et Lipsiae 2:1–214. https://doi.org/10.5962/bhl.title.46299

Oheimb PV v, Albrecht C, Riedel F et al (2011) Freshwater biogeography and limnological evolution of the Tibetan Plateau – insights from a plateau-wide distributed gastropod taxon (*Radix* spp.). PLoS One 6:e26307. https://doi.org/10.1371/journal.pone.0026307

Økland J (1990) Lakes and snails. Environment and Gastropoda in 1,500 Norwegian lakes, ponds and rivers. Universal Book Services / Dr. W. Backhuys Publ, Oegstgeest

Over HJ (1962) A method of determining the liver fluke environment by means of the vegetation type. Bulletin de l'Office International des Epizooties 58:297–304

Over HJ (1967) Ecological biogeography of *Lymnaea truncatula* in The Netherlands. PhD Thesis, University of Utrecht, The Netherlands

Patterson CM, Burch JB (1978) Chromosomes of pulmonate molluscs. In: Fretter V, Peake J (eds) Pulmonates, vol 2A. Academic Press, London, pp 171–217

Patzer HE (1927) Beiträge zur Biologie der Leberegelschnecke, *Galba* (*Limnaea*) *truncatula* Müll. Zool Jahrb Allg Zool 53:321–372

Pchelintsev VF, Korobkov IA (eds) (1960) Fundamentals of Paleontology: a manual for Paleontologists and geologists of the U.S.S.R. Molluscs – Gastropoda. Academy of Sciences of the USSR Press, Moscow-Leningrad. [in Russian]

Pecheur M (1974) Lutte stratégique contre la distomatose. Comptes-Rendus de Recherches, Travaux du Centre de Recherches sur les Maladies Parasitaires des Animaux Domestiques. IRSIA, Bruxelles 38:85–150

Pereira AE, Uribe N, Pointier JP (2020) Lymnaeidae from Santander and bordering departments of Colombia: morphological characterization, molecular identification and natural infection with *Fasciola hepatica*. Vet Parasitol Reg Stud Rep 20:100408. https://doi.org/10.1016/j.vprsr.2020.100408

Perrot JL, Perrot M (1938) Note sur les chromosomes de cinq espèces de limnées. Compte-rendu des séances de la Société de physique et d'histoire naturelle de Genève 53:92–93

Peters BG (1938a) Biometrical observations on shells of *Limnaea* species. J Helminthol 16(4): 181–212. https://doi.org/10.1017/S0022149X00018587

Peters BG (1938b) Habitats of *Limnaea truncatula* in England and Wales during dry seasons. J Helminthol 16(4):213–260. https://doi.org/10.1017/S0022149X00018599

Pfenninger M, Cordellier M, Streit B (2006) Comparing the efficacy of morphological and DNA-based taxonomy in the freshwater gastropod genus *Radix* (Basommatophora, Pulmonata). BMC Evol Biol 6:100. https://doi.org/10.1186/1471-2148-6-100

Piechocki A, Wawrzyniak-Wydrowska B (2016) Guide to freshwater and marine Mollusca of Poland. Bogucki Wydawnictwo Naukowe, Poznań

Pointier JP (ed) (2015) Freshwater molluscs of Venezuela and their medical and veterinary importance. Conchbooks, Harxheim

Pointier JP, Vàzquez AA (2020) Lymnaeoidea, Lymnaeidae. In: Damborenea C, Rogers DC, Thorp JH (eds) Thorp and Covich's freshwater invertebrates, Volume 5: Keys to Neotropical and Antarctic Fauna. Academic Press, London, pp 281–290

Ponder WF, Waterhouse J (1997) A new genus and species of Lymnaeidae from the lower Franklin River, South-Western Tasmania. J Molluscan Stud 63:441–468. https://doi.org/10.1093/mollus/63.3.441

Puslednik L, Ponder WF, Dowton M et al (2009) Examining the phylogeny of the Australasian Lymnaeidae (Heterobranchia: Pulmonata: Gastropoda) using mitochondrial, nuclear and morphological markers. Mol Phylogenet Evol 52(3):643–659. https://doi.org/10.1016/j.ympev.2009.03.033

Remigio EA (2002) Molecular phylogenetic relationships in the aquatic snail genus *Lymnaea*, the intermediate host of the causative agent of fascioliasis: insight from broader taxon sampling. Parasitol Res 88:687–696. https://doi.org/10.1007/s00436-002-0658-8

Remigio EA, Blair D (1997a) Molecular systematics of the freshwater snail family Lymnaeidae (Pulmonata: Basommatophora) utilising mitochondrial ribosomal DNA sequences. J Molluscan Stud 63(2):173–185. https://doi.org/10.1093/mollus/63.2.173

Remigio EA, Blair D (1997b) Relationships among problematic North American stagnicoline snails (Pulmonata: Lymnaeidae) reinvestigated using nuclear ribosomal DNA internal transcribed spacer sequences. Can J Zool 75:1540–1545. https://doi.org/10.1139/z97-779

Roberts EW (1950) Studies on the life-cycle of *Fasciola hepatica* (Linnaeus) and of its snail host, *Limnaea (Galba) truncatula* Müller in the field and under controlled conditions. Ann Trop Med Parasit 44:187–206. https://doi.org/10.1080/00034983.1950.11685441

Rondelaud D, Barthe D (1982) Les générations rédiennes de *Fasciola hepatica* L. chez *Lymnaea truncatula* Müller. Pluralité des schémas de développement. Ann Parasit Hum Comp 57:639–642

Rondelaud D, Vignoles P, Dreyfuss G et al (2006) The control of *Galba truncatula* (Gastropoda: Lymnaeidae) by the terrestrial snail *Zonitoides nitidus* on acid soils. Biol Control 39:290–299

Rondelaud D, Belfaiza M, Vignoles P et al (2009) Redial generations of *Fasciola hepatica*: a review. J Helminthol 83:245–254. https://doi.org/10.1017/s0022149x09222528

Roszkowski W (1914a) Contribution à l'étude des Limnées du lac Léman. Rev Suisse Zool 22(15): 457–539

Roszkowski W (1914b) Note sur l'appareil génital de *Limnaea auricularia* L. et de *Limnaea ovata* Drap. Zool Anz 44(4):175–179

Roszkowski W (1925) Contributions to the knowledge of the family Lymnaeidae. II. Some new data to the knowledge of the Altai *Lymnaea*. Annales Zoologici Musei Polonici Historiae Naturalis 4(4):277–281

Roszkowski W (1928) The distribution of Lymnaeids in Europe and in North America, with relation to Wegener's theory. Annales Zoologici Musei Polonici Historiae Naturalis 7(2–3):64–97

Roszkowski W (1929) Contributions to the study of the family Lymnaeidae. I. On the systematic position and the geographical distribution of the genus *Myxas* J. Sowerby. Annales Zoologici Musei Polonici Historiae Naturalis 8(2):64–97

Russell-Hunter WD (1964) Physiological aspects of ecology in non-marine mollusks. In: Wilbur KM, Yonge CM (eds) Physiology of Mollusca. Academic Press, London, pp 83–125

Russell-Hunter WD (1978) Ecology of freshwater pulmonates. In: Fretter V, Peake J (eds) Pulmonates, vol 2A. Academic Press, London, pp 335–384

Saadi AJ, Davison A, Wade CM (2020) Molecular phylogeny of freshwater snails and limpets (Panpulmonata: Hygrophila). Zool J Linnean Soc 190(2):518–531. https://doi.org/10.1093/zoolinnean/zlz177

Saito T, Hirano T, Ye B et al (2021) A comprehensive phylogeography of the widespread pond snail genus *Radix* revealed restricted colonization due to niche conservatism. Ecol Evol 11(24): 18446–18459. https://doi.org/10.1002/ece3.8434

Say T (1817) Conchology. In: Nicholson W (ed) American edition of the British Encyclopedia, or, dictionary of arts and sciences comprising an accurate and popular view of the present improved state of human knowledge, 1st edn, 2:A3–C6 [= 1–20], pp [1–4]
Say T (1821) Description of univalve shells of the United States. J Acad Natl Sci Phila 2(1): 149–179
Schniebs K, Glöer P, Vinarski M et al (2011) Intraspecific morphological and genetic variability in *Radix balthica* (Linnaeus 1758) (Gastropoda: Basommatophora: Lymnaeidae) with morphological comparison to other European *Radix* species. J Conchol 40(6):657–678
Schniebs K, Glöer P, Vinarski M et al (2013) Intraspecific morphological and genetic variability in the European freshwater snail *Radix labiata* (Rossmaessler, 1835) (Gastropoda: Basommatophora: Lymnaeidae). Contrib Zool 82(1):55–68. https://doi.org/10.1163/18759866-08201004
Schrank von Paula F (1803) Fauna Boica. Durchgedachte Geschichte der in Baiern einheimischen und zahmen Thiere. Philipp Krull: Landshut 3(2):1–372. https://doi.org/10.5962/bhl.title.44923
Schröter JS (1779) Die Geschichte der Flußconchylien mit vorzüglicherRücksicht auf diejenigenwelche in den thüringischen Wassern leben.Gebauer, Halle
Seba A (1758) Locupletissimi rerum naturalium thesauri accuratadescriptio, et iconibus artificiosissimis expressio, per universamphysices historium. Tomus III. J. Wetsten, W. Smith & Jansson-Waesbergios, Amsterdam
Sindou P, Cabaret J, Rondelaud D (1991) Survival of snails and characteristics lesions of *Fasciola hepatica* infection in four European species of *Lymnaea*. Vet Parasitol 40:47–58
Sitnikova TY, Kijashko PV, Sysoev AV (2012) Species names of J.-R. Bourguignat and their application in current taxonomy of freshwater gastropods of the Russian fauna. The Bulletin of the Russian Far East Malacological Society 15(16):87–116
Sitnikova TY, Sysoev AV, Prozorova LA (2014) Types of freshwater gastropods described by Ya.I. Starobogatov, with additional data on the species: family Lymnaeidae. Zoologicheskie Issledovania 16:7–37
Smith G (1981) A three-year study of *Lymnaea truncatula* habitats, disease foci of fascioliasis. Brit Vet J 17:329–342. https://doi.org/10.1016/S0007-1935(17)31638-X
Sowerby GB II (1872) Monograph of the genus *Limnaea*. Conchologia iconica, or, illustrations of the shells of molluscous animals, vol 18. Reeve and Co, London, pp 1–15. unpaginated text
Stadnichenko AP (2004) Pond snails and limpet snails (Lymnaeidae and Acroloxidae) of Ukraine. Tsentr uchebnoi literatury, Kiev. [in Russian]
Stadnichenko AP (2006) Lymnaeidae and Acroloxidae of Ukraine: sampling and studying methods, biology, ecology, their practical importance. Ruta Press, Zhytomyr. [in Russian]
Starobogatov YI (1967) On the systematization of freshwater pulmonate molluscs. Trudy Zoologicheskogo Instituta AN SSSR 42:280–304. [in Russian]
Starobogatov YI (1970) Molluscan fauna and zoogeographic zonation of continental freshwater bodies of the world. Nauka, Leningrad. [in Russian]
Stefanski W (1959) Occurrence and ecology of *Galba truncatula* in Poland. Wiad Parazytol 5:311–314. [in Polish]
Strong EE, Gargominy O, Ponder WP et al (2008) Global diversity of gastropods (Gastropoda; Mollusca) in freshwater. Hydrobiologia 595:149–166. https://doi.org/10.1007/s10750-007-9012-6
Sturtevant AH (1923) Inheritance of direction of coiling in *Limnaea*. Science 58:269–270. https://doi.org/10.1126/science.58.1501.269
Subba Rao MV (1989) Handbook: freshwater molluscs of India. Zoological Survey of India, Calcutta
Swammerdam J (1738) Bybel der natuure of historie der insecten, tot zeekere zoorten gebracht: door voorbeelden, ontleedkundige onderzoekingen van veelerhande kleine gediertens, als ook door kunstige kopere plaaten opgeheldert. Verrykt met ontelbaare waarnemingen van nooit ontdekte zeldzaamheden in de natuur. Alles in de hollandse, des auteurs moedertaale, beschreven. Hier by komt een voorreeden, waar in het leven van den auteur beschreven is

door Herman Boerhaave. De Latynsche overzetting heeft bezorgt Hieronimus David Gaubius. II. deelen. I. Severinus, B. van der Aa and P. van der Aa, Leiden, 2:551–990. https://doi.org/10.5962/bhl.title.119987

Taktakishvili IG (1967) The historical development of the family Valenciennidae. Metsniereba, Tbilisi. [in Russian]

Taylor DW (1988) Aspects of freshwater mollusc ecological biogeography. Palaeogeogr Palaeoclim 62:511–576. https://doi.org/10.1016/0031-0182(88)90071-5

Taylor DW, Sohl NF (1962) An outline of gastropod classification. Malacologia 1(1):7–32

Thiele J (1931) Handbuch der Systematischen Weichtierkunde, vol 2. Gustav Fischer Verlag, Jena, pp 377–778

Thomas AP (1882) Second report of experiments on the development of the liver fluke (*Fasciola hepatica*). J Roy Agr Soc Engl 18:439–455

Thompson FG (2011) An annotated checklist and bibliography of the land and freshwater snails of Mexico and Central America. Bull Fla Mus Nat Hist 50(1):1–299

Throckmorton LH (1968) Biochemistry and taxonomy. Ann Rev Entomol 13:99–114. https://doi.org/10.1146/annurev.en.13.010168.000531

Troschel FH (1839) Ueber die Gattung *Amphipeplea* Nilss. Archiv für Naturgeschichte 5(1): 177–184

van Beneden PJ (1838) Mémoire sur le *Limneus glutinosus*. Nouveaux mémoires de l'Académie Royale des Sciences et Belles-Lettres de Bruxelles 11(22):3–16

Vázquez AA, Pilar A, Lounnas M et al (2018) Lymnaeid snails hosts of *Fasciola hepatica* and *Fasciola gigantica* (Trematoda: Digenea): a worldwide review. CAB Rev Perspect Agric Vet Sci Nutr Nat Resour 13:1–15. https://doi.org/10.1079/PAVSNNR201813062

Vázquez AA, Vargas M d, Alba A et al (2019) Reviewing *Fasciola hepatica* transmission in the West Indies and novel perceptions from experimental infections of sympatric vs. allopatric snail/fluke combinations. Vet Parasitol 275:108955. https://doi.org/10.1016/j.vetpar.2019.108955

Vinarski MV (2013) One, two, or several? How many lymnaeid genera are there? Ruthenica Russian Malacol J 23(1):41–58

Vinarski MV (2014) The birth of malacology: when and how? Zoosyst Evol 90(1):1–5. https://doi.org/10.3897/zse.90.7008

Vinarski MV (2015) Conceptual shifts in animal systematics as reflected in the taxonomic history of a common aquatic snail species (*Lymnaea stagnalis*). Zoosyst Evol 91:91–103. https://doi.org/10.3897/zse.91.4509

Vinarski MV (2016) Annotated type catalogue of lymnaeid snails (Mollusca, Gastropoda) in the collection of the Natural History Museum, Berlin. Zoosyst Evol 92(1):131–152. https://doi.org/10.3897/zse.92.8107

Vinarski MV (2019) 11. Class Gastropoda. In: Rogers DC, Thorp JH (eds) Thorp and Covich's freshwater invertebrates. Vol. IV. Keys to Palaearctic Fauna. Academic Press, London, pp 310–345

Vinarski MV, Kantor YI (2016) Analytical catalogue of fresh and brackish water molluscs of Russia and adjacent countries. A.N. Severtsov Institute of Ecology and Evolution of RAS, Moscow

Vinarski M, Schniebs K, Glöer P et al (2011) The taxonomic status and phylogenetic relationships of the genus *Aenigmomphiscola* Kruglov et Starobogatov, 1981 (Gastropoda: Pulmonata: Lymnaeidae). J Nat Hist 45(33–34):2049–2068. https://doi.org/10.1080/00222933.2011.574800

Vinarski MV, Schniebs K, Glöer P et al (2012) The steppe relics: taxonomic study on two lymnaeid species endemic to the former USSR (Gastropoda: Pulmonata: Lymnaeidae). Arch Molluskenkd 141(1):67–85. https://doi.org/10.1127/arch.moll/1869-0963/141/067-085

Vinarski MV, Nekhaev IO, Glöer P et al (2013) Type materials of freshwater gastropod species described by C.A. Westerlund and accepted in current malacological taxonomy: a taxonomic and nomenclatorial study. Ruthenica Russian Malacol J 23(2):79–108

Vinarski MV, Aksenova OV, Bespalaya YV et al (2016) *Radix dolgini*: the integrative taxonomic approach supports the species status of a Siberian endemic snail (Mollusca, Gastropoda, Lymnaeidae). C R Biol 339(1):24–36. https://doi.org/10.1016/j.crvi.2015.11.002

Vinarski MV, Clewing C, Albrecht C (2019) Family Lymnaeidae. In: Lydeard C, Cummings KS (eds) Freshwater molluscs of the world: a distribution atlas. Johns Hopkins University Press, Baltimore, MD, pp 158–162

Vinarski MV, Aksenova OV, Bolotov IN (2020) Taxonomic assessment of genetically-delineated species of radicine snails (Mollusca, Gastropoda, Lymnaeidae). Zoosyst Evol 96(2):577–608. https://doi.org/10.3897/zse.96.52860

Wagner J (1927) Nouvelles contributions a l'anatomie de l'appareil génital des limnees de Hongrie. Állattani Közlemények 24(1–2):29–39. [in Hungarian]

Wagner H (1929) Zur Variation von *Limnaea* und biometrische Untersuchungen an *Planorbis*. Zool Anz 80(7/9):183–193

Walter HJ (1969) Illustrated biomorphology of the "*angulata*" lake form of the basommatophoran snail *Lymnaea catascopium* Say. Malacol Rev 2:1–102

Welter-Schultes F (2012) European non-marine molluscs: a guide for species identification. Planet Poster Editions, Göttingen

Wenz W (1923) Gastropoda extramarina tertiaria, vol 4. W. Junk, Berlin, pp 1069–1420

Wenz W, Zilch A (1959–1960) Gastropoda. Teil 2. Euthyneura. In: Schindewolf OH (ed) Handbuch der Paläozoologie. Gebrüder Borntraeger, Berlin. 6(2):ii+1–834

Westerlund CA (1885) Fauna der in der Paläarktischen Region (Europa, Kaukasien, Sibirien, Turan, Persien, Kurdistan, Armenien, Mesopotamien, Kleinasien, Syrien, Arabien, Egypten, Tripolis, Tunisien, Algerien und Marocco) lebenden Binnenconchylien. V. Fam. Succineidae, Auriculidae, Limnaeidae, Cyclostomidae und Hydrocenidae. H. Ohlsson, Lund

Wright CA (1971) Flukes and snails. George Allen and Unwin, London

Yen TC (1939) Die Chinesische Land- und Süsswasser-Gastropoden in Natur-Museums Senckenberg. Abhandlungen der Senckenbergischen naturforschenden Gesellschaft 444:1–233

Zhadin VI (1923) The variation of *Limnaea stagnalis* in waterbodies of the vicinities of Murom. Russkiy Gidrobiologicheskiy Zhurnal 2(97–106):173–178. [in Russian]

Zhadin VI, Pankratova VA (1931) Studies on biology of mollusks – vectors of fasciolosis, and the development of measures against them. Raboty Okskoi Biologicheskoi Stantsii 6:79–157. [in Russian]

Chapter 2
General Characteristics of the Family Lymnaeidae

Maxim V. Vinarski and Jean-Pierre Pointier

Abstract The article provides a brief introduction to the Lymnaeidae and contains short reviews of taxonomy, the fossil record, morphology, biogeography, and ecology of these snails, with an emphasis on the extant representatives of the family.

The aim of this chapter is to provide the reader with a brief and general overview to the family Lymnaeidae, in other words to give an introductory reading to the subjects discussed in the rest of the book.

2.1 Taxonomic Position of the Family

Lymnaeidae Rafinesque, 1815 is a relatively speciose family of aquatic snails belonging to a large group informally known as the "pulmonate snails" (Pulmonata). The main diagnostic feature (synapomorphy) of the subclass Pulmonata was thought to be their mode of respiration based on the utilization of atmospheric air and the use of lungs as an organ of breathing. Since 1816, when Georges Cuvier proposed a new system of the Gastropoda based on morphology of the respiratory system (Cuvier 1816), a taxon named Pulmonata figured in almost each version of classification, though the rank of this group might have been varied. Cuvier (1816) and many of his followers ranked it as an order (e.g., Fischer 1880–1887; Pelseneer 1905), whereas in the twentieth century systematics the Pulmonata have usually been treated as a

M. V. Vinarski (✉)
Laboratory of Macroecology & Biogeography of Invertebrates, Saint-Petersburg State University, Saint-Petersburg, Russia

S.I. Vavilov Institute for the History of Science and Technology, St. Petersburg Branch, Russian Academy of Sciences, Saint-Petersburg, Russia

J.-P. Pointier
PSL Research University, USR 3278 CNRS–EPHE, CRIOBE Université de Perpignan, Perpignan, France
e-mail: pointier@univ-perp.fr

M. V. Vinarski, A. A. Vázquez (eds.), *The Lymnaeidae*, Zoological Monographs 7,
https://doi.org/10.1007/978-3-031-30292-3_2

subclass of Gastropoda (Prosobranchia and Opisthobranchia constituted the two other subclasses) [Thiele 1929, 1931 and many subsequent taxonomists]. An alternative, albeit somewhat less popular, approach was to build the system of Gastropoda on the gross morphology of the nervous system. The use of this criterion led Spengel (1881) to distribute gastropods between two subclasses—Streptoneura (with *chiastoneury*, i.e. crossing of the cerebrovisceral commissures caused by the torsion process) and Euthyneura (without chiastoneury), and such a division was accepted by some workers of the early-mid-twentieth century (Grobben 1910; Ivanov 1940; Taylor and Sohl 1962).

Most often, the subclass (or order) Pulmonata was divided into two subgroupings—Basommatophora (chiefly aquatic forms) and Stylommatophora (almost exclusively terrestrial). In this classical scheme, the Lymnaeidae, alongside with such families as Acroloxidae, Planorbidae, Physidae, and several else, constituted Basommatophora (Thiele 1931; Hubendick 1978; Salvini-Plawen 1980). This classification has been challenged in the 1970s–1980s (Golikov and Starobogatov, 1975, 1988; Minichev and Starobogatov 1979; Salvini-Plawen 1980; Haszprunar 1988; Ponder and Lindberg 1997; reviewed in Ponder and Lindberg 2008), and now the class Gastropoda is divided into several subclasses, none of which is identical to Pulmonata of the older authors. The current system of Gastropoda is grounded chiefly on the results of molecular phylogenetic research, according to which Pulmonata is a part of a larger group called Heterobranchia (Ponder and Lindberg 2008). It is ranked as a subclass by Bouchet et al. (2017). Heterobranchia is the taxon embracing virtually all representatives of the former subclasses Opisthobranchia and Pulmonata. Thus, traditional “Pulmonata” has become an abandoned concept. According to Schrödl (2013, p. 162), “traditional Pulmonata most likely has failed from a phylogenetic point of view, and thus should be avoided in a systematic, taxonomic and classificatory context.” However, it does not prevent from using the informal term “pulmonate snails.” Basommatophora is also an outdated concept; in the current taxonomy, the higher group to which the Lymnaeidae belongs is Hygrophila (Box 2.1). Historically, Hygrophila constituted a subset of the old “Basommatophora,” a taxon uniting all families of freshwater basommatophoran gastropods (Thiele 1931).

2.2 Evolutionary Origin of the Family and Its Fossil Record

The direct ancestors of the family Lymnaeidae are unknown. According to Kruglov (2005), the origin and initial diversification of this group should be dated to the Triassic Period, and possible progenitors of the lymnaeids were some *Chilina*-like gastropods. It correlates well with the basal position of Chilinidae within Hygrophila revealed in both morphology- and molecular-based phylogenies (Hubendick 1978; Nordsieck 1992; Jörger et al. 2010; Saadi et al. 2020).

The oldest known lymnaeid remains are dated as early Jurassic and come from the western USA. Paleontologists classify these fossil species as belonging to extant genera *Galba* and *Lymnaea* (Robinson 1915; Yen and Reeside 1946; Tracey et al. 1993), which, however, must be treated with great caution. As it has been revealed during recent molecular phylogenetic studies on the living Lymnaeidae, shell characters are not a good source of phylogenetic signal; phylogenetically distant genera may demonstrate convergent similarity in their shell shape and proportions (see below). The estimates of the geological age of particular lymnaeid genera obtained by independent approaches can be quite different. For example, Martinson (1961) described three fossil species of the genus *Radix* from the Upper Cretaceous of Siberia. On the other hand, a fossil-calibrated phylogeny presented by Aksenova et al. (2018) dates the origin of this genus to the mid-Eocene, and the mean age of divergence between *Radix* and its sister clade, which includes eight lymnaeid genera, is 58.2 Ma, which corresponds to the Paleocene. However, this is just the mean age, the confidence interval for this estimate is quite broad and ranges from ca. 45 to 73 Ma, which involves also the latest Cretaceous.

As of 01.10.2022, 360 fossil lymnaeid species are accepted as valid (MolluscaBase 2022). Around 15 extinct genera of the Lymnaeidae have been described, including those forming a conchologically aberrant subfamily, Valencienniinae Gorjanović-Kramberger, 1923 (sometimes ranked as a family; see Taktakishvili 1967), known from the Pliocene of southeast Europe (Fig. 2.1).

2.3 Diversity and Taxonomic Structure of the Extant Lymnaeidae

According to Hubendick (1951) and Kruglov and Starobogatov (1981, 1993), the extant Lymnaeidae are distributed among only two genera—*Lymnaea* and *Lanx* Clessin, 1882 (Hubendick 1951), or *Lymnaea* and *Aenigmomphiscola* (Kruglov and Starobogatov, 1993). Today, most taxonomists tend to split the family Lymnaeidae into a series of genera, the actual number of which is, however, unknown since the integrative revision of the taxonomic structure of this family has not been completed yet. The most recent published estimates counted 20 to 25 extant lymnaeid genera (Ponder and Waterhouse 1997; Vinarski 2013; Vinarski et al. 2019), but in the course of a molecular phylogenetic study (Aksenova et al. 2018), the number of valid genera increased, mainly due to elevation of rank of some taxa previously treated as subgenera (*Ampullaceana*, *Kamtschaticana*, and some other). Two new lymnaeid genera, *Idaholanx* Clark, Campbell & Lydeard, 2017, and *Tibetoradix* Bolotov, Vinarski & Aksenova, 2018, have recently been described. Therefore, for the time being, it is reasonable to estimate the generic richness of living lymnaeids to lie between 30 and 35 valid genera.

To estimate the lymnaeid species richness in the recent fauna is even more difficult. The estimates available from the literature vary from only 40 (Hubendick 1951) to about 100 (Strong et al. 2008) and even more than 200 valid species worldwide (Kruglov 2005). As of 01.11.2021, MolluscaBase (2022) lists 161 species

Fig. 2.1 A selection of fossil lymnaeid shells. (**a**) *Stagnicola longiscatus* (Brongniart, 1810), Eocene of England. (**b**) *Lymnaea caudata* F.E. Edwards, 1852, Eocene of England. (**c**) *L. megarensis* Gaudry in Gaudry et Fischer, 1867, Pliocene of Greece. (**d**) *Lymnaea bogatschevi* Vinarski et Frolov, 2017, late Miocene of South Russia. (**e**) *Radix kobelti* (Brusina, 1884), Pliocene of Croatia. (**f**) *Corymbina bicarinata* (Fuchs, 1887), Pliocene of Greece. (**g, h**) *Valenciennius reussi* Neumayr in Neumayr et Paul, 1875, Pliocene of Croatia. Two different specimens in two views. All shells, except E, are from the collection of the Natural History Museum in Vienna. E – from collection of the Zoological Institute, Russian Academy of Sciences, St. Petersburg. Scale bars 5 mm. Photos: Maxim V. Vinarski

of Lymnaeidae flagged as "accepted." In order to understand this number properly, one needs to keep in mind that a great portion of these taxa were described in the nineteenth to the first half of the twentieth centuries, when shell characters served as the virtually sole basis for species delineation. The validity of some of the accepted species still requires to be confirmed.

The synonymy rate among living Lymnaeidae is rather high. According to MolluscaBase, there are 1050 unaccepted species names, which is 6.5 times higher than the number of accepted species. This synonymy rate would increase substantially, if all unaccepted taxa of the "variety" rank, introduced in the past, are taken into account (it would be impossible to make such a calculation here since a large portion of these "varieties" has still not been indexed in MB).

On the other hand, recent integrative taxonomic studies (Campbell et al. 2017; Aksenova et al. 2018, 2019; Mahulu et al. 2019) have resulted in description of several new species of lymnaeid snails (or resurrection of the old-forgotten ones), and there are all reasons to believe that this process is not finished yet. Furthermore, instances of the so-called taxonomic crypsis have been detected among lymnaeids (Correa et al. 2010, 2011; Aksenova et al. 2018; Alda et al. 2018, 2021), which means that at least some "traditional" species of this family may, in fact, represent complexes of several distinct species. Since 2005, a series of new lymnaeid species, still not supported by molecular data, has been described (e.g., Glöer and Yildirim 2006; Glöer and Pešić 2008; Vinarski 2011; Qian et al. 2012). Therefore, an estimate of 200–225 living lymnaeid species in the world's fauna seems realistic.

The internal taxonomic structure of the extant Lymnaeidae has been considered differently by various authors. Traditionally, most systematists did not divide it into subfamilies (Hubendick 1951; Starobogatov 1967; Kruglov and Starobogatov 1993; Jackiewicz 1998; Ponder and Waterhouse 1997). The most controversial was the taxonomic status and rank of a peculiar North American group of limpet lymnaeids (the genera *Lanx* and *Fisherola* Hannibal, 1912). Both conchologically and anatomically, these limpets seem to be rather distinct from the rest of the Lymnaeidae. For example, this group is characterized by the presence of two prostates (H.B. Baker 1925; Starobogatov 1967). Hannibal (1914) proposed to erect a separate subfamily, Lancinae, whose rank was subsequently elevated to the family (Pilsbry 1925; Thiele 1931). Later on, the family status for this group was accepted by Taylor and Sohl (1962) and Starobogatov (1967).

On the other hand, Walter (1969, p. 5), in his thorough morphological study of the North American lymnaeids, concluded that "the limpet-like lymnaeids are more closely related to some species of '*Stagnicola*' than are some species of '*Stagnicola*' to each other" and therefore the author proposed "to reject the taxa Lancinae and *Lanx*." His point of view, however, was so radical that no malacologist shared it.

Recently, Campbell et al. (2017) showed that this group does not constitute a separate family within the Lymnaeoidea. Though the phylogenetic relationships of this group to other Lymnaeidae are not resolved yet, it represents a monophyletic and morphologically divergent clade that may be characterized by some synapomorphies (Campbell et al. 2017). Campbell et al. (2017) proposed to retain it as a subfamily of the Lymnaeidae, and their opinion was followed by Bouchet et al. (2017) and Vinarski et al. (2019).

The most recent phylogenetic study by Saadi et al. (2020) did not resolve the internal taxonomic structure of the Lymnaeidae. The authors showed that the subfamily Lymnaeinae may be not monophyletic, but the set of genera included in their study is too limited to warrant a thorough conclusion on the phylogenetic relationships among Lymnaeidae.

The current system of the living Lymnaeidae, as it is accepted by Bouchet et al. (2017) and MolluscaBase (2022), is summarized in Box 2.2. The contributors to this book, in the main, used this classification, however, in some cases they followed their own views on the volume and taxonomic position of particular genera and subgenera. Aksenova et al. (2023) propose an updated and more detailed version of this system, whose reliability and workability will be evaluated in the nearest future.

2.4 Shell and Soft Body Morphology Features

The vast majority of freshwater species of Gastropoda, including lymnaeids, are shell-bearing animals. Following Burch (1989), one can divide the overall diversity of shell morphologies observed among freshwater snails into four general categories. *Neritiform* shells vary in shape from hemispherical to globose; their spires are very low and small and sometimes become virtually invisible (Fig. 2.2a). *Ancyliform* (*limpet-shaped*) shells are uncoiled, have a shape of cap or obtuse cone (see Fig. 2.2c). *Planorbiform* (*discoidal*) shells, with sunken spire, are coiled in the same plane (see Fig. 2.2b), whereas *turbospiral* (*conic*) shells have a more or less raised spire. Both spire height and whorl number in turbospiral shells are very variable, which creates a plethora of conchological variants even within a single family. Most lymnaeid snails possess turbospiral shells that may take various shapes, from slender and subulate to subglobose (Fig. 2.3). The shell of the lymnaeid genus *Lantzia* may be described as neritiform (see Fig. 2.3l), while auriculate shells occupy an intermediate position in the shell morphospace, being a transition from conic to neritiform (see Fig. 2.3k). The limpet-shaped forms among living lymnaeids are known exclusively from North America where three genera, forming the subfamily Lancinae, occur (see Fig. 2.2c). In the late Miocene and Pliocene, ancyliform lymnaeids were living in large lacustrine basins of south-eastern Europe; most limpets of this group have been placed into the genus *Valenciennius* Rousseau, 1842 (see Fig. 2.1g,h) and other, closely allied, genera (Gorjanović-Kramberger 1923; Moos 1944; Taktakishvili 1967; Marinescu 1969; Harzhauser et al. 2016). As Lancinae and Valencienniinae belong to two distant phylogenetic lineages, separated both in space and time, it is clear that the limpet-shaped form among lymnaeids evolved convergently from coiled ancestors (which was a common situation within the class Gastropoda, see Vermeij 2017).

Fig. 2.2 Examples of neritiform (a), planorbiform (b), and ancyliform (c) shell. (**a**) *Theodoxus pallasi* (Neritidae). (**b**) *Anisus leucostoma* (Planorbidae). (**c**) *Lanx subrotundatus* (Lymnaeidae: Lancinae). This figure is not in a scale. Photos: Maxim V. Vinarski

The absolute shell size within the discussed family varies considerably. Its largest representatives in the recent fauna belong to the genus *Lymnaea*; their shells may reach 60–65 mm height. Shells of fossil Valencienniinae may have been even larger (up to 132 mm in diameter; see Taktakishvili 1967). On the other hand, shell height of species of the genus *Galba* rarely exceeds 10 mm (in most cases, the adult shells are 5–8 mm height). The majority of extant pond snails are, however, animals of medium size; their shell heights lie between 10 and 20–22 mm.

It should be remembered that the absolute shell size is a very plastic variable, and a number of abiotic and biotic factors are known to affect size. For example, the so-called *parasitic castration* may lead to development of unusually large individuals of pond snails (Wilson and Denison 1980). An opposite trend is *dwarfism,* which has been discovered in various species of the family. Dwarf morphs are known in lymnaeid populations inhabiting geothermal springs (Bolotov et al. 2012; Aksenova et al. 2017; Vinarski et al. 2023), in deep alpine lakes (Geyer 1923, 1929), and, occasionally, may represent a result of starvation (*Hungerformen* of Kobelt 1871). Typically, a dwarf individual is 1.5–2.0 times smaller than its congeners from populations living under "normal" conditions (Fig. 2.4). Most probably, this "thermal dwarfism" must be explained by the shortening of the life cycle of snails and their shift to neotenic reproduction (Vinarski et al. 2023). Large-

Fig. 2.3 Diversity of lymnaeid shell shapes. (**a**) Subulate (*Acella haldemani*). (**b**) Cylindric (*Omphiscola glabra*). (**c**) Turriculate (*Lymnaea taurica*). (**d, e**) High-conical (two morphotypes of *Lymnaea stagnalis*). (**f–h**) Ovate-conical (*Bulimnea megasoma*, *Pseudosuccinea columella*, *Ampullaceana lagotis*). (**i**) Ovoid (*Ampullaceana balthica*). (**j**) Ear-shaped, of auriculate (*Radix auricularia*). (**k**) Globose (*Myxas glutinosa*). (**l**) Neritiform (*Lantzia carinata*). (**a–g**) belong to the subfamily Lymnaeinae; (**h–l**) – to the subfamily Amphipepleinae. Scale bars 2 mm. Photos: Maxim V. Vinarski

scale spatial trends in shell size have been revealed in some widespread lymnaeid species. Generally, the absolute shell size tends to decrease in the south-north direction, which contradicts the predictions of Bergmann's rule (Vinarski 2012a, b; Vinarski and Kramarenko 2019). This relationship between shell size and latitude has been explained by reduction of the growth season (i.e., a warm period when shell growth is possible) in high latitudes. This reduction leads to formation of a climatically-driven "converse Bergmann" cline of shell size found in lymnaeids and some other aquatic pulmonates (Vinarski and Kramarenko 2019).

However, some clear exceptions from this general size gradient can be mentioned. The largest recent lymnaeid species, *Lymnaea stagnalis,* is native to the temperate belt of the North Hemisphere and is almost totally absent from the tropic latitudes (Hubendick 1951). On the other hand, the fauna of the Lymnaeidae of Hawaii, the islands lying at 20^0 N, consists of small-sized species, whose mean shell length is between 5 and 10 mm. Only rare individuals of these species can reach 15 mm height (Hubendick 1952). Possibly, this is a consequence of the so-called island rule that predicts that the insular endemics are generally diminutive in absolute size as compared to their relatives of the mainlands (Lomolino 2005; McClain et al. 2006; Lomolino et al. 2013; and references therein).

The lymnaeids are notoriously known as a group demonstrating enormous variability in shell shape and proportions, both at the intrapopulational and intraspecific level. Many shell characters of these snails are subject to local adaptation, and the observed conchological variability is typically merely direct effects of the ambient conditions on a plastic phenotype (though some distinct conchological

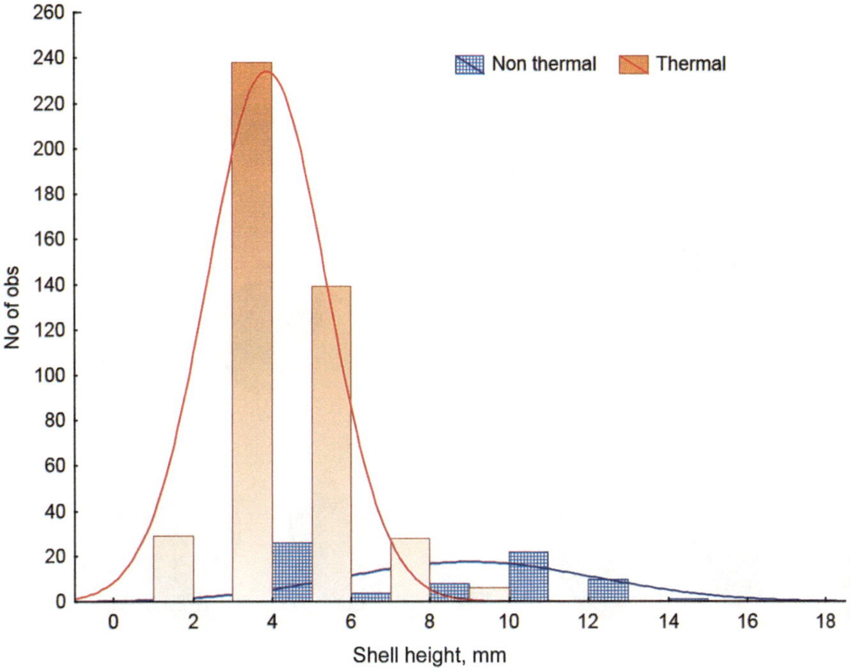

Fig. 2.4 Shell height distribution in *Ampullaceana balthica* individuals from thermal (water temperature range + 15... + 38 ° C) and non-thermal (water temperature range + 5... + 7 ° C) waterbodies of Island. (after Vinarski et al. 2023; modified)

variants may well have a hereditary basis). In a large body of water, like Lake Konstanz (= Bodensee) in Germany and Switzerland, the extent of intraspecific variation may be colossal (Fig. 2.5). In smaller habitats, the conchological disparity within a population is usually less prominent, though some patterns of variation may be discerned. Typically, shells can be arranged into a continuous series of phenotypes, varying from relatively slender and high-spired to more inflated and low-spired shapes (Fig. 2.6). In the past, it was not an uncommon situation when shells corresponding to the opposite extremes of variation were classified as belonging to two distinct varieties or even species. The biometrical analysis usually shows that all individual shells actually belong to a continuum, and the values of the measured traits follow the normal distribution pattern.

The variability in shell surface coloration and sculpture can also be very prominent at the intraspecific level. As a rule, it lacks any taxonomic value, though there are data that the pattern of shell microsculpture may be species-specific in some lymnaeids (Jackiewicz and Koralewska-Batura 1995; Jackiewicz 1998).

Examples of the *convergent* and *parallel* shell evolution are widespread among lymnaeid snails. The genera *Aenigmomphiscola* and *Omphiscola* represent one of the most salient instances. The structure of the copulatory apparatus in these genera is strikingly different, whereas their shells are almost indistinguishable (Kruglov and

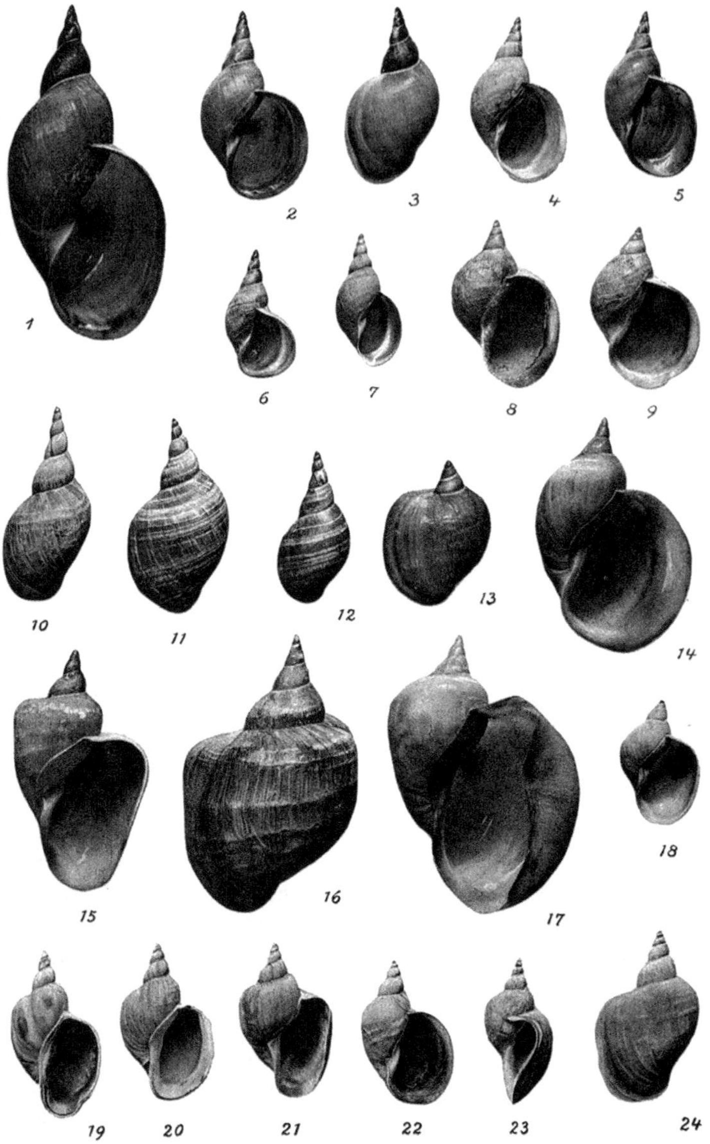

Fig. 2.5 Conchological disparity of *Lymnaea stagnalis* from Lake Konstanz. A "typical" form (= *L. stagnalis* f. *normalis perfecta*) is given under No. 1. After Geyer (1929), with modifications

Starobogatov 1981). A molecular phylogenetic study has brought evidence of the sister relationship between the two genera (Vinarski et al. 2011); most probably, they share the same shell morphology which was inherited from their common ancestor. Similar shell morphologies have been developed in lymnaeid species belonging to phylogenetically distant genera and living in different continents (Figs. 2.7 and 2.8).

Fig. 2.6 Intrapopulation shell variation in two lymnaeid species: *Stagnicola palustris* (A) and *Radix auricularia* (B). (**a**) Russia, Tula Region, the Yasskoye Lake. (**b**) Mongolia, the Khar-Us-Nuur Lake. Scale bars 2 mm. Photos: Maxim V. Vinarski

Fig. 2.7 Neritiform and subneritiform shells independently developed in phylogenetically and geographically distant lymnaeid species. (**a**) *Erinna newcombi*, Hawaii. (**b**) *Lantzia carinata*, Reunion Island. (**c**) *Radix onychia*, Japan, Biwa Lake. (**d**) *Peregriana subonychia*. Russia, Neogene of the Altay Mts. Scale bars 2 mm. Photos: Maxim V. Vinarski

Fig. 2.8 Auriculate shells independently developed in phylogenetically and geographically distant lymnaeid species. (**a**) *Radix obliquata*, Kyrgyzstan, Issyk-Kul' Lake. (**b**) *Ladislavella emarginata* f. *mighelsi*, USA, Square Lake, Maine. (**c**) *Lymnaea stagnalis* f. *lacustris*, Russia, Karelia, Svyatozero Lake. (**d**) *Ampullaceana ampla*, Sweden, Östergotland. Scale bars 5 mm. Photos: Maxim V. Vinarski

The vast majority of the lymnaeid species have dextral (= right-coiled) shells, however, for a small group known as the subgenus *Pseudisidora* (*Pseudobulinus* Kruglov & Starobogatov, 1993), sinistrality is a diagnostic trait. It is endemic to Hawaii and, most probably, includes only one valid species—*Pseudisidora* (*Pseudobulinus*) *reticulata* (Gould, 1848) [Fig. 2.9]. In other lymnaeid species, sinistral (= left-coiled) individuals may occasionally arise as a consequence of a single-gene mutation (see Freeman and Lundelius 1982; Pierce 1996; Asami et al. 2008; Davison et al. 2009; Kuroda and Abe 2020 for reviews of sinistrality in the Lymnaeidae). Some recent results hint that the sinistrality can be the ancestral state for the family of pond snails (Davison et al. 2016), though the recent *Pseudobulinus*, with their narrow endemicity, do not seem a good candidature for the lymnaeid ancestral group and, most probably, their sinistrality is a secondary trait. Unfortunately, the fossil record gives us no data on the assumed left-coiled lymnaeid ancestors.

Within the Hygrophila, the direction of shell coiling typically (but not always) corresponds to the type of asymmetry of the soft body. Species of the families

Fig. 2.9 Sinistral individuals of lymnaeid snails. (**a**) *Pseudisidora reticulata*, Hawaii. (**b, c**) *Stagnicola saridalensis*, Russia, Western Siberia (sinistral and dextral specimens). (**d**) *Lymnaea stagnalis*. Russia, vicinity of St. Petersburg. (**e**) *Radix auricularia*. England, Northamptonshire. Scale bars 2 mm. Photos: Maxim V. Vinarski (**a, c-d**), Peter Glöer (**b**)

Chilinidae, Latiidae, Lymnaeidae, and Acroloxidae are conchologically and anatomically dextral, so that the genital, anal, and respiratory openings are on the right side of the body. In contrast, the families Bulinidae, Physidae, and Planorbidae include anatomically sinistral species, and in some of them the shell is right-handed, but in this case, it is designated as *pseudodextral*.

The body plan of the Lymnaeidae represents a specific case of the Hygrophila body plan; the latter has been described in many sources (Ivanov 1940; Duncan 1960; Starobogatov 1967; Franc 1968; Hubendick 1978; Nordsieck 1992; Ponder et al. 2020) that allows us to omit a detailed anatomical characterization of the family (see also anatomical accounts presented in Hubendick 1951, 1978; Jackiewicz 1954, 1998; Walter 1969; Stadnichenko, 2004; Kruglov 2005). Walter's paper (1969) especially merits to be mentioned here since it represents a minutely and richly

Fig. 2.10 A crawling individual of the mammoth pond snail, *Bulimnea megasoma* (Say, 1824). Canada, Ontario Province, the Cart Lake near Cobalt Town. Photo: Maxim V. Vinarski

illustrated descriptive monograph of the anatomy of a certain lymnaeid species, *Lymnaea* (*Ladislavella* of the current system) *catascopium* (Say, 1817).

The body of a lymnaeid snail divides into three main parts—the head, the foot, and the visceral mass covered by the mantle. In a living animal, the whole visceral mass is hidden within its shell (Fig. 2.10). In some groups, like the genus *Myxas* (the gelatinous lymnaeid snails), the mantle is greatly extended and can be reflected over the shell, covering it except for a small oval space on the upper side of the last whorl. The head bears a pair of flat tentacles, triangular in outline, and a pair of eyes located at the tentacle bases. The foot is a muscular organ allowing snails to crawl over a substrate. The coloration of both foot and head is usually homogeneous and varies from pale yellow or light-grayish to almost black, while the mantle surface may exhibit a more complex pattern of coloration, with light spots or small "freckles" against a dark background, and vice versa (Falniowski 1980a, b; Jackiewicz 1993; Schniebs et al. 2011). Within a species, some, more or less distinct, "types" of pigmentation can be distinguished (e.g., in *Ampullaceana balthica*; see Schniebs et al. 2011). Some authors reported some stable intraspecific differences in the mantle pigmentation pattern (Roszkowski 1914; Falniowski 1980a, b). For example, the mantle pigmentation pattern between isolates of Cuban *Pseudosuccinea columella* susceptible and resistant to *Fasciola hepatica* infection was analyzed by Gutiérrez et al. (2003). These authors showed that the distribution pattern of the spots was clearly different between resistant and susceptible isolates, whereas some other researchers questioned the usefulness of this character as a diagnostic one (Schniebs et al. 2011).

The mantle cavity in the pulmonate snails has been transformed into an unpaired "lung," an adaptation to air-breathing. In all lymnaeids, except *Pseudobulinus*, it is located on the right side of the body and opens via the *pneumostome* (*mantle pore*). The atmospheric air is usually used as the oxygen source, however, in some situations, it is not easy for pond snails (for example, for those living in large lakes of tectonic origin, at considerable depths) to rise to the water surface for breathing. A classic example of this is the Lake Geneva lymnaeids, living at depths of 40–200 m (Russell-Hunter 1964). Under such circumstances, the mantle cavity becomes filled by water and thus acts similarly to the "adaptive gill" (*pseudobranch*) of some other Hygrophila belonging to Acroloxidae, Bulinidae, and Planorbidae (Starobogatov 1967). The cutaneous respiration is another—not alternative to the described above—mechanism of oxygen uptake (Russell-Hunter 1978). This acts especially effectively in species (like those of *Myxas*) with the mantle reflected over the shell (Stadnichenko 2004).

The lung also functions as a hydrostatic apparatus. A change in the volume of air in the lung leads to a change in the buoyancy of the animal, and also allows it to hold the shell in a certain position in relation to the substrate (Checa and Jiménez-Jiménez, 1997).

Figure 2.11 depicts the general topography of the internal organs of a lymnaeid mollusc.

The alimentary system starts from a relatively small mouth, which is located ventrally in the frontal part of the head. The mouth leads to the oral cavity containing brown-colored *jaws* made of conchiolin. Their function is to grind up large pieces of food. The *buccal mass*, or *pharynx* (see Fig. 2.11), incorporates the *radula*, which is a protrusile tonguelike structure serving as a device for getting the food by means of substrate scraping (see Ivanov 1990 for a detailed review on the functional morphology and evolution of the molluscan radula). The radula tears or rasps food particles and then pulls them into the mouth. Solem (1974) describes in detail the biomechanics of this process in herbivorous pulmonate snails. The radula is mounted on a cartilaginous *odontophore* and consists of hundreds to thousands of minute teeth.

Hubendick (1978, p. 13) described the lymnaeid radula (like the radulas of other Hygrophila) as being "of a primitive type with numerous small and comparatively simple teeth of fairly similar shape in each cross row" (Fig. 2.12). The teeth of the lymnaeid radula form transversal rows, up to several dozens, each of them has a more or less identical structure. The *central* (*rachidian*) cusp is situated medially and divides a row into two virtually symmetrical halves containing approximately the same number of individual teeth. The rachidial tooth (see Fig. 2.12) is smaller than the *lateral teeth* located on either side of the latter. The *marginal teeth* are situated externally of the lateral ones. Some authors (e.g., Stadnichenko 2004) distinguish the third group of teeth, the *interstitial* ones, positioned between the lateral and marginal ones.

The rachidian teeth bear a single cusp or two cusps (*monocuspid* or *bicuspid* tooth), while the lateral ones have either two or three (*tricuspid*) well-developed cusps. Typically, these cusps are of uneven size, with the central cusp being the most

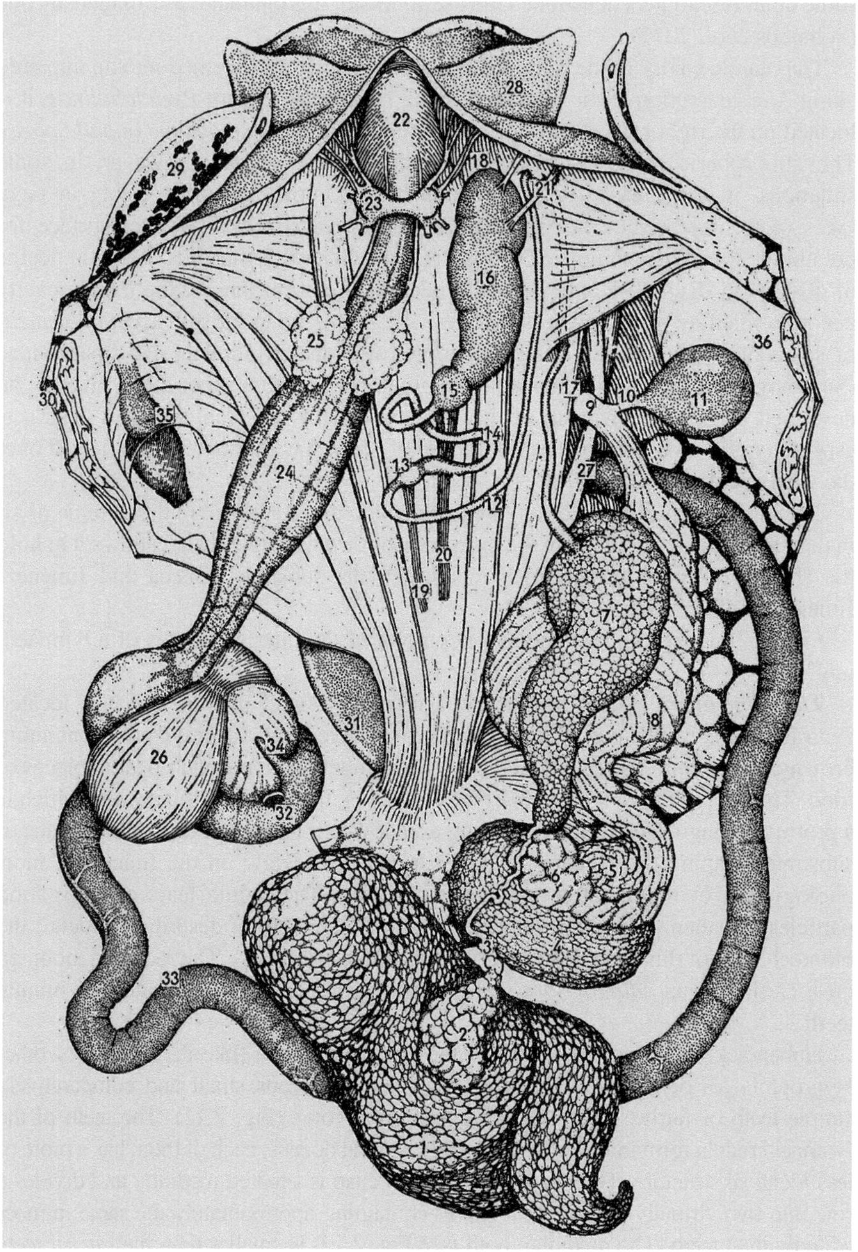

Fig. 2.11 Internal anatomy of *Ampullaceana balthica* (Poland, vicinities of Poznań). After Jackiewicz (1954), slightly modified. Explanation of numbers: 1 – liver; 2 – hermaphroditic gland; 3 – hermaphroditic duct with the seminal vesicles; 4 – albumen gland; 5 – nidamental gland; 6 – uterus; 7 – prostate; 8 – pyriform body; 9 – vagina; 10 – spermathecal duct; 11 – spermatheca (*bursa copulatrix*); 12 – *vas deferens*; 13 – bulbous termination of the penis sheath; 14 – penis sheath; 15 – bulbous termination of the praeputium; 16 – praeputium; 17 – female gonopore; 18 – male gonopore; 19 – retractor of the penis sheath; 20 – retractor of the praeputium; 21 – protractor; 22 – buccal mass (*pharynx*); 23 – cerebral ganglia; 24 – esophagus; 25 – salivary

developed. The marginal teeth cusps are weakly developed, their number varies from 1 to 6, rarely more.

Each individual radula may be described by means of a simple formula indicating the number of lateral and marginal teeth and the number of cusps which are bearing. Jackiewicz (1998, p. 15) provided an example of a radula studied in a specimen of *Stagnicola corvus*. "The length and the width of this radula are 4.27 mm and 1.93 mm, respectively. There are 87 longitudinal and 118 transversal rows. The shell height and width were 31.83 and 14.00 mm, respectively. Radula size is directly proportional to the specimen size." The radular formula of this specimen was (where M—marginal teeth; L—lateral teeth):

$$\begin{array}{ccccc} (M) & (L) & (C) & (L) & (M) \\ \underline{32} & \underline{11} & 1 & \underline{11} & \underline{32} \\ 3\text{-}5 & 2\text{-}3 & & 2\text{-}3 & 3\text{-}5 \end{array} \quad = \quad 43\text{-C-}43, \ \text{where}$$

the numbers in the middle line are equal to the numbers of teeth in a half-row, and the numbers in the lower line correspond to the number of cusps.

The morphology of the radula is subject to some ontogenetic changes; some abnormalities of the radular structure are sometimes occurring (Jackiewicz 1998; Stadnichenko 2004).

The significance of the lymnaeid radular morphology for the classification of the family had long been considered low if not negligible (Annandale and Rao 1925; Roszkowski 1929; Hubendick 1951; Jackiewicz 1998; but see Yakchali and Deilamy 2012). In the late 1950s, Berrie (1959) demonstrated biometrically that the radular morphology may vary intraspecifically, and two separate populations of the same species may be so different in this respect that "they could legitimately be established as separate subspecies" (Berrie 1959, p. 403). On the other hand, the intraspecific differences are sometimes not prominent and not always stable, and, in some studied cases, the revealed distinctions in the radular morphology do not fit the branching pattern of a phylogenetic tree (e. g., see Vinarski et al. 2022). At the family level, the radular morphology of the lymnaeids has been characterized as "strikingly uniform" (Hubendick 1951, p. 38). The lack of the taxonomic signal in the radular morphology was reported in other groups of Pulmonata (e.g., terrestrial family Helicidae; see Schileyko 1972).

Three main parts of the alimentary system lie posterior to the pharynx—esophagus, stomach, and intestine. The *salivary glands* ducts open laterally into the pharynx, somewhat anterior to the esophagus. The *esophagus* is a muscular tube-like organ, usually rather long and dark-pigmented. The lymnaeid *stomach* is a relatively large and well-muscled organ, typically of rounded shape, consisting of three parts (crop, stomach s.str., and gizzard). The gizzard has a blind pocket called

Fig. 2.11 (continued) glands; 26 – stomach; 27 – anal opening; 28 – tentacle; 29 – mantle collar; 30 – mantle over the mantle cavity; 31 – foot; 32 – liver duct outlet; 33 – intestine; 34 – cecum; 35 – heart; 36 – lung

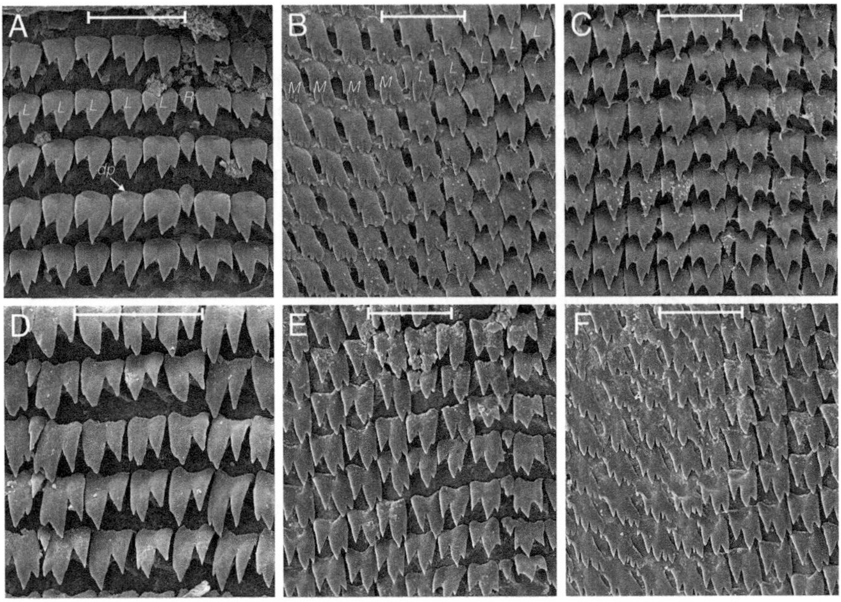

Fig. 2.12 Fragments of radulae of *Tibetoradix kruglovi* (**a–c**) and *T. imitator* (**d–f**). Abbreviations: *dp* – distal part of tooth, *L* – lateral tooth, *M* – marginal tooth, *R* – rachidian (central) tooth. Scale bars = 40 μm. Photos: Ivan O. Nekhaev

the *cecum*. The stomach is followed by the *intestine*, a long and thin duct of almost the same length as the esophagus. The posterior end of the intestine forms the *rectum*; the *anal opening* is located on the right side of the soft body, close to the head (see Fig. 2.11).

A large well-developed digestive gland called the *hepatopancreas*, or, expressing less technically, the liver, occupies the upper part of the visceral mass located within the upper whorls of the shell. This is the largest internal organ in most (if not all) snails. The lymnaeid hepatopancreas is bilobate, dark-brown colored; its ducts open into the gizzard. The main function of the hepatopancreas is the intracellular digestion of fats and proteins, whereas the digestion of carbohydrates takes place in the gizzard (Kruglov 2005).

The lymnaeids are potentially omnivorous snails, but vegetative food dominates in their diet. Larger species (like *L. stagnalis*) may feed on dead corpses of other animals, including representatives of their own species (Baker 1911; Frömming 1956; Tsikhon-Lukanina 1987; Stadnichenko 2004). The instances of their feeding on egg masses of other snails are known (Frömming 1956). In some representatives of the family, the detritivory becomes the dominant, if not obligate, feeding mode. A typical lymnaeid snail is a grazer, scraping its food from the substrate using the radula.

The circulatory system is open, of the lacunar type. It consists of the heart, blood vessels, the sinus system, and the lacunae. There is a well-developed arterial system, consisting of vessels with developed walls (aorta, arteries), but venous vessels, as a rule, lack their own walls. The *heart* is a relatively simple bicameral organ serving as a blood pump. It is located on the left side of the soft body between the lung and the kidney and is surrounded by the *pericardium* filled with fluid. The heart *ventricle* is followed by the aorta which further divides into two arteries, the anterior and the posterior ones, both subdivide into arteries of the second order. The arteries bring the *hemolymph* (the body fluid analogous to blood of vertebrates), through the sinus system, to the various internal organs. Next, the hemolymph returns to the lung through the weakly branched pulmonary veins. In the lung, it is oxygenated and returns to the heart.

The hemolymph of the Lymnaeidae has two main components—the hemolymph plasma, containing the oxygen-carrying protein hemocyanin, and the hemocytes (amoebocytes). The latter are responsible for diverse aspects of molluscan life such as immune response, biomineralization, shell formation, and regeneration processes. Recently, Machałowski and Jesionowski (2021) published a detailed review article on the mollusc hemolymph and its constituents, which we recommend as a further reading on this topic.

The central organ of ***the excretory system*** is the *kidney*, or *nephridium*, which is described as a large unpaired sac-like organ of spongy structure (Jackiewicz 1998). This is situated together with the lung and the pericardium in the last whorl of the visceral mass. The kidney is tripartite, its duct (ureter) opens as a pore on the right side of the snail's body, near the pneumostome and opposite the anal opening.

The nervous system is of scattered-nodular type, and it tends to concentrate into a cerebral ring around the esophagus, which consists of separate ganglia. The ganglia are connected by transverse *commissures* and longitudinal *connectives*, the latter being relatively short, as a result of which the ganglia are located close to each other. The tendency of the fusion of these ganglia leading to the further concentration of the nervous system is observed among the Hygrophila (Hubendick 1978). In the Lymnaeidae, the central nervous system consists of paired brain or *cerebral ganglia*, foot or *pedal ganglia*, lateral or *pleural ganglia*, *parietal ganglia* and *esophageal* or *buccal* ganglia as well as solitary *visceral (or abdominal) ganglion*. The large cerebral ganglia, located just behind the pharynx on the dorsal side of the body, innervate the front of the head, the copulatory apparatus, and sensory organs. The pleural ganglia innervate the mantle, the pedal ganglia innervate the leg muscles, the pallial ganglia innervate the edge of the mantle, osphradium, pneumatic bridge, etc. Finally, the abdominal ganglion, which occupies the extreme posterior position in the cerebral ring, innervates almost all organs of the visceral sac. A pair of very small buccal ganglia lies under the esophagus and innervates the pharynx, esophagus, and salivary glands.

The sensory organs of the Lymnaeidae are represented by sensitive cells scattered over the body surface and especially numerous on the tentacles, along the edges of the mouth, mantle, and foot. These cells perform tactile, olfactory, and gustatory functions. The head tentacles, of triangular shape, are a specialized organ of touch and smell. In the mantle cavity lies a specialized organ of chemical sense—*osphradium*, which determines the composition of water supplied to the lung, perceiving fluctuations in osmotic pressure and the dissolved oxygen concentration. There are paired organs of balance and seismic and audio reception—*statocysts* (*otocysts*), located in the front of the foot in the immediate vicinity of the pedal ganglia. They are hollow vesicles lined with epithelial ciliary cells that perform a receptor function. The cavity of the statocyst is filled with liquid statolymph, in are located *statoconia*—small auditory stones, which, when the equilibrium is disturbed, change their position and press on the receptor cells. In larger species, like *Lymnaea stagnalis*, the number of statoconia can reach 700. The organs of vision are represented by a pair of eyes located at the base of the tentacles. The eyes are responsible for photoreception, but they are not able to create visual images; therefore, the lymnaeid snails cannot distinguish the shape of objects. The dermal (non-ocular) photoreceptor system was described in *Lymnaea stagnalis* (Chono et al. 2002).

At the classical, i.e., pre-molecular, stage of the lymnaeid taxonomy, the ***reproductive system*** was considered the most important source of the taxonomic and phylogenetic signal in the family (Baker 1911; Hubendick 1951; Walter 1969; Paraense 1976, 1982, 1984, 1994, 1995; Kruglov and Starobogatov 1981; Jackiewicz 1998; Kruglov 2005; Pointier 2015; Pointier and Vázquez 2020). Though the recent "integrative" studies have shown that the significance of the reproductive anatomy is not as prominent as it was thought in the last century, the features of the lymnaeid genitals remain an important tool for species identification, at least in some genera (i.e., *Stagnicola*, *Galba*).

Figure 2.13a illustrates the general morphology of the lymnaeid reproductive system.

All freshwater pulmonates, not excluding the lymnaeids, are dioecious snails, and their reproductive system is based on a common *Bauplan* (Duncan 1960; Starobogatov 1967; Hubendick 1978; Jordaens et al. 2007).

An unpaired hermaphroditic organ—the gonad (*hermaphroditic gland*, *ovotestis*) is located apically in the visceral sac and in most cases is immersed in the tissues of the hepatopancreas. The gonad is divided into separate lobes—*diverticulae* opening into the gonadial ducts, and the latter into the *hermaphroditic duct*, which forms more or less numerous swellings—seminal vesicles (see Fig. 2.13). At the end of the hermaphroditic duct, the reproductive system is divided into the male and female ducts, emerging from a small formation called the *quadrivium*. The hermaphroditic duct and the duct of the albuminoid gland open into the quadrivium, which forms a special extension—the fertilizing pocket.

The proximal part of the male genital duct is represented by the *prostate* (see Fig. 2.13c)—a sac-like gland that produces seminal fluid. The lancine snails have two prostates of different structure. There may be one, two, or many folds inside the prostate, formed by the prostate wall, usually in its widened part. In some genera (*Omphiscola*, *Pseudosuccinea*), the prostate lacks any folds (Paraense 1983; Kruglov and Starobogatov 1993; Jackiewicz 1998). The prostate is followed by the *vas deferens*, which, near the female genital opening, plunges into the integument of the body and reaches the male genital opening, where it leaves the muscle tissue and opens into the *copulatory apparatus* (see Fig. 2.13b), the morphology of which is relatively uniform within the family (but see Kruglov and Starobogatov 1981). The copulatory apparatus consists of a tubular *penis* (see Fig. 2.13d,e), embedded into the *penis sheath*, and a *praeputium* opening outward with the male genital opening. In the normal state,

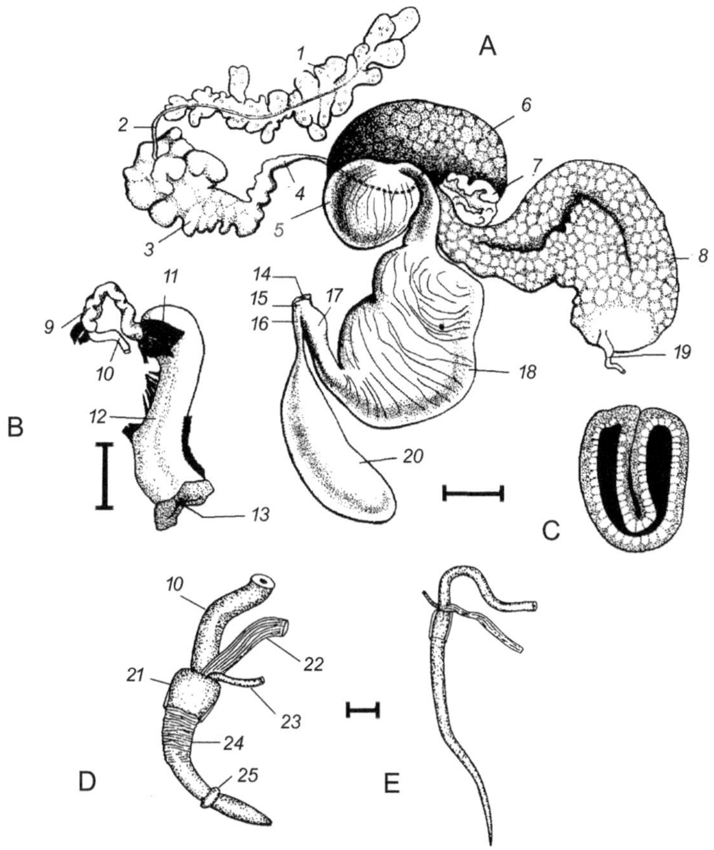

Fig. 2.13 The reproductive anatomy of the lymnaeid snails: The distal part of the reproductive system (**a**), the copulatory apparatus (**b**), and cross-section through the prostate (**c**) of *Ampullaceana fontinalis* (**a–c**), and the penises of *Lymnaea stagnalis* (**d**) and *Myxas glutinosa* (**e**). After Beriozkina and Starobogatov (1988) and Andreeva et al. (2010), slightly modified. Scale bars 0.5 mm (D, E), 1 mm (A–C). Explanation of numbers: 1 – hermaphroditic gland; 2 – the distal part of the hermaphroditic duct; 3 – seminal vesicles; 4 – the proximal part of the hermaphroditic duct; 5 – nidamental gland; 6 – albumen gland; 7 – the labyrinth of the oviduct; 8 – prostate; 9 – penis sheath; 10 – the distal part of the spermathecal duct; 11 – the praeputium retractor; 12 – praeputium; 13 – male gonopore; 14 – female gonopore; 15 – vagina; 16 – spermathecal duct; 17 – provagina; 18 – uterus; 19 – the proximal part of the spermathecal duct; 20 – spermatheca (*bursa copulatrix*); 21 – the base of the penis; 22 – the penis sheath retractor; 23 – the penial nerve; 24 – penis; 25 – the fixative swelling (present not in all species)

the copulatory apparatus is immersed inside the soft body, and it turns outward during copulation. Two annular folds—*velum* (outer) and *sarcobellum* (inner)—are situated on the border of the penis sheath and the praeuputium. Various lymnaeid genera differ from each other in relation to the degree of development of these folds; sometimes (like in the genus *Omphiscola*) the velum can completely disappear (Kruglov and Starobogatov 1981; Jackiewicz 1998).

The most peculiar penial morphology is observed in a small group of lymnaeids separated in a genus of its own, *Aenigmomphiscola* Kruglov et Starobogatov, 1981. Two extant species of this genus are endemic to Russia and Kazakhstan (Vinarski and Grebennikov 2012). In these snails, the velum is asymmetrically enlarged and forms a so-called *praeputial organ* (Kruglov and Starobogatov 1981; Vinarski et al. 2011). According to Kruglov (2005), it plays the role of a sucker that holds the individuals during copulation. A praeputial organ performing a similar function is also present in species of the genus *Planorbarius* Dumeril, 1806 and *Segmentina* Fleming, 1818 of the family Planorbidae (Stadnichenko 1990; Maksimova and Yakovleva 1991). In some groups that do not have a praeputial organ, the fixative swelling, or ring, develops on the penis (Fig. 2.13d).

In the proximal part of the penis sheath, there are glandular deepenings called the *proximal chambers*. The penis sheath becomes noticeably swollen at the proximal end when the chambers are well-developed. Several muscle retractors approach the copulatory apparatus from the outside.

The female reproductive tract in the Lymnaeidae is represented by the *oviduct*, the enlarged part of which is called the *uterus*. The uterus begins after the point of confluence of the nidamental gland duct into the oviduct and passes into the *provagina*, which opens outward (see Fig. 2.13a). The proximal part of the provagina, near the female genital opening, receives the seminal duct.

The *chromosomal diversity* within the family Lymnaeidae has been studied intensively during the past 50–60 years (Inaba 1969; Patterson and Burch 1978; Garbar and Korniushin 2003; Garbar et al. 2004). This diversity appeared relatively low as compared to that found in some other Hygrophila families. Most lymnaeid species possess 17 or 18 pairs of chromosomes, and the extant subfamilies can be characterized using the chromosome number. Thus, representatives of the subfamily Amphipepleinae (= Radicinae) typically have 17 (sometimes 16) chromosome pairs, whereas in the Lymnaeidae the typical haploid number is 18 (some species have $n = 19$) [Vinarski 2013]. The low chromosome number is characteristic for the most Hygrophila families (Burch 1960; Burch et al. 1960; Patterson and Burch 1978; Thiriot-Quiévreux 2003; Garbar et al. 2004) and thus can be considered a plesiomorphic character (Nordsieck 1992). The exception is some polyploid species of the families Bulinidae and Planorbidae (the genera *Ancylus*, *Bulinus*, *Gyraulus*). In these groups, tetraploid, hexaploid and even octoploid species are known (Burch et al. 1960; Goldman et al. 1983; Burch and Jung 1993).

Though the chromosome number as such cannot be used as a unique taxonomic characteristic of a genus or species in the Lymnaeidae, some interspecific differences in other karyological characters can be reported. For example, the European species of the genus *Stagnicola* can be distinguished on the basis of their NF (= *nombre fundamentale*), i.e., the number of the chromosome arms. Thus, *Stagnicola palustris* and *S. corvus* share the same chromosome number ($n = 18$) but their NFs are different, equal to 72 and 60, respectively (Garbar et al. 2004). Some morphological features of the chromosomes allow one to distinguish two conchologically indistinguishable genera, *Stagnicola* and *Catascopia* (= *Ladislavella* of the current system) [Garbar et al. 2004].

The life cycles of the Lymnaeidae are relatively uniform. These molluscs are typical *r*-strategists characterized by relatively short life spans, high fecundity, increased rate of food consumption, and fast generation turnover. Under natural conditions, the life cycle duration in the family varies between one and 2 years (Russell-Hunter 1978; Beriozkina and Starobogatov 1988). The number of generations per year is also variable and depends on local conditions.

The dioecy of the Lymnaeidae and their ability to reproduce by self-fertilization have often been considered as an adaptation to life in non-permanent waterbodies as well as to rapid colonization of novel habitats (Boss 1978; Beriozkina and Starobogatov 1992; Heller 1993; Jarne et al. 1993; Jordaens et al. 2007; Vinarski et al. 2021). Despite the simultaneous hermaphroditism of freshwater Pulmonata, self-fertilization (*autogamy*) is not obligatory for these animals (Jarne et al. 1993). In most cases, cross-fertilization (*allogamy*) prevails, in which one individual plays the role of a "male" and the other as a "female," but the possibility of autogamy remains though its frequency varies among different species (Beriozkina and Starobogatov 1988; Jordaens et al. 2007). The ability to self-fertilize also determines the phenomenon of *aphally*—the absence of the copulatory apparatus, which is known in some representatives of Hygrophila (Harutyunova 1977; Starobogatov et al. 1989).

In the vast majority of aquatic pulmonates, embryonic development occurs outside the body of the parent. In the oviduct, a fertilized egg is covered by a series of protective membranes that form an egg capsule (Beriozkina 2002). Then, individual egg capsules combine under a common membrane secreted by the glands of the distal part of the oviduct, and form a separate clutch, called *an egg mass* (or *egg cluster*, or *egg capsule*, or *syncapsule*). The number of egg capsules in an egg mass varies considerably and, as a rule, depends on the body size of the molluscs. In large species of Lymnaeidae, such as *Lymnaea stagnalis*, the number of egg capsules per egg mass can exceed 100 (Beriozkina and Starobogatov 1988; Kruglov and Starobogatov 1992), and in some representatives of the genus *Radix* it reaches 250 (Prozorova 1992). On the other hand, species of the genus *Galba* are characterized by the small size of egg masses and, as a consequence, the small number of egg capsules (Kruglov and Starobogatov 1992). Nutrients are concentrated outside the egg capsules, usually in the protein matrix of the egg mass, which is eaten by juveniles before leaving the syncapsule (Beriozkina 2002).

The lymnaeid egg masses demonstrate a marked diversity in their size and structure within the family (Fig. 2.14). The data on the morphology of egg masses of different representatives of the Lymnaeidae can be found in the literature (Nekrasov 1927, Piechocki 1979; Beriozkina and Starobogatov 1981, 1988; Kruglov and Starobogatov 1985, 1992; Prozorova 1992; Jackiewicz 1998). Some authors have considered them species-specific. In the 1980s, there were attempts in the USSR to use the structure of the egg masses as a diagnostic tool for the identification of the lymnaeid (and other freshwater pulmonate) species of Eastern Europe (Beriozkina and Starobogatov 1981, 1988). A diagnostic key for the lymnaeids based on size and some qualitative features of the egg masses was even proposed (Beriozkina & Starobogatov 1988, p. 232–234).

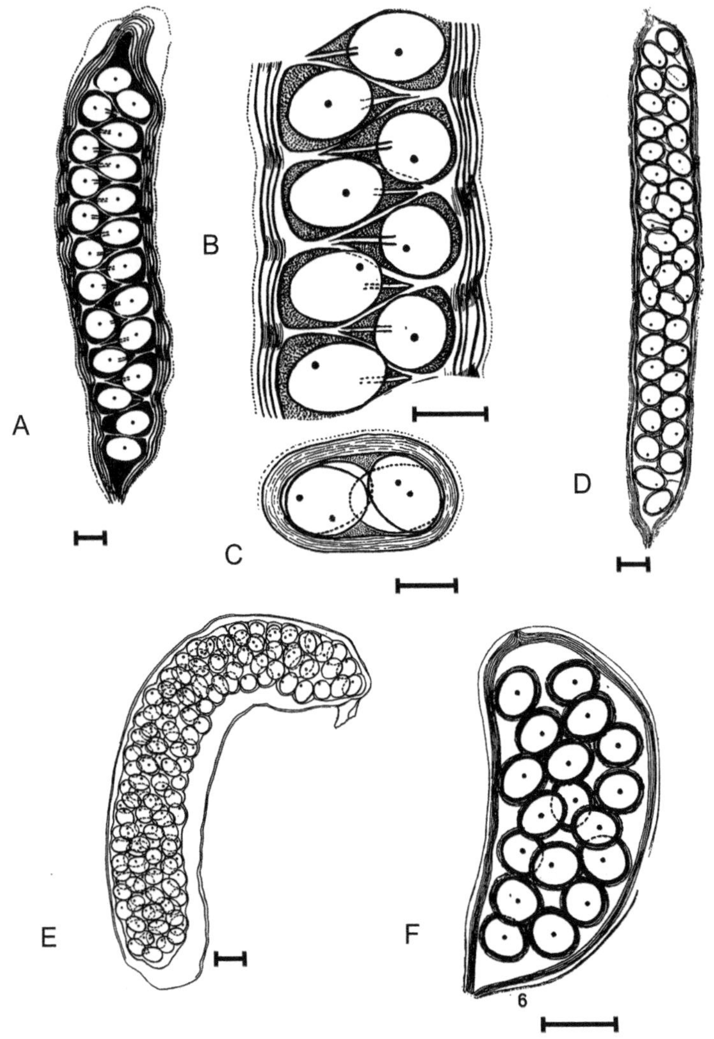

Fig. 2.14 The structure of the egg masses in *Myxas glutinosa* (**a–c**), *Lymnaea stagnalis* (**d**), *Radix auricularia* (**e**), and *Stagnicola palustris* (**f**). After Beriozkina and Starobogatov (1988), slightly modified. Scale bars 1 mm. (**a, d–f**) General views. (**b**). The middle region of an egg mass (enlarged). (**c**). Cross-section through an egg mass

However, the latest research has revealed that the idea of the species specificity of the egg masses morphology was incorrect. In one of the latest publications, Beriozkina (2019) explicitly revised her earlier views and stated that "the high degree of variability [of egg masses of pulmonate snails] questions the use of their morphology as a species criterion. At the same time, this morphology retains its significance as a characteristic of the subgenera" (Beriozkina 2019, p. 40).

2.5 Elements of Biogeography and Ecology

The modern range of the family can be classified as "subcosmopolitan." The lymnaeids are distributed in waterbodies of all continents, except for the Antarctic (Starobogatov 1970; Banarescu 1990; Strong et al. 2008; Vinarski et al. 2019). The most species-rich faunas of this family are those of North America, Central Asia, South-East Asia, and Europe, whereas Africa and South America maintain a relatively low species richness of the Lymnaeidae. In the Northern hemisphere, the family is represented far north of the Polar circle, though these snails are not found in many Arctic islands (Vinarski et al. 2021). It has been thought that the Lymnaeidae obey the global species richness gradient, i.e. the pattern of decline in species richness toward the poles (Hubendick 1962; Starobogatov 1970; Vinarski et al. 2012), however, actually, the form of the relationship between the number of lymnaeid species and the latitude is not linear. In the equatorial parts of Africa and South America, the species richness of the family is considerably lower than in the temperate latitudes. It is also characteristic of tropical Asia. In the Northern hemisphere, the Lymnaeidae tend to concentrate in the latitudes lying between 30 and 60 degrees. In the Southern hemisphere, the highest number of lymnaeid species is observed in Australia.

The altitudinal range of the family is rather wide. The lymnaeids are found in some of the deepest continental depressions lying below the sea level (i.e., Turfan Depression in Western China; Vinarski, pers. observations). The highest occurrences of the family are known in the Himalaya and Tibet plateau. The genus *Tibetoradix* Bolotov, Vinarski & Aksenova, 2018, is a unique example of a freshwater mollusc genus adapted to live under high-altitude conditions. In Tibet Plateau, the species of this genus are distributed vertically between 3544 and 4749 m a.s.l. (Vinarski et al. 2022). Another example is *Galba cousini*, endemic to the northern part of the Andes Cordillera (Venezuela, Colombia, and Ecuador), which occurs at altitudes ranging from 2500 to 4200 m a.s.l. (Pointier 2015).

The lymnaeids (and other freshwater pulmonates) use different mechanisms of their dispersal, ranging from the "natural" vectors (birds, fish, large flying insects) to dissemination with humans' transport vehicles and aquarium trade (Malone 1965; Starobogatov 1970; Boag 1986; Madsen and Frandsen 1989; Brown 2007; Kappes and Haase 2012; Ng et al. 2016; Patoka et al. 2017). Many species use rivers and streams for their passive dispersal, drifting downstream (Starobogatov 1970; Shikov 1977). Large rivers flowing in latitudinal direction have long been thought to serve as the main vectors of dispersal of continental molluscs in the south-north direction (Vinarski et al. 2012, 2021).

Some representatives of the family are effective invaders, distributed today far outside their native range. The most widespread of them is *Pseudosuccinea columella*, a North American species, which is currently distributed in Europe, South America, South Africa, Australia, New Zealand, and some other countries (Lounnas et al. 2017; Vinarski et al. 2019). *Lymnaea stagnalis*, a Holarctic species, whose range lies mainly within the Northern Hemisphere temperate zone, has been introduced to Australia, New Zealand, and Cameroon (Hubendick 1951; Climo and Pullan 1972; Tchakonté et al. 2014). *Radix auricularia*, another Eurasian species, is known now from North America and New Zealand (Hubendick 1951; Burch 1989; Mills et al. 1993). This list of invasive lymnaeids is far from being complete.

The main factors limiting the dispersal of freshwater pulmonates are the chemical composition of water, temperature, and oxygen regimes of waterbodies (Aho 1978a, b; Beriozkina and Starobogatov 1988; Pip 1993; Økland 1990; Økland and Økland 1997; Dillon 2000). One of the most important chemical elements for molluscs is calcium. A number of studies returned strong positive correlations between the abundance and/or diversity of freshwater molluscs and calcium concentration (Dillon 2000). Calcium ions absorbed by animals from the water in the form of carbonate salts are used to build the shell (Russell-Hunter 1964; Økland 1990). Many lymnaeid species are calciphiles and do not occur in habitats with a low content of this element (below 20 mg/l). According to Briers (2003), such species as *Lymnaea stagnalis*, *Radix auricularia*, and *Myxas glutinosa* belong to this ecological group.

It has been experimentally established that a decrease in the concentration of calcium ions to 1.04 mg/l leads to its negative balance in the body and the death of snails (Kruglov 2005). The number of lymnaeid species in waterbodies poor in calcium is reduced in comparison with habitats where there is no deficiency of this element (Russell-Hunter 1964; Aho 1978a, b; Økland 1990).

The optimal hydrogen-ion concentration (pH) values for freshwater Pulmonata are 6.9–8.0. When it decreases to 6.0–6.5, it results in in vivo destruction of the shell material and high embryonic mortality (Kruglov 2005). Under pH $= 4.7$–4.9, significantly lower adult growth rates and reduced gross fecundity are observed (Hunter 1990). Freshwater molluscs are totally absent from waterbodies with pH < 4.0 (Salazkin 1969). At the same time, the reactions of molluscs to the effect of this factor are often species- and genus-specific (Salazkin 1969; Økland 1990; Kruglov 2005). For example, the gelatinous pond snail, *Myxas glutinosa*, exhibits low resistance to water acidification which was revealed under experimental conditions (Berezina 2001). On the other hand, such species as *Lymnaea stagnalis* and *Galba truncatula* in Norway were found to dwell in a wide pH range, from 4.4 to 8.8. Both species demonstrated increased frequency of occurrence in waterbodies with pH values $= 7.4$–8.8 (Økland 1990).

The presence of oxygen dissolved in water is necessary for the normal embryonic development of the Lymnaeidae (Stadnichenko 2006), therefore, these snails cannot survive anoxic conditions, despite their possibility of breathing atmospheric air.

The Lymnaeidae is a group of ecologically plastic snails able to dwell in various aquatic environments, whose range embraces almost the entire spectrum of waterbody types (large lakes and ponds, reservoirs, rivers and streams, brooks, ephemeral pools, etc.). Certain representatives of the genera *Galba* and *Orientogalba* are, in fact, amphibiotic snails, living on various wet surfaces like shores of rivers and streams, moistened stones near waterfalls, and other similar habitats. On the other hand, only a few lymnaeid species are known to inhabit the giant ancient lakes (Ohrid, Baikal, Lugu) [Schütt 1974; Albrecht et al. 2008;

Vinarski 2018; Wiese et al. 2020], and generally the Lymnaeidae do not form in such waterbodies endemic species flocks as it is characteristic for other families of freshwater snails—Planorbidae, Acroloxidae, Thiaridae, Viviparidae (Boss 1978). Possibly the only freshwater environment not conquered by lymnaeid snails is the subterranean waterbodies, including the cave streams and lakes.

Lymnaeids overwinter buried in the ground, often among the roots of aquatic plants. The activity of molluscs in winter is sharply reduced, they, as a rule, stop feeding, the intensity of gas exchange also decreases, and their lung is filled with water. Such a state resembling anabiosis is regarded as the "hibernation" (Ivanov 1940), "dormancy" (Storey 1972), or, to use a term coined by Boss (1974), "oblomovism" of molluscs. If the waterbody does not freeze to the bottom, then some snails can remain motile and feed under the ice (Russell-Hunter 1964; Stadnichenko 2006).

In relatively large waterbodies, the lymnaeids, similar to other freshwater pulmonates, tend to concentrate in the shallow zone near the shore, well warmed up and rich with vegetation and light. However, in large lakes, these snails may occur at depths 250–280 m (Russell-Hunter 1978). Most representatives of the family are phytophilous snails though potentially may live on different substrata (submerged stones, deadwood, plastic and metal objects of anthropogenic origin, and so on). The lymnaeids inhabiting non-permanent habitats are adapted to summer aestivation that helps them survive when their waterbody is dried up. The snails are waiting out the dry period in moist soil or among dead leaves and stems of aquatic plants (Cheatum 1934; McMahon 1983; Alyakrinskaya 2004). According to different authors, the duration of stay in an anhydrous environment varies significantly among freshwater pulmonates, ranging from 1 month to 390 days (Alyakrinskaya 2004). Despite these adaptations, the survival rate of molluscs during the unfavorable dry season is very low.

Though this family is commonly classified among freshwater molluscs, some Lymnaeidae demonstrate a considerable salinity tolerance. A classical example is the lymnaeids found in brackish water environments, for instance, in the Baltic Sea, under salinity around 2‰ or even higher (Clessin 1878; Dregol'skaya et al. 1988). In the Issyk-Kul Lake in Central Asia (Kyrgyzstan), the lymnaeid *Radix obliquata* lives under salinity 5.8–5.9‰ (Pavlova 1983). In Northern Kazakhstan, several representatives of the family were found in mineralized lakes with salinity equal to 4.95‰ (Andreeva et al. 2010). According to Taktakishvili (1967), *Valenciennius*, the giant Neogene lymnaeid of south-east Europe (see above), was essentially a marine mollusc adapted to rather deep-sea conditions. Taktakishvili argued that the unique morphological characteristics of this genus, including their gigantism and limpet-like shape, evolved as an adaptation to living under increased salinity and at considerable depths.

The family demonstrate some instances of profound thermal adaptation, i.e. living in the hot waterbodies (lakes, ponds, streams) situated in geothermal sites scattered throughout the world—in North America (Alaska, Yellowstone Park), Iceland, Kamchatka Peninsula, Tibet Plateau, Pamir, etc. (Starmühlner 1957; von Oheimb et al. 2011; Aksenova et al. 2016, 2017; Vinarski et al. 2023). Hydrothermal habitats are characterized by high temperature (typically $>$20–25 ° C) that is maintained the year round, without drastic seasonal fluctuations. From the biological point of view, they represent a very peculiar environment, and animals need to develop special adaptations in order to be able to conquer it (Mitchell 1974; Bolotov et al. 2012). Formation of ecological ("thermal") races in such habitats is a common outcome of their colonization by various lymnaeid species (Vinarski et al. 2023). In earlier times, such races often were described as separate species (Boubee 1833; Kruglov and Starobogatov 1989), but the recent molecular studies show that, in most cases, they should be classified merely as ecotypes of widespread species (Aksenova et al. 2017). In total, 17 species of the family representing several genera and all extant subfamilies were found to form populations in geothermal sites around the world (Table 2.1).

The practical significance of the family is multifarious. The public health research is focused on the lymnaeids as intermediate hosts of trematode larvae, many species of which cause serious diseases in humans, domestic and wild animals. The classical instance of this is the host–parasite relationships between *Galba truncatula* and *Fasciola hepatica*, the main causative agent of fasciolosis. Except for *G. truncatula*, a number of other lymnaeid species are found to transmit the larvae of *F. hepatica* and other fasciolids, however, the suitability of different species for the larvae is unequal (Bargues and Mas-Coma 2005; Kruglov 2005; Malcicka 2015; Vázquez et al. 2018). Such opportunism in host choice is characteristic of many other trematode species. For example, the obligate hosts of the trematode *Echinostoma revolutum* Looss, 1899 are molluscs of two families (Lymnaeidae, Planorbidae), and additional hosts are several other families of freshwater molluscs (Bithyniidae, Viviparidae, and some families of bivalves) and even dragonfly larvae (Sudarikov et al. 2002).

And, vice versa, a single lymnaeid species may serve as a host for many trematode larvae belonging to different genera and families (Stadnichenko 2004; Kruglov 2005; Żbikowska 2011). For example, in the northern and central parts of Poland *Lymnaea stagnalis* hosts 14 species of Trematoda (Żbikowska 2007), in Ukraine, this snail harbors 21 species of parasites (Stadnichenko 2004), whereas the full spectrum of trematode species parasitizing this snail species throughout its range is 35 (Kruglov 2005). It is assumed (Faltýnková et al. 2008) that the most important factors determining the species richness of trematode larvae in molluscs are the host body size (the larger the snail, the greater the number of trematode species infect a given host species) and the density of mollusc populations. The number of cercariae produced by a mollusk directly depends on the size of the animal, temperature, season, and other factors. According to Shigin (1971), a large individual of *Lymnaea stagnalis* is capable of producing up to 13.5 million *Diplostomum spathaceum* (Rudolphi, 1819) cercariae during its life.

Many species of lymnaeids play a significant role in the aquatic food webs, being the destructors of vegetation and detritus and serving as the food item for many species of fish, waterfowl, and predatory macroinvertebrates. These snails are used as biological indicators of water pollution and ecosystem health. The shells of fossil lymnaeids are useful for the purposes of biostratigraphy.

The great pond snail, *Lymnaea stagnalis*, is actively used as a standard model object for the studies in neurophysiology, ecotoxicology, developmental biology, biomineralization, and some other branches of science (Meshcheryakov 1990;

Table 2.1 A list of lymnaeid species inhabiting the geothermal sites in various continents

Species	Country, region	Source of data
Ampullaceana balthica (Linnaeus, 1758)	Iceland, French Pyrenees	Bolotov et al. 2017
A. lagotis (Schrank, 1803)	Northeast European Russia; East Kazakhstan	Vinarski et al. 2023
Austropeplea tomentosa (L. Pfeiffer, 1855)	New Zealand	Winterbourn 1968
Galba truncatula (O.F. Müller, 1774)	Austria, Iceland, Hungary	Pinter 1979; Hafner et al. 1986
Kamtschaticana kamtschatica (Middendorff, 1851)	Russian Far East	Aksenova et al. 2016
Ladislavella catascopium (Say, 1817)	USA (Yellowstone Park), Kamchatka Peninsula, Canada (British Columbia and Yukon Provinces)	Brues 1927; Vinarski et al. 2016, 2023
Ladislavella elodes (Say, 1821)	USA, Nevada, Yellowstone Park	Brues 1928, 1932
Lanx patelloides (I. Lea, 1856)	USA, California	Brues 1932
Lymnaea stagnalis (Linnaeus, 1758)	Austria	Hafner et al. 1986
Peregriana peregra (O.F. Müller, 1774)	Croatia	Krkać 1979
Pseudosuccinea columella (Say, 1817)	Austria	Hafner et al. 1986
Radix alticola (Izzatullaev, Kruglov & Starobogatov, 1983)	Tajikistan, Nepal	Aksenova et al. 2018
R. auricularia (Linnaeus, 1758)	Europe (Austria), Russia (Baikal area, Kamchatka), Afghanistan; Tibet	Weber 1910; Likharev and Starobogatov 1967; Hafner et al. 1986; Aksenova et al. 2016, 2017; von Oheimb et al. 2016
R. rubiginosa (Michelin, 1831)	Indonesia, Sulawesi and Sumatra Islands	Brues 1939
R. rufescens (Gray, 1822)[a]	India, Himalaya	Chatterjee et al., 2022
Stagnicola palustris (O.F. Müller, 1774)	Austria, Hungary	Pinter 1979; Hafner et al. 1986
Tibetoradix kozlovi Vinarski, Bolotov et Aksenova, 2018	China, Tibetan Plateau	Vinarski et al. 2022

[a]Referred to as *Lymnaea acuminata* (Lamarck, 1822) [see Chatterjee et al., 2022]

Amorim et al. 2019; Fodor et al. 2020; Kuroda and Abe 2020; Rivi et al. 2021). The use of other lymnaeid species as laboratory animals is much less widespread, though one can cite classical experiments on the heredity of shell sinistrality made by Boycott and his colleagues (Boycott and Diver 1923, 1927; Boycott et al. 1930). The model object of their research was another species, *Peregriana peregra* (Freeman and Lundelius 1982).

Most aspects of the Lymnaeidae biology, ecology, morphology, and distribution, reviewed in this chapter only cursory, will be discussed in detail in the following chapters of this book.

Box 2.1 Taxonomic Position of the Lymnaeidae Within the Class Gastropoda, According to Opinions of Different Systematists

Thiele (1931)	Taylor and Sohl (1962)
Class: Gastropoda Cuvier, 1795	**Class:** Gastropoda Cuvier, 1795
Subclass: Pulmonata Cuvier, 1817	**Subclass**: Euthyneura Spengel, 1881
Order: Basommatophora Keferstein, 1865	**Order**: Basommatophora Keferstein, 1865
Superfamily: Lymnaeoidea Rafinesque, 1815	**Superfamily**: Lymnaeacea Rafinesque, 1815
Family: Lymnaeidae Rafinesque, 1815	**Family**: Lymnaeidae Rafinesque, 1815
Hubendick (1978)	Nordsieck (1992)
Class: Gastropoda Cuvier, 1795	**Class:** Gastropoda Cuvier, 1795
Subclass: Pulmonata Cuvier, 1817	**Subclass**: Pulmonata Cuvier, 1817
Order: Basommatophora Keferstein, 1865	**Superorder**: Basommatophora Keferstein, 1865
Superfamily: Lymnaeacea Rafinesque, 1815	**Order**: Hygrophila A. Férussac, 1822
Family: Lymnaeidae Rafinesque, 1815	**Infraorder**: Branchiopulmonata Minichev & Starobogatov, 1975
	Superfamily: Lymnaeacea Rafinesque, 1815
	Family: Lymnaeidae Rafinesque, 1815
Golikov and Starobogatov (1988)	Bouchet et al. (2017)
Class: Gastropoda Cuvier, 1795	**Class:** Gastropoda Cuvier, 1795
Subclass: Pulmonata Cuvier, 1817	**Subclass**: Heterobranchia Burmeister, 1837
Superorder: Lymnaeiformii A. Férussac, 1822	**Cohort**: Tectipleura Schrödl et al., 2011
Order: Lymnaeiformes A. Férussac, 1822	**Subcohort**: Panpulmonata Jörger et al., 2010
Family: Lymnaeidae Rafinesque, 1815	**Superorder**: Hygrophila A. Férussac, 1822
	Superfamily: Lymnaeoidea Rafinesque, 1815
	Family: Lymnaeidae Rafinesque, 1815

Box 2.2 The Taxonomic Structure of the Extant Lymnaeidae Accepted in MolluscaBase (2022)[a]

Family Lymnaeidae Rafinesque, 1815[b]		
Subfamily Lymnaeinae	Subfamily Amphipepleinae Pini, 1877 (= Radicinae Vinarski 2013)	Subfamily Lancinae Hannibal 1914
Acella Haldeman, 1841 *Aenigmomphiscola* Kruglov et Starobogatov, 1981 *Bulimnea* Haldeman, 1841 *Erinna* H. & A. Adams, 1855 *Galba* Schrank, 1803 (subgenera *Galba* s.str., *Bakerilymnaea* Weyrauch, 1964; *Pseudogalba* F.C. Baker, 1913; *Sibirigalba* Kruglov et Starobogatov, 1985)[c] *Hinkleyia* F.C. Baker, 1928 *Ladislavella* B. Dybowski, 1913 *Lymnaea* Lamarck, 1799 *Omphiscola* Rafinesque, 1819 *Pectinidens* Pislbry, 1911 *Pseudisidora* Thiele, 1931 *Pseudosuccinea* F.C. Baker, 1908 *Stagnicola* Jeffreys, 1830 *Walterigalba* Kruglov et Starobogatov, 1985	*Ampullaceana* Servain, 1882 *Austropeplea* Cotton, 1942 (subgenera *Austropeplea* s. str., *Kutikina* Ponder et Waterhouse, 1997) *Bullastra* Bergh, 1901 *Kamtschaticana* Kruglov et Starobogatov, 1984 *Lantzia* Jousseaume, 1872 *Limnobulla* Kruglov et Starobogatov, 1985 *Myxas* G.B. Sowerby I, 1822 *Orientogalba* Kruglov et Starobogatov, 1985 *Peregriana* Servain, 1882 *Racesina* Vinarski et Bolotov, 2018 *Radix* Montfort, 1810 (subgenera *Radix* s.str. and *Exsertiana* Bourguignat, 1883) *Tibetoradix* Bolotov, Vinarski et Aksenova, 2018	*Idaholanx* Clark, Campbell et Lydeard, 2017 *Fisherola* Hannibal, 1912 *Lanx* Clessin, 1882

[a]The genera are given in alphabetical order which does not reflect their phylogenetic relationships. Please note that this classification is somewhat different from an updated version proposed in this book (Aksenova et al. 2023). In particular, Aksenova et al. (2023) reject the subfamily Lancinae and rank it as a tribe Lancini within the Lymnaeinae.
[b]The fourth subfamily, Valencienniinae, is completely extinct
[c]The subgeneric classification of *Galba* requires a revision

Acknowledgments MVV is thankful for the financial support of his lymnaeid studies to the Russian Scientific Fund (project No. 19-14-00066/P). Peter Glöer (Hetlingen, Germany) and Ivan Nekhaev (St. Petersburg, Russia) are acknowledged for sending us some photos used in this chapter.

References

Aho J (1978a) Regional variation in the diversity of freshwater gastropods in Southern and Western Finland. Publ Univ Joensuu BII(8):1–10

Aho J (1978b) Freshwater snail populations and the equilibrium theory of island biogeography. II. Relative importance of chemical and spatial variables. Ann Zool Fenn 15:155–164
Aksenova OV, Vinarski MV, Bolotov IN et al (2016) An overview of *Radix* species of the Kamchatka Peninsula (Gastropoda: Lymnaeidae). Bulletin of the Russian Far Eastern Malacological Society 20:5–27
Aksenova O, Vinarski M, Bolotov I et al (2017) Two *Radix* spp. (Gastropoda: Lymnaeidae) endemic to thermal springs around Lake Baikal represent ecotypes of the widespread *Radix auricularia*. J Zool Syst Evol Res 55(4):298–309. https://doi.org/10.1111/jzs.12174
Aksenova OV, Bolotov IN, Gofarov MY et al (2018) Species richness, molecular taxonomy and biogeography of the radicine pond snails (Gastropoda: Lymnaeidae) in the Old World. Sci Rep-UK 8:11199. https://doi.org/10.1038/s41598-018-29451-1
Aksenova OV, Vinarski MV, Kondakov AV et al (2019) A new Radix species from Qinling Mountains, China (Gastropoda: Lymnaeidae). Ecol Montenegrina 26:137–146. https://doi.org/10.37828/em.2019.26.10
Aksenova MV, Bolotov IN, Kondakov AV, Vinarski MV (2023, this volume) Phylogeny and taxonomy of the family Lymnaeidae. In: Vinarski MV, Vázquez A (eds) The Lymnaeidae. A handbook on their natural history and parasitological significance. Springer, Cham, pp 67–101
Albrecht C, Wolff C, Glöer P et al (2008) Concurrent evolution of ancient sister lakes and sister species: the freshwater gastropod genus *Radix* in lakes Ohrid and Prespa. Hydrobiologia 615:157–167. https://doi.org/10.1007/978-1-4020-9582-5_11
Alda P, Lounnas M, Vázquez AA et al (2018) A new multiplex PCR assay to distinguish among three cryptic *Galba* species, intermediate hosts of *Fasciola hepatica*. Vet Parasitol 251:101–105. https://doi.org/10.1016/j.vetpar.2018.01.006
Alda P, Lounnas M, Vázquez AA et al (2021) Systematics and geographical distribution of *Galba* species, a group of cryptic and worldwide freshwater snails. Molec Phyl Evol 157:107035. https://doi.org/10.1016/j.ympev.2020.107035
Alyakrinskaya IO (2004) The resistance to desiccation in freshwater molluscs. Izvestiya RAN Seriya biologicheckaya 3:367–374. [in Russian]
Amorim J, Abreu I, Rodrigues P et al (2019) *Lymnaea stagnalis* as a freshwater model invertebrate for ecotoxicological studies. Sci Total Environ 669:11–28. https://doi.org/10.1016/j.scitotenv.2019.03.035
Andreeva SI, Andreev NI, Vinarski MV (2010) Key to freshwater gastropods of Western Siberia (Mollusca: Gastropoda). V. 1. Gastropoda: Pulmonata. Fasc. 1. Families Acroloxidae and Lymnaeidae. The authors, Omsk. [in Russian]
Annandale N, Rao HS (1925) Materials for a revision of the recent Indian Lymnaeidae (Mollusca, Pulmonata). Rec Ind Mus 27(3):137–189
Asami T, Gittenberger E, Falkner G (2008) Whole-body enantiomorphy and maternal inheritance of chiral reversal in the pond snail *Lymnaea stagnalis*. J Hered 99:552–557. https://doi.org/10.1093/jhered/esn032
Baker FC (1911) The Lymnaeidae of North and Middle America. Special Publication of the Chicago Academy of Sciences 3:1–539. https://doi.org/10.5962/bhl.title.20500
Baker HB (1925) Anatomy of *Lanx*, a limpet-like Lymnaeid Mollusk. Proc Calif Acad Sci Fourth Series 14(8):143–169
Banarescu P (1990) Zoogeography of fresh waters. Vol. 1. General distribution and dispersal of freshwater animals. AULA-Verlag, Wiesbaden
Bargues MD, Mas-Coma S (2005) Reviewing lymnaeid vectors of fascioliasis by ribosomal DNA sequence analyses. J Helminthol 79:257–267. https://doi.org/10.1079/JOH2005297
Berezina NA (2001) Influence of ambient pH on freshwater invertebrates under experimental conditions. Russ J Ecol 32:343–351. https://doi.org/10.1023/A:1011978311733
Beriozkina GV (2002) Some trends in the adaptive evolution of egg clutches in freshwater Gastropoda. Visnik Zhitomirs'kogo pedagogichnogo universitetu 10:55–57. [in Russian]

Beriozkina GV (2019) The normal intraspecific variability of the egg masses morphology in the Lymnaeiformes. Biologicheskie nauki v shkole i vuze 20:35–41. [in Russian]. https://elibrary.ru/jrsbrz
Beriozkina GV, Starobogatov YI (1981) Morphology of the egg capsules of some molluscs of the genus *Lymnaea* (Gastropoda, Pulmonata). Zool Zh 60(12):1756–1768. [in Russian]
Beriozkina GV, Starobogatov YI (1988) Reproductive ecology and egg clusters of freshwater Pulmonata. Trudy Zoologicheskogo Instituta AN SSSR 174:1–306. [in Russian]
Beriozkina GV, Starobogatov YI (1992) ("1991"). The systems of reproduction of freshwater Pulmonata. Trudy Zoologicheskogo Instituta AN SSSR 228:130–139. [in Russian]
Berrie AD (1959) Variation in the radula of the freshwater snail *Lymnaea peregra* (Müller, 1774) from northwestern Europe. Arkiv för Zoologi, series 2 12(27):391–404
Boag DA (1986) Dispersal in pond snails: potential role of waterfowl. Can J Zool 64:904–909. https://doi.org/10.1139/z86-136
Bolotov IN, Bespalaya YV, Usacheva OV (2012) Ecology and evolution of hydrobionts in hot springs of the Subarctic and Arctic: Formation of similar assemblages, adaptation of species, and microevolutionary processes. Biol Bull Rev 2:340–348. https://doi.org/10.1134/S2079086412040020
Bolotov IN, Aksenova OV, Bespalaya YV et al (2017) Origin of a divergent mt DNA lineage of a freshwater snail species, *Radix balthica*, in Iceland: cryptic glacial refugia or a postglacial founder event? Hydrobiologia 787:73–98. https://doi.org/10.1007/s10750-016-2946-9
Boss KJ (1974) Oblomovism in the Mollusca. Trans Amer Micros Soc 93:460–481. https://doi.org/10.2307/3225152
Boss KJ (1978) On the evolution of gastropods in ancient lakes. In: Fretter V, Peake J (eds) Pulmonates, vol 2A. Academic Press, London, pp 385–428
Boubee SSN (1833) Nouveaux gisemens, en France, de Mollusques et Zoophytes. Bulletin d'histoire naturelle de France: pour servir à la statistique, à la géographie naturelle et à l'étude géologique de cette contrée 1(3):3–24
Bouchet P, Rocroi JP, Hausdorf B et al (2017) Revised classification, nomenclator and typification of gastropod and monoplacophoran families. Malacologia 61(1–2):1–526. https://doi.org/10.4002/040.061.0201
Boycott AE, Diver C (1923) On the inheritance of sinistrality in *Limnaea peregra*. P Roy Soc Lond B Bio 95:207–213. https://doi.org/10.1098/rspb.1923.0033
Boycott AE, Diver C (1927) The origin of an albino mutation in *Limnaea peregra*. Nature 119:9. https://doi.org/10.1038/119009a0
Boycott AE, Diver C, Garstang SL et al (1930) The inheritance of sinistrality in *Limnaea peregra* (Mollusca, Pulmonata). Philos T Roy Soc B 219:51–131
Briers RA (2003) Range size and environmental calcium requirements of British freshwater gastropods. Glob Ecol Biogeogr 12:47–51. https://doi.org/10.1046/j.1466-822X.2003.00316.x
Brown RJ (2007) Freshwater mollusks survive fish gut passage. Arctic 60(2):124–128. https://doi.org/10.14430/arctic237
Brues CT (1927) Animal life in hot springs. Q Rev Biol 2(2):181–203. https://doi.org/10.1086/394272
Brues CT (1928) Studies on the fauna of hot springs in the Western United States and the biology of thermophilous animals. Proc Am Acad Arts Sci 63:139–228
Brues CT (1932) Further studies of the fauna of North American hot springs. Proc Am Acad Arts Sci 67:185–303
Brues CT (1939) Studies on the fauna of some thermal springs in the Dutch East Indies. Proc Am Acad Arts Sci 73(4):71–95
Burch JB (1960) Chromosome studies of aquatic pulmonate snails. Nucleus 3(2):177–208
Burch JB (1989) North American freshwater snails. Malacological Publications, Hamburg, MI
Burch JB, Jung Y (1993) Polyploid chromosome numbers in the *Torquis* group of the freshwater snail genus *Gyraulus* (Mollusca: Pulmonata: Planorbidae). Cytologia 58:145–149

Burch JB, Basch PF, Bush LL (1960) Chromosome numbers in ancylid snails. Revista Portuguesa de Zoologia e Biologia geral 2(3/4):199–204

Campbell DC, Clark SA, Lydeard C (2017) Phylogenetic analysis of the Lancinae (Gastropoda, Lymnaeidae) with a description of the U.S. federally endangered Banbury Springs lanx. Zookeys 663:107–132. https://doi.org/10.3897/zookeys.663.11320

Chatterjee N, Gupta AK, Tiwari SK et al (2022) Small size gastropod fauna from the Matli Geothermal Spring, Bhagirathi Valley, Garhwal Himalaya, Uttarakhand: Ecological implications. Jour Geol Soc India 98:47–52. https://doi.org/10.1007/s12594-022-1926-0

Cheatum EP (1934) Limnological investigations on respiration, annual migratory cycle, and other related phenomena in fresh-water pulmonate snails. T Am Misrosc Soc 53(4):348–407. https://doi.org/10.2307/3222383

Checa AG, Jiménez-Jiménez AP (1997) Regulation of spiral growth in planorbid gastropods. Lethaia 30:257–269. https://doi.org/10.1111/j.1502-3931.1997.tb00468.x

Chono K, Fujita Y, Ito I (2002) Non-ocular dermal photoreception in the pond snail *Lymnaea stagnalis*. Brain Res 951(1):107–112. https://doi.org/10.1016/S0006-8993(02)03143-8

Clessin S (1878) Nordschwedische Mollusken. Malakozoologische Blätt 25:67–79

Climo FM, Pullan NB (1972) A taxonomic review of the family Lymnaeidae (Mollusca: Gastropoda) in New Zealand. J Roy Soc New Zeal 2:5–13. https://doi.org/10.1080/03036758.1972.10423300

Correa AC, Escobar JC, Durand P et al (2010) Bridging gaps in the molecular phylogeny of the Lymnaeidae (Gastropoda: Pulmonata), vectors of fascioliasis. BMC Evol Biol 10:381. https://doi.org/10.1186/1471-2148-10-381

Correa AC, Escobar JS, Noya O (2011) Morphological and molecular characterization of Neotropic Lymnaeidae (Gastropoda: Lymnaeoidea), vectors of fasciolosis. Infect Genet Evol 11:1978–1988. https://doi.org/10.1016/j.meegid.2011.09.003

Cuvier G (1816) "1817" Le règne animal distribué d'après son organisation. Tome 2 contenant les reptiles, les poissons, les mollusques, les annélides. Deterville, Paris. https://doi.org/10.5962/bhl.title.41460

Davison A, Barton NH, Clarke B (2009) The effect of coil phenotypes and genotypes on the fecundity and viability of *Partula suturalis* and *Lymnaea stagnalis*: implications for the evolution of sinistral snails. J Evol Biol 22(8):1624–1635. https://doi.org/10.1111/j.1420-9101.2009.01770.x

Davison A, McDowell GS, Holden JM et al (2016) Formin is associated with left-right asymmetry in the pond snail and the frog. Curr Biol 26(5):654–660. https://doi.org/10.1016/j.cub.2015.12.071

Dillon RT (2000) The ecology of freshwater molluscs. Cambridge University Press, Cambridge etc. https://doi.org/10.1017/CBO9780511542008

Dregol'skaya IN, Korotneva NV, Starobogatov YI (1988) Molluscs of the eastern part of the Finnish Gulf and their cytoecological study. Zool Zh 67(11):1631–1637. [in Russian]

Duncan CJ (1960) The genital systems of the freshwater Basommatophora. Proc Zool Soc London 135(3):339–356. https://doi.org/10.1111/j.1469-7998.1960.tb05851.x

Falniowski A (1980a) Pigmentation of the mantle border in Polish representatives of the subgenus *Radix* (Lymnaeidae, Basommatophora, Gastropoda). Basteria 44:3–8

Falniowski A (1980b) Podrodzaj *Radix* s.str. (Gastropoda, Basommatophora) w Polsce. 1. Pigmentacja i anatomia. Opis *Lymnaea peregra roszkowskiana* subsp. nov. Zeszyty Naukowe Uniwersytetu Jagiellonskiego. Prace Zoologiczne 26:67–108

Faltýnková A, Našincová V, Kablásková L (2008) Larval trematodes (Digenea) of planorbid snails (Gastropoda: Pulmonata) in Central Europe: a survey of species and key to their identification. Syst Parasitol 69:155–178. https://doi.org/10.1007/s11230-007-9127-1

Fischer P (1880–1887) Manuel de conchyliologie et de paleontologie conchyliologique ou, Histoire naturelle des mollusques vivants et fossiles. V. Savy, Paris. https://doi.org/10.5962/bhl.title.13213

Fodor I, Hussein AAA, Benjamin PR et al (2020) The unlimited potential of the great pond snail, *Lymnaea stagnalis*. eLife 9:e56962. https://doi.org/10.7554/eLife.56962.sa2

Franc A (1968) Sous-classe des pulmonés. In: Grassé PP (ed) Traité de Zoologie. Anatomie, Systematique, Biologie, vol 5(3). Masson et Cie, Paris, pp 325–607

Freeman G, Lundelius JW (1982) The developmental genetics of dextrality and sinistrality in the gastropod *Lymnaea peregra*. W Roux Arch Dev Biol 191:69–83. https://doi.org/10.1007/BF00848443

Frömming E (1956) Biologie der mitteleuropäischen Süßwasserschnecken. Dunker-Humblot, Berlin

Garbar AV, Korniushin AV (2003) Karyotypes of European species of *Radix* (Gastropoda: Pulmonata: Lymnaeidae) and their relevance to species distinction in the genus. Malacologia 45(1):141–148

Garbar AV, Manilo VV, Korniushin AV (2004) Probable directions of evolution of karyotypes of the European Lymnaeidae (Mollusca, Gastropoda, Pulmonata) in the light of modern concepts of phylogeny of the family. Vestnik zoologii 38(2):29–37. [in Russian]

Geyer D (1923) Zwergschnecken. Archiv für Molluskenkunde 55(6):231–235

Geyer D (1929) Die Molusken des Bodenseestrandes. Zoologischer Jahrbucher. Abteilung für Systematik, Ökologie und Geographie der Tiere 58:135–172

Glöer P, Pešić V (2008) *Radix skutaris* n.sp., a new species from Montenegro (Gastropoda: Lymnaeidae). Mollusca (Dresden) 26(1):83–88

Glöer P, Yildirim MZ (2006) *Stagnicola* records from Turkey with the description of two new species, *Stagnicola tekecus* n. sp. and *S. kayseris* n. sp. (Gastropoda: Lymnaeidae). J Conchol 39(1):85–89

Goldman MA, LoVerde PT, Chrisman CL (1983) Hybrid origin of polyploidy in freshwater snails of the genus *Bulinus* (Mollusca, Planorbidae). Evolution 37(3):592–600. https://doi.org/10.1111/j.1558-5646.1983.tb05576.x

Golikov AN, Starobogatov YI (1975) Systematics of prosobranch gastropods. Malacologia 15(1):185–232

Golikov AN, Starobogatov YI (1988) Problems of phylogeny and system of the prosobranchiate gastropods. Trudy Zoologicheskogo Instituta AN SSSR 187:4–87. [in Russian]

Gorjanović-Kramberger C (1923) Die Valenciennesiden und einige anderen Limnaeiden der pontischen Stufe des unteren Pliocaens in ihrer stratigraphischen und genetischen Bedeutung. Glasnik hrvatskoga prirodoslovnoga društva 35:87–114

Grobben K (1910) Claus-Grobben Lehrbuch der Zoologie. N.G. Elwert, Marburg

Gutiérrez A, Pointier JP, Yong M et al (2003) Evidence of phenotypic differences between resistant and susceptible isolates of *Pseudosuccinea columella* (Gastropoda: Lymnaeidae) to *Fasciola hepatica* (Trematoda: Digenea) in Cuba. Parasitol Res 90:129–134. https://doi.org/10.1007/s00436-002-0824-z

Hafner W, Honsig-Erlemburg W, Mildner P (1986) Faunistischer Bericht über die Thermen in Warmbad Villach. Carinthia II(176/96Jg):231–239

Hannibal HA (1914) Note on the classification of Ancylidae. The Nautilus 28(2):23–24

Harutyunova LD (1977) Aphally in a population of *Radix* (Gastropoda; Lymnaeidae) of Armenia. Biologicheskiy Zhurnal Armenii 30(3):89–90. [in Russian]

Harzhauser M, Mandic O, Neubauer TA, Georgopoulou E, Hassler A (2016) Disjunct distribution of the Miocene limpet-like freshwater gastropod genus *Delminiella*. J Mollus Stud 82:129–136. https://doi.org/10.1093/mollus/eyv040

Haszprunar G (1988) On the origin and evolution of major gastropod groups, with special reference to the Streptoneura. J Mollus Stud 54:367–441. https://doi.org/10.1093/mollus/54.4.367

Heller J (1993) Hermaphroditism in molluscs. Biol J Linn Soc 48:19–42. https://doi.org/10.1111/j.1095-8312.1993.tb00874.x

Hubendick B (1951) Recent Lymnaeidae. Their variation, morphology, taxonomy, nomenclature and distribution. Kungliga Svenska Vetenskapsakademiens Handlingar. Fjärde Serien 3(1):1–223

Hubendick B (1952) Hawaiian Lymnaeidae. Occasional papers of Bernice P. Bishop Museum, Honolulu, Hawaii 20(19):307–328

Hubendick B (1962) Aspects of diversity of fresh-water fauna. Oikos 13:249–261

Hubendick B (1978) Systematics and comparative morphology of the Basommatophora. In: Fretter V, Peake J (eds) Pulmonates, vol 2A. Academic Press, London, pp 1–47

Hunter RD (1990) Effects of low pH and low calcium concentration on the pulmonate snail *Planorbella trivolvis*, a laboratory study. Can J Zool 68:1578–1583. https://doi.org/10.1139/z90-233

Inaba A (1969) Cytotaxonomic studies of lymnaeid snails. Malacologia 7(2–3):143–168

Ivanov AV (1940) Class Gastropoda. In: Dogiel VA, Zenkevich LA (eds) A guide on zoology. Invertebrates, vol 2. Annelida. Mollusca. The Soviet Academy of Sciences Press, Moscow-Leningrad, pp 323–455. [in Russian]

Ivanov DL (ed) (1990) Evolutionary morphology of the Mollusca (the peculiarities of the morphofunctional alterations of the radular apparatus). The Moscow State University Press, Moscow. [in Russian]

Jackiewicz M (1954) Z badań anatomiczno-porównawczych nad niektórymi gatunkami z rodzaju *Radix* Montfort na terenie Wielkopolski. Prace Komisji Biologicznej, Poznańskie Towarzystwo Przyjaciół Nauk 15(3):1–20

Jackiewicz M (1993) Die Mantelpigmentation als Diagnosemerkmal bei Schlammschnecken (Gastropoda, Pulmonata, Lymnaeidae). Malakologische Abhandlungen / Museum für Tierkunde Dresden 16:165–172

Jackiewicz M (1998) European species of the family Lymnaeidae (Gastropoda, Pulmonata, Basommatophora). Genus 9(1):1–93

Jackiewicz M, Koralewska-Batura E (1995) The shell-surface sculpture of Lymnaeidae (Gastropoda, Pulmonata, Basommatophora). Malakologische Abhandlungen / Museum für Tierkunde Dresden 17(2):191–197

Jarne P, Vianey-Liaud M, Delay B (1993) Selfing and outcrossing in hermaphrodite freshwater gastropods (Basommatophora): where, when and why. Biol J Linn Soc 49:99–125. https://doi.org/10.1111/j.1095-8312.1993.tb00893.x

Jordaens K, Dillen L, Backeljau T (2007) Effects of mating, breeding system and parasites on reproduction in hermaphrodites: pulmonate gastropods. Anim Biol 57(2):137–195. https://doi.org/10.1163/157075607780377965

Jörger KM, Stöger I, Kano Y et al (2010) On the origin of Acochlidia and other enigmatic euthyneuran gastropods, with implications for the systematics of Heterobranchia. BMC Evol Biol 8:323. https://doi.org/10.1186/1471-2148-10-323

Kappes H, Haase P (2012) Slow, but steady: dispersal of freshwater molluscs. Aquat Sci 74:1–14. https://doi.org/10.1007/s00027-011-0187-6

Kobelt W (1871) Zur Kenntniss der europäischen Limnäen. II. *L stagnalis* Linné. Malakozoologischer Blätt 18:108–119

Krkać N (1979) Temperature and reproductive cycle relations in *Radix peregra* O.F. Müller. Malacologia 18:227–232

Kruglov ND (2005) Lymnaeid snails of Europe and Northern Asia. Smolensk State Pedagogical University Press, Smolensk. [in Russian]

Kruglov ND, Starobogatov YI (1981) A new genus of the Lymnaeidae and taxonomy of the subgenus *Omphiscola* (*Lymnaea*) (Pulmonata, Gastropoda). Zool Zh 60(7):965–977. [In Russian]

Kruglov ND, Starobogatov YI (1985) Methods of experimental hybridization and some results of its applications in the taxonomy of Lymnaeidae. Malacol Rev 18:21–35

Kruglov ND, Starobogatov YI (1989) Morphology and taxonomy of the molluscs from the subgenus *Radix* of the genus *Lymnaea* (Gastropoda, Pulmonata, Lymnaeidae) from Siberia and Far East of the USSR. Zool Zh 68(5):17–30. [in Russian]

Kruglov ND, Starobogatov YI (1992) ("1991") The unknown egg masses of molluscs of the genus *Lymnaea* (Gastropoda Pulmonata Lymnaeidae). Trudy Zoologicheskogo Instituta Akademii Nauk SSSR 228:111–129. [in Russian]

Kruglov ND, Starobogatov YI (1993) Annotated and illustrated catalogue of species of the family Lymnaeidae (Gastropoda Pulmonata Lymnaeiformes) of Palaearctic and adjacent river drainage areas. Part I Ruthenica Russian Malacol J 3(1):65–92

Kuroda R, Abe M (2020) The great pond snail *Lymnaea stagnalis*. EvoDevo 11:24. https://doi.org/10.1186/s13227-020-00169-4

Likharev IM, Starobogatov YI (1967) On the molluscan fauna of Afghanistan. Trudy Zoologicheskogo Instituta Akademii Nauk SSSR 42:159–197. [in Russian]

Lomolino MV (2005) Body size evolution in insular vertebrates: generality of the Island rule. J Biogeogr 32:1683–1699. https://doi.org/10.1111/j.1365-2699.2005.01314.x

Lomolino MV, van der Geer AA, Lyras GA et al (2013) Of mice and mammoths: generality and antiquity of the Island rule. J Biogeorg 40:1427–1439. https://doi.org/10.1111/jbi.12096

Lounnas M, Correa AC, Vázquez AA et al (2017) Self-fertilization, long-distance flash invasion and biogeography shape the population structure of *Pseudosuccinea columella* at the worldwide scale. Mol Ecol 26:887–903. https://doi.org/10.1111/mec.13984

Machałowski T, Jesionowski T (2021) Hemolymph of molluscan origin: from biochemistry to modern biomaterials science. Applied Physics, series A 127:3. https://doi.org/10.1007/s00339-020-04166-1

Madsen H, Frandsen F (1989) The spread of freshwater snails including those of medical and veterinary importance. Acta Trop 46:139–146. https://doi.org/10.1016/0001-706x(89)90030-2

Mahulu A, Clewing C, Stelbrink B et al (2019) Cryptic intermediate snail host of the liver fluke *Fasciola hepatica* in Africa. Parasite Vectors 12:573. https://doi.org/10.1186/s13071-019-3825-9

Maksimova TI, Yakovleva ZA (1991) Morphofunctional aspects of physiology of copulation in molluscs of the genus *Planorbarius* (Gastropoda Pulmonata Bulinidae). Trudy Zoologicheskogo Instituta Akademii Nauk SSSR 228:49–60. [in Russian]

Malcicka M (2015) Life history and biology of *Fascioloides magna* (Trematoda) and its native and exotic hosts. Ecol Evol 5(7):1381–1397. https://doi.org/10.1002/ece3.1414

Malone CR (1965) Dispersal of aquatic gastropods via the intestinal tract of water birds. The Nautilus 78:135–139

Marinescu F (1969) *Velutinellus*, nouveau genre fossile de la famille des Lymneidae, et ses relations avec *Velutinopsis* et *Valenciennius*. Malacologia 9:313–325

Martinson GG (1961) Mesozoic and Cenozoic Mollusca, Non-marine Sedimentary Succession of Sibirican platform, Baikalia and Mongolia. Trudy Baikal'skoi Limnologicheskoi Stantsii 19:1–332. [in Russian]

McClain CR, Boyer A, Rosenberg G (2006) The Island rule and the evolution of body size in the deep sea. J Biogeogr 33:1578–1584. https://doi.org/10.1111/j.1365-2699.2006.01545.x

McMahon RF (1983) Physiological ecology of freshwater pulmonates. In: Russell-Hunter WD (ed) The Mollusca, vol 6. Academic Press, Orlando etc, pp 360–430

Meshcheryakov VN (1990) The common pond snail, *Lymnaea stagnalis*. In: Dettlaff TA, Wassetsky SG (eds) Animal species for developmental studies: Invertebrates. Plenum Publishing Corp, New York, pp 69–132

Mills EL, Leach JH, Carlton JT et al (1993) Exotic species in the Great Lakes: a history of biotic crises and anthropogenic introductions. J Great Lakes Res 19(1):1–54. https://doi.org/10.1016/S0380-1330(93)71197-1

Minichev YS, Starobogatov YI (1979) The subclasses of Gastropoda and their phylogenetic relations. Zool Zh 58:293–305. [in Russian]

Mitchell R (1974) The evolution of thermophyly in hot springs. Q Rev Biol 49(3):229–242. https://doi.org/10.1086/408084

MolluscaBase eds. (2022) MolluscaBase. Accessed at http://www.molluscabase.org on 2022-10-01. https://doi.org/10.14284/448

Moos A (1944) Neue Funde von Lymnaeiden, insbesondere von Valenciennesiiden im Pannon Kroatiens. Vjestnik Hrvatskog državnog geološkog zavoda i Hrvatskog državnog geološkog muzeja 2(3):341–390

Nekrasov AD (1927) The observations on the egg clutches of freshwater animals. II. The comparative morphology of the egg clutches of freshwater Gastropoda. Russkiy Zoologicheskiy Zhurnal 7(4):153–179. [in Russian]

Ng TH, Tan SK, Wong WH et al (2016) Molluscs for sale: Assessment of freshwater gastropods and bivalves in the Ornamental Pet Trade. PLoS One 11(8):e0161130. https://doi.org/10.1371/journal.pone.0161130

Nordsieck H (1992) Phylogeny and system of the Pulmonata. Arch Molluskenk 121(1/6):31–52

Økland J (1990) Lakes and snails. Environment and Gastropoda in 1,500 Norwegian lakes, ponds and rivers. Universal Book Services / Dr. W. Backhuys Publ, Oegstgeest

Økland J, Økland KA (1997) Distribution limits of freshwater molluscs in Northern Europe – a survey of edge populations in Norway and adjacent areas. Heldia 4(sonderheft 5):78–93

Paraense WL (1976) *Lymnaea viatrix*: a study of topotypic specimens (Mollusca: Lymnaeidae). Rev Bras Biol 36(2):419–428

Paraense WL (1982) *Lymnaea rupestris* sp. n. from southern Brazil (Pulmonata: Lymnaeidae). Mem I Oswaldo Cruz 77(4):437–443. https://doi.org/10.1590/S0074-02761982000400011

Paraense WL (1983) *Lymnaea columella* in northern Brazil. Mem I Oswaldo Cruz 78(4):477–482. https://doi.org/10.1590/S0074-02761983000400011

Paraense WL (1984) *Lymnaea diaphana*: a study of topotypic specimens (Pulmonata: Lymnaeidae). Mem I Oswaldo Cruz 79:75–81. https://doi.org/10.1590/S0074-02761984000100009

Paraense WL (1994) A Mexican population of *Lymnaea elodes* (Gastropoda: Lymnaeidae). Malacol Rev 27:5–11

Paraense WL (1995) *Lymnaea cousini* Jousseaume, 1887, from Ecuador (Gastropoda: Lymnaeidae). Mem I Oswaldo Cruz 90(5):605–609. https://doi.org/10.1590/S0074-02761995000500011

Patoka J, Kopecký O, Vrabec V et al (2017) Aquarium molluscs as a case study in risk assessment of incidental freshwater fauna. Biol Invasions 19:2039–2046. https://doi.org/10.1007/s10530-017-1412-6

Patterson CM, Burch JB (1978) Chromosomes of pulmonate molluscs. In: Fretter V, Peake J (eds) Pulmonates, vol 2A. Academic Press, London, pp 171–217

Pavlova MV (1983) Biology and productivity of *Limnaea* (*Radix*) *auricularia* (L.) var. *obliquata* Martens in the Lake Issyk-Kul. In: Konurbaev AO (ed) Limnologicheskiye issledovaniya v Kirgizii. Ilim, Frunze, pp 34–46. [in Russian]

Pelseneer P (1905) Mollusca. In: Ray Lankester E (ed) A treatise on zoology, vol V. Adam & Charles Black, London

Piechocki A (1979) Mięczaki (Mollusca). Ślimaki (Gastropoda). In: Fauna słodkowodna Polski, vol 7. Państwowe wydawnictwo naukowe, Warszawa-Poznań, pp 1–188

Pierce HG (1996) On sinistral coiling among fossil North American Lymnaeidae. The Veliger 39(3):220–225

Pilsbry HA (1925) The family Lancidae distinguished from the Ancylidae. The Nautilus 38:73–75

Pinter I (1979) A Hévízi-tó és a környező berek csigafaunája. Allattani Közlemények 66(1/4): 117–130

Pip E (1993) The distribution of freshwater gastropods in central North America in relation to water temperature. Heldia 2(1/2):21–27

Pointier JP (ed) (2015) Freshwater molluscs of Venezuela and their medical and veterinary importance. Conchbooks, Harxheim

Pointier JP, Vàzquez AA (2020) Lymnaeoidea, Lymnaeidae. In: Damborenea C, Rogers DC, Thorp JH (eds) Thorp and Covich's Freshwater Invertebrates, Keys to Neotropical and Antarctic Fauna, vol 5. Academic Press, London, pp 281–290

Ponder WF, Lindberg DR (1997) Towards a phylogeny of gastropod molluscs: an analysis using morphological characters. Zool J Linnean Soc 119:83–265. https://doi.org/10.1111/j.1096-3642.1997.tb00137.x
Ponder WF, Lindberg DR (eds) (2008) Phylogeny and evolution of the Mollusca. University of California Press, Berkeley etc
Ponder WF, Waterhouse J (1997) A new genus and species of Lymnaeidae from the lower Franklin River, South Western Tasmania. J Molluscan Stud 63:441–468. https://doi.org/10.1093/mollus/63.3.441
Ponder WF, Lindberg DR, Ponder JM (2020) Biology and evolution of the Mollusca, vol 2. CRC Press, Boca Raton
Prozorova LA (1992) ("1991") Morphology of gastropod egg-clusters of the Primorye Region. Trudy Zoologicheskogo Instituta Akademii Nauk SSSR 228:74–110. [in Russian]
Qian Z, Yang J, Lu Y, He J (2012) Description of three freshwater species (Gastropoda) from China. Shell Discoveries 1(1):30–31
Rivi V, Benatti C, Lukowiak K et al (2021) What can we teach *Lymnaea* and what can *Lymnaea* teach us? Biol Rev 96(4):1590–1602. https://doi.org/10.1111/brv.12716
Robinson WI (1915) Two new fresh-water gastropods from the Mesozoic of Arizona. Am J Sci, fourth series 40(240):649–651
Roszkowski W (1914) Note sur l'appareil génital de *Limnaea auricularia* L. et de *Limnaea ovata* Drap. Zool Anz 44(4):175–179
Roszkowski W (1929) Contributions to the study of the family Lymnaeidae. I. On the systematic position and the geographical distribution of the genus *Myxas* J. Sowerby. Annales Zoologici Musei Polonici Historiae Naturalis 8(2):64–97
Russell-Hunter WD (1964) Physiological aspects of ecology in non-marine mollusks. In: Wilbur KM, Yonge CM (eds) Physiology of Mollusca. Academic Press, London, pp 83–125
Russell-Hunter WD (1978) Ecology of freshwater pulmonates. In: Fretter V, Peake J (eds) Pulmonates, vol 2A. Academic Press, London, pp 335–384
Saadi AJ, Davison A, Wade CM (2020) Molecular phylogeny of freshwater snails and limpets (Panpulmonata: Hygrophila). Zool J Linnean Soc 190(2):518–531. https://doi.org/10.1093/zoolinnean/zlz177
Salazkin AA (1969) On some pecularities of distribution of gastropods in lakes of different type in the humid zone of Western Siberia. In: Johanzen BG (ed) Voprosy malakologii Sibiri. Tomsk State University, Tomsk, pp 57–60. [in Russian]
Salvini-Plawen L (1980) A reconsideration of systematics in the Mollusca (phylogeny and higher classification). Malacologia 19:249–258
Schileyko AA (1972) The radular morphology in Helicidae and the taxonomic significance of the radular characters. Sbornik trudov Zoologicheskogo muzeya Moskovskogo universiteta 12: 157–168. [in Russian]
Schniebs K, Glöer P, Vinarski M et al (2011) Intraspecific mophological and genetic variability in *Radix balthica* (Linnaeus 1758) (Gastropoda: Basommatophora: Lymnaeidae) with morphological comparison to other European *Radix* species. J Conchol 40(6):657–678
Schrödl M (2013) Time to say "bye-bye" Pulmonata? Spixiana 37(2):161–164
Schütt H (1974) Zwei neue reliktäre Süßwassermollusken der Dinariden. Annalen des Naturhistorischen Museums in Wien 78:473–480
Shigin AA (1971) The role of molluscs in the biology of trematodes of the genus *Diplostomum*. Molluscs, the ways, methods and results of their study. In: All-Union scientific meeting on the study of molluscs. Abstracts of communications. Nauka, Leningrad, pp 129–130. [in Russian]
Shikov EV (1977) On dispersal of terrestrial snails during spring floods. Zool Zh 56(3):361–367. [in Russian]
Solem A (1974) The shell makers. Introducing mollusks. John Wiley & Sons, New York etc
Spengel JW (1881) Die Geruchsorgane und des Nervensystem der Mollusken. Z Wiss Zool 35(3): 333–383

Stadnichenko AP (1990) Lymnaeiformes (physids, bulinids, planorbids). Fauna of Ukraine in 40 volumes 29(4):1–282. [in Russian]
Stadnichenko AP (2004) Pond snails and limpet snails (Lymnaeidae and Acroloxidae) of Ukraine. Tsentr uchebnoi literatury, Kiev. [in Russian]
Stadnichenko AP (2006) Lymnaeidae and Acroloxidae of Ukraine: sampling and studying methods, biology, ecology, their practical importance. Ruta Press, Zhytomyr. [in Russian]
Starmühlner F (1957) Zur Individuendichte und Formänderung von *Lymnaea peregra* Müller in isländischen Thermalbiotopen. Ergebnisse der Österreichischen Island-Expedition 1955. Sitzungsberichte der Akademie der Wissenschaften in Wien mathematisch-naturwissenschaftlichen Klasse (Abt. 1) 166:331–384
Starobogatov YI (1967) On the systematization of freshwater pulmonate molluscs. Trudy Zoologicheskogo Instituta AN SSSR 42:280–304. [in Russian]
Starobogatov YI (1970) Molluscan fauna and zoogeographic zonation of continental freshwater bodies of the world. Nauka, Leningrad. [in Russian]
Starobogatov YI, Prozorova LA, Zatravkin MN (1989) Species composition of the family Physidae (Gastropoda Pulmonata Lymnaeiformes) of Siberia and Far East of the USSR (with notes on European physids). Byulleten' Moskovskogo Obshchestva Ispytateley Pripody, otdel biologicheskiy 94(1):62–78. [in Russian]
Storey R (1972) Dormancy in *Lymnaea peregra* (Müller) during periods of dryness. J Conchol 27: 377–386
Strong EE, Gargominy O, Ponder WF, Bouchet P (2008) Global diversity of gastropods (Gastropoda; Mollusca) in freshwater. Hydrobiologia 595:149–166. https://doi.org/10.1007/s10750-007-9012-6
Sudarikov VE, Shigin AA, Kurochkin YV et al (2002) The trematode metacercariae – parasites of freshwater hydrobionts of Central Russia. Nauka, Moscow. [in Russian]
Taktakishvili IG (1967) The historical development of the family Valenciennidae. Metsniereba, Tbilisi. [in Russian]
Taylor DW, Sohl NF (1962) An outline of gastropod classification. Malacologia 1(1):7–32
Tchakonté S, Ajeagah GA, Diomandé C et al (2014) Diversity, dynamic and ecology of freshwater snails related to environmental factors in urban and suburban streams in Douala-Cameroon (Central Africa). Aquat Ecol 48:379–395. https://doi.org/10.1007/s10452-014-9491-2
Thiele J (1929) Handbuch der Systematischen Weichtierkunde, vol 1. Gustav Fischer Verlag, Jena, pp 1–376
Thiele J (1931) Handbuch der Systematischen Weichtierkunde, vol 2. Gustav Fischer Verlag, Jena, pp 377–778
Thiriot-Quiévreux K (2003) Advances in chromosomal studies of gastropod molluscs. J Molluscan Stud 69:187–201. https://doi.org/10.1093/mollus/69.3.187
Tracey S, Todd JA, Erwin DH (1993) Mollusca: Gastropoda. In: Benton MJ (ed) Fossil Record 2. Chapman & Hall, London etc, pp 131–167
Tsikhon-Lukanina EA (1987) Trophology of aquatic molluscs. Nauka, Moscow. [in Russian]
Vázquez AA, Alda P, Lounnas M et al (2018) Lymnaeid snails hosts of *Fasciola hepatica* and *Fasciola gigantica* (Trematoda: Digenea): A worldwide review. CAB Rev Perspect Agric Vet Sci Nutr Nat Resour 13:1–15. https://doi.org/10.1079/PAVSNNR201813062
Vermeij GJ (2017) The limpet form in gastropods: evolution, distribution, and implications for the comparative study of history. Biol J Linn Soc 120(1):22–37. https://doi.org/10.1111/bij.12883
Vinarski MV (2011) A new species of stagnicoline snails (Mollusca: Gastropoda: Lymnaeidae) from the extreme North of Western Siberia. Zootaxa 2817:55–58. https://doi.org/10.11646/zootaxa.2817.1.2
Vinarski MV (2012a) Geographical variability in freshwater mollusks. Biol Bull Rev 2:390–399. https://doi.org/10.1134/S2079086412050088
Vinarski MV (2012b) The problem of geographical variation in freshwater gastropods. In: Fedorov A, Yakovlev H (eds) Mollusks: morphology, behavior and ecology. Nova Science Publishers, Hauppauge, NY, pp 181–208

Vinarski MV (2013) One, two, or several? How many lymnaeid genera are there? Ruthenica Russian Malacol J 23(1):41–58

Vinarski MV (2018) The family Lymnaeidae in ancient lakes: A review and synthesis of data on extant and fossil taxa. In: SIAL 8 – Speciation in Ancient Lakes: Celebrating 25 Years and Moving Towards the Future, 29 July–3 August 2018, Entebbe (Uganda). Conference Program and Abstracts, p 66

Vinarski MV, Grebennikov ME (2012) An overview of species of the genus *Aenigmomphiscola* Kruglov et Starobogatov, 1981 (Gastropoda: Pulmonata: Lymnaeidae). Ruthenica Russian Malacol J 22(2):159–170. [in Russian]

Vinarski MV, Kramarenko SS (2019) Scale-dependence in geographic variation in a freshwater gastropod across the Palearctic. Molluscan Res 39(2):159–170. https://doi.org/10.1080/13235818.2018.1497570

Vinarski M, Schniebs K, Glöer P, Hundsdoerfer A (2011) The taxonomic status and phylogenetic relationships of the genus *Aenigmomphiscola* Kruglov and Starobogatov, 1981 (Gastropoda: Pulmonata: Lymnaeidae). J Nat Hist 45(33–34):2049–2068. https://doi.org/10.1080/00222933.2011.574800

Vinarski MV, Andreyev NI, Andreyeva SI et al (2012) Latitudinal changes in the diversity of freshwater gastropods (Mollusca: Gastropoda) in waterbodies of Western Siberia. Inland Water Biology 5(1):83–90. https://doi.org/10.1134/S1995082911040213

Vinarski MV, Aksenova OV, Bespalaya YV et al (2016) *Ladislavella tumrokensis*: The first molecular evidence of a Nearctic clade of lymnaeid snails inhabiting Eurasia. Syst Biodivers 14(3):276–287. https://doi.org/10.1080/14772000.2016.1140244

Vinarski MV, Clewing C, Albrecht C (2019) Family Lymnaeidae. In: Lydeard C, Cummings KS (eds) Freshwater mollusks of the world: a distribution atlas. John Hopkins University Press, Baltimore, pp 158–162

Vinarski MV, Bolotov IN, Aksenova OV et al (2021) Freshwater Mollusca of the Circumpolar Arctic: A review on their taxonomy, diversity and biogeography. Hydrobiologia 848:2891–2918. https://doi.org/10.1007/s10750-020-04270-6

Vinarski MV, von Oheimb PV, Aksenova OV et al (2022) Trapped on the roof of the world: taxonomic diversity and evolutionary patterns of Tibetan Plateau endemic freshwater snails (Gastropoda: Lymnaeidae: *Tibetoradix*). Integr Zool 17:825848. https://doi.org/10.1111/1749-4877.12600

Vinarski MV, Aksenova OV, Bespalaya YV et al (2023) How an ecological race is forming: morphological and genetic disparity among thermal and non-thermal populations of aquatic Lymnaeid Snails (Gastropoda: Lymnaeidae). Diversity 15(4):548. https://doi.org/10.3390/d15040548

von Oheimb PV, Albrecht C, Riedel F et al (2011) Freshwater biogeography and limnological evolution of the Tibetan Plateau – Insights from a plateau-wide distributed gastropod taxon (*Radix* spp.). PLoS One 6:e26307. https://doi.org/10.1371/journal.pone.0026307

von Oheimb PV, Landler L, von Oheimb CM (2016) Cold snails in hot springs: Observations from Patagonia and the Tibetan plateau. Malacologia 59:313–320. https://doi.org/10.4002/040.059.0209

Walter HJ (1969) Illustrated biomorphology of the "*angulata*" lake form of the basommatophoran snail *Lymnaea catascopium* Say. Malacol Rev 2:1–102

Weber A (1910) Über Binnenmollusken aus Ost-Turkestan, Kaschmir und West-Tibet. Zoologische Jahrbücher. Abteilung für Systematik, Geographie und Biologie der Thiere 29:299–312

Wiese R, Clewing C, Albrecht C et al (2020) How ancient is Lake Lugu (Yunnan, China)? The gastropods' viewpoint with focus on *Radix* (Lymnaeidae). J Great Lakes Res 46:1099–1112. https://doi.org/10.1016/j.jglr.2020.06.003

Wilson RA, Denison J (1980) The parasitic castration and gigantism of *Lymnaea truncatula* infected with the larval stages of *Fasciola hepatica*. Z Parasitenk 61:109–119. https://doi.org/10.1007/bf00925458

Winterbourn MJ (1968) The faunas of thermal waters in New Zealand. Tuatara 16(2):111–122

Yakchali M, Deilamy LJ (2012) Radular variations in freshwater snails of the family Lymnaeidae (Mollusca: Gastropoda: Pulmonata) from northwestern Iran. Am Malac Bull 30(2):323–326. https://doi.org/10.4003/006.030.0212

Yen TC, Reeside JBJ (1946) Fresh-water mollusks from the Morrison Formation (Jurassic) of Sublette County, Wyoming. J Paleontol 20(1):52–58

Żbikowska E (2007) Digenea species in chosen populations of freshwater snails in northern and central part of Poland. Wiad Parazitol 53(4):301–308

Żbikowska E (2011) One snail – three Digenea species, different strategies in host-parasite interaction. Anim Biol 61:1–19. https://doi.org/10.1163/157075511X554383

Chapter 3
Phylogeny and Taxonomy of the Family Lymnaeidae

**Olga V. Aksenova, Ivan N. Bolotov, Irina S. Khrebtova,
Alexander V. Kondakov, and Maxim V. Vinarski**

Abstract A new three-locus (COI + 16S rRNA +28S rRNA) phylogeny of the living representatives of the family Lymnaeidae, which covers the majority of recent genera and subgenera, is proposed. Two living subfamilies (Lymnaeinae Rafinesque, 1815 and Amphipepleinae Pini, 1877), each divided into a series of tribes, are recognized. Four tribes (Austropepleini, Omphiscolini, Peregrianini, and Tibetoradicini), are described as new for science, however, some lymnaeid genera have not been assigned to a tribe due to scarcity of available information. An annotated list of all accepted lymnaeid taxa above the species level is provided, with short synonymies, morphological diagnoses, distributional and taxonomic remarks (when appropriate).

3.1 Introduction

Both phylogeny and taxonomic structure of the family Lymnaeidae have been a matter of hot debates during the last century (e.g., Baker 1911; Hubendick 1951; Starobogatov 1967, 1970, 1976; Inaba 1969; Jackiewicz 1993, 1998; Garbar et al. 2004; Kruglov 2005; Ponder and Waterhouse 1997; Vinarski 2013; and many others). Until the 2010s, the vast majority of such works relied on the interpretations of morphological features (those of shell, radula, reproductive anatomy) and, more rarely, available cytogenetic information. Vinarski (2013) discussed the previous attempts to reconstruct the lymnaeid phylogeny and to develop a working taxonomy based on it. According to this author, the living lymnaeids must be divided into two

O. V. Aksenova · I. N. Bolotov · I. S. Khrebtova · A. V. Kondakov
N. Laverov Federal Center for Integrated Arctic Research, The Ural Branch of the Russian Academy of Sciences, Arkhangelsk, Russia

M. V. Vinarski (✉)
Laboratory of Macroecology & Biogeography of Invertebrates, Saint-Petersburg State University, Saint-Petersburg, Russia

S.I. Vavilov Institute for the History of Science and Technology, St. Petersburg Branch, Russian Academy of Sciences, Saint-Petersburg, Russia

M. V. Vinarski, A. A. Vázquez (eds.), *The Lymnaeidae*, Zoological Monographs 7,
https://doi.org/10.1007/978-3-031-30292-3_3

subfamilies—Lymnaeinae Rafinesque, 1815 and Radicinae Vinarski, 2013 (which appeared to be a junior synonym of Amphipepleinae Pini, 1877; see Bouchet et al. 2017). The phylogenetic relationships and taxonomic position of the extinct lymnaeids, including the fossil subfamily Valencienninae Gorjanović-Kramberger, 1923, were discussed neither by Vinarski (2013) nor by most other experts in the field (see, for example, Jackiewicz 1998; Kruglov 2005). In this publication, we avoid any questions concerning the fossil representatives of the family as well, focusing instead on the reconstruction of the lymnaeid phylogeny based on a molecular genetic study of the recent pond snails. Since 2010, genetic methods have become the mainstream in the lymnaeid taxonomy and phylogeny, and all most recent attempts to reveal the taxonomic structure for this family heavily relied on molecular phylogenetic information (Puslednik et al. 2009; Correa et al. 2010; Campbell et al. 2017; Aksenova et al. 2018; Saadi et al. 2020).

According to all phylogenetic reconstructions of the recent time, the Lymnaeidae form a well-resolved monophyletic clade among the Hygrophila (Dayrat et al. 2011; Saadi et al. 2020) that evolved, presumably, in the mid-Mesozoic. This clade includes both Lymnaeidae s. str. and Lancidae (ranked as a separate family by Taylor and Sohl 1962 and followed by Starobogatov 1967, 1970). Despite having a set of morphological features absent in the rest of the Lymnaeidae, the latter group proved to be nested within the subfamily Lymnaeinae in our phylogenetic reconstruction (see below) and cannot be ranked even a subfamily (contrary to Campbell et al. 2017). Starobogatov (1976) and Kruglov (2005) supposed that the family evolved from certain "*Chilina*-like" ancestors (Fig. 3.1) though the fossil forms which would support this hypothesis are unknown. The nearest living relatives of the Lymnaeidae belong to the clade comprising the families Bulinidae P. Fischer & Crosse, 1880, Burnupiidae Albrecht, 2017, and Planorbidae Rafinesque, 1815 (= Planorboidea auct.) [Saadi et al. 2020].

This chapter consists of two main parts. The first one provides a new phylogenetic hypothesis for the living representatives of the family, based on a multi-locus approach (COI, 16S rRNA, and 28S rRNA). This hypothesis then becomes a backbone for a new taxonomic structure of the living Lymnaeidae proposed here for the first time. As compared with the previous versions of the system (Vinarski 2013; Campbell et al. 2017), this one is more complicated, with the two recent subfamilies each divided into several tribes. The rank and phylogenetic position of the lancine group (= Lancinae Hannibal, 1914; = Lancidae auct.) is clarified. The second (systematic) part of the chapter provides an annotated list of all recent taxa of the Lymnaeidae between the family and species rank (i.e., subfamilies, tribes, genera, subgenera), with remarks on their taxonomic content, distribution, ecology, etc.

It should be noted that our phylogenetic reconstruction cannot be seen as an ultimate one. Firstly, it does not comprise the extinct taxa, and, that is more important, we still lack available genetic data on some lymnaeid groups of high phylogenetic interest (the genera *Acella*, *Pseudisidora*, *Lantzia*, and some others). Inevitably, the future phylogenetic work, based on an extended set of taxa and more advanced technologies (such as mitogenomics) will shed new light on some

Fig. 3.1 Shells of living *Chilina* that possibly resemble shells of the direct ancestors of the Lymnaeidae (see text for explanation). (**a**) *Chilina parchappii* (D'Orbigny, 1835). (**b**) *Ch. ampullacea* G.B. Sowerby I, 1838. Both shells from the collection of the Natural History Museum, London, UK. Scale bars 5 mm. Photos: Maxim Vinarski

unresolved problems in the Lymnaeidae phylogeny, and the system of the family will change accordingly.

3.2 Material and Methods

Our phylogenetic reconstruction is based on DNA sequences of 124 lymnaeid specimens, representing 27 genera and 116 species, and two outgroup taxa, belonging to the family Physidae (Table 3.1). Here, we used a combined alignment with partial sequences of the mitochondrial *cytochrome c oxidase subunit I* (*COI*), *small ribosomal RNA* (*16S rRNA*), and the nuclear *large ribosomal RNA* (*28S rRNA*) genes. New sequences were generated using the sequencing approach and primer pairs as described in our earlier work (Aksenova et al. 2018). Each gene sequence dataset was aligned separately using the MUSCLE algorithm of MEGA7 (Kumar et al. 2016). Absent sites were treated as missing data. The maximum likelihood phylogeny was reconstructed with IQ-TREE v. 1.6.12 (Nguyen et al. 2015). The analysis was run using an automatic identification of the best evolutionary models for each partition (Kalyaanamoorthy et al. 2017) and an ultrafast bootstrap algorithm

Table 3.1 List of the *COI*, *16S rRNA*, and *28S rRNA* gene sequences used in the multi-locus phylogenetic reconstruction of the Lymnaeidae

Genus	Species	Region	Code	*COI*	*16S rRNA*	*28S rRNA*
Racesina Vinarski & Bolotov, 2018	*R. luteola* (Lamarck, 1822)	Nepal	RacLut	JN794496	JN794322	N/A
	R. oxiana (Boettger, 1889)	Tajikistan	RacOxi	MH189935	**ON620112**	MH168046
	R. siamensis (Sowerby, 1873)	Myanmar	RacSia	MH190023	**ON620118**	MH168050
Radix Montfort, 1810	*R. rubiginosa* (Michelin, 1831)	Indonesia	RadRub	MH189925	**ON620106**	MH168042
	R. brevicauda (Sowerby, 1873)	China: Tibet	RadBre	JN794435	JN794210	N/A
	R. auricularia (Linnaeus, 1758)	Russia: Kamchatka	RadAu1	MH189923	**ON620105**	**ON620444**
	R. auricularia (Linnaeus, 1758)	China: Tibet	RadAu2	MH189863	**ON620099**	MH168033
	R. makhrovi Bolotov, Vinarski & Aksenova, 2018	China: Tibet	RadMak	MH189861	**ON620098**	MH168032
	R. plicatula (Benson, 1842)	China: Tibet	RadPl1	MH190049	**ON620130**	**ON620458**
	R. plicatula (Benson, 1842)	Russian Far East	RadPl2	**ON603571**	**ON620155**	N/A
	R. cf. *swinhoei* (Adams, 1866)	Laos	RadSwi	MT344026	MT345558	N/A
	R. sp. Lake Lugu A	China: Lake Lugu	RadSpA	MT344013	MT345554	N/A
	R. sp. Lake Lugu B	China: Lake Lugu	RadSpB	MT344008	MT345557	N/A
	R. sp. Lake Lugu C	China: Lake Lugu	RadSpC	MT344011	MT345556	N/A
	R. sp. Lake Lugu D	China: Lake Lugu	RadSpD	MT344012	MT345555	N/A
	R. coreana (Martens, 1886)	South Korea	RadCor	**ON603572**	**ON620156**	N/A
	R. japonica (Jay, 1857)	Japan	RadJap	BFMN059–18	N/A	N/A
	R. hamadai Habe, 1968 [=*Radix* sp. Ra-03 Ohari et al., 2020]	Japan	RadShi	LC360961	**ON552481**	**ON552516**
	R. sp. Japan	Japan	RadHon	**ON603567**	**ON620151**	N/A
	R. alticola (Izzatullaev, Kruglov & Starobogatov, 1983)	Tajikistan	RadAlt	MH189949	**ON620113**	MH168047

	R. sp. Trichonis Lake	Greece	RadTri	EU818805	N/A	N/A
	R. euphratica (Mousson, 1874)	Tajikistan	RadEuf	MH189976	**ON620114**	MH168048
	R. dgebuadzei Aksenova, Vinarski, Bolotov & Kondakov, 2019	China: Gansu	RadDge	MN718571	**ON620131**	MN719901
	R. sp. Turkey	Turkey	RadTur	**ON603537**	**ON620108**	ON620445
	R. exsertus (Martens, 1866)	Uganda: Nile Basin	RadExs	**ON603553**	**ON620137**	ON620464
	R. debaizei (Bourguignat, 1887)	Uganda: Lake Victoria	RadDeb	**ON603552**	**ON620136**	ON620463
	R. natalensis (Krauss, 1848)	Cabo Verde	RadNat	HG977206	N/A	N/A
	R. rufescens (Gray, 1822)	Myanmar	RadRuf	MH190025	**ON620119**	MH168051
Ampullaceana Servain, 1881	*A. ampla* (Hartmann, 1821)	European Russia	AmpAmp	MH190044	**ON620124**	MH168052
	A. balthica (Linnaeus, 1758)	European Russia	AmpBal	MH190000	**ON620115**	**ON620447**
	A. sp. Italy	Italy	AmpIta	**ON603562**	**ON620146**	ON620472
	A. intermedia (Lamarck, 1822)	France	AmpInt	KP242511	N/A	N/A
	A. lagotis (Schrank, 1803)	European Russia	AmpLag	MH189858	**ON620096**	MH168030
	A. relicta (Poliński, 1929)	Macedonia	AmpRel	EU818821	JN794307	N/A
	A. fontinalis (Studer, 1820)	European Russia	AmpFon	MH189903	**ON620104**	MH168040
	A. sp. Ohrid	Albania	AmpSp1	EU818833	N/A	N/A
	A. cf. *dipkunensis* (Gundrizer & Starobogatov, 1979)	European Russia	AmpDip	MH189854	**ON620095**	MH168029
Peregriana Servain, 1881	*P. dolgini* (Gundrizer & Starobogatov, 1979)	European Russia	PerDol	MH189886	**ON620101**	MH168036
	P. peregra (O.F. Müller, 1774)	Slovakia	PerPer	MH189931	**ON620109**	MH168044
Kamtschaticana Kruglov & Starobogatov, 1984	*K. kamtschatica* (Middendorff, 1850)	Russia: Kamchatka	KamKam	**ON603574**	**ON620158**	MH168041
	K. sp. Japan [=*Radix* sp. Ra-c2 Ohari et al., 2020]	Japan	KamNip	**LC360970**	**ON552506**	**ON552540**
	K. sp. Kamchatka.	Russia: Kamchatka	KamSp1	**ON603575**	**ON620159**	ON620483

(continued)

Table 3.1 (continued)

Genus	Species	Region	Code	*COI*	*16S rRNA*	*28S rRNA*
Myxas G. B. Sowerby I, 1822	*M. glutinosa* (O. F. Müller, 1774)	European Russia	MyxGlu	**ON603541**	**ON620121**	**ON620450**
Tibetoradix Bolotov, Vinarski & Aksenova, 2018	*T. hookeri* (Reeve, 1850)	China: Tibet	TibHoo	MH189865	**ON620100**	MH168034
	T. kozlovi Vinarski, Bolotov & Aksenova, 2018	China: Tibet	TibKoz	MH190045	**ON620129**	**ON620457**
	T. kruglovi Bolotov, Vinarski & Oheimb, 2021	China: Tibet	TibSp1	JN794395	JN794169	N/A
	T. imitator Vinarski, Bolotov & Oheimb, 2021	China: Tibet	TibSp2	JN794441	JN794216	N/A
	T. khamensis Oheimb, Vinarski & Bolotov, 2021	China: Tibet	TibSp3	JN794384	JN794158	N/A
	T. transhimalayensis Oheimb, Bolotov & Vinarski, 2021	China: Tibet	TibSp4	JN794436	JN794211	N/A
Orientogalba Kruglov & Starobogatov, 1985	*O. ollula* (Gould, 1859)	China: Tibet	OriOll	**ON603547**	**ON620128**	ON620456
	O. bowelli (Preston, 1909)	China: Tibet	OriBow	JN794473	JN794248	N/A
	O. viridis (Quoy & Gaimard, 1833)	Indonesia	OriVir	MH189927	**ON620107**	MH168043
	O. sp. Hokkaido [=*Austropeplea ollula* Ohari et al., 2020]	Japan	OriHok	**LC360950**	**ON552480**	**ON552515**
	O. sp. Uzbekistan	Uzbekistan	OriUzb	**ON603548**	**ON620132**	ON620459
Austropeplea Cotton, 1942	*A. tomentosa* (L. Pfeiffer, 1855)	Australia	AusTom	AY227365	AF485645	HQ156217
	A. hispida (Ponder & Waterhouse, 1997)	Tasmania	AusHis	N/A	EU556268	N/A
Bullastra Bergh, 1901	*B. lessoni* (Deshayes, 1830)	Australia	BulLes	N/A	EU556252	N/A
	B. cumingiana (L. Pfeiffer, 1855)	Philippines	BulCum	N/A	U82068	N/A
Lymnaea Lamarck, 1799	*L. stagnalis* (Linnaeus, 1758) [Asian subclade: European Russia]	European Russia	LymSt1	MH189887	**ON620102**	MH168037
	L. stagnalis (Linnaeus, 1758) [Asian subclade: European Russia]	European Russia	LymSt2	**ON603550**	**ON620134**	ON620461

	L. fragilis (Linnaeus, 1758) Siberia	Russia: Siberia	LymFra	HG932266	N/A	N/A
	L. sorensis B. Dybowski, 1912 [Asian subclade: Western China, Kamchatka, Krasnoyarsk, Kemerovo]	China: Xinjiang	LymEx1	**ON603546**	**ON620127**	ON620455
	L. sorensis B. Dybowski, 1912 [Asian subclade: Western China (Tarim basin), Kamchatka, Krasnoyarsk, Kemerovo]	Russia: East Siberia (Krasno-yarsk Region)	LymEx2	**ON603570**	**ON620154**	ON620479
	L. sp. France [Asian subclade: Iberia]	France	LymAlt	HG932252	N/A	N/A
	L. appressa Say, 1821 [Nearctic subclade: North America]	Canada	LymApp	**ON603557**	**ON620141**	ON620467
	L. sp. Kazakhstan [Asian subclade: Kazakhstan]	Kazakhstan	LymKaz	**ON603559**	**ON620143**	ON620469
	L. sp. Rostov [Nearctic subclade: Rostov]	European Russia	LymRos	**ON603560**	**ON620144**	ON620470
	L. sp. Baltic Sea [Nearctic subclade: Novgorod]	European Russia	LymNov	**ON614717**	**ON620111**	ON620446
	L. sp. Arkhangelsk [East European subclade: Arkhangelsk, Karelia, NAO, KhMAO]	European Russia	LymArk	**ON603554**	**ON620138**	ON620465
	L. sp. Crimea [East European subclade: Crimea]	Crimea	LymCri	**ON603539**	**ON620117**	**ON620448**
Kazakhlymnaea Kruglov & Starobogatov, 1984	*K. taurica* (Clessin, 1880)	Russia:Western Siberia	KazTau	HG932240	N/A	N/A
	K. sp. Italy	Italy	KazIta	HG932245	N/A	N/A
Stagnicola Jeffreys, 1830	*S. fuscus* (C. Pfeiffer, 1821)	Germany	StaFus	HG932234	N/A	N/A
	S. corvus (Gmelin, 1791)	European Russia	StaCo1	MH189932	**ON620110**	MH168045
	S. palustris (O. F. Müller, 1774)	European Russia	StaPal	MH189888	**ON620103**	MH168038
	S. montenegrinus Glöer & Pešič, 2009	Ukraine	StaMon	**ON603555**	ON620139	N/A
		European Russia	StaArc	**ON603540**	**ON620120**	**ON620449**

(continued)

Table 3.1 (continued)

Genus	Species	Region	Code	*COI*	*16S rRNA*	*28S rRNA*
	S. archangelica (Kruglov & Starobogatov, 1986)					
	S. sp. Italy	Italy	StaIta	**ON603563**	**ON620147**	ON620473
	S. sp. Rostov	European Russia	StaRos	**ON603551**	**ON620135**	ON620462
	S. sp. Kyrgyzstan	Russia: Krasnodar Region	StaKyr	**ON603576**	**ON620160**	ON620484
	S. sp. Krasnodar	Russia: Krasnodar Region	StaKra	**ON603579**	**ON620163**	ON620487
	S. turricula (Held, 1836)	Poland	StaTur	KP070778	N/A	N/A
Lanx Clessin, 1880	*L. alta* (Tryon, 1865)	USA: Oregon	LanAlt	HM230361	N/A	HM230318
	L. patelloides (Lea, 1856)	USA: California	LanPat	HM230363	KT267276	HM230322
Idaholanx Clark, Campbell & Lydeard, 2017	*I. fresti* Clark, Campbell & Lydeard, 2017	USA: Idaho	IdaFre	HM230356	KT267273	HM230308
Bulimnea Haldeman, 1841	*B. megasoma* (Say, 1824)	Canada	BulMeg	**ON603556**	**ON620140**	**ON620466**
Hinkleyia Baker, 1928 [=*Sphaerogalba* Kruglov & Starobogatov, 1985 **syn. nov.**]	*H. caperata* (Say, 1829)	USA: New Mexico	HinCap	MF962229	MF962327	N/A
	H. bulimoides (Lea, 1841) **comb. nov.**	USA: Colorado	SphBul	EU038362	EU038315	N/A
Walterlymnaea Starobogatov & Budnikova, 1976	*W. catascopium* (Say, 1817)	USA: Yellowstone National Park	LadCa1	**ON603543**	**ON620123**	**ON620452**
	W. catascopium (Say, 1817)	Russia: Kamchatka	LadCa3	KP830105	**ON620097**	**ON620443**
	W. emarginata (Say, 1821)	USA	LadEma	N/A	U82081	AY465069
	W. bonnevillensis (Call, 1884)	USA: Utah	LadBon	N/A	AF485655	N/A
	W. elrodi (Baker & Henderson, 1933)	USA: Montana	LadElr	N/A	AF485656	N/A
	W. exilis (Lea, 1834)	Canada	LadExi	MG421226	N/A	N/A
	W. sp. Ontario	Canada	LadOnt	MG421900	N/A	N/A

	W. arctica (Lea, 1864)	USA: Alaska	LadArc	**ON603545**	**ON620126**	ON620454
Ladislavella B. Dybowski, 1913	*L. liogyra* (Westerlund, 1897)	Russia: Khanka Lake	LadLio	**MH190007**	**ON620116**	MH168049
	L. terebra (Westerlund, 1885)	Russia: Western Siberia	LadTer	**ON603568**	**ON620152**	ON620477
	L. sp. Altai [*terebra*-group]	Russia: Altai Mts.	LadAlt	**ON603577**	**ON620161**	ON620485
	L. occulta (Jackiewicz, 1959)	Poland	LadOcc	KP070796	N/A	N/A
Dallirhytis Kruglov & Starobogatov, 1989 **stat. rev.**	*D. atkaensis* (Dall, 1884) **comb. rev.**	Russia: Chukchi Peninsula	DalAt1	**ON603573**	**ON620157**	ON620481
	D. atkaensis (Dall, 1884) **comb. rev.**	USA: Alaska	DalAt2	**ON603544**	**ON620125**	**ON620453**
Galba Schrank, 1803	*G. obrussa* (Say, 1825)	North America	GalObr	N/A	AF485658	N/A
	G. parva (Lea, 1841)	Canada	GalPar	KM612176	N/A	N/A
	G. truncatula (O. F. Müller, 1774)	Georgia	GalTru	**ON603566**	**ON620150**	**ON620476**
	G. neotropica (Bargues, Artigas, Mera y Sierra, Pointier & Mas-Coma, 2007)	Uruguay	GalNeo	KX781342	KX712144	N/A
	G. cubensis (L. Pfeiffer, 1839)	USA: South Carolina	GalCub	FN182205	FN182204	N/A
	G. meridensis (Bargues, Artigas, Khoubbane & Mas-Coma, 2011)	Venezuela	GalMer	JN614389	HQ283237	N/A
	G. sp. Cyprus	Cyprus	GalCyp	**ON603565**	**ON620149**	ON620475
	G. humilis (Say, 1822)	USA: Yellow-stone National Park	GalHum	**ON603542**	**ON620122**	**ON620451**
	G. schirazensis (Küster, 1862) [=*Galba* sp. Gt-14 Ohari et al., 2020]	Iran	GalSch	JF272607	JF272605	N/A
	G. viator (Orbigny, 1835)	Argentina	GalVia	JN872451	JN872461	N/A
	G. mweruensis (Connolly, 1929)	Ethiopia	GalMwe	MN601410	MN602707	N/A

(continued)

Table 3.1 (continued)

Genus	Species	Region	Code	*COI*	*16S rRNA*	*28S rRNA*
	G. cousini (Jousseaume, 1887)	Columbia	GalCou	KJ495741	N/A	N/A
	G. sp. Japan. [=*Galba* sp. Gt-c1 Ohari et al., 2020]	Japan	GalPa1	**LC360899**	**ON552492**	**ON552527**
	G. sp. Japan. [=*Galba* sp. Gt-c1 Ohari et al., 2020]	Russia: Kamchatka	GalPa2	**ON603549**	**ON620133**	**ON620460**
Omphiscola Rafinesque, 1819	*O. glabra* (O. F. Müller, 1774)	France	OmpGla	**ON603564**	**ON620148**	**ON620474**
Aenigmomphiscola Kruglov & Starobogatov, 1981	*A. europaea* Kruglov & Starobogatov, 1981	European Russia	AenEur	**ON603569**	**ON620153**	ON620478
	A. cf. *kazakhstanica* Kruglov & Starobogatov, 1981	Russia: Altai Mts.	AenAlt	**ON603578**	**ON620162**	ON620486
Pectinidens Pilsbry, 1911	*P. diaphanus* (King, 1832)	Argentina	PecDia	JN872456	JN872475	N/A
Pseudosuccinea F. C. Baker, 1908	*P. columella* (Say, 1817)	Canada	PseCo1	**ON603561**	**ON620145**	ON620471
	P. columella (Say, 1817)	Canada	PseCo2	**ON603558**	**ON620142**	ON620468
Fisherola Hannibal, 1912	*F. nuttallii* (Haldeman, 1841)	USA: Idaho	FisNut	HM230359	HM230355	HM230315
Erinna H. Adams & A. Adams, 1855	*E. aulacospira* (Ancey, 1899)	USA: Hawaii	EriAul	AY150091	N/A	N/A
Physella Haldeman, 1843[a]	*P. acuta* (Draparnaud, 1805)[a]		PhyAcu	AY651174	AY651213	EF489368
Aplexa J. Fleming, 1820[a]	*A. elongata* (Say, 1821)[a]		AplElo	EU038377	EU038330	AY465071

[a]Outgroup (Physidae). N/A – not available. New sequences generated in this study are in bold

(Hoang et al. 2017) through an online server (http://iqtree.cibiv.univie.ac.at) (Trifinopoulos et al. 2016).

Morphological studies of various lymnaeid snails have been conducted in a number of zoological depositories of Russia and Western Europe that are listed below. During this work, we were able to familiarize ourselves with representatives of all living genera and subgenera of the Lymnaeidae except for *Idaholanx* Clark, Campbell & Lydeard, 2017 and *Kutikina* Ponder and Waterhouse, 1997.

Museum Acronyms:
LMBI – Laboratory of Macroecology & Biogeography of Invertebrates, Saint-Petersburg State University (Saint-Petersburg, Russia); MHNH – Museum national d'Histoire naturelle (France, Paris); NHMUK – Natural History Museum of the United Kingdom (London, UK); NHMW – Natural History Museum of Vienna, Austria; SMF – The Naturmuseum Senckenberg, Frankfurt am Main, Germany; USNM – National Museum of Natural History, Smithsonian Institution, Washington D.C., USA; ZIN – Zoological Institute of the Russian Academy of Sciences (Saint-Petersburg, Russia); ZMB – Natural History Museum of Berlin (Germany).

3.3 Phylogenetic Relationships Within the Family Lymnaeidae

A new multi-locus phylogenetic hypothesis obtained as a result of this study is presented in Fig. 3.2. Though not all recovered branches have equally high statistical support, the basic division of the living Lymnaeidae into two large monophyletic clades corresponding to the two subfamilies delineated by Vinarski (2013)—Lymnaeinae and Amphipepleinae—receives full support. The lancine snails of North America (genera *Lanx*, *Idaholanx*, *Fisherola*) do not constitute a separate clade and, instead, are clustered within the subfamily Lymnaeinae. It corresponds to the earlier findings of Aksenova et al. (2018) and Saadi et al. (2020).

The deep split of the family into two large clades corresponding to subfamilies is based mainly on the molecular phylogenetic data. The only non-molecular character allowing to distinguish between the two taxa is their chromosome numbers: 16 or (most often) 17 chromosome pairs in Amphipepleinae and 18 (rarely 19) pairs in Lymnaeinae (Vinarski, 2013).

Each lymnaeid subfamily comprises several internal clades of different rank and volume (see Fig. 3.2). Within the subfamilies, we recovered a number of groups of closely related genera that may be characterized geographically. Each of these groups, that we ranked here as tribes, is distributed in a more or less well-defined region of the world which can be rather small (e. g., the Tibetan Plateau). The practice of division of a large and speciose family (or subfamily) in a series of tribes has been applied to such families of freshwater molluscs as Unionidae (Lopes-Lima et al. 2017; Bolotov et al. 2018; Froufe et al. 2020) and Planorbidae (Albrecht et al. 2007; Saadi et al. 2020). Unfortunately, a substantial portion of genera within the

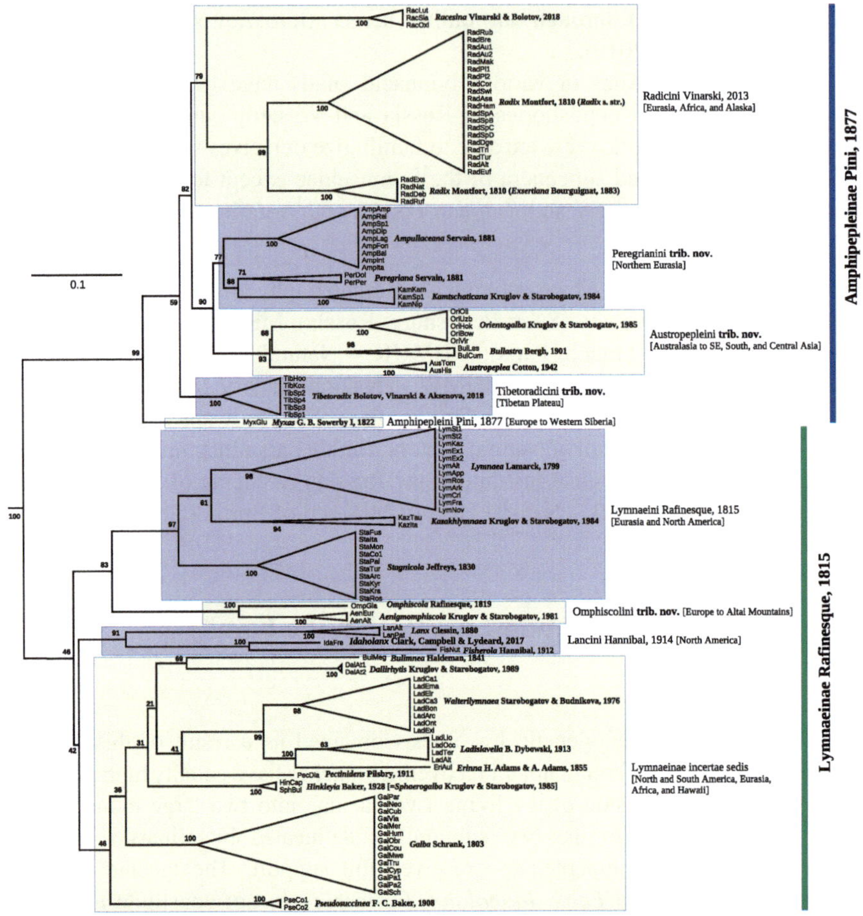

Fig. 3.2 Maximum likelihood consensus phylogeny of the family Lymnaeidae recovered from IQ-TREE analysis and obtained for the complete data set of mitochondrial and nuclear sequences (five partitions: three codons of *COI* + *16S rRNA* + *28S rRNA*). Black numbers near nodes are ultrafast bootstrap support values. The species branches are collapsed to the genus-level clades. The tribe-level clades are highlighted in color. The scale bar indicates the branch lengths. The codes, sequence accession numbers, and sampling regions for each species are given in Table 3.1

Lymnaeinae remains unassigned to tribes owing to the insufficiently high statistical support of the respective clades (see Fig. 3.2). The intergeneric relationships within this subfamily require further research with an expanded set of included genera and taxa. Thus, only three tribes may be surely defined within the subfamily: Lymnaeini Rafinesque, 1815; Omphiscolini **trib. nov.**, and Lancini Hannibal, 1914. Five tribes, including three having no available names, were recovered within the Amphipepleinae (see Fig. 3.2).

As a conclusion, we accept among the recent Lymnaeidae two subfamilies, eight tribes, and 30 genera, which are briefly characterized in the following section.

3.4 Taxonomy (Systematic Part)

The earlier morphology-based works on the taxonomy of the recent Lymnaeidae include, among others, those of W. Dybowski (1903), Baker (1911, 1915), Hubendick (1951), Jackiewicz (1993, 1998), Kruglov and Starobogatov (1993a, b), Ponder and Waterhouse (1997), Kruglov (2005), and Vinarski (2013). Since the late-2000s, most works aiming to construct a system of the family rely chiefly on the molecular genetic evidence (see Bargues et al. 2006; Puslednik et al. 2009; Correa et al. 2010; Campbell et al. 2017; Aksenova et al. 2018, and some others). The working taxonomic system for the living lymnaeids based primarily on the multi-locus phylogeny presented above may be summarized as follows.

Class Gastropoda Cuvier, 1875
Subclass Heterobranchia Burmeister, 1837
Infraclass Euthyneura Spengel, 1881
Subterclass Tectipleura Schrödl, Jörger, Klussmann-Kolb & N. G. Wilson, 2011
Superorder Hygrophila Férussac, 1822 (= Lymnaeiformes Minichev & Starobogatov, 1975)
Superfamily Lymnaeoidea Rafinesque, 1815
Family Lymnaeidae Rafinesque, 1815

(Rafinesque 1815, p. 144)

[= Limnophysidae W. Dybowski, 1903; = Acellinae Haldeman, 1912; = Fossariinae B. Dybowski, 1913].

Type genus: *Lymnaea* Lamarck, 1799.

The family comprises two extant subfamilies—Lymnaeinae Rafinesque, 1815 and Amphipepleinae Pini, 1877.

Subfamily Lymnaeinae Rafinesque, **1815**
Diagnosis. Shell helicoid or patelliform, of different size (shell height varies from 5–6 to almost 70 mm) and shape (from subulate or turriculate to ovate conical, auriculate, or even neritiform). Prostate with one or several (2–10) internal folds or unfolded. Haploid chromosome number 18–19. Copulatory apparatus simple or with a so-called preputial organ. Spermathecal duct long, its length several times exceeds that of spermatheca.

The distribution of this subfamily is almost cosmopolitan, with an exception to Australia and other large Pacific islands (excluding Hawaii), where no aboriginal representatives of the Lymnaeinae are known.

Tribe Lymnaeini Rafinesque, **1815**
Diagnosis. Shell helicoid, large or medium-sized (up to 70 mm height), typically more or less slender, with 5 to 7 whorls and relatively small aperture. Prostate with one or several (2–10) internal folds (branched or unbranched). Copulatory apparatus simple. Penis with or without fixatory ring-like swelling.

1. ***Lymnaea* Lamarck**, 1799 (Fig. 3.3a)

(Lamarck 1799, p. 75)

[= *Helix* Linnaeus, 1758 (partim); = *Lymnaeus* Cuvier, 1816; = *Limnaeus* Pfeiffer, 1821; = *Limnea* Sowerby, 1822 = *Limneus* Sandberger, 1875; = *Limnoturgida* Chiamenti, 1899; = *Limnaea* Germain, 1903, non Poli, 1791 (Bivalvia); = *Eulimneus* Sandberger, 1875; = *Psiliana* Servain, 1882; = *Stagnaliana* Servain, 1882; = *Limnoturgida* Chiamanti, 1899; = *Omphalolimnus* W. Dybowski, 1903; = *Lymnoëa* Suter, 1913; = *Kobeltilymnaea* Kruglov & Starobogatov, 1993].[1]

Type species. *Helix stagnalis* Linnaeus, 1758 = *Lymnaea stagnalis* [M].[2]

Taxonomic content. Two subgenera have been assigned to this genus (Vinarski et al. 2012): *Lymnaea* s. str., and *Kazakhlymnaea* Kruglov and Starobogatov, 1984; the latter is considered below as a separate genus. Traditionally, a single recent species, *Lymnaea stagnalis* (Linnaeus, 1758), is assigned to *Lymnaea* (Glöer 2002; Welter-Schultes 2012; but see Kruglov and Starobogatov 1985c, 1993a; Glöer 2019). The latest molecular genetic data (Aksenova et al., submitted), however, reveals a cryptic genetic diversity within *Lymnaea stagnalis* s. lato, which represents a complex of more than 10 species, one of them—*Lymnaea appressa* (Say, 1821)—is endemic to North America (note that these putative species do not correspond to those delineated by Kruglov and Starobogatov 1985c, 1993a).

Distribution and ecology. The native range of *Lymnaea* lies in the Holarctic. *L. stagnalis* s. lato was introduced into Australia, New Zealand, and tropical Africa (Hubendick 1951; Tchakonté et al. 2014). *L. stagnalis* is a generalist snail inhabiting a very wide range of biotopes and demonstrating an enormous conchological variability (see Vinarski 2015 for review and discussion).

2. *Kazakhlymnaea* **Kruglov and Starobogatov, 1984** (Fig. 3.3b)

(Kruglov and Starobogatov 1984b, p. 66)

Type species. *Lymnaea palustris kazakensis* Mozley, 1934 = *Kazakhlymnaea taurica kazakensis* [OD].

Taxonomic content. The genus includes the type species, with two subspecies (Vinarski et al. 2012; Vinarski and Kantor 2016), and one undescribed species in South Europe.

Distribution and ecology. South Europe to North Kazakhstan and southwest Siberia. *Kazakhlymnaea* snails are living in ephemeral steppe waterbodies (Beriozkina and Starobogatov 1988; Vinarski et al. 2012).

3. *Stagnicola* **Jeffreys,** 1830 (Fig. 3.3c)

(Jeffreys 1830, p. 376)

---

[1]This list of generic synonyms is, most probably, not exhaustive.

[2]Here and below, the following abbreviations are used. M – type species designated by monotypy; OD – type species designated in the original description; SD – subsequent designation of the type species; SM – designated by subsequent monotypy.

Fig. 3.3 Shells of representatives of genera and species of the Lymnaeinae. (**a**) *Lymnaea stagnalis*. Central Kazakhstan, Sholak Lake [LMBI]. (**b**) *Kazakhlymnaea taurica kazakensis* [a syntype of *Lymnaea palustris kazakensis* Mozley, 1934; USNM]. (**c**) *Stagnicola corvus*. Ukraine, Zhitomir Region, Romanov settlement [LMBI]. (**d**) *Omphiscola glabra*. Germany, Nassau [ZIN] (**e**) *Aenigmomphiscola uvalievae* [holotype; ZIN]. (**f**) *Lanx patelloides*. USA, Oregon [SMF]. (**g**) *Fisherola nuttalli*. USA, Oregon, Deschutes River [NHMUK]. (**h**) *Acella haldemani*. USA, Illinois,

[= *Limnophysa* Fitzinger, 1833; = *Fenziana* Servain, 1882; = *Ligericiana* Servain, 1882; = *Palustrisiana* Servain, 1882; = *Costolymnaea* B. Dybowski, 1913; = *Entochilius* Sandberger, 1880; = *Berlaniana* Kruglov et Starobogatov, 1986; = *Kuesterilymnaea* Vinarski, 2003].

Type species. *Limneus communis* Jeffreys, 1830 = *Stagnicola palustris* (O.F. Müller, 1774) [M].

Taxonomic content. Two morphologically-defined subgenera: *Stagnicola* s. str., and *Corvusiana* Servain, 1882 (type species *Helix corvus* Gmelin, 1791) have been delineated (Vinarski 2013; Vinarski and Kantor 2016). Though the latter subgenus has some unique cytogenetic and anatomic differences from *Stagnicola* s. str. (Kruglov and Starobogatov 1984b; Garbar et al. 2004; Vinarski 2013), our results, as well as the results obtained by other workers (Pieńkowska et al. 2015; Pieńkowska and Lesicki 2018), give no strong evidence of its independent phylogenetic position. Possibly, the latter taxon should be treated as a full synonym of *Stagnicola*. Kruglov and Starobogatov (1986, 1993a) and Burch (1989) listed dozens of nominal species of *Stagnicola* living in North Eurasia and North America, but the actual number of valid species is not known yet since some nominal taxa remain unassessed genetically. Perhaps, there are no more than 10–15 living members in this genus.

Distribution and ecology. Palearctic. Numerous species of *Stagnicola* recorded from North America (Burch 1989; Johnson et al. 2013) belong to either *Hinkleyia* or *Ladislavella*. Most species of *Stagnicola* inhabit minor waterbodies as well as shallow zones of large lakes and streams.

Tribe Omphiscolini Bolotov, Vinarski & Aksenova *trib. nov.*

Diagnosis. Shell slender, turriculate, small to medium-sized (up to 20 mm height), with 6 to 8 whorls and diminutive aperture. Prostate with a single internal fold or unfolded. Copulatory apparatus simple or with a so-called preputial organ. Penis without fixatory ring-like swelling.

Type genus. *Omphiscola* Rafinesque, 1819.

4. ***Omphiscola* Rafinesque,** 1819 (Fig. 3.3d)

(Rafinesque 1819, p. 423)

[= *Leptolimnea* Swainson, 1840; = *Glabriana* Servain, 1882]

Type species: *Buccinum glabrum* O.F. Müller, 1774 = *Omphiscola glabra* [SM].

Fig. 3.3 (continued) Lake Cedar [LMBI]. (**i**) *Bulimnea megasoma*. Canada, Ontario, Cart Lake [LMBI]. (**j**) *Dallirhytis atkaensis*. Canada, Yukon Territory [ZIN]. (**k**) *Erinna newcombi*. Hawaii Islanda, Kenda River [NHMUK]. (**l**) *Galba viator*. Peru, Huaura River [SMF]. (**m**) *Ladislavella (Ladislavella) liogyra*. Russia, Nikol'sk-Ussuriysky [ZIN]. (**n**) *Ladislavella (Walterigalba) catascopium*. Greenland [LMBI]. (**o**) *Pectinidens diaphana*. Magellan Strait, King Cape [holotype, NHMUK]. (**p**) *Pseudosuccinea columella*. Germany, Tübingen, invasive [SMF]. (**q**) *Hinkleyia caperata*. USA, Colorado [ZIN]. (**r**) *Pseudisidora (Pseudisidora) rubella* (the lectotype of *Limnaea oahuensis* Souleyet, 1852]. Hawaii, Oahu [MHNH]. (**s**) *Pseudisidora (Pseudobulinus) reticulata*, Hawaii, Kauai [ZMB]. Scale bars 2 mm (D, E, H, K–L, N–S), 5 mm (A–C, F–G, I–J, M). Photos: Maxim Vinarski. See "Material & methods" section for museum acronyms

Taxonomic content, distribution, and ecology: A monotypic genus restricted to northwestern Europe in its distribution (Welter-Schultes 2012). *Omphiscola glabra* inhabits shallow, still waters with dense vegetation (Jackiewicz 1998). Jackiewicz (1998, p. 38) characterizes it as "a rare, even very rare species."

5. ***Aenigmomphiscola* Kruglov and Starobogatov**, 1981 (Fig. 3.3e)

(Kruglov and Starobogatov 1981, p. 966)

Type species: *Aenigmomphiscola europaea* Kruglov and Starobogatov, 1981 [OD].

Taxonomic content, distribution, and ecology: Two species endemic to European Russia, Western Siberia, and Northern Kazakhstan (Vinarski et al. 2011; Vinarski and Grebennikov 2012). The snails typically live in ephemeral habitats such as wetland and floodplain pools, wet meadows (Beriozkina and Starobogatov 1988; Vinarski and Grebennikov 2012).

Tribe Lancini Hannibal, **1914**
(Hannibal, 1914, p. 24)

Type genus. *Lanx* Clessin, 1882.

Diagnosis. Shell patelliform, small to medium-sized (shell width up to 20 mm), aperture broad, ovoid. Two prostatic glands of different structure and the secondarily enlarged columellar muscle (C-shaped or almost circular) constitute the anatomical synapomorphies of this tribe (Hubendick 1951; Campbell et al. 2017).

Remarks. This group of patelliform lymnaeids was ranked as a separate family within Hygrophila by some authors (Pilsbry 1925; Taylor and Sohl 1962; Starobogatov 1967; Gray 1988), while other systematists treated it as a subfamily of the family Lymnaeidae (Burch 1989; Bouchet et al. 2017; Campbell et al. 2017) or even as merely a genus of lymnaeid snails (Hubendick 1951). According to Campbell et al. (2017), the subfamily includes only three genera and four extant species distributed in the Pacific Northwest USA and Canada. The finding of lancine snails in the Cretaceous deposits of Nevada (MacNeil 1939) superficially indicates it is a very ancient taxon, but one cannot exclude the repeated origin of the limpet form in independent clades of lymnaeids as a result of convergent evolution (Vermeij 2017). The absence of a profound genetic distance between the recent lancines and the rest of the Lymnaeinae assumes a relatively young geological age of this tribe. Ecologically, these snails may be characterized as dwellers of well-oxygenated lotic waterbodies such as rivers and streams, living under and on the sides of submerged stones (Gray 1988; Campbell et al. 2017). The genus *Lanx* is also occurring in lakes (Gray 1988).

6. ***Lanx* Clessin,** 1880 (Fig. 3.3f)

(Clessin, 1880, p. 10)

[= *Ancylus* O.F. Müller, 1774 (partim); = *Walkerola* Hannibal, 1912]

Type species. *Ancylus patelloides* I. Lea, 1856 = *Lanx patelloides* [SD, Hubendick, 1951, p. 114].

Taxonomic content and distribution. Two extant species, inhabiting the western USA (California and Oregon).

7. ***Fisherola*** **Hannibal,** 1912 (Fig. 3.3g)

(Hannibal, 1912, p. 10)
[= *Ancylus* O.F. Müller, 1774 (partim)]
Type species. *Fisherola lancides* Hannibal, 1912 = *Fisherola nuttallii* (Haldeman, 1841) [OD].
Taxonomic content and distribution. The genus is known from the west of the USA (California, Oregon) and Canada (British Columbia) [Burch 1989; Campbell et al. 2017].

8. ***Idaholanx*** **Clark, Campbell & Lydeard, 2017**

(Campbell et al. 2017, p. 121)
Type species. *Idaholanx fresti* Clerk, Campbell & Lydeard, 2017 [OD].
Taxonomic content and distribution. A monotypic genus found exclusively in Idaho, USA.

Genera of the Lymnaeinae still unassigned to tribes

The genera listed below in alphabetic order have been unassigned to a particular tribe owing either to the low statistical support of corresponding clades or to the lack of available molecular information on their representatives.

9. ***Acella*** **Haldeman,** 1841 (Fig. 3.3h)

(Haldeman 1841, p. 6)
Type species. *Lymnaea gracilis* Jay, 1839 = *Acella haldemani* (Binney, 1867) [M].
Taxonomic content, distribution, and ecology. A single extant species, the spindle pond snail, *Acella haldemani* (Binney, 1867), is included. It is distributed in the USA and south of Canada (Baker 1911; Burch 1989; Johnson et al. 2013), lives in large perennial lakes and streams (Baker 1911; Taylor et al. 1963).

10. ***Bulimnea*** **Haldeman,** 1841 (Fig. 3.3i)

(Haldeman 1841, p. 6)
Type species. *Lymnaea megasoma* Say, 1824 = *Bulimnea megasoma* [M].
Taxonomic content, distribution, and ecology. A monotypic genus endemic to North America: inhabits Great Lakes and St. Lawrence river drainage area and parts of the Canadian Interior Basin (Burch 1989). Lives in large perennial rivers and lakes (Taylor et al. 1963), registered from sloughs and ponds (Baker 1911).

11. ***Dallirhytis*** **Kruglov & Starobogatov,** 1989 (Fig. 3.3j)

(Kruglov and Starobogatov, 1989, p. 15)
Type species. *Lymnaea petersi* Dall, 1905 = *Dallirhytis atkaensis* (Dall, 1884) [OD].

Taxonomic remark. Established as a "section" of the subgenus *Polyrhytis* Meek, 1876 of the genus *Lymnaea* s. lato (Kruglov and Starobogatov 1989, 1993a). The use of the generic name *Polyrhytis*, with a fossil *Lymnaea kingi* Meek, 1876 as its type species, is, in our opinion, unwarranted especially since we have no data on the real phylogenetic relationship between *L. kingi* and the recent lymnaeids of North America (see Vinarski 2012 for further discussion of the meaning and applicability of the name *Polyrhytis*). Walter (cited after Clarke 1973, p. 310) once characterized the type species of this genus as "a chimaera having marked advanced and primitive stagnicoline anatomical features, and in this it looks like a hybrid between *Lymnaea stagnalis* and *L. catascopium*." The independent generic position of this species proposed here allows us to explain this observation.

Taxonomic content, distribution, and ecology. A single species included, *Dallirhytis atkaensis*, is a "glacial" relict of the Beringian freshwater malacofauna (Clarke 1973); now it is distributed in Alaska, north-western Canada, and eastern Chukotka (Clarke 1973, 1981; Kruglov and Starobogatov 1993a). The ecological information on this snail is relatively scarce. In British Columbia it lives in "clear, cold, oligotrophic lakes" (Clarke 1981, p. 128); in Chukotka it is registered from lakes and "large lentic waterbodies" (Kruglov 2005, p. 177).

12. ***Erinna* H. Adams & A. Adams in A. Adams,** 1855 (Fig. 3.3**k**)

(A. Adams 1855, p. 120)
[= *Pelagolimnaea* Germain, 1928]
Type species. *Erinna newcombi* H. & A. Adams, 1855 [M].

Taxonomic content, distribution, and ecology. This taxon includes two species endemic to Hawaii (Johnson et al. 2013). One of these species, *E. aulacospira* (Ancey, 1899), is, probably, extinct now (Cowie et al. 2017). *Erinna* snails were reported from rivers and waterfalls, as well as from the wet surface rocks of a precipice (Baker 1911; Hubendick 1952).

Remark. Our phylogeny recovered *Erinna* as a part of *Ladislavella*, which makes the latter a paraphyletic taxon (see Fig. 3.2). Possibly, *Ladislavella* s. lato must be divided into several genera.

13. ***Galba* Schrank,** 1803 (Fig. 3.3l)

(Schrank 1803, p. 262)
[= *Buccinum* O.F. Müller, 1774 (partim); = *Truncatuliana* Servain, 1882; = *Fossaria* Westerlund, 1885; = *Microlimnaea* W. Dybowski, 1908; = *Palustria* W. Dybowski, 1908; = *Turrilimnaea* W. Dybowski, 1908; = *Simpsonia* F.C. Baker, 1911; = *Pseudogalba* F.C. Baker, 1913; = *Nasonia* F.C. Baker, 1928; = *Montigalba* Izzatullaev, Kruglov et Starobogatov, 1983; = *Afrogalba* Kruglov et Starobogatov, 1985].

Type species. *Buccinum truncatulum* O.F. Müller, 1774 = *Galba truncatula* [SD: ICZN, 1998].

Taxonomic content and nomenclature. Vinarski (2013) considered this genus as including three morphologically-defined subgenera: *Galba* s.str., *Bakerilymnaea* Weyrauch, 1964 (type species *Limnaea cubensis* Pfeiffer, 1839), and *Sibirigalba*

Kruglov & Starobogatov, 1985 (type species *Limnaea truncatula* var. *sibirica* Westerlund, 1885). Our phylogenetic analysis recovered *Galba* as a compact monophyletic clade not subdivided into subgenera (see also Alda et al. 2021). The actual number of valid species in the genus is difficult to define, since many nominal species (especially numerous representatives of *Galba* occurring in North America; see Burch 1989; Johnson et al. 2013; Alda et al. 2021) have yet not been studied molecularly. Possibly, there are 15–20 valid species in this genus. Some recent malacologists, especially those working beyond Europe, prefer to use the generic name *Fossaria* instead of *Galba* (Burch 1989; Ponder and Waterhouse 1997). The reason is that the taxonomic identity of the type species of *Galba* has been thought to be unidentifiable (Ponder and Waterhouse 1997). In this case, however, the generic name *Truncatuliana* Servain, 1882 takes precedence before *Fossaria*.

Distribution and ecology. Most species of *Galba* s. str. inhabit the Americas (Burch 1989; Artigas et al. 2011; Standley et al. 2013), two representatives of the genus, *G. truncatula* (O.F. Müller, 1774) and *G. schirazensis* (Küster, 1862), are broadly distributed in the Palearctic (van Damme 1984; Kruglov 2005; Bargues et al. 2011; Mahulu et al. 2019). Possibly, both of them are of American origin and were introduced to the Old World in the historical time (see Correa et al. 2010; Lounnas et al. 2018), however, this conclusion has been disputed (Artigas et al. 2011; Bargues et al. 2011; Mahulu et al. 2019). The fossil records of *G. truncatula* (or closely related species) in the Neogene and Quaternary deposits of Eastern Europe (Danilovsky 1955; Ložek 1964; Sanko 2007), Siberia (Popova 1981), and the Late Pleistocene and Holocene of North Africa (van Damme 1984) contradict the hypothesis on the recent introduction of this species from the New World. *G. truncatula* has been introduced to New Zealand (Climo and Pullan 1972). Recently, Mahulu et al. (2019) demonstrated that *G. mweruensis* (Connolly, 1929), which is widely distributed through Sub-Saharan Africa, is a distinct species, possibly sister to *G. truncatula*; the status of *G. robusta* Vinarski, 2018 (type locality situated in Yemen) is unclear because of the lack of molecular data.

Most representatives of *Galba* are semi-amphibious or truly amphibious snails living in seasonal pools, ditches, wet shores of lakes, streams and springs, on wet surfaces of stones (Frömming 1956; Beriozkina and Starobogatov 1988; Jackiewicz 1998).

14. ***Hinkleyia* F.C. Baker,** 1928 (Fig. 3.3q)

(Baker 1928, p. 259)

[= *Sphaerogalba* Kruglov & Starobogatov, 1985]

Type species. *Lymnaeus caperatus* Say, 1829 = *Hinkleyia caperata* [OD].

Taxonomic content, distribution, and ecology. Three species of *Hinkleyia* were listed by North American malacologists as being distributed in the Nearctic (Taylor et al. 1963; Burch 1989). They may live in seasonal bodies of water (sloughs, irrigation ditches, shallow ponds) as well as in small springs and mountain streams (Taylor et al. 1963). Taylor et al. (1963) and Burch (1989) treated *Hinkleyia* as a subgenus of *Stagnicola*. Kruglov and Starobogatov (1985b) separated *H. montanensis* in a subgenus of its own, *Walterigalba* Kruglov and Starobogatov,

1985. Due to the lack of molecular data, it is difficult to decide if *Walterigalba* represents a valid subgenus, or it is merely a synonym of *Hinkleyia*. The genus *Sphaerogalba* Kruglov & Starobogatov, 1985 (type species *Lymnea bulimoides* I. Lea, 1841), with more than 10 species distributed in North and South America (Kruglov and Starobogatov 1985b), according to our phylogenetic hypothesis (see Fig. 3.2) must be treated as a junior synonym of *Hinkleyia*. At least one of the species included by Kruglov and Starobogatov (1985b) to *Sphaerogalba*, the South American *Lymnaea viator* (d'Orbighy, 1835), belongs to the genus *Galba* (Artigas et al. 2011). The habitats of *Sphaerogalba* are very similar to these of *Hinkleyia* (Kruglov and Starobogatov 1985b).

15. ***Ladislavella* B. Dybowski,** 1913 (Fig. 3.3m, n)

(B. Dybowski 1913, p. 179)

[= *Polyrhytis* Meek, 1876 sensu Kruglov et Starobogatov, 1989 (partim); = *Catascopia* Meier-Brook et Bargues et, 2002].

Type species. *Ladislavella sorensis* B. Dybowski, 1913 = *Ladislavella terebra* (Westerlund, 1885) [SD, Hubendick 1951, p. 116].

Taxonomic content and nomenclature. Vinarski (2012) assigned two subgenera to this genus: *Ladislavella* s. str. (two species), and *Walterlymnaea* Starobogatov & Budnikova, 1976 (type species: *Lymnaea catascopium* Say, 1817). The latter group corresponds to the *L. emarginata/catascopium* and *L. elodes* species groups (sensu Burch 1989). Our multi-locus phylogeny recovered *Walterigalba* as a sister group to *Ladislavella* s. str. + *Erinna* clade. Possibly, *Walterlymnaea* has to be assigned to a genus of its own, but the absence of molecular data on two lymnaeid taxa which are endemic to Hawaii (i.e., *Pseudisidora* s.str. and *Pseudobulinus*) do not allow us to resolve this question here. *Ladislavella* sensu Vinarski (2012) embraces approximately 20 nominal taxa (Burch 1989; Johnson et al. 2013), but it is not clear yet how many valid species are there in *Walterlymnaea*.

Distribution and ecology. *Ladislavella* s. str. is a Palearctic subgenus broadly distributed in Central and Eastern Europe as well as in Asiatic Russia (Vinarski 2012); it includes not less than three recent species. Another subgenus is confined almost exclusively to North America and Greenland, with a single subspecies in the Asiatic part of Beringia (Vinarski et al. 2016b, 2017). *Ladislavella* may occupy a wide range of habitats, from perennial large lakes to seasonal pools and thermal springs (Vinarski et al. 2016b, 2023).

16. ***Pectinidens* Pilsbry,** 1911 (Fig. 3.3o)

(Pilsbry 1911, p. 522)

Type species. *Lymnaea diaphana* King, 1832 = *Pectinidens diaphanus* [OD].

Taxonomic content, distribution, and ecology. This genus, whose validity has recently been supported by a molecular study (Bargues et al. 2012), includes 1–2 species inhabiting southernmost parts of South America and the Falkland Islands (Paraense 1984; Bargues et al. 2012). The snails dwell in small waterbodies—ponds, ditches, in wetlands (Paraense 1984; Bargues et al. 2012).

17. ***Pseudisidora* Thiele,** 1931 (Fig. 3.3r,s)

(Thiele 1931, p. 476)

Type species: *Lymnaea rubella* Lea, 1844 = *Pseudisidora rubella* [M].

Taxonomic content, distribution, and ecology: A small group comprising a few (2–3) species distributed in Hawaii. Kruglov and Starobogatov (1989, 1993a) included three species described from the Kamchatka Peninsula (eastern Asiatic Russia) there, but this opinion is very doubtful from the zoogeographic point of view and needs a confirmation. Two subgenera are recognized (Vinarski 2013): *Pseudisidora* s. str., and *Pseudobulinus* Kruglov & Starobogatov, 1993 (type species—*Physa reticulata* Gould, 1847). The snails of the genus have been reported from streams, artificial pools, waterfalls, ditches, and springs (Hubendick 1952).

18. ***Pseudosuccinea* F.C. Baker,** 1908 (Fig. 3.3p)

(Baker, 1908, p. 943)

Type species: *Lymnaea columella* Say, 1817 = *Pseudosuccinea columella* [OD].

Taxonomic content, distribution, and ecology: The genus is distributed in both Americas (Baker 1911; Roszkowski 1927; Burch 1989). From 1927 onward, *Pseudosuccinea columella* has been reported from glasshouses and aquaria of different countries of Europe where it is broadly distributed now (Schlesch 1930; Madsen and Frandsen 1989; Lounnas et al. 2017); also it has been registered as a non-indigenous species in Sub-Saharan Africa (van Eeden and Brown 1966; Brown 1994; Tchakonté et al. 2014), South America (Davies et al. 2014; Martín et al. 2016), and in the Pacific Region—Hawaii, Australia, and New Zealand (Climo and Pullan 1972; Madsen and Frandsen 1989). All indigenous species of China and India classified by some authors as belonging to *Pseudosuccinea* (Yen 1939; Subba Rao 1989) are, actually, members of the genus *Radix*. *P. columella* is a generalist species with the highest invasive potential. It has been reported from waterbodies of a different sort (Roszkowski 1927; Welter-Schultes 2012), though it clearly prefers lotic habitats and is rarely registered from streams and brooks (Baker, 1911).

Subfamily Amphipepleinae Pini, **1877**
(Pini, 1877, p. 174)

[? = Valencienninae Kramberger-Gorjanovic, 1923]

Type genus. *Amphipeplea* Nillson, 1822 = *Myxas* G.B. Sowerby I, 1822.

Diagnosis. Shell helicoid, small to medium-sized (shell height varies from 5–6 to around 35 mm), of various shape (from ovate-conical to auriculate or neritiform). Prostate with one internal fold or unfolded. Haploid chromosome number 16–17. Copulatory apparatus simple. Penis without fixatory ring-like swelling. The spermathecal duct may be long, short, or virtually absent.

Remark. The only synapomorphy characteristic of this subfamily is the reduced haploid chromosome number (n = 16–17) [Vinarski 2013]. The Amphipepleinae are distributed almost worldwide, with most genera and species native to the Old World (Starobogatov 1970; Kruglov and Starobogatov 1993a; Vinarski 2013; Aksenova et al. 2018; Vinarski et al. 2020).

Tribe Amphipepleini Pini, **1877**

Diagnosis. Shell of moderate size, broad, almost ovoid, very fragile, with extremely expanded body whorl and almost invisible spire. The mantle of a living animal is reflected over its shell. Prostate with a single internal fold. The spermathecal duct is relatively short, its length is roughly equal to that of spermatheca or slightly exceeds it.

The tribe is endemic to Europe.

19. ***Myxas* G.B. Sowerby I, 1822** (Fig. 3.4a)

(Sowerby 1822: part VI)

[= *Amphipeplea* Nilsson, 1822; = *Lutea* T. Brown, 1827; = *Cyclolimnea* Dall, 1905].

Type species. *Buccinum glutinosum* O.F. Müller, 1774 = *Myxas glutinosa* [M].

Taxonomic content, distribution, and ecology. 1–2 species distributed in the north-western part of Palearctic westward to south-western Siberia (Jackiewicz 1998; Glöer 2002, 2019; Welter-Schultes 2012; Vinarski and Kantor 2016). Other *Myxas*-similar lymnaeids of the world have been classified within separate taxa showing no close affinity to this genus (Kruglov and Starobogatov 1985a; Vinarski et al. 2021). *Myxas* usually inhabit stagnant, permanent, and well-oxygenated waterbodies (Feliksiak 1939; Jackiewicz, 1998; Glöer 2002).

Tribe Austropepleini Bolotov, Vinarski & Aksenova trib. nov.

Type genus. *Austropeplea* Cotton, 1942.

Diagnosis. Shell of moderate size, conical to almost round. Penis simple, prostate with a single unbranched fold. The spermathecal duct relatively short. This tribe has no morphological synapomorphies and is delineated solely on the basis of molecular evidence.

20. ***Austropeplea* Cotton,** 1942 (Fig. 3.4e)

(Cotton 1942, p. 80)

[= *Glacilimnea* Iredale, 1943; = *Simlimnea* Iredale, 1943]

Type species: *Lymnaea aruntalis* Cotton & Godfrey, 1938 = *Austropeplea papyracea* (Tate, 1880) [OD].

Taxonomic content, distribution, and ecology. The genus comprises a few (6–7) species distributed in Australia, New Guinea, Tasmania, and New Zealand (Starobogatov 1970; Ponder and Waterhouse 1997). Two subgenera are recognized (Ponder et al. 2020): *Austropeplea* s. str., and *Kutikina* Ponder and Waterhouse, 1997, with *Kutikina hispida* Ponder and Waterhouse, 1997 as its type species. Ecologically, the representatives of this genus may be characterized as dwellers of ponds, springs, and wetlands (Boray 1964).

21. ***Bullastra* Bergh,** 1901 (Fig. 3.4f)

(Bergh 1901, p. 254)

[=? *Peplimnea* Iredale, 1943]

Type species. *Bullastra velutinoides* Bergh, 1901 = *Bullastra cumingiana* (L. Pfeiffer, 1855) [M].

Taxonomic content, distribution, and ecology. Five species, inhabiting Australia, the Philippines, and the Moluccas (Kruglov and Starobogatov, 1985a; Ponder et al. 2020; Vinarski et al. 2020). The ecology of *Bullastra* is similar to that of the previous genus.

22. ***Orientogalba* Kruglov & Starobogatov, 1985** (Fig. 3.4g)

(Kruglov and Starobogatov 1985b, p. 28)

[= *Lenagalba* Kruglov & Starobogatov, 1985; = *Viridigalba* Kruglov & Starobogatov, 1985]

Type species. *Lymnaea heptapotamica* Lazareva, 1967 = *Orientogalba heptapotamica* [OD].

Taxonomic content, distribution, and ecology. A small group consisting of 3–4 species widely distributed in Asia (except for its western and northern parts) and in some Pacific islands (Kruglov and Starobogatov 1993b; Aksenova et al. 2018; Vinarski et al. 2020), introduced into southwest Europe (*O. viridis*; see Schniebs et al. 2017). Most species are semi-amphibious and inhabit small ponds, pools, and wetlands (Kruglov and Starobogatov 1985b).

Tribe Peregrianini Bolotov, Vinarski & Aksenova trib. nov.

Type genus. *Peregriana* Servain, 1882.

Diagnosis. Shell of moderate size (shell height up to 20 mm), conical or ovate-conical. Penis simple, prostate with a single unbranched fold. The spermathecal duct is typically short or virtually absent. This tribe has no morphological synapomorphies and is delineated solely on the basis of molecular evidence.

23. ***Peregriana* Servain,** 1882 (Fig. 3.4j)

(Servain 1882, p. 56)

Type species. *Buccinum peregrum* O.F. Müller, 1774 = *Peregriana peregra* [SD; Kobelt 1883, p. 14].

Taxonomic content, distribution, and ecology. According to Aksenova et al. (2018), the genus comprises two species. The range of *Peregriana* covers Europe and Siberia (Vinarski et al. 2016a; Aksenova et al. 2018). The snails of this genus are inhabitants of non-permanent waterbodies of various types; sometimes may occur in small permanent wetlands (Vinarski et al. 2016a).

24. ***Ampullaceana* Servain,** 1882 (Fig. 3.4i)

(Servain 1882, p. 53)

[= *Biformiana* Servain 1882; = *Bouchardiana* Servain, 1882; = *Caenisiana* Servain 1882; = *Effusiana* Servain 1882; = *Limosiana* Servain, 1882; = *Nivalisiana* Servain 1882; = *Ohridlymnaea* Kruglov & Starobogatov 1993]

Type species. *Limnaeus ampullaceus* Rossmäßler, 1835 =? *Ampullaceana balthica* (Linnaeus, 1758), SD. According to Kruglov and Starobogatov (1993b), *L. ampullaceus* is a valid species allied to *A. balthica*.

Fig. 3.4 Shells of representatives of genera and species of the Amphipepleinae. (**a**) *Myxas glutinosa*. North Kazakhstan, the Tobol River floodplain [ZIN]. (**b**) *Radix* (*Radix*) *makhrovi*. Tibet, west of the Lhasa River mouth [holotype, ZIN]. (**c**) *Radix* (*Exsertiana*) *rufescens*. Myanmar, Yeto River [LMBI]. (**d**) *Racesina oxiana*. Tajikistan, Kurban-Shakhid [LMBI]. (**e**) *Austropeplea papyracea*. Australia, Torrens River at Adelaida [MHNH]. (**f**) *Bullastra lessoni*. Australia, Brisbane [ZMB]. (**g**) *Orientogalba viridis*, Guam Island [syntype, MHNH]. (**h**) *Kamtschaticana kamtschatica*. Russia, Kamchatka, Esso River [ZIN]. (**i**) *Ampullaceana ampla*. Ukraine, Transkarpatian Region [LMBI]. (**j**) *Peregriana peregra*. Austria, South Tirol [NHMW]. (**k**) *Tibetoradix imitator*, Tibet, pond near Drongpa Tradun [holotype, ZIN]. (**l**) *Limnobulla peculiaris*. Falkland Islands [holotype, damaged; NHMUK]. (**m**) *Lantzia carinata*. Réunion Island [NHMW]. Scale bars 2 mm (B, C, E, G, H, J, L, M), 5 mm (A, D, F, I, K). Photos: Maxim Vinarski. See "Material & methods" section for museum acronyms

Taxonomic content, distribution, and ecology. Aksenova et al. (2018) include in this genus seven genetically defined species, some of them, such as *A. lagotis* (Schrank, 1803), are widely distributed throughout Europe, Siberia, and Central Asia. The species *A. relicta* (Poliński, 1929), with its two subspecies, is endemic to large Balkan lakes, Ohrid and Prespa (Aksenova et al. 2018). The native range of the genus is Palearctic; at least one species, *A. balthica*, has recently been introduced to eastern Canada (Vinarski et al. 2022b).

Most representatives of *Ampullaceana* live in small stagnant waterbodies and slowly running streams, though some species are known to inhabit the shallow zones of large lakes or thermal springs and pools (Kruglov 2005; Bolotov et al. 2017; Vinarski et al. 2020). A series of nominal species belonging to this genus remains unassessed genetically, and their validity, as well as proper generic position, are unknown.

25. ***Kamtschaticana* Kruglov & Starobogatov, 1984** (Fig. 3.4h)

(Kruglov & Starobogatov 1984a, p. 30)
[= *Lymnaea* (*Pacifimyxas*) Kruglov & Starobogatov, 1985]
Type species. *Limnaeus kamtschaticus* Middendorff, 1850 [OD].

Taxonomic content, distribution, and ecology. A single species, *K. kamtschatica* (Middendorff, 1850), is included, and two more species living in Kamchatka and Japan are not formally described yet (Aksenova, Bolotov, Vinarski, unpublished). *K. kamtschatica* inhabits the northeastern part of Eurasia and has not been found outside Russia, except for Alaska (Kruglov and Starobogatov 1984a, 1993b; Aksenova et al. 2018; Vinarski et al. 2021). *K. kamtschatica* lives in waterbodies of different type, including wet shores of lakes and rivers, and the geothermal sites (e.g., the Valley of Geysers, Kamchatka), where it can form sustainable populations in warm water up to +39.9 °C (Aksenova et al. 2016).

Tribe Radicini Vinarski, 2013
(Vinarski, 2013, p. 51)

Type genus. *Radix* Montfort, 1810.

Diagnosis. Shell of medium size (shell height up to 35 mm), high-conical to broadly ear-shaped. Penis simple, prostate one-folded or (in *Racesina*) with several internal folds. Spermathecal duct long. This tribe has no morphological synapomorphies and is delineated solely on the basis of molecular data.

26. ***Radix* Montfort,** 1810 (Fig. 3.4b,c)

(Montfort, 1810, p. 266)

[= *Helix* Linnaeus, 1758 (partim); = *Xymorus* Gebler, 1829; = *Gulnaria* Turton, 1831; = *Neritostoma* Adams & Adams, 1855; = *Cerasina* Kobelt, 1881; = *Auriculariana* Servain, 1882; = *Raffrayana* Bourguignat, 1883; = *Desertiradix* Kruglov & Starobogatov, 1989; = *Iraniradix* Kruglov & Starobogatov, 1989; = *Okhotiradix* Kruglov & Starobogatov, 1989; = *Pamiriradix* Kruglov & Starobogatov, 1989; = *Thermoradix* Kruglov & Starobogatov, 1989; = *Ussuriradix* Kruglov & Starobogatov, 1989].[3]

[3]The full synonymy of this genus is too extensive to be given here in full.

Type species. *Radix auriculatus* Montfort, 1810 = *Radix auricularia* (Linnaeus, 1758) [OD].

Taxonomic content. The genus contains two subgenera: *Radix* s.str. and *Exsertiana* Bourguignat, 1883 [type species *Limnaeus natalensis* var. *exsertus* Martens, 1866 = *Radix natalensis* (Krauss, 1848)]. The latter subgenus in its distribution is restricted to the tropical regions of the Old World (Vinarski et al. 2020). According to different authors, the genus includes 2 (Jackiewicz 1998) to more than 40 (Kruglov 2005) species, but the validity of many nominal species of *Radix* accepted by the latter author has recently been rejected as a result of the integrative taxonomic studies (Bolotov et al. 2014; Aksenova et al. 2016, 2017). The exact number of valid species of *Radix* is unknown, it may be 15–20.

Distribution and ecology: *Radix* is widely distributed in Eurasia and Africa but absent from Australia and South America (Starobogatov 1970; Aksenova et al. 2018; Vinarski et al. 2020). One species (*R. auricularia*) was introduced into North America (Burch 1989). Most representatives of *Radix* live in streams and permanent lakes, reservoirs, and quarries. They are phytophilic snails, usually restricted to shallow zones of waterbodies (Jackiewicz 1998).

27. ***Racesina* Vinarski & Bolotov,** 2018 (Fig. 3.4d)

(Vinarski and Bolotov 2018, p. 332)

Type species: *Lymnaea luteola* Lamarck, 1822 [OD].

Taxonomic content, distribution, and ecology: The genus contains not less than 3 species distributed in Central and South Asia, from Tajikistan and Uzbekistan southwards to Ceylon, Thailand, and south China (Brandt 1974; Kruglov 2005; Aksenova et al. 2018; Vinarski and Bolotov 2018; Vinarski et al. 2020). The molluscs dwell in small bodies of water, often in non-permanent ones (Subba Rao 1989; Kruglov 2005).

Remark: Three species included in *Racesina* by Vinarski and Bolotov (2018) were previously treated as members of the (sub-) genus *Cerasina* Kobelt, 1881 (Subba Rao 1989; Kruglov 2005; Aksenova et al. 2018). As Vinarski and Bolotov (2018) have shown, the genus *Cerasina* sensu Kobelt is a junior synonym of *Radix*, and a new generic name was proposed by these authors for *Cerasina* sensu auct. non Kobelt.

Tribe Tibetoradicini Bolotov, Vinarski & Aksenova trib. nov.

Type genus. *Tibetoradix* Bolotov, Vinarski and Aksenova, 2018.

Diagnosis. Shell of medium size (shell height up to 23 mm), high-conical to almost ear-shaped. Penis simple, prostate with a single unbranched fold. Spermathecal duct long. This tribe has no morphological synapomorphies and is delineated solely on the basis of molecular evidence.

A single genus is included. This tribe is endemic to the Tibetan Plateau and thus is characterized by the narrowest range among all lymnaeid tribes.

28. ***Tiberoradix* Bolotov, Vinarski & Aksenova, 2018** (Fig. 3.4k)

(Aksenova et al., 2018, p. 11)

Type species. *Lymnaea hookeri* Reeve, 1850 [OD].

Taxonomic content. The genus comprises not less than six species (Aksenova et al. 2018; Vinarski et al. 2022a). *Tibetoradix* is endemic to the Tibetan Plateau in China, where the crown group of this genus evolved in the Miocene (Vinarski et al. 2022a). The snails of this genus are known to live in streams, wetlands, and floodplain waterbodies. *Tibetoradix* is a unique genus among freshwater snails completely consisting of high-mountain species adapted to live at heights between 3500 and 5000 m a.s.l.

Genera of the Amphipepleinae still unassigned to tribes

Two nominal genera, on which no molecular genetic information is accessible, are provisionally classified here within the Amphipepleinae, however, their placement to particular tribes stays unclear.

29. ***Lantzia* Jousseaume,** 1872 (Fig. 3.4m)

(Jousseaume 1872, p. 5)

Type species. *Lantzia carinata* Jousseaume, 1872 [M].

Taxonomic content, distribution, and ecology. The monotypic genus which is found exclusively in the Réunion Island lying in the Indian Ocean (Brown 1994; Ponder and Waterhouse 1997). Lives in waterfalls (Brown 1994). Cowie et al. (2017) mention *Lantzia carinata* as probably an extinct species.

30. ***Limnobulla* Kruglov & Starobogatov, 1985** (Fig. 3.4l)

(Kruglov and Starobogatov 1985a, p. 71)

Type species. *Lymnaea peculiaris* Hubendick, 1951 = *Limnobulla peculiaris* [OD].

Taxonomic content, distribution, and ecology. A single species of *Limnobulla* inhabits the Falkland Islands. It is unknown both anatomically and ecologically (Kruglov and Starobogatov 1985a; Ponder and Waterhouse 1997) and has been provisionally included in the Amphipepleinae.

Acknowledgments The authors thank the Russian Science Foundation for the financial support of our research on the phylogeny and systematics of the Lymnaeidae (projects Nos. 19-14-00066/P and 21-74-10155).

References

Adams A (1855) Descriptions of two new genera and several new species of Mollusca, from the collection of Hugh Cummings, Esq. P Zool Soc Lond 23:119–124

Aksenova OV, Vinarski MV, Bolotov IN et al (2016) An overview of *Radix* species of the Kamchatka Peninsula (Gastropoda: Lymnaeidae). Bulletin of the Russian Far Eastern Malacological Society 20:5–27

Aksenova O, Vinarski M, Bolotov I et al (2017) Two *Radix* spp. (Gastropoda: Lymnaeidae) endemic to thermal springs around Lake Baikal represent ecotypes of the widespread *Radix auricularia*. J Zool Syst Evol Res 55(4):298–309. https://doi.org/10.1111/jzs.12174

Aksenova OV, Bolotov IN, Gofarov MY et al (2018) Species richness, molecular taxonomy and biogeography of the radicine pond snails (Gastropoda: Lymnaeidae) in the Old World. Sci Rep-UK 8:11199. https://doi.org/10.1038/s41598-018-29451-1

Albrecht C, Kuhn K, Streit B (2007) A molecular phylogeny of Planorboidea (Gastropoda: Pulmonata): insights from enhanced taxon sampling. Zool Scr 36:27–39. https://doi.org/10.1111/j.1463-6409.2006.00258.x

Alda P, Lounnas M, Vázquez AA et al (2021) Systematics and geographical distribution of *Galba* species, a group of cryptic and worldwide freshwater snails. Molec Phyl Evol 157:107035. https://doi.org/10.1016/j.ympev.2020.107035

Artigas P, Bargues MD, Mera y Sierra RL et al (2011) Characterisation of fascioliasis lymnaeid intermediate hosts from Chile by DNA sequencing, with emphasis on *Lymnaea viator* and *Galba truncatula*. Acta Trop 120:245–257. https://doi.org/10.1016/j.actatropica.2011.09.002

Baker FC (1908) Suggestions for a natural classification of the family Lymnæidæ. Science 27(703): 942–943

Baker FC (1911) The Lymnaeidae of North and Middle America. Special publication of the Chicago Academy of Sciences 3:1–539. https://doi.org/10.5962/bhl.title.20500

Baker FC (1915) On the classification of the Lymnaeids. The Nautilus 29(2):20–24

Baker FC (1928) The fresh water Mollusca of Wisconsin. Part I. Gastropoda. Bulletin of the Wisconsin Geological and Natural History Survey 70(1):xx+1–x507

Bargues MD, Mera y Sierra RL, Artigas P (2012) DNA multigene sequencing of topotypic specimens of the fascioliasis vector *Lymnaea diaphana* and phylogenetic analysis of the genus *Pectinidens* (Gastropoda). Mem I Oswaldo Cruz 107(1):111–124. https://doi.org/10.1590/s0074-02762012000100016

Bargues MD, Artigas P, Jackiewicz M et al (2006) Ribosomal DNA ITS-1 sequence analysis of European stagnicoline Lymnaeidae (Gastropoda). Heldia 6(1–2):57–68

Bargues MD, Artigas P, Khoubbane M et al (2011) *Lymnaea schirazensis*, an overlooked snail distorting fascioliasis data: genotype, phenotype, ecology, worldwide spread, susceptibility, applicability. PLoS One 6:e24567. https://doi.org/10.1371/journal.pone.0024567

Bergh R (1901) Bullacea. In: Reisen in Archipel der Philippinen von Dr. C. Semper. C.W. Kreidel, Wiesbaden 7(4):209–256

Beriozkina GV, Starobogatov YI (1988) Reproductive ecology and egg clusters of freshwater Pulmonata. Trudy Zoologicheskogo Instituta AN SSSR 174:1–306. [in Russian]

Bolotov I, Bespalaya Y, Aksenova O et al (2014) A taxonomic revision of two local endemic *Radix* spp. (Gastropoda: Lymnaeidae) from Khodutka geothermal area, Kamchatka, Russian Far East. Zootaxa 3869:585–593. https://doi.org/10.11646/zootaxa.3869.5.9

Bolotov IN, Aksenova OV, Bespalaya YV et al (2017) Origin of a divergent mtDNA lineage of a freshwater snail species, *Radix balthica*, in Iceland: cryptic glacial refugia or a postglacial founder event? Hydrobiologia 787:73–98. https://doi.org/10.1007/s10750-016-2946-9

Bolotov IN, Pfeiffer JM, Konopleva ES et al (2018) A new genus and tribe of freshwater mussel (Unionidae) from Southeast Asia. Sci Reps-UK 8:10030. https://doi.org/10.1038/s41598-018-28385-y

Boray JC (1964) Studies on the ecology of *Lymnaea tomentosa*, the intermediate host of *Fasciola hepatica*: I. History, geographical distribution, and environment. Aust J Zool 12(2):217–230. https://doi.org/10.1071/ZO9640217

Bouchet P, Rocroi JP, Hausdorf B et al (2017) Revised classification, nomenclator and typification of gastropod and monoplacophoran families. Malacologia 61(1–2):1–526. https://doi.org/10.4002/040.061.0201

Brandt RAM (1974) The non-marine aquatic Mollusca of Thailand. Arch Molluskenkd 105:1–423

Brown DS (1994) Freshwater snails of Africa and their medical importance. Taylor and Francis, London. https://doi.org/10.1201/9781482295184

Burch JB (1989) North American freshwater snails. Malacological Publications, Hamburg, MI

Campbell DC, Clark SA, Lydeard C (2017) Phylogenetic analysis of the Lancinae (Gastropoda, Lymnaeidae) with a description of the U.S. federally endangered Banbury Springs lanx. Zookeys 663:107–132. https://doi.org/10.3897/zookeys.663.11320

Clarke AH (1973) The freshwater molluscs of the Canadian Interior Basin. Malacologia 13:1–509

Clarke AH (1981) The freshwater molluscs of Canada. National Museum of Natural Sciences, Ottawa

Clessin S (1880) Die Familie der Ancylinen. Systematisches Conchylien-Cabinet von Martini und Chemnitz. Bauer und Raspe, Nürnberg 1(6):1–40

Climo FM, Pullan NB (1972) A taxonomic review of the family Lymnaeidae (Mollusca: Gastropoda) in New Zealand. J Roy Soc New Zeal 2:5–13. https://doi.org/10.1080/03036758.1972.10423300

Correa AC, Escobar JC, Durand P et al (2010) Bridging gaps in the molecular phylogeny of the Lymnaeidae (Gastropoda: Pulmonata), vectors of fascioliasis. BMC Evol Biol 10:381. https://doi.org/10.1186/1471-2148-10-381

Cotton BC (1942) Some australian freshwater Gasteropoda. Trans R Soc S Aust 66(1):75–82

Cowie RH, Régnier C, Fontaine R et al (2017) Measuring the sixth extinction: what do mollusks tell us? The Nautilus 131(1):3–41

Danilovsky IV (1955) Scandinavian glaciation deposits key site of the Russian plain and index quaternary molluscs. Gosgeoltekh Publishers, Moscow. [in Russian]

Davies D, Nieva L, Choke LA et al (2014) First record of *Pseudosuccinea columella* (Say, 1817) from Salta province, Northwest Argentina (Mollusca: Gastropoda: Lymnaeidae). CheckList 10(3):597–599. https://doi.org/10.15560/10.3.597

Dayrat B, Conrad M, Balayan S et al (2011) Phylogenetic relationships and evolution of pulmonate gastropods (Mollusca): new insights from increased taxon sampling. Mol Phylogenet Evol 59: 425–437. https://doi.org/10.1016/j.ympev.2011.02.014

de Montfort PD (1810) Conchyliologie Systématique et classification Méthodique des coquilles, vol 2. Coquilles Univalves, Non Cloisonées. F. Schoell, Paris, pp 1–676. https://doi.org/10.5962/bhl.title.10571

Dybowski W (1903) Bemerkungen über die gegenwärtige Systematik der Süsswasserschnecken. Nachrichtsblatt der Deutschen Malakozoologischen Gesellschaft 35(9–10):130–145

Dybowski B (1913) Bemerkungen und Zusatze zu der Arbeit von Dr. W. Dybowski "Molluscen aus der Uferregion des Baicalsees". Ezhegodnik Zoologicheskogo muzeya Imperatorskoy Akademii nauk 17:165–218

Feliksiak S (1939) Über Biologie und Morphologie der Mantelschnecke, *Radix glutinosa* (O.F. Müller). Zoologische Jahrbücher, Abteilung für Systematik, Ökologie und Geografie der Tiere 72:17–70

Frömming E (1956) Biologie der mitteleuropäischen Süßwasserschnecken. Dunker-Humblot, Berlin

Froufe E, Bolotov I, Aldridge DC et al (2020) Mesozoic mitogenome rearrangements and freshwater mussel (Bivalvia: Unionoidea) macroevolution. Heredity 124:189–196. https://doi.org/10.1038/s41437-019-0242-y

Garbar AV, Manilo VV, Korniushin AV (2004) Probable directions of evolution of karyotypes of the European Lymnaeidae (Mollusca, Gastropoda, Pulmonata) in the light of modern concepts of phylogeny of the family. Vestnik zoologii 38(2):29–37. [in Russian]

Glöer P (2002) Die Sußwassergastropoden Nord- und Mitteleuropas: Bestimmungschlussel, Lebenweise, Verbreitung. Die Tierwelt Deutschlands. Conchbooks, Hackenheim 73:1–327

Glöer P (2019) The freshwater gastropods of the west-Palaearctis. Volume 1. Fresh- and brakish waters except spring and subterranean snails. Identification key, anatomy, ecology, distribution. The author, Hetlingen

Gray J (1988) Evolution of the freshwater ecosystem: the fossil record. Palaeogeogr Palaeocl 62:1–214. https://doi.org/10.1016/0031-0182(88)90054-5

Haldeman SS (1841) A monograph of the Limniades and other fresh-water univalve shells of North America. J Dobson, Philadelphia 3:1–16. https://doi.org/10.5962/bhl.title.11226

Hannibal HA (1912) A synopsis of the recent and tertiary freshwater Mollusca of the California Province, based upon an ontogenetic classification. Proc Malac Soc London 10(2):112–165. https://doi.org/10.1093/oxfordjournals.mollus.a063488

Hannibal HA (1914) Note on the classification of Ancylidae. The Nautilus 28(2):23–24

Hoang DT, Chernomor O, von Haeseler A et al (2017) UFBoot2: improving the ultrafast bootstrap approximation. Mol Biol Evol 35:518–522. https://doi.org/10.1093/molbev/msx281

Hubendick B (1951) Recent Lymnaeidae. Their variation, morphology, taxonomy, nomenclature and distribution. Kungliga Svenska Vetenskapsakademiens Handlingar Fjärde Serien 3(1): 1–223

Hubendick B (1952) Hawaiian Lymnaeidae. Occasional Papers of Bernice P, vol 20(19). Bishop Museum, Honolulu, Hawaii, pp 307–328

ICZN (1998) Opinion 1896. *Galba* Schrank, 1803 (Mollusca: Gastropoda): *Buccinum truncatulum* designated as the type species. Bull Zool Nomencl 55(2):123

Inaba A (1969) Cytotaxonomic studies of lymnaeid snails. Malacologia 7(2/3):143–168

Jackiewicz M (1993) Phylogeny and relationships within the European species of the family Lymnaeidae. Folia Malacologica 5:61–95

Jackiewicz M (1998) European species of the family Lymnaeidae (Gastropoda, Pulmonata, Basommatophora). Genus 9(1):1–93

Jeffreys JG (1830) A synopsis of the testaceous pneumobranchous Mollusca of Great Britain. Trans Linn Soc Lond 16:323–392

Johnson PD, Bogan AE, Brown KM et al (2013) Conservation status of freshwater gastropods of Canada and the United States. Fisheries 38(6):247–282

Jousseaume F (1872) Description de quatre mollusques nouveaux. Revue et Magasin de Zoologie Pure et Appliquée, series 2(23):5–15

Kalyaanamoorthy S, Minh BQ, Wong TKF et al (2017) ModelFinder: fast model selection for accurate phylogenetic estimates. Nat Methods 14:587–589. https://doi.org/10.1038/nmeth.4285

Kobelt W (1883) Erster Nachtrag zur zweiten Auflage des Catalogs der im europäischen Faunengebiet lebenden Binnenconchylien. Nachrichtsblatt der Deutschen Malakozoologischen Gesellschaft 15(1/2):1–25

Kruglov ND (2005) Lymnaeid snails of Europe and Northern Asia. Smolensk State Pedagogical University Press, Smolensk. [in Russian]

Kruglov ND, Starobogatov YI (1981) A new genus of the Lymnaeidae and taxonomy of the subgenus *Omphiscola* (*Lymnaea*) (Pulmonata, Gastropoda). Zool Zh 60(7):965–977. [in Russian]

Kruglov ND, Starobogatov YI (1984a) A contribution to the morphology and taxonomy of the subgenus *Peregriana* (*Lymnaea,* Gastropoda, Pulmonata) of the Asiatic part of the USSR and adjacent regions. Zool Zh 63:22–33. [in Russian]

Kruglov ND, Starobogatov YI (1984b) Morphological peculiarities and taxonomy of mollusks of the subgenus *Corvusiana* of the genus *Lymnaea* (Gastropoda, Pulmonata, Lymnaeidae). Byulleten' Moskovskogo Obshchestva Ispytateley Prirody, otdel biologicheskiy 89(2):58–70. [in Russian]

Kruglov ND, Starobogatov YI (1985a) *Myxas*-similar Lymnaeidae (Gastropoda, Pulmonata), their origin and specific composition. Byulleten' Moskovskogo Obshchestva Ispytateley Prirody, otdel biologicheskiy 90(2):69–78. [in Russian]

Kruglov ND, Starobogatov YI (1985b) The volume of the subgenus *Galba* and of other similar subgenera of the genus *Lymnaea* (Gastropoda, Pulmonata). Zool Zh 64(1):24–35. [in Russian]

Kruglov ND, Starobogatov YI (1985c) Methods of experimental hybridization and some results of its applications in the taxonomy of Lymnaeidae. Malacol Rev 18:21–35

Kruglov ND, Starobogatov YI (1986) Mollusks of the subgenus *Stagnicola* of the genus *Lymnaea* of the USSR fauna. Byulleten' Moskovskogo Obshchestva Ispytateley Prirody, otdel biologicheskiy 91(2):59–72. [in Russian]

Kruglov ND, Starobogatov YI (1989) Mollusks of the subgenus *Polyrhytis* of the genus *Lymnaea* in the fauna of the USSR (Pulmonata, Lymnaeidae). Zool Zh 68(3):14–20. [In Russian]

Kruglov ND, Starobogatov YI (1993a) Annotated and illustrated catalogue of species of the family Lymnaeidae (Gastropoda Pulmonata Lymnaeiformes) of Palaearctic and adjacent river drainage areas. Part I Ruthenica: Russian Malacol J 3(1):65–92

Kruglov ND, Starobogatov YI (1993b) Annotated and illustrated catalogue of species of the family Lymnaeidae (Gastropoda Pulmonata Lymnaeiformes) of Palaearctic and adjacent river drainage areas. Part II Ruthenica: Russian Malacol J 3(1):161–180

Kumar S, Stecher G, Tamura K (2016) MEGA7: molecular evolutionary genetics analysis version 7.0 for bigger datasets. Mol Biol Evol 33(7):1870–1874. https://doi.org/10.1093/molbev/msw054

Lamarck JB (1799) Prodrome d'une nouvelle classification des coquilles, comprenant une rédaction appropriée des caractères génériques, et l'établissement d'un grand nombre de genres nouveaux. Lu à l'Institut national, le 21 frimaire an VII. Mémoires de la Société d'Histoire Naturelle de Paris 1:63–91

Lopes-Lima M, Froufe E, Do VT et al (2017) Phylogeny of the most species-rich freshwater bivalve family (Bivalvia: Unionida: Unionidae): defining modern subfamilies and tribes. Mol Phylogenet Evol 106:174–191. https://doi.org/10.1016/j.ympev.2016.08.021

Lounnas M, Correa AC, Vázquez AA et al (2017) Self-fertilization, long-distance flash invasion and biogeography shape the population structure of *Pseudosuccinea columella* at the worldwide scale. Mol Ecol 26:887–903. https://doi.org/10.1111/mec.13984

Lounnas M, Correa AC, Alda P et al (2018) Population structure and genetic diversity in the invasive freshwater snail *Galba schirazensis*. Can J Zool 96(5):425–435. https://doi.org/10.1139/cjz-2016-0319

Ložek V (1964) Quartärmollusken der Tschechoslowakei. Rozpravy Ústředního Ústavu Geologického 31:1–374

MacNeil FS (1939) Fresh-water invertebrates and land plants of Cretaceous age from Eureka, Nevada. J Paleontol 13:355–360

Madsen H, Frandsen F (1989) The spread of freshwater snails including those of medical and veterinary importance. Acta Trop 46:139–146. https://doi.org/10.1016/0001-706x(89)90030-2

Mahulu A, Clewing C, Stelbrink B et al (2019) Cryptic intermediate snail host of the liver fluke *Fasciola hepatica* in Africa. Parasite Vectors 12:573. https://doi.org/10.1186/s13071-019-3825-9

Martín PR, Ovando XMC, Seuffert ME (2016) First record of the freshwater snail *Pseudosuccinea columella* (Gastropoda: Lymnaeidae) in southern Pampas (Argentina) and assessment of future spread. Mollusc Res 36(3):213–221. https://doi.org/10.1080/13235818.2015.1128602

Nguyen LT, Schmidt HA, Haeseler VA et al (2015) IQ-TREE: a fast and effective stochastic algorithm for estimating maximum-likelihood phylogenies. Mol Biol Evol 32:268–274. https://doi.org/10.1093/molbev/msu300

Paraense WL (1984) *Lymnaea diaphana*: a study of topotypic specimens (Pulmonata: Lymnaeidae). Mem I Oswaldo Cruz 79:75–81. https://doi.org/10.1590/S0074-02761984000100009

Pieńkowska JR, Lesicki A (2018) A note on the status of *Galba occulta* Jackiewicz, 1959 (Gastropoda: Hygrophila: Lymnaeidae). Folia Malacologica 26(4):231–247. https://doi.org/10.12657/folmal.026.02

Pieńkowska JR, Rybska E, Banasiak J et al (2015) Taxonomic status of *Stagnicola palustris* (O.F. Müller, 1774) and *S. turricula* (Held, 1836) (Gastropoda: Pulmonata: Lymnaeidae) in view of new molecular and chorological data. Folia Malacologica 23(1):3–18. https://doi.org/10.12657/folmal.023.003

Pilsbry HA (1911) Non-marine Mollusca of Patagonia. Reports of the Princeton University expeditions to Patagonia, 1896–1899. Princeton, New York, The University; Stuttgart, Schweizerbart 3(5):513–633. https://doi.org/10.5962/bhl.title.12486

Pilsbry HA (1925) The family Lancidae distinguished from the Ancylidae. The Nautilus 38:73–75

Pini N (1877) Molluschi terrestri e d'acque dolce viventi nel territorio d'Esino. Bullettino della Società Malacologica Italiana 2(2):67–205

Ponder WF, Waterhouse J (1997) A new genus and species of Lymnaeidae from the lower Franklin River, south western Tasmania. J Mollus Stud 63:441–468. https://doi.org/10.1093/mollus/63.3.441

Ponder WF, Hallan A, Shea ME et al (2020) Australian freshwater molluscs. Revision 1. https://keys.lucidcentral.org/keys/v3/freshwater_molluscs/

Popova SM (1981) The Cenozoic continental malacofauna of southern Siberia and adjacent territories. Nauka Publishers, Moscow. [in Russian]

Puslednik L, Ponder WF, Dowton M et al (2009) Examining the phylogeny of the Australasian Lymnaeidae (Heterobranchia: Pulmonata: Gastropoda) using mitochondrial, nuclear and morphological markers. Mol Phylogenet Evol 52(3):643–659. https://doi.org/10.1016/j.ympev.2009.03.033

Rafinesque CS (1815) Analyse de la nature ou Tableau de l'Univers et des corps organisés. de l'Auteur, Palerme. https://doi.org/10.5962/bhl.title.106607

Rafinesque CS (1819) Prodrome. De 70 nouveaux genres d'animaux découverts dans l'intérieur des Etats-Unis d'Amérique, durant l'année 1818. Journal de Physique, de Chimie, d'Histoire Naturelle et des Arts 88:417–429

Roszkowski W (1927) Contributions to the study of the family Lymnaeidae VIII. The genus *Pseudosuccinea* from South Brazil. Annales Zoologici Musei Polonici Historiae Naturalis 6(1):1–33

Saadi AJ, Davison A, Wade CM (2020) Molecular phylogeny of freshwater snails and limpets (Panpulmonata: Hygrophila). Zool J Linnean Soc 190(2):518–531. https://doi.org/10.1093/zoolinnean/zlz177

Sanko AF (2007) Quaternary freshwater mollusks of Belarus and neighboring regions of Russia, Lithuania, Poland (field guide). Institute of Geochemistry and Geophysics, National Academy of Sciences, Belarus, Minsk. [in Russian]

Schlesch H (1930) *Pseudosuccinea peregrina* Clessin, espèce d'Amérique du Sud, introduite dans les jardins des plantes de l'Europe. Bulletin de la Sociéte Zoologique de France 55(5):425–426

Schniebs K, Glöer P, Vinarski MV et al (2017) A new alien species in Europe: first record of *Austropeplea viridis* (Quoy & Gaimard, 1833) in Spain. J Conchol 42(5):357–370

Schrank von Paula F (1803) Fauna Boica. Durchgedachte Geschichte der in Baiern einheimischen und zahmen Thiere, vol 3(2). Philipp Krull, Landshut, pp 1–372

Servain G (1882 ["1881"]). Histoire malacologique du lac Balaton en Hongrie. S. Lejay et Co., Poissy. https://doi.org/10.5962/bhl.title.10458

Sowerby GBI (1822) The genera of recent and fossil shells, for the use of students in conchology and geology, vol 1. G.B. Sowerby, London, pp 1–274. https://doi.org/10.5962/bhl.title.86281

Standley CJ, Prepelitchi L, Pietrokovsky SM et al (2013) Molecular characterization of cryptic and sympatric lymnaeid species from the *Galba/Fossaria* group in Mendoza Province, Northern Patagonia, Argentina. Parasit Vectors 6:304. https://doi.org/10.1186/1756-3305-6-304

Starobogatov YI (1967) On the systematization of freshwater pulmonate molluscs. Trudy Zoologicheskogo Instituta AN SSSR 42:280–304. [in Russian]

Starobogatov YI (1970) Molluscan fauna and zoogeographic zonation of continental freshwater bodies of the world. Nauka, Leningrad. [in Russian]

Starobogatov YI (1976) The system and phylogeny of the Lymnaeidae. Problemy Zoologii. Zoologicheskiy Institut AN SSSR, p 79–81 [in Russian]

Subba Rao MV (1989) Handbook: freshwater molluscs of India. Zoological Survey of India, Calcutta

Taylor DW, Sohl NF (1962) An outline of gastropod classification. Malacologia 1(1):7–32

Taylor DW, Walter HJ, Burch JB (1963) Freshwater snails of the subgenus *Hinkleyia* (Lymnaeidae: Stagnicola) from the Western United States. Malacologia 1(2):237–281

Tchakonté S, Ajeagah GA, Diomandé C et al (2014) Diversity, dynamic and ecology of freshwater snails related to environmental factors in urban and suburban streams in Douala-Cameroon (Central Africa). Aquat Ecol 48:379–395. https://doi.org/10.1007/s10452-014-9491-2

Thiele J (1931) Handbuch der Systematischen Weichtierkunde, vol 2. Gustav Fischer Verlag, Jena, pp 377–778

Trifinopoulos J, Nguyen LT, von Haeseler A et al (2016) W-IQ-TREE: a fast online phylogenetic tool for maximum likelihood analysis. Nucleic Acids Res 44:W232–W235. https://doi.org/10.1093/nar/gkw256

van Damme D (1984) The freshwater Mollusca of northern Africa: distribution, biogeography and paleoecology. Dev Hydrobiol 25:1–164

van Eeden JA, Brown DS (1966) Colonization of fresh waters in the Republic of South Africa by *Lymnaea columella* Say (Mollusca: Gastropoda). Nature 210:1172–1173

Vermeij G (2017) The limpet form in gastropods: evolution, distribution, and implications for the comparative study of history. Biol J Linn Soc 120:22–37. https://doi.org/10.1111/bij.12883

Vinarski MV (2012) The lymnaeid genus *Catascopia* Meier-Brook et Bargues, 2002 (Mollusca: Gastropoda: Lymnaeidae), its synonymy and species composition. Invertebr Zool 9(2):91–104

Vinarski MV (2013) One, two, or several? How many lymnaeid genera are there? Ruthenica Russian Malacol J 23(1):41–58

Vinarski MV (2015) Conceptual shifts in animal systematics as reflected in the taxonomic history of a common aquatic snail species (*Lymnaea stagnalis*). Zoosyst Evol 91(2):91–103. https://doi.org/10.3897/zse.91.4509

Vinarski MV, Bolotov IN (2018) *Racesina*, a new generic name for a group of Asian lymnaeid snails (Gastropoda: Hygrophila: Lymnaeidae). Zoosystematica Rossica 27:328–333. https://www.zin.ru/journals/zsr/content/2018/zr_2018_27_2_Vinarski.pdf

Vinarski MV, Grebennikov ME (2012) An overview of species of the genus *Aenigmomphiscola* Kruglov et Starobogatov, 1981 (Gastropoda: Pulmonata: Lymnaeidae). Ruthenica The Russian Malacol J 22(2):159–170. [in Russian]

Vinarski MV, Kantor YI (2016) Analytical catalogue of fresh and brackish water molluscs of Russia and adjacent countries. A.N. Severtsov Institute of Ecology and Evolution of RAS, Moscow

Vinarski M, Schniebs K, Glöer P et al (2011) The taxonomic status and phylogenetic relationships of the genus *Aenigmomphiscola* Kruglov and Starobogatov, 1981 (Gastropoda: Pulmonata: Lymnaeidae). J Nat Hist 45(33–34):2049–2068. https://doi.org/10.1080/00222933.2011.574800

Vinarski M, Schniebs K, Glöer P et al (2012) The steppe relics: taxonomic study on two lymnaeid species endemic to the former USSR (Gastropoda: Pulmonata: Lymnaeidae). Arch Molluskenk 141(1):67–85. https://doi.org/10.1127/arch.moll/1869-0963/141/067-085

Vinarski MV, Aksenova OV, Bespalaya YV et al (2016a) *Radix dolgini*: the integrative taxonomic approach supports the species status of a Siberian endemic snail (Mollusca, Gastropoda, Lymnaeidae). C R Biol 339(1):24–36. https://doi.org/10.1016/j.crvi.2015.11.002

Vinarski MV, Aksenova OV, Bespalaya YV et al (2016b) *Ladislavella tumrokensis*: the first molecular evidence of a Nearctic clade of lymnaeid snails inhabiting Eurasia. Syst Biodivers 14(3):276–287. https://doi.org/10.1080/14772000.2016.1140244

Vinarski MV, Bolotov IN, Schniebs K et al (2017) Endemics or strangers? The integrative re-appraisal of taxonomy and phylogeny of the Greenland Lymnaeidae (Mollusca: Gastropoda). C R Biol 340:541–557. http://dx.doi.org/10.1016/j.crvi.2017.09.005

Vinarski MV, Aksenova OV, Bolotov IN (2020) Taxonomic assessment of genetically-delineated species of radicine snails (Mollusca, Gastropoda, Lymnaeidae). Zoosyst Evol 96(2):577–608. https://doi.org/10.3897/zse.96.52860

Vinarski MV, Aksenova OV, Bespalaya YV et al (2021) One Beringian genus less: a re-assessment of *Pacifimyxas* Kruglov & Starobogatov, 1985 (Mollusca: Gastropoda: Lymnaeidae) questions the current estimates of Beringian biodiversity. J Zool Syst Evol Res 59:44–59. https://doi.org/10.1111/jzs.12411

Vinarski MV, von Oheimb PV, Aksenova OV et al (2022a) Trapped on the roof of the world: taxonomic diversity and evolutionary patterns of Tibetan plateau endemic freshwater snails

(Gastropoda: Lymnaeidae: *Tibetoradix*). Integr Zool 17:825–848. https://doi.org/10.1111/1749-4877.12600

Vinarski MV, Aksenova OV, Bolotov IN (2022b) A new alien snail *Ampullaceana balthica* for the Canadian fauna, with an overview of transatlantic malacofaunal exchange in the Anthropocene. Aquat Invasions 17(1):21–35. https://doi.org/10.3391/ai.2022.17.1.02

Vinarski MV, Aksenova OV, Bespalaya YV et al (2023) How an ecological race is forming: Morphological and genetic disparity among thermal and non-thermal populations of aquatic lymnaeid snails (Gastropoda: Lymnaeidae). Diversity 15:548. https://doi.org/10.3390/d15040548

Welter-Schultes F (2012) European non-marine molluscs: a guide for species identification. Planet Poster Editions, Göttingen

Yen TC (1939) Die Chinesische Land- und Süsswasser-Gastropoden in Natur-Museums Senckenberg. Abhandlungen der Senckenbergischen naturforschenden Gesellschaft 444:1–233

Chapter 4
Conchological and Anatomical Identification of the Lymnaeid Snails

Maxim V. Vinarski and Jean-Pierre Pointier

Abstract This chapter aims to familiarize the readers with terminology and basic concepts applied to conchological and anatomical characters that have been widely used for the identification of lymnaeid snails. The main diagnostic traits of the lymnaeid shell and soft body anatomy are reviewed and illustrated.

For a very long time, shell characters served as the sole source of information for the classification and identification of lymnaeid snails. Shell size, shape, proportions, sculpture, coloration were viewed as the main source of taxonomic signal even though all these characters are subject to enormous variation, both at the intra- and interpopulation level. As Falniowski (1980, p. 327) stated, the shell variation among lymnaeids is "probably the largest one in all of the freshwater gastropods." Today, in the epoch of the molecular taxonomy predominance, the significance of conchological information for the lymnaeid identification is lowered but the shells and their properties did not lose completely their usefulness. This is especially evident in paleontology, where shells represent virtually the only material available to researchers. Another field where conchology still keeps its importance is working with historical (museum) samples, including type specimens of lymnaeid species described in the mid-eighteenth–mid-twentieth centuries. In the majority of cases, the pond snail specimens collected till the dawn of the last century are represented in museum collections by dried shells only. At last, for amateurs and "citizen scientists," the examination of shells of lymnaeids and other families of freshwater molluscs is virtually the only method of their fast identification, which may be

M. V. Vinarski (✉)
Laboratory of Macroecology & Biogeography of Invertebrates, Saint-Petersburg State University, Saint-Petersburg, Russia

S.I. Vavilov Institute for the History of Science and Technology, St. Petersburg Branch, Russian Academy of Sciences, Saint-Petersburg, Russia

J.-P. Pointier
PSL Research University, USR 3278 CNRS–EPHE, CRIOBE Université de Perpignan, Perpignan, France
e-mail: pointier@univ-perp.fr

M. V. Vinarski, A. A. Vázquez (eds.), *The Lymnaeidae*, Zoological Monographs 7,
https://doi.org/10.1007/978-3-031-30292-3_4

Fig. 4.1 The main parts of a lymnaeid shell (as exemplified by shells of *Omphiscola glabra*, left, and *Myxas glutinosa*, right). Abbreviations: *ap*—apex; *apr*—aperture; *bm*—basal margin; *bw*—body whorl; cf—columellar fold; *cm*—columellar margin; *pam*—parietal margin; *plm*—palatal margin; *sp*—spire; *st*—suture

important for monitoring of invasive species. Below, we provide an overview of the lymnaeid shells, with emphasis on structures and characters most frequently used in identification keys (see, for example, Glöer 2002; Kruglov 2005; Khokhutkin et al. 2009; Andreeva et al. 2010; Piechocki and Wawrzyniak-Wydrowska 2016; Vinarski 2019; Pointier and Vàzquez 2020).

This chapter aims to familiarize the readers with terminology and basic concepts applied to conchological and anatomical characters that have been widely used for the identification of lymnaeid snails. One can find similar accounts in works by previous authors (Jackiewicz 1998; Stadnichenko 2004; Kruglov 2005; Andreeva et al. 2010; Vinarski 2019; Pointier and Vàzquez 2020). An overview of molecular techniques used for the taxonomic identification of lymnaeid snails can be found in Alda et al. (2023).

Whereas most of the recent lymnaeid snails possess turbospiral shells, a small group of limpets (sometimes separated as a distinct subfamily, the Lancinae) is known within the family. A turbospiral lymnaeid shell comprises several parts, illustrated in Fig. 4.1. The uppermost point of a shell, where coiling starts, is named *apex*. The *aperture* (or *mouth*), the opening through which the soft body extends, is situated on the opposite part of a shell tube formed during the coiling. In all pulmonate aquatic snails, including Lymnaeidae, the aperture lacks an *operculum* (characteristic for the gill-breathing freshwater Gastropoda) as well as other protective structures. The *spire* and the *body (ultimate) whorl* represent two more shell parts important for identification.

In addition, some other terms related to gastropod shell morphology should be explained here (see the full glossary in Cox 1960):

The *shell axis* is an imaginary line through the shell apex, about which whorls are coiled (not applicable to limpets). It is surrounded by a solid or hollow pillar, not visible from outside, which is formed by adaxial (this term means “toward shell axis inward”; Cox 1960) walls of whorls. This pillar is called the *columella*.

The *suture* is a continuous line adjoining two consequent whorls. This line can be deep or relatively shallow, straight, or oblique. Its state depends on the degree of

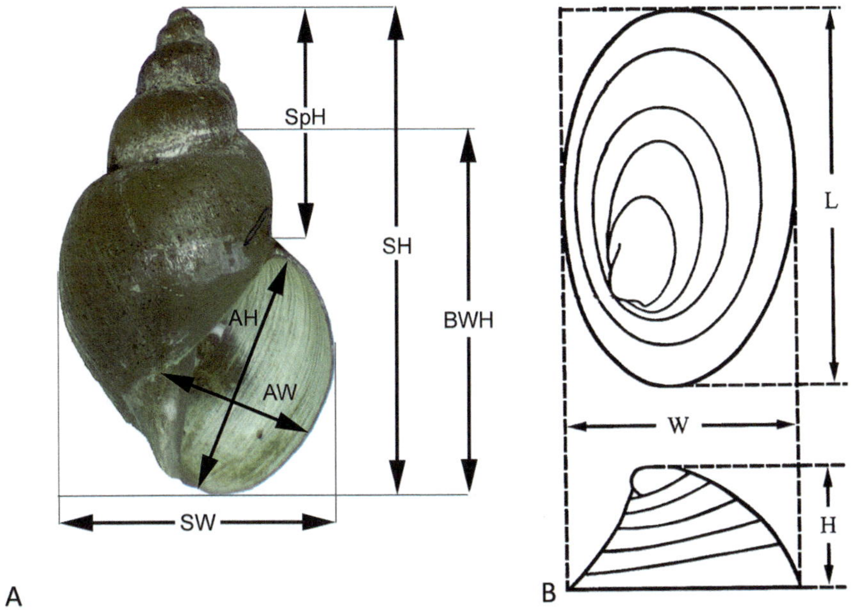

Fig. 4.2 (**a**) A scheme of linear measurements of a turbospiral shell (as exemplified by a shell of the dwarf pond snail, *Galba truncatula*). Abbreviations: SH—shell height; SW—shell width; SpH—spire height; BWH—body whorl height; AH—aperture height; AW—aperture width. (**b**) A scheme of linear measurements of an ancyliform (limpet) shell. Abbreviations: H—shell height; L—shell length; W—aperture width

shell whorl convexity. In most lymnaeid species, shell whorls are slightly convex or flattened, though some pond snails are characterized by strongly convex, sometimes almost stepped, whorls [for example, *Galba truncatula* (O.F. Müller, 1774); see Fig. 4.2].

The *umbilicus* is a cavity or depression formed around the shell axis inside the body whorl and situated at the base of the shell. In most species of the discussed family, it is absent or looks like a very thin slit covered by the columellar lip.

The *sculpture* is a relief pattern on the shell surface. It may be represented by ribs or riblets, thin striae, small depressions, ridges, and other structures. Some lymnaeid shells (for example, that of the gelatinous pond snail, *Myxas glutinosa*) virtually lack any prominent sculpture, being smooth and glossy. If sculptural elements are oriented parallel to the shell axis, then these are designated as *radial sculpture*. The *spiral sculpture* is formed by sculptural elements oriented transversely to the axis. The latter type of sculpture is quite rare among lymnaeids.

The variability in shell surface sculpture (and coloration) can be very prominent at the intraspecific level but, as a rule, it lacks any taxonomic value though there are data that the pattern of shell sculpture may be species-specific in some lymnaeids (Jackiewicz and Koralewska-Batura 1995; Jackiewicz 1998).

A keel (or *carina*) is a prominent spiral ridge of a shell. It may be solid or relatively soft, in the latter case, the keel may look like a thin and fragile fringe on the shell edge. This structure is rarely manifested among freshwater pulmonate snails, except for some genera of the family Planorbidae. Within Lymnaeidae, the keeled species are known among extinct species (see Vinarski and Pointier 2023).

A special nomenclature for the designation of parts of the shell aperture has been developed (see Fig. 4.1). The margin of the aperture adjacent to the body whorl is called *parietal*; the free lateral margin is *palatal*, and the lower free margin is *basal*. The *columellar* margin is adjacent to the shell columella. In some species of the family, there is a more or less prominent depression situated at the junction of the columellar and parietal margins. Often, it is called the *columellar fold*.

The columellar margin in lymnaeids is usually reflected over the adjacent part of the shell which forms the *columellar lip*, more or less broad. The coloration of this lip is usually lighter than the body whorl surface. In most lymnaeid species, the columellar lip is rather thin; usually, it completely covers the umbilicus, which makes the latter virtually invisible. In other cases, the umbilicus looks like a very narrow slot.

The absolute shell size is very plastic in most taxa of the Lymnaeidae and therefore is generally avoided in identification keys (though it may be useful for fast identification of some genera). There are many factors influencing shell size, including age (ontogenic variation), ambient temperature, food supply, parasitism, and so on (Abel 1920; Hubendick 1951; Calow 1981; Lam and Calow 1988; Lakowitz et al. 2008; Vinarski 2012, 2013; Whelan 2021; Vinarski and Pointier 2023). The absolute shell size is, thus, of a low significance or taxonomy, however, most genera and subgenera may be characterized by a more or less clear range of body sizes, which may be attained by their representatives. There are genera of large-bodied lymnaeids (*Bulimnea*, *Lymnaea*) and genera of much smaller size (*Galba, Orientogalba*). In some cases, the congeneric species may differ by their absolute shell sizes. For example, shells of *Stagnicola corvus* (Gmelin, 1791) are generally larger than those of *S. palustris* (O.F. Müller, 1774) or *S. turricula* (Held, 1836); shells of *Radix auricularia* (Linnaeus, 1758) are more voluminous as compared to shells of *R. euphratica* (Mousson, 1874) and *R. natalensis* (Krauss, 1848).

The use of the so-called morphometric indices, like the ratio between shell width (SW) and shell height (SH), has been served as a standard means for quantitative expression and analysis of these interspecific differences, though today it is gradually replacing by more advanced statistical techniques such as geometric morphometry. These indices are simple to calculate, and therefore they are still in wide use. Their use aims to give the simplest quantitative characteristics of proportions of a given shell or a particular shell sample.

Shell proportions are thought to be a rather more reliable source of taxonomically relevant information. In order to use shell proportions for species identification, different conchometric indices are calculated based on plain linear measurements. For example, one may use the ratio between shell width and shell height (SW/SH; Fig. 4.2) for characterization of relative slenderness of a snail shell, and so on. These ratios, however, are also prone to change with age, which is explained by the allometric growth (Fig. 4.3). Consequently, the shell habitus is also subject to ontogenic alterations.

Though the use of conchometric indices has been a long tradition in the lymnaeid studies, we have to stress that only in a minority of cases do these indices represent a reliable way of distinguishing between closely affined species. As a rule, their values greatly overlap even among the most tightly related taxa. Such an overlap of values is characteristic for comparisons based on multivariate statistics as well (Fig. 4.4).

The shell measurement scheme (see Fig. 4.2a) includes six basic measurements however it by no means is standard; the number of linear measurements can be increased depending on the task of a particular research (see, for example, Samadi et al. 2000). For example, it may be needed to introduce supplementary measurements like the body whorl width above the aperture, the height and width of the penultimate whorl, and so on. In the case of limpet shells, the measurement scheme undergoes corresponding changes (see Fig. 4.2b).

In addition to linear measurements, it is sometimes helpful to count the whorl number, since many genera and subgenera of the Lymnaeidae differ from each other by the maximum number of shell whorls. The scheme of whorl counting is given in Fig. 4.5. This parameter is frequently used as a proxy of the absolute age of snails. Another useful parameter is the shell *tangent line*, i.e. an imaginary line that "just touches" the surface of all or several of the whorls constituting the spire. It may be straight or almost straight, convex or concave (see 4.5b). The form of this line may be species-specific and in certain cases allows one to distinguish between closely allied and conchologically similar species.

Ever since Baker (1911, 1915) and Roszkowski (1914a, b, 1922, 1926), it was believed that the internal organs of the lymnaeids, as being not greatly modified by external influences, represent a more reliable criterion for purposes of classification as compared to conchological characters. The structure of the reproductive system was thought to be the most important source of taxonomic and phylogenetic signals. Most consistently, this idea was developed in seminal works of Hubendick (1951), Jackiewicz (1959, 1993, 1998), and Paraense (1976, 1982, 1983, 1984, 1994, 1995), who relied heavily on anatomical data in their search for a "natural" system of the Lymnaeidae and the reliable identification of these snails. The extensive use of the soft body characters for the identification of the species and higher taxa of the Lymnaeidae has also been advocated by Falniowski (1980) and Kruglov (2005).

The Lymnaeidae are hermaphroditic snails, with the male and female reproductive organs co-existing in the same animal (see for example Fig. 4.6). Not all of these organs are equally important for taxonomy. The size and proportions of the copulatory organ, the shape of the spermatheca, and the relative length of its duct as well as the internal structure of the prostate are the characters that have most frequently been used by practitioners for the sake of taxonomic identification of lymnaeids.

The copulatory organ in the Lymnaeidae typically consists of two large structures named the *praeputium* and the *penis sheath* (Fig. 4.6). The ratio of their lengths (known as the "index of the copulatory organ," ICA) has been used for species delimitation in different genera of Lymnaeidae (Jackiewicz 1959, 1996, 1998; Falniowski 1980; Kruglov and Starobogatov 1985; Vinarski 2011a; Standley et al.

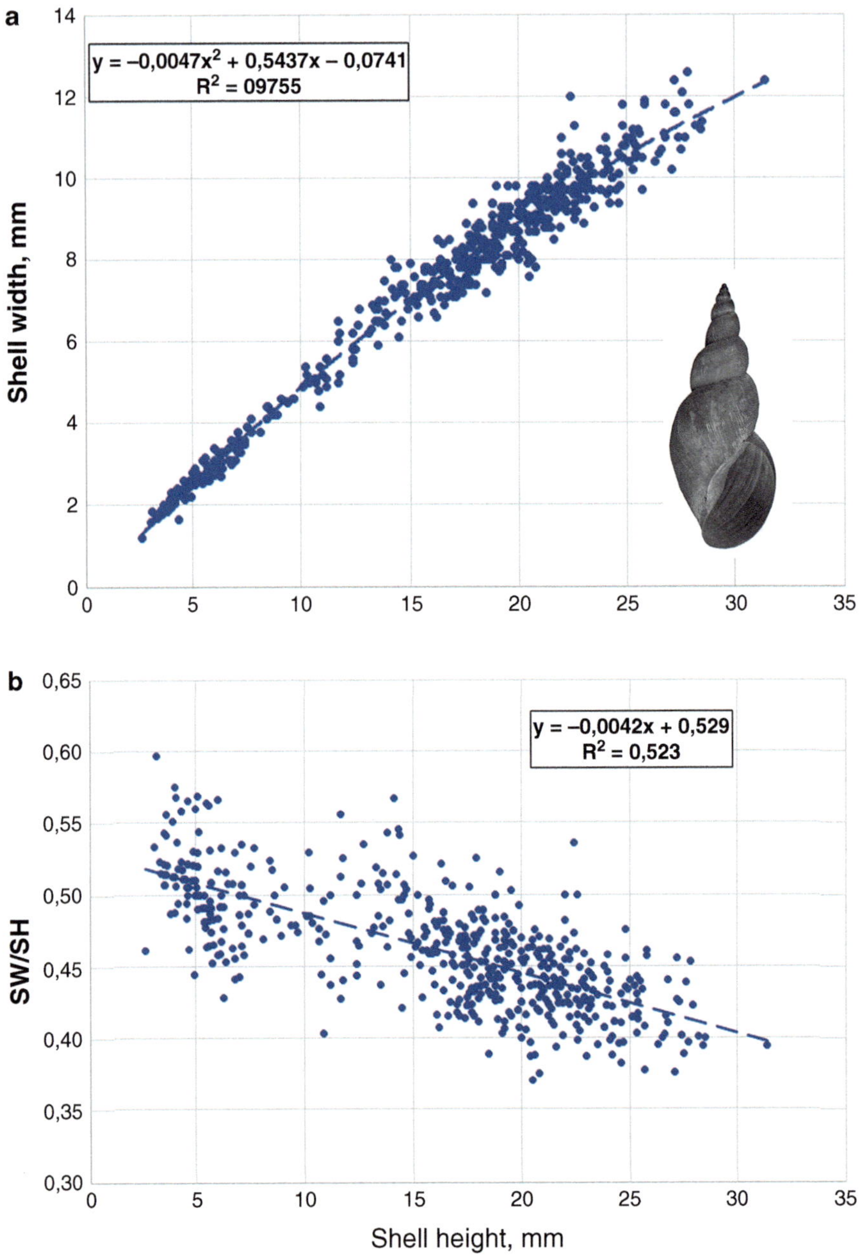

Fig. 4.3 Ontogenic changes in shell proportions in a sample of *Lymnaea taurica kazakensis* (Central Kazakhstan, Sary-Kopa Lake, n = 548). (**a**) The relationship between shell height and shell width. (**b**) Ontogenic change in the values of SW/SH index. SH is used as a proxy for snail age. Regression equations and trend lines are given. Based on unpublished data of M. Vinarski

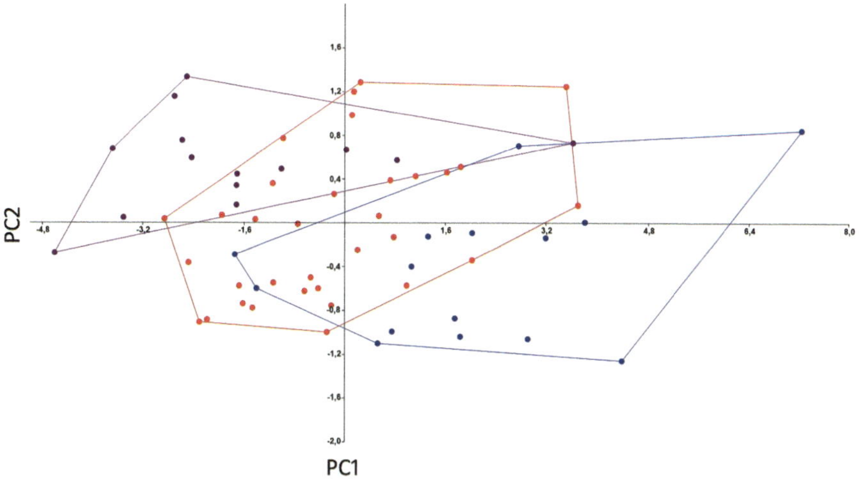

Fig. 4.4 The overlap of three samples of two sister lymnaeid species (*Ampullaceana fontinalis*, *A. lagotis*) in the plane of the first two principal components. The analysis is based on the six standard shell measurements with addition of whorl number. The two first PCs explain 96.6% of variation. Red dots—*A. fontinalis* from Seversky Donets River, Russia. Blue dots—*A. fontinalis* from Veselovskoye Reservoir, Russia. Violet dots—*A. lagotis* from the Panj River floodplain, Tajikistan. All three samples were identified genetically. The primary data taken from Vinarski et al. (2020)

2013; Pointier 2015). It has been revealed, though, that both intra- and interspecific variation of ICA values is significant that limits its direct use for lymnaeid identification (Schniebs et al. 2011, 2013; Vinarski 2011b). It should also be noted that in certain species of the genus *Radix* this ratio may be polymorphic within a population, and conchologically indistinguishable individuals may differ drastically from each other in the penis sheath length (Vinarski 2011b). Another source of uncertainty in using this ratio is its ontogenic variation (Beriozkina and Starobogatov 1988); in some species, the spatial (geographical) variation of this index is observed (Vinarski 2009).

A statistical analysis of the intrapopulation variability of ICA has shown that, in most populations, this variability is rather moderate, with the coefficient of variation (*Cv*) being lower than 20% (Table 4.1). The data of Table 4.1 reveals the phylogenetically allied species show virtually no significant differences in this index (contrary to the opinion of Beriozkina and Starobogatov 1988; Kruglov 2005), which greatly limits the usefulness of this ratio for identification purposes. However, some contrary instances can be reported here. In the genus *Stagnicola*, two sister and conchologically cryptic species, *S. palustris* and *S. turricula*, demonstrate a statistically significant difference in their mean ICA values (Falkner 1985; Glöer 2002, 2019). In *S. turricula*, the praeputium is nearly twice shorter than the penis sheath, which results in the ICA values between 0.34–0.63 (Falkner 1985; Vinarski, unpublished). In another species, the praeputium length is subequal to the penis sheath length (Fig. 4.7) and sometimes slightly exceeds the latter, and the ICA lies in

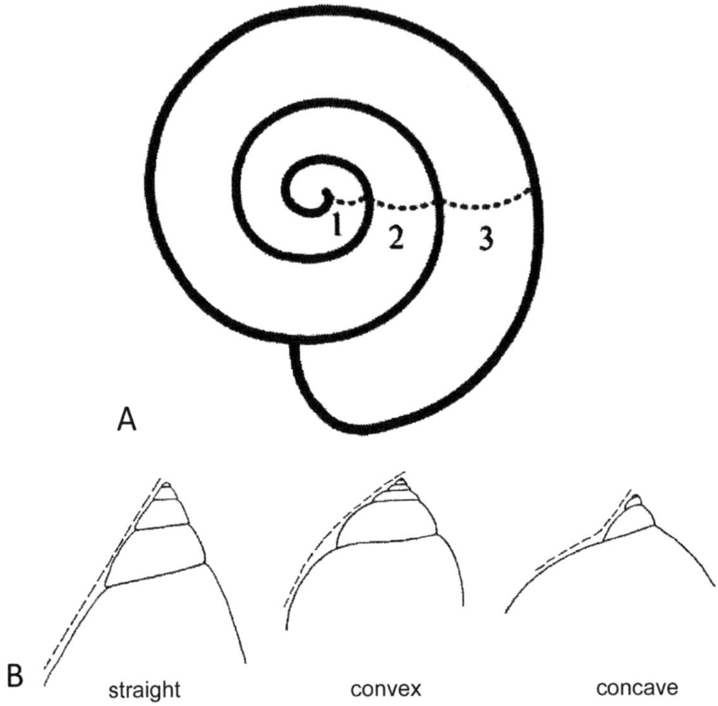

Fig. 4.5 (**a**) A scheme of the whorl number counting. (**b**) Three types of the tangent line. After Vinarski (2019), slightly modified

the range 0.60–1.18 (Vinarski, unpublished data). On the other hand, some overlap in the ICA values in the two species is observed (see Fig. 4.7), therefore the diagnostic significance of this ratio is not absolute. However, in the vast majority of cases, the two stagnicoline species can be distinguished through this index. Though it is almost impossible to differentiate *S. palustris* and *S. turricula* based on their shell traits (Jackiewicz 1998), the molecular evidence of their distinct species status has recently been provided (Pieńkowska et al. 2015). There is another stagnicoline species, *S. saridalensis* (Mozley, 1934), distributed in Siberia, whose ICA ratio is less than 0.25 (a very short praeputium and extremely long and narrow penis sheath; see Lazareva 1967; Kruglov 2005; Vinarski 2014) [see Table 4.1].

In South America, conchologically cryptic species of the genus *Galba* can be distinguished based on the proportions of their copulatory organs. The ICA was reported as the only morphological character allowing the distinction between *G. viator* (d'Orbigny, 1835), *G. truncatula*, and *G. cubensis* (L. Pfeiffer, 1839) [Standley et al. 2013; Pointier 2015; Pointier and Vàzquez 2020]. However, as it was cautiously noted, this index "has no practical interest for the identification of a single snail collected in the field" (Pointier 2015, p. 109), which is explained by the overlap in the distribution of the values of this parameter.

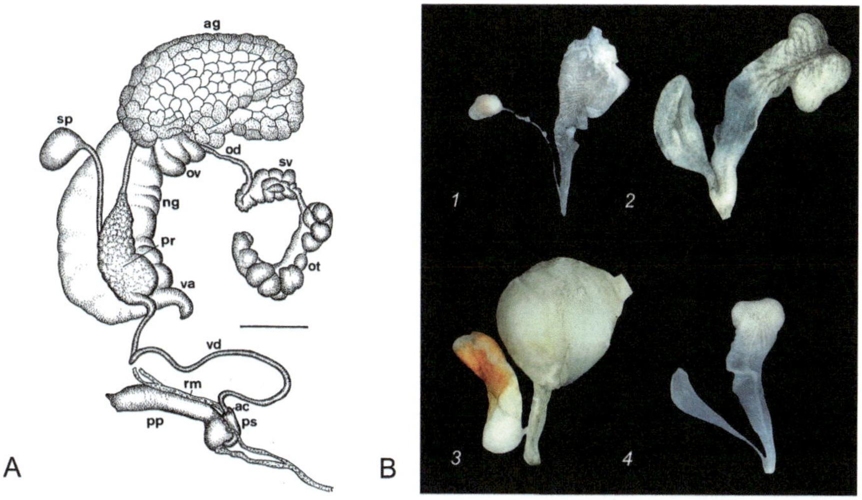

Fig. 4.6 (**a**) Anatomy of the reproductive system of *Galba truncatula* from France: ac = apical chamber; ag = albumen gland; ng = nidamental gland; od = ovispermiduct; ot = ovotestis; ov = oviduct; pr = prostate; pp. = preputium; ps = penis sheath; rm. = retractor muscle of penial complex; sd = spermiduct; sp. = spermatheca; va = vagina; scale bar = 1 mm. (**b**) Variability of the form of the spermatheca and the length of the spermathecal duct in selected species of radicine snails: 1 = *Radix auricularia*; 2 = *Peregriana peregra*; 3 = *Ampullaceana balthica*; 4 = *A. lagotis*. Photos are taken from Schniebs et al. (2011) and modified

A small group of Palearctic Lymnaeidae described as the genus *Aenigmomphiscola* Kruglov & Starobogatov, 1981 demonstrates a more advanced structure of the copulatory organ with the so-called praeputial organ within the praeputium that is regarded to be asymmetrically enlarged velum (Kruglov and Starobogatov 1981; Vinarski et al. 2011). In all other respects, however, these snails are quite similar to the rest of lymnaeids.

The relative length and proportions (long and narrow vs short and thick) of the spermathecal duct as well as the determination of the number of internal prostate folds (see Fig. 4.8) are useful for discrimination among genera and subgenera of Lymnaeidae. For example, the species of the genus *Radix* are typically characterized by a compact globose (or sac-like) spermatheca (or *bursa copulatrix*) sitting on a relatively long and narrow spermathecal duct. The representatives of the genera *Ampullaceana* and *Peregriana*, having similar shell proportions, can be distinguished from the genus *Radix* by the very short spermathecal duct (which sometimes is virtually absent) and oblong spermatheca (see Fig. 4.6b). In most instances, these differences are prominent enough to differentiate between species of *Radix* and other, conchologically similar, genera. However, as Schniebs et al. (2011, 2013, 2019) reported, there are cases, when this distinguishing character fails to help. Among specimens of *Ampullaceana ampla* (W. Hartmann, 1821), for example,

Table 4.1 Index of copulatory apparatus (ICA) in samples of selected lymnaeid species[a, b]

Species, sampling site	Number of specimens	ICA Variation limits	ICA Mean ± standard deviation	R	Cv %
Aenigmomphiscola europaea					
Moscow Province, Oka River floodplain	12	0.53–0.75	0.63 ± 0.07	1.42	11.1
Ampullaceana ampla					
Chelyabinsk Province, Miass River	21	0.68–1.20	0.92 ± 0.15	1.78	8.3
Ampullaceana fontinalis					
Rostov Province, Seversky Donets River	17	0.88–1.20	1.08 ± 0.10	1.36	9.3
Rostov Province, Veselovskoye Reservoir	12	1.04–1.47	1.20 ± 0.16	1.41	13.3
Ampullaceana balthica					
Altai Territory, Sklyuikha River	20	1.05–1.88	1.38 ± 0.23	1.79	21.0
Murmansk Province, Pirenga Lake	13	0.81–1.54	1.15 ± 0.22	1.91	19.1
Germany, Hamburg City, ditch Marschlande	10	0.69–1.24	1.00 ± 0.16	1.80	6.0
Ampullaceana lagotis					
Omsk Province, Irtysh River floodplain in Omsk City	23	1.15–2.11	1.47 ± 0.24	1.84	16.3
Tajikistan, Panj River floodplain	12	1.29–1.77	1.51 ± 0.13	1.32	8.6
Bulimnea megasoma					
Canada, Ontario Province, vicinities of Cobalt Town	27	0.93–0.12	1.70 ± 0.11	1.70	12.9
Galba truncatula					
Tyumen Province, Kochu-Uriy oxbow	34	2.44–3.33	2.94 ± 0.22	1.37	7.5
Yamal Peninsula, Shchuchya River floodplain	31	2.56–3.18	2.89 ± 0.19	1.24	6.6
Kamtschaticana kamtschatica					
Kamchatka Peninsula, the Geyser Valley	12	0.97–1.35	1.24 ± 0.09	1.39	7.9
Ladislavella catascopium vahlii					
Greenland, Ferguson Lake	15	1.33–1.95	1.63 ± 0.18	1.47	11.0
Ladislavella liogyra					
Khabarovsk Territory, Bochinka River floodplain	15	1.34–2.06	1.64 ± 0.18	1.54	11.0
Ladislavella terebra					
Altai Territory, floodplain marsh of Kulunda River	31	1.00–1.68	1.29 ± 0.22	1.68	17.1
Altai Republic, Teletskoye Lake	30	0.91–1.50	1.10 ± 0.14	1.65	12.7
Novosibirsk Province, pool near Kuibyshev – Severnoye road	24	0.82–1.20	1.01 ± 0.13	1.46	12.9
Omsk Province, marsh at Krivoye Lake	29	0.83–1.36	1.05 ± 0.13	1.64	12.4

(continued)

Table 4.1 (continued)

Species, sampling site	Number of specimens	ICA			
		Variation limits	Mean ± standard deviation	R	Cv %
Omsk Province, lake at Cherlak settlement	23	1.04–1.84	1.31 ± 0.23	1.72	17.6
Omsk Province, small lake near Malye Bodachi Lake	29	0.77–1.38	1.03 ± 0.16	1.79	15.6
Omsk Province, marsh in Omsk City	36	0.95–1.55	1.19 ± 0.12	1.63	10.1
Omsk Province, Lake at Kirgap village	31	0.87–1.35	1.13 ± 0.13	1.55	11.5
Tomsk Province, floodplain marsh of Ob' River near Nikolskoye settlement	30	0.88–1.35	1.08 ± 0.13	1.53	12.0
Tyumen Province, vicinity of Labytnangi, lake at Vylposl Channel	30	0.78–1.32	1.06 ± 0.14	1.69	13.2
Tyumen Province, marsh near Labytnangi	37	0.81–1.30	1.05 ± 0.13	1.61	12.4
Lymnaea stagnalis					
Altai Territory, swamp in floodplain of Kulunda River	26	3.52–5.91	4.83 ± 0.64	1.67	13.3
Altai Territory, Valovoye Lake	30	3.77–5.90	4.73 ± 0.56	1.56	11.8
Altai Territory, Kuznetsovo Lake	30	3.48–5.74	4.56 ± 0.57	1.65	12.5
Altai Territory, Noven'koye Lake	25	3.83–5.93	4.63 ± 0.61	1.55	13.2
Altai Territory, stream at Uglovskoye settlement	20	3.70–6.33	4.59 ± 0.74	1.71	16.1
Bryansk Province, drainage channel near Maloye Kozhanovskoye Lake	25	4.04–5.62	4.76 ± 0.48	1.39	10.1
Omsk Province, Atachka River	22	3.45–5.48	4.37 ± 0.51	1.59	11.7
Omsk Province, Omsk City, lake in Moskovka District	26	4.23–6.64	5.19 ± 0.65	1.57	12.5
Tyumen Province, floodplain lake of Irtysh River. near Tobol'sk Town	22	3.95–5.75	4.74 ± 0.45	1.46	9.5
Lymnaea taurica kazakensis					
Altai Territory, floodplain marsh of Kulunda River	31	2.06–2.88	2.48 ± 0.22	1.39	8.9
Orientogalba heptapotamica					
Mongolia, Khovd River floodplain	34	0.88–1.38	1.15 ± 0.12	1.57	10.4
Peregriana dolgini					
Omsk Province, lake close to Ananyevskoye Lake	20	0.75–1.27	1.03 ± 0.13	1.69	12.6
Peregriana peregra					
Armenia, Agartskin stream	11	1.03–1.33	1.15 ± 0.11	1.29	9.6
Radix alticola					
Tajikistan, a thermal spring near Djavshangoz village	16	0.81–1.18	0.97 ± 0.11	1.46	11.3
Radix auricularia					
Altai Territory, Kuznetsovo Lake	25	0.95–1.29	1.09 ± 0.09	1.36	8.3

(continued)

Table 4.1 (continued)

Species, sampling site	Number of specimens	ICA Variation limits	Mean ± standard deviation	R	Cv %
Mongolia, Khar-Us-Nuur Lake	42	0.67–1.29	0.88 ± 0.15	1.93	17.0
Novosibirsk Province, Novosibirsk Reservoir (form with long praeputium)	18	0.91–1.31	1.06 ± 0.10	1.44	9.4
Novosibirsk Province, Novosibirsk Reservoir (form with short praeputium)	7	0.34–0.61	0.52 ± 0.09	1.79	17.3
Omsk Province, Krivoye Lake	21	1.02–1.36	1.12 ± 0.09	1.36	8.0
Tomsk Province, lake at Novomikhailovka village	30	0.93–1.39	1.10 ± 0.12	1.49	10.9
Radix euphratica					
Tajikistan, ditch near Kurban-Shakhid village	15	0.87–1.69	1.25 ± 0.20	1.94	16.0
Radix natalensis					
Uganda, crater lake Kyamwiga	20	0.84–1.13	0.96 ± 0.07	1.35	7.3
Radix parapsilia					
Tomsk Province, lake at Novomikhailovka village	30	0.70–0.96	0.86 ± 0.06	1.37	7.0
Stagnicola palustris					
Pskov Province, Yasskoye Lake	34	0.59–1.07	0.77 ± 0.12	1.81	15.6
Arkhangelsk Province, Uyemlyanka River floodplain	30	0.73–1.18	0.96 ± 0.10	1.62	10.4
Stagnicola saridalensis					
Novosibirsk Province, Fadikha Lake	32	0.08–0.18	0.12 ± 0.03	2.25	25.0
Novosibirsk Province, floodplain marsh of Kargat River	40	0.09–0.22	0.12 ± 0.03	2.44	25.0
Omsk Province, floodplain marsh of Irtysh River in Omsk City	78	0.12–0.24	0.17 ± 0.03	2.33	17.7
Omsk Province, Trauly Lake	37	0.10–0.21	0.14 ± 0.02	2.10	14.2
Omsk Province, Kabankul' Lake	54	0.10–0.23	0.16 ± 0.03	2.30	18.8
Sverdlovsk Province, Murzinka River	28	0.11–0.24	0.18 ± 0.04	2.18	22.2
Tyumen Province, vicinity of Labytnangi Town, Vylposl Channel	35	0.10–0.19	0.14 ± 0.02	1.90	14.3
Stagnicola zebrella					
Yamal Peninsula, Talbey-Yakha River floodplain	12	0.88–1.15	1.00 ± 0.11	1.31	11.0

[a]All dissections and measurements were made by the same person (MVV) according to the standard protocol (see Vinarski 2011b). All samples collected in Russia (unless otherwise stated)
[b]R—the ratio of the maximum to the minimum ICA values in a given sample; Cv—the coefficient of variation

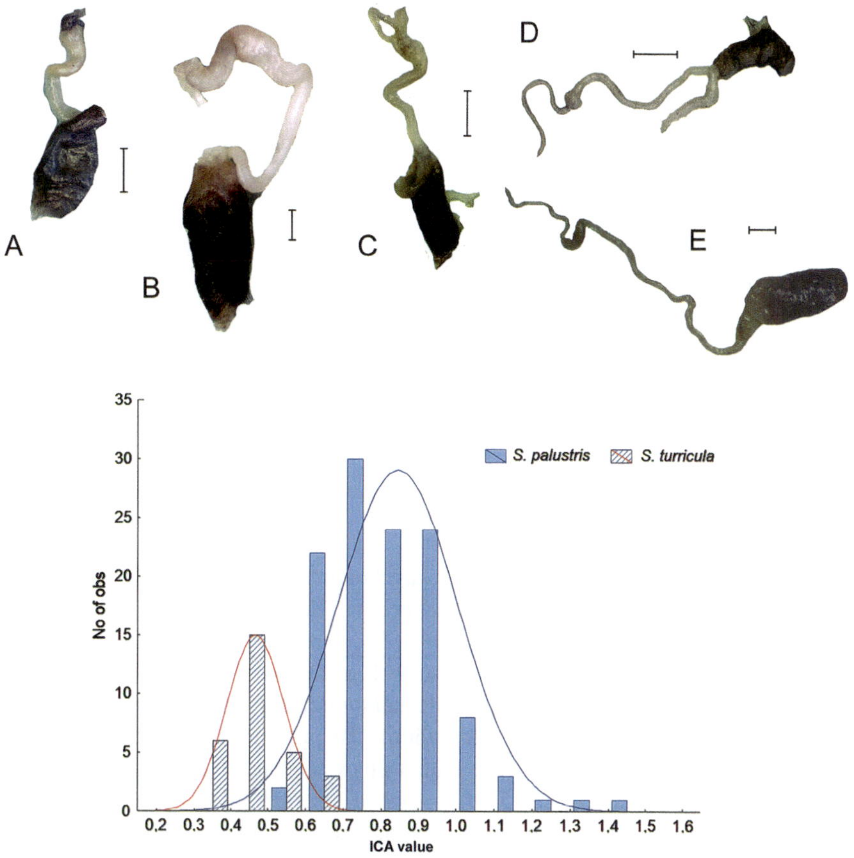

Fig. 4.7 The copulatory organs of *Stagnicola palustris* (a–c) and *S. turricula* (d–e), with a diagram showing an overlap of the distributions of their ICA values. (**a**) Germany, Brandenburg, Kesselsee Lake. (**b**) Russia, Pskov Region, Yasskoye Lake. (**c**) Russia, Murmansk Region, Gorelyi Island. (**d**) Germany, Bavaria, Danube River near Oberalteich. (**e**) Azerbaijan, a lake adjoining the Caspian Sea near Narimanabad Town. Scale bars 1 mm. After Vinarski (unpublished data)

some individuals with very long spermathecal duct were found (see Fig. 5 in Schniebs et al. 2019), which makes these individuals anatomically indistinguishable from *Radix auricularia* (note: the identity of these specimens as belonging to *A. ampla* was confirmed by molecular methods).

It may be concluded, on the basis of the abovementioned facts, that neither particular anatomical character allows unequivocal differentiation between two lymnaeid individuals belonging to phylogenetically related species. The absolutely correct identification of a single snail is possible by means of genetic tools (Schniebs et al. 2011, 2013). Both qualitative and quantitative anatomical characters are, thus, only workable if whole samples, not single snails, are compared (see Vinarski 2011b).

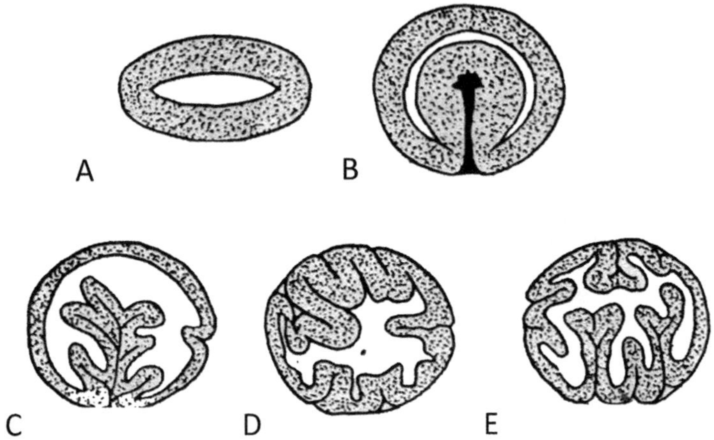

Fig. 4.8 The five structural types of the lymnaeid prostate delineated by Kruglov (2005). Explanation in the text. Modified after Kruglov (2005)

The internal structure of the prostate was thought to provide useful information for distinguishing between the lymnaeid genera (Starobogatov 1976; Jackiewicz 1993, 1998; Kruglov and Starobogatov 1993; Kruglov 2005). Kruglov (2005) delineated the five main types of the prostate (Fig. 4.8): (A) simple prostate without any internal fold (genera *Omphiscola*, *Pseudobulinus*, *Pseudosuccinea*); (B) one (rarely two) internal unbranched folds within the prostate (most recent lymnaeid genera fall to this group); (C) prostate with a single branched fold (genus *Racesina*); (D) many-folded prostate with 5–10 branched internal folds (genus *Lymnaea*); (E) many-folded prostate with several branched folds (some representatives of *Stagnicola*). The distribution of these character states among the living Lymnaeidae reveals that the five structural types lack any phylogenetical meaning, since the same structural type might have been developed independently in different lineages of the family. The loss of the internal folds of the prostate, for instance, has happened at least thrice in the lymnaeid evolution, and the unfolded prostate is found in the European genus *Omphiscola*, the North American *Pseudosuccinea*, and in the genus *Pseudobulinus*, which is endemic to Hawaii (Kruglov and Starobogatov 1993).

The use of molecular techniques has revealed that the *Bauplan* of the reproductive anatomy within the Lymnaeidae is extremely conservative (except for the subfamily Lancinae having two prostates of different structure; see Vinarski and Pointier 2023). At least several genetically divergent genera and subgenera of the subfamily Amphipepleinae exhibit the same structure of the copulatory apparatus and the prostate (Aksenova et al. 2018; Vinarski et al. 2020). The results of the most recent "integrative" studies have confirmed the enormous level of intra- and interspecific variability in the lymnaeid snails, which makes the genetic methods the solely reliable tool for identification of the pond snails. However, even popular molecular approaches like DNA barcoding can sometimes fail in the lymnaeid identification (Schniebs et al. 2016), therefore we advocate here the integrative approach to taxonomic diagnostic of the lymnaeid snails which combines the data provided by conchological, anatomical, and molecular genetic methods.

Acknowledgments MVV thanks the Russian Scientific Fund for financial support of his research on the lymnaeid taxonomy and phylogeny (project No. 19-14-00066-P).

References

Abel O (1920) Lehrbuch der Paläozoologie. G. Fischer Verlag, Jena

Aksenova OV, Bolotov IN, Gofarov MY et al (2018) Species richness, molecular taxonomy and biogeography of the radicine pond snails (Gastropoda: Lymnaeidae) in the Old World. Sci Rep 8:11199. https://doi.org/10.1038/s41598-018-29451-1

Alda P, Bonel N, Alba A et al (2023, this volume) Molecular techniques for the study of ecological and evolutionary processes in lymnaeids. In: Vinarski MV, Vázquez A (eds) The Lymnaeidae. A handbook on their natural history and parasitological significance. Springer, Cham, pp 121–146

Andreeva SI, Andreev NI, Vinarski MV (2010) Key to freshwater gastropods of Western Siberia (Mollusca: Gastropoda). V. 1. Gastropoda: Pulmonata. Fasc. 1. Families Acroloxidae and Lymnaeidae. The authors, Omsk [in Russian]

Baker FC (1911) The Lymnaeidae of North and Middle America. Special publication of the Chicago Academy of Sciences 3:1–539. https://doi.org/10.5962/bhl.title.20500

Baker FC (1915) On the classification of the Lymnaeids. The Nautilus 29(2):20–24

Beriozkina GV, Starobogatov YI (1988) Reproductive ecology and egg clusters of freshwater Pulmonata. Trudy Zoologicheskogo Instituta AN SSSR 174:1–306. [in Russian]

Calow P (1981) Adaptational aspects of growth and reproduction in *Lymnaea peregra* (Gastropoda: Pulmonata) from exposed and sheltered aquatic habitats. Malacologia 21:5–13

Cox LM (1960) Gastropoda – general characteristics of Gastropoda. In: Moore RC (ed) Treatise on invertebrate Paleontology. Part I. Mollusca 1. University of Kansas Press, Lawrence, pp 84–169

Falkner G (1985) *Stagnicola turricula* (Held) – eine selbständige Art neben *Stagnicola palustris* (O.F. Müller). Heldia 1(2):47–50

Falniowski A (1980) The anatomical determination of Polish Lymnaeidae (Mollusca, Basommatophora). Acta Hydrobiologica 22(3):327–335

Glöer P (2002) Die Sußwassergastropoden Nord- und Mitteleuropas: Bestimmungschlussel, Lebenweise, Verbreitung. Die Tierwelt Deutschlands. Conchbooks, Hackenheim 73:1–327

Glöer P (2019) The freshwater gastropods of the West-Palaearctis. Volume 1. Fresh- and brakish waters except spring and subterranean snails. Identification key, anatomy, ecology, distribution. The author, Hetlingen

Hubendick B (1951) Recent Lymnaeidae. Their variation, morphology, taxonomy, nomenclature and distribution. Kungliga Svenska Vetenskapsakademiens Handlingar Fjärde Serien 3(1): 1–223

Jackiewicz M (1959) Badania nad zmiennością i stanowiskiem systematycznym *Galba palustris* O. F Müll. Prace Komisji Biologicznej, Poznańskie Towarzystwo Przyjaciół Nauk 19(3):1–86

Jackiewicz M (1993) Phylogeny and relationships within the European species of the family Lymnaeidae. Folia Malacologica 5:61–95. https://doi.org/10.12657/folmal.005.003

Jackiewicz M (1996) Concerning Hungarian populations of "*Galba palustris*" studied by Kilias (1992) (Gastropoda; Pulmonata: Lymnaeidae). Malakologische Abhandlungen / Museum für Tierkunde Dresden 18:165–172

Jackiewicz M (1998) European species of the family Lymnaeidae (Gastropoda, Pulmonata, Basommatophora). Genus 9(1):1–93

Jackiewicz M, Koralewska-Batura E (1995) The shell-surface sculpture of Lymnaeidae (Gastropoda, Pulmonata, Basommatophora). Malakologische Abhandlungen / Museum für Tierkunde Dresden 17(2):191–197

Khokhutkin IM, Vinarski MV, Grebennikov ME (2009) Molluscs of the Urals and the adjacent areas. The family Lymnaeidae (Gastropoda, Pulmonata, Lymnaeiformes). Fasc. 1. Goshchitskiy Publishers, Yekaterinburg. [in Russian]

Kruglov ND (2005) Lymnaeid snails of Europe and Northern Asia. Smolensk State Pedagogical University Press, Smolensk. [in Russian]

Kruglov ND, Starobogatov YI (1981) A new genus of the Lymnaeidae and taxonomy of the subgenus *Omphiscola* (*Lymnaea*) (Pulmonata, Gastropoda). Zool Zh 60(7):965–977. [in Russian]

Kruglov ND, Starobogatov YI (1985) Methods of experimental hybridization and some results of its applications in the taxonomy of Lymnaeidae. Malacol Rev 18:21–35

Kruglov ND, Starobogatov YI (1993) Annotated and illustrated catalogue of species of the family Lymnaeidae (Gastropoda Pulmonata Lymnaeiformes) of Palaearctic and adjacent river drainage areas. Part I Ruthenica Russian Malacol J 3(1):65–92

Lakowitz T, Brönmark C, Nyström P (2008) Tuning in to multiple predators: conflicting demands for shell morphology in a freshwater snail. Freshw Biol 53:2184–2191. https://doi.org/10.1111/j.1365-2427.2008.02045.x

Lam PKS, Calow P (1988) Differences in the shell shape of *Lymnaea peregra* (Müller) (Gastropoda: Pulmonata) from lotic and lentic habitats; environmental or genetic variance? J Mollus Stud 54:197–207. https://doi.org/10.1093/mollus/54.2.197

Lazareva AI (1967) On the systematics of freshwater snails of Kazakhstan from the group *Lymnaea palustris* Müller (Gastropoda, Pulmonata). Zool Zh 46(9):1340–1349. [in Russian]

Paraense WL (1976) *Lymnaea viatrix*: a study of topotypic specimens (Mollusca: Lymnaeidae). Rev Bras Biol 36(2):419–428

Paraense WL (1982) *Lymnaea rupestris* sp. n. from southern Brazil (Pulmonata: Lymnaeidae). Mem I Oswaldo Cruz 77(4):437–443. https://doi.org/10.1590/S0074-02761982000400011

Paraense WL (1983) *Lymnaea columella* in Northern Brazil. Mem I Oswaldo Cruz 78(4):477–482. https://doi.org/10.1590/S0074-02761983000400011

Paraense WL (1984) *Lymnaea diaphana*: a study of topotypic specimens (Pulmonata: Lymnaeidae). Mem I Oswaldo Cruz 79:75–81. https://doi.org/10.1590/S0074-02761984000100009

Paraense WL (1994) A Mexican population of *Lymnaea elodes* (Gastropoda: Lymnaeidae). Malacol Rev 27:5–11

Paraense WL (1995) *Lymnaea cousini* Jousseaume, 1887, from Ecuador (Gastropoda: Lymnaeidae). Mem I Oswaldo Cruz 90(5):605–609. https://doi.org/10.1590/S0074-02761995000500011

Piechocki A, Wawrzyniak-Wydrowska B (2016) Guide to freshwater and marine Mollusca of Poland. Bogucki Wydawnictwo Naukowe, Poznań

Pieńkowska JR, Rybska E, Banasiak J et al (2015) Taxonomic status of *Stagnicola palustris* (O.F. Müller, 1774) and *S. turricula* (Held, 1836) (Gastropoda: Pulmonata: Lymnaeidae) in view of new molecular and chorological data. Folia Malacologica 23(1):3–18. https://doi.org/10.12657/folmal.023.003

Pointier JP (ed) (2015) Freshwater molluscs of Venezuela and their medical and veterinary importance. Conchbooks, Harxheim

Pointier JP, Vàzquez AA (2020) Lymnaeoidea, Lymnaeidae. In: Damborenea C, Rogers DC, Thorp JH (eds) Thorp and Covich's freshwater invertebrates, Volume 5: Keys to Neotropical and Antarctic Fauna. Academic Press, London, pp 281–290

Roszkowski W (1914a) Contribution à l'étude des Limnées du lac Léman. Rev Suisse Zool 22(15): 457–539

Roszkowski W (1914b) Note sur l'appareil génital de *Limnaea auricularia* L. et de *Limnaea ovata* Drap. Zool Anz 44(4):175–179

Roszkowski W (1922) Przyczynki do poznania rodziny Lymnaeidae. VI. Aparat płciowy *Galba truncatula* (Müll.). Archivum Nauk Biologicznych Towarzystwa Naukowego Warszawskiego 1(5):1–6

Roszkowski W (1926) Contributions to the study of the family Lymnaeidae. VII. The structure of prostate of the Lymnaeidae. Annales Zoologici Musei Polonici Historiae Naturalis 5(1):1–14

Samadi S, Roumégoux A, Bargues MD et al (2000) Morphological studies of lymnaeid snails from the human fascioliasis endemic zone of Bolivia. J Mol Stud 66:31–44. https://doi.org/10.1093/mollus/66.1.31

Schniebs K, Glöer P, Vinarski M et al (2011) Intraspecific mophological and genetic variability in *Radix balthica* (Linnaeus 1758) (Gastropoda: Basommatophora: Lymnaeidae) with morphological comparison to other European *Radix* species. J Conchol 40(6):657–678

Schniebs K, Glöer P, Vinarski M et al (2013) Intraspecific morphological and genetic variability in the European freshwater snail *Radix labiata* (Rossmaessler, 1835) (Gastropoda: Basommatophora: Lymnaeidae). Contrib Zool 82(1):55–68. https://doi.org/10.1163/18759866-08201004

Schniebs K, Glöer P, Vinarski M et al (2016) A barcode pitfall in Palaearctic *Stagnicola* specimens (Mollusca: Lymnaeidae): incongruence of mitochondrial genes, a nuclear marker and morphology. North West J Zool 12(2):239–254

Schniebs K, Glöer P, Vinarski M et al (2019) Intraspecific morphological and genetic variability in the Palaearctic freshwater snail *Radix ampla* (Hartmann, 1821) (Gastropoda: Basommatophora: Lymnaeidae). J Conchol 43(3):245–267

Stadnichenko AP (2004) Pond snails and limpet snails (Lymnaeidae and Acroloxidae) of Ukraine. Tsentr uchebnoi literatury, Kiev. [in Russian]

Standley CJ, Prepelitchi L, Pietrokovsky SM (2013) Molecular characterization of cryptic and sympatric lymnaeid species from the *Galba/Fossaria* group in Mendoza Province, Northern Patagonia, Argentina. Parasit Vectors 6:304. http://www.parasitesandvectors.com/content/6/1/304

Starobogatov YI (1976) The system and phylogeny of the Lymnaeidae. Problemy Zoologii. Zoologicheskiy Institut AN SSSR: 79–81 [in Russian]

Vinarski MV (2009) Geographical variability in the male genitalia in two stagnicoline species (Gastropoda: Pulmonata: Lymnaeidae). Mollusca (Dresden) 27(2):157–166

Vinarski MV (2011a) A new species of stagnicoline snails (Mollusca: Gastropoda: Lymnaeidae) from the extreme north of Western Siberia. Zootaxa 2817:55–58. https://doi.org/10.11646/zootaxa.2817.1.2

Vinarski MV (2011b) The "index of the copulatory apparatus" and its application to the systematics of freshwater pulmonates (Mollusca: Gastropoda: Pulmonata). Zoosystematica Rossica 20(1): 11–27

Vinarski MV (2012) Geographical variability in freshwater mollusks. Biol Bull Rev 2:390–399. https://doi.org/10.1134/S2079086412050088

Vinarski MV (2013) The variation in freshwater pulmonate mollusks (the taxonomic aspect). Omsk State Pedagogical University Press, Omsk. [In Russian]

Vinarski MV (2014) *Lymnaea likharevi* Lazareva, 1967 is a junior synonym of *Lymnaea saridalensis* Mozley, 1934 (Gastropoda: Pulmonata: Lymnaeidae). Ruthenica. Russian Malacol J 24(1):35–44

Vinarski MV (2019) 11. Class Gastropoda. In: Rogers DC, Thorp JH (eds) Thorp and Covich's freshwater invertebrates, vol IV. Keys to Palaearctic Fauna. Academic Press, London, pp 310–345

Vinarski MV, Pointier JP (2023, this volume) General characteristics of the family Lymnaeidae. In: Vinarski MV, Vàzquez AA (eds) The Lymnaeidae: a handbook. Springer, Cham, pp 25–66

Vinarski MV, Schniebs K, Glöer P, Hundsdoerfer A (2011) The taxonomic status and phylogenetic relationships of the genus *Aenigmomphiscola* Kruglov and Starobogatov, 1981 (Gastropoda:

Pulmonata: Lymnaeidae). J Nat Hist 45(33–34):2049–2068. https://doi.org/10.1080/00222933.2011.574800
Vinarski MV, Aksenova OV, Bolotov IN (2020) Taxonomic assessment of genetically-delineated species of radicine snails (Mollusca, Gastropoda, Lymnaeidae). Zoosyst Evol 96(2):577–608. https://doi.org/10.3897/zse.96.52860
Whelan N (2021) Phenotypic plasticity and the endless forms of freshwater gastropod shells. Freshw Mollusk Biol Conserv 24(2):87–103. https://doi.org/10.31931/fmbc-d-20-00015

Chapter 5
Molecular Techniques for the Study of Ecological and Evolutionary Processes in Lymnaeids

Pilar Alda, Nicolás Bonel, Annia Alba, Olga V. Aksenova, Ivan N. Bolotov, Alexander V. Kondakov, and Philippe Jarne

Abstract In this chapter, we review the molecular approaches that have been applied to the study of ecological and evolutionary processes in lymnaeids. First, we describe the most common molecular techniques used to fix, preserve, dissect, collect, extract, and analyze lymnaeid samples. Second, we give some examples by mentioning studies in which these techniques have been applied in lymnaeids. We group the reviewed literature according to its experimental purpose: species identification, systematics, and phylogenetics; population genetics and phylogeography; mating system and reproductive behavior; and biotic and abiotic interactions. Finally, we introduce and discuss our view on the trends and perspectives on molecular procedures and techniques that will bring light to our understanding of the ecology and evolution of lymnaeids.

P. Alda (✉)
Genética y Ecología Evolutiva (CERZOS–CCT–CONICET Bahía Blanca), Bahía Blanca, Argentina

MIVEGEC, University of Montpellier, CNRS, IRD, Montpellier, France

N. Bonel
Centro de Recursos Naturales Renovables de la Zona Semiárida (CERZOS, CCT, UNS-CONICET Bahía Blanca), Bahía Blanca, Argentina

Consejo Nacional de Investigaciones Científicas y Técnicas (CONICET), Buenos Aires, Argentina

A. Alba
Laboratory of Malacology, Institute of Tropical Medicine "Pedro Kourí", Havana, Cuba

O. V. Aksenova · I. N. Bolotov · A. V. Kondakov
N. Laverov Federal Center for Integrated Arctic Research, The Ural Branch of the Russian Academy of Sciences, Arkhangelsk, Russia

P. Jarne
Centre d'Ecologie Fonctionnelle et Evolutive, CNRS, Université de Montpellier, Montpellier, France

M. V. Vinarski, A. A. Vázquez (eds.), *The Lymnaeidae*, Zoological Monographs 7,
https://doi.org/10.1007/978-3-031-30292-3_5

5.1 Introduction

The advent of molecular technologies in the biological sciences has radically increased our knowledge about living organisms. Applying and developing tools to study the molecular basis of biological activity in and between cells became essential for understanding living organisms and their vital processes. At the beginning of the 1960s, researchers started to apply molecular tools for studying ecological and evolutionary processes (Lewontin 1974). Today, the new molecular approaches based on "omics" technologies (genomics, transcriptomics, proteomics, and metabolomics) allow researchers for characterizing and quantifying pools of molecules to study ecological or evolutionary responses of lymnaeids from an integrative point of view (Seppälä et al. 2021).

In this chapter, we show how molecular methodologies have been applied to the study of ecological and evolutionary processes in lymnaeids. Thus, we do not cover biochemical technologies used in embryology, physiology, and neuroscience without an ecological or evolutionary perspective. Some of those studies have been reviewed elsewhere (Fodor et al. 2020; Rivi et al. 2020) (Box 5.1). In the first section, we describe the most common molecular techniques used to fix, preserve, dissect, collect, extract, and analyze lymnaeid samples. In the second section, we give some examples by mentioning studies in which these techniques have been applied in lymnaeids. We group the reviewed literature according to its experimental purpose. Our review does not pretend to be either systematic, or exhaustive, but rather to explore and illustrate the molecular technologies that researchers usually apply in lymnaeids. Finally, we introduce and discuss our view on the trends and perspectives on molecular procedures and techniques that will bring light to our understanding of the ecology and evolution of lymnaeids.

5.2 How to Study Lymnaeid Molecules?

The answer depends on the targeted molecule, which is tightly related to the scientific purpose of any given study. Techniques for investigating DNA, RNA, enzymes, nutrients, heavy metals, and pollutants have specific protocols. In the following section, we will mention and briefly describe the most common techniques used to fix, preserve, dissect, collect, extract, and analyze molecules from lymnaeids. These snail samples can belong to natural or experimental populations, depending on the purpose of the study.

5.2.1 Fixation and Preservation of Samples

Fixation and preservation prevent sample decay due to autolysis or putrefaction. First, fixation terminates any ongoing biochemical reactions and, then, preservation maintains sample stability by preventing reduction or loss for a given period of time. The fixation and preservation method depends on the targeted molecule, the purpose of the study, and the sample used. It usually includes immersing the sample in a fixative solution for stopping metabolic processes and preventing further molecular degradation. For instance, a common practice has been to fix and preserve lymnaeid specimens in formaldehyde or Bouin's fluid. Both fixative agents are appropriate for morphological studies, but hinder tissue use for most molecular analyses. They cause molecular cross-links and extensive fragmentations seriously hampering the possibility to analyze molecules from those samples (Schander 2003).

Tissue fixation that enables both a proper study of lymnaeid internal anatomy and good DNA quality, for example, can be achieved by placing snails in water at 70 °C for 30–45 s shortly after collection. Snails relax and do not contract inside their shells as when they are directly placed in a freezer at −20 °C or in ethanol within a tube. Heating individuals before immersion in ethanol promotes relaxation and prevents the activation of nucleases that are very aggressive and can rapidly degrade significant amounts of DNA. The body is then carefully withdrawn from the shell using forceps and stored in 70% ethanol at 4 °C (Alda et al. 2021). This procedure allows both fixation of individuals without contraction of soft parts (necessary for morphological studies) and rapid penetration of ethanol in all snail organs (necessary for DNA studies). Appropriate conservation shortly after snail death ensures the success of any molecular analysis (Schiller et al. 2014). Another option for obtaining good DNA quality is to directly place snails in a freezer at −80 °C or in a liquid nitrogen container.

RNA is more labile than DNA. High RNA quality is only achieved if endogenous RNAses are inactivated immediately upon snail death to prevent RNA degradation. Thus, for transcriptomics studies, snails are snap-frozen in liquid nitrogen and stored at low temperature (−80 °C) or shipped in dry ice (Tills et al. 2015). This is often not feasible in remote fieldwork locations. In these cases, samples can be directly placed in a solution such as RNA*later*®—a commercial solution that stabilizes and protects RNA from degradation, allowing their proper preservation at a temperature higher than −80 °C (Alba et al. 2019a; Seppälä et al. 2021).

Other approaches to fix and preserve lymnaeid molecules, like proteins, for instance, include freezing or lyophilization (Alba et al. 2020, 2019a). For some studies, like those estimating enzyme activity, the use of fresh samples is recommended (Tripathi et al. 2004). In these studies, snails are maintained alive in aquaria until dissection and tissue collection.

5.2.2 Dissection and Tissue Collection

Dissection and tissue collection must not only ensure fidelity, quality, and quantity of the recovered molecule, but also must minimize as much as possible snail suffering. Van Iersel et al. (2014) provided a video protocol to show how to euthanize snails with an injection of magnesium chloride solution before removing the snail soft tissue from the shell. Sometimes the targeted molecule is extracted from the whole snail individual or even a set of individuals. Other molecules are isolated from shells or specific organs or tissues (van Iersel et al. 2014). For extracting DNA, for instance, the whole snail individual is used. However, when individuals are large, DNA is extracted from foot tissue because it does not require cracking the shell, and it prevents contamination with foreign DNA from the digestive tract or gonads (Alba et al. 2019b). However, gonads and digestive glands may be targeted when one is interested in studying parasites, like larval trematodes, that occupy these organs. Additionally, by doing so, snail and parasite DNA is extracted in a single procedure (Alba et al. 2015).

Dissection may sometimes require careful and complex manipulation (van Iersel et al. 2014). When collecting the hemolymph, for instance, the foot sole is gently pushed forcing the snail to retract deeply into its shell and thereby extruding a drop of hemolymph which is collected with a micropipette and placed in a tube with an anticoagulant (Ali et al. 2015). In smaller individuals or when higher quantities of hemolymph are needed, hemolymph is collected by puncturing the pericardial cavity (Alba et al. 2020).

Lymnaeid molecules can also be extracted from water (Thomsen and Willerslev 2015). The environmental DNA (eDNA) approach is based on the analysis of free DNA molecules released by individuals in water (Box 5.2). Researchers collected snail eDNA by water filtering using a vacuum pump and a sterile filtration device including a membrane on which eDNA and other molecules accumulate (Jones et al. 2018; Mulero et al. 2021). Membranes are immediately stored in a sterile buffer until conventional DNA extraction methods. Alternatively, instead of filtering water, Rathinasamy et al. (2018) isolated snail eDNA from water by using a silica matrix where eDNA adheres.

5.2.3 Molecule Extraction

The choice of the appropriate technique for extracting or isolating lymnaeid molecules depends on study goal, equipment, amount of sample available, desired purity, molecule, and cost. DNA isolation technique, for instance, should be carefully selected before proceeding to the extraction. Following earlier work in other freshwater snails, one of the first DNA extraction methods used in lymnaeids is based on phenol-chloroform and proteinase K (Bargues et al. 2007). The yield and quality of DNA obtained by this protocol are very good, but the technique requires toxic organic solvents (e.g., phenol, chloroform) and is time-consuming (multiple tube transfers, a minimum of three centrifugations, overnight incubation). Non-toxic techniques based on commercial kits are currently widely used to extract lymnaeid DNA. They should be used if high purity levels are required.

These commercial kits are, however, rather expensive and their use is also, as the phenol-chloroform approach, time-consuming if a large number of samples must be analyzed. Caron et al. (2011) developed and applied a DNA extraction protocol that is cheap, simple, fast (5 h) and does not require toxic organic solvents or tube transfers. This procedure uses a Chelex® solution that prevents DNA degradation by trapping degradative enzymes and potential contaminants and leaving DNA in solution. Despite the low DNA concentration and the suspended impurities compared to the phenol-chloroform technique and commercial DNA extraction kits, the Chelex solution is excellent when subsequent analyses are based on PCR and sequencing specific DNA targets (Alda et al. 2021; Lounnas et al. 2018). Whatever the DNA extraction technique, if the suspension is not used in the short term, it must be stored at −20 °C, or below, in order to preserve DNA quality. Long storage results, however, in the degradation of DNA, making recovery more difficult with long periods of storage. Thus, when performing long-read DNA sequencing, fresh or recently fixed material is recommended.

When studying the structure, function, and evolution of the whole genome (genomics), DNA quality, quantity, and purity are a priority. Previously mentioned DNA extraction kits may not be the best option for performing snail genomics. Specific mollusc DNA extraction kits or modified protocols that remove polysaccharides while ensuring a high molecular DNA weight have been used in molluscs (Gomes-dos-Santos et al. 2020). For small species, as many lymnaeid species, obtaining high quantities of DNA from a single individual may not be possible. Therefore, pooling several individuals is the most frequent option. However, this strategy may introduce errors during subsequent analysis, given that it may increase polymorphism within the pooled data. When performing genomics on small lymnaeid species that reproduce by cross-fertilization, polymorphism can be reduced by experimentally inbreeding individuals. However, this strategy is limited to those species that sustain long-term inbreeding and that are easily cultured at the laboratory (Gomes-dos-Santos et al. 2020).

For transcriptomics and other RNA-based methods, tissues are sometimes crushed in liquid nitrogen using a mortar and pestle (Seppälä et al. 2021). Commercial RNA extraction kits are efficient for isolating high-quality total RNA while removing the majority of genomic DNA. However, some applications such as gene expression analysis by qPCR may require more complete removal of residual contaminating DNA (Nakadera et al. 2019). After isolating DNA or RNA, sample quality and purity are usually measured before further molecular analysis using commercial kits and a fluorometer or a spectrophotometer (Seppälä et al. 2021).

Some snail molecules do not need to be isolated before analysis. Enzyme activity has been, for example, directly measured in the snail nervous tissue (Tripathi et al. 2004). Separation and isolation of metabolites and proteins from snail tissues for downstream use in metabolomics and proteomics analyses require, however, a very detailed protocol (Giusti et al. 2013; Tufi et al. 2015). Heavy metals and other pollutants are also isolated from snail tissues following various and different steps depending on the targeted molecule (Gonçalves et al. 2016; Ng et al. 2011; Reátegui-Zirena et al. 2017).

5.2.4 *Molecular Analyses*

5.2.4.1 DNA-Based Techniques

Most DNA studies require a DNA amplification phase to rapidly generate millions of copies of the DNA marker. PCR conditions such as enzyme and primers used, reagent concentration, number of cycles, annealing temperature, and duration of each step vary depending on the technique, the targeted gene, and the lymnaeid species. Multiplex PCR has been developed in lymnaeids enabling the amplification of more than one DNA marker using multiple sets of primers with a single PCR mixture (Alba et al. 2015; Caron et al. 2011). Another type of PCR is the qPCR that monitors and quantifies a targeted DNA molecule during the whole PCR amplification (Duffy et al. 2009). TaqMan probes are designed to increase the sensitivity and specificity of qPCR (Schweizer et al. 2007).

PCR amplification followed by Sanger sequencing of conserved genes has been the most common DNA-based technique applied to study lymnaeid phylogenetic relationships. The mitochondrial Cytochrome c oxidase subunit I (COI) and the nuclear ITS2 genes have generally been used because they evolve rapidly enough to show differences between species, but not too fast to show huge differences within species. The primers LCO1490 and HCO2198 (Folmer et al. 1994) and NEWS2 and ITS2-RIXO (Almeyda-Artigas et al. 2000) are among the most used for COI and ITS2 amplification, respectively.

Polymorphic markers have also been applied to study lymnaeid snails. For instance, in the Random Amplification of Polymorphic DNA (RAPD) several arbitrary, short primers are used in a PCR to randomly amplify DNA. A RAPD reaction can generate distinctive molecular patterns between individuals (Calienes et al. 2004). However, this technique has too many drawbacks (e.g., unstable variation patterns, unsure homology of same-sized bands). Microsatellite genotyping is a much more efficient approach for evaluating polymorphism. A microsatellite is a repetitive DNA segment in which certain DNA motifs (ranging in length from one to six or more base pairs) are repeated, typically 5–50 times. Microsatellites occur at thousands of locations within genomes. They have a higher mutation rate than other areas of DNA, generating large polymorphism, and therefore facilitating analyses of genetic diversity among and within populations (Jarne and Lagoda 1996). Microsatellite genotyping targets specific loci and therefore allows easy genotypic identification. The targeted locus is PCR-amplified with fluorescently labeled primers (previously described and tested) and alleles are stained and sized by capillary

Table 5.1 Polymorphic microsatellites of lymnaeid species that have been described and tested so far

Species	Number of polymorphic microsatellites	Reference
Ampullaceana balthica	8	Salinger and Pfenninger (2009)
Galba cubensis	15	Lounnas et al. (2017b)
Galba schirazensis	13	Lounnas et al. (2018)
Galba truncatula	9	Trouvé et al. (2000)
Lymnaea stagnalis	9	Knott et al. (2003)
Pseudosuccinea columella	15	Nicot et al. (2008)
Radix auricularia	2	de Boer (2004)

electrophoresis. Microsatellites of seven lymnaeid species have been described and tested so far (Table 5.1).

More recently, Next Generation Sequencing (NGS) technologies like Illumina and Pyrosequencing have been developed allowing one to determine the order of nucleotides—as Sanger technology does but much faster. However, while the Sanger method only sequences a single DNA fragment at a time, NGS massively sequences millions of fragments simultaneously per run. The major difference between Sanger and NGS is sequencing volume. The high throughput of NGS opened a new era in genomics research, with rapidly falling prices for DNA sequencing allowing whole-genome sequencing at affordable prices (Gomes-dos-Santos et al. 2020). However, it is still limited for researchers with access to proper and affordable infrastructure and resources.

Once millions of long DNA sequences have been obtained, genomic analysis begins. First, DNA fragments are assembled to reconstruct full-length base sequences. Then, gene locations are identified within the sequences and the structure and function of these genes are determined—the annotation step. A wide range of powerful analytical tools has been developed for genome assembly and annotation. Thanks to pipelines, bioinformatics combines algorithms, parameter setting methods, and input formats through semi-automated, iterative, and high-throughput analyses to obtain whole genomes (Gomes-dos-Santos et al. 2020). The use of these pipelines usually requires, however, high bioinformatic expertise in a fast-evolving field. This is probably the reason why only the mitochondrial genomes of *Galba pervia*, *Pseudosuccinea columella*, *Ampullaceana balthica*, *A. intermedia* [= *Radix* MOTU3], *Peregriana peregra* [= *Radix* MOTU5], and *Radix plicatula*, and the whole genome of *R. auricularia* have been described so far (Feldmeyer et al. 2010, 2015; Liu et al. 2012; Qin et al. 2019; Schell et al. 2017).

5.2.4.2 RNA-Based Techniques

RNA can also be sequenced or quantified. Targeted RNA sequencing is a highly accurate technique that selects and sequences RNA sequences of interest. However, one may need to only quantify RNA in a given sample. Quantitative reverse transcription PCR (RT-qPCR) is an efficient, simple, and low-cost technique used to detect the increase or decrease in the expression of a particular gene or set of genes of organisms under different conditions (Alba et al. 2020; Nakadera et al. 2019). RT-qPCR first converts RNA into complementary DNA (cDNA) by reverse transcriptase from total or messenger RNA and, then, amplifies and quantifies cDNA by qPCR.

Researchers may need to measure or sequence the sum of all the RNA transcripts of a given tissue under, for instance, different conditions or at different life times (transcriptomics). RNA sequencing generates a large volume of raw sequence reads that requires complex analysis to yield useful information (Seppälä et al. 2021). Data analysis needs a combination of bioinformatics software tools that vary according to data and goals. NGS technologies and the subsequent assembly and annotation are quite similar when analyzing DNA and RNA, only the targeted molecule and the interpretation of the results change.

5.2.4.3 Enzymes and Other Proteins

Enzymes can be identified, purified, or quantified by different biochemical methods: electrophoresis, western blots, ELISAs, chromatography, mass spectrometry, through in-house protocols and commercial kits (Burgess and Deutscher 2009). They can also be studied through the estimation of their biological activity [e.g., Ellman's colorimetric acetylcholinesterase assay (Ellman et al. 1961)]. In this case, parameters such as substrate/ligand concentration, pH and buffer type, ionic strength, and temperature must be tightly controlled. Lymnaeid enzymes like cytochrome oxidase, succinic dehydrogenase, lactic dehydrogenase, and acid and alkaline phosphatases have been studied through spectrophotometry, fluorescence, and chromatography (Ali et al. 2014; Tripathi and Singh 2004). The function and dynamic of other proteins involved in different signaling pathways such as receptors and cytokines have also been elucidated in lymnaeid snails (Seppälä et al. 2021). As in the case of other omics, proteomics also allows to simultaneously study the complexity of the protein network in a given sample. For instance, Alba et al. (2019a) applied two-dimensional electrophoresis followed by mass spectrometry to identify snail proteins.

Enzymes have also been for a long time a valuable target to evaluate genetic variation within and among species (Lewontin 1974), including lymnaeid species (Jabbour-Zahab et al. 1997; Jarne and Delay 1990). Multilocus Enzyme Electrophoresis (MLEE) indeed offers a rapid and inexpensive approach to test for genetic polymorphism by observing different banding patterns of enzyme mobility in an electrophoretic gel. Their use, however, has vanished over the last three decades. MLEE has been replaced by DNA methods, like microsatellites, for several reasons: limited loci polymorphism, the potential effect of selection on variants, and detection of only a fraction of the variation (Avise, 2004; Rowe et al. 2017).

5.2.4.4 Metabolites, Toxins, and Pollutants

Colorimetric-based assays—in which a measurable color change occurs in the presence of the targeted molecule—have been widely used to quantify amino acids, lipids, carbohydrates, and minerals in lymnaeids (Ali et al. 2014; Tripathi and Singh, 2004). More accurate techniques like chromatography, especially high-performance liquid chromatography and spectrophotometry were used for separating, identifying, and quantifying these molecules and metabolites (Brix et al. 2011; Hiripi et al. 1998; Reátegui-Zirena et al. 2016). Nowadays, commercial assay kits provide the reagents required to perform analytical tests for the measurement of these molecules. More recently, metabolomics—the simultaneous detection and quantification of small-molecule metabolites—has been applied on lymnaeids (Tufi et al. 2015).

To reveal and predict the effects of pollution, researchers measure toxic chemicals on organisms. Heavy metals and many other pollutants like pesticides, herbicides, fertilizers, or pharmaceuticals have been measured in lymnaeids (Ali et al. 2012). Heavy metals, for instance, are quantified by high-sensitivity spectroscopy where complex equipment is used to measure the spectra produced by heavy metal interacting with or emitting electromagnetic radiation. Different spectroscopical technologies have been applied on lymnaeids (Gonçalves et al. 2016; Leung et al. 2003; Ng et al. 2011; Reátegui-Zirena et al. 2017).

5.3 Why Use Molecular Technologies in Lymnaeids?

In the following section, we intend to illustrate how the previously mentioned technologies have been applied for developing specific investigations in lymnaeids. We will briefly describe the results of some of the studies that use the most common molecular tools. For the sake of clarity, we group the studies reviewed according to research themes and field, rather than by technical approach.

5.3.1 Species Identification, Systematics, and Phylogenetics

Lymnaeids snails exhibit great diversity in shell morphology due to both genetic variation and phenotypic plasticity, but they show rather homogeneous anatomical traits (Correa et al. 2010; Vinarski et al. 2020). Thus, species identification and

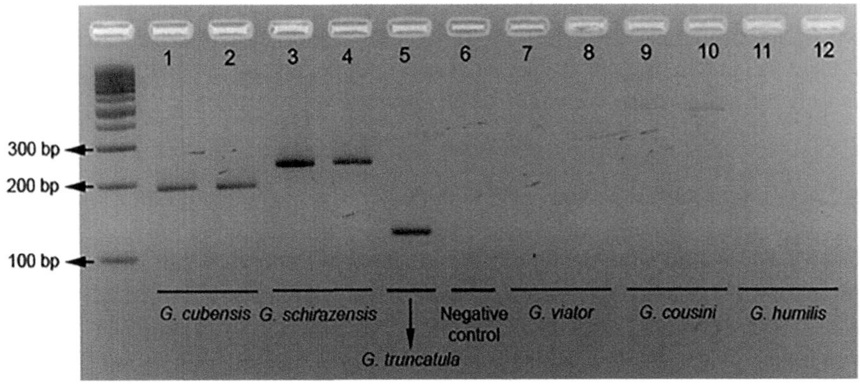

Fig. 5.1 Agarose gel electrophoresis of multiplex PCR products of individuals of the six *Galba* species (extracted from Alda et al. 2018)

systematics have almost always been challenging in this snail group (Schniebs et al. 2011, 2013, 2019; Vinarski 2013). This is a critical issue since accurate species identification is required, especially with regard to invasive or disease-transmitting species. When species are morphologically similar, molecular biology becomes compulsory to distinguish them (Alda et al. 2021). Sequencing conserved gene markers has almost exclusively been the only technique applied to molecularly identify lymnaeid species and to study their systematics and phylogenetic relationships. Twenty years ago, researchers used one or two genes and a few individuals per species to describe the phylogeny of a genus or a group of lymnaeids living in a given region (e.g., Bargues et al. 2001). More recent comprehensive phylogenetic studies, however, include more genes and at least ten populations per species. For instance, Correa et al. (2010) constructed the phylogeny of the family Lymnaeidae using three genes and 50 taxa, the most complete dataset to date at the family level (see also Aksenova et al. 2023). Similarly, Aksenova et al. (2018) conducted a phylogenetic study within the subfamily Amphipepleinae using 2602 sequences, and Alda et al. (2021) did it within the genus *Galba* using 796 sequences. Furthermore, Aksenova et al. (2018) developed the first comprehensive DNA barcode library for the reliable molecular identification of 35 species of the radicine lymnaeids from the Old World using the COI barcode fragment.

Cheaper techniques have also been developed for rapid molecular discrimination of lymnaeid species. We provide here the example of MLEE and multiplex PCR. MLEE was successfully used within and among species of the genera *Galba*, *Lymnaea*, *Pseudosuccinea*, and *Omphiscola* (Durand et al. 2002; Hurtrez-Boussès et al. 2005; Jabbour-Zahab et al. 1997; Ward et al. 1997). Multiplex PCR has been used to discriminate cryptic species in one step (Alda et al. 2018) (Fig. 5.1). Alternatively, multiplex PCR can be used to detect snail DNA and parasite DNA in the same sample. Caron et al. (2011) developed a multiplex PCR that amplifies a highly repeated DNA *Fasciola* sequence and the ITS2 of *Galba truncatula*—a technique that has been useful for surveying parasite transmission through one of its main snail hosts.

5.3.2 *Population Genetics and Phylogeography*

DNA sequencing of single and conserved markers (ITS2, ITS1, COI, and 16S) has also been applied to differentiate populations of lymnaeids. The resulting sequences can be used to construct haplotype networks that visually represent genealogical relationships, biogeography, and evolutionary history among populations. For instance, this approach was used to demonstrate that several endemic taxa, previously considered as separate endemic species, are in fact synonyms of a widespread Holarctic species, *Radix auricularia*, and represent ecological morphs adapted to living in thermal springs of Lake Baikal and Kamchatka (Bolotov et al. 2014; Aksenova et al. 2017; Vinarski et al. 2023) (Fig. 5.2). An approximate Bayesian computation approach using the COI sequence data revealed that a distant lineage of *Ampullaceana balthica* in Iceland originated via postglacial colonization followed by rapid divergence under an elevated mutation rate (Bolotov et al. 2017).

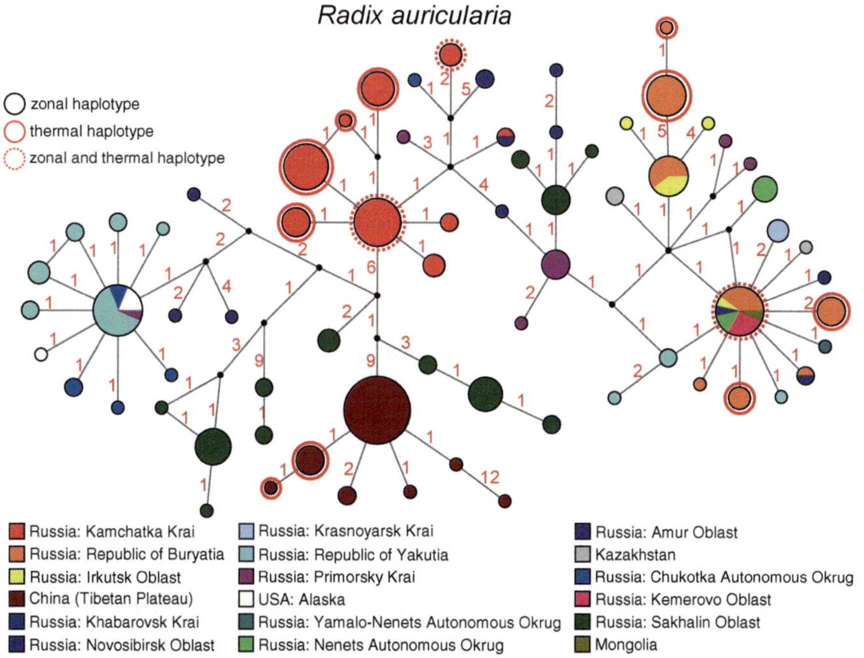

Fig. 5.2 Haplotype network of the COI sequences of *Radix auricularia* specimens from European Russia, China, Kazakhstan, Mongolia, Siberia, and the Russian Far East (n = 239). The circle size is proportional to the number of available sequences belonging to a certain haplotype. Red numbers near branches indicate the number of nucleotide substitutions between haplotypes

Combining markers with contrasted mutation rates has been very useful for studying the phylogeography of some lymnaeid species (Kopp et al. 2012; Lounnas et al. 2018, 2017a; Pfenninger et al. 2011). Microsatellites can detect recent expansion events, while conserved markers like COI can reveal ancient population splits. This is because microsatellite loci have high mutation rates and tens of generations can generate mutations at least at one locus while, for example, several thousands of years are needed to generate one base substitution at the COI gene (Lounnas et al. 2018). In the case of *Pseudosuccinea columella*, the analyses of eight nuclear microsatellites and two mitochondrial genes revealed that populations from North America, the putative origin area, harbored much more genetic variability than invasive populations whereas a unique genotype has been repeatedly introduced in South America and all over the world (Lounnas et al. 2017a). This result clearly supports the use of conserved and polymorphic molecular markers to unveil lymnaeid phylogeography.

5.3.3 Mating System and Reproductive Behavior

Like other simultaneous hermaphrodite snails, lymnaeids can reproduce by self- and cross-fertilization. Studies based on microsatellites have shown, however, that the selfing rate can vary among lymnaeid species (Escobar et al. 2011) and sometimes among natural populations within species (Bürkli et al. 2017; Chapuis et al. 2007; Lounnas et al. 2018, 2017a, b; Meunier et al. 2004). Microsatellites and MLEE have also been used for measuring selfing rates in experimental populations (Wiehn et al. 2002). For instance, by genotyping a single microsatellite marker, Nakadera et al. (2014) experimentally investigated the pattern of cross-fertilization over time after a single mating in *L. stagnalis*. They found that offspring can be produced from stored mating partner's sperm for a little over 2 months, which indicates a prolonged paternity gain.

Functional genomics has been used in the model species *Lymnaea stagnalis* to investigate reproduction, focusing on seminal (prostatic) fluid proteins secreted by accessory reproductive glands. These accessory gland proteins (ACPs) have been shown to be involved in sexual selection and partner manipulation in species such as the fruit fly *Drosophila melanogaster* (Sirot et al. 2011). Experimental approaches have been conducted in *Lymnaea stagnalis* in order to evaluate their impact under different social conditions or in intravaginally injected focal individuals (van Iersel et al. 2014). Individuals strategically alter seminal fluid production in response to mate availability and sperm competition (Nakadera et al. 2019). Specifically, these authors found that snails plastically elevated ACP expression in the presence of at least one mating partner. The transfer of ACPs also temporarily depresses egg production in recipient snails in *L. stagnalis*, and similar results have been obtained in other lymnaeids, even in selfing species with limited copulation activity (Swart et al. 2020). More generally, ACPs are involved in the regulation of post-copulatory processes in *L. stagnalis* with a marked effect on fitness (Nakadera et al. 2019; Swart et al. 2019).

An approach that has rarely been mobilized for studying lymnaeid mating system and reproductive behavior is transcriptomics. Koene and colleagues (Nakadera et al. 2019; Swart et al. 2019) followed such an approach to characterize the general expression of proteins in the prostate in *Lymnaea stagnalis* (see above). A more general approach has been used by Burgarella et al. (2015) to address molecular evolution in outcrossing species (*Physa acuta*) vs. selfing ones (*Galba truncatula*) based on whole-genome transcriptomics. All results predicted in selfers from theory, such as strongly reduced genetic diversity or slower rate of adaptive evolution, were detected in *G. truncatula*. It, therefore, remains unclear how *Galba* managed to adapt over its long history of selfing.

5.3.4 Abiotic Interactions

The simplicity of breeding various lymnaeid species in the laboratory, the occurrence of these species in anthropogenic habitats, and the knowledge about the normal metabolism of *Lymnaea stagnalis* have turned lymnaeids into a widely used animal model to investigate how the environment affects metabolism and behavior. Physiological stress responses have been investigated based on enzymatic activity, nutrient concentration, and the bioaccumulation of heavy metals and toxins. For instance, studies have tested the effect of pesticides, herbicides, fertilizers, pharmaceuticals, and other pollutants on the activity of acetylcholinesterase on the nerve tissue of the Indian snail *Radix rufescens* [= *Lymnaea acuminata*] (Singh and Agarwal, 1984) and the pond snail *Lymnaea stagnalis* (Tufi et al. 2016).

Industrial activities such as paint manufacture, printing, battery and tire manufacture, mining, and water run-off from shooting ranges are a source of heavy metals and other pollutants that find their way into rivers through effluent discharge. Heavy metals (e.g., aluminum, iron, copper, zinc, lead, and cadmium) accumulate in different tissues decreasing snail feeding, growth, fertility, and fecundity (Ng et al. 2011; Reátegui-Zirena et al. 2017; Walton et al. 2010). Likewise, pollutants like zinc oxide nanoparticles present in hazardous materials such as chlorinated solvents and pesticides cause DNA damage in snails through oxidative stress (Ali et al. 2012). Other environmental stressors that have been found and broadly studied in lymnaeids are microcystins that are toxins produced by certain freshwater cyanobacteria during algal blooms jeopardizing the life of freshwater gastropods and higher consumers through food webs (Lance et al. 2014; Zurawell et al. 2006).

Instead of measuring a single or a group of molecules, modern toxicology uses metabolomics to understand toxicological effects in lymnaeids (Boisseaux et al. 2017). Transcriptomics and proteomics have also been applied to study how pharmaceutical and other industrial compounds and by-products in wastewater may affect the expression of genes involved in the nervous regulation of feeding, digestion, reproduction, and locomotion (Giusti et al. 2013). Similarly, gene expression assays revealed that pharmaceutical pollutants changed the mRNA expression of some genes involved in the innate immune response of *L. stagnalis* threatening snail defenses against infectious agents like parasites, bacteria, and viruses (Gust et al. 2013).

Most assays have been performed in a single natural or experimental population. However, Côte et al. (2015) showed that genetically different natural populations widely varied in their copper-induced mortality, demonstrating that species-level extrapolations from single population studies are highly unreliable. In this sense, ecotoxicological studies should incorporate the analysis of population genetic diversity in order to refine environmental risk assessment.

5.3.5 Biotic Interactions

Molecular technologies have also been applied in studies investigating the interactions among lymnaeids, and between lymnaeids and other organisms. The majority of these studies investigate snail physiological responses to parasites. The trematodes *Fasciola hepatica* and *Trichobilharzia ocellata* and snails from the genera *Pseudosuccinea*, *Galba*, and *Lymnaea* have been the most common parasite-lymnaeid snail models used for evaluating parasite-induced physiological studies (Alba et al. 2020; de Lange et al. 2001; Southgate 1970).

Similar to other invertebrates, lymnaeid snails have an innate immune system that inactivates and eliminates penetrating parasites. One of the methods used to evaluate the innate immune system is to measure the formation of toxic reactive oxygen species and the peroxidase activity during snail defense reactions against trematodes (Gornowicz et al. 2013; Vorontsova et al. 2019). An increase in the activity of other proteins in response to parasites and other pathogens has also been reported (Plows et al. 2005; Seppälä et al. 2021; Skála et al. 2014). Likewise, parasites can increase the concentration of metallic ions, carbohydrates, lipids, proteins, and free amino acids. Infected individuals show, for instance, an increase of zinc, copper, and potassium concentration in the soft tissues and a decrease in calcium levels in the soft parts and shells (Mostafa 2008).

Gene expression, transcriptomics, and proteomics have also been applied to investigate the molecular and immunological bases of the resistance of *Pseudosuccinea columella* to *Fasciola hepatica* prior to and during experimental challenges with the parasite (Alba et al. 2019a, 2020). For instance, Alba et al. (2020) compared the whole-snail transcriptome and the proteome of the albumen gland between resistant and susceptible populations of *Pseudosuccinea columella* and found that resistant snails have constitutively a greater abundance of transcripts and proteins associated with immune defense and stress response.

The microbiota has important implications for how the host's immune system is structured to accommodate them and for how it responds to pathogens (Van Horn et al. 2012). Indigenous microbial communities interact with their host to assist in obtaining nutrients and detoxifying secondary compounds from food, and to protect the host from invasion by pathogenic exogenous microbes. Lymnaeid snails, which live primarily in freshwater environments and feed on algae, decaying macrophytes and detritus, may have a significantly different gut and hemolymph flora than other freshwater organisms, but specific knowledge of microbiota in lymnaeid snails is very limited (Dar et al. 2017). For example, the first biochemical study of bacterial strains isolated from snail intestines was conducted for *Lymnaea stagnalis* from India (Sayyad and Shaikh 2015), but without the use of molecular methods. Owing to the development of NGS, there has been a recent increase in research on the microbiome based on high-throughput sequencing of the V3/V4 region of the 16S rRNA gene for freshwater gastropods (Van Horn et al. 2012; Dar et al. 2017; Chevalier et al. 2020; Allan et al. 2018; Huot et al. 2020). Recently, the intestinal bacterial communities of *Radix auricularia* were characterized using 16S rRNA amplicon sequencing, and then the intestinal bacteria were compared at different stages of snail's growth by researchers from China (Hu et al. 2018). Later, Horton et al. (2020) investigated how microplastics may affect inherent microbiome diversity and community composition of *Lymnaea stagnalis*. The microbiota may influence the response of lymnaeids to metazoan pathogens such as schistosomes and related parasites or changed environmental conditions. It is unknown, this response fully independent of the accommodations made to the gut flora. The first data on the composition of microbial communities of pond snails emphasize the importance of conducting further research to better understand the mechanisms of interaction between microorganisms and the host mollusk.

5.4 Trends and Perspectives in Ecological and Evolutionary Studies

In this chapter, we briefly show the steps that should be followed to study any targeted molecule in lymnaeids (Fig. 5.3). We also tried to illustrate how molecular methodologies have been applied to study ecological and evolutionary processes on lymnaeids. We intend now to propose how some previously mentioned and other technologies might improve our knowledge on the ecology and evolution of lymnaeids. Moreover, lymnaeids can be used as animal models to answer broader questions about the ecology and evolution of other living beings (Box 5.1). Some of these answers can only be obtained by applying a molecular approach, while others are enriched by applying it. One of the biggest challenges in future ecological and evolutionary studies will be to consider the analysis of intraspecific diversity. Most ecotoxicological studies, for instance, base their conclusions on experiments performed in single natural or experimental populations (Côte et al. 2015). Future studies should perform experiments in genetically differentiated populations in order to obtain more accurate conclusions.

Comparing genomic polymorphism between native and invasive lymnaeid populations—by applying, for instance, double digest restriction-site associated

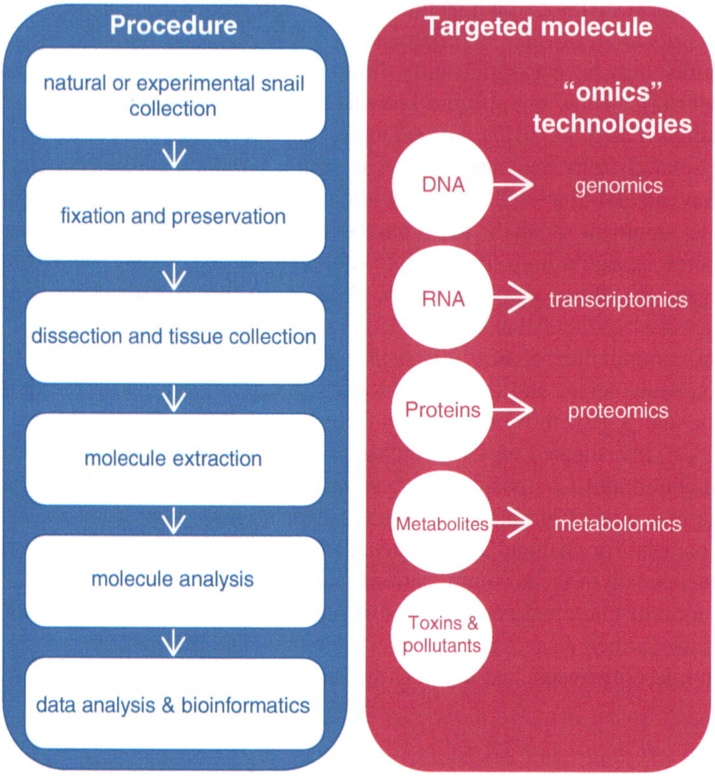

Fig. 5.3 Steps (in blue) that should be followed to study any targeted molecule (in pink) in lymnaeids

sequencing—can help us to reconstruct routes of invasions, to unveil the genetic consequences in snail invaders, and to understand the spread of infectious diseases transmitted by invaders. Compared to native populations, introduced populations usually display less genetic variation as a result of bottlenecks and founding events (Cristescu 2015). This effect enhances when selfing is the preferred mating system (Lounnas et al. 2018, 2017a). It is widely admitted that genetically homogenous host populations (resulting from recent invasions or selfing) are more vulnerable to infection than genetically diverse populations (King and Lively 2012). For instance, the rapid expansion of the selfing snail *Biomphalaria pfeifferi*, an intermediate host for human schistosomiasis, resulted in the establishment of genetically homogenous populations in Zimbabwe that were more susceptible to the parasite, promoting the resurgence of the disease in this country (Campbell et al. 2010). As most invasive host populations have no genetic variability to evolve resistance mechanisms, the parasite could freely expand without opposition, representing a health threat. Future genomic studies on lymnaeids could help us to reveal the mechanisms and processes behind the interaction between native and introduced host populations and parasites.

Molecular tools should be routinely used to prevent the (re)emergence of infectious diseases transmitted by lymnaeids. An integrated control program on fasciolosis, for instance, can combine the administration of antihelminths with the implementation of an eDNA approach to early detect the arrival of an invasive lymnaeid species to a given irrigation or drainage canal and immediately react by preventing herds from drinking water at these sites (Jones et al. 2018; Sabourin et al. 2018). Molecular species identification should also be used to construct updated and dynamic maps of the distribution of native and invasive snails that transmit parasites to humans, cattle, or wildlife. Mapping is essential for determining the risk of infectious diseases.

Our knowledge of the lymnaeid phylogeny and biogeography is still scarce compared to other snail groups. Integrative taxonomy has proved to be an appropriate and accurate approach for describing and delimiting species (Pante et al. 2015; Vinarski et al. 2016a, b, 2021; Aksenova et al. 2018). To do so, we need to bring together traditional taxonomy with NGS, advanced computational tools, and mating experiments. This last approach is only feasible in lymnaeid species that prefer to reproduce by cross-fertilization. Geographically wider sampling and a higher number of genes and individuals that better represent intra- and interspecific variability are essentials for obtaining more precise and informative phylogenies and species distribution models. Non-morphological or genetic data should also be included in phylogenetic studies. Lymnaeid species significantly differ, for instance, in their habitat preference, and their amphibious behavior. While some species prefer permanent, larger, and deeper water bodies which they leave only sporadically, others live on mud or wet sites (Bargues et al. 2011). Among these last species, some are usually found below the water while others have a terrestrial behavior (Bargues et al. 2011). Including behavioral traits in phylogenetic studies and species delimitation methods could enhance our knowledge about lymnaeids systematics. All this morphological, geographic, environmental, and genetic data should be available in online free data repositories. This extensive database would be in line with FAIR data (https://en.wikipedia.org/wiki/FAIR_data), which is an initiative that makes scientific research transparent and accessible to society, amateurs, or professionals. Broad development of DNA barcode libraries is necessary for accurate molecular identification of lymnaeid species, many of which represent cryptic taxa that can hardly be distinguished by traditional morphology-based approaches (Aksenova et al. 2018).

Whole genomes of good quality are compulsory for addressing many issues in evolutionary biology and ecology. At this point, we only have a few genomes in mollusks, and not all are of good-enough quality. This is especially true in lymnaeids, and genome projects are called for. They could begin by sequencing at least one species per genus or species group. Whole-genome sequencing can be nowadays performed at affordable prices (Gomes-dos-Santos et al. 2020). However, the posterior genome analyses (assembly and annotation) are time-consuming and require qualified staff with expertise in bioinformatic databases and tools. Genome evolution and adaptation is one question that would benefit from such an approach. For example, comparing genomes of preferentially selfing and preferentially outcrossing lymnaeid species could help to understand the effect of the mating system on genome structure and function. A species that reproduces by selfing for many generations (i.e., long-term) shows a genomic degradation that translates into a low adaptive potential, which is why selfing is considered to be an evolutionary dead-end (Noël et al. 2017). In other words, this detrimental genomic effect is predicted to drive selfing lineages to extinction in the long term. Transcriptomic analysis showed that *Galba truncatula*, a preferentially selfing lymnaeid species, present decreased efficacy of purifying selection, a slower rate of adaptive evolution, and weakened codon usage bias/GC-biased gene conversion than the preferentially outcrossing species *Physa acuta* (Burgarella et al. 2015). Transcriptomic coupled with functional genomic analysis could be performed to study coding sequence polymorphism, divergence, and expression levels between preferentially selfing and preferentially outcrossing *Galba* species to search for signatures of selection.

The "omics" approaches could also enhance our knowledge about the epidemiological triad "lymnaeids, parasites, and the environment" (Alba et al. 2019a). One current challenge in biological science is to accurately predict species response to environmental changes and species interactions. Some pioneer articles have been published in lymnaeids. For example, an eco-immunological study performed in *L. stagnalis* showed that snails exposed to high temperatures had a weaker immune defense, which potentially predisposes them to infections (Seppälä and Jokela 2011). Future ecotoxicological and eco-immunological studies should integrate genomics, transcriptomic, proteomic, and metabolomic information to provide a better understanding of the cellular response to biotic and abiotic stressors.

Box 5.1 The Great Pond Snail, *Lymnaea stagnalis*, as a Model Organism

One of the most versatile model organisms used in the history of medical and biological science is the pond snail *Lymnaea stagnalis*, a lymnaeid species found in many lakes and ponds throughout the Holarctic region. This freshwater snail was first an excellent model for studying the basic and conserved mechanisms of central nervous system physiology and pathology (Rivi et al. 2020). Among the main reasons, we can highlight: (1) unlike other lymnaeid species, it is easy to breed it under experimental conditions; (2) its large, readily accessible, and identified neurons offer a large easily manipulated amount of biological material for molecular analyses; and (3) its well-known and characterized anatomy, histology, physiology, life cycle, and behavior make this organism an excellent model to test how an external pressure like environmental stressors and parasites can alter biological functions (Fodor et al. 2020; Rivi et al. 2020). *Lymnaea stagnalis* has been extensively used for addressing fundamental mechanisms in neurobiology (Fodor et al. 2020) focusing, for instance, in learning and memory responses (Rivi et al. 2020) that could be extrapolated to humans and other mammals. Following the

(continued)

Box 5.1 (continued)
neurobiological boom in the 1970s, *L. stagnalis* became a popular model in other medical and biological fields as diverse as host–parasite interactions, ecotoxicology, evolution, genome editing, and human disease modeling (Fodor et al. 2020). The current vast knowledge about the physiology of *L. stagnalis* is the result of 50 years of intense molecular research (Fodor et al. 2020).

Box 5.2 Environmental DNA
Aiming at monitoring and measuring biodiversity, this approach relies on sampling a complex mixture of genomic DNA from many different organisms found in an environmental sample (Mulero et al. 2021). Species detection is based on PCR amplification: barcoding uses specific primers targeting a given species and metabarcoding uses fewer specific targets that can amplify a group of species including the snails of interest and many other species. Studies based on eDNA, especially metabarcoding, highly depend on the existence and reliability of currently available databases, which sometimes prevents the identification of a sample to the species level (Mulero et al. 2021). Environmental variables such as pH, water flow, and temperature could affect the accuracy of detecting a given species in a water sample (Mulero et al. 2021; Nicholson et al. 2020). However, the great advantage of eDNA is that very rare species or species at low abundance can be detected, even when they are not detected in classical malacological survey (Mulero et al. 2021).

References

Aksenova O, Vinarski M, Bolotov I et al (2017) Two *Radix* spp. (Gastropoda: Lymnaeidae) endemic to thermal springs around Lake Baikal represent ecotypes of the widespread *Radix auricularia*. J Zool Syst Evol Res 55:298–309. https://doi.org/10.1111/jzs.12174

Aksenova OV, Bolotov IN, Gofarov MY et al (2018) Species richness, molecular taxonomy and biogeography of the radicine pond snails (Gastropoda: Lymnaeidae) in the Old World. Sci Rep-UK 8:11199. https://doi.org/10.1038/s41598-018-29451-1

Aksenova OV, Bolotov IN, Kondakov AV et al (2023, this volume) Phylogeny and taxonomy of the family Lymnaeidae. In: Vinarski MV, Vázquez A (eds) The Lymnaeidae. A handbook on their natural history and parasitological significance. Springer, Cham, pp 67–101

Alba A, Vázquez AA, Hernández H et al (2015) A multiplex PCR for the detection of *Fasciola hepatica* in the intermediate snail host *Galba cubensis*. Vet Parasitol 211:195–200. https://doi.org/10.1016/j.vetpar.2015.05.012

Alba A, Tetrau G, Chaparro C et al (2019a) Field-occurring resistance to *Fasciola hepatica* (Trematoda) in *Pseudosuccinea columella* snails: a review from literature and insights from comparative "omic" analyses. Dev Comp Immunol 101:103463. https://doi.org/10.1016/j.dci.2019.103463

Alba A, Vázquez AA, Sánchez J et al (2019b) Patterns of distribution, populations genetics and ecological requirements of field-occurring resistant and susceptible *Pseudosuccinea columella* snails to *Fasciola hepatica* in Cuba. Sci Rep-UK 9:14359. https://doi.org/10.1038/s41598-019-50894-7

Alba A, Duval D, Sánchez J et al (2020) The immunobiological interplay between *Pseudosuccinea columella* resistant/susceptible snails with *Fasciola hepatica*: hemocytes in the spotlight. Dev Comp Immunol 102:103485. https://doi.org/10.1016/j.dci.2019.103485

Alda P, Lounnas M, Vázquez AA et al (2018) A new multiplex PCR assay to distinguish among three cryptic *Galba* species, intermediate hosts of *Fasciola hepatica*. Vet Parasitol 251:101–105. https://doi.org/10.1016/j.vetpar.2018.01.006

Alda P, Lounnas M, Vázquez AA et al (2021) Systematics and geographical distribution of *Galba* species, a group of cryptic and worldwide freshwater snails. Molec Phyl Evol 157:107035. https://doi.org/10.1016/j.ympev.2020.107035

Ali D, Alarifi S, Kumar S et al (2012) Oxidative stress and genotoxic effect of zinc oxide nanoparticles in freshwater snail *Lymnaea luteola* L. Aquat Toxicol 124–125:83–90. https://doi.org/10.1016/j.aquatox.2012.07.012

Ali D, Yadav PG, Kumar S et al (2014) Sensitivity of freshwater pulmonate snail *Lymnaea luteola* L., to silver nanoparticles. Chemosphere 104:134–140. https://doi.org/10.1016/j.chemosphere.2013.10.081

Ali D, Ali H, Alarifi S et al (2015) Impairment of DNA in a freshwater gastropod (*Lymnea luteola* L.) after exposure to titanium dioxide nanoparticles. Arch Environ Contam Toxicol 68:543–552. https://doi.org/10.1007/s00244-015-0132-0

Allan ERO, Tennessen JA, Sharpton TJ et al (2018) Allelic variation in a single genomic region alters the microbiome of the snail *Biomphalaria glabrata*. J Hered 109:604–609. https://doi.org/10.1093/jhered/esy014

Almeyda-Artigas RJ, Bargues MD, Mas-Coma S (2000) ITS-2 rDNA sequencing of *Gnathostoma* species (Nematoda) and elucidation of the species causing human gnathostomiasis in the Americas. J Parasitol 86:537–544. https://doi.org/10.1645/0022-3395(2000)086[0537,IRSOGS]2.0.CO;2

Avise JC (2004) Molecular markers, natural history, and evolution, 2nd edn. Sinauer Associates, Sunderland, MA

Bargues MD, Vigo M, Horák P et al (2001) European Lymnaeidae (Mollusca: Gastropoda), intermediate hosts of trematodiases, based on nuclear ribosomal DNA ITS-2 sequences. Infect Genet Evol 1:87–107. https://doi.org/10.1016/S1567-1348(01)00019-3

Bargues MD, Artigas P, Mera y Sierra R et al (2007) Characterisation of *Lymnaea cubensis*, *L. viatrix* and *L. neotropica* n. sp., the main vectors of *Fasciola hepatica* in Latin America, by analysis of their ribosomal and mitochondrial DNA. Ann Trop Med Parasit 101:621–641. https://doi.org/10.1179/136485907X229077

Bargues MD, Artigas P, Khoubbane M et al (2011) *Lymnaea schirazensis*, an overlooked snail distorting fascioliasis data: genotype, phenotype, ecology, worldwide spread, susceptibility, applicability. PLoS One 6:e24567. https://doi.org/10.1371/journal.pone.0024567

Boisseaux P, Noury P, Thomas H et al (2017) Immune responses in the aquatic gastropod *Lymnaea stagnalis* under short-term exposure to pharmaceuticals of concern for immune systems: diclofenac, cyclophosphamide and cyclosporine A. Ecotox Environ Safe 139:358–366. https://doi.org/10.1016/j.ecoenv.2017.02.003

Bolotov I, Bespalaya Y, Aksenova O et al (2014) A taxonomic revision of two local endemic *Radix* spp. (Gastropoda: Lymnaeidae) from Khodutka geothermal area, Kamchatka, Russian Far East. Zootaxa 3869(5):585–593. https://doi.org/10.11646/zootaxa.3869.5.9

Bolotov IN, Aksenova OV, Bespalaya YV et al (2017) Origin of a divergent mtDNA lineage of a freshwater snail species, *Radix balthica*, in Iceland: cryptic glacial refugia or a postglacial founder event? Hydrobiologia 787:73–98. https://doi.org/10.1007/s10750-016-2946-9

Brix KV, Esbaugh AJ, Grosell M (2011) The toxicity and physiological effects of copper on the freshwater pulmonate snail, *Lymnaea stagnalis*. Comp Biochem Phys C 154:261–267. https://doi.org/10.1016/j.cbpc.2011.06.004

Burgarella C, Gayral P, Ballenghien M et al (2015) Molecular evolution of freshwater snails with contrasting mating systems. Mol Biol Evol 32:2403–2416. https://doi.org/10.1093/molbev/msv121

Burgess RR, Deutscher MP (2009) Guide to protein purification, Methods in enzymology, 2nd edn. Elsevier/Academic Press, Amsterdam-Boston, MA

Bürkli A, Sieber N, Seppälä K et al (2017) Comparing direct and indirect selfing rate estimates: when are population-structure estimates reliable? Heredity 118:525–533. https://doi.org/10.1038/hdy.2017.1

Calienes AF, Fraga J, Pointier JP et al (2004) Detection and genetic distance of resistant populations of *Pseudosuccinea columella* (Mollusca: Lymnaeidae) to *Fasciola hepatica* (Trematoda: Digenea) using RAPD markers. Acta Trop 92:83–87. https://doi.org/10.1016/j.actatropica.2004.05.014

Campbell G, Noble LR, Rollinson D et al (2010) Low genetic diversity in a snail intermediate host (*Biomphalaria pfeifferi* Krauss, 1848) and schistosomiasis transmission in the Senegal River basin. Mol Ecol 19:241–256. https://doi.org/10.1111/j.1365-294X.2009.04463.x

Caron Y, Righi S, Lempereur L et al (2011) An optimized DNA extraction and multiplex PCR for the detection of *Fasciola* sp. in lymnaeid snails. Vet Parasitol 178:93–99. https://doi.org/10.1016/j.vetpar.2010.12.020

Chapuis E, Trouve S, Facon B et al (2007) High quantitative and no molecular differentiation of a freshwater snail (*Galba truncatula*) between temporary and permanent water habitats. Mol Ecol 16:3484–3496. https://doi.org/10.1111/j.1365-294X.2007.03386.x

Chevalier FD, Diaz R, McDew-White M et al (2020) The hemolymph of *Biomphalaria* snail vectors of schistosomiasis supports a diverse microbiome. Env Microbiol 22(12):5450–5466. https://doi.org/10.1111/1462-2920.15303

Correa AC, Escobar JC, Durand P et al (2010) Bridging gaps in the molecular phylogeny of the Lymnaeidae (Gastropoda: Pulmonata), vectors of fascioliasis. BMC Evol Biol 10:381. https://doi.org/10.1186/1471-2148-10-381

Côte J, Bouétard A, Pronost Y et al (2015) Genetic variation of *Lymnaea stagnalis* tolerance to copper: a test of selection hypotheses and its relevance for ecological risk assessment. Environ Pollut 205:209–217. https://doi.org/10.1016/j.envpol.2015.05.040

Cristescu ME (2015) Genetic reconstructions of invasion history. Mol Ecol 24:2212–2225. https://doi.org/10.1111/mec.13117

Dar MA, Pawar KD, Pandit RS (2017) Gut microbiome analysis of snails: a biotechnological approach. In: Ray S (ed) Organismal and molecular malacology. Intech Open, pp 189–217. https://doi.org/10.5772/68133

de Boer MG (2004) Two new highly polymorphic microsatellite loci and inadvertent minisatellite loci for *Lymnaea auricularia*. J Mollus Stud 70:95–96. https://doi.org/10.1093/mollus/70.1.95

de Lange RPJ, Moorer-van Delft CM, de Boer PACM et al (2001) Target-dependent differentiation and development of molluscan neurons and neuroendocrine cells: use of parasitisation as a tool. Neuroscience 103:289–299. https://doi.org/10.1016/S0306-4522(00)00556-X

Duffy T, Kleiman F, Pietrokovsky S et al (2009) Real-time PCR strategy for rapid discrimination among main lymnaeid species from Argentina. Acta Trop 109:1–4. https://doi.org/10.1016/j.actatropica.2008.08.003

Durand P, Pointier JP, Escoubeyrou K et al (2002) Occurrence of a sibling species complex within neotropical lymnaeids, snail intermediate hosts of fascioliasis. Acta Trop 83:233–240. https://doi.org/10.1016/S0001-706X(02)00128-6

Ellman GL, Courtney KD, Andres V et al (1961) A new and rapid colorimetric determination of acetylcholinesterase activity. Biochem Pharmacol 7:88–95. https://doi.org/10.1016/0006-2952(61)90145-9

Escobar JS, Auld JR, Correa AC et al (2011) Patterns of mating-system evolution in hermaphroditic animals: correlations among selfing rate, inbreeding depression, and the timing of reproduction: mating-system evolution in hermaphroditic animals. Evolution 65:1233–1253. https://doi.org/10.1111/j.1558-5646.2011.01218.x

Feldmeyer B, Hoffmeier K, Pfenninger M (2010) The complete mitochondrial genome of *Radix balthica* (Pulmonata, Basommatophora), obtained by low coverage shot gun next generation sequencing. Mol Phylogenet Evol 57:1329–1333. https://doi.org/10.1016/j.ympev.2010.09.012

Feldmeyer B, Greshake B, Funke E et al (2015) Positive selection in development and growth rate regulation genes involved in species divergence of the genus *radix*. BMC Evol Biol 15(1):1–11. https://doi.org/10.1186/s12862-015-0434-x

Fodor I, Hussein AAA, Benjamin PR et al (2020) The unlimited potential of the great pond snail, *Lymnaea stagnalis*. eLife 9:e56962. https://doi.org/10.7554/eLife.56962.sa2

Folmer O, Black M, Hoeh W et al (1994) DNA primers for amplification of mitochondrial cytochrome c oxidase subunit I from diverse metazoan invertebrates. Mol Marine Biol Biotechnol 3:294–299

Giusti A, Leprince P, Mazzucchelli G et al (2013) Proteomic analysis of the reproductive organs of the hermaphroditic gastropod *Lymnaea stagnalis* exposed to different endocrine disrupting chemicals. PLoS One 8:e81086. https://doi.org/10.1371/journal.pone.0081086

Gomes-dos-Santos A, Lopes-Lima M, Castro LFC et al (2020) Molluscan genomics: the road so far and the way forward. Hydrobiologia 847:1705–1726. https://doi.org/10.1007/s10750-019-04111-1

Gonçalves S, Bennett M, Raab A et al (2016) Sub-lethal cadmium exposure increases phytochelatin concentrations in the aquatic snail *Lymnaea stagnalis*. Sci Total Environ 568:1054–1058. https://doi.org/10.1016/j.scitotenv.2016.06.149

Gornowicz D, Dmochowska K, Żbikowska E et al (2013) Total antioxidative status and the activity of peroxidase and superoxide dismutase in the haemolymph of *Lymnaea stagnalis* (L.) naturally infected with digenean trematodes. J Mollus Stud 79:225–229. https://doi.org/10.1093/mollus/eyt019

Gust M, Fortier M, Garric J et al (2013) Immunotoxicity of surface waters contaminated by municipal effluents to the snail *Lymnaea stagnalis*. Aquat Toxicol 126:393–403. https://doi.org/10.1016/j.aquatox.2012.09.001

Hiripi L, Vehovszky Á, Juhos S et al (1998) An octopaminergic system in the CNS of the snails, *Lymnaea stagnalis* and *Helix pomatia*. Phil Trans R Soc Lond B 353:1621–1629. https://doi.org/10.1098/rstb.1998.0314

Horton AA, Newbold LK, Palacio-Cortés AM et al (2020) Accumulation of polybrominated diphenyl ethers and microbiome response in the great pond snail *Lymnaea stagnalis* with exposure to nylon (polyamide) microplastics. EcotoxEnviron Safe 188:109882. https://doi.org/10.1016/j.ecoenv.2019.109882

Hu Z, Chen X, Chang J et al (2018) Compositional and predicted functional analysis of the gut microbiota of *Radix auricularia* (Linnaeus) via high-throughput Illumina sequencing. Peer J 6: e5537. https://doi.org/10.7717/peerj.5537

Huot C, Clerissi C, Gourbal B et al (2020) Schistosomiasis vector snails and their microbiota display a Phylosymbiosis pattern. Front Microbiol 10:3092. https://doi.org/10.3389/fmicb.2019.03092

Hurtrez-Boussès S, Pendino A, Barnabé C et al (2005) Comparison between shell morphology and genetic diversity in two sympatric lymnaeid snails, vectors of fasciolosis. Can J Zool 83:1643–1648. https://doi.org/10.1139/z05-150

Jabbour-Zahab R, Pointier JP, Jourdane J et al (1997) Phylogeography and genetic divergence of some lymnaeid snails, intermediate hosts of human and animal fascioliasis with special reference to lymnaeids from the Bolivian Altiplano. Acta Trop 64:191–203. https://doi.org/10.1016/S0001-706X(96)00631-6

Jarne P, Delay B (1990) Inbreeding depression and self-fertilization in *Lymnaea peregra* (Gastropoda: Pulmonata). Heredity 64:169–175. https://doi.org/10.1038/hdy.1990.21

Jarne P, Lagoda PJL (1996) Microsatellites, from molecules to populations and back. Trends Ecol Evol 11:424–429. https://doi.org/10.1016/0169-5347(96)10049-5

Jones RA, Brophy PM, Davis CN et al (2018) Detection of *Galba truncatula*, *Fasciola hepatica* and *Calicophoron daubneyi* environmental DNA within water sources on pasture land, a future tool for fluke control? Parasite Vector 11:342. https://doi.org/10.1186/s13071-018-2928-z

King KC, Lively CM (2012) Does genetic diversity limit disease spread in natural host populations? Heredity 109:199–203. https://doi.org/10.1038/hdy.2012.33

Knott KE, Puurtinen M, Kaitala V (2003) Primers for nine microsatellite loci in the hermaphroditic snail *Lymnaea stagnalis*. Mol Ecol Notes 3:333–335. https://doi.org/10.1046/j.1471-8286.2003.00444.x

Kopp KC, Wolff K, Jokela J (2012) Natural range expansion and human-assisted introduction leave different genetic signatures in a hermaphroditic freshwater snail. Evol Ecol 26:483–498. https://doi.org/10.1007/s10682-011-9504-8

Lance E, Petit A, Sanchez W et al (2014) Evidence of trophic transfer of microcystins from the gastropod *Lymnaea stagnalis* to the fish *Gasterosteus aculeatus*. Harmful Algae 31:9–17. https://doi.org/10.1016/j.hal.2013.09.006

Leung KMY, Lewis JW, Ibrahim H et al (2003) Concentrations of Metallothionein-like proteins and heavy metals in the freshwater snail *Lymnaea stagnalis* exposed to different levels of waterborne cadmium. B Environ Contam Tox 71:1084–1090. https://doi.org/10.1007/s00128-003-8856-3

Lewontin RC (1974) The genetic basis of evolutionary change. Columbia University Press, New York

Liu GH, Wang SY, Huang WY et al (2012) The complete mitochondrial genome of *Galba pervia* (Gastropoda: Mollusca), an intermediate host snail of *Fasciola* spp. PLoS One 7:e42172. https://doi.org/10.1371/journal.pone.0042172

Lounnas M, Correa AC, Vázquez AA et al (2017a) Self-fertilization, long-distance flash invasion and biogeography shape the population structure of *Pseudosuccinea columella* at the worldwide scale. Mol Ecol 26:887–903. https://doi.org/10.1111/mec.13984

Lounnas M, Vázquez AA, Alda P (2017b) Isolation, characterization and population-genetic analysis of microsatellite loci in the freshwater snail *Galba cubensis* (Lymnaeidae). J Mollus Stud 83:63–68. https://doi.org/10.1093/mollus/eyw041

Lounnas M, Correa AC, Alda P et al (2018) Population structure and genetic diversity in the invasive freshwater snail *Galba schirazensis*. Can J Zool 96(5):425–435. https://doi.org/10.1139/cjz-2016-0319

Meunier C, Hurtrez-Boussès S, Jabbour-Zahab R et al (2004) Field and experimental evidence of preferential selfing in the freshwater mollusc *Lymnaea truncatula* (Gastropoda, Pulmonata). Heredity 92:316–322. https://doi.org/10.1038/sj.hdy.6800410

Mostafa OMS (2008) Effects of *Fasciola gigantica* experimental infection on some inorganic elements in the snail host *Lymnaea natalensis*. Exp Parasitol 118:468–471. https://doi.org/10.1016/j.exppara.2007.10.009

Mulero S, Toulza E, Loisier A et al (2021) Malacological survey in a bottle of water: a comparative study between manual sampling and environmental DNA metabarcoding approaches. Glob Ecol Conserv 25:e01428. https://doi.org/10.1016/j.gecco.2020.e01428

Nakadera Y, Blom C, Koene JM (2014) Duration of sperm storage in the simultaneous hermaphrodite *Lymnaea stagnalis*. J Mollus Stud 80:1–7. https://doi.org/10.1093/mollus/eyt049

Nakadera Y, Giannakara A, Ramm SA (2019) Plastic expression of seminal fluid protein genes in a simultaneously hermaphroditic snail. Behav Ecol 30(4):904–913. https://doi.org/10.1093/beheco/arz027

Ng TYT, Pais NM, Wood CM (2011) Mechanisms of waterborne cu toxicity to the pond snail *Lymnaea stagnalis*: physiology and cu bioavailability. Ecotox Environ Safe 74:1471–1479. https://doi.org/10.1016/j.ecoenv.2011.03.010

Nicholson A, McIsaac D, MacDonald C et al (2020) An analysis of metadata reporting in freshwater environmental DNA research calls for the development of best practice guidelines. Environ DNA 2:343–349. https://doi.org/10.1002/edn3.81

Nicot A, Dubois MP, Debain C et al (2008) Characterization of 15 microsatellite loci in the pulmonate snail *Pseudosuccinea columella* (Mollusca, Gastropoda). Mol Ecol Resour 8: 1281–1284. https://doi.org/10.1111/j.1755-0998.2007.02065.x

Noël E, Jarne P, Glémin S et al (2017) Experimental evidence for the negative effects of self-fertilization on the adaptive potential of populations. Curr Biol 27:237–242. https://doi.org/10.1016/j.cub.2016.11.015

Pante E, Schoelinck C, Puillandre N (2015) From integrative taxonomy to species description: one step beyond. Syst Biol 64:152–160. https://doi.org/10.1093/sysbio/syu083

Pfenninger M, Salinger M, Haun T et al (2011) Factors and processes shaping the population structure and distribution of genetic variation across the species range of the freshwater snail *Radix balthica* (Pulmonata, Basommatophora). BMC Evol Biol 11:135. https://doi.org/10.1186/1471-2148-11-135

Plows LD, Cook RT, Davies AJ et al (2005) Carbohydrates that mimic schistosome surface coat components affect ERK and PKC signalling in *Lymnaea stagnalis* haemocytes. Int J Parasitol 35:293–302. https://doi.org/10.1016/j.ijpara.2004.11.012

Qin DM, Huang XC, Yang LM et al (2019) Complete mitochondrial genome of the radicine pond snail *Radix plicatula* (Gastropoda: Lymnaeidae). Mitochondr DNA B 4(2):2861–2862. https://doi.org/10.1080/23802359.2019.1661300

Rathinasamy V, Hosking C, Tran L et al (2018) Development of a multiplex quantitative PCR assay for detection and quantification of DNA from *Fasciola hepatica* and the intermediate snail host, *Austropeplea tomentosa*, in water samples. Vet Parasitol 259:17–24. https://doi.org/10.1016/j.vetpar.2018.06.018

Reátegui-Zirena EG, Fidder BN, Salice CJ (2016) A cost or a benefit? Counterintuitive effects of diet quality and cadmium in *Lymnaea stagnalis*. Ecotoxicology 25:1771–1781. https://doi.org/10.1007/s10646-016-1720-0

Reátegui-Zirena EG, Fidder BN, Olson AD et al (2017) Transgenerational endpoints provide increased sensitivity and insight into multigenerational responses of *Lymnaea stagnalis* exposed to cadmium. Environ Pollut 224:572–580. https://doi.org/10.1016/j.envpol.2017.02.040

Rivi V, Benatti C, Colliva C et al (2020) *Lymnaea stagnalis* as model for translational neuroscience research: from pond to bench. Neurosci Biobehav R 108:602–616. https://doi.org/10.1016/j.neubiorev.2019.11.020

Rowe G, Sweet M, Beebee TJC (2017) An introduction to molecular ecology, 3rd edn. Oxford University Press, Oxford-New York

Sabourin E, Alda P, Vázquez A et al (2018) Impact of human activities on fasciolosis transmission. Trends Parasitol 34:891–903. https://doi.org/10.1016/j.pt.2018.08.004

Salinger M, Pfenninger M (2009) Highly polymorphic microsatellite markers for *Radix balthica* (Linnaeus 1758). Mol Ecol Resour 9:1152–1155. https://doi.org/10.1111/j.1755-0998.2009.02588.x

Sayyad MN, Shaikh TT (2015) Biochemical study of bacterial strains isolated from snail gut collected from Rauza Bagh, Maulana Azad College, Aurangabad, Maharashtra, India. J Environ Res Dev 9(3):577–584

Schander C (2003) DNA, PCR and formalinized animal tissue? A short review and protocols. Org Divers Evol 3:195–205. https://doi.org/10.1078/1439-6092-00071

Schell T, Feldmeyer B, Schmidt H et al (2017) An annotated draft genome for *Radix auricularia* (Gastropoda, Mollusca). Genome Biol Evol 9:585–592. https://doi.org/10.1093/gbe/evx032

Schiller E, Haring E, Däubl B et al (2014) Ethanol concentration and sample preservation considering diverse storage parameters: a survey of invertebrate wet collections of the Natural History Museum Vienna. Annalen des Naturhistorischen museums in Wien. Serie B Für Botanik Und Zoologie 116:41–68

Schniebs K, Glöer P, Vinarski M et al (2011) Intraspecific morphological and genetic variability in *Radix balthica* (Linnaeus 1758) (Gastropoda: Basommatophora: Lymnaeidae) with morphological comparison to other European *radix* species. J Conchol 40(6):657–678

Schniebs K, Glöer P, Vinarski M et al (2013) Intraspecific morphological and genetic variability in the European freshwater snail *Radix labiata* (Rossmaessler, 1835) (Gastropoda: Basommatophora: Lymnaeidae). Contrib Zool 82(1):55–68. https://doi.org/10.1163/18759866-08201004

Schniebs K, Glöer P, Vinarski M et al (2019) Intraspecific morphological and genetic variability in the palaearctic freshwater snail *Radix ampla* (Hartmann, 1821)(Gastropoda: Basommatophora: Lymnaeidae). J Conchol 43:245–267

Schweizer G, Meli ML, Torgerson PR et al (2007) Prevalence of Fasciola hepatica in the intermediate host *Lymnaea truncatula* detected by real time TaqMan PCR in populations from 70 Swiss farms with cattle husbandry. Vet Parasitol 150:164–169. https://doi.org/10.1016/j.vetpar.2007.08.006

Seppälä O, Jokela J (2011) Immune defence under extreme ambient temperature. Biol Lett 7:119–122. https://doi.org/10.1098/rsbl.2010.0459

Seppälä O, Walser JC, Cereghetti T et al (2021) Transcriptome profiling of *Lymnaea stagnalis* (Gastropoda) for ecoimmunological research. BMC Genom 22:144. https://doi.org/10.1186/s12864-021-07428-1

Singh DK, Agarwal RA (1984) Correlation of the anticholinesterase and Molluscicidal activity of the latex of *Euphorbia royleana* on the snail *Lymnaea acuminata*. J Nat Prod 47:702–705. https://doi.org/10.1021/np50034a023

Sirot LK, Wolfner MF, Wigby S (2011) Protein-specific manipulation of ejaculate composition in response to female mating status in *Drosophila melanogaster*. P Natl Acad Sci USA 108:9922–9926. https://doi.org/10.1073/pnas.1100905108

Skála V, Černíková A, Jindrová Z et al (2014) Influence of *Trichobilharzia regenti* (Digenea: Schistosomatidae) on the defence activity of *Radix lagotis* (Lymnaeidae) haemocytes. PLoS One 9:e111696. https://doi.org/10.1371/journal.pone.0111696

Southgate VR (1970) Observations on the effect of the rediae of *Fasciola hepatica* on the lipid composition of the hepatopancreas of *Lymnaea truncatula*. Parasitology 61:293–299. https://doi.org/10.1017/S0031182000041123

Swart EM, Davison A, Ellers J et al (2019) Temporal expression profile of an accessory-gland protein that is transferred via the seminal fluid of the simultaneous hermaphrodite *Lymnaea stagnalis*. J Mollus Stud 85:177–183. https://doi.org/10.1093/mollus/eyz005

Swart EM, Starkloff NC, Ypenburg S et al (2020) The effect of mating on female reproduction across hermaphroditic freshwater snails. Invertebr Biol 139(1):e12275. https://doi.org/10.1111/ivb.12275

Thomsen PF, Willerslev E (2015) Environmental DNA – an emerging tool in conservation for monitoring past and present biodiversity. Biol Conserv 183:4–18. https://doi.org/10.1016/j.biocon.2014.11.019

Tills O, Truebano M, Rundle S (2015) An embryonic transcriptome of the pulmonate snail *Radix balthica*. Mar Genom 24:259–260. https://doi.org/10.1016/j.margen.2015.07.014

Tripathi PK, Singh A (2004) Toxic effects of cypermethrin and alphamethrin on reproduction and oxidative metabolism of the freshwater snail, *Lymnaea acuminata*. Ecotox Environ Safe 58:227–235. https://doi.org/10.1016/j.ecoenv.2003.12.006

Tripathi SM, Singh VK, Singh S et al (2004) Enzyme inhibition by the molluscicidal agent *Punica granatum* Linn. Bark and *Canna indica* Linn. Root. Phytother Res 18:501–506. https://doi.org/10.1002/ptr.1168

Trouvé S, Degen L, Meunier C et al (2000) Microsatellites in the hermaphroditic snail, *Lymnaea truncatula*, intermediate host of the liver fluke, *Fasciola hepatica*. Mol Ecol 9:1662–1664. https://doi.org/10.1046/j.1365-294x.2000.01040-2.x

Tufi S, Stel JM, de Boer J et al (2015) Metabolomics to explore Imidacloprid-induced toxicity in the central nervous system of the freshwater snail *Lymnaea stagnalis*. Environ Sci Technol 49: 14529–14536. https://doi.org/10.1021/acs.est.5b03282

Tufi S, Wassenaar PNH, Osorio V et al (2016) Pesticide mixture toxicity in surface water extracts in snails (*Lymnaea stagnalis*) by an *in vitro* acetylcholinesterase inhibition assay and metabolomics. Environ Sci Technol 50:3937–3944. https://doi.org/10.1021/acs.est.5b04577

Van Horn DJ, Garcia JR, Loker ES et al (2012) Complex intestinal bacterial communities in three species of planorbid snails. J Mollus Stud 78(1):74–80. https://doi.org/10.1093/mollus/eyr038

van Iersel S, Swart EM, Nakadera Y et al (2014) Effect of male accessory gland products on egg laying in gastropod molluscs. J Vis Exp 88:e51698. https://doi.org/10.3791/51698

Vinarski MV (2013) One, two, or several? How many lymnaeid genera are there? Ruthenica Russian Malacol J 23(1):41–58

Vinarski MV, Aksenova OV, Bespalaya YV et al (2016a) *Radix dolgini*: the integrative taxonomic approach supports the species status of a Siberian endemic snail (Mollusca, Gastropoda, Lymnaeidae). C R Biol 339(1):24–36. https://doi.org/10.1016/j.crvi.2015.11.002

Vinarski MV, Aksenova OV, Bespalaya YV et al (2016b) *Ladislavella tumrokensis*: the first molecular evidence of a Nearctic clade of lymnaeid snails inhabiting Eurasia. Syst Biodivers 14(3):276–287. https://doi.org/10.1080/14772000.2016.1140244

Vinarski MV, Aksenova OV, Bolotov IN (2020) Taxonomic assessment of genetically-delineated species of radicine snails (Mollusca, Gastropoda, Lymnaeidae). Zoosyst Evol 96(2):577–608. https://doi.org/10.3897/zse.96.52860

Vinarski MV, Aksenova OV, Bespalaya YV et al (2021) One Beringian genus less: a re-assessment of *Pacifimyxas* Kruglov & Starobogatov, 1985 (Mollusca: Gastropoda: Lymnaeidae) questions the current estimates of Beringian biodiversity. J Zool Syst Evol Res 59(1):44–59. https://doi.org/10.1111/jzs.12411

Vinarski MV, Aksenova OV, Bespalaya YV et al (2023) How an ecological race is forming: morphological and genetic disparity among thermal and non-thermal populations of aquatic Lymnaeid snails (Gastropoda: Lymnaeidae). Diversity 15(4):548. https://doi.org/10.3390/d15040548

Vorontsova YL, Slepneva IA, Yurlova NI et al (2019) The effect of trematode infection on the markers of oxidative stress in the offspring of the freshwater snail *Lymnaea stagnalis*. Parasitol Res 118:3561–3564. https://doi.org/10.1007/s00436-019-06494-5

Walton RC, White KN, Livens F et al (2010) The suitability of gallium as a substitute for aluminum in tracing experiments. Biometals 23:221–230. https://doi.org/10.1007/s10534-009-9280-x

Ward PI, Goater CP, Mikos M (1997) Shell variation in sympatric freshwater *Lymnaea peregra* and *L. ovata* (Gastropoda: Lymnaeidae). Biol J Linn Soc 61:139–149. https://doi.org/10.1111/j.1095-8312.1997.tb01782.x

Wiehn J, Kopp K, Rezzonico S et al (2002) Family-level covariation between parasite resistance and mating system in a hermaphroditic freshwater snail. Evolution 56:1454–1461. https://doi.org/10.1111/j.0014-3820.2002.tb01457.x

Zurawell RW, Holmes CFB, Prepas EE (2006) Elimination of the cyanobacterial Hepatotoxin microcystin from the freshwater Pulmonate snail *Lymnaea stagnalis jugularis* (Say). J Toxicol Env Health A 69:303–318. https://doi.org/10.1080/15287390500227498

Chapter 6
The Fossil Record of the Lymnaeidae: Revisiting a 200-Myr-Long Story of Success

Thomas A. Neubauer

Abstract Here, I present a review of the global pre-Pleistocene fresh- to brackish-water fossil record of the gastropod family Lymnaeidae based on a thorough literature survey of over 450 scientific works. I discuss the putative origin of the family, assess diversity development through geological time (based on the fossil records of Europe and North America), and illustrate the family's geographic distribution over the past 200 Myr using paleogeographic maps. The following section deals with potential dispersal mechanisms to explain the family's disjunct fossil distribution. A final part is devoted to peculiar cases of morphological evolution toward limpet-like and/or strongly sculptured shells.

6.1 Introduction

Today, the Lymnaeidae are a diverse family in the world's fresh water (e.g., Pyron and Brown 2015; Aksenova et al. 2018; Vinarski et al. 2019; and other articles in this volume). The family is near globally distributed, whereas the northern hemisphere maintains a much higher diversity (Vinarski et al. 2019). Also morphologically, they are highly diverse and comprise a great variety of shell shapes (e.g., Baker 1911; Hubendick 1951; Correa et al. 2011; Campbell et al. 2017; Aksenova et al. 2018;

Supplementary Information The online version contains supplementary material available at https://doi.org/10.1007/978-3-031-30292-3_6.

T. A. Neubauer (✉)
Department of Animal Ecology and Systematics, Systematics and Biodiversity Group, Justus Liebig University Giessen, Gießen, Germany

Naturalis Biodiversity Center, Leiden, The Netherlands

Present Address: SNSB - Bavarian State Collection for Paleontology and Geology, Munich, Germany
e-mail: neubauer@snsb.de

M. V. Vinarski, A. A. Vázquez (eds.), *The Lymnaeidae*, Zoological Monographs 7,
https://doi.org/10.1007/978-3-031-30292-3_6

Vinarski et al. 2019; Alda et al. 2023). Moreover, Lymnaeidae play a vital role in the transmission of liver fluke and are thus of medical importance (e.g., Correa et al. 2010; Mahulu et al. 2019).

Since lymnaeids occur in a great range of habitats, they are a common component of fossil freshwater communities and are often used as paleoecological indicators (e.g., Good 1987; Carobene et al. 2018; Salvador et al. 2018a; Esu and Girotti 2020; Kadolsky 2020; Rasser et al. 2020). Numerous species have been described from fossil deposits all around the globe (e.g., Wenz 1923). Yet, in spite of their high diversity and relevance to paleoecology, a thorough review of the family's diversity and distribution through geological time is absent.

Here, I review the literature to provide an overview over the global fossil record of lymnaeids and their geographic and stratigraphic distribution. Given the number of publications documenting lymnaeids in fossil deposits, the overview focuses on the pre-Pleistocene record. Although not including every single record in the literature, I sought to cover the complete geographic range of the family through time. I review the oldest records of fossil species attributed to the Lymnaeidae and discuss the family's potential origin. Based on the available data for Europe and North America, which contain particularly rich records, I reconstruct continental diversity through time. Lymnaeid distribution through time and space is illustrated via paleogeographic maps. Two further chapters deal with potential dispersal mechanisms and morphological evolution.

6.2 Material and Methods

Information for European and North American faunas was retrieved in the frame of a project dealing with diversification dynamics of freshwater gastropods over the past 200 Myr. This project aligns with a continuous effort to gather and complete an inventory of all fossil freshwater gastropods worldwide. Most of the taxonomic information has already been made available via MolluscaBase (MolluscaBase eds 2021). Data were mostly obtained directly from the literature. Additional information were derived from various online sources, but in all cases these data were checked for accuracy and correctness against the primary literature. In total, the data were derived from 454 scientific works (Supplementary Table S1).

Systematically, I include in this review all species currently classified in the Lymnaeidae. Since fossil (as well as recent) lymnaeids have been regularly mixed up with members of the terrestrial families Succineidae (e.g., Marshall 1926), Bulimulidae (e.g., Salvador et al. 2018b), and Oleacinidae (e.g., Wenz 1923; Harzhauser et al. 2014b), I follow the latest revisions to account for systematic changes. This concerns also some genera formerly placed in the family Clivunellidae (*Delminiella, Neoclivunella*, and *Neodelminiella*), now a monotypic family restricted to the Miocene of the Balkans (Harzhauser et al. 2016). *Neoclivunella* and *Neodelminiella* comprise limpet-shaped shells and are classified tentatively in the lymnaeid subfamily Valencienniinae (Harzhauser et al. 2016). That taxon is still poorly defined and its genus- and species-level taxonomy is not well resolved and depends on the author (e.g., Gorjanović-Kramberger 1923; Moos 1944; Taktakishvili 1967; Marinescu 1970). Because of their descent from a species belonging to the genus *Radix*, Bouchet et al. (2017) treated the Valencienniinae as a synonym of Amphipepleinae. Here, I follow the traditional concept and treat the subfamily as distinct. In addition to the two abovementioned genera, I include here *Velutinopsis*, *Undulotheca*, *Hiscerus*, *Velutinellus*, *Provalenciennesia*, and *Valenciennius*. Despite the morphological similarity to *Delminiella* and the Valenciennünae, the family Clivunellidae is believed to be closely related to Planorbidae (Harzhauser et al. 2016) and is thus not treated herein. The Mesozoic genus *Proauricula* Huckriede, 1967, which was considered a basal Basommatophora and probably predecessor of the Lymnaeidae by Bandel (1991), was lately reclassified in the Ellobiidae by Yu et al. (2021).

Stratigraphically, the review is limited to the pre-Pleistocene fossil record. With few exceptions (e.g., the Caspian Sea; Andrusov 1923), the Early Pleistocene faunas resemble in species composition the modern fauna (e.g., Taylor 1960, 1966; La Rocque 1963; Esu and Girotti 1975; Schlickum and Puisségur 1978; Dehm 1979). Only the overview of the number of fossil species discovered per year and decade is based on all fossil species. Data were derived from a MolluscaBase query for all fossil-only species (including nomina dubia, taxa inquirenda and temporary names, e.g., junior homonyms without replacement name; queried 7 November 2021; Supplementary Table S2).

Age classifications of the localities are based on the most recent assessments of the fossil-bearing strata. For most European sites, age data are available from Neubauer et al. (2015a, b, c, 2021). Stratigraphic data on formations in China were derived mostly from Zhang (2009). Based on the ages, paleocoordinates were estimated using the GPlates Portal under the PALEOMAP model (http://portal.gplates.org/service/d3_demo/?view=points).

The statistical analyses were performed using R v. 4.0.3 (R Core Team 2020). Diversity reconstructions are made only for Europe (including Turkey) and North America because of the paucity of data for the other continents as well as the uncertain stratigraphy or taxonomy of many of those records. Subspecies, doubtful records or identifications ("cf.", "aff."), nomina nuda and nomina dubia were not considered. Taxa inquirenda and temporary names were included. Range-through species richness binned over stratigraphic stages was calculated using the package divDyn v. 0.8.0 (Kocsis et al. 2019). In order to avoid a biased decrease of range-through richness toward the Pleistocene by assuming that all taxa went extinct at the end of the Pliocene, I included species occurring today in a separate Holocene time bin for the calculation. Data for the modern species was taken from MolluscaBase. Stage boundaries follow the International Chronostratigraphic Chart (July 2021; https://stratigraphy.org/chart).

Since all fossil time series are inevitably skewed by uneven preservation (Alroy 2010), I applied a subsampling approach to assess whether any observed diversity pattern is the result of sampling. For this purpose, I used the shareholder quorum subsampling (SQS) method of Alroy (2010) as implemented and advanced in the package divDyn using 1000 iterations and a quorum level varying between 0.3 and

0.7 at increments of 0.1. The SQS approach was only applied to the European dataset, since not enough data are available for other continents.

Genus-level diversity is not discussed herein due to the problems associated with genus classification in lymnaeids. While modern lymnaeid systematics rely on anatomical characters and results from molecular phylogenetics (e.g., Vinarski 2013; Aksenova et al. 2018), any fossil species classification is inevitably based on shell characters, which are often of limited value (see also Vinarski 2013 for an in-depth discussion of the subject). Most fossil species are placed in "standard" genera like *Lymnaea* or *Galba*, whereas probably very few of them actually belong in these genera.

Institutional abbreviations. MBFSZ – Mining and Geological Survey of Hungary, Budapest; NHMW – Natural History Museum Vienna; ZMBH – Zemaljski Muzej Bosne i Hercegovine, Sarajevo.

6.3 Diversity Through Time

The first fossil lymnaeids were described by Alexandre Brongniart in 1810 (Brongniart 1810). He introduced nine new species from Eocene and Oligocene deposits of the Paris Basin in France; all of them are still considered as valid. Since then, the number of species descriptions per year has almost linearly increased (Fig. 6.1a). Noteworthy publications advancing the number of species above average were contributed by Deshayes (1863), who described 15 species also from the Paleogene of the Paris Basin species, and Gorjanović-Kramberger (1901), who introduced eleven species for the Valencienniinae of Lake Pannon. Overall, with 18 described species each, these two authors introduced the most (still valid) fossil lymnaeid species (Fig. 6.1b).

The curve of new species descriptions is not saturated yet. Only since the year 2000, twenty new species of Lymnaeidae have been described from fossil freshwater deposits (Macaleţ 2000; Pierce and Constenius 2001; Kovalenko 2004, 2017; Prysjazhnjuk et al. 2006; Pan and Zhu 2007; Harzhauser et al. 2012, 2014a, 2016; Neubauer et al. 2014, 2015d; Vinarski and Frolov 2017; Yu et al. 2021).

Altogether, 379 valid accepted species have been described over the 210 year-long history of fossil lymnaeid taxonomy (Supplementary Table S2). Of these, pre-Pleistocene distribution data are available for 376 species in 32 genera (of which 15 are only known from fossil species) and four subfamilies (Lymnaeinae, Amphipepleinae, Lancinae, and Valencienniinae) (Supplementary Table S1).

The species richness reconstructions showed similar tendencies for Europe and North America, yet in quite different magnitudes (Fig. 6.2). While the pre-Pleistocene European fossil record comprised altogether 262 species, only 56 species were recorded for North America. Species richness remained low in both continents during the Mesozoic and earliest Paleogene. No notable diversity decline was observed at the Cretaceous–Paleogene boundary, which contrasts a

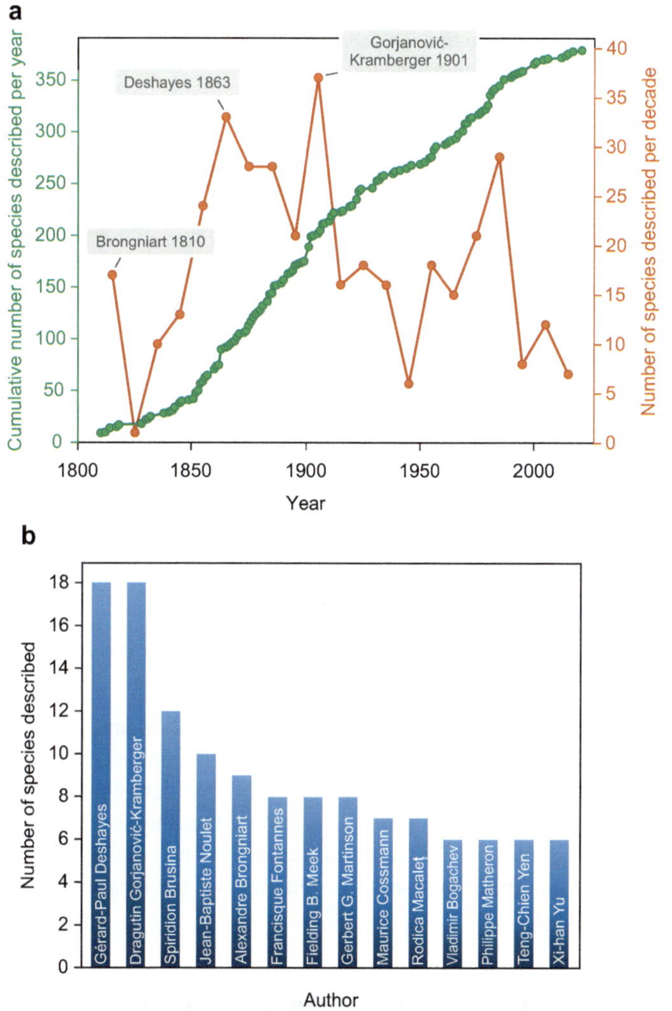

Fig. 6.1 (**a**) Number of lymnaeid species described per year (cumulative, green) and decade (orange). Indicated are additionally the three publications that have contributed the most to the number of species. (**b**) Authors that described more than five fossil lymnaeid species. Invalid or unavailable names (e.g., synonyms, nomina nuda) were not considered in any of the two plots

massive breakdown of the overall freshwater gastropod diversity in Europe (Neubauer et al. 2021).

Diversification speeded up about coevally in both continents in the Lutetian (Middle Eocene). In Europe, a diversity peak was reached with 45 species in the Rupelian. Diversity dropped in both continents toward the Oligocene–Miocene boundary and increased again during the Miocene. The all-time species richness peak was reached in both continents during the Tortonian. In Europe, this peak is

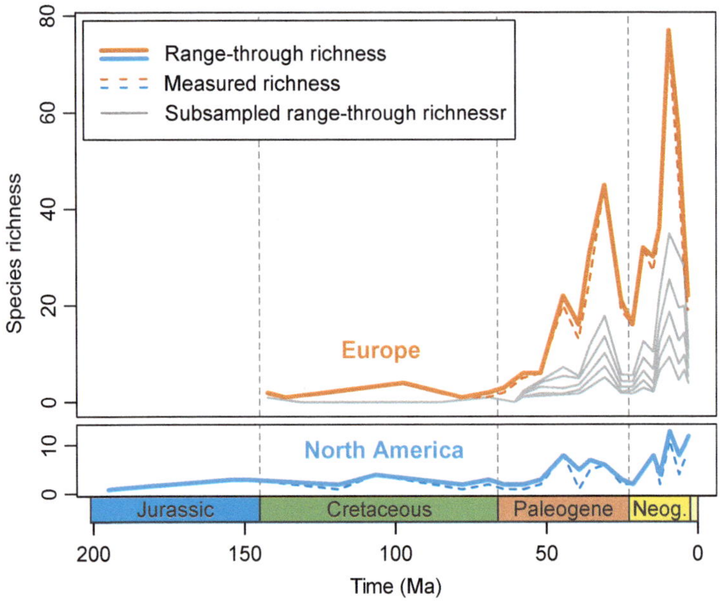

Fig. 6.2 Lymnaeidae species richness through geological time for Europe and North America, including subsampled richness for Europe (using quorum levels of 0.3, 0.4, 0.5, 0.6, and 0.7)

partly owed to major diversification events in Lake Pannon and the Dacian Basin. Hence, not all of the 77 species recorded for the Tortonian lived simultaneously.

The diversity trend for Europe derived from the raw data was confirmed by the subsampling (SQS) approach (Fig. 6.2), suggesting that the observed pattern is not biased by uneven sampling. There is unfortunately not enough data for North America to use subsampling to assess the reliability of the measured trend.

6.4 Geographic Distribution Through Time and the Origins of Lymnaeidae

In this chapter, I give an overview of the geographic distribution of Lymnaeidae from their first occurrence in the fossil record to the Pliocene, illustrated by a series of paleographic maps (Figs. 6.3, 6.4, 6.5, 6.6). Given the high number of records in some stratigraphic intervals, these maps are not exhaustive and do not contain all records as single points but are rather intended to cover the entire geographic range of lymnaeids in each interval. Also, the lymnaeid fossil record is clearly geographically biased toward Europe and North America (Supplementary Fig. S1). The spotty occurrences on other continents, such as Asia, where lymnaeids are common today (Vinarski et al. 2019), suggest a strong preservation, sampling, and/or research bias.

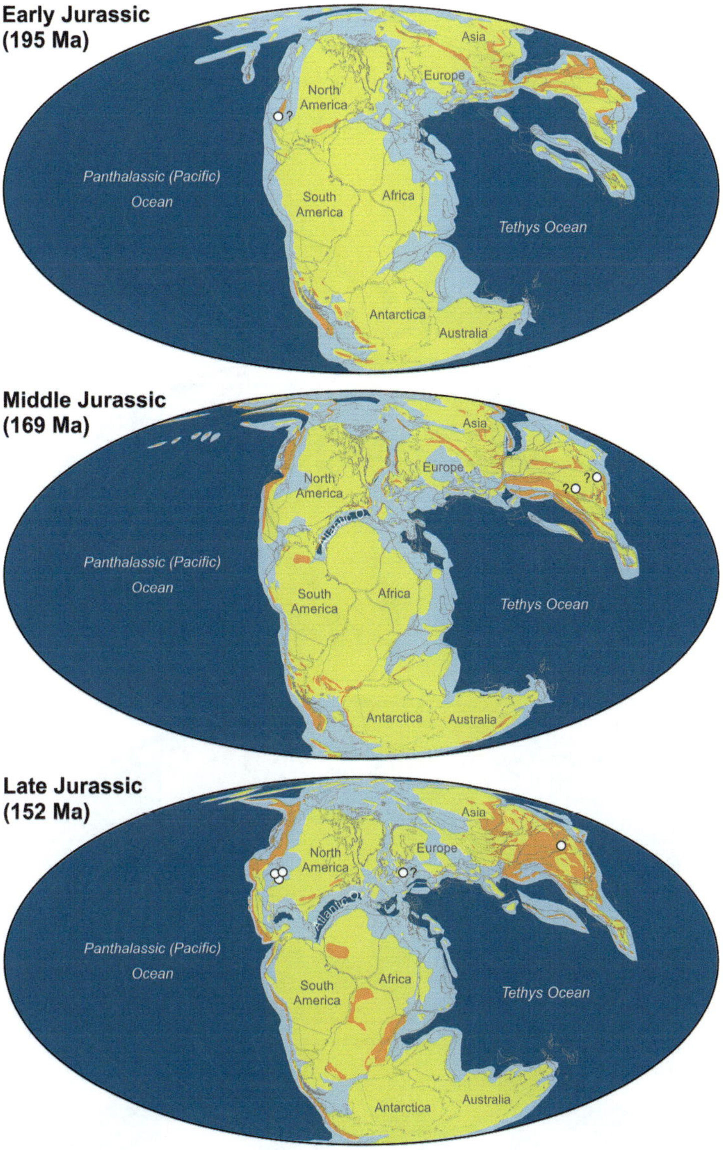

Fig. 6.3 Geographic distribution of Lymnaeidae in the Jurassic. Close-by localities are represented by single points. Question marks indicate species of uncertain lymnaeid affinity. Base map from Cao et al. (2017) used under Creative Commons Attribution 3.0 (CC BY 3.0) license. Yellow areas mark land masses, orange indicates mountain ranges

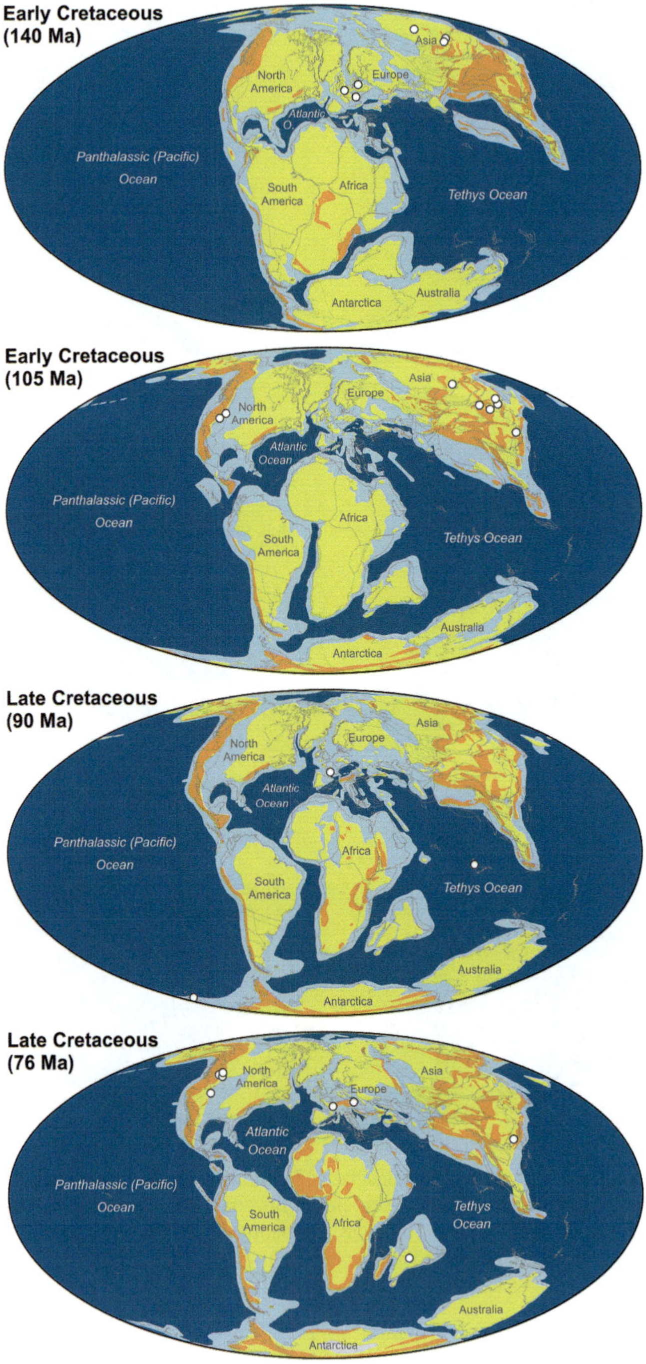

Fig. 6.4 Geographic distribution of Lymnaeidae in the Cretaceous. Close-by localities are represented by single points. Base map from Cao et al. (2017)

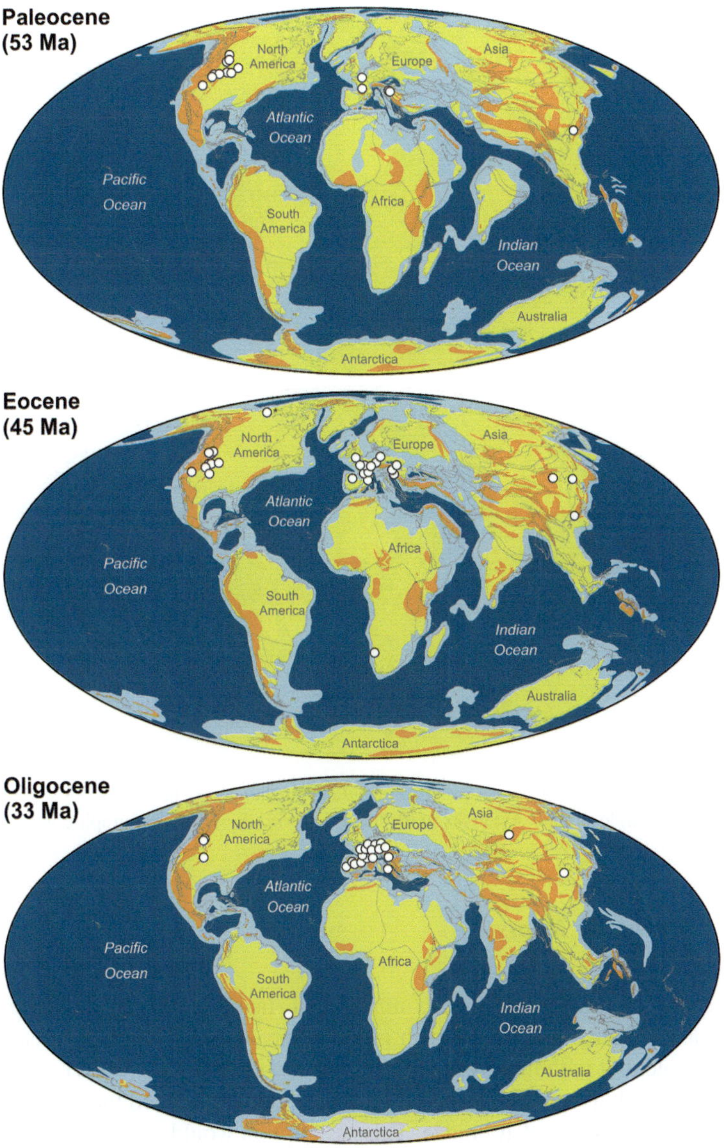

Fig. 6.5 Geographic distribution of Lymnaeidae in the Paleogene. Close-by localities are represented by single points. Asterisks mark localities with doubtful stratigraphic age. Base map from Cao et al. (2017)

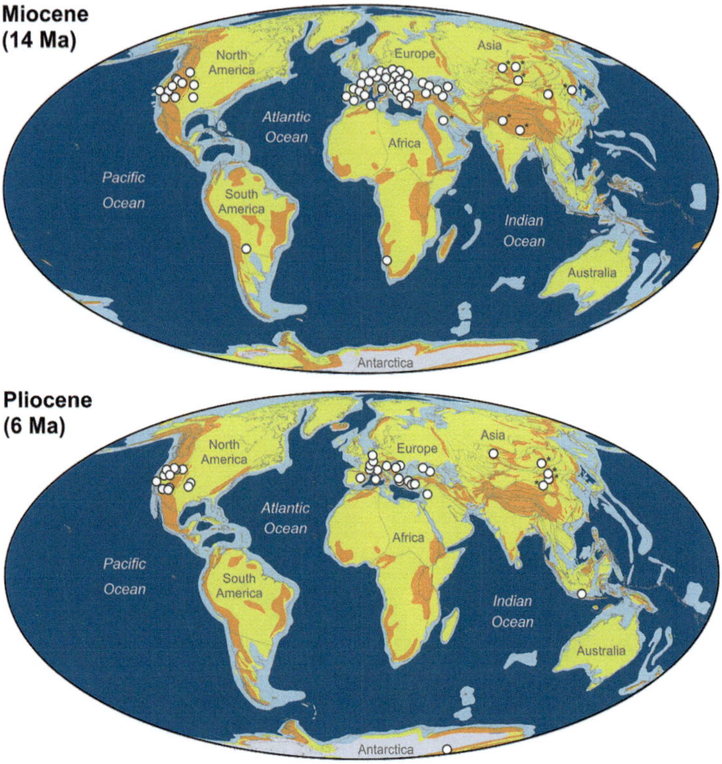

Fig. 6.6 Geographic distribution of Lymnaeidae in the Neogene. Close-by localities are represented by single points. Asterisks mark localities with doubtful stratigraphic age. Base map from Cao et al. (2017)

6.4.1 Mesozoic Roots

The ancestry of the Lymnaeidae is unresolved. Inaba (1969) hypothesized a late "pro-lymnaeid" ancestor, from which modern lineages arose starting already in the late Paleozoic, but without any evidence from fossils. Kruglov (2005) suggested an origin and initial diversification in the Triassic and hypothesized that the ancestor of Lymnaeidae is related to Chilinidae. This assumption would at least match the basal position of Chilinidae within the Hygrophila as indicated by phylogenies based on morphology and molecular data (Hubendick 1978; Nordsieck 1992; Jörger et al. 2010), although their fossil record is comparably sparse and young (Dayrat et al. 2011). However, the latest molecular phylogeny by Saadi et al. (2020) suggests that the ancestor of Lymnaeidae is more closely related with Physidae. Yet, the fossil record of that family only starts in the Late Jurassic (Huckriede 1967).

A Triassic origin of Lymnaeidae as supposed by Kruglov (2005) predates the fossil record of the family. The earliest putative mention of lymnaeids in the fossil record derives from the United States (Fig. 6.3). Robinson (1915) described the two species *Valvata gregorii* and *Limnaea hopii* from "Painted Desert beds" in Arizona, which he attributed to the Morrison Formation, suggesting a Late Jurassic age for the fossils. Yen (1951) later attributed these findings erroneously to the Upper Triassic Chinle Formation (Harshbarger et al. 1957). The stratigraphic mystery was finally solved by Harshbarger et al. (1957), who correlated the fossil-bearing strata with the silty facies of the Kayenta Formation, pointing toward an Early Jurassic (Pliensbachian) age (Marsh and Rowe 2018). However, the type material has apparently not been studied since the first description (see, e.g., Yen 1952, p. 27), and the illustrations and description provided by Robinson (1915) do not unequivocally ascertain placement in the Lymnaeidae.

The next oldest records of supposed Lymnaeidae come from the Middle Jurassic of China (Fig. 6.3). Pan (1977) described *Galba lufengensis* and *Galba yunnanensis* from the Zhanghe Formation and Hepingxiang Formation of Yunnan, and Yü and Pan (1980) recorded *Galba tongshangensis* from the Tongshan Group of Zhejiang. All these stratigraphic units are presently attributed to the Middle Jurassic (Zhang 2009). The preservation of these fossil species is poor (Yü and Pan 1980) and, as for Robinson's material above, the attribution to Lymnaeidae cannot be confirmed with certainty. In fact, *Galba lufengensis* resembles terrestrial succineids.

A careful revision of these earliest fossils is required to assess the origin of the Lymnaeidae. At least from the Late Jurassic onward, the number of lymnaeid records increases and they come from various continents (Fig. 6.3). This includes species from the Morrison Formation in the USA (White 1886; Yen 1952; Solem and Yochelson 1979; Gray 1988; Evanoff et al. 1998), the Tithonian of Shandong in China (Pan 1983) and, possibly, the Kimmeridgian of Lower Saxony in Germany (Huckriede 1967), involving several species undoubtedly belonging in the family Lymnaeidae.

Records from the Jurassic–Cretaceous boundary were derived from the Tithonian–Valanginian Purbeck Formation in southern England (Arkell 1941) and its stratigraphic equivalents (mostly of Berriasian age) in Switzerland (De Loriol and Jaccard 1865) and Germany (Dunker 1846) (Fig. 6.4). Early Cretaceous lymnaeids were documented from the Valanginian–Hauterivian and the Aptian–Albian of Siberia (Reis 1910; Martinson 1956, 1961), the Guantou, Yehe, Banlashan, Yixian and Xiazhuang formations of Zhejiang, Jilin, Liaoning and Beijing in China (Yü and Pan 1980; Zhu 1980; Yü 1987; Pan and Zhu 2007) and the Aptian–Albian Cloverly Formation (Yen 1946c, 1951) and the Albian Bear River Formation (White 1876; Yen 1954) in the USA. This also involves findings of Yen and Reeside (1946) and Yen (1952), who formerly attributed their fossils to the Morrison Formation, but Evanoff et al. (1998) identified the strata as belonging to the Cloverly Formation.

The Late Cretaceous (potentially) saw the first appearance of Lymnaeidae on the southern hemisphere (Fig. 6.4). Beu et al. (2014) reported an unidentified species from the Cenomanian of New Zealand, which they tentatively referred to the genus *Austropeplea*. Additional early Late Cretaceous records come from the Cenomanian of France (Repelin 1902; age after Fabre-Taxy 1948) and Myanmar (Yu et al. 2021). After a long stratigraphic gap, end-Cretaceous taxa appear in the Campanian–

Maastrichtian St. Mary River Formation (Whiteaves 1885; Tozer 1956), the Campanian–Danian Edmonton Group (Tozer 1956) and the Maastrichtian lower Willow Creek Formation (Tozer 1956) of Canada, the Maastrichtian Laramie Formation in the USA (Meek 1873), the Maastrichtian Nanxiong Formation of Guangdong, China (Yü 1977), the Maastrichtian of India (Sowerby 1840; Hislop 1860; Hartman et al. 2008), the Campanian–Maastrichtian ("Rognacien") of France (Matheron 1843; Fabre-Taxy 1959), and the Maastrichtian of Haţeg in Romania (Pană et al. 2002). (Note that Pană et al. (2002) introduced a new genus and three new species names for Lymnaeidae but without describing them or designating type material, rendering all these names unavailable.)

6.4.2 The Paleogene Diversification

Rich Paleocene faunas containing Lymnaeidae have been documented from the USA and Canada, including the Green River Formation (White 1880; La Rocque 1960), Paskapoo Formation (Whiteaves 1885; Russell 1926; Tozer 1956), Devils Basin Formation (Love 1989) and Fort Union Formation (Meek and Hayden 1856; Yen 1948b; Bickel 1977) (Fig. 6.5). European records were derived from the "Liburnian" (Danian) of Croatia (Sandberger 1870–1875; Stache 1889) and the Thanetian of France, i.e., the "Sables de Bracheux" (Cossmann 1889; Glibert 1962) and the "Calcaire de Rilly" in the Paris Basin (Bayan 1870) and the "Calcaire à *Physa prisca*" in southern France (Noulet 1854; Rey and Villatte 1971). Only a single Paleocene record comes from Asia, namely from the Shanghu Formation of Guangdong, China (Yü 1977).

The Eocene diversification is represented by numerous localities in Europe and North America (Fig. 6.5). The majority of records come from the Paris Basin in France (e.g., Brongniart 1810; Deshayes 1863; Leriche 1899; Cossmann 1902; Jodot and Feugueur 1953; Glibert 1962) and the Hampshire Basin in southern England (e.g., Sowerby 1826–1829; Edwards 1852; Glibert 1962; Paul 1989; Munt 2014), as well as from the Upper Rhine Graben in Germany and France (e.g., Sandberger 1870–1875; Rollier 1910), the "Calcaires de Mas-Saintes-Puelles", "Calcaire de Castres" and "Molasse de Castelnaudary" in southern France (Serres 1844; Noulet 1854, 1857; Jodot and Rey 1956), and the Anjou (Glibert 1962), Matelles (Rey 1962b) and Aix-en-Provence (Matheron 1843; Roule 1886; Rey 1962a) basins, amongst several others. In North America, several species have been reported from the Middle–Late Eocene of the Kishenehn Basin (Russell 1952, 1955; Pierce and Constenius 2001, 2014). Additional records come from the Upper Eocene Florissant Formation (Cockerell 1906, 1908), the Upper Eocene to Lower Oligocene White River Formation (Evans and Shumard 1856; Meek 1876) and several more localities in the mid-west (e.g., Yen 1946b, 1948a; Good 1987; Hartman and Roth 1998). Only three Asian regions have yielded Eocene lymnaeids. They are located in China (Guangdong and Hebei provinces; Yü and Pan 1982; Yü and Zhang 1982) and Korea (Suzuki 1949). Finally, Lymnaeidae first appeared in the fossil record of

Africa during the Eocene, with two species from the Lutetian of Namibia (Pickford et al. 2008).

During the Oligocene, Lymnaeidae maintain a diverse fossil record in France, specifically in the Paris Basin (e.g., Brongniart 1810; Deshayes 1863; Dollfus 1906, 1920; Glibert 1962; Lozouet and Maestrati 1981), the "Calcaires d'Albi" in southern France (e.g., Fontannes 1884), and the Limagne Basin (e.g., Rey 1965) (Fig. 6.5). Numerous findings have been reported from the Chattian "*Ramondi* Beds" and "Formation du Gypse d'Aix" in France, Switzerland, and Germany (e.g., Fontannes 1881, 1884; Douxami 1904; Jodot 1954; Rey 1965). Additional important records come from the Upper Rhine Graben in France (e.g., Sandberger 1870–1875; Gillet 1953; Jodot 1954), the Mainz Basin (e.g., Thomä 1845; Mödden et al. 2000), and the Hessian Depression in Germany (e.g., Dunker 1853) and Limburg in Belgium (Glibert and Heinzelin de Braucourt 1954; Janssen 1980; Marquet et al. 2008). Only few records come from outside central Europe; these include localities in Portugal (Roman 1907; Pais 2012), Spain (Almera 1894; Royo Gómez 1922; Esu 1984), and Italy (Sacco 1886; Esu and Girotti 2010). Comparably little is known about North American Oligocene Lymnaeidae. Only three species have been documented from the Renova Formation (Pierce 1993) and the Antero Formation (Russell 1938) in the USA. Similarly, only few Asian localities have yielded lymnaeids. Popova (1964, 1981) described a few taxa from the Upper Oligocene–Lower Miocene (?) Bayanday Formation of Siberia, and Yen (1969) documented one species from Shandong Province, China. Finally, an unidentified lymnaeid was found in the Upper Oligocene Tremembé Formation of São Paulo State, Brazil (Salvador et al. 2018b), being the oldest fossil Lymnaeidae of South America.

6.4.3 The Neogene Diversity Peak

The Miocene record of Lymnaeidae is by far the richest (Figs. 6.2 and 6.6), and the sheer number of records and works mentioning lymnaeid fossils makes a detailed listing impractical (Supplementary Table S1). The European Miocene fossil record includes 158 species from over 1340 records from 26 countries, spanning from Portugal in the West to the Caspian Sea in the East (Fig. 6.6). The majority of records come from Germany and France, particularly from the Aquitaine Basin (e.g., Grateloup 1828; d'Orbigny 1852; Degrange-Touzin 1892; Peyrot 1933; Wenz 1936), the Upper Freshwater Molasse (e.g., Klein 1853; Sandberger 1870–1875; Clessin 1885; Jooss 1913; Wenz 1935; Schlickum 1976; Salvador and Rasser 2016; Salvador et al. 2016), and the Mainz Basin (e.g., Thomä 1845; Braun 1851; Kadolsky 1989). Paleogeographically, most records were derived from the Late Miocene Lake Pannon. This huge long-lived brackish-water lake alone yielded over 50 species of Lymnaeidae, most of which are endemic to the lake (see also chap. 6.1). Important contributions were made by Brusina (1874, 1902), Gorjanović-Kramberger (1901, 1923), Pavlović (1927), Strausz (1942), Moos (1944), Stevanović (1951, 1955, 1990b, c), Bartha (1955, 1971), Papp (1956),

Kochansky-Devidé and Pikija (1976), Sremac (1981), Marinescu (1973), Lennert et al. (1999), Vrsaljko (1999), and Botka et al. (2019), amongst many others. Further rich faunas were reported from the Dacian Basin in Romania (Krejci-Graf and Wenz 1932; Wenz 1942; Marinescu 1970; Hanganu 1972; Huică 1977; Marinescu and Papaianopol 1990; Macaleț 2000) and the Bakony Mountains in Hungary (Kókay 2006).

Important North American faunas yielding Miocene lymnaeids were described by Hanna (1923), Yen (1946a, 1947), Taylor (1954, 1957, 1966, 1983), Carr and Trimble (1963), Firby (1966), and Pierce (1993). A South American record comes from the Middle Miocene San José Formation of Argentina (Morton and Herbst 2003; Salvador et al. 2018b). Asian taxa were recorded by Suzuki (1949) from Korea, Willis (1967) from Iraq, and Yen (1969) from China. Additional reports of undefined Neogene age were made by Martinson (1954), Popova (1981), and Yü (1982) from Russia, Mongolia, and Tibet. Pickford (2008) reported two species from the Early Miocene of Namibia.

The Pliocene distribution reflects more or less the Miocene one, with many records in Europe and North America, a few in Asia and a single one in South America (Fig. 6.6). The center of European diversity is slightly shifted to the east compared to the Miocene, with the Dacian Basin holding the most diverse fauna (e.g., Macarovici 1940; Hanganu 1972; Macaleț 2000; Papaianopol et al. 2003; Van Baak et al. 2015). Additionally, the Transylvanian Basin in Romania (Jekelius 1932) and the Lower Rhône Basin in France (Serres 1853; Michaud 1855; Fontannes 1879, 1883) have yielded faunas rich in lymnaeids. The North American faunas mostly come from the Blancan deposits and were summarized by Taylor (1966). Additional records were listed by Henderson and Rodeck (1934), Nations (1974), Taylor and Smith (1981), and Love (1989). In Asia, taxa were documented from the Kyzylgir Formation in Siberia (Popova et al. 1970; Popova 1981), the Late Pliocene of Java (van Benthem Jutting 1937) and several localities in China (Ping 1931; Suzuki 1949; Zhu 1985; note that some of these records may range into the Early Pleistocene). Finally, an unidentified lymnaeid species was documented from the Pliocene Meyer Desert Formation of Antarctica by Ashworth and Preece (2003), representing the sole fossil record of that continent as yet.

6.5 Dispersal in the Fossil Record

When viewing the distribution patterns of Lymnaeidae in the fossil record and their spread from the presumed origin in the Early Jurassic of North America to nearly all over the world, the most obvious question is: how did they get there? Already during the Jurassic, the early Atlantic Ocean separated North America and Eurasia (Cao et al. 2017; Fig. 6.3). In fact, during most of the Mesozoic and Cenozoic, broad seaways isolated the continents, and only short intermittent phases of terrestrial pathways are known (Brikiatis 2014, 2016; Cao et al. 2017). However, these land bridges existed only between selected continents and it is unlikely that any of them offered hydrological connections allowing direct dispersal. A prime example of dispersal with an obvious lack of land connection is posed by the Pliocene occurrence of a lymnaeid on Antarctica (Ashworth and Preece 2003).

Dispersal in freshwater snails today knows active and passive pathways. Snails move actively along streams or are passively dispersed via ingestion and defecation by fish or waterbirds or via attachment to birds, mammals, amphibians, or even larger insects (e.g., Walther et al. 2008; Kappes and Haase 2012; van Leeuwen et al. 2012, 2013; Kolenda et al. 2017). Dispersal by both fish and waterfowl has already been witnessed for Lymnaeidae previously (Baker 1911; Kappes and Haase 2012; van Leeuwen and van der Velde 2012). Especially long-distance aerial dispersal is invoked to explain the pan-continental distribution of the family and disjunct occurrences on islands or in distant water bodies (Aksenova et al. 2018; Vinarski et al. 2021). Only Lancinae may form an exception here. In contrast to most lymnaeids, lancines have a poorly developed lung and probably do not survive transport outside of the water for long, constraining the possibility for dispersal by birds or mammals over long distances (Campbell et al. 2017).

As concerns fossil representatives, aerial dispersal has been proposed to explain the disjunct occurrence of *Delminiella* in Austria, separated from the remaining records in the Dinaride Lake System by a marine connection between the Mediterranean and Paratethys seas (Harzhauser et al. 2016). Similarly, Esu and Girotti (2018) suggested transport by waterfowl as likely mechanism to explain the occurrence of *Corymbina*, otherwise restricted to the Aegean region, in Lower Pliocene deposits of Italy. The Antarctic Pliocene record of a lymnaeid is likely a result of bird-mediated dispersal as well (Ashworth and Preece 2003) and the same might be true for many a disjunct occurrence in the Late Cretaceous or Cenozoic (Taylor 1988). However, waterfowl did not originate before the Late Cretaceous (Claramunt and Cracraft 2015). The mechanisms for dispersal prior to their evolution remain enigmatic.

Migrating land vertebrates, amphibians, or insects may have constituted viable options to reach water bodies not hydrologically connected, but it is unclear how lymnaeids could disperse across large distances and seaways. Early Cretaceous birds show similar anatomical features as modern migratory birds, but it is uncertain if they also underwent long-distance migration (Falk 2011). A possible yet so far entirely unsupported hypothesis involves pterosaurs (Yu et al. 2021). Some species of pterosaurs were likely capable of long-distance travels (Witton and Habib 2010), and some of them probably contributed to angiosperm seed dispersal during the Cretaceous (Fleming and Lips 1991). A few species probably fed on mollusks (Bestwick et al. 2018). Perhaps, freshwater mollusks could have been dispersed via ingestion and defecation or via attachment to the feet, just like in modern birds. However, to support this hypothesis a joint fossil record of gastropod and pterosaur species would be necessary.

Another aerial dispersal alternative may be offered by strong winds and storms. Several examples are known of terrestrial and freshwater mollusks and even fishes to be transported via storms (Rees 1965; Ożgo et al. 2016). Although the number of records is still anecdotal, Ożgo et al. (2016) considered storms a possible means of long-distance dispersal, at least for terrestrial gastropods. However, the chance that storms pick up lymnaeids and deliver them safely to new freshwater environments might be comparatively low.

6.6 Exploring the Limits of Morphological Disparity

The typical lymnaeid shell, both today and in the fossil record, is smooth and turbiniform. However, a number of "special cases" of morphological evolution happened over the long evolutionary history of the Lymnaeidae. Four of these are detailed in this chapter.

6.6.1 *Valenciennіinae*

Probably the most outstanding among those cases is the evolution of the Valenciennіinae. This subfamily originated during the Early Tortonian (Late Miocene) in early Lake Pannon from a small ancestor species currently attributed to the genus *Radix* (Gorjanović-Kramberger 1923; Moos 1944; Taktakishvili 1967; Marinescu 1970; Neubauer et al. 2016). During the c. 7-Myr-lifetime of the long-lived Lake Pannon, several evolutionary lineages evolved from that ancestor exhibiting widely expanded apertures and/or limpet-shaped shells (Fig. 6.7a–c, f, i). This also involved a massive size increase over time: starting from about 20 mm in height (Gorjanović-Kramberger 1890), some species developed shell sizes of up to 132 mm (Taktakishvili 1967). As such, they are among the largest non-marine aquatic snails ever recorded. Some of the species also developed a peculiar siphonal fold (Fig. 6.7c) and/or sculpture in the form of strong concentric ribs (Fig. 6.7a–c, f), both of which are rare features among lymnaeids in general.

While starting in Lake Pannon, these lineages were not confined to it. Several species have been reported from the Upper Miocene and Pliocene of the Dacian Basin (Krejci-Graf and Wenz 1932; Marinescu 1970; Hanganu 1972; Huică 1977; Marinescu and Papaianopol 1990; Papaianopol and Marinescu 1995) and Black Sea Basin (Rousseau 1842; Sinzov 1875; Davitashvili 1930; Taktakishvili 1967; Özsayar 1977). During the Pontian (latest Miocene), the subfamily expanded as far as the Kura Basin in Azerbaijan in the East (Davitashvili 1931). The last known occurrence is *Valenciennius kujalnicus* Taktakishvili, 1962 from the lower Kuyalnikian (latest Pliocene to earliest Pleistocene) of the Gurian region in Georgia (Taktakishvili 1967). ("?*Valenciennesia metochiana*" Pavlović, 1933 from the Middle Miocene Peć Series in Kosovo is probably not related with the Valenciennіinae; a revision is required.)

Because of the sequential appearance of (morpho-)species of these evolutionary lineages, the individual species have been used as biostratigraphic markers (Moos 1944; Taktakishvili 1967; Botka et al. 2019). In the Black Sea Basin, stratigraphers

Fig. 6.7 Examples of lymnaeid species with extraordinary morphological adaptations. (**a**) *Undulotheca halavatsi* (Gorjanović-Kramberger, 1901), Guşteriţa, Romania, Late Miocene, coll. Botka (no number) (from Botka et al. 2019). (**b**) *Provalenciennesia arthaberi* (Gorjanović-Kramberger, 1901), Beočin, Serbia, Late Miocene, MBFSZ Pl. 2545. (**c**) *Valenciennius reussi* Neumayr in Neumayr and Paul, 1875, Tirol, Romania, Late Miocene, NHMW 1900/IX/5 (from Vermeij 2017). (**d**) *Delminiella* cf. *soklici*, Podhum, Bosnia, and Herzegovina, Early Miocene, ZMBH Mg3657. (**e**) *Velutinopsis velutina* (Deshayes, 1838), Mihalţ, Romania, Late Miocene, coll. Botka. (**f**) *Valenciennius reussi*, Okrugljak near Zagreb, Croatia, Late Miocene, NHMW 1888/XIV/2. (**i**) *Radix kobelti* (Brusina, 1884), Okrugljak near Zagreb, Croatia, Late Miocene, NHMW 1888/XIV/9. (**g**) *Delminiella soklici* Kochansky-Devidé and Slišković, 1972, Vučipolje near Tomislavgrad, Bosnia, and Herzegovina, Early Miocene, holotype, ZMBH Mg3631 (from Harzhauser et al. 2016). (**h**) *Delminiella excentrica* Kochansky-Devidé and Slišković, 1972, Eminovo selo near Tomislavgrad, Bosnia and Herzegovina, Early Miocene, holotype, ZMBH Mg3623 (from Harzhauser et al. 2016). (**j**) *Corymbina bicarinata* (Fuchs, 1877), Livanates, Greece, Early Pleistocene, holotype, NHMW 1878/XX/3. (**k, l**) *Corymbina elegans* (Cantraine, 1841), Livanates, Greece, Early Pleistocene, NHMW 1878/XX/1 (label reads "*Lymnaeus Adelinae*", a

even used to distinguish lower and upper "*Valenciennius* clays" in Pontian strata (Seninski 1905; Stevanović 1990a; Popov et al. 2016).

Little is known about the causes for that peculiar morphological diversification. In Lake Pannon, the Valencienniinae seem to be constricted to deep-water (sublittoral and profundal) environments (Harzhauser and Mandic 2008; Cziczer et al. 2009). One hypothesis is that the broad, flattened shell is an adaptation to life on mudgrounds in deeper settings (Marinescu 1970; Müller et al. 1999). Cziczer et al. (2009) proposed, in turn, that the limpet-like shape serves floating or swimming in the water column.

6.6.2 *Delminiella*

A similar case of morphological evolution toward a limpet-like shell is found in the Early Miocene genus *Delminiella* (Fig. 6.7d, g, h). Unlike for Valenciennninae, no ancestor–descendant relationship is known for that genus, but a lymnaeid ancestor has been proposed (Harzhauser et al. 2016). The genus is known by three species only; two occur in the Dinaride Lake System in Croatia and Bosnia and Herzegovina (Kochansky-Devidé and Slišković 1972), the third one is endemic to the paleo-lake Lavant in southern Austria (Harzhauser et al. 2016). They are all much smaller than Valenciennninae and range around 15 mm in maximum diameter. All species have widely expanded apertures and concentric growth lines or riblets, and all retain a small, coiled apex (Kochansky-Devidé and Slišković 1972; Harzhauser et al. 2016; Fig. 6.7d, g, h).

In the Dinaride Lake System, their appearance in the fossil record parallels the evolution of the similarly limpet-shaped but presumably unrelated family Clivunellidae. In the Livno and Tomislavgrad basins in Bosnia and Herzegovina, *Clivunella* and *Delminiella* even co-occur in the same strata (Kochansky-Devidé and Slišković 1972). In contrast to the alleged deep-water Valenciennninae, *Delminiella* (as well as Clivunellidae) are suggested to be more generalist species that lived in a greater variety of paleohabitats (Harzhauser et al. 2016). They occurred in muddy lake bottoms in deeper water with intermittent low oxygen levels (De Leeuw et al. 2011) and in littoral settings alike (Bulić and Jurišić-Polšak 2009).

Fig. 6.7 (continued) junior objective synonym of *C. elegans*). (**m, n**) *Lanx kirbyi* Hanna and Gester, 1963, Dorris, California, Early Pleistocene, paratype, CASG 34807.02 (formerly CASG 12451). Photos by D. Botka (a, b), CAS (m, n), O. Mandic (d, g, h), T.A. Neubauer (j, l), A. Schumacher (c), M. Vinarski (f, i, k). Scale bars = 10 mm (a–c, f), 5 mm (d, e, g–n)

6.6.3 Lancinae

Similar to Valencienniinae and *Delminiella*, the subfamily Lancinae comprises limpet-like shells (Vermeij 2017). Species of that group have completely lost coiling and are the only true patelliform lymnaeids (Campbell et al. 2017; Fig. 6.7m, n). Shells are smooth aside from concentric growth lines. Because of their close resemblance of "true" freshwater limpets, species of Lancinae have been previously placed in the planorbid subfamily Ancylinae (Campbell et al. 2017).

Lancinae are restricted to North America and include the three genera *Lanx*, *Fisherola,* and *Idaholanx* (Campbell et al. 2017), of which only *Lanx* is known to have a fossil record. Four fossil species are recognized and their appearance in the fossil record is patchy, which might indicate that not all of them belong to the same genus. This is especially true for the oldest record of the genus and subfamily, *Lanx nevadensis* MacNeil, 1939 from the Lower Cretaceous of Nevada (MacNeil 1939). Given that molecular data indicate a close relationship between Lancinae and Lymnaeinae (Saadi et al. 2020), it is unlikely that the subfamily originated already in the Cretaceous. Probably, *Lanx nevadensis* does not belong in Lancinae; its systematic position still needs to be assessed.

The stratigraphically next occurrence of (probably true) Lancinae derives from Middle Miocene strata of California, over 100 Myryounger (Firby 1966). Further occurrences have been documented from the Late Miocene, Pliocene, and Pleistocene of the western USA (Hannibal 1912; Hanna 1922; Yen 1944, 1947; Hanna and Gester 1963; Taylor 1966, 1981; Fig. 6.7m, n). Today, Lancinae dwell primarily in cool, flowing, well-oxygenated water, often in rivers, springs or spring-influenced areas of the Pacific drainage region, where they are commonly found under stones or attached to them (Taylor 1981; Burch 1982; Campbell et al. 2017).

6.6.4 Corymbina

A different type of morphospace expansion is exhibited by the fossil genus *Corymbina* (syn. *Adelina* Cantraine, 1841, non Dejean, 1835; *Adelinella* Wenz, 1922). The genus includes five species that are characterized by the presence of strong axial ribs, sometimes accompanied by a spiral keel (Fig. 6.7j–l). Most species have a classical lymnaeid-type shell, but some forms developed a detached aperture or an even uncoiled body whorl (Willmann 1981). The genus is mostly confined to the Aegean region (Greece and Turkey), where it occurs in Middle Miocene to Middle Pleistocene strata (Oppenheim 1919; Willmann 1981; Schütt and Kavuşan 1984; Esu and Girotti 2020). Two exceptions of disjunct occurrences were found in the Sarmatian (late Middle Miocene) of Romania (Marinescu 1992) and the early Pliocene of Italy (Esu and Girotti 2018). Because of the limited occurrence in space and time, the genus has been used for biostratigraphic correlations between Greek and Turkish Neogene strata (Willmann 1981, 1982; Böger 1983).

6.7 Synthesis

The fossil record of Lymnaeidae is patchy but indicates a widespread distribution over long periods of time of the family's evolutionary history. Given the paucity of freshwater deposits over large parts of the globe, especially in the Mesozoic, we know little about the family's origin, early history, and distribution. The earliest supposed member of the family occurred during the Early Jurassic in North America. Already in the course of the Jurassic the Lymnaeidae spread to Europe and Asia. Diversity remained low until the early Paleogene until the family started to diversify during the Middle Eocene—approximately 150 Myr after its (presumed) origin. The all-time peak of lymnaeid diversity was reached in the Late Miocene and coincided with major morphological diversification of the subfamily Valencienniinae. The resulting species developed partly huge and strongly ribbed, limpet-like shells and were probably adapted to dysoxic deep-water settings. The associated evolutionary lineages evolved in long-lived lakes, which are widely considered natural laboratories of evolution (e.g., Michel et al. 1992; Harzhauser et al. 2013; Van Damme and Gautier 2013; Jenny et al. 2020). A similar yet smaller event of morphological diversification happened during the Early Miocene in central to southeastern Europe (genus *Delminiella*). The earliest supposed member of the North American Lancinae, a small subfamily of truly patelliform species that still exists today, dates back to the Early Cretaceous.

The disjunct distribution of Lymnaeidae through geological time is strongly shaped by long-distance dispersal. During the Late Cretaceous to Cenozoic, waterfowl probably constituted the main vector, suggested by numerous examples of isolated occurrences. In fact, the diversification of waterfowl and development of migratory routes may have contributed markedly to the diversification of Lymnaeidae. Conversely, the mechanisms for long-distance dispersal prior to waterfowl evolution remain doubtful and potentially involve Mesozoic ornithurine birds, pterosaurs, or storms. However, the patchy Mesozoic fossil record and the uncertainties regarding the systematic classification of early alleged lymnaeids further complicate the reconstruction of dispersal pathways.

Acknowledgments I am grateful to Maxim Vinarski for the invitation to contribute to this book. Dániel Botka, Christine Garcia, Oleg Mandic, Alice Schumacher, and Maxim Vinarski kindly provided photographs. Heinz Kollmann, Simon Schneider, and Maxim Vinarski helped retrieving rare pieces of literature. Maxim Vinarski and Mathias Harzhauser provided helpful comments on an early draft of the manuscript. The study contributes to the project "Unraveling drivers of species diversification—an integrative deep-time approach on continental aquatic biota" funded by the German Research Foundation (DFG grant no. NE 2268/2-1).

References

Aksenova OV, Bolotov IN, Gofarov MY et al (2018) Species richness, molecular taxonomy and biogeography of the radicine pond snails (Gastropoda: Lymnaeidae) in the Old World. Sci Rep 8:11199. https://doi.org/10.1038/s41598-018-29451-1

Alda P, Bonel N, Alba A et al (2023, this volume) Molecular techniques for the study of ecological and evolutionary processes in lymnaeids. In: Vinarski MV, VÃzquez A (eds) The Lymnaeidae. A handbook on their natural history and parasitological significance. Springer, Cham, pp 121–146

Almera J (1894) Descripción de los depósitos pliocénicos de la cuenca del Bajo Llobregat y llano de Barcelona. Mem R Acad Cien Art Barcelona 3:1–355

Alroy J (2010) Geographical, environmental and intrinsic biotic controls on Phanerozoic marine diversification. Palaeontology 53(6):1211–1235. https://doi.org/10.1111/j.1475-4983.2010.01011.x

Andrusov N (1923) Apsheronskiy yarus. Trudy Geologicheskoy Komissii novaya seriya 110:1–294. [in Russian]

Arkell WJ (1941) The gastropods of the Purbeck beds. Quart Journ Geol Soc London 97:79–128. https://doi.org/10.1144/GSL.JGS.1941.097.01-04.04

Ashworth AC, Preece RC (2003) The first freshwater molluscs from Antarctica. J Molluscan Stud 69(1):89–92. https://doi.org/10.1093/mollus/69.1.89

Baker FC (1911) The Lymnaeidae of North and Middle America, recent and fossil. The Chicago Academy of Sciences, Chicago, IL. https://doi.org/10.5962/bhl.title.20500

Bandel K (1991) Gastropods from brackish and fresh water of the Jurassic – Cretaceous transition (a systematic reevaluation). Berliner geowiss Abh A 134:9–55

Bartha F (1955) A várpalotai pliocén puhatestü fauna biosztratigrafiai vizsgálata. Magyar All Földt Intezet Evk 43(2):275–359

Bartha F (1971) A magyarországi pannon biosztratigráfiai vizsgálata. In: Góczán F, Benkö J (eds) A Magyarországi pannonkori képzödmények kutatásai. Akadémiai Kiadó, Budapest, pp 9–172

Bayan F (1870) Études faites dans la collection de l'École des Mines sur des fossiles nouveaux ou mal connus. Premier fascicule. Mollusques tertiaires. F. Savy, Paris

Bestwick J, Unwin DM, Butler RJ et al (2018) Pterosaur dietary hypotheses: a review of ideas and approaches. Biol Rev 93(4):2021–2048. https://doi.org/10.1111/brv.12431

Beu AG, Marshall BA, Reay MB (2014) Mid-Cretaceous (Albian–Cenomanian) freshwater Mollusca from the Clarence Valley, Marlborough, New Zealand, and their biogeographical significance. Cretac Res 49:134–151. https://doi.org/10.1016/j.cretres.2014.02.011

Bickel D (1977) A new genus and species of freshwater gastropod from the Paleocene Tongue River formation of North Dakota. J Paleontol 54(1):123–130

Böger H (1983) Stratigraphische und tektonische Verknüpfungen kontinentaler Sedimente des Neogens im Ägäis-Raum. Geol Rundsch 72(3):771–814

Botka D, Magyar I, Csoma V et al (2019) Integrated stratigraphy of the Guşteriţa clay pit: a key section for the early Pannonian (late Miocene) of the Transylvanian Basin (Romania). Austrian J Earth Sci 112(2):221–247. https://doi.org/10.17738/ajes.2019.0013

Bouchet P, Rocroi J-P, Hausdorf B et al (2017) Revised classification, nomenclator and typification of gastropod and monoplacophoran families. Malacologia 61(1–2):1–526. https://doi.org/10.4002/040.061.0201

Braun A (1851) Darstellung der geognostischen Verhältnisse des Mainzer Beckens und seiner fossilen Fauna und Flora. In: Walchner FA (ed) Handbuch der Geognosie zum Gebrauche bei seinen Vorlesungen und zum Selbststudium mit besonderer Berücksichtigung der geognostischen Verhältnisse des Grossherzogthums Baden. Christian Theodor Gross, Karlsruhe, pp 1121–1169

Brikiatis L (2014) The De Geer, Thulean and Beringia routes: key concepts for understanding early Cenozoic biogeography. J Biogeogr 41(6):1036–1054. https://doi.org/10.1111/jbi.12310

Brikiatis L (2016) Late Mesozoic North Atlantic land bridges. Earth-Sci Rev 159:47–57. https://doi.org/10.1016/j.earscirev.2016.05.002
Brongniart A (1810) Sur des Terrains qui paroissent avoir été formés sous l'eau douce. Ann Mus Natl Hist Nat 15:357–405
Brusina S (1874) Fossile Binnen-Mollusken aus Dalmatien. Kroatien und Slavonien nebst einem Anhange, Actienbuchdruckerei, Agram
Brusina S (1902) Iconographia Molluscorum Fossilium in tellure tertiaria Hungariae, Croatiae, Slavoniae, Dalmatiae, Bosniae, Herzegovinae, Serbiae and Bulgariae inventorum. Officina Soc, Typographicae, Agram
Bulić J, Jurišić-Polšak Z (2009) Macropalaeontology and stratigraphy of lacustrine Miocene deposits at Crnika Beach on the Island of Pag (Croatia). Geol Croat 62(3):135–156. https://doi.org/10.4154/gc.2009.16
Burch JB (1982) Freshwater snails (Mollusca: Gastropoda) of North America. In: U.-S. environmental protection agency, office of research and development, environmental monitoring and support laboratory, Cincinnati, OH
Campbell DC, Clark SA, Lydeard C (2017) Phylogenetic analysis of the Lancinae (Gastropoda, Lymnaeidae) with a description of the U.S. federally endangered Banbury Springs lanx. Zookeys 663:107–132. https://doi.org/10.3897/zookeys.663.11320
Cantraine F (1841) Malacologie méditerranéenne et littorale, ou description des mollusques qui vivent dans la Méditerranée ou sur le continent de l'Italie, ainsi que des coquilles qui se trouvent dans les terrains tertiaires italiens, avec des observations sur leur anatomie, leurs moeurs, leur analogie et leur gisement. Ouvrage servant de faune malacologique italienne et de complément à la Conchiologia fossile subapennina de Brocchi. Nouv Mem Acad R Sci Bruxelles 13:1–173
Cao W, Zahirovic S, Flament N et al (2017) Improving global paleogeography since the late Paleozoic using paleobiology. Biogeosciences 14(23):5425–5439. https://doi.org/10.5194/bg-14-5425-2017
Carobene D, Harzhauser M, Mandic O et al (2018) Taxonomy and palaeoecology of continental Gastropoda (Mollusca) from the Late Pleistocene mammoth-bearing site of Bullendorf in NE Austria. Riv Ital Paleontol Stratigr 124(3):509–534. https://doi.org/10.13130/2039-4942/10616
Carr WJ, Trimble DE (1963) Geology of the American Falls quadrangle, Idaho. U.S. Geol Surv Bull 1121-G:1–44
Claramunt S, Cracraft J (2015) A new time tree reveals Earth history's imprint on the evolution of modern birds. Sci Adv 1(11):e1501005. https://doi.org/10.1126/sciadv.1501005
Clessin S (1885) Die Conchylien der Obermiocaenen Ablagerungen von Undorf. Malakozool Bl Neue Folge 7:71–95
Cockerell TDA (1906) The fossil Mollusca of Florissant, Colorado. Bull Am Mus Nat Hist 22:459–462
Cockerell TDA (1908) The Miocene species of *Lymnaea*. The Nautilus 22(6):69–70
Correa AC, Escobar JC, Durand P et al (2010) Bridging gaps in the molecular phylogeny of the Lymnaeidae (Gastropoda: Pulmonata), vectors of fascioliasis. BMC Evol Biol 10:381. https://doi.org/10.1186/1471-2148-10-381
Correa AC, Escobar JS, Noya O (2011) Morphological and molecular characterization of Neotropic Lymnaeidae (Gastropoda: Lymnaeoidea), vectors of fasciolosis. Infect Genet Evol 11:1978–1988. https://doi.org/10.1016/j.meegid.2011.09.003
Cossmann M (1889) Catalogue illustré des coquilles fossils de l'Éocène des environs de Paris. Ann Soc R zool malacolog Belg 24:3–376
Cossmann M (1902) Appendice No. 3 au Catalogue illustré des coquilles fossils de l'Éocène des environs de Paris. Ann Soc R zool malacolog Belg 36:9–110
Cziczer I, Magyar I, Pipík R et al (2009) Life in the sublittoral zone of long-lived Lake Pannon: paleontological analysis of the Upper Miocene Szák Formation, Hungary. Int J Earth Sci 98: 1741–1766. https://doi.org/10.1007/s00531-008-0322-3
d'Orbigny A (1852) Prodrome de Paléontologie. Stratigraphique universelle des animaux mollusques et rayonnés faisant suitre au cours élémentaire de paléontologie et de géologie

stratigraphique. Troisième volume, Victor Masson, Paris. https://www.biodiversitylibrary.org/page/31730040

Davitashvili LS (1930) Kimmeriyskiy yarus. In: Archangelsky AD, Davitashvili LS (eds) Rukovodyashchiye iskopayemyye neftyenosnykh rayonov Krymsko-Kavkazskoy oblasti, VIII, vol 6. Trudy Gosudarstvennogo Issledovatel'skogo Neftyanogo Instituta, pp 1–44. [in Russian]

Davitashvili LS (1931) Ponticheskiy yarus. In: Archangelsky AD, Davitashvili LS (eds) Rukovodyashchiye iskopayemyye neftyenosnykh rayonov Krymsko-Kavkazskoy oblasti, VII. Trudy Gosudarstvennogo Issledovatel'skogo Neftyanogo Instituta, pp 1–56. [in Russian]

Dayrat B, Conrad M, Balayan S et al (2011) Phylogenetic relationships and evolution of pulmonate gastropods (Mollusca): new insights from increased taxon sampling. Mol Phylogenet Evol 59(2):425–437. https://doi.org/10.1016/j.ympev.2011.02.014

De Leeuw A, Mandic O, Krijgsman W et al (2011) A chronostratigraphy for the Dinaride Lake System deposits of the Livno-Tomislavgrad Basin: the rise and fall of a long-lived lacustrine environment. Stratigraphy 8(1):29–43

De Loriol P, Jaccard A (1865) Étude géologique et paléontologique de la formation d'eau douce infracrétacée du Jura et en particulier de Villers-le-Lac. Mem Soc phys hist nat Genève 18(1): 63–128

Degrange-Touzin A (1892) Étude sur la faune terrestre, lacustre et fluviatile de l'Oligocène supérieur et du Miocène dans le Sud-Ouest de la France et principalement dans la Gironde. Actes Soc Linn Bordeaux 45:125–230

Dehm R (1979) Artenliste der altpleistozänen Molluskenfauna vom Uhlenberg bei Dinkelscherben. Geol Bavarica 80:123–125

Dejean PFMA (1833–1836) Catalogue des coléoptères de la collection de M. le comte Dejean. Méquignon-Marvis Père et Fils, Paris

Deshayes GP (1861–1864) Description des animaux sans vertèbres découverts dans le bassin de Paris pour servir de supplément a la description des coquilles fossiles des environs de Paris comprenant une revue générale de toutes les espèces actuellement connues. Tome deuxième. – Texte. Mollusques Acéphalés Monomyairew et Brachiopodes. Mollusques céphalés. Première Partie. J.-B. Baillière et Fils, Paris. https://www.biodiversitylibrary.org/page/35623270

Dollfus GF (1906) Bassin de Paris. Feuille de Bourges au 320.000e (révision des faunes continentales). In: Anonymous (ed.) Comptes rendus des collaborateurs pour la campagne de 1905. Bull Serv Carte Géol Fr 16(110):284–304

Dollfus GF (1920) Bassin de Paris. Feuille de Paris au 80.000e. In: Anonymous (Ed.) Comptes rendus des collaborateurs pour la campagne de 1919. Bull Serv Carte Géol Fr 24(140):1–13

Douxami H (1904) Étude sur la molasse rouge. Ann Soc Linn Lyon nouv ser 51:1–30. https://www.biodiversitylibrary.org/page/54584266

Dunker W (1846) Monographie der Norddeutschen Wealdenbildung. Ein Beitrag zur Geognosie und Naturgeschichte der Vorwelt, Oehme und Müller, Braunschweig

Dunker W (1853) Ueber die in der Braunkohlenformation von Großalmerode in neuerer Zeit entdeckten Süßwasser-Mollusken. Programm der höheren Gewerbschule in Cassel zu Michaelis 1853:3–18

Edwards FE (1852) Part II. Pulmonata. In: Edwards FE, Wood SV (eds) A monograph of the Eocene Mollusca, or descriptions of shells from the older Tertiaries of England. Paleontographical Society, London, pp 57–122. https://www.biodiversitylibrary.org/page/12894018

Esu D (1984) Gasteropodi dei bacini continentali terziari eocenico-oligocenici dell'isola di Maiorca (Baleari). Thalassia Salent 14:85–99. https://doi.org/10.1285/i15910725v14p85

Esu D, Girotti O (1975) La malacofauna continentale del Plio-Pleistocene dell'Italia centrale. I Paleontologia Geol Rom 13:203–294

Esu D, Girotti O (2010) The Late Oligocene molluscan fauna from Otranto (Apulia, southern Italy): an example of alternating freshwater, lagoonal and emerged environments. Palaeontology 53(1): 137–174. https://doi.org/10.1111/j.1475-4983.2009.00923.x

Esu D, Girotti O (2018) *Valvata mathiasi* n. sp. (Gastropoda: Heterobranchia: Valvatidae) from the lower Pliocene of the Val di Pesa (Tuscany, Central Italy). Arch Molluskenkd 147(1):49–54. https://doi.org/10.1127/arch.moll/147/049-054

Esu D, Girotti O (2020) Updating a late Early – Middle Pleistocene non-marine molluscan fauna from Achaia (Greece). Systematics and palaeoecological remarks. Boll Malacol 56:59–83

Evanoff E, Good SC, Hanley JH (1998) An overview of the freshwater mollusks from the Morrison Formation (Upper Jurassic, Western Interior, USA). Mod Geol 22:423–450

Evans J, Shumard BF (1856) Descriptions of new fossil species from the fresh water Tertiary formation of Nebraska, collected by the North Pacific railroad expedition, under Gov. J. J Stevens. Proc Acad Nat Sci Philad 7:164–165. https://www.biodiversitylibrary.org/page/26299385

Fabre-Taxy S (1948) Faunes lagunaires et continentales du Crétacé Supérieur de Provence. I – Le Turonien saumâtre. Ann Paléont 34:63–95

Fabre-Taxy S (1959) Faunes lagunaires et continentales du Crétacé Supérieur de Provence. III – Le Maestrichtien et le Danien. Ann Paléont 45:55–124

Falk AR (2011) Tracking Mesozoic birds across the world. J Syst Palaeontol 9(1):85–90. https://doi.org/10.1080/14772019.2010.512616

Firby JR (1966) New non-marine Mollusca from the Esmeralda formation, Nevada. Proc Calif Acad Sci 4th ser 33(14):453–480. https://www.biodiversitylibrary.org/page/26236670

Fleming TH, Lips KR (1991) Angiosperm endozoochory: were pterosaurs Cretaceous seed dispersers? Am Nat 138(4):1058–1065. https://doi.org/10.1086/285269

Fontannes F (1879) Note sur la découverte d'un gisement de marne a Limnées a Celleneuve, près Montpellier. Revue Sci Nat 8(64–75):159

Fontannes F (1881) Les terrains tertiaires du bassin de Crest. Ann Soc Agr Hist Nat Arts Utiles Lyon 5th ser 3:827–1046. http://gallica.bnf.fr/ark:/12148/bpt6k6154617d/f856.image

Fontannes F (1883) Diagnoses d'espèces et de variétés nouvelles des terrains tertiaires du bassin du Rhône. Mougin-Rusand, Lyon

Fontannes F (1884) Description sommaire de la faune malacologique des formations saumâtres et d'eau douce du groupe d'Aix (Bartonien-Aquitanien) dans le Bas-Languedoc, la Provence et le Dauphiné. Georg, F. Savy, Lyon, Paris. http://gallica.bnf.fr/ark:/12148/bpt6k55007367

Gillet S (1953) Les marnes à Cyrènes de l'Oligocène d'Alsace. Rev Inst Francais Pétrole 8(8): 395–422

Glibert M (1962) Euthyneura et Pulmonata fossiles du Cénozoïque étranger des collections de l'Institut Royal des Sciences Naturelles de Belgique. Mem Inst R Sci nat Belgique 2nd ser 70:1–140. www.vliz.be/imisdocs/publications/252233.pdf

Glibert M, Heinzelin de Braucourt J (1954) Le gîte des vertébrés tongriens de Hoeleden. Bull Inst R Sci nat Belgique 30(1):1–14

Good SC (1987) Mollusc-based interpretations of lacustrine paleoenvironments of the Sheep Pass Formation (latest Cretaceous to Eocene) of east central Nevada. PALAIOS 2:467–478. https://doi.org/10.2307/3514618

Gorjanović-Kramberger C (1890) Die praepontischen Bildungen des Agramer Gebirges. Glasnik Hrvatsk Nar Društva 5:151–163. https://www.biodiversitylibrary.org/page/11105025

Gorjanović-Kramberger K (1901) Über die Gattung *Valenciennesia* und einige unterpontische Limnaeen. Ein Beitrag zur Entwicklungsgeschichte der Gattung *Valenciennesia* und ihr Verhältnis zur Gattung *Limnaea*. Beitr Paläont Österr-Ung Orient 13(3):121–140. https://www.biodiversitylibrary.org/page/14512804

Gorjanović-Kramberger C (1923) Die Valenciennesiden und einige anderen Limnaeiden der pontischen Stufe des unteren Pliocaens in ihrer stratigraphischen und genetischen Bedeutung. Glasnik Hrvatsk Prir Društva 35:87–114

Grateloup J-PS (1828) Tableau des coquilles Fossiles qu'on rencontre dans des terrains calcaires tertiaires (faluns) des environs de Dax, dans le département des Landes (Articles 1–3). Bull Hist Nat Soc linn Bordeaux 2:72–109. 123–158, 192–204. https://www.biodiversitylibrary.org/page/35729945

Gray J (1988) Evolution of the freshwater ecosystem: the fossil record. Palaeogeogr Palaeoclimatol Palaeoecol 62:1–214. https://doi.org/10.1016/0031-0182(88)90054-5

Hanganu E (1972) Des espèces de *Valenciennius* au Pontien du Bassin dacique. Revue Roum Geol Geophys Geogr 16(1):21–39

Hanna GD (1922) Fossil freshwater mollusks from Oregon, contained in the Condon Museum of the University of Oregon. Univ Oregon Publ 1(12):1–22. https://www.biodiversitylibrary.org/page/20825593

Hanna GD (1923) Upper Miocene lacustrine mollusks from Sonoma County, California. Proc Calif Acad Sci 4th ser 12(3):31–41. https://www.biodiversitylibrary.org/page/15974899

Hanna GD, Gester GC (1963) Pliocene lake beds near Dorris, California. Occas Pap Calif Acad Sci 43:1–17

Hannibal H (1912) A synopsis of the recent and tertiary freshwater Mollusca of the Californian province, based upon an ontogenetic classification. Proc Malacol Soc London 10(2–3): 112–165. 167–211. https://www.biodiversitylibrary.org/page/3158751

Harshbarger JW, Repenning CA, Irwin JH (1957) Stratigraphy of the uppermost Triassic and the Jurassic rocks of the Navajo country. US Geol Surv Prof Paper 291:1–74. https://pubs.usgs.gov/pp/0291/report.pdf

Hartman JH, Roth B (1998) Late Paleocene and Early Eocene nonmarine molluscan faunal change in the Bighorn Basin, northwestern Wyoming and south-central Montana. In: Aubrey MP, Lucas SG, Berggren WA (eds) Late Paleocene–Early Eocene biotic and climatic events in the marine and terrestrial records. Columbia University Press, pp 323–379

Hartman JH, Erickson DN, Bakken A (2008) Stephen Hislop and his 1860 Cretaceous continental molluscan new species descriptions in Latin from the Deccan Plateau. India. Palaeontology 51(6):1225–1252. https://doi.org/10.1111/j.1475-4983.2008.00807.x

Harzhauser M, Mandic O (2008) Neogene lake systems of Central and South-Eastern Europe: faunal diversity, gradients and interrelations. Palaeogeogr Palaeoclimatol Palaeoecol 260(3–4): 417–434. https://doi.org/10.1016/j.palaeo.2007.12.013

Harzhauser M, Neubauer TA, Mandic O et al (2012) A Middle Miocene endemic freshwater mollusc assemblage from an intramontane Alpine lake (Aflenz Basin, Eastern Alps, Austria). Palaeontol Z 86(1):23–41. https://doi.org/10.1007/s12542-011-0117-x

Harzhauser M, Mandic O, Kern AK et al (2013) Explosive demographic expansion by dreissenid bivalves as a possible result of astronomical forcing. Biogeosciences 10:8423–8431. https://doi.org/10.5194/bg-10-8423-2013

Harzhauser M, Neubauer TA, Gross M et al (2014a) The early Middle Miocene mollusc fauna of Lake Rein (Eastern Alps, Austria). Palaeontogr Abt A 302(1–6):1–71

Harzhauser M, Neubauer TA, Georgopoulou E et al (2014b) The Early Miocene (Burdigalian) mollusc fauna of the North Bohemian Lake (Most Basin). Bull Geosci 89(4):819–908. https://doi.org/10.3140/bull.geosci.1503

Harzhauser M, Mandic O, Neubauer TA et al (2016) Disjunct distribution of the Miocene limpet-like freshwater gastropod genus *Delminiella*. J Molluscan Stud 82(1):129–136. https://doi.org/10.1093/mollus/eyv040

Henderson J, Rodeck HG (1934) New species of Pliocene Mollusca from Eastern Oregon. J Paleontol 8(3):264–269

Hislop S (1860) On the Tertiary deposits, associated with trap-rock, in the East Indies. Quart Journ Geol Soc London 16:154–182. http://archive.org/stream/quarterlyjournal161860ge#page/154/mode/2up

Hubendick B (1951) Recent Lymnaeidae. Their variation, morphology, taxonomy, nomenclature and distribution. Kungliga Svenska Vetenskapsakademiens Handlingar Fjärde Serien 3(1): 1–223

Hubendick B (1978) Systematics and comparative morphology of the Basommatophora. In: Fretter V, Peake J (eds) Pulmonates. Volume 2A – systematics, evolution and ecology. Academic Press, London, pp 1–47

Huckriede R (1967) Molluskenfaunen mit limnischen und brackischen Elementen aus Jura, Serpulit und Wealden NW-Deutschlands und ihre paläogeographische Bedeutung. Geol Jb Beih 67:1–263

Huică IV (1977) Studiul geologic al depozitelor miocene şi pliocene dintre Valea Sohodol şi Valea Blahniţa, judeţul Gorj (depresiunea getica). Anuarul Institutului de Geologie şi Geofizică 51:5–68

Inaba A (1969) Cytotaxonomic studies of lymnaeid snails. Malacologia 7(2–3):143–168. https://www.biodiversitylibrary.org/page/13122332

Janssen AW (1980) A mollusc-fauna with '*Pseudamnicola*' *helicella* (Braun) from the Atuatuca formation (Oligocene) at St.-Truiden (Belgium, province of Limburg). Meded Werkgr Tert Kwart Geol 17(2):43–55

Jekelius E (1932) Fauna Neogenă a României. Die Molluskenfauna der dazischen Stufe des Beckens von Braşov. Memoriile Institutului geologic al României 2:1–118

Jenny J-P, Anneville O, Arnaud F et al (2020) Scientists' warning to humanity: rapid degradation of the world's large lakes. J Gt Lakes Res 46(4):686–702. https://doi.org/10.1016/j.jglr.2020.05.006

Jodot P (1954) Mollusques de petites dimensions des formations continentales ludienne, sannoisienne et stampienne du Jura méridional et de la Haute-Savoie. Bull Soc Geol Fr 6th ser 4(7–9):537–556

Jodot P, Feugueur L (1953) Le passage du Lutétien au Bartonien a Montagny-en-Vexin (Oise). Présence d'un calcaire lacustre a faune bartonienne subordonnée aux couches à *Potamides lapidum*. Bull Soc Geol Fr 6th ser 3(9):933–940

Jodot P, Rey R (1956) Observations stratigraphiques et malacologiques sur les bassins lacustres de Saint-Alban-sur-Limagnole (Lozère) et de Massiac (Cantal). Bull Soc Geol Fr 6th ser 6(7–9): 937–968

Jooss CH (1913) Ueber *Limnaea* (*Limnaea* s.str.) *turrita* Klein emend. Jooss. Centralbl Min Geol Paläont 1913(2):58–64. https://www.biodiversitylibrary.org/page/49054066

Jörger KM, Stöger I, Kano Y et al (2010) On the origin of Acochlidia and other enigmatic euthyneuran gastropods, with implications for the systematics of Heterobranchia. BMC Evol Biol 8:323. https://doi.org/10.1186/1471-2148-10-323

Kadolsky D (1989) Stratigraphie und Molluskenfaunen von "Landschneckenkalk" und "Cerithienschichten" im Mainzer Becken (Oberoligozän bis Untermiozän?). Stratigraphische, paläogeographische und paläoökologische Ergebnisse. Geol Jb A 110:69–133

Kadolsky D (2020) A remarkable non-marine mollusc fauna of early Eocene age from a fissure infill in Karsdorf quarry (Sachsen-Anhalt, Germany). Geol Saxonica 65:31–76. https://doi.org/10.26049/GEOLSAX65-66-2019-2020-04

Kappes H, Haase P (2012) Slow, but steady: dispersal of freshwater molluscs. Aquat Sci 74(1): 1–14. https://doi.org/10.1007/s00027-011-0187-6

Klein A (1853) Conchylien der Süsswasserkalkformation Württembergs. Jahresh Ver vaterl Naturk Württemberg 9:203–223. https://www.biodiversitylibrary.org/page/7983357

Kochansky-Devidé V, Pikija M (1976) Panonske Clivunellidae (Gastropoda) sjeverne Hrvatske. Geol Vjesn 29:397–407

Kochansky-Devidé V, Slišković T (1972) Revizija roda *Clivunella* Katzer, 1918 i *Delminiella* n. gen. (Gastropoda). Geol glasn 16:47–70

Kocsis AT, Alroy J, Reddin CJ et al (2019) divDyn: diversity dynamics using fossil sampling data. R package version 0.8.0. http://cran.r-project.org/package=divDyn. Accessed 1 November 2020

Kókay J (2006) Nonmarine mollusc fauna from the Lower and Middle Miocene, Bakony Mts., W Hungary. Geol Hung Ser Palaeont 56:1–196

Kolenda K, Najbar A, Kusmierek N et al (2017) A possible phoretic relationship between snails and amphibians. Folia Malacol 25(4):281–285. https://doi.org/10.12657/folmal.025.019

Kovalenko V (2004) Lymnaeidae iz mestonahoždeni Trijebine i Vračević, Serbia. Bull Acad Serbe Sci Arts 42:327–339

Kovalenko VA (2017) Presnovodnyye mollyuski (Gastropoda, Pulmonata, Lymnaeidae) v sarmatskikh otlozheniyakh yuga Ukrainy. Heoloho-mineralohichnyy visnyk Kryvoriz'koho natsional'noho universytetu 1(37):17–31. [in Russian]

Krejci-Graf K, Wenz W (1932) Stratigraphie und Paläontologie des Obermiozäns und Pliozäns der Muntenia (Rumänien). Z Dtsch Geol Ges 83:65–163

Kruglov ND (2005) Lymnaeid snails of Europe and Northern Asia. Smolensk State Pedagogical University Press, Smolensk. [in Russian]

La Rocque A (1960) The molluscan faunas of the Flagstaff Formation of central Utah. Geol Soc Am Mem 78:1–100. https://doi.org/10.1130/MEM78-p1

La Rocque A (1963) Late Cenozoic non-marine molluscan associations in eastern North America. Sterkiana 11:1–50

Lennert J, Szónoky M, Szuromi-Korecz A et al (1999) The Lake Pannon fossils of the Bȋtaszék brickyard. Acta Geol Hung 42(1):67–88

Leriche M (1899) Description de la Faune d'eau douce sparnacienne de Cuvilly (Oise). Ann Soc geol Nord 28:95–104. http://iris.univ-lille1.fr/bitstream/handle/1908/1774/DP11_28.pdf?sequence=2

Love JD (1989) Names and descriptions of new and reclassified formations in northwestern Wyoming (geology of the Teton-Jackson Hole region, northwestern Wyoming). US Geol Surv Prof Paper 932-C:1–45. https://pubs.usgs.gov/pp/0932c/report.pdf

Lozouet P, Maestrati P (1981) Sables de Vauroux-St.-Antoine. Afzett Werkgr Tert Kwart Geol 3(2):61–71

Macaleț R (2000) New *Radix* species identified in the Neogene deposits of the Dacic Basin. Acta Palaeont Romaniae 2:251–259

Macarovici N (1940) Recherches géologiques et paléontologiques dans la Bessarabie méridionale (Roumanie). Ann sci Univ Jassy 36(1):177–422

MacNeil FS (1939) Fresh-water invertebrates and land plants of cretaceous age from Eureka, Nevada. J Paleontol 13:355–360

Mahulu A, Clewing C, Stelbrink B et al (2019) Cryptic intermediate snail host of the liver fluke *Fasciola hepatica* in Africa. Parasit Vectors 12:573. https://doi.org/10.1186/s13071-019-3825-9

Marinescu F (1970) *Velutinellus*, nouveau genre fossile dc la famille des Lymnaeidae, et ses relations avec *Velutinopsis* et *Valenciennius*. Malacologia 9(2):313–325

Marinescu F (1973) Les mollusques pontiens de Tirol (Banat Roumain). Mem Inst Geol Geophys 18:7–56

Marinescu F (1992) *Radix* (*Adelinella*) *coronatus* n. sp. (Mollusca, Gastropoda) dans le Sarmatien du Bassin de Borod. Rom J Palaeont 75:9–10

Marinescu F, Papaianopol I (1990) Faciostratotypes et stratotypes de limite del'avantpays Carpatho-Balkanique et du Bassin Dacique (Roumanie). In: Stevanović P, Nevesskaya LA, Marinescu F et al (eds) Chronostratigraphie und Neostratotypen. Neogen der Westlichen ("Zentrale") Paratethys, Bd. VIII, Pl1. Pontien. Verlag der Jugoslawischen Akademie der Wissenschaften und Künste und der Serbischen Akademie der Wissenschaften und Künste, Zagreb, Beograd, pp 398–410

Marquet R, Lenaerts J, Karnekamp C et al (2008) The molluscan fauna of the Borgloon formation in Belgium (Rupelian, Early Oligocene). Palaeontos 12:1–100

Marsh AD, Rowe TB (2018) Anatomy and systematics of the sauropodomorph *Sarahsaurus aurifontanalis* from the early Jurassic Kayenta formation. PLoS One 13(10):e0204007. https://doi.org/10.1371/journal.pone.0204007

Marshall WB (1926) New fossil fresh-water mollusks from Florida. Proc U S Nat Mus 68(2612): 1–4. https://www.biodiversitylibrary.org/page/7610769

Martinson GG (1954) Nekotoryye presnovodnyye bryukhonogiye mollyuski iz neogenovykh otlozheniy Irkutskogo amfiteatra. Trudy Baikal'skoy Limnologicheskoy Stantzii Akademii Nauk SSSR 13:108–121. [in Russian]

Martinson GG (1956) Opredelitel' mezozoyskikh i kaynozoyskikh presnovodnykh mollyuskov Vostochnoy Sibiri. Izdatel'stvo Akademii Nauk SSSR, Moskva, Leningrad. [in Russian]
Martinson GG (1961) Mezozoiskie i Kainozoiskie Molliuski kontinentalnykh otlozhenii Sibirskoi Platformy Zabaikalia i Mongolii. Trudy Baikal'skoy Limnologicheskoy Stantzii Akademii Nauk SSSR 19:1–332. [in Russian]
Matheron P (1843) Catalogue méthodique et descriptif des corps organisés fossiles du départment des Bouches-du-Rhône et lieux circonvoisins; précédé D'un Mémoire sur les terrains supérieurs au Grès Bigarré du S. E. de la France. Rep Trav Soc stat Marseille 6:1–269
Meek FB (1873) Preliminary paleontological report, consisting of lists and descriptions of fossils, with remarks on the ages of the rocks in which they were found, etc., etc. US Geol Geogr Surv Terr Ann rep 6:431–541. https://biodiversitylibrary.org/page/15677973
Meek FB (1876) A report on the invertebrate Cretaceous and Tertiary fossils of the upper Missouri country. Rep US Geol Surv Terr 9:1–629. https://www.biodiversitylibrary.org/page/2024294
Meek FB, Hayden FV (1856) Descriptions of new species of Acephala and Gastropoda, from the Tertiary formations of Nebraska territory, with some general remarks on the geology of the country about the sources of the Missouri River. Proc Acad Nat Sci Philad 8:111–126. https://www.biodiversitylibrary.org/page/1935165
Michaud G (1855) Descriptions de coquilles fossiles découvertes dans les environs de Hauterive (Drôme). Ann Soc linn Lyon nouv ser 2:33–64. https://www.biodiversitylibrary.org/page/3487714
Michel AE, Cohen AS, West K et al (1992) Large African Lakes as natural laboratories for evolution: examples from the endemic gastropod fauna of Lake Tanganyika. Mitt Intern Ver Theor Angew Limnol 23:85–99
Mödden C, Schäfer P, Reichenbacher B et al (2000) Säugetiere, Fisch-Otolithen, Ostracoden, Mollusken und Charophyten aus den Süßwasser-Schichten (Oligozän) von Wolfsheim im Mainzer Becken. Palaeontol Z 74(3):343–361. https://doi.org/10.1007/BF02988106
MolluscaBase (eds) (2021) MolluscaBase. https://www.molluscabase.org. Accessed 7 Nov 2021
Moos A (1944) Neue Funde von Lymnaeiden, insbesondere von Valenciennesiiden im Pannon Kroatiens. Vjestnik Hrvatskog državnog geološkog zavoda i Hrvatskog državnog geološkog muzeja 2(3):341–390
Morton LS, Herbst R (2003) Moluscos dulceacuícolas de las Formaciones San José y Chiquimil (Mioceno) del Valle de Santa María (Catamarca y Tucumán). Argentina Ameghiniana 40(2): 205–216
Müller P, Geary DH, Magyar I (1999) The endemic molluscs of the Late Miocene Lake Pannon: their origin, evolution, and family-level taxonomy. Lethaia 32:47–60
Munt MC (2014) Mollusca from the Insect Limestone (Bembridge Marls Member: Bouldnor Formation: Solent Group), Palaeogene, Isle of Wight, southern England. Earth Env Sci Trans Roy Soc Edinb 104:263–274. https://doi.org/10.1017/S1755691014000048
Nations JD (1974) Paleontology, biostratigraphy, and paleoecology of the Verde Formation of late Cenozoic age, north-central Arizona. In: TNV K, Swann GA, Eastwood RL (eds) Geology of northern Arizona with notes on archaeology and paleoclimate. Part II – Area studies and field guides. Geological Society of America, Rocky Mountain Section Meeting, Flagstaff, Arizona. Geological Society of America, Flagstaff, AZ, pp 611–629
Neubauer TA, Harzhauser M, Kroh A et al (2014) Replacement names and nomenclatural comments for problematic species-group names in Europe's Neogene freshwater Gastropoda. Part 2. ZooKeys 429:13–46. https://doi.org/10.3897/zookeys.429.7420
Neubauer TA, Harzhauser M, Kroh A et al (2015a) A gastropod-based biogeographic scheme for the European Neogene freshwater systems. Earth-Sci Rev 143:98–116. https://doi.org/10.1016/j.earscirev.2015.01.010
Neubauer TA, Harzhauser M, Georgopoulou E et al (2015b) Tectonics, climate, and the rise and demise of continental aquatic species richness hotspots. Proc Natl Acad Sci U S A 112(37): 11478–11483. https://doi.org/10.1073/pnas.1503992112

Neubauer TA, Georgopoulou E, Kroh A et al (2015c) Synopsis of European Neogene freshwater gastropod localities: updated stratigraphy and geography. Palaeontol Electron 18(1):3T. https://doi.org/10.26879/478

Neubauer TA, Harzhauser M, Pipík R (2015d) Upper Miocene endemic lacustrine gastropod fauna of the Turiec Basin: addressing taxonomic, paleobiogeographic and stratigraphic issues. Geol Carpathica 66(2):139–156. https://doi.org/10.1515/geoca-2015-0016

Neubauer TA, Georgopoulou E, Harzhauser M et al (2016) Predictors of shell size in long-lived lake gastropods. J Biogeogr 43(10):2062–2074. https://doi.org/10.1111/jbi.12777

Neubauer TA, Hauffe T, Silvestro D et al (2021) Current extinction rate in European freshwater gastropods greatly exceeds that of the late Cretaceous mass extinction. Commun Earth Environ 2:97. https://doi.org/10.1038/s43247-021-00167-x

Nordsieck H (1992) Phylogeny and system of the Pulmonata (Gastropoda). Arch Molluskenkd 121: 31–52

Noulet J-B (1854) Mémoires sur les coquilles fossiles des terrains d'eau douce du Sud-Ouest de la France. Victor Masson, Paris

Noulet J-B (1857) Coquilles fossiles nouvelles des terrains d'eau douce du Sud-Ouest de la France. Editions Masson, Paris

Oppenheim P (1919) Das Neogen in Kleinasien. Z Dtsch Geol Ges 70:1–210. https://www.biodiversitylibrary.org/page/43778447

Ożgo M, Örstan A, Kirschenstein M et al (2016) Dispersal of land snails by sea storms. J Molluscan Stud 82(2):341–343. https://doi.org/10.1093/mollus/eyv060

Özsayar TY (1977) Karadeniz kiyi bölgesindeki Neojen formasyonlari ve bunlarin mollusk faunasinin incelenmesi. Karadeniz Teknik Üniversitesi Yayin 79:1–80

Pais J (ed) (2012) The Paleogene and Neogene of Western Iberia (Portugal): a Cenozoic record in the European Atlantic domain. Springer, Berlin

Pan HZ (1977) Mesozoic and Cenozoic fossil Gastropoda from Yunnan. In: Nanjing Institute of Geology and Palaeontology (ed) Mesozoic Fossils from Yunnan. 2. Science Press, Beijing, pp 83–152

Pan HZ (1983) Jurassic-cretaceous non-marine gastropod from Shandong Province. Acta Palaeontol Sin 22(2):210–218

Pan HZ, Zhu XG (2007) Early Cretaceous non-marine gastropods from the Xiazhuang Formation in North China. Cretac Res 27:215–224. https://doi.org/10.1016/j.cretres.2006.12.001

Pană I, Grigorescu D, Csiki Z et al (2002) Paleo-ecological significance of the continental gastropod assemblages from the Maastrichtian dinosaur beds of the Haţeg Basin. Acta Palaeont Romaniae 3:337–343

Papaianopol I, Marinescu F (1995) Faune de mollusques daciens du Bassin Dacique. In: Marinescu F, Papaianopol I (eds) Chronostratigraphie und Neostratotypen. Neogen der Zentrale Paratethys, Bd. IX, Pl1. Dacien. Editura Academiei Române, Bucuresti, pp 130–267

Papaianopol I, Marinescu F, Macaleț R (2003) Les coupes représentatives (stratotypes, faciostratotypes, stratotypes de limite). In: Papaianopol I, Marinescu F, Krstić N et al (eds) Chronostratigraphie und Neostratotypen. Neogen der Zentrale Paratethys, Bd. X, Pl2. Romanien. Editura Academiei Române, Bucuresti, pp 133–173

Papp A (1956) Paläontologische Beobachtungen im Pannon von Podsused bei Zagreb (Kroatien). Geol Vjesn 8–9:67–79

Paul CRC (1989) The molluscan faunal succession in the Hatherwood Limestone Member (Upper Eocene), Isle of Wight, England. Tert Res 10(4):147–162

Pavlović PS (1927) Donjopontiski mekušci iz okoline Beograda (s narocitim obzirom na fosilnu faunu okoline sela Vrcina). Sprska Akademija nauka, posebna izdanja 66(Prirodnjački i matematički spisi 17):1–121

Pavlović PS (1933) O fosilnoj fauni mekušaca iz okoline Peci. Glas Srpske Kraljevske Akademije 158(78):75–91

Peyrot A (1933) Conchologie néogènique de l'Aquitaine. Gastropodes Actes Soc Linn Bordeaux 84(2):129–288. 11–18. https://www.biodiversitylibrary.org/page/47871688

Pickford M (2008) Freshwater and terrestrial Mollusca from the Early Miocene deposits of the northern Sperrgebiet, Namibia. Mem Geol Surv Namibia 20:65–74
Pickford M, Senut B, Morales J et al (2008) Fossiliferous Cainozoic carbonates of the northern Sperrgebiet. Mem Geol Surv Namibia 20:25–42
Pierce HG (1993) The nonmarine mollusks of the Late Oligocene-Early Miocene Cabbage Patch fauna of western Montana III Aquatic mollusks and conclusions. J Paleont 67(6):980–993
Pierce HG, Constenius KN (2001) Late Eocene-Oligocene nonmarine mollusks of the northern Kishenehn Basin, Montana and British Columbia. Ann Carnegie Mus 70(1):1–112. https://www.biodiversitylibrary.org/page/52462132
Pierce HG, Constenius KN (2014) Terrestrial and aquatic mollusks of the Eocene Kishenehn Formation, Middle Fork Flathead River, Montana. Ann Carnegie Mus 82(4):305–329. https://doi.org/10.2992/007.082.0401
Ping C (1931) Tertiary and Quaternary non-marine gastropods of North China. Palaeont Sin Ser B 6(6):1–39
Popov SV, Rostovtseva YV, Fillippova NY et al (2016) Paleontology and stratigraphy of the Middle–Upper Miocene of the Taman Peninsula: Part 1. Description of key sections and benthic fossil groups. Paleont J 50(10):1039–1206. https://doi.org/10.1134/S0031030116100014
Popova SM (1964) K poznaniyu paleogenovykh i neogenovykh presnovodnykh mollyuskov Pribaykal'ya i yuga Sovetskogo Dal'nego Vostoka. In: Popova SM, Martinson GG (eds) Stratigrafiya i paleontologiya mezozoyskikh i kaynozoyskikh otlozheniy Vostochnoy Sibiri i Dal'nego Vostoka. Trudy Limnologicheskogo instituta 4(24):151–271. [in Russian]
Popova SM (1981) Kaynozoyskaya kontinental'naya malakofauna yuga Sibiri i sopredel'nykh territoriy. Nauka, Moskva. [in Russian]
Popova SM, Devyatkin YV, Starobogatov YI (1970) Mollyuski kyzylgirskoy svity Gornogo Altaya. Nauka, Moskva. [in Russian]
Prysjazhnjuk V, Kovalenko V, Górka M et al (2006) The fresh-water gastropods from the Sarmatian (Lymnaeidea, Bulinidae, Planorbiidae) of Zwierzyniec (Central Poland). In: Gozhik PF (ed) Problemy paleontologii ta biostratigrafii proterozoyu i fanerozoyu Ukraini. Natsional'naya Akademiya Nauk Ukraini, Kiev, pp 256–264
Pyron M, Brown KM (2015) Introduction to Mollusca and the class Gastropoda. In: Thorp JH, Rogers DC (eds) Thorp and Covich's freshwater invertebrates: ecology and general biology, vol 1. Elsevier, London, pp 383–421
R Core Team (2020) R: A language and environment for statistical computing. Version 4.0.3. R Foundation for Statistical Computing, Vienna. http://www.R-project.org. Accessed 1 November 2020
Rasser MW, Höltke O, Salvador RB (2020) Gastropod palaeohabitats of Miocene Lake Randeck Maar and its hinterland defined by an actualistic genus-level approach. Lethaia 53(2):229–241. https://doi.org/10.1111/let.12353
Rees WJ (1965) The aerial dispersal of Mollusca. J Molluscan Stud 36(5):269–282. https://doi.org/10.1093/oxfordjournals.mollus.a064955
Reis O (1910) Fauna rybnykh slantsev Zabaykal'skoy oblasti. Geologicheskiye issledovaniya i razvedyvatel'nyye raboty po linii Sibirskoy zheleznoy dorogi 29:117–126. [in Russian]
Repelin J (1902) Description des faunes et des gisements du Cénomanien saumâtre ou d'eau douce du Midi de la France. Ann Mus hist nat Marseille Sect Geol 7:1–133. https://www.biodiversitylibrary.org/page/11806897
Rey R (1962a) Remarques sur les faunes malacologiques de l'Éocène inférieur de Provence. C r somm seanc Soc geol Fr 1962a(8):235–236
Rey R (1962b) Sur l'importance paléontologique de faunes malacologiques continentales de très petites dimensions, comme celles de la région de Montpellier. C r somm seanc Soc geol Fr 1962b(10):312–313
Rey R (1965) L'Oligocène et le Miocène inférieur de la Limagne bourbonnaise (Notes de paléontologie malacologique). Rev sci Bourbonnais Cent 1964:56–81

Rey R, Villatte J (1971) Gastéropodes des calcaires lacustres du Thanétien sous-pyrénéen entre les vallées du Douctouyre et de l'Aude. Bull Soc Hist nat Toulouse 107(3):414–422. http://gallica.bnf.fr/ark:/12148/bpt6k6556471 3/f66.item

Robinson WI (1915) Two new fresh-water gastropods from the Mesozoic of Arizona. Am J Sci 4th ser 40(240):649–651. https://doi.org/10.2475/ajs.s4-40.240.649

Rollier L (1910) Troisième supplément à la description géologique de la partie jurassienne de la feuille VII de la carte géologique de la Suisse au 1:100,000. Mat Carte Geol Suisse Nouv ser 25: i–vii:1–230

Roman F (1907) 1re partie – Paléontologie. In: Roman F, Torres A (eds) Le Néogène continental dans la basse vallée du Tage (rive droite). Imprimerie de l'Académie Royale des Sciences, Lisbonne, pp 1–88

Roule L (1886) Nouvelle recherches sur les mollusques du terrain lacustre inférieur de Provence. Ann Malacolog 2:195–228. https://www.biodiversitylibrary.org/page/16301303

Rousseau L (1842) Description des principaux fossiles de la Crimée. In: Demidov A (ed) Voyage dans la Russie méridionale et la Crimée, par la Hongrie, la Valachie et la Moldavie, exécuté en 1837, sous la direction de M. Anatole de Démidoff, par MM. de Sainson, Le Play, Huot, Léveillé, Raffet, Rousseau, de Nordmann et Du Ponceau. Tome 2. Ernest Bourdin et Ce., Paris, pp 781–823

Royo Gómez J (1922) El Mioceno continental ibérico y su fauna malacológica. Com invest paleontol prehist 30:1–230

Russell LS (1926) Mollusca of the Paskapoo Formation in Alberta. Trans R Soc Can 3rd ser Geol Sci 20:207–220

Russell LS (1938) New species of Gastropoda from the Oligocene of Colorado. J Paleontol 12(5): 505–507

Russell LS (1952) Molluscan fauna of the Kishenehn Formation, Southeastern British Columbia. In: Anonymous (Ed.) Annual report of the National Museum of Canada for the fiscal year 1950–51. Natl Mus Can Bull 126:120–141

Russell LS (1955) Additions to the molluscan fauna of the Kishenehn formation, Southeastern British Columbia and adjacent Montana. In: Anonymous (Ed.) Annual report of the National Museum of Canada for the fiscal year 1953–54. Natl Mus Can Bull 136:102–119

Saadi AJ, Davison A, Wade CM (2020) Molecular phylogeny of freshwater snails and limpets (Panpulmonata: Hygrophila). Zool J Linnean Soc 190(2):518–531. https://doi.org/10.1093/zoolinnean/zlz177

Sacco F (1886) Nuove species terziarie di molluschi terrestri e d'aqua dolce e salmastra del Piemonte. Atti Soc Ital Sci Nat 29:427–476. https://www.biodiversitylibrary.org/page/16237730

Salvador RB, Rasser MW (2016) Fossil land and freshwater gastropods from the Middle Miocene of Bechingen and Daugendorf, southwestern Germany. Arch Molluskenkd 145(1):111–124. https://doi.org/10.1127/arch.moll/1869-0963/145/111-124

Salvador RB, Höltke O, Rasser MW et al (2016) Annotated type catalogue of the continental fossil gastropods in the Staatliches Museum für Naturkunde Stuttgart, Germany. Palaeodiversity 9: 15–70. https://doi.org/10.18476/pale.v9.a3

Salvador RB, Tütken T, Tomotani BM et al (2018a) Paleoecological and isotopic analysis of fossil continental mollusks of Sandelzhausen (Miocene, Germany). Palaeontol Z 92:395–409. https://doi.org/10.1007/s12542-017-0400-6

Salvador RB, Cabrera F, Martínez S et al (2018b) Annotated catalogue of the fossil Hygrophila and Eupulmonata (Mollusca: Gastropoda) from South America (cretaceous – Neogene). N Jb Geol Paläont Abh 289(3):249–280. https://doi.org/10.1127/njgpa/2018/0760

Sandberger CLF (1870–1875) Die Land- und Süßwasser-Conchylien der Vorwelt. C. W. Kreidel, Wiesbaden. https://www.biodiversitylibrary.org/bibliography/112296

Schlickum WR (1976) Die in der pleistozänen Gemeindekiesgrube von Zwiefaltendorf a. d. Donau abgelagerte Molluskenfauna der Silvanaschichten. Arch Molluskenkd 107(1–3):1–31

Schlickum WR, Puisségur J-J (1978) Die Molluskenfauna der Schichten mit *Viviparus burgundinus* und *Pyrgula nodotiana* von Montagny-les-Beaune (Dép. Côte-d'Or). Arch Molluskenkd 109(1/3):1–26
Schütt H, Kavuşan G (1984) Mollusken der miozänen Süßwasserablagerungen in der Umgebung von Harmancık bei Kütahya-Bursa in Nordwestanatolien. Arch Molluskenkd 114(4/6):217–229
Seninski K (1905) Novyya Dannyya o neogenovych' plastach' Yugozapadnogo Zakavkaz'ya. Trudy Obshchestva Estestvoispytateley pri Imperatorskom Yuryevskom Universitete 16:1–80. [in Russian]
Serres M (1844) Notice sur les terrains d'eau douce du Bassin émergé de Castelnaudary (Aude). Ann Sci Nat Zool 3rd ser 2:168–190. https://www.biodiversitylibrary.org/page/2272726
Serres M (1853) Note sur les dépôts diluviens, les sables et les marnes teriaires d'eau douce mis à découvert lors des fondations du Palais-de-Justice de Montpellier, où l'on a rencontré des débris de Singes fossiles. Revue Mag Zool 2nd ser 5:446–461. 557–565. https://www.biodiversitylibrary.org/page/13711970
Sinzov I (1875) Opisaniye novykh i maloissledovannykh form rakovin iz tretichnykh obrazovaniy Novorossii. Zapiski Novorossiiskogo obshchestva estestvoispytatelei 3:1–59. [in Russian]
Solem A, Yochelson EL (1979) North American Paleozoic land snails, with a summary of other Paleozoic nonmarine snails. US Geol Surv Prof Paper 1072:1–42. https://pubs.usgs.gov/pp/1072/report.pdf
Sowerby JdC (1826–1829) The mineral conchology of Great Britain; or, Coloured figures and descriptions of those remains of testaceous animals or shells, which have been preserved at various times and depths in the earth. Vol. VI. Privately published by author, London. https://www.biodiversitylibrary.org/page/14386101
Sowerby JC (1840) Organic remains collected by Mr. Malcolmson and described by Mr. James de Carle Sowerby. In: Malcolmson JG. On the fossils of the eastern portion of the Great Basaltic District of India. Trans Geol Soc London Series 2 5(3):550–551. https://www.biodiversitylibrary.org/page/36236116
Sremac J (1981) Neke nove i manje poznate vrste mekušaca Croatica-naslaga i Banatica-naslaga sjeverne Hrvatske. Geol Vjesn 33:107–121
Stache G (1889) Die Liburnische Stufe und deren Grenzhorizonte. Eine Studie über die Schichtenfolgen der cretacisch-eocänen oder protocänen Landbildungsperiode im Bereiche der Küstenländer von Österreich-Ungarn mit einer einleitenden Uebersicht der geologischen Verhältnisse dieses Gebietes. Erste Abtheilung: Geologische Uebersicht und Beschreibung der Faunen und Floren-Reste. Abh k k Geol R-A 13:1–170
Stevanović PM (1951) Pontische Stufe im engeren Sinne – obere Congerienschichten Serbiens und der angrenzenden Gebiete. Srpska Akademija nauka, posebna izdanja 187(Geološki institut 2): 1–361
Stevanović PM (1955) Neue Beiträge zur Kenntnis der kaspibrackischen Fazies des Portaferrian (O.-Pont s. str.) in Serbien. Ann Geol Penins Balk 23:45–65
Stevanović P (1990a) Discussion on the Pontian in the Pannonian Basin of the Western ("Central") Paratethys. In: Stevanović P, Nevesskaya LA, Marinescu F et al (eds) Chronostratigraphie und Neostratotypen. Neogen der Westlichen ("Zentrale") Paratethys, Bd. VIII, Pl1. Pontien. Verlag der Jugoslawischen Akademie der Wissenschaften und Künste und der Serbischen Akademie der Wissenschaften und Künste, Zagreb, Beograd, pp 31–38
Stevanović P (1990b) Pontien südlich von der Sava und Donau in Serbien und Bosnien. In: Stevanović P, Nevesskaya LA, Marinescu F et al (eds) Chronostratigraphie und Neostratotypen. Neogen der Westlichen ("Zentrale") Paratethys, Bd. VIII, Pl1. Pontien. Verlag der Jugoslawischen Akademie der Wissenschaften und Künste und der Serbischen Akademie der Wissenschaften und Künste, Zagreb, Beograd, pp 119–153
Stevanović P (1990c) Pontien nördlich von der Sava und Donau, in Syrmien, Bačka und Banat. In: Stevanović P, Nevesskaya LA, Marinescu F et al (eds) Chronostratigraphie und Neostratotypen. Neogen der Westlichen ("Zentrale") Paratethys, Bd. VIII, Pl1. Pontien. Verlag der

Jugoslawischen Akademie der Wissenschaften und Künste und der Serbischen Akademie der Wissenschaften und Künste, Zagreb, Beograd, pp 195–211
Strausz L (1942) Das Pannon des mittleren Westungarns. Ann hist nat Mus Natl Hung Pars Min Geolg Palaeont 35:1–102
Suzuki K (1949) Development of the fossil non-marine molluscan faunas in Eastern Asia. Japan. J Geol Geogr Trans 21(1–4):91–133
Taktakishvili IG (1962) Novyye dannyye o stratigraficheskom rasprostranenii roda *Valenciennius* Rousseau. Soobshcheniya Akademii nauk Gruzinskoy SSR 29(3):301–306. [in Russian]
Taktakishvili IG (1967) Istoricheskoye razvitiye semeystva valensiyenniid. Metsniyereba, Tbilisi. [in Russian]
Taylor DW (1954) Nonmarine mollusks from Barstow formation of southern California. US Geol Surv Prof Paper 254-C:67–80. https://doi.org/10.3133/pp254C
Taylor DW (1957) Pliocene fresh-water mollusks from Navajo County, Arizona. J Paleont 31(1): 654–661
Taylor DW (1960) Late Cenozoic molluscan faunas from the High Plains. US Geol Surv Prof Paper 337:1–94
Taylor DW (1966) Summary of North American Blancan nonmarine mollusks. Malacologia 4(1): 1–172. https://www.biodiversitylibrary.org/page/13147234
Taylor DW (1981) Freshwater mollusks of California: a distributional checklist. Calif Fish Game 67(3):140–163
Taylor DW (1983) Late Tertiary mollusks from the lower Colorado River valley. Contr Mus Paleont Univ Michigan 26(13):289–298
Taylor DW (1988) Aspects of freshwater mollusc ecological biogeography. Palaeogeogr Palaeoclimatol Palaeoecol 62(1–4):511–576. https://doi.org/10.1016/0031-0182(88)90071-5
Taylor DW, Smith GR (1981) Pliocene molluscs and fishes from northeastern California and northwestern Nevada. Contr Mus Paleont Univ Michigan 25(18):339–413
Thomä C (1845) Fossile Conchylien aus den Tertiärschichten bei Hochheim und Wiesbaden gesammelt und im naturhistorischen Museum zu Wiesbaden ausgestellt. Jb Ver Naturk Nassau 2:125–162
Tozer ET (1956) Uppermost Cretaceous and Paleocene nonmarine molluscan faunas of western Alberta. Geol Surv Can Mem 280:1–125. https://doi.org/10.4095/101507
Van Baak CGC, Mandic O, Lazăr I et al (2015) The Slanicul de Buzau section, a unit stratotype for the Romanian stage of the Dacian Basin (Plio-Pleistocene, Eastern Paratethys). Palaeogeogr Palaeoclimatol Palaeoecol 440:594–613. https://doi.org/10.1016/j.palaeo.2015.09.022
van Benthem Jutting T (1937) Non marine Mollusca from fossil horizons in Java with special reference to the Trinil fauna. Zool Meded 20(11):83–180. https://www.repository.naturalis.nl/document/149951
Van Damme D, Gautier A (2013) Lacustrine mollusc radiations in the Lake Malawi Basin: experiments in a natural laboratory for evolution. Biogeosciences 10:5767–5778. https://doi.org/10.5194/bg-10-5767-2013
van Leeuwen CHA, van der Velde G (2012) Prerequisites for flying snails: external transport potential of aquatic snails by waterbirds. Freshw Sci 31(3):963–972. https://doi.org/10.1899/12-023.1
van Leeuwen CHA, van der Velde G, van Lith B et al (2012) Experimental quantification of long distance dispersal potential of aquatic snails in the gut of migratory birds. PLoS One 7(3): e32292. https://doi.org/10.1371/journal.pone.0032292
van Leeuwen CHA, Huig N, van der Velde G et al (2013) How did this snail get here? Several dispersal vectors inferred for an aquatic invasive species. Freshw Biol 58:88–99. https://doi.org/10.1111/fwb.12041
Vermeij GJ (2017) The limpet form in gastropods: evolution, distribution, and implications for the comparative study of history. Biol J Linn Soc 120(1):22–37. https://doi.org/10.1111/bij.12883
Vinarski MV (2013) One, two, or several? How many lymnaeid genera are there? Ruthenica 23(1): 41–58

Vinarski MV, Frolov PD (2017) A new late Miocene *Lymnaea* with aberrant suture structure unique for the family (Gastropoda: Pulmonata: Lymnaeidae). Hist Biol 29(4):480–487. https://doi.org/10.1080/08912963.2016.1192618
Vinarski MV, Clewing C, Albrecht C (2019) Lymnaeidae Rafinesque, 1815. In: Lydeard C, Cummings KS (eds) Freshwater mollusks of the world. A distribution atlas. Johns Hopkins University Press, Baltimore, MD, pp 158–162
Vinarski MV, Bolotov IN, Aksenova OV et al (2021) Freshwater Mollusca of the circumpolar Arctic: a review on their taxonomy, diversity and biogeography. Hydrobiologia 848:2891–2918. https://doi.org/10.1007/s10750-020-04270-6
Vrsaljko D (1999) The Pannonian palaeoecology and biostratigraphy of molluscs from Kostanjek-Medvednica Mt., Croatia. Geol Croat 52:9–27
Walther AC, Benard MF, Boris LP et al (2008) Attachment of the freshwater limpet *Laevapex fuscus* to the hemelytra of the water bug *Belostoma flumineum*. J Freshw Ecol 23(2):337–339. https://doi.org/10.1080/02705060.2008.9664207
Wenz W (1922) Zur Nomenklatur tertiärer Land- und Süßwassergastropoden. IV. Senckenbergiana 4:5–7. https://www.biodiversitylibrary.org/page/30068315
Wenz W (1923) Fossilium Catalogus I: Animalia. Gastropoda extramarina tertiaria. Vol. IV. W. Junk, Berlin. https://www.biodiversitylibrary.org/page/40593222
Wenz W (1935) Weitere Beiträge zur Land- und Süßwasser-Molluskenfauna der subalpinen Molasse des Pfändergebietes. Senckenbergiana 17:223–225
Wenz W (1936) Die Molluskenfauna der Mergel von Paulhiac (Lot-et-Garonne). Arch Molluskenkd 68(6):228–238
Wenz W (1942) Die Mollusken des Pliozäns der rumänischen Erdöl-Gebiete als Leitversteinerungen für die Aufschluß-Arbeiten. Senckenbergiana 24:1–293
White CA (1876) Invertebrate paleontology of the plateau province, together with notice of a few species from localities beyond its limits in Colorado. In: Powell JW (ed) Report on the geology of the eastern portion of the Uinta Mountains and a region of country adjacent thereto. With atlas. U. S. Geological and Geographic Survey of the Territories, Washington, pp 74–135. https://www.biodiversitylibrary.org/page/19391191
White CA (1880) Descriptions of new invertebrate fossils from the Mesozoic and Cenozoic rocks of Arkansas, Wyoming, Colorado, and Utah. Proc US Nat Mus 3:157–162. https://www.biodiversitylibrary.org/page/14917257
White CA (1886) On the fresh-water invertebrates of the North American Jurassic. US Geol Surv Bull 29:691–725. https://doi.org/10.3133/b29
Whiteaves JF (1885) Report on the Invertebrata of the Laramie and Cretaceous rocks of the vicinity of the Bow and Belly Rivers and adjacent localities in the North-west Territory. Contrib Can Palaeont 1(1):1–89. https://www.biodiversitylibrary.org/page/38502645
Willis RP (1967) Geology of the Arabian peninsula. Bahrain. US Geol Surv Prof Paper 560-E:1–54. https://pubs.usgs.gov/pp/0560e/report.pdf
Willmann R (1981) Evolution, Systematik und stratigraphische Bedeutung der neogenen Süßwassergastropoden von Rhodos und Kos/Ägäis. Palaeontogr Abt A 174:10–235
Willmann R (1982) Biostratigraphisch wichtige Süßwassergastropoden (Prosobranchia, Hydrobiidae) aus dem Neogen des Ägäis-Raumes. N Jb Geol Paläont Abh 162(3):304–331
Witton MP, Habib MB (2010) On the size and flight diversity of giant pterosaurs, the use of birds as pterosaur analogues and comments on pterosaur flightlessness. PLoS One 5(11):e13982. https://doi.org/10.1371/journal.pone.0013982
Yen TC (1944) Notes on fresh-water mollusks of Idaho formation at Hammett, Idaho. J Paleont 18(1):101–108
Yen TC (1946a) Late Tertiary fresh-water mollusks from southeastern Idaho. J Paleontol 20(5): 485–494
Yen TC (1946b) Eocene nonmarine gastropods from Hot Spring County, Wyoming. J Paleont 20(5):495–500

Yen TC (1946c) On Lower Cretaceous fresh-water mollusks of Sage Creek, Wyoming. Not Nat Acad Nat Sci 166:1–13
Yen TC (1947) Pliocene fresh-water mollusks from northern Utah. J Paleontol 21(3):268–277
Yen TC (1948a) Eocene fresh-water Mollusca from Wyoming. J Paleontol 22(5):634–640
Yen TC (1948b) Paleocene fresh-water mollusks from southern Montana. US Geol Surv Prof Paper 214-C:35–50
Yen TC (1951) Fresh-water mollusks of Cretaceous age from Montana and Wyoming. Part 1: a fluviatile fauna from the Kootenai formation near Harlowton, Montana. US Geol Surv Prof Paper 233-A:1–9. https://pubs.usgs.gov/pp/0233a/report.pdf
Yen TC (1952) Molluscan fauna of the Morrison formation. US Geol Surv Prof Paper 233-B:21–51. https://doi.org/10.3133/pp233B
Yen TC (1954) Nonmarine mollusks of Late Cretaceous age from Wyoming, Utah and Colorado. Part 1: a fauna from western Wyoming. US Geol Surv Prof Paper 254-B:45–59. https://pubs.usgs.gov/pp/0254b/report.pdf
Yen JTC (1969) Fossile nicht-marine Mollusken-Faunen aus Nordchina. Sitzber Österr Akad Wiss 177(1/3):21–64. http://www.zobodat.at/stable/pdf/SBAWW_177_0021-0064.pdf
Yen TC, Reeside JB Jr (1946) Fresh-water mollusks from the Morrison formation (Jurassic) of Sublette County, Wyoming. J Paleont 20(1):52–58
Yü W (1977) Cretaceous and early Tertiary non-marine gastropods from South China with their stratigraphical significance. Acta Palaeontol Sin 16(2):191–213
Yü W (1982) Some fossil gastropods from Xizang. In: Anonymous (ed) The series of the comprehensive scientific expedition to the Qinghai-Xizang plateau, Palaeontology of Xizang (Tibetan) plateau, vol 4. Science Press, Beijing, pp 255–281
Yü XH (1987) Late Jurassic and Early Cretaceous fresh water gastropods (Mollusca) from western Liaoning Province, China. In: Yu X, Wang W, Liu X et al (eds) Mesozoic stratigraphy and palaeontology of Western Liaoning, vol 3. Geological Publishing House, Beijing, pp 29–116
Yü W, Pan HZ (1980) Mesozoic non-marine gastropods from Zhejiang and South Anhui. In: Anonymous (ed) Divisions and correlation of the Mesozoic volcano-sedimentary rocks in Zhejiang and Anhui provinces, China. Academia Sinica, Nanking Institute of Geology and Paleontology, Nanking, pp 135–172
Yü W, Pan HZ (1982) Eocene non-marine Gastropoda from Zhuo Xian, Hebei. Bull Nanjing Inst Geol Palaeont 4:189–212
Yü W, Zhang X-Q (1982) Late Cretaceous and early Tertiary non-marine gastropods from Sanshui Basin, Guangdong. Mem Nanjing Inst Geol Palaeont Acad Sin 17:37–84
Yu T, Neubauer TA, Jochum A (2021) First freshwater gastropod preserved in amber suggests long-distance dispersal during the Cretaceous. Geol Mag 158(7):1327–1334. https://doi.org/10.1017/S0016756821000285
Zhang SX (2009) Geological formation names of China (1866–2000), vol 1. Springer, Dordrecht
Zhu GX (1980) Mollusca. In: Shenyang Institute of Geology and Mineral Resources (ed) Paleontological Atlas of Northeast China II. Mesozoic and Cenozoic volume. Geological Publishing House, Beijing, pp 8–59
Zhu XG (1985) Late Cenozoic non-marine gastropods from Xili Basin in Gansu. Acta Palaeontol Sin 24(6):672–679

Chapter 7
Biogeography of the Living Lymnaeidae

Maxim V. Vinarski, Olga V. Aksenova, Ivan N. Bolotov, Antonio A. Vázquez, Pilar Alda, Jean-Pierre Pointier, and Sylvie Hurtrez-Boussès

Abstract The distribution of the recent Lymnaeidae across continents and zoogeographical regions of the world is reviewed, with a brief characteristic of the geographical distribution of each extant lymnaeid genus. The quantitative estimates of the similarity between the lymnaeid faunas of different continents and the largest zoogeographical regions are provided and discussed. It is shown that the highest taxonomic level of endemism in the family is observed in the Nearctic region, and the overall faunal similarity is highest among the Palaearctic and Oriental region,

M. V. Vinarski (✉)
Laboratory of Macroecology & Biogeography of Invertebrates, Saint-Petersburg State University, Saint-Petersburg, Russia

S.I. Vavilov Institute for the History of Science and Technology, St. Petersburg Branch, Russian Academy of Sciences, Saint-Petersburg, Russia

O. V. Aksenova · I. N. Bolotov
N. Laverov Federal Center for Integrated Arctic Research, The Ural Branch of the Russian Academy of Sciences, Arkhangelsk, Russia

A. A. Vázquez
Laboratory of Malacology, Institute of Tropical Medicine "Pedro Kourí,", Havana, Cuba

MIVEGEC, University of Montpellier, CNRS, IRD, Montpellier, France

P. Alda
Genética y Ecología Evolutiva (CERZOS–CCT–CONICET Bahía Blanca), Bahía Blanca, Argentina

MIVEGEC, University of Montpellier, CNRS, IRD, Montpellier, France

J.-P. Pointier
PSL Research University, USR 3278 CNRS–EPHE, CRIOBE Université de Perpignan, Perpignan, France
e-mail: pointier@univ-perp.fr

S. Hurtrez-Boussès
MIVEGEC, University of Montpellier, CNRS, IRD, Montpellier, France

Département de Biologie-Ècologie, Faculté des Sciences, Université de Montpellier, Montpellier, France
e-mail: sylvie.hurtrez@ird.fr

M. V. Vinarski, A. A. Vázquez (eds.), *The Lymnaeidae*, Zoological Monographs 7,
https://doi.org/10.1007/978-3-031-30292-3_7

whereas the Australian fauna is the most peculiar. A special section of the article is devoted to a review of the invasive lymnaeid species and the possible consequences of their invasions to the non-native areas.

7.1 Introductory Remarks

A biogeographical characterization of a family of animals or plants can be provided following two different, but not mutually exclusive, approaches. First, one can compare lists of species and genera of various large regions of the world (continents, countries, zoogeographical domains, or provinces) and thereby not only to obtain a picture of the global distribution of the studied group but also potentially reveal some interesting distribution patterns. The second approach is to focus on particular taxa (genus, subgenus) instead of territories, depicting the distribution of all genera within a family one after another. In the text below, the two approaches are followed. However, some preliminary remarks are needed before one can proceed to them.

Meier-Brook once stated that the family Lymnaeidae "is difficult to discuss" in the biogeographic context (Meier-Brook 1984, p. 35) since this taxon needs a revision and the status of many nominal species and genera of this family is uncertain. Today, almost 40 years later, the situation is essentially the same. Only a portion of nominal species of the family, accepted as valid in the most authoritative databases (MolluscaBase, IUCN Red List), has been studied molecularly, and their species status still lacks genetic support. Thus, the available species lists include taxa of two categories: *euspecies* (whose validity is supported morphologically and/or genetically) and *morphospecies*. The latter category embraces nominal species delineated based on their morphological (conchological, anatomical) distinctness, but the genetic data able to clarify their species status are unavailable. The current integrative research in the lymnaeid taxonomy has revealed that, if in some instances, the morphospecies appear to be *euspecies* and thus are resurrected from the limbo of synonyms (see, for example, Vinarski et al. 2016a; Aksenova et al. 2018; Mahulu et al. 2019), in other cases the nominal species described in the nineteenth–first half of the twentieth century do not constitute real taxa (Vinarski et al. 2016b, 2021a; Aksenova et al. 2017; Schniebs et al. 2019). Combining the two categories into a single one (species sensu lato) generates some degree of taxonomic uncertainty, affecting all biogeographical inferences made on the basis of the analyses of species lists. Though, it would be nearly impossible to estimate the extent of this uncertainty, we have to acknowledge this problem here.

The third category, *candidate species* (or *putative species*), are those that have been delineated using the molecular techniques but not described formally and lacking the Latin binomial names. Though, there are all chances that such entities will once be formalized taxonomically. Currently, they form a sort of "dark matter" of taxonomy and nomenclature (Minelli 2019) and their use in a biogeographical analysis is problematic. For instance, Wiese et al. (2020) delineated four candidate species of the lymnaeid genus *Radix* inhabiting Lake Lugu in China, but none of them was described under the guidelines of the International Code of Zoological Nomenclature. All the four species remain unnamed and thus cannot be indexed by either MolluscaBase or IUCN Red List.

Wallace (1894, p. 613) gave a classical definition of the zoogeographical regions which are "primary divisions of the earth's surface of approximately continental extent, which are characterized by distinct assemblages of animal types". In the Anthropocene epoch, the distinctness of such assemblages is becoming blur due to the homogenization of biota caused by human-mediated extinctions, biological invasions, intentional and unintentional introductions of non-indigenous species (including freshwater molluscs) to new areas (Bernardo-Madrid et al. 2019). The progress of taxonomy is also affecting the zoogeographical region delineation and the form and strength of macroecological patterns revealed during analyses of species distributions (Vinarski and Kramarenko 2015; Vavalidis et al. 2019).

To conclude, any biogeographical hypothesis, reconstruction, or regionalization scheme is a product of the time when it was created and as such is of limited significance, being destined to be greatly altered (if not totally rejected) with time (e.g., Cox 2001). Therefore, we avoid here an overview of biogeographical hypotheses on the Lymnaeidae given by previous authors (Fischer 1880–1887; Baker 1911; Roszkowski 1928; Hubendick 1951; Inaba 1969; Starobogatov 1970; Russell-Hunter 1978; Meier-Brook 1984; Banarescu 1990, 1992; Kruglov and Starobogatov 1993a; Kruglov 2005). We acknowledge, however, that the picture which is given in this chapter also reflects the current taxonomy of the family. We realize that it will change in the near future, following the novelties in taxonomy, but, hopefully, the achieved state of knowledge on the lymnaeid taxonomy allows us to reveal some real patterns, not depending on taxonomic uncertainty. The discussion below is restricted to the extant lymnaeid taxa only; Neubauer (2023) considers some questions of the historical biogeography of this family. Starobogatov (1970) and Vinarski et al. (2019) published brief overviews of the global distribution of the family.

Given the considerations above, some conventions were adopted before carrying out zoogeographical analyses and calculations.

1. The lists of lymnaeid species found in different regions are compiled following MolluscaBase (2021). As of 01.11.2021, it lists 161 recent species of Lymnaeidae flagged as "accepted" though not all of them have molecular support. We do not discuss the taxonomic validity of these species and use all of them conventionally as having the same status.
2. The occurrences of the invasive lymnaeid snails outside their native ranges are not considered. In other words, we try to reveal the "natural" patterns of the lymnaeid distribution as they were before the Anthropocene epoch.
3. For the sake of comparability, the scheme of delineation of the world's main zoogeographical regions proposed by Strong et al. (2008, Fig. 7.3) is followed here. This scheme, however, is not the standard one and neither the single available; in some details it disagrees with other authors' schemes (see below).

7.2 Distribution of the Lymnaeid Snails Across Continents and Zoogeographical Regions

The primary data for the analysis were collected from published sources and—to a much lesser extent—from our unpublished observations made in the field. We tried to use the most comprehensive and taxonomically sound sources of data; the taxonomy follows that of MolluscaBase (2021). The main sources of information for compiling the species lists were as follows:

- For Europe, Northern Asia, and Northern Africa: Kruglov and Starobogatov (1993a, b); Brown (1994); Glöer (2002, 2019); Welter-Schultes (2012); Vinarski and Kantor (2016); Aksenova et al. (2018); Mabrouki et al. (2020); Vinarski et al. (2020).
- For Central, South, and East Asia: Likharev and Starobogatov (1967); Brandt (1974); Liu et al. (1979); Subba Rao (1989); Neubert (1998); Yildirim et al. (2006); Nesemann et al. (2007); Glöer and Pešić (2012); Glöer and Bößneck (2013); Vinarski et al. (2017).
- For North America: Clarke (1981); Burch (1989); Johnson et al. (2013).
- For Central and South America: Thompson (2011); Pointier (2015); Pointer and Vázquez (2020).
- For Afrotropical region: Van Damme (1984); Brown (1994); Appleton and Miranda (2015); Mahulu et al. (2019).
- For Australasia and Pacific Islands: Hubendick (1952); Climo and Pullan (1972); Starmühlner (1976); Ponder and Waterhouse (1997), Ponder et al. (2020); Christensen et al. (2021).

The possible approaches to choosing the primary geographical units for the analysis are numerous. In some cases, it is more appropriate to use the units delineated on the basis of non-biological factors–continents, drainage basins separated by watersheds, geopolitical and administrative regions (countries, states, provinces). Another approach is to use the specific biogeographical units of various ranks, like realms, domains, and provinces. Such commonly used terms as Palaearctic, Nearctic, Neotropica, and others refer to the highest-rank biogeographical units. At least two attempts of the global biogeographical regionalization of the freshwaters have been undertaken during the last 50 years – these of Starobogatov in the USSR and Banarescu in Romania (Starobogatov 1970, 1986; Banarescu 1992; Kruglov and Starobogatov 1993a), one of them (Starobogatov's) being based exclusively on the distribution of Mollusca. The alternative solution is to divide the Earth's surface into a large number of ecoregions, which have the same rank and do not form a hierarchical system (e.g., Laske et al. 2019). Abell et al. (2008, p. 404) define an ecoregion as "a large area encompassing one or more freshwater systems with a distinct assemblage of natural freshwater communities and species." In the global regionalization scheme proposed by Abell et al. (2008), 830 freshwater ecoregions have been delineated. At last, a conventionally chosen latitudinal, longitudinal, or altitudinal span (for example, whole area lying between 10 and 20, 20 and

30°) may be used as the unit for the primary analysis. This approach is especially common in studies focused on the revealing of macroecological gradients, like the global species richness gradient.

In this section of the article, we follow the first two approaches and compare the genus and species diversity of the Lymnaeidae among continents and the seven highest-rank zoogeographical regions of the world (sensu Strong et al. 2008). These seven regions are Afrotropical (AT), Australasian (AU), Nearctic (NA), Neotropical (NT), Oriental (OL), Pacific Islands (PAC), and Palaearctic (PA). There is an eighth region, the Antarctic, but its modern fauna is totally devoid of the Lymnaeidae and will not be considered further. Due to the current state of our knowledge on the distribution of the family, we were unable to compile reliable species lists for all zoogeographical provinces delineated by Starobogatov (1970, 1986) and even less for all freshwater ecoregions delimited by Abell et al. (2008). There are more than a few regions of the world whose lymnaeid fauna is studied incompletely (and in some cases is not studied at all).

For a quantitative comparison between the faunal lists, Ochiai's coefficient of similarity (Bolton 1991) was used. According to Malyshev (2001), among the plethora of similarity indices proposed to date, Ochiai's coefficient belongs to the most informative (especially in those cases when two faunal lists of drastically different lengths are compared). Earlier, it was applied by Vinarski and Kantor (2016) and Vinarski et al. (2021b).

It is calculated as:

$$I_{OCH} = \frac{a}{\sqrt{(a+b)(a+c)}}, \qquad \text{where}$$

a–the number of species common between two compared lists; b and c–the number of unique species for the first and second list, respectively.

A value of 0 of this coefficient corresponds to two completely non-overlapping species lists; a value of 1 means that the two compared faunas have the same species content. The calculations of Ochiai's coefficient and subsequent cluster analysis were performed using PAST 2.00 software (Hammer et al. 2001).

The map (Fig. 7.1) visualizes the differences in lymnaeid species richness among continents and countries. The global distribution of the family departs considerably from the universally accepted concept of the latitudinal gradient of species richness (Willig et al. 2003; Hillebrand 2004)—the "Holy Grail" of modern ecology and biogeography (Adams 2009). The most species-rich assemblages of Lymnaeidae are found in the temperate latitudes of the Northern Hemisphere, with North America and North Eurasia maintaining the highest number of the pond snail species. The equatorial zone malacofauna, especially in Africa and South America, is impoverished in respect of Lymnaeidae. In sub-Saharan Africa, only two native species of the family are known (Brown 1994; Mahulu et al. 2019), though North Africa, belonging to the Palaearctic, has at least five species (van Damme 1984; Brown 1994; Lotfy and Lotfy 2015). The lymnaeid generic richness in tropical regions is also significantly decreased (Table 7.1).

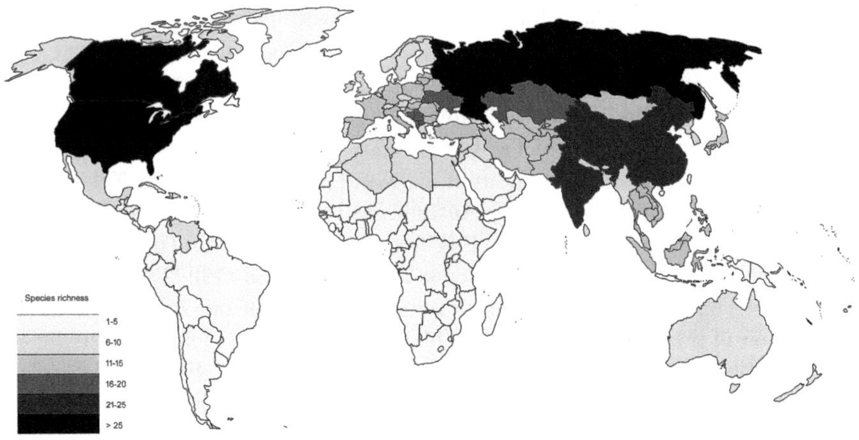

Fig. 7.1 Distribution of the lymnaeid species diversity across continents and countries. See text for the sources of primary data. Invasive species are not counted

Table 7.1 Genus and species richness of the Lymnaeidae arranged by main zoogeographical region, with the percentage of endemism

Region	Number of genera		Number of species	
	Total[a]	% of endemic	Total[b]	% of endemic
Afrotropical (AT)	3(1)	33.3	3 (2)	66.7
Australasian (AU)	2(1)	50.0	9 (9)	100.0
Nearctic (NA)	12 (8)	66.7	61 (54)	84.4
Neotropical (NT)	3 (2)	66.7	10 (6)	60.0
Oriental (OR)	6 (0)	0	19 (11)	57.9
Pacific Islands (PAC)	3(2)	66.7	5 (4)	80.0
Palearctic (PA)	13 (7)	53.9	69 (58)	84.1

[a]In brackets—number of endemic genera;
[b]In brackets—number of endemic species

The high level of genus- and species-level endemism in the seven zoogeographical regions shows that the lymnaeids rarely cross the boundaries between continents, and an "average" lymnaeid species is confined to a single region. (This rule was broken during the historical epoch when some species of snails started to "travel" with humans' help). The highest taxonomic level of endemism is observed in the Nearctic, with several genera and a single subfamily (Lancinae) being unique to this region. This fact, alongside a high species richness, allowed some authors to hypothesize that North America is the "center of origin" of this family (Meier-Brook 1984). However, the oldest known fossil member of the Lymnaeidae, "*Lymnaea*" *hopii* (Robinson, 1915), which originates from the Early Jurassic Morrison Formation of Arizona, was described so insufficiently (Robinson 1915) that its placement in the Lymnaeidae raises some doubts (Neubauer, 2023). The findings of the fossil lymnaeids of the Late Jurassic age come from different continents, then constituted parts of Laurasia, and the "motherland" of the family remains unknown (Neubauer, 2023). It may well be North America. Starobogatov (1970) considered North America (with Europe, which was then a part of the latter) and Southeast Asia as two main centers of evolutionary diversification of this family. This corresponds well with the current distribution of the lymnaeid species richness, which is more concentrated in the Northern Hemisphere (see Fig. 7.1). Thus, the Lymnaeidae could be considered a group of Laurasian origin.

However, a cautionary note needs to be placed here. The high level of both species richness and endemism in North America (and, to a lesser extent, in Northern Eurasia) arises partially due to the unresolved taxonomy of their lymnaeid snails. For example, Johnson et al. (2013) list 21 species of the genus *Galba* and 25 species of "*Stagnicola*" (= *Ladislavella* of the current systematics) in the fauna of Canada and the USA. The genetic data are not available for the vast majority of these nominal species. On the other hand, the existing molecular genetic information indicates that even the three species of *Ladislavella*, most common and widespread in North America (i.e., *Ladislavella catascopia*, *L. elodes*, and *L. emarginata*), may be synonyms (Remigio and Blair 1997; Bargues et al. 2003). The same is also applicable to some nominal Nearctic species of *Galba* (Alda et al. 2021). In our opinion, the current estimates of the lymnaeid species richness in North America (Burch 1989; Johnson et al. 2013) may be greatly exaggerated, and an integrative revision is required to bring order into this taxonomic mess. Thus, the status of North America as a global hotspot of lymnaeid diversity may be reconsidered in the future.

The most peculiar, albeit not rich, lymnaeid fauna is that of Australia, where all six native lymnaeids are endemic. The fauna of the Pacific islands, with the exclusion of Hawaii, is extremely poor in lymnaeids, with probably one native representative of this group (*Orientogalba viridis* described from Guam Island). The Hawaiian Islands, on the other hand, have a small group of endemic genera (*Erinna*, *Pseudisidora*), which includes the only lymnaeid taxon—subgenus *Pseudisidora* (*Pseudobulinus*)—with obligately sinistral shell (Hubendick 1952; Kruglov and Starobogatov 1993a, b; Christensen et al. 2021). The oceanic islands of the Atlantic and Indian oceans harbor their own endemic species (*Radix mauritiana* of Mauritius) and even genera (*Limnobulla* of the Falkland Islands and *Lantzia* of the Reunion Island).

The Palaearctic, being the greatest of the zoogeographical regions, has, predictably, the richest fauna of the Lymnaeidae. Of course, not all regions of the Palaearctic maintain species-rich faunas. Thus, that part of the Arctic zone which belongs to the Palaearctic is poor in lymnaeids as well as in other families of freshwater snails and bivalves (Vinarski et al. 2021b). An analysis carried out at the subcontinental scale reveals the latitudinal gradient of species richness that can be traced between 50 and 72° N (Fig. 7.2). Remarkably, the form of the relationship between the number of lymnaeid species and latitude is not linear, the highest species richness is observed in the temperate regions of the Northern Hemisphere lying between 40 and 60° N, whereas the southern Palearctic.

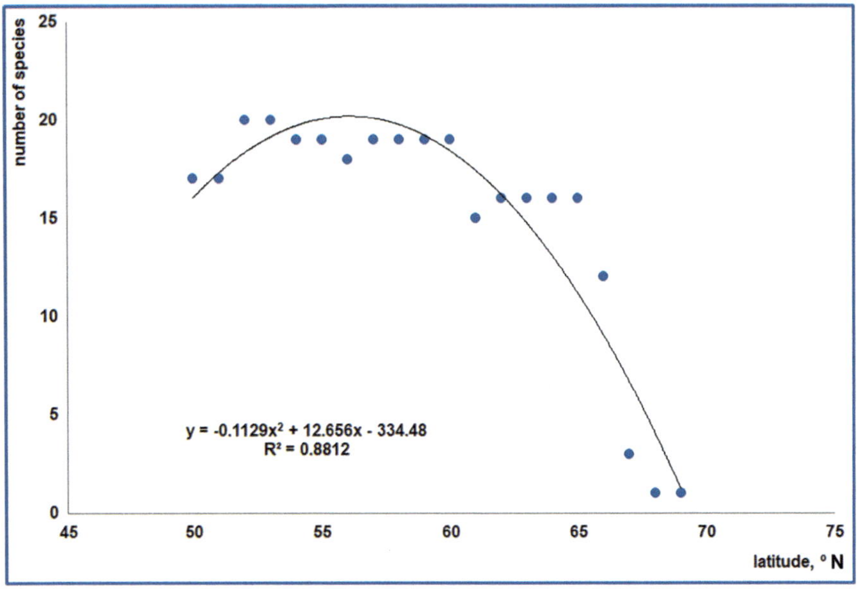

Fig. 7.2 Latitudinal changes in the lymnaeid species richness in Western Siberia (Asiatic Russia) between 50 and 70 degrees of northern latitude. Regression equations and approximating power indices (R^2) are shown. After Vinarski (unpublished data)

Such a non-linear relationship is characteristic not only for the Lymnaeidae but also for the fauna of freshwater Gastropoda as a whole (Hubendick 1962; Starobogatov 1970; Vinarski et al. 2012a; Vinarski and Kramarenko 2015).

The overall (Ochiai-based) similarity between the lymnaeid faunas is highest among the Palaearctic and Oriental region, whereas the Australian fauna is the most peculiar (Fig. 7.3; Table 7.2). The lymnaeid faunas from the Northern Hemisphere continents are, in general, more similar to each other than to the faunas of the southern continents. Possibly, it reflects the geological history of the family, which evolved in Laurasia and only subsequently colonized Gondwana.

The faunal similarity between North America and Eurasia is relatively low in spite of the existence of several land bridges between the two mainlands in the geological past (Brikiatis 2014). Thus, the concept of the Holarctic should be rejected, at least if the family Lymnaeidae is chosen as a taxon on which the regionalization is based (see also Vinarski et al. 2021b).

The size of ranges of different lymnaeid genera is very unequal (Figs. 7.4 and 7.5). Some genera (*Galba, Lymnaea, Radix*) are distributed over more than one continent, while some others (*Erinna*, *Lantzia*, *Pseudisidora*, *Tibetoradix*) have very restricted ranges, being endemic to particular archipelagoes or mountain countries. The ranges of such taxa as *Ampullaceana*, *Myxas*, *Peregriana* can be classified as "subcontinental" (see below). The same is characteristic for particular species. One of the most broadly distributed lymnaeid snails of the world fauna is the ear pond snail, *Radix auricularia* (Linnaeus, 1758), whose native range covers almost the

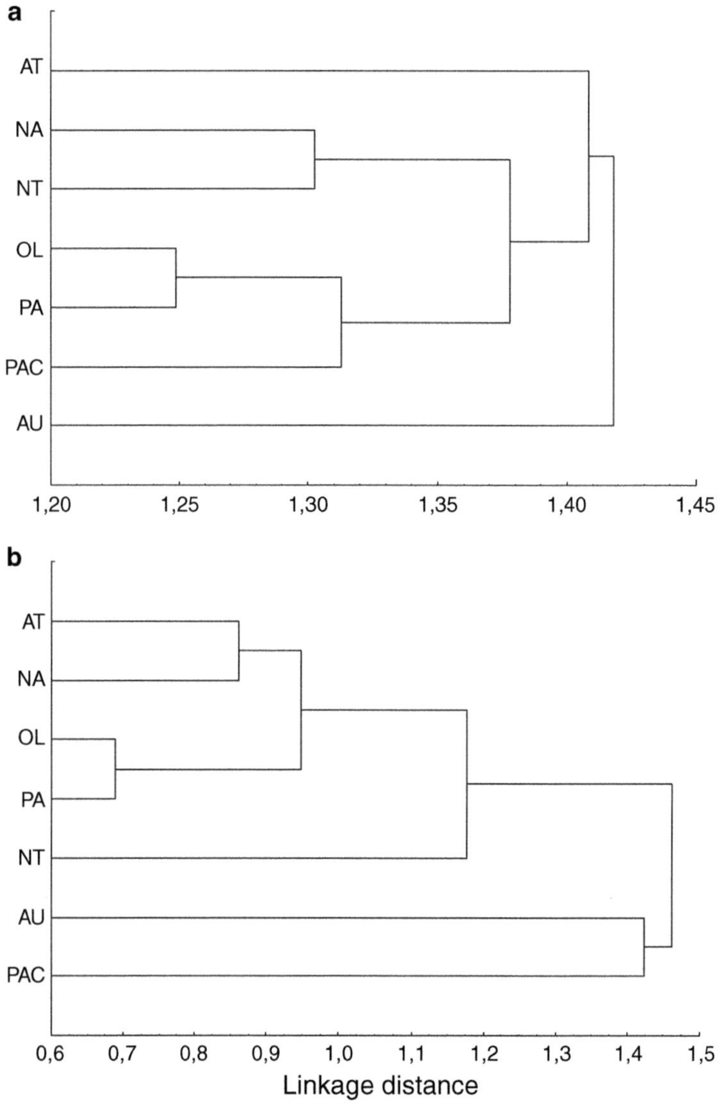

Fig. 7.3 Dendrograms of the UPGMA based on the lymnaeid species (**a**) and genera similarity (**b**) between the zoographical regions. For the primary data, see Table 7.2

entire Eurasia, northern Africa, and the extreme north-west of North America (Hubendick 1951; Kruglov 2005; Vinarski et al. 2020). Such narrow-ranged species as *Ampullaceana relicta* (Poliński, 1929), restricted to the large lakes of the Balkan peninsula, or *Radix onychia* (Westerlund, 1883), found only in the Biwa Lake in Japan, give examples of endemic species with very local distribution.

Table 7.2 The faunal similarity between the seven zoogeographical regions (based on Ochiai's coefficient of similarity)[a]

	AT	AU	NA	NT	OL	PA	PAC
AT	**1.00**	0.00	0.41	0.33	0.47	0.39	0.00
AU	0.00	**1.00**	0.00	0.00	0.29	0.00	0.00
NA	0.00	0.00	**1.00**	0.17	0.35	0.32	0.00
NT	0.00	0.00	0.08	**1.00**	0.24	0.16	0.00
OL	0.00	0.00	0.03	0.07	**1.00**	0.57	0.24
PA	0.06	0.00	0.09	0.04	0.12	**1.00**	0.16
PAC	0.00	0.00	0.00	0.00	0.10	0.05	**1.00**

[a] The values above the diagonal are based on lists of genera, below the diagonal—on species lists. Abbreviations as in Table 7.1

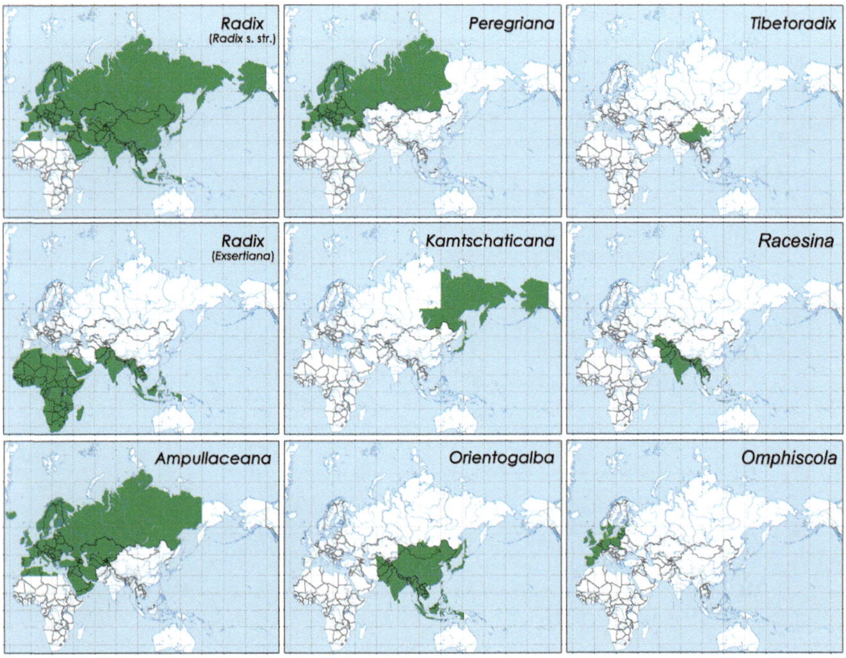

Fig. 7.4 The native ranges of selected genera and subgenera of the Lymnaeidae. After Aksenova et al. (2018), corrected and modified

Some representatives of the family appeared to be extremely efficient anthropogenic invaders, muddling our knowledge of species distribution (Alda et al. 2021). Upon our estimate, not less than 12 lymnaeid species belonging to different genera and subfamilies can be placed under this category (Table 7.3), though the degree of their invasive success is variable. Such species as *Galba truncatula*, *G. cubensis*, *Radix auricularia*, and, especially, *Pseudosuccinea columella* are able to cross the oceans and to gain a foothold in areas very remote from their native ranges (Lounnas

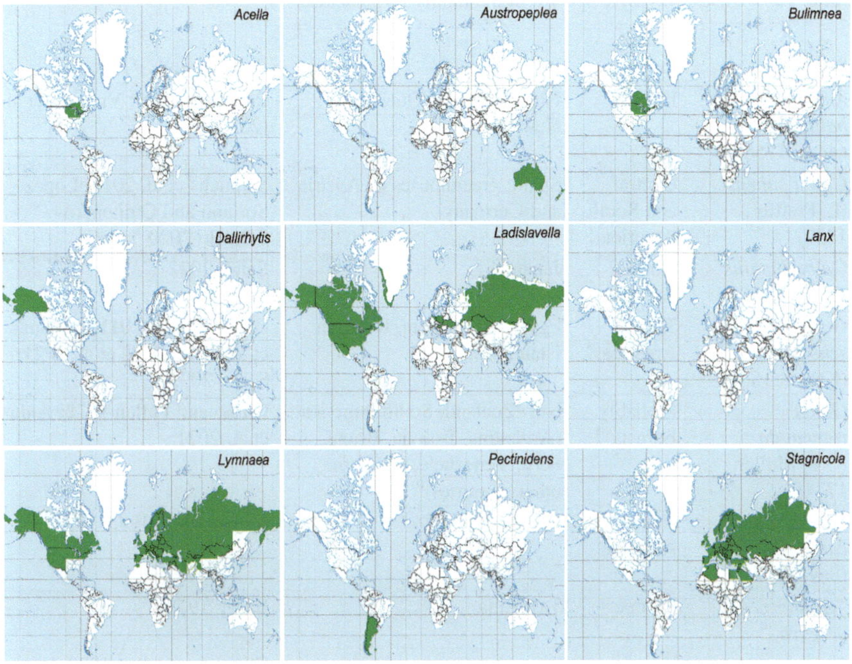

Fig. 7.5 The native ranges of selected genera and subgenera of the Lymnaeidae. (Continuation)

et al. 2017; Schniebs et al. 2018; Alda et al. 2021; Saito 2022; Vinarski et al. 2022a). Evidently, these snails used humans' transportation vehicles as the mean of their dispersal. For example, the dwarf pond snail (*G. truncatula*) managed to colonize such countries as New Zealand, Venezuela, Brazil, Argentina, and some others (see Table 7.3). The current invasive range of *P. columella* may be considered almost pantropical. This snail is also common in aquaria and greenhouses of Europe. On the other hand, cases of "failed" invasions are known. For instance, the Nearctic species *Ladislavella catascopium* was discovered in 1929 in Leith, Scotland, in a warm engine pond at 27 ° C. The population of this species remained viable at least until 1940 when it was studied by Kevan (1943). The author believed that the snails were "introduced to the timber yard at Leith with logs of birch and elm imported from Eastern Canada, for these on arrival are immediately floated in the engine pond" (Kevan 1943, p. 431). The later authors writing on the UK malacofauna (Kerney 1976; Anderson 2005) did not mention this species (except Macan 1949); it is absent from the recent pan-European checklists (Falkner et al. 2001) and atlases (Welter-Schultes 2012; Glöer 2019). Upon Kevan's information, the pond in Leith was "repopulated annually after cleaning by the emergence of snails from the overflow pipes" (Kevan 1943, p. 459). It means that, besides the high water temperature in the pond, this population was highly dependent on human activity and, probably, could not sustain without it. One may hypothesize that once the water regime in this habitat

Table 7.3 A list of invasive species in the family Lymnaeidae

Species	Native range	Invasive range	Reference(s)
Ampullaceana balthica (Linnaeus, 1758)	Europe	Israel; Canada (Quebec, Newfoundland)	Roll et al. 2009; Vinarski et al. (2022a)
Galba cubensis (L. Pfeiffer, 1839)	Central and South America	Spain; greenhouses of Austria and Germany	Schniebs et al. 2018; López-Soriano and Quiñonero-Salgado 2020
Galba humilis (Say, 1822)	North America	Japan	Saito, 2022
Galba schirazensis (Küster, 1862)	The Americas[a]	Egypt, Japan, Iran, Reunion Island, Spain[a]	Bargues et al. 2011; Lounnas et al. 2018; Alda et al. 2021; Saito, 2022
Galba truncatula (O.F. Müller, 1774)	Probably Eurasia)[b]	New Zealand, South America (Argentina, Bolivia, Brazil, Peru, Venezuela, and some other countries)[b]	Climo and Pullan 1972; Alda et al. 2021
Ladislavella catascopium (Say, 1817)	North America	UK (Scotland)[c]	Kevan 1941, 1943
"Lymnaea peregra"[d]	Europe	Tasmania	Boray 1978
Lymnaea stagnalis (Linnaeus, 1758)	Eurasia	Australia, Cameroon, New Zealand	Climo and Pullan 1972; Boray 1978; Ponder et al. 2020
Orientogalba viridis (Qiuy et Gaimard, 1833)	South and East Asia	Australia; Israel; Spain	Boray 1978; Roll et al. 2009; Schniebs et al. 2017; López-Soriano and Quiñonero-Salgado 2020; Ponder et al. 2020
Pseudosuccinea columella (Say, 1817)	North America	In tropical and subtropical countries of all continents; Europe (chiefly in greenhouses)	Pointier et al. 2007; Lounnas et al. 2017; Glöer 2019
Radix auricularia (Linnaeus, 1758)	Palearctic	Israel; North America (patchily distributed through much of the USA; reported from Mexico)	Dundee 1974; Burch 1989; Naranjo-García and Castillo-Rodríguez 2017; Roll et al. 2009
R. rubiginosa (Michelin, 1831)	Southeast Asia	Australia; Israel; South Africa; in greenhouses and tropical aquaria: Austria,[e] England; Ireland; Russia	Boray 1978; Anderson 2005; Leiss and Reischütz 1996; Roll et al. 2009; Appleton and Miranda 2015; Ponder et al. 2020

[a]The place of origin of this species remains unknown. Bargues et al. (2011) argued it is an Old World species recently invaded the Americas, whereas Lounnas et al. (2018, p. 434) are inclined to think that *G. schirazensis* "has an American origin with recent colonization of the Old World by a genetically uniform strain related to populations from Venezuela and Colombia";
[b]Correa et al. (2010) hypothesized, however, that *G. truncatula* belongs to a branch of a clade of American origin and invaded the Old World in the eighteenth century. Alda et al. (2021) considered it as a species alien to South America;
[c]A failed invasion; the invasive population went extinct presumably after WWII;
[d]The taxonomic identity of this record remains unclear. Probably, it refers to *Ampullaceana balthica*;
[e]Referred to as *Radix javanica* (see Leiss and Reischütz 1996)

was changed drastically, it caused the disappearance of *L. catascopium* from the pond.

Apart from muddling the current biogeographical regionalization schemes, the invasions of the alien lymnaeids have a significant practical aspect. Some of these species are vectors of trematodes (like *Fasciola hepatica* and *F. gigantica*), and their arrival to new regions may lead to the emergence of the foci of fascioliasis there. This process has been especially well studied in South America (Bargues et al. 2011; Pointier 2015; Lounnas et al. 2018; Vázquez et al. 2018; Alda et al. 2021).

7.3 Biogeography of Particular Subfamilies and Genera of the Lymnaeidae

In this section, we present a concise overview of geographical distribution of particular subtaxa of the family Lymnaeidae, with emphasis on genera and subgenera.

Subfamily Lymnaeinae. The range of the nominative subfamily almost coincides with the range of the entire family. The species of the Lymnaeinae are represented in all continents, except Australia which, probably, is due to its older (as compared to the rest of the subfamilies) geological age.

The genus ***Lymnaea*** is a taxon of broadly Holarctic distribution (see Fig. 7.5). The *Lymnaea stagnalis* species complex is distributed throughout Eurasia (except its south and southeast parts), North Africa, and North America. It is absent from the Far East. These snails can be found in lowlands and highlands (*Lymnaea kashmirensis* described from the Himalaya), able to live under low salinity (like in the Finnish Gulf of the Baltic Sea). *Lymnaea stagnalis* s. lato has been reported, as an alien species, from Cameroon, Australia, and New Zealand (see Table 7.3). "*Lymnaea*" *rupestris* (Paraense, 1982), described from the Santa Catarina State of Brazil (Paraense 1982), most probably belongs to another lymnaeid genus.

Acella and ***Bulimnea*** are two monotypic genera endemic to North America (see Fig. 7.5). The distribution of *Acella* is restricted to the central part of North America, from southern Ontario and Quebec in Canada to Illinois, Vermont, and Ohio (Burch 1982, 1989; Johnson et al. 2013). The range of *Bulimnea* is rather similar; Burch (1989, p. 170) describes the distribution of this genus as "Great Lakes and St. Lawrence River drainage area, parts of the Albany, Winnipeg, and Nelson River systems in the Canadian Interior Basin."

Hinkleyia is another genus endemic to the Nearctic region. Its range extends from the Yukon Territory and Quebec in Canada to California and Colorado (Taylor et al. 1963; Burch 1982, 1989). *Hinkleyia montanensis*, which is sometimes separated in a special genus ***Walterigalba*** (see Vinarski 2013 for details), occurs in the western USA, including Idaho, Montana, Nevada, Utah, and Wyoming (Burch 1989), as well as in western Canada: Alberta and British Columbia (Johnson et al. 2013).

Genus ***Dallirhytis***, with its single species, *D. atkaensis* (Dall, 1884), is a Beringian taxon (see Fig. 7.5) known from the northwestern part of North America (Alaska, northwestern Canada) and Chukotka peninsula of Northern Asia (Clarke 1981; Kruglov and Starobogatov 1989; Johnson et al. 2013; Vinarski and Kantor 2016).

Genus ***Kazakhlymnaea*** is endemic to Palearctic; its range extends from the southeast Europe in the west to the northern and central parts of Kazakhstan in the east (Vinarski et al. 2012b; Vinarski and Kantor 2016). The known occurrences of snails of this genus are lying between 40 and 55 degrees of northern latitude.

A pair of vicariant genera, ***Aenigmomphiscola*** and ***Omphiscola***, occupies the western part of the Palearctic. The former genus, still very poor studied, is known from a handful of reliable recordings made in the central part of European Russia, middle and southern Urals, south-west Siberia, and Northern Kazakhstan (Kruglov and Starobogatov 1981; Vinarski et al. 2011; Vinarski and Grebennikov 2012). This genus is thus endemic to the former USSR territory. *Omphiscola* is restricted in its distribution to Europe (see Fig. 7.4). Its range extends from southern Scandinavia in the north to Spain in the south (Økland 1990; Kruglov and Starobogatov 1993a; Arconada et al. 2019; Glöer 2019). It is absent from many countries of Central and South Europe (Glöer 2019).

The range of the genus ***Stagnicola*** (see Fig. 7.5) was often described as Holarctic, since many species of Nearctic lymnaeids had been classified within this genus (Baker 1911; Burch 1989; Johnson et al. 2013). According to the current taxonomy, no true *Stagnicola* is inhabiting North America, and all species of the Nearctic fauna once placed to this genus must be re-classified as belonging to the genus *Ladislavella*. Thus, the range of *Stagnicola* s. str. is restricted to Europe and Northern Asia, where it extends eastwards to the Baikal region and Yakutia (Vinarski and Kantor 2016). *S. palustris* has been reported from Morocco, Algeria, and even Zaire (as an introduced species); a single finding of this species from Saudi Arabia is known (Brown 1994; Neubert 1998). The taxonomic identity of "*Stagnicola*" *tungabhadraensis* (Ray, 1967), described from India, Andhra Pradesh state (Ray 1967; Subba Rao 1989), remains unknown due to absence of anatomical and/or genetic data. It is very doubtful, though, this taxon is actually a member of *Stagnicola*.

Ladislavella (see Fig. 7.5) consists of two subgenera—*Ladislavella* s. str. is distributed in Northern Eurasia, and its range is disjunctive. One species (*L. occulta*) lives in Central Europe, while the rest of the species of this subgenus occur in Siberia and the Russian Far East (Vinarski 2012, 2020; Vinarski and Kantor 2016). The subgenus *Ladislavella* (*Walterlymnaea*) Starobogatov et Budnikova, 1976 is widely distributed in North America and Greenland (Vinarski et al. 2017); one species has recently been found in Kamchatka Peninsula (Vinarski et al. 2016b). In North America, the southern boundary of the genus range lies in Mexico. Possibly, this continent, where more than 20 nominal species of "*Stagnicola*" are known (Burch 1989; Johnson et al. 2013), is the center of origin of this genus, whereas the Eurasian part of its range was only relatively recently occupied by species of *Ladislavella*.

The monotypic genus ***Pectinidens*** is the only lymnaeid genus endemic to South America (see Fig. 7.5). It inhabits the southern part of the continent being distributed in Chile, Argentina, and Falkland (Malvine) Islands (Pointer and Vázquez 2020).

The Hawaiian Islands have two endemic lymnaeid genera—***Erinna*** and ***Pseudisidora***, with two species each (Christensen et al. 2021).

Galba is probably the genus with the largest geographical distribution among the lymnaeids. *Galba* species are common inhabitants of unstable freshwater habitats worldwide like ditches and ponds. In line with a probable American origin, most species are found in the Americas but with some having been introduced into other continents such as *Galba schirazensis* and *Galba cubensis/viator* (Correa et al. 2011; Alda et al. 2021). Two *Galba* species have a much more restricted distribution: *Galba cousini* lives in high-altitude lakes in Ecuador, Colombia, and Venezuela and *Galba humilis* live in plains of North America (Alda et al. 2021). Uncertainty has prevailed for decades in *Galba* biogeography mainly because most of the nominal species share a similar shell morphology and internal anatomy, as well as plasticity in shell, anatomy, and life-history traits (Samadi et al. 2000; Correa et al. 2010). Recent molecular studies have shaded some light, but regions such as North America and Africa still need to be extensively explored (Correa et al. 2010; Alda et al. 2021).

Genus ***Pseudosuccinea*** only accounts for the species *P. columella*, showing a relatively recent multi-continent invasion that probably began with the late nineteenth–early twentieth century globalization. This species is believed to have a North American origin, spreading out initially to Central America (including the Caribbean countries), reaching out to South America and secondarily other continents. Higher genetic diversity found in its native range in the USA and a reduced variation in populations sampled elsewhere endorse such assumption (Lounnas et al. 2017). Introductions of *P. columella* have been of rather epidemiological interest given its role in liver fluke transmission worldwide. Out of the Americas, its presence is confirmed from open fields in Africa (Brown 1994), Western Europe (Pointier et al. 2007), Argentina (Martín et al. 2016), Australia (Molloy and Anderson 2006), New Zealand (Boray 1978), and several oceanic islands including Hawaii, La Réunion, and French Polynesia (Lounnas et al. 2017). However, the species is known to occur in many artificial ponds in botanical gardens or zoos from several European countries (Glöer 2019) with the aquarium plant-trade pointed as the main path for introductions worldwide (Vázquez et al. 2018).

Subfamily Lancinae,[1] with its three genera, is endemic to northwestern North America and is unknown outside the USA and western Canada. Campbell et al. (2017, p. 108) define the range of the Lancinae as "from the upper Sacramento and Pit Rivers of northern California, north to the Columbia River system in the states of Idaho, Oregon, Washington and Montana in the United States and the province of British Columbia, Canada." The genus ***Lanx*** is endemic to western USA (see Fig. 7.5). It occurs in California, Oregon, and Idaho (Johnson et al. 2013). The genus ***Idaholanx*** is very limited in its distribution, being found exclusively in cold

[1] Treated as a tribe by Aksenova et al. (2023).

water springs flowing into the eastern side of an 8 km section of the Snake River, in Gooding County, Idaho (Campbell et al. 2017). ***Fisherola***, the third extant genus of the subfamily, is the most widely distributed. Apart from British Columbia, it occurs in Idaho, Montana, Oregon, Utah, Washington, and Wyoming (Johnson et al. 2013).

Subfamily Amphipepleinae is essentially an Old World group, whose center of origin was, probably, situated in Central Asia (Aksenova et al. 2018). At least 8 genera of the Amphipepleinae are unknown beyond the Old World.

The genus ***Radix*** has, perhaps, the largest native range among the lymnaeid genera except for *Galba*. The nominative subgenus is distributed throughout Eurasia and occurs also in northwestern North America (see Fig. 7.4). Another subgenus, *Radix* (*Exsertiana*), is a tropical group inhabiting sub-Saharan Africa, Arabia, and South Asia eastward to Indonesia (see Fig. 7.4). The highest species richness of *Radix* is observed in Central Asia (Aksenova et al. 2018; Vinarski et al. 2020), with a series of species endemic to this area (*R. brevicauda*, *R. makhrovi*, *R. dgebuadzei*, and some others).

Tibetoradix is a unique example of the lymnaeid evolutionary radiation that took place in a high-altitude plateau at altitudes exceeding 3500 m a.s.l. (Aksenova et al. 2018; Vinarski et al. 2022b). All 6–7 species of *Tibetoradix* are unknown from regions other than the Tibet Plateau (see Fig. 7.4).

The range of the genus ***Ampullaceana*** almost completely coincides with the boundaries of the Palearctic region (see Fig. 7.4). *Ampullaceana lagotis*, the most widely distributed member of this group, is distributed in Europe, Northern and Central Asia eastwards to Yakutia in Eastern Siberia. It is also found in the High Asia (Vinarski et al. 2020). On the other hand, *A. relicta* is endemic to two ancient lakes of the Balkan Peninsula—Ohrid and Prespa (Vinarski et al. 2020). Most species of *Ampullaceana* occur in Europe, two of them are registered from Northern Africa (Brown 1994; Mabrouki et al. 2020).

The distribution of the genus ***Peregriana*** resembles that of the preceding genus, but *Peregriana* is absent from Central Asia as well as from some regions of northeastern Asia. This genus includes two vicariant species, one of which (*P. peregra*) is distributed chiefly in Europe, while the second (*P. dolgini*) occupies Siberia and can be found in Europe in the Pechora River basin only (Vinarski et al. 2020).

Like the genus *Dallirhytis* (Lymnaeinae), the genus ***Kamtschaticana*** is a Beringian group, whose range covers both northwestern North America and northeastern Asia (Vinarski et al. 2021a; see Fig. 7.4). *Kamtschaticana* is widely distributed in Eastern Siberia and the Russian Far East; a single species of this genus has recently been discovered in Japan (Aksenova et al. 2023).

Myxas, a morphologically peculiar lymnaeid genus, is a European taxon, with a few reliable findings in Western Siberia and Northern Kazakhstan (Vinarski et al. 2013, 2020; Glöer 2019). According to some data, the eastern extension of its range lies in the Lower Yenisei drainage basin (Kruglov and Starobogatov 1993b; Kruglov 2005). It is virtually absent from the southern regions of Europe and seems to be more abundant in the northern latitudes.

Table 7.4 Global distribution of the Lymnaeidae genera: a summary (based on native ranges only)

Type of distribution	Genera
Subcosmopolitan	*Galba*
Holarctic	*Ladislavella, Lymnaea, Radix*
Palaearctic	*Aenigmomphiscola, Ampullaceana, Kazakhlymnaea, Myxas, Omphiscola, Peregriana, Stagnicola*
Nearctic	*Acella, Bulimnea, Fisherola, Hinkleyia, Idaholanx, Lanx, Pseudosuccinea, Walterigalba*
Beringian	*Dallirhytis, Kamtschaticana*
Central Asian	*Tibetoradix*
Central-East Asian	*Orientogalba*[a], *Racesina*
South American	*Pectinidens*
Australasian	*Austropeplea, Bullastra*
Endemics of oceanic islands	*Erinna, Lantzia, Limnobulla, Pseudisidora*

[a]Found also in Australasia

The next two lymnaeid genera, ***Orientogalba*** and ***Racesina*** (= *Cerasina* auct.), are widely distributed in East, Southeast, and Central Asia. For example, the range of *Racesina* covers a vast territory of Asia, from Turkmenistan and Tajikistan in the north to Thailand and, possibly, Malaysia, in the southeast (Vinarski and Bolotov 2018; Vinarski et al. 2020; see Fig. 7.4). The native range of *Orientogalba* is even broader; besides Central and South Asia, it embraces the Far East and some Pacific Islands (but absent from Australia). One species of this genus has recently been registered as non-indigenous in South Europe (see Table 7.3).

The Australasian distribution is characteristic for the genera ***Austropeplea*** and ***Bullastra***, the first of them being endemic to Australia and New Zealand (see Fig. 7.5). The species of the genus *Bullastra* occur in Australia as well as in Indonesia, New Guinea, and the Philippine Islands; a single species, *B. buruana*, is endemic to the Buru Island of Indonesia (Haas 1913; Starmühlner 1976; Ponder et al. 2020; Vinarski et al. 2020).

The monotypic genus ***Limnobulla***, tentatively placed into the subfamily Lymnaeinae, occurs in the Falkland Islands only. Another island genus with very limited distribution is ***Lantzia*** which inhabits Réunion Island in the Indian Ocean (Brown 1994).

A summary of the geographical distribution of the Lymnaeidae genera is presented in Table 7.4.

Acknowledgments MVV and OVA are grateful for the financial support of their lymnaeid studies to the Russian Science Foundation (projects No. 19-14-00066/P and 21-74-10155).

References

Abell R, Thieme M, Revenga C (2008) Freshwater ecoregions of the world: A new map of biogeographic units for freshwater biodiversity conservation. Bioscience 58:403–414. https://doi.org/10.1641/B580507

Adams J (2009) Species richness: patterns in the diversity of Life. Praxis Publishing Ltd, Chichester

Aksenova O, Vinarski M, Bolotov I et al (2017) Two *Radix* spp. (Gastropoda: Lymnaeidae) endemic to thermal springs around Lake Baikal represent ecotypes of the widespread *Radix auricularia*. J Zool Syst Evol Res 55(4):298–309. https://doi.org/10.1111/jzs.12174

Aksenova OV, Bolotov IN, Gofarov MY et al (2018) Species richness, molecular taxonomy and biogeography of the radicine pond snails (Gastropoda: Lymnaeidae) in the Old World. Sci Rep-UK 8:11199. https://doi.org/10.1038/s41598-018-29451-1

Aksenova OV, Bolotov IN, Kondakov AV et al (2023, this volume) Phylogeny and taxonomy of the family Lymnaeidae. In: Vinarski MV, Vázquez A (eds) The Lymnaeidae. A handbook on their natural history and parasitological significance. Springer, Cham, pp 67–101

Alda P, Lounnas M, Vázquez AA et al (2021) Systematics and geographical distribution of *Galba* species, a group of cryptic and worldwide freshwater snails. Molec Phyl Evol 157:107035. https://doi.org/10.1016/j.ympev.2020.107035

Anderson R (2005) An annotated list of the non-marine Mollusca of Britain and Ireland. J Conchol 38:607–638

Appleton CC, Miranda NAF (2015) Two Asian freshwater snails newly introduced into South Africa and an analysis of alien species reported to date. Afr Invertebr 56(1):1–17. https://doi.org/10.5733/afin.056.0102

Arconada B, De Andrés J, Araujo R (2019) *Omphiscola glabra* (Müller, 1774) (Gastropoda: Lymnaeidae) en la Península Ibérica. Iberus 37(2):209–217

Baker FC (1911) The Lymnaeidae of North and Middle America. Special publication of the Chicago Academy of Sciences 3:1–539. https://doi.org/10.5962/bhl.title.20500

Banarescu P (1990) Zoogeography of fresh waters. Vol. 1. General distribution and dispersal of freshwater animals. AULA-Verlag, Wiesbaden

Banarescu P (1992) Zoogeography of fresh waters. Vol. 2. Distribution and dispersal of freshwater animals in North America and Eurasia. AULA-Verlag, Wiesbaden

Bargues MD, Horák P, Patzner RA et al (2003) Insights into relationships of Palearctic and Nearctic lymnaeids (Mollusca: Gastropoda) by rDNA ITS-2 sequencing and phylogeny of stagnicoline intermediate host species of *Fasciola hepatica*. Parasite 10:243–255. https://doi.org/10.1051/parasite/2003103243

Bargues MD, Artigas P, Khoubbane M et al (2011) *Lymnaea schirazensis*, an overlooked snail distorting fascioliasis data: Genotype, phenotype, ecology, worldwide spread, susceptibility, applicability. PLoS One 6:e24567. https://doi.org/10.1371/journal.pone.0024567

Bernardo-Madrid R, Calatayud J, González-Suárez M et al (2019) Human activity is altering the world's zoogeographical regions. Ecol Lett 22:1297–1305. https://doi.org/10.1111/ele.13321

Bolton HC (1991) On the mathematical significance of the similarity index of Ochiai as a measure for biogeographical habitats. Austr J Zool 39:143–156. https://doi.org/10.1071/ZO9910143

Boray JC (1978) The potential impact of exotic *Lymnaea* spp. on fascioliasis in Australasia. Vet Parasitol 4:127–141. https://doi.org/10.1016/0304-4017(78)90004-3

Brandt RAM (1974) The non-marine aquatic Mollusca of Thailand. Arch Molluskenkd 105:1–423

Brikiatis L (2014) The De Geer, Thulean and Beringia routes: key concepts for understanding early Cenozoic biogeography. J Biogeogr 41:1036–1054. https://doi.org/10.1111/jbi.12310

Brown DS (1994) Freshwater snails of Africa and their medical importance. Taylor and Francis, London. https://doi.org/10.1201/9781482295184

Burch JB (1982) North American freshwater snails. Walkerana 4:217–365

Burch JB (1989) North American freshwater snails. Malacological publications, Hamburg, MI

Campbell DC, Clark SA, Lydeard C (2017) Phylogenetic analysis of the Lancinae (Gastropoda, Lymnaeidae) with a description of the U.S. federally endangered Banbury Springs lanx. Zookeys 663:107–132. https://doi.org/10.3897/zookeys.663.11320

Christensen CC, Hayes KA, Yeung NW (2021) Taxonomy, conservation, and the future of native aquatic snails in the Hawaiian Islands. Diversity 13:215. https://doi.org/10.3390/d13050215

Clarke AH (1981) The freshwater molluscs of Canada. National Museum of Natural Sciences, Ottawa

Climo FM, Pullan NB (1972) A taxonomic review of the family Lymnaeidae (Mollusca: Gastropoda) in New Zealand. J Roy Soc New Zeal 2:5–13. https://doi.org/10.1080/03036758.1972.10423300

Correa AC, Escobar JC, Durand P et al (2010) Bridging gaps in the molecular phylogeny of the Lymnaeidae (Gastropoda: Pulmonata), vectors of fascioliasis. BMC Evol Biol 10:381. https://doi.org/10.1186/1471-2148-10-381

Correa AC, Escobar JS, Noya O et al (2011) Morphological and molecular characterization of Neotropic Lymnaeidae (Gastropoda: Lymnaeoidea) vectors of fasciolosis. Inf Gen Evol 11(8): 1978–1988. https://doi.org/10.1016/j.meegid.2011.09.003

Cox CB (2001) The biogeographic regions reconsidered. J Biogeogr 28:511–523. https://doi.org/10.1046/j.1365-2699.2001.00566.x

Dundee DS (1974) Catalogue of introduced mollusks of eastern North America (north of Mexico). Sterkiana 55:1–37

Falkner G, Bank RA, Proschwitz T (2001) Check-list of the non-marine molluscan species-group taxa of the states of Northern, Atlantic and Central Europe (CLECOM I). Heldia 4:1–76

Fischer P (1880–1887) Manuel de conchyliologie et de paleontologie conchyliologique ou, Histoire naturelle des mollusques vivants et fossiles. V. Savy, Paris. https://doi.org/10.5962/bhl.title.13213

Glöer P (2002) Die Sußwassergastropoden Nord- und Mitteleuropas: Bestimmungschlussel, Lebenweise, Verbreitung. Die Tierwelt Deutschlands. Conchbooks, Hackenheim 73:1–327

Glöer P (2019) The freshwater gastropods of the West-Palaearctis. Volume 1. Fresh- and brackish waters except spring and subterranean snails. Identification key, anatomy, ecology, distribution. The author, Hetlingen

Glöer P, Bößneck U (2013) Freshwater molluscs from Nepal and North India with the description of seven new species. Arch Molluskenkd 142:137–156. https://doi.org/10.1127/arch.moll/1869-0963/142/137-156

Glöer P, Pešić V (2012) The freshwater snails (Gastropoda) of Iran, with descriptions of two new genera and eight new species. Zookeys 219:11–61. https://doi.org/10.3897/zookeys.219.3406

Hammer Ø, Harper DAT, Ryan PD (2001) PAST: PAleontological STatistics software package for education and data analysis. Palaeontol Electron 4(1):1–9

Haas F (1913) Neue Süsswasserschnecke aus Central Buru. Nachrichtsblatt der Deutschen Malakozoologischen Gesellschaft. 45:184. https://biodiversitylibrary.org/page/15108159

Hillebrand H (2004) On the generality of the latitudinal diversity gradient. Am Nat 163:192–211. https://doi.org/10.1086/381004

Hubendick B (1951) Recent Lymnaeidae. Their variation, morphology, taxonomy, nomenclature and distribution. Kungliga Svenska Vetenskapsakademiens Handlingar Fjärde Serien 3(1): 1–223

Hubendick B (1952) Hawaiian Lymnaeidae. Occasional Papers of Bernice P. Bishop Museum, Honolulu, Hawaii 20(19):307–328

Hubendick B (1962) Aspects on the diversity of the freshwater fauna. Oikos 13:249–261

Inaba A (1969) Cytotaxonomic studies of lymnaeid snails. Malacologia 7(2–3):143–168

Johnson PD, Bogan AE, Brown KM et al (2013) Conservation status of freshwater gastropods of Canada and the United States. Fisheries 38(6):247–282. https://doi.org/10.1080/03632415.2013.785396

Kerney MP (1976) Atlas of the non-marine Mollusca of the British Isles. Institute of Terrestrial Ecology, Cambridge

Kevan DK (1941) A recent addition to the British Fauna. J Conchol 21(9):279
Kevan DK (1943) Study of an introduced North American freshwater mollusc, *Stagnicola catascopium* (Say). Proc R Soc Edinb Sect B 61:430–461
Kruglov ND (2005) Lymnaeid snails of Europe and Northern Asia. Smolensk State Pedagogical University Press, Smolensk. [in Russian]
Kruglov ND, Starobogatov YI (1981) A new genus of the Lymnaeidae and taxonomy of the subgenus *Omphiscola* (*Lymnaea*) (Pulmonata, Gastropoda). Zool Zh 60:965–977. [in Russian]
Kruglov ND, Starobogatov YI (1989) Mollusks of the subgenus *Polyrhytis* of the genus *Lymnea* in the fauna of the USSR (Pulmonata, Lymnaeidae). Zool Zh 68:14–20. [in Russian]
Kruglov ND, Starobogatov YI (1993a) Annotated and illustrated catalogue of species of the family Lymnaeidae (Gastropoda Pulmonata Lymnaeiformes) of Palaearctic and adjacent river drainage areas. Part I Ruthenica Russian Malacol J 3(1):65–92
Kruglov ND, Starobogatov YI (1993b) Annotated and illustrated catalogue of species of the family Lymnaeidae (Gastropoda Pulmonata Lymnaeiformes) of Palaearctic and adjacent river drainage areas. Part II Ruthenica Russian Malacol J 3(1):161–180
Laske SM, Amundsen PA, Christoffersen KS et al (2019) Circumpolar patterns of Arctic freshwater fish biodiversity: A baseline for monitoring. Freshw Biol 67(1):176–193. https://doi.org/10.1111/fwb.13405
Leiss A, Reischütz P (1996) Ein Beitrag zur Kenntnis der Molluskenfauna der Gewächshäuser in Wien und Niederösterreich. Wissenschaftliche Mitteilungen aus dem Niederösterreichischen Landesmuseum 9:173–184
Likharev IM, Starobogatov YI (1967) On the molluscan fauna of Afghanistan. Trudy Zoologicheskogo Instituta Akademii Nauk SSSR 42:159–197. [in Russian]
Liu YY, Wang YX, Zhang WZ (1979) Freshwater Mollusks. Economic Fauna of China. Science Press, Beijing. [in Chinese]
López-Soriano J, Quiñonero-Salgado S (2020) Status of the populations of two alien species of Lymnaeidae in the Ebro Delta (NE Iberian Peninsula). Elona: Revista de Malacología Ibérica 2: 17–23
Lotfy WM, Lotfy LM (2015) Synopsis of the Egyptian freshwater snail fauna. Folia Malacologica 23:19–40. https://doi.org/10.12657/folmal.023.002
Lounnas M, Correa AC, Vázquez AA et al (2017) Self-fertilization, long-distance flash invasion and biogeography shape the population structure of *Pseudosuccinea columella* at the worldwide scale. Mol Ecol 26:887–903. https://doi.org/10.1111/mec.13984
Lounnas M, Correa AC, Alda P et al (2018) Population structure and genetic diversity in the invasive freshwater snail *Galba schirazensis*. Can J Zool 96(5):425–435. https://doi.org/10.1139/cjz-2016-0319
Mabrouki Y, Fouzi Taybi A, Glöer P (2020) New additions to the freshwater gastropod fauna (Gastropoda: Hydrobiidae, Lymnaeidae) of Morocco. Ecologica Montenegrina 31:40–44. https://doi.org/10.37828/em.2020.31.8
Macan TT (1949) A key to the British fresh- and brackish-water gastropods with notes on their ecology. Freshwater Biological Association, Scientific Publication No. 13, Ambleside, UK
Mahulu A, Clewing C, Stelbrink B et al (2019) Cryptic intermediate snail host of the liver fluke *Fasciola hepatica* in Africa. Parasit Vectors 12:573. https://doi.org/10.1186/s13071-019-3825-9
Malyshev LI (2001) The syndromes in the comparative floristics. Evolyutsionnaya Biologiya, vol 1. Tomsk State University Press, Tomsk, pp 190–206. [in Russian]
Martín PR, Ovando XMC, Seuffert ME (2016) First record of the freshwater snail *Pseudosuccinea columella* (Gastropoda: Lymnaeidae) in southern Pampas (Argentina) and assessment of future spread. Molluscan Res 36(3):213–221. https://doi.org/10.1080/13235818.2015.1128602
Meier-Brook C (1984) A preliminary biogeography of freshwater pulmonate gastropods. In: Solem A, van Bruggen AC (eds) World-wide snails. Biogeographical studies on non-marine Mollusca, Brill, Leiden, pp 23–37

Minelli A (2019) The galaxy of the non-Linnaean nomenclature. Hist Philos Life Sci 41:31. https://doi.org/10.1007/s40656-019-0271-0
Molloy JB, Anderson GF (2006) The distribution of *Fasciola hepatica* in Queensland, Australia, and the potential impact of introduced snail intermediate hosts. Vet Parasitol 137(1–2):62–66. https://doi.org/10.1016/j.vetpar.2005.12.024
MolluscaBase eds. (2021) MolluscaBase. http://www.molluscabase.org. Accessed 01 May 2021. https://doi.org/10.14284/448
Naranjo-García E, Castillo-Rodríguez ZG (2017) First inventory of the introduced and invasive mollusks in Mexico. The Nautilus 131(2):107–126
Nesemann H, Sharma S, Sharma G et al (2007) Class Gastropoda. In: Nesemann H, Sharma S, Sharma G et al (eds) Aquatic invertebrates of the Ganga river system (Mollusca, Annelida, Crustacea [in part]), vol 1. H. Nesemann, Hofheim-Lorsbach, pp 57–102
Neubauer TA (2023, this volume) The fossil record of the Lymnaeidae – revisiting a 200-myr-long story of success. In: Vinarski MV, Vàzquez AA (eds) The Lymnaeidae: a handbook on their natural history and parasitological significance. Springer, Cham, pp 147–181
Neubert E (1998) Annotated checklist of the terrestrial and freshwater molluscs of the Arabian Peninsula with descriptions of new species. Fauna of Arabia 17:333–461
Økland J (1990) Lakes and snails. Environment and Gastropoda in 1,500 Norwegian lakes, ponds and rivers. Universal Book Services; Dr. W. Backhuys Publ., Oegstgeest
Paraense WL (1982) *Lymnaea rupestris* sp. n. from southern Brazil (Pulmonata: Lymnaeidae). Mem I Oswaldo Cruz 77(4):437–443. https://doi.org/10.1590/S0074-02761982000400011
Pointer JP, Vázquez AA (2020) Phylum Mollusca: Lymnaeoidea, Lymnaeidae. In: Rogers CD, Thorp J, Damborenea C (eds) Thorp and Covich's freshwater invertebrates, Volume 5: Keys to neotropical and Antarctic fauna, Fourth Edition. Academic Press, London etc, pp 281–289
Pointier JP (ed) (2015) Freshwater molluscs of Venezuela and their medical and veterinary importance. Conchbooks, Harxheim
Pointier JP, Coustau C, Rondelaud D et al (2007) *Pseudosuccinea columella* (Say 1817) (Gastropoda, Lymnaeidae), snail host of *Fasciola hepatica*: first record for France in the wild. Parasitol Res 101(5):1389–1392. https://doi.org/10.1007/s00436-007-0656-y
Ponder WF, Waterhouse J (1997) A new genus and species of Lymnaeidae from the lower Franklin River, south western Tasmania. J Molluscan Stud 63:441 468. https://doi.org/10.1093/mollus/63.3.441
Ponder WF, Hallan A, Shea M et al (2020) Australian freshwater Molluscs. Revision 1. http://keys.lucidcentral.org/keys/v3/freshwater_molluscs/
Ray HC (1967) A new or little known species of the genus *Stagnicola* Leach, 1830 from India (Mollusca: Gastropoda: family Lymnaeidae). Journal de Conchyliologie 106(1):3–7
Remigio EA, Blair D (1997) Relationships among problematic North American stagnicoline snails (Pulmonata: Lymnaeidae) reinvestigated using nuclear ribosomal DNA internal transcibed spacer sequences. Can J Zool 75:1540–1545. https://doi.org/10.1139/z97-779
Robinson WI (1915) Two new fresh-water gastropods from the Mesozoic of Arizona. Am J Sci, fourth series 40(240):649–651
Roll U, Dayan T, Simberloff D et al (2009) Non-indigenous land and freshwater gastropods in Israel. Biol Invasions 11(8):1963–1972. https://doi.org/10.1007/s10530-008-9373-4
Roszkowski W (1928) The distribution of lymnaeids in Europe and North America, with relation to Wegener's theory. Annales Zoologici Musei Polonici Historiae Naturalis 7:85–94
Russell-Hunter WD (1978) Ecology of freshwater pulmonates. In: Fretter V, Peake J (eds) Pulmonates, vol 2A. Academic Press, London, pp 335–384
Saito T (2022) First record of the non-indigenous freshwater snail *Galba humilis* (Say, 1822) (Mollusca: Hygrophila:Lymnaeidae) in Japan. Bioinvasion Records 11(2):428–439. https://doi.org/10.3391/bir.2022.11.2.16
Samadi S, Roumégoux A, Bargues MD et al (2000) Morphological studies of Lymnaeid snails from the human fasciolosis endemic zone of Bolivia. J Mollus Stud 66:31–44. https://doi.org/10.1093/mollus/66.1.31

Schniebs K, Glöer P, Vinarski MV et al (2017) A new alien species in Europe: first record of *Austropeplea viridis* (Quoy & Gaimard, 1833) in Spain. J Conchol 42(5):357–370

Schniebs K, Glöer P, Quiñonero-Salgado S et al (2018) The first record of *Galba cubensis* (L. Pfeiffer, 1839) (Mollusca: Gastropoda: Lymnaeidae) from open fields of Europe. Folia Malacologica 26(1):3–15. https://doi.org/10.12657/folmal.026.002

Schniebs K, Glöer P, Vinarski M et al (2019) Intraspecific morphological and genetic variability in the Palaearctic freshwater snail *Radix ampla* (Hartmann, 1821) (Gastropoda: Basommatophora: Lymnaeidae). J Conchol 43(3):245–267

Starmühlner F (1976) Beiträge zur Kenntnis der Süßwasser-Gastropoden pazifischer Inseln. Ergebnisse der Österreichischen Indopazifik-Expedition des 1. Zoologischen Institutes der Universität Wien. Annalen des Naturhistorischen Museum in Wien 80:473–656

Starobogatov YI (1970) Molluscan fauna and zoogeographic zonation of continental freshwater bodies of the world. Nauka, Leningrad. [in Russian]

Starobogatov YI (1986) The fauna of lakes as a source of information about their history. In: Kvasov DD (ed) The general patterns of origin and development of lakes. Method of studies of the history of lakes. Gidrometeoizdat, Moscow, pp 33–50. [in Russian]

Strong EE, Gargominy O, Ponder WF, Bouchet P (2008) Global diversity of gastropods (Gastropoda; Mollusca) in freshwater. Hydrobiologia 595:149–166. https://doi.org/10.1007/s10750-007-9012-6

Subba Rao MV (1989) Handbook: freshwater molluscs of India. Zoological Survey of India, Calcutta

Taylor DW, Walter HJ, Burch JB (1963) Freshwater snails of the subgenus *Hinkleyia* (Lymnaeidae: Stagnicola) from the Western United States. Malacologia 1(2):237–281

Thompson FG (2011) An annotated checklist and bibliography of the land and freshwater snails of Mexico and Central America. Bull Fla Mus Nat His 50(1):1–299

van Damme D (1984) The freshwater Mollusca of Northern Africa: Distribution, biogeography and paleoecology. Dev Hydrobiol 25:1–164

Vavalidis T, Zogaris S, Economou AC et al (2019) Changes in fish taxonomy affect freshwater biogeographical regionalisations: insights from Greece. Water 11:1743. https://doi.org/10.3390/w11091743

Vázquez AA, Alda P, Lounnas M et al (2018) Lymnaeid snails hosts of *Fasciola hepatica* and *Fasciola gigantica* (Trematoda: Digenea): A worldwide review. CAB Rev Perspect Agric Vet Sci Nutr Nat Res 13:1–15. https://doi.org/10.1079/PAVSNNR201813062

Vinarski MV (2012) The lymnaeid genus *Catascopia* Meier-Brook et Bargues, 2002 (Mollusca: Gastropoda: Lymnaeidae), its synonymy and species composition. Invertebr Zool (Moscow) 9(2):91–104

Vinarski MV (2013) One, two, or several? How many lymnaeid genera are there? Ruthenica: Russian Malacol J 23(1):41–58

Vinarski MV, Bolotov IN (2018) *Racesina*, a new generic name for a group of Asian lymnaeid snails (Gastropoda: Hygrophila:Lymnaeidae). Zoosystematica Rossica 27(2):328–333. https://doi.org/10.31610/zsr/2018.27.2.328

Vinarski MV, Grebennikov ME (2012) A review of the species of the genus *Aenigmomphiscola* Kruglov et Starobogatov, 1981 (Gastropoda: Pulmonata: Lymnaeidae). Ruthenica: Russian Malacol J 22(2):159–170. [in Russian]

Vinarski MV, Kantor YI (2016) Analytical catalogue of fresh and brackish water molluscs of Russia and adjacent countries. A.N. Severtsov Institute of Ecology and Evolution of RAS, Moscow

Vinarski MV, Kramarenko SS (2015) How does the discrepancies among taxonomists affect macroecological patterns? A case study of freshwater snails of Western Siberia. Biodivers Conserv 24(8):2079–2091. https://doi.org/10.1007/s10531-015-0934-4

Vinarski M, Schniebs K, Glöer P et al (2011) The taxonomic status and phylogenetic relationships of the genus *Aenigmomphiscola* Kruglov et Starobogatov, 1981 (Gastropoda: Pulmonata:

Lymnaeidae). J Nat Hist 45(33–34):2049–2068. https://doi.org/10.1080/00222933.2011.574800
Vinarski MV, Andreyev NI, Andreyeva SI et al (2012a) Latitudinal changes in the diversity of freshwater gastropods (Mollusca: Gastropoda) in waterbodies of Western Siberia. Inland Water Biol 5(1):83–90. https://doi.org/10.1134/S1995082911040213
Vinarski MV, Schniebs K, Glöer P et al (2012b) The steppe relics: taxonomic study on two lymnaeid species endemic to the former USSR (Gastropoda: Pulmonata: Lymnaeidae). Arch Molluskenkd 141(1):67–85. https://doi.org/10.1127/arch.moll/1869-0963/141/067-085
Vinarski MV, Grebennikov ME, Shishkoedova OS (2013) Past and present distribution of *Myxas glutinosa* (O.F. Müller, 1774) in the waterbodies of the Urals and Siberia. J Limnol 72(2): 336–342. https://doi.org/10.4081/jlimnol.2013.e27
Vinarski MV, Aksenova OV, Bespalaya YV et al (2016a) *Radix dolgini*: The integrative taxonomic approach supports the species status of a Siberian endemic snail (Mollusca, Gastropoda, Lymnaeidae). C R Biol 339(1):24–36. https://doi.org/10.1016/j.crvi.2015.11.002
Vinarski MV, Aksenova OV, Bespalaya YV et al (2016b) *Ladislavella tumrokensis*: The first molecular evidence of a Nearctic clade of lymnaeid snails inhabiting Eurasia. Syst Biodivers 14(3):276–287. https://doi.org/10.1080/14772000.2016.1140244
Vinarski MV, Bolotov IN, Schniebs K (2017) Endemics or strangers? The integrative re-appraisal of taxonomy and phylogeny of the Greenland Lymnaeidae (Mollusca: Gastropoda). C R Biol 340:541–557. https://doi.org/10.1016/j.crvi.2017.09.005
Vinarski MV, Clewing C, Albrecht C (2019) Family Lymnaeidae. In: Lydeard C, Cummings KS (eds) Freshwater mollusks of the world: a distribution atlas. John Hopkins University Press, Baltimore, MD, pp 158–162
Vinarski MV (2020) *Ladislavella occulta* (Jackiewicz 1959) – a species of aquatic snails new for Hungary with remarks on its distribution in Central and Eastern Europe Acta Zoologica Academiae Scientiarum Hungaricae 66(4):369–377. https://doi.org/10.17109/AZH.66.4.369.2020
Vinarski MV, Aksenova OV, Bolotov IN (2020) Taxonomic assessment of genetically-delineated species of radicine snails (Mollusca, Gastropoda, Lymnaeidae). Zoosyst Evol 96(2):577–608. https://doi.org/10.3897/zse.96.52860
Vinarski MV, Aksenova OV, Bespalaya YV et al (2021a) One Beringian genus less: a re-assessment of *Pacifimyxas* Kruglov & Starobogatov, 1985 (Mollusca: Gastropoda: Lymnaeidae) questions the current estimates of Beringian biodiversity. J Zool Syst Evol Res 59:44–59. https://doi.org/10.1111/jzs.12411
Vinarski MV, Bolotov IN, Aksenova OV et al (2021b) Freshwater Mollusca of the Circumpolar Arctic: a review on their taxonomy, diversity and biogeography. Hydrobiologia 848:2891–2918. https://doi.org/10.1007/s10750-020-04270-6
Vinarski MV, Aksenova OV, Bolotov IN et al (2022a) A new alien snail *Ampullaceana balthica* for the Canadian fauna, with an overview of Transatlantic malacofaunal exchange in the Anthropocene. Aquat Invasions 17(1):21–35. https://doi.org/10.3391/ai.2022.17.1.02
Vinarski MV, von Oheimb PV, Aksenova OV et al (2022b) Trapped on the Roof of the World: Taxonomic diversity and evolutionary patterns of Tibetan Plateau endemic freshwater snails (Gastropoda: Lymnaeidae: *Tibetoradix*). Integr Zool 17(5):825–848. https://doi.org/10.1111/1749-4877.12600
Wallace AR (1894) What are zoological regions? Nature 49:610–613. https://doi.org/10.1038/049610a0
Welter-Schultes F (2012) European non-marine molluscs: a guide for species identification. Planet Poster Editions, Göttingen
Wiese R, Clewing C, Albrecht C et al (2020) How ancient is Lake Lugu (Yunnan, China)? The gastropods' viewpoint with focus on *Radix* (Lymnaeidae). J Great Lakes Res 46:1099–1112. https://doi.org/10.1016/j.jglr.2020.06.003

Willig MR, Kaufman DM, Stevens RD (2003) Latitudinal gradients of biodiversity: pattern, process, scale, and synthesis. Annu Rev Ecol Evol Syst 34:273–309. https://doi.org/10.1146/annurev.ecolsys.34.012103.144032

Yildirim MZ, Gümüş BA, Kebapçi Ü et al (2006) The basommatophoran pulmonate species (Mollusca: Gastropoda) of Turkey. Turk J Zool 30:445–458

Chapter 8
Patterns and Processes of Speciation in Lymnaeidae

Christian Albrecht, Björn Stelbrink, and Catharina Clewing

Abstract In the present chapter, we aimed to summarize the current knowledge of patterns and processes of speciation in the gastropod family Lymnaeidae. Using selected case studies, we evaluated biodiversity and speciation patterns at both large and small spatial scale. We outlined adaptations to extreme and isolated environments and examined the studies dealing with speciation processes in lymnaeid gastropods.

Despite being widespread globally, the family is not very speciose and species flocks or small-scale radiations are an absolute exception. The family obtained a nearly cosmopolitan but mainly Northern Hemispheric distribution, with both widespread and narrow-range taxa. Adaptation to special or extreme habitats illustrates the importance of ecology in lymnaeid speciation. Whereas the prevailing geographical mode is allopatry, with likely cases of parapatry, no evidence currently exists for sympatric speciation in Lymnaeidae. Niche conservatism seems to be pronounced in some lineages and supports the assumption that speciation by adapting to different selective regimes is a common phenomenon. Mechanisms involved in speciation include biotic interactions such as in host–parasite pairs, leading to co-evolution in various species.

Lymnaea stagnalis, and increasingly some species of *Radix*, serve as model systems in developmental biology and neurophysiology and are thus among the gastropods spearheaded for genomic and transcriptomic analyses, allowing for a genome-level perspective on various biological aspects. All these new developments in the "omics" era will, in the near future, improve our understanding of speciation mechanisms in lymnaeids and beyond.

C. Albrecht (✉) · B. Stelbrink · C. Clewing
Department of Animal Ecology and Systematics, Systematics and Biodiversity Group, Justus Liebig University Giessen, Giessen, Germany
e-mail: Christian.Albrecht@allzool.bio.uni-giessen.de; Bjoern.Stelbrink@allzool.bio.uni-giessen.de; Catharina.Clewing@allzool.bio.uni-giessen.de

M. V. Vinarski, A. A. Vázquez (eds.), *The Lymnaeidae*, Zoological Monographs 7,
https://doi.org/10.1007/978-3-031-30292-3_8

8.1 Introduction

Lymnaeidae represent a major group of predominantly freshwater pulmonate gastropods (Hygrophila) that reached a factual cosmopolitan distribution throughout their roughly 200 million years of evolution (Zilch 1960; Taylor 1988; Tracey et al. 1993). While being present on all continents, formerly including Antarctica as suggested by Pliocene fossils (Ashworth and Preece 2003), members of this group fulfill several ecological functions (Taylor 1988; Vinarski et al. 2023 Chap. 7 of this book). Moreover, many species have biomedical importance since they serve as intermediate hosts for diseases relevant for both animals and humans (Brown 1994), of which fascioliasis is the most well-known, affecting millions of people worldwide (Vázquez et al. 2023, this book).

Lymnaeidae have been recognized for their pronounced shell-shape plasticity. This morphological variability together with an anatomical uniformity has led to very different approaches and estimates of both species and generic diversity in both extant (Hubendick 1951, 1978; Kruglov 2005; Vinarski 2013) and extinct taxa (Zilch 1960; Walter 1968; Neubauer 2023, this book). Given that most nominal species in the family were delineated on a strictly conchiological basis, it is thus not surprising that estimates of global species richness of Lymnaeidae vary from only 40 (Hubendick 1951) to about 100 (Strong et al. 2008) or even not less than 200 species (Kruglov 2005). However, there has been some progress in recent years when studies on lymnaeid systematics are increasingly phylogeny-based and conducted in an integrative taxonomic framework (e.g., Remigio and Blair 1997; Bargues et al. 2003, 2007, 2013; Correa et al. 2010; Aksenova et al. 2018; Vinarski et al. 2020; Vinarski et al. 2022). Nevertheless, controversy exists about the generic-level composition of the family. Depending on the underlying assumptions, authors have used systems recognizing a single genus *Lymnaea* (e.g., Walter 1968), two genera (Hubendick 1951; Kruglov 2005), or far more than these (e.g., Burch 1989; Correa et al. 2010; Vinarski 2013). However, in recent years, the latter approach became more and more dominating (see, e.g., Puslednik et al. 2009; von Oheimb et al. 2011; Vinarski 2013; Aksenova et al. 2018; Vinarski et al. 2019).

Following a multi-genus concept with 32 recognized recent genera, an analysis of their occurrences across 22 zoogeographical freshwater regions (Fig. 8.1) highlights the unequal distribution patterns of generic diversity of worldwide Lymnaeidae (Fig. 8.2, Table 8.1) that contrasts a general latitudinal diversity gradient (Hillebrand 2004). The majority of genera are found in the Northern Hemisphere, i.e., the Holarctic region. Here, eastern North America, western North America, the Euro-Mediterranean subregion, and Siberia are particular biodiversity hotspots with 9 or 10 genera being present. Regions of intermediate diversity are found both in large regions such as northern North America, East and Central Asia but also comparatively small ones such as western Mongolia and Lake Baikal (Fig. 8.1). In sharp

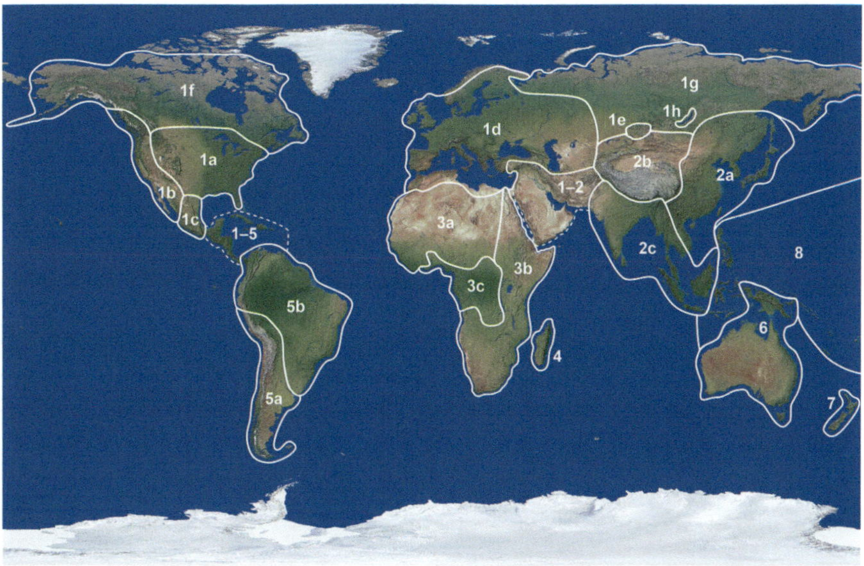

Fig. 8.1 Accepted zoogeographical freshwater regions after Bănărescu (1990) and Schäfer (1997). 1 – Holarctic region: 1a – Eastern North American subregion, 1b – Western North American subregion, 1c – Central Mexican subregion, 1d – Euro-Mediterranean subregion, 1e – Western Mongolian subregion, 1f – Arctic North American subregion, 1 g – Siberian subregion, 1 h – Baikal subregion; 2 – Sino-Indian region: 2a – East Asian subregion, 2b – High Asian subregion, 2c – South Asian subregion; 1–2: Western Asian transitional area; 3 – Ethiopian (African) region: 3a – West Africa, 3b – South East Africa, 3c – Central Africa; 4 – Malagasy region; 5 – Neotropical (South American) region: 5a – Andean-Patagonian subregion, 5b – Guayana-Brazilian subregion; 1–5: Middle American/Antillean transitional area; 6 – Australian region; 7 – New Zealand region; 8 – Indo-West Pacific region

contrast are those, mostly southern continental, regions that are inhabited by only one or two genera of Lymnaeidae, namely Africa, the Australian, and New Zealandian regions. Note, however, that there are other non-indigenous genera present in Africa (Appleton and Miranda 2015), and depending on taxonomic treatment, there might be more genera recognized for Australia (Puslednik et al. 2009).

Whereas the biogeography of Lymnaeidae has been summarized elsewhere (Vinarski et al. 2019; Vinarski et al. 2023, this book Chap. 7), the major aim of the present chapter is to review prominent patterns of speciation in selected case studies and the underlying mechanisms that led to the disparately distributed biodiversity of these gastropods. Given the wealth of literature that is currently available, such a review is naturally biased and not intended to be complete, but rather to provide an overview.

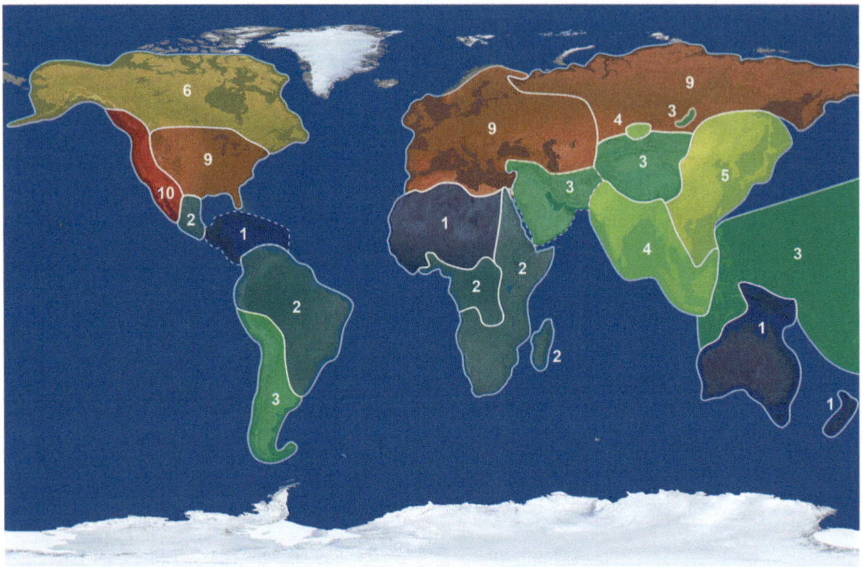

Fig. 8.2 Generic-level global diversity of Lymnaeidae (modified map of Fig. 8.1). Note the latitudinal gradient showing a generally decreasing diversity towards the South. Genera accounted for the respective freshwater zoogeographical regions are listed in Table 8.1

8.2 Speciation Patterns

8.2.1 Large-Scale Patterns

Global distribution patterns of generic biodiversity (Fig. 8.2 and Table 8.1) are also reflected on the species level (Hubendick 1951). Some species occupy very large ranges, some of them constantly extending their distribution anthropogenically, such as *Pseudosuccinea columella* (e.g., Lounnas et al. 2017). In contrast, small range endemics exist, especially in special and extreme habitats (see below). Moreover, the latitudinal biodiversity gradient observed (Fig. 8.2) is also present below the genus level, i.e., the total number of recognized species decreases toward the South (Vinarski et al. 2019). Given their biological characteristics and particularly their dispersal capacity, reproductive mode, and ecological flexibility, it has long been recognized that Lymnaeidae tend to speciation comparatively less often than caenogastropods (e.g., Boss 1978; Strong et al. 2008). In fact, there are no conspicuous species flocks known among the 32 recognized genera (but see below).

A superspecies concept has often been applied for widespread species such as *Radix auricularia* (Hubendick 1951) and *Lymnaea stagnalis* (Vinarski 2015). However, a detailed examination of seemingly widespread lymnaeid species has repeatedly revealed the existence of cryptic species or species complexes. A typical example is the endemic genus *Tibetoradix*, which originally was considered to belong to *Radix* and to be represented by only a single species, namely *Lymnaea*

Table 8.1 Distribution table of the 32 recognized lymnaeid genera across 22 accepted zoogeographical freshwater regions (see Fig. 8.1 and text for details). Note that Madagascar (region **4**) includes surrounding archipelagos such as the Mascarene Islands and that the Hawaiian Islands and other Pacific isolated archipelagos are subsumed under Indo-West-Pacific (region **8**). The nomenclature largely follows Vinarski et al. (2019), with additions from Vinarski et al. (2020) for radicine taxa

	Zoogeographical freshwater regions																						
Genus	1a	1b	1c	1d	1e	1f	1 g	1 h	2a	2b	2c	1–2	3a	3b	3c	4	5a	5b	1–5	6	7	8	Total regions per genus
Lymnaea	1	1		1	1	1	1		1														**7**
Acella	1																						**1**
Aenigmomphiscola				1			1																**2**
Bakerilymnaea	1	1				1																	**3**
Bulimnea	1	1																					**2**
Erinna																						1	**1**
Galba				1			1					1		1	1		1	1					**7**
Hinkleyia	1	1				1											?	?					**3**
Ladislavella	1	1	1		1	1	1	1	1														**8**
Omphiscola				1																			**1**
Pseudoisidora																						1	**1**
Pseudosuccinea	1		1			1												1	1				**5**
Sibirigalba					1		1		1														**3**
Sphaerogalba	1	1																					**2**
Stagnicola				1			1	1			1	1											**5**
Walterigalba	1	1				1																	**3**
Lanx		1																					**1**
Fisherola		1																					**1**
Idaholanx		1																					**1**
Myxas				1																			**1**
Ampullaceana										1													**1**

(continued)

Table 8.1 (continued)

Genus	Zoogeographical freshwater regions																						
	1a	1b	1c	1d	1e	1f	1 g	1 h	2a	2b	2c	1–2	3a	3b	3c	4	5a	5b	1–5	6	7	8	Total regions per genus
Austropeplea																				1	1		**2**
Bullastra											1											1	**2**
Kamchaticana							1																**1**
Lantzia																1							**1**
Limnobulla																	1						**1**
Orientogalba									1		1												**2**
Pectitidens																	1						**1**
Peregriana				1			1																**2**
Racesina				1							1	1											**3**
Radix				1	1		1	1	1	1			1	1	1	1							**10**
Tibetoradix										1													**1**
Generic diversity	**9**	**10**	**2**	**9**	**4**	**6**	**9**	**3**	**5**	**3**	**4**	**3**	**1**	**2**	**2**	**2**	**3**	**2**	**1**	**1**	**1**	**3**	

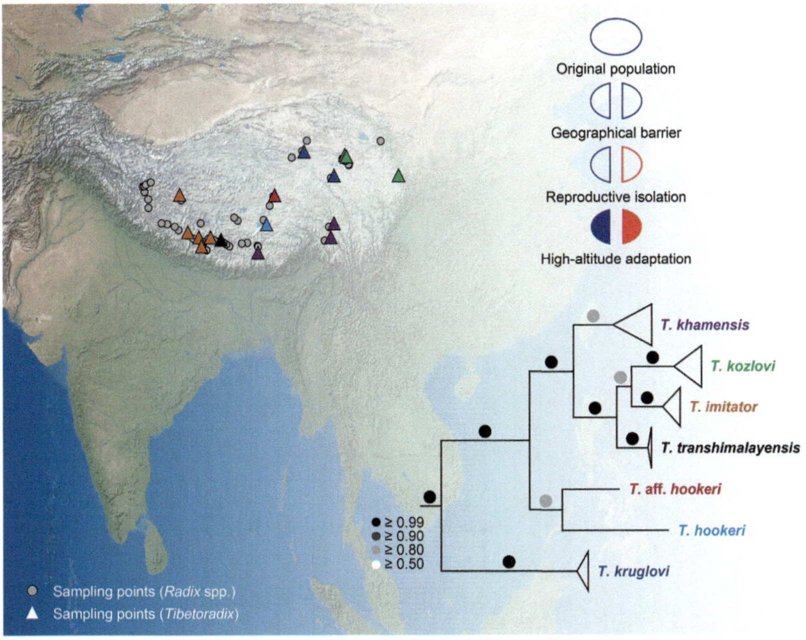

Fig. 8.3 *Tibetoradix* and other lymnaeids on the Tibetan Plateau. The molecular phylogeny (bottom-right) and the respective sampling (triangles in map) were redrawn after Vinarski et al. (2022). The remaining lymnaeid sampling points (gray circles in map) were compiled from the literature (von Oheimb et al. 2011; Clewing et al. 2016). For *Tibetoradix*, a process of allopatric speciation and high-altitude adaptation has been assumed (scheme on top-right), which resulted in a comparatively high number of species that are "trapped" in their environment (see Vinarski et al. 2022 for a discussion)

hookeri. However, new sampling from the Tibetan Plateau resulted in the recurrent number of six known species endemic to the plateau and small adjacent regions (Vinarski et al. 2022). Allopatric speciation, competitive exclusion, and high-altitude adaptation are interpreted for the pattern observed on the Tibetan Plateau (Fig. 8.3). Thereby, competitive exclusion likely prohibits the co-occurrence of several lineages (von Oheimb et al. 2011; Clewing et al. 2016; Vinarski et al. 2022), while adaptation to high altitudes probably reduces the chances of dispersal into lower regions (Vinarski et al. 2022).

Another prime example for misinterpreted widespread species relates to *Galba truncatula*, a species that turned out to hide several congeners, among them *G. schirazenzis*, an overlooked species (Bargues et al. 2013) that is even invasive (Lounnas et al. 2018). In Africa, *G. truncatula* has been confused with *G. mweruensis*, a species occurring in high-altitude mountains of eastern and South Africa (Fig. 8.4). According to Mahulu et al. (2019), sub-tropical Africa has been colonized only once in either the Pliocene or possibly as early as the Miocene. From here on, diversification within *G. mweruensis* started finally in the

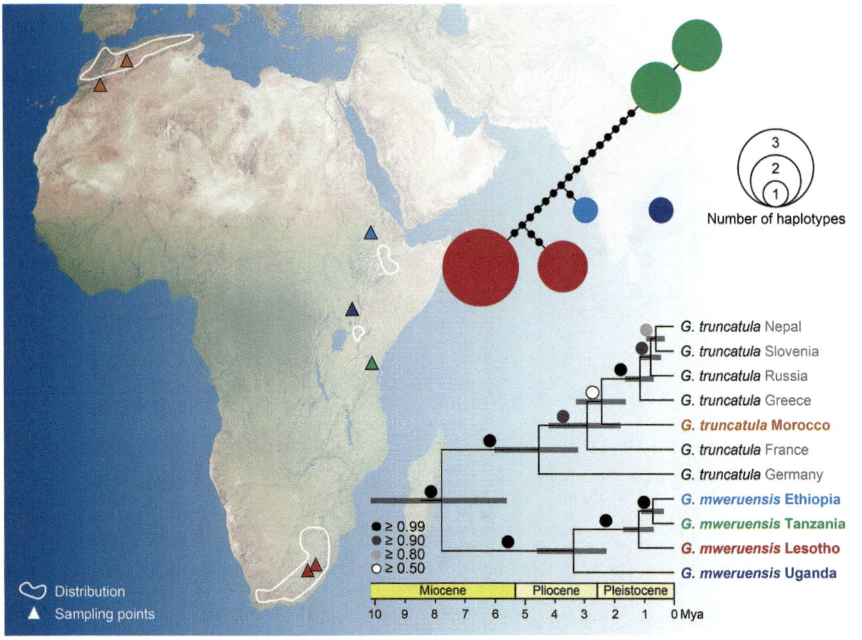

Fig. 8.4 Diversification of *Galba* on the African continent. The time-calibrated molecular phylogeny and the respective sampling (triangles in map) were redrawn after Mahulu et al. (2019). Both the phylogenetic relationships and the haplotype network (top-right) show that all African mountain populations (except those from Morocco) belong to a different species other than *G. truncatula* and that their genetic distinctiveness is probably a result of their Plio-Pleistocene origin

Plio-Pleistocene and is probably ongoing, leading to potentially even more cryptic species in high-altitude areas of Africa as exemplified by the genetic distinctness of the population from Mt. Elgon, Uganda (Fig. 8.4). The widely disjunct populations of *G. mweruensis* are indicative of allopatric speciation. However, it remains unclear whether allopatry or an "evolutionary trap" scenario as proposed for *Tibetoradix* or competitive exclusion by the widespread *Radix natalensis*, which actually represents a species complex (C. Albrecht & C. Clewing, unpubl. data), is responsible for the quite restricted distribution of that species.

Cryptic species have been found in several other lymnaeid genera, namely in *Tibetoradix* (Vinarski et al. 2022) and *Galba* (Bargues et al. 1997, 2013; Standley et al. 2013; Mahulu et al. 2019). Originally, cryptic species were assumed to be less common among Lymnaeidae (Vinarski et al. 2019). However, more recent studies on smaller geographical scales continue to reveal more diversity that is either morphostatic (Schniebs et al. 2013) or obscured by phenotypic plasticity (Vinarski et al. 2020). In a recent comprehensive and geographically large-scale study, American representatives of the genus *Galba* were studied with respect to the presence of cryptic species (Fig. 8.5). It was concluded that morphostatic speciation in widely distributed and genetically diverse populations accounted for at least six cryptic

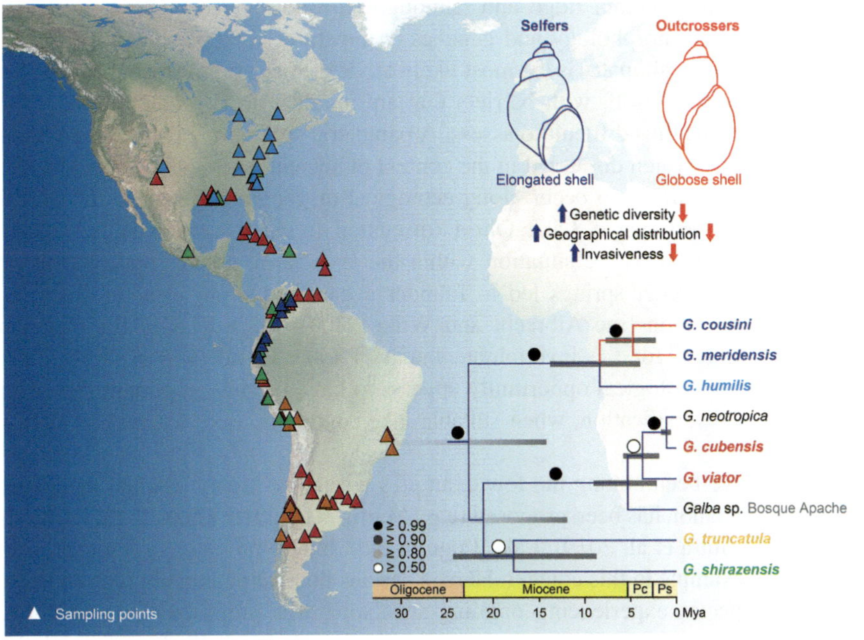

Fig. 8.5 Diversification of *Galba* on the American continent. The time-calibrated molecular phylogeny and the respective sampling (triangles in map) were redrawn after Alda et al. (2021). In selfing lineages (blue branches), shells are elongated, while genetic diversity, geographical distribution, and invasiveness are high (scheme on top-right). In contrast, outcrossing species (red branches) possess a globose shell and are less diverse and invasive and only occupy small ranges

species and that more detailed studies are needed to understand the roles of both ecology and mating system evolution (Alda et al. 2021).

8.2.2 Small-Scale Patterns

Surprisingly, only few studies addressed speciation in lymnaeids directly at smaller scales, i.e., in field studies from natural populations. However, some observational evidence for diversification patterns of sister species or diverging populations within a single species in *Radix* s.l., i.e., encompassing all radicine species, exists from confined areas. Isolated ecosystems, and specifically ancient lakes, can serve here again. Allopatric speciation has often been proposed, but should involve strict geographical barriers, preventing or reducing gene flow considerably (see, e.g., Bolnick and Fitzpatrick 2007).

A study on *Radix* including *R. relicta* and *R. pinteri* from lakes Ohrid and Prespa on the Balkans showed the difficulty in elucidating what (geographical) type of speciation is present in these sister lakes. The two species are geographically separated by a mountain ridge and, although still closely related, can be distinguished by shell morphology and genetics (Albrecht et al. 2008, but see Vinarski et al. 2020). Here, allopatry is the most likely mode of speciation, which may also act in the lake proper itself, with barriers that are intralacustrine, both vertically and horizontally, and thus difficult to assess. A parapatric speciation process (see Coyne and Orr 2004) is often discussed in the context of ancient lakes (Schön and Martens 2004) and is assumed to occur along ecological or geographical clines or habitat fragmentation settings in Lake Ohrid (discussed in Albrecht and Wilke 2008). However, less clear is the situation within the lake basin, where small peripheral populations in feeder springs led to founder events and strong genetic drift, i.e., peripatric speciation (see Albrecht and Wilke 2008). Regardless of the different modes presented, for the invertebrate fauna of Lake Ohrid (and possibly other ancient lakes), ecological opportunity appears to have been an important factor for the onset of diversification, when suitable lake conditions were reached (Stelbrink et al. 2020).

Lakes have been among the few examples where the highly debated sympatric mode of speciation has been proposed (e.g., Martin et al. 2015; Foote 2018; Richards et al. 2019; Sutra et al. 2019). Lake Lugu in the Chinese province of Yunnan is an interesting example in this context. Here, a species flock of potentially four endemic lymnaeid species experiencing ongoing speciation was proposed (Fig. 8.6). The lake's specific environment gave rise to neritid-like *Radix* shells three times independently (Wiese et al. 2020). However, syntopic occurrence, i.e., living in the same lake, is not synonymous with sympatric speciation (Bolnick and Fitzpatrick 2007). This type of speciation involves hybridization and reproductive barriers within a single, initially randomly mating population (Futuyma 2013). However, such processes have not been studied for any lymnaeid species assemblage and thus render a discussion premature. To date, microallopatry seems more likely responsible for population divergences at the intralacustrine scale. Moreover, either ecophenotypic shell-shape changes or morphostasis makes the detection and delimitation of diversification and speciation events challenging (see discussion on cryptic species above).

8.2.3 Adaptation in Extreme and Isolated Habitats: The Role of Ecology

Ecophenotypy in Lymnaeidae has been discussed for decades, and it is undoubted that the immediate environmental conditions of these gastropods play a role in shell size and shape parameters (Vinarski & Pointier, 2023, this book, Chap. 2). However, to what extent such plasticity is manifested through selective forces is less well-known. A look at the adaptations to extreme environments can serve here to identify potential cases, where ecological speciation (see Rundle and Nosil 2005; Nosil 2012) might be involved in species formation. Occurrence in extreme altitudes

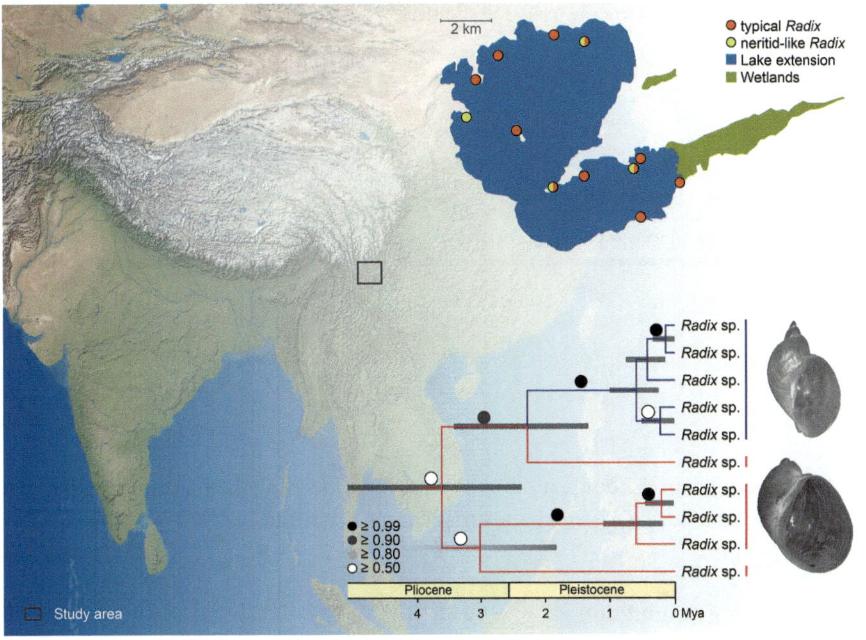

Fig. 8.6 *Radix* diversification in Lake Lugu (Yunnan, China). The time-calibrated molecular phylogeny was redrawn after Wiese et al. (2020). The phylogenetic relationships together with the genetic distances between the lineages suggest that neritid-like forms (lower shell in the phylogeny) evolved at least three times independently (red lineages) together with regular shell forms (blue lineages and upper shell in the phylogeny)

(up to and sometimes above 4000 m a.s.l.) have been reported for several *Galba* species across the Altiplano in South America (e.g., Alda et al. 2021) and for *G. mweruensis* in Afro-montane regions (Mahulu et al. 2019). Although high-altitude freshwater habitats provided unexplored niches, adaptations to these harsh conditions (freezing over extended periods of time, scarcity of food, etc.) are hypothesized to have led to "evolutionary dead ends" for the high-mountain fauna (Vinarski et al. 2022). On the other hand, such extreme habitats may also have acted as evolutionary reservoirs during severe environmental conditions. Populations of *Radix* from the Lake Donggi Cona drainage system (north-eastern Tibetan Plateau) were examined in order to study the impact of glaciation-related changes on the demographic history of this group (Clewing et al. 2016). Phylogenetic, phylogeographical, and demographic analyses suggested that not only a single refugium was present but also that isolated sub-refugia in and outside the lake proper existed and provided suitable habitats for ancestral *Radix* populations during the Last Glacial Maximum (Fig. 8.7).

Ancient lakes are known to host an extraordinary range of biodiversity and endemics (e.g., Brooks 1950; Martens 1997; Albrecht et al. 2020). Lymnaeidae are only showing moderate endemism in such lakes, which is apparently restricted to

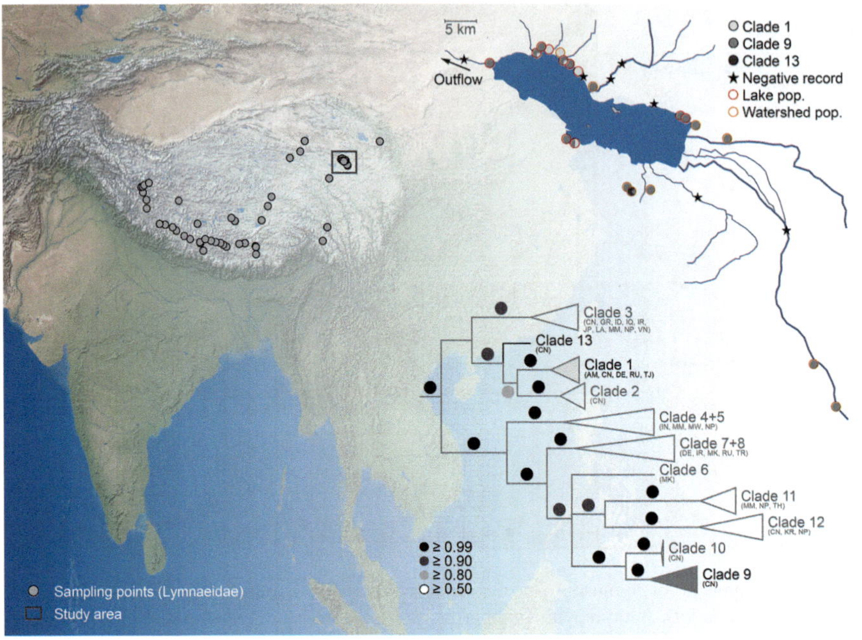

Fig. 8.7 Radicine lymnaeids on the eastern Tibetan Plateau. The molecular phylogeny and the Lake Donggi Cona and watershed sampling were adopted from Clewing et al. (2016). The remaining lymnaeid sampling points (grey circles in map) were compiled from the literature (von Oheimb et al. 2011; Clewing et al. 2016). The occurrence of genetically distinct lineages suggests the presence of sub-refugia during the Last Glacial Maximum that served as suitable habitats for ancestral populations

the radicine clade *Radix* s.l. (Vinarski et al. 2017, but see Neubauer, 2023, this book). A key ecological feature is the adaptation to sublittoral and even profundal habitats in these often very deep lakes. An impressive example is *Radix relicta* (Polinski, 1929) that occurs in depths up to 30 m owing to special physiological modifications related to respiration and tissue vascularization (Kaiser 1959; see Albrecht et al. 2008). Similar to isolated insular-like ecosystems such as ancient lakes, true islands sometimes host endemic forms of Lymnaeidae that are distinguished on the genus level, for example *Pseudisidora* and *Erinna* from the Hawaiian Islands. Species of these genera are only scarcely explored, and in addition, potentially cryptic forms exist (Christensen et al. 2021). These lymnaeids live in special habitats such as waterfalls and seeps, a situation also existent at La Reunion Island in the Indian Ocean, where a single limpet-shaped species, *Lantzia carinata*, is known only from very few waterfalls. Although these island taxa have not yet been included in lymnaeid phylogenetic studies, convergent evolution has potentially led to the limpet or neritid-like shell forms in the particular environment these gastropods are confined to.

Another factor presumably driving ecological speciation in lymnaeids is temperature. Here, both cold and warm-stenotherm forms have been described (von Oheimb et al. 2016; Vinarski et al. 2021), and in some cases, a closer examination revealed that supposedly endemic forms are rather ecotypes of widespread species (Vinarski et al. 2017; Vinarski et al. 2020).

8.3 Speciation Processes

Given the patterns observed at different geographical, temporal, and taxonomic scales, the actual underlying processes leading to the morphospace disparity and generally to diversification in Lymnaeidae should also be discussed. Following the mechanisms of initial divergence proposed by Schluter (2001), all four suggested categories can be discussed for the Lymnaeidae. (1) Polyploid speciation due to chromosomal rearrangements does not play a major role, unlike in other Hygrophila, especially the Planorbidae and Bulinidae (see Albrecht et al. 2019a, 2019b). In Lymnaeidae, chromosome numbers are rather uniform and range from the plesiomorphic haploid $N = 18$ to $N = 16$ and $N = 17$ in derived clades (Inaba 1969; Correa et al. 2010). To what extent hybridization is involved in species formation in Lymnaeidae is currently not well understood. (2) Speciation by genetic drift requires effective long-time isolation and dispersal barriers to allow for drift-based speciation in allopatry. Whereas long-term isolation is very plausible in insular systems as described above, it is more difficult to assume dispersal limitations given the colonization potential of most lymnaeids (Hubendick 1951). (3) Maybe more likely is ecological speciation due to adaptations to different selective regimes. This idea has been proposed earlier for lymnaeid taxa (Pfenninger et al. 2006) and is now supported by a recent study, which suggests that niche conservatism and varying climatic conditions (mainly temperature) most likely act as different selective regimes to *Radix* lineages (Saito et al. 2021). (4) Another category encompasses cases where differential adaptation to a similar selective regime occurs (divergence under uniform selection). However, such an initial cause of speciation is not easy to prove and might not be very common among lymnaeids.

Despite environmental factors driving speciation, biotic interactions can also lead to adaptations and even co-evolution. The best-studied system in this context is the interaction of lymnaeid intermediate hosts with trematode parasites (see, e.g., Vázquez et al. 2023 this book). More and more genomic approaches help to disentangle host–parasite relationships on the organismic level through simultaneous genotyping of gastropods and infecting trematode parasites using high-throughput amplicon sequencing (Hammoud et al. 2021). Moreover, the immunobiological interplay between gastropod and parasite (Alba et al. 2019, 2020) are starting to be understood on the molecular level (Juhász and Lawton 2022). Here, transcriptomes of the model species *Lymnaea stagnalis* revealed highly varying factors that may be the key determinants of differences in parasite resistance among individuals both within and between natural host populations (Seppälä et al. 2021).

The steadily growing number of mitogenomes, (annotated) genomes, and transcriptomes of lymnaeid species (e.g., Feldmeyer et al. 2010; Tills et al. 2015; Schell et al. 2017; Qin et al. 2019) will revolutionize our abilities to study genetic mechanisms involved in important life-history traits such as developmental time, age and size at maturity, chirality (Kuroda and Abe 2020; Davison and Neiman 2021), and even reproductive systems as a whole. For example, left–right asymmetry has been extensively studied (Davison 2020), but mainly in *L. stagnalis* (e.g., Davison et al. 2016; Kuroda et al. 2016). Chirality is a topic of general interest in developmental biology and has implications to speciation (Abe and Kuroda 2019). A debate around chirality-determining single genes in *L. stagnalis* has arisen, suggesting that chirality is maternally inherited by the actin-related gene *Lsdia1* (Kuroda and Abe 2020). To what extent this is directly related to speciation remains a question, with a "single-gene speciation"—though possible in theory—considered unlikely (Davison 2020). Other developmental traits relate to the reproductive system and the mode of reproduction. The evolution of hermaphroditism and its consequences for sexual selection is of particular interest here. Most recently, sexual selection has been shown to exist in *L. stagnalis*, something that used to be considered less important in hermaphroditic animals (Fodor et al. 2020). Moreover, selfing or outcrossing as the predominant mode of hermaphroditic reproduction has been supposed to be a key factor in diversification of American *Galba* (e.g., Alda et al. 2021; Fig. 8.5).

Radix balthica (= *Ampullaceana balthica*) emerges as a new model system to study developmental plasticity of functional traits that are partly genetically based (Tills et al. 2013a, 2013b). The relative importance of genetic drift vs. selective processes during the embryonic development can now be evaluated on the transcriptome level for both juvenile (Tills et al. 2015) and adult stages (Feldmeyer et al. 2011). Given the fast progress in the "omics" era, that is not reflected yet in mollusc and specifically gastropod studies (see Klein et al. 2019; Gomes-dos-Santos et al. 2020; Yang et al. 2020), very few studies have set such datasets in a specific speciation framework studying life-history traits directly related to fitness. An exception is the study by Feldmeyer et al. (2015) on European *Radix*, showing that species divergence is driven by positive selection on life-history traits such as development and growth-rate regulated genes, which are thought to be advantageous under contemporary climates with distinct climate niches occupied by the different species. Preadaptation to persisting climate regimes is also thought to be a key component in the invasion success of alien species. The North American *Pseudosuccinea columella* is a self-fertilizing global invader, in which new genotypes are found in many of the colonized regions, despite an otherwise very low genetic variability (Lounnas et al. 2017). Such cases provide ideal models for studying potentially rapid speciation in its initial stages (Fig. 8.8).

Fig. 8.8 Examples of live specimens of Lymnaeidae and selected extreme and isolated habitats. (**a**) *Radix relicta*, Lake Ohrid, North Macedonia; (**b**) *Lymnaea stagnalis*; (**c**) *Stagnicola* sp.; (**d**) Ancient Lake Ohrid, North Macedonia and Albania; (**e**) Cascade Maniquet, La Réunion, type locality of *Lantzia carinata*; (**f**) Jackson Pool on Mt. Elgon Uganda, highest known occurrence of *Galba mweruensis*; (**g**) High-altitude habitat at Lake Donggi Cona (Tibetan Plateau); (**h**) Hot spring on the eastern Tibetan Plateau where radicine specimens were found; (**i**) Wetland in the highlands of Lesotho, habitat of *Galba mweruensis*. Photo credit (Christian Albrecht: d, e, i; Catharina Clewing: g; Daniel Engelhard: f; Jan Hamrsky: b, c; Parm Viktor von Oheimb: h; Thomas Wilke: a)

References

Abe M, Kuroda R (2019) The development of CRISPR for a mollusc establishes the formin Lsdia1 as the long-sought gene for snail dextral/sinistral coiling. Development 146:dev175976. https://doi.org/10.1242/dev.175976

Aksenova OV, Bolotov IN, Gofarov MY et al. (2018) Species richness, molecular taxonomy and biogeography of the radicine pond snails (Gastropoda: Lymnaeidae) in the Old World. Sci Rep 8:11199. https://doi.org/10.1038/s41598-018-29451-1

Alba A, Vázquez AA, Sánchez J et al. (2019) Patterns of distribution, population genetics and ecological requirements of field-occurring resistant and susceptible *Pseudosuccinea columella* snails to *Fasciola hepatica* in Cuba. Sci Rep 9:14359. https://doi.org/10.1038/s41598-019-50894-7

Alba A, Duval D, Sánchez J et al. (2020) The immunobiological interplay between *Pseudosuccinea columella* resistant/susceptible snails with *Fasciola hepatica*: hemocytes in the spotlight. Dev Comp Immunol 102:103485. https://doi.org/10.1016/j.dci.2019.103485

Albrecht C, Wilke T (2008) Ancient Lake Ohrid: biodiversity and evolution. Hydrobiologia 615: 103–140. https://doi.org/10.1007/s10750-008-9558-y

Albrecht C, Wolff C, Glöer P, Wilke T (2008) Concurrent evolution of ancient sister lakes and sister species: the freshwater gastropod genus *Radix* in lakes Ohrid and Prespa. Hydrobiologia 615: 157–167. https://doi.org/10.1007/s10750-008-9555-1
Albrecht C, Stelbrink B, Clewing C (2019a) Planorbidae Rafinesque, 1815. In: Lydeard C, Cummings KS (eds) Freshwater mollusks of the world: a distribution atlas. Johns Hopkins University Press, Baltimore, pp 181–186
Albrecht C, Stelbrink B, Clewing C (2019b) Bulinidae P. Fischer & Cross, 1880. In: Lydeard C, Cummings KS (eds) Freshwater mollusks of the world: a distribution atlas. Johns Hopkins University Press, Baltimore, pp 167–171
Albrecht C, Salzburger W, Umba Tolo C, Stelbrink B (2020) Speciation in Ancient Lakes 8 – celebrating 25 years and moving towards the future. J Great Lakes Res 46:1063–1066. https://doi.org/10.1016/j.jglr.2020.08.009
Alda P, Lounnas M, Vázquez AA et al. (2021) Systematics and geographical distribution of *Galba* species, a group of cryptic and worldwide freshwater snails. Mol Phylogenet Evol 157:107035. https://doi.org/10.1101/647867
Appleton CC, Miranda NAF (2015) Two Asian freshwater snails newly introduced into South Africa and an analysis of alien species reported to date. African Invertebr 56:1–17. https://doi.org/10.5733/afin.056.0102
Ashworth AC, Preece RC (2003) The first freshwater molluscs from Antarctica. J Molluscan Stud 69:89–92
Bănărescu P (1990) Zoogeography of fresh waters. Volume 1. General distribution and dispersal of freshwater animals. Aula-Verlag, Wiesbaden
Bargues MD, Mangold AJ, Munoz-Antoli C et al. (1997) SSU rDNA characterization of lymnaeid snails transmitting human fascioliasis in south and Central America. J Parasitol 83:1086–1092
Bargues MD, Horák P, Patzner RA et al. (2003) Insights into the relationships of Palearctic and Nearctic lymnaeids (Mollusca: Gastropoda) by rDNA ITS-2 sequencing and phylogeny of stagnicoline intermediate host species of *Fasciola hepatica*. Parasite 10:243–255. https://doi.org/10.1051/parasite/2003103243
Bargues MD, Artigas P, Mera y Sierra RL et al. (2007) Characterisation of *Lymnaea cubensis, L. viatrix* and *L. neotropica* n. sp., the main vectors of *Fasciola hepatica* in Latin America, by analysis of their ribosomal and mitochondrial DNA. Ann Trop Med Parasitol 101:621–641. https://doi.org/10.1179/136485907X229077
Bargues MD, Artigas P, Khoubbane M et al. (2013) *Lymnaea schirazensis*, an overlooked snail distorting fascioliasis data: genotype, phenotype, ecology, worldwide spread, susceptibility, applicability. PLoS One 6:e24567. https://doi.org/10.1371/journal.pone.0024567
Bolnick DI, Fitzpatrick BM (2007) Sympatric speciation: models and empirical evidence. Annu Rev Ecol Evol Syst 38:459–487. https://doi.org/10.1146/annurev.ecolsys.38.091206.095804
Boss KJ (1978) On the evolution of gastropods in ancient lakes. In: Fretter V, Peake JF (eds) Pulmonates. Systematics, evolution and ecology. Academic Press, London, pp 385–428
Brooks JL (1950) Speciation in ancient lakes. Q Rev Biol 25(30–60):131–176
Brown DS (1994) Freshwater snails of Africa and their medical importance. Taylor and Francis, London
Burch JB (1989) North American freshwater snails. Malacological Publications, Hamburg, Michigan
Christensen CC, Hayes KA, Yeung NW (2021) Taxonomy, conservation, and the future of native aquatic snails in the Hawaiian islands. Diversity 13:215. https://doi.org/10.3390/d13050215
Clewing C, Albrecht C, Wilke T (2016) A complex system of glacial sub-refugia drives endemic freshwater biodiversity on the Tibetan Plateau. PLoS One 11:e0160286. https://doi.org/10.1371/journal.pone.0160286
Correa AC, Escobar JC, Durand P et al. (2010) Bridging gaps in the molecular phylogeny of the Lymnaeidae (Gastropoda: Pulmonata), vectors of Fascioliasis. BMC Evol Biol 10:381. https://doi.org/10.1186/1471-2148-10-381

Coyne JA, Orr HA (2004) Speciation. Sinauer, Sunderland, MA
Davison A (2020) Flipping shells! Unwinding LR asymmetry in mirror-image molluscs. Trends Genet 36:189–202. https://doi.org/10.1016/j.tig.2019.12.003
Davison A, Neiman M (2021) Mobilizing molluscan models and genomes in biology. Philos Trans R Soc London B 376:20200163. https://doi.org/10.1098/rstb.2020.0163
Davison A, McDowell GS, Holden JM et al. (2016) Formin is associated with left-right asymmetry in the pond snail and the frog. Curr Biol 26:654–660. https://doi.org/10.1016/j.cub.2015.12.071
Feldmeyer B, Hoffmeier K, Pfenninger M (2010) The complete mitochondrial genome of *Radix balthica* (Pulmonata, Basommatophora), obtained by low coverage shot gun next generation sequencing. Mol Phylogenet Evol 57:1329–1333
Feldmeyer B, Wheat CW, Krezdorn N et al. (2011) Short read Illumina data for the *de novo* assembly of a non-model snail species transcriptome (*Radix balthica*, Basommatophora, Pulmonata), and a comparison of assembler performance. BMC Genomics 12:317. https://doi.org/10.1186/1471-2164-12-317
Feldmeyer B, Greshake B, Funke E et al (2015) Positive selection in development and growth rate regulation genes involved in species divergence of the genus Radix. BMC Evol Biol 15:164. https://doi.org/10.1186/s12862-015-0434-x
Fodor I, Hussein AAA, Benjamin PR et al. (2020) The unlimited potential of the great pond snail, *Lymnaea stagnalis*. eLife 9:e56962. https://doi.org/10.7554/eLife.56962
Foote AD (2018) Sympatric speciation in the genomic era. Trends Ecol Evol 33:85–95. https://doi.org/10.1016/j.tree.2017.11.003
Futuyma DJ (2013) Evolution. Sinauer Associates, Sunderland, Mass
Gomes-dos-Santos A, Lopes-Lima M, Castro LFC, Froufe E (2020) Molluscan genomics: the road so far and the way forward. Hydrobiologia 847:1705–1726. https://doi.org/10.1007/s10750-019-04111-1
Hammoud C, Mulero S, Van Bocxlaer B et al. (2021) Simultaneous genotyping of snails and infecting trematode parasites using high-throughput amplicon sequencing. Mol Ecol Resour 22: 567–586. https://doi.org/10.1111/1755-0998.13492
Hillebrand H (2004) On the generality of the latitudinal diversity gradient. Am Nat 163:192–211. https://doi.org/10.1086/381004
Hubendick B (1951) Recent Lymnaeidae. Their variation, morphology, taxonomy, nomenclature, and distribution. K Sven Vetenskapsakademiens Handl 3:1–223
Hubendick B (1978) Systematics and comparative morphology of the Basommatophora. In: Fretter V, Peake J (eds) Pulmonates, Systematics, evolution and ecology, vol 2A. Academic Press, London, pp 1–47
Inaba A (1969) Cytotaxonomic studies of lymnaeid snails. Malacologia 7:143–168
Juhász A, Lawton SP (2022) Toll like receptors and their evolution in the lymnaeid freshwater snail species *Radix auricularia* and *Lymnaea stagnalis*, key intermediate hosts for zoonotic trematodes. Dev Comp Immunol 127:104297. https://doi.org/10.1016/j.dci.2021.104297
Kaiser P (1959) Über die Atmung von *Radix relicta* Pol. aus dem Ohrid-See. Recl des Trav / Stn Hydrobiol 7:1–5
Klein AH, Ballard KR, Storey KB et al. (2019) Multi-omics investigations within the phylum Mollusca, class Gastropoda: from ecological application to breakthrough phylogenomic studies. Brief Bioinform 18:377–394. https://doi.org/10.1093/bfgp/elz017
Kruglov ND (2005) Molluscs of the family Lymnaeidae (Gastropoda, Pulmonata) in Europe and northern Asia [in Russian]. Smolensk State Pedagogical University Press, Smolensk
Kuroda R, Abe M (2020) The pond snail *Lymnaea stagnalis*. Evodevo 11:24. https://doi.org/10.1186/s13227-020-00169-4
Kuroda R, Fujikura K, Abe M et al. (2016) Diaphanous gene mutation affects spiral cleavage and chirality in snails. Sci Rep 6:34809. https://doi.org/10.1038/srep34809

Lounnas M, Correa AC, Vázquez AA et al. (2017) Self-fertilization, long-distance flash invasion and biogeography shape the population structure of *Pseudosuccinea columella* at the worldwide scale. Mol Ecol 26:887–903. https://doi.org/10.1111/mec.13984

Lounnas M, Correa AC, Alda P et al. (2018) Population structure and genetic diversity in the invasive freshwater snail *Galba schirazensis* (Lymnaeidae). Can J Zool 96:425–435. https://doi.org/10.1139/cjz-2016-0319

Mahulu A, Clewing C, Stelbrink B et al. (2019) Cryptic intermediate snail host of the liver fluke *Fasciola hepatica* in Africa. Parasit Vectors 12:573. https://doi.org/10.1186/s13071-019-3825-9

Martens K (1997) Speciation in ancient lakes. Trends Ecol Evol 12:177–182. https://doi.org/10.1016/S0169-5347(97)01039-2

Martin CH, Cutler JS, Friel JP et al. (2015) Complex histories of repeated gene flow in Cameroon crater lake cichlids cast doubt on one of the clearest examples of sympatric speciation. Evolution 69:1406–1422. https://doi.org/10.1111/evo.12674

Neubauer TA (2023) The fossil record of the Lymnaeidae: revisiting a 200-Myr-long story of success. In: Vinarski MV, Vázquez AA (eds) The Lymnaeidae: a handbook on their natural history and parasitological significance. Springer, Cham

Nosil P (2012) Ecological speciation. Oxford Univ Press, Oxford

Pfenninger M, Cordellier M, Streit B (2006) Comparing the efficacy of morphologic and DNA-based taxonomy in the freshwater gastropod genus *Radix* (Basommatophora, Pulmonata). BMC Evol Biol 6:100

Puslednik L, Ponder WF, Dowton M, Davis AR (2009) Examining the phylogeny of the Australasian Lymnaeidae (Heterobranchia: Pulmonata: Gastropoda) using mitochondrial, nuclear and morphological markers. Mol Phylogenet Evol 52:643–659. https://doi.org/10.1016/j.ympev.2009.03.033

Qin D-M, Huang X-C, Yang L-M et al. (2019) Complete mitochondrial genome of the radicine pond snail *Radix plicatula* (Gastropoda: Lymnaeidae). Mitochondrial DNA Part B 4:2861–2862. https://doi.org/10.1080/23802359.2019.1661300

Remigio EA, Blair D (1997) Molecular systematics of the freshwater snail family Lymnaeidae (Pulmonata: Basommatophora) utilising mitochondrial ribosomal DNA sequences. J Molluscan Stud 63:173–185. https://doi.org/10.1093/mollus/63.2.173

Richards EJ, Servedio MR, Martin CH (2019) Searching for sympatric speciation in the genomic era. BioEssays 41:1900047. https://doi.org/10.1002/bies.201900047

Rundle HD, Nosil P (2005) Ecological speciation. Ecol Lett 8:336–352. https://doi.org/10.1111/j.1461-0248.2004.00715.x

Saito T, Hirano T, Ye B et al. (2021) A comprehensive phylogeography of the widespread pond snail genus *Radix* revealed restricted colonization due to niche conservatism. Ecol Evol 11:18446–18459. https://doi.org/10.1002/ece3.8434

Schäfer A (1997) Biogeographie der Binnengewässer: eine Einführung in die biogeographische Areal- und Raumanalyse in limnischen Ökosystemen. Teubner, Stuttgart

Schell T, Feldmeyer B, Schmidt H et al. (2017) An annotated draft genome for *Radix auricularia* (Gastropoda, Mollusca). Genome Biol Evol 9:585–592. https://doi.org/10.1093/gbe/evx032

Schluter D (2001) Ecology and the origin of species. Trends Ecol Evol 16:372–380

Schniebs K, Peter G, Vinarski MV et al. (2013) Intraspecific morphological and genetic variability in the European freshwater snail *Radix labiata* (Rossmaessler, 1835) (Gastropoda: Basommatophora: Lymnaeidae). Contrib to Zool 82:55–68. https://doi.org/10.1163/18759866-08201004

Schön I, Martens K (2004) Adaptive, pre-adaptive and non-adaptive components of radiations in ancient lakes: a review. Org Divers Evol 4:137–156. https://doi.org/10.1016/j.ode.2004.03.001

Seppälä O, Walser J-C, Cereghetti T et al. (2021) Transcriptome profiling of *Lymnaea stagnalis* (Gastropoda) for ecoimmunological research. BMC Genomics 22:144. https://doi.org/10.1186/s12864-021-07428-1

Standley CJ, Prepelitchi L, Pietrokovsky SM et al. (2013) Molecular characterization of cryptic and sympatric lymnaeid species from the *Galba/Fossaria* group in Mendoza Province, Northern Patagonia, Argentina. Parasit Vectors 6:304. https://doi.org/10.1186/1756-3305-6-304

Stelbrink B, Wilke T, Albrecht C (2020) Ecological opportunity enabled invertebrate radiations in ancient Lake Ohrid. J Great Lakes Res 46:1156–1161. https://doi.org/10.1016/j.jglr.2020.06.012

Strong EE, Gargominy O, Ponder WF, Bouchet P (2008) Global diversity of gastropods (Gastropoda; Mollusca) in freshwater. Hydrobiologia 595:149–166. https://doi.org/10.1007/s10750-007-9012-6

Sutra N, Kusumi J, Montenegro J et al. (2019) Evidence for sympatric speciation in a Wallacean ancient lake. Evolution 73:1898–1915. https://doi.org/10.1111/evo.13821

Taylor DW (1988) Aspects of freshwater mollusc ecological biogeography. Palaeogeogr Palaeoclimatol Palaeoecol 62:511–576

Tills O, Bitterli T, Culverhouse P et al. (2013a) A novel application of motion analysis for detecting stress responses in embryos at different stages of development. BMC Bioinformatics 14:37. https://doi.org/10.1186/1471-2105-14-37

Tills O, Rundle SD, Spicer JI (2013b) Parent–offspring similarity in the timing of developmental events: an origin of heterochrony? Proc R Soc London B 280:20131479. https://doi.org/10.1098/rspb.2013.1479

Tills O, Truebano M, Rundle S (2015) An embryonic transcriptome of the pulmonate snail *Radix balthica*. Mar Genomics 24:259–260. https://doi.org/10.1016/j.margen.2015.07.014

Tracey S, Todd JA, Erwin DH (1993) Mollusca: Gastropoda. In: Benton MJ (ed) The fossil record 2. Chapman and Hall, London, pp 131–167

Vázquez AA, Alba A, Alda P, Vittecoq M, Chapuis E, Faugère D, Pointier J-P, Hurtrez-Boussès S (2023) Lymnaeid snails and the transmission of fasciolosis: understanding the differential risks from local to global scale. In: Vinarski MV, Vázquez AA (eds) The Lymnaeidae: a handbook on their natural history and parasitological significance. Springer, Cham

Vinarski MV (2013) One, two, or several? How many lymnaeid genera are there? Ruthenica Russian Malacol J 23:41–58

Vinarski MV (2015) Conceptual shifts in animal systematics as reflected in the taxonomic history of a common aquatic snail species (*Lymnaea stagnalis*). Zoosystematics Evol 91:91–103. https://doi.org/10.3897/zse.91.4509

Vinarski MV, Pointier J-P (2023) General characteristics of the family Lymnaeidae. In: Vinarski MV, Vázquez AA (eds) The Lymnaeidae: a handbook on their natural history and parasitological significance. Springer, Cham

Vinarski MV, Bolotov IN, Schniebs K et al. (2017) Endemics or strangers? The integrative re-appraisal of taxonomy and phylogeny of the Greenland Lymnaeidae (Mollusca: Gastropoda). C R Biol 340:541–557. https://doi.org/10.1016/j.crvi.2017.09.005

Vinarski MV, Clewing C, Albrecht C (2019) Lymnaeidae Rafinesque, 1815. In: Lydeard C, Cummings KS (eds) Freshwater mollusks of the world: a distribution atlas. Johns Hopkins University Press, Baltimore, pp 158–162

Vinarski MV, Aksenova OV, Bolotov IN (2020) Taxonomic assessment of genetically-delineated species of radicine snails (Mollusca, Gastropoda, Lymnaeidae). Zoosystematics Evol 96:577–608. https://doi.org/10.3897/zse.96.52860

Vinarski MV, Aksenova OV, Bespalaya YV et al. (2021) One Beringian genus less: a re-assesment of Pacifimyxas Kruglov & Starobogatov, 1985 (Mollusca: Gastropoda: Lymnaeidae) questions the current estimates of Beringian biodiversity. J Zool Syst Evol Res 59:44–59. https://doi.org/10.1111/jzs.12411

Vinarski MV, von Oheimb PV, Aksenova OV et al. (2022) Trapped on the roof of the world: taxonomic diversity and evolutionary patterns of Tibetan Plateau endemic freshwater snails (Gastropoda: Lymnaeidae: *Tibetoradix*). Integr Zool 17(5):825–848. https://doi.org/10.1111/1749-4877.12600

Vinarski MV, Aksenova OV, Bolotov IN, Vázquez AA, Alda P, Pointier J-P, Hurtrez-Boussès S (2023) Biogeography of the living Lymnaeidae. In: Vinarski MV, Vázquez AA (eds) The Lymnaeidae: a handbook on their natural history and parasitological significance. Springer, Cham

von Oheimb PV, Albrecht C, Riedel F et al. (2011) Freshwater biogeography and limnological evolution of the Tibetan plateau – insights from a plateau-wide distributed gastropod taxon (*Radix* spp.). PLoS One 6:e26307. https://doi.org/10.1371/journal.pone.0026307

von Oheimb PV, Landler L, von Oheimb KCM (2016) Cold snails in hot springs: observations from Patagonia and the Tibetan Plateau. Malacologia 59:313–320. https://doi.org/10.4002/040.059.0209

Walter HJ (1968) Evolution, taxonomic revolution, and zoogeography of the Lymnaeidae. Bull Am Malacol Union 34:18–20

Wiese R, Clewing C, Albrecht C et al. (2020) How ancient is Lake Lugu (Yunnan, China)? The gastropods' viewpoint with focus on *Radix* (Lymnaeidae). J Great Lakes Res 46:1099–1112. https://doi.org/10.1016/j.jglr.2020.06.003

Yang Z, Zhang L, Hu J et al. (2020) The evo-devo of molluscs: insights from a genomic perspective. Evol Dev 22:409–424. https://doi.org/10.1111/ede.12336

Zilch A (1960) Gastropoda (Euthyneura). In: Schindewolf OH (ed) Handbuch der Paläozoologie. Gebrüder Borntraeger, Berlin, pp 91–102

Chapter 9
Ecology of Lymnaeid Snails

Maxim V. Vinarski and Olga V. Aksenova

Abstract The ecology of the lymnaeid snails, including their habitat preferences, resistance to abiotic factors, and interactions with other species of freshwater organisms, is briefly reviewed and accompanied by numerous illustrations. A special section deals with the patterns of life cycles, reproduction, and fecundity of the pond snails.

The family Lymnaeidae belongs to a taxonomic group (superorder) called Hygrophila—the "moisture lovers" in literal translation from Latin. These aquatic or, less commonly, semiaquatic (amphibious) snails are typical dwellers of inland waterbodies and watercourses of different sorts—large and small lakes, rivers, streams, brooks, and springs, ponds and ephemeral pools, swamps, and marshes (Figs. 9.1 and 9.2) as well as artificially created or transformed by humans reservoirs (Fig. 9.3). Some representatives of the Hygrophila can form sustainable populations in less typical habitats like brackish water sea gulfs and hydrothermal springs (Fig. 9.4); on very rare occasions, these snails can be found in cave waterbodies (e.g., Vinarski and Palatov 2018). The ecology and lifestyle of the lymnaeid snails are characteristic of the Hygrophila as a whole, and the main aim of this chapter is to outline the lymnaeid ecology, their habitat preferences, tolerance to various abiotic factors, and biotic interactions with other organisms with which they share their habitats. However, the different aspects of the relationship between the lymnaeids and a certain group of their parasites—trematode larvae—are omitted in this chapter as this matter is reviewed in detail in other chapters of this book (Rondelaud

M. V. Vinarski (✉)
Laboratory of Macroecology & Biogeography of Invertebrates, Saint-Petersburg State University, Saint-Petersburg, Russia

S.I. Vavilov Institute for the History of Science and Technology, St. Petersburg Branch, Russian Academy of Sciences, Saint-Petersburg, Russia

O. V. Aksenova
N. Laverov Federal Center for Integrated Arctic Research, The Ural branch of the Russian Academy of Sciences, Arkhangelsk, Russia

M. V. Vinarski, A. A. Vázquez (eds.), *The Lymnaeidae*, Zoological Monographs 7,
https://doi.org/10.1007/978-3-031-30292-3_9

Fig. 9.1 Examples of large permanent waterbodies maintaining populations of lymnaeid snails. (**a**) Russia, Lake Baikal, Peschanaya Bay (*Radix auricularia*). (**b**) Kyrgyzstan, Lake Issyk-Kul' in Cholpon-Ata (*Radix obliquata*). (**c**) Uganda, Lake Victoria (*Radix natalensis*). (**d**) Russia, Western Siberia, Mrassu River (*Ampullaceana* sp., *Radix auricularia*). (**e**) Russia, Yakutsk, Lena River (*Lymnaea stagnalis, Ampullaceana lagotis*). (**f**) Russia, Republic of Karelia, Lake Onega (*Stagnicola palustris, Ampullaceana fontinalis*). (**g**) Switzerland, Geneva, Lake Geneva (*Ampullaceana balthica, Radix auricularia, Stagnicola* sp.). (**h**) Russia, Chukotka Peninsula, Iultinsky District, Chirovoye Lake (*Dallirhytis atkaensis*). Photos: Maxim V. Vinarski (A–D) & Olga V. Aksenova (E–H)

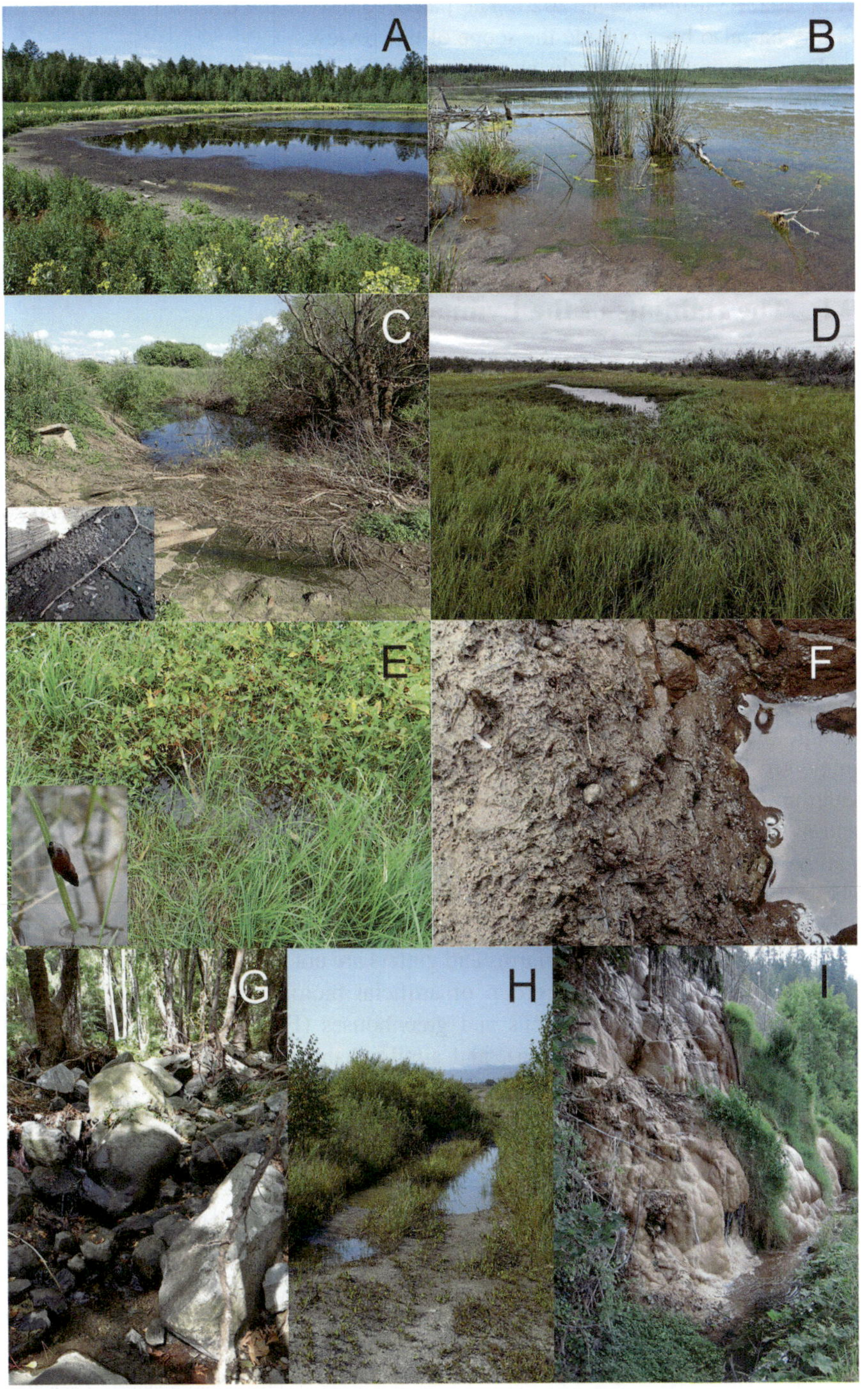

Fig. 9.2 Small permanent and non-permanent habitats of various lymnaeid snails. (**a**) Russia, Republic of Yakutia, Yakutsk City, a small eutrophic lakelet (*Lymnaea stagnalis*, *Ampullaceana lagotis*). (**b**) Canada, Yukon Territory, a small swamp (*Ladislavella catascopium*). (**c**) Russia, Jewish Autonomous Region, ephemeral pool near the road and dried snail shells nearby (*Radix*

et al. 2023; Vázquez et al. 2023. The paleoecology of the extinct members of the family has also been outside the scope of this review.

For a more detailed discussion of the lymnaeid ecology, the reader is recommended to consult special monographs and review articles dealing with the subject (Boycott 1936; Wesenberg-Lund 1939; Boettger 1944; Frömming 1956; Russell-Hunter 1964, 1978; McMahon 1983; Beriozkina and Starobogatov 1988; Økland 1990; Dillon 2000; Stadnichenko 2006).

9.1 The Habitats of the Lymnaeid Snails

If we consider the family Lymnaeidae as a whole, it demonstrates a very wide range of the habitats where its representatives can dwell, and therefore can be considered as ecologically very plastic. The subcosmopolitan distribution of the living Lymnaeidae also corresponds to this characteristic. However, an "average" lymnaeid species is a dweller of small or medium-sized lentic waterbodies restricted to their inshore zones—shallow, relatively warm, and rich in light and vegetation. The typical depth of their occurrence is less than 1 meter (Beriozkina and Starobogatov 1988). Populations of lymnaeid snails can be occasionally found in small- and medium-sized rivers, where the molluscs are living on and under submerged stones (see Fig. 9.1d).

Mostly herbivorous animals, the lymnaeids are found mainly on the surface of aquatic plants—their stems or leaves, as well as on the surface of various objects immersed in water—stones, logs, man-made rubbish (Fig. 9.5). The pieces of the latter are often used by lymnaeids both for scraping microalgae from their surface and as a substrate for the egg-laying (Fig. 9.5h–m). At least some species of the family are rather tolerant to the anthropogenic garbage and can become relatively abundant even in waterbodies heavily polluted by household and industrial waste. The artificial habitats such as channels, dammed reservoirs, road ditches, tanks, and even drinking troughs at cattle farms and gutters are not always deprived of the pond snails (Fig. 9.3). A peculiar group of artificial biotopes is composed of heated waterbodies of botanical gardens and greenhouses (Fig. 9.3f), cooling ponds of thermal power plants (Fig. 9.3g), and similar habitats whose thermal regime is drastically altered as compared with that of "natural" waterbodies. These unusually

---

Fig. 9.2 (continued) auricularia). (**d**) Russia, Republic of Yakutia, Indigirka River basin, overflow of a small stream (*Radix* sp.). (**e**) Russia, Primorsky Territory, a small puddle on the shore of the Utinoye Lake (*Ladislavella liogyra*). (**f**) Russia, Sakhalin Island, vicinity of Nogliki settlement, a wet clay on the Jimdan River bank (? *Kamtschaticana kamtschatica*). (**g**) Cyprus, Troodos Mountains, Kalidonia River basin, mountain creek (*Galba* sp.). (**h**) Russia, Primorsky Territory, Partizanskaya River basin, puddles in a rut on a dirt road (*Ladislavella liogyra*). (**i**) Canada, British Columbia, Ainsworth Thermal Springs, warm stream (*Ladislavella catascopium*). Photos: Olga V. Aksenova

Fig. 9.3 Examples of artificial habitats maintaining populations of lymnaeid snails. (**a**) Tajikistan, Dushanbe City, a fountain near the President's Palace (*Radix euphratica*). (**b**) India, Uttarakhand State, a gutter in Purola Town (*Galba* cf. *truncatula*). (**c**) Canada, Quebec Province, a small waterbody formed by a dam built at Lake d'Anguille (*Ampullaceana balthica*, invasive). (**d**) Russia, Rostov Region, the Volga-Don canal (*Lymnaea stagnalis*, *Radix auricularia*, *Stagnicola palustris*). (**e**) Myanmar, Loikaw District, rice field (*Radix* sp.). (**f**) Russia, Krasnodarsky Territory,

warm habitats, characterized by very slight temperature fluctuations throughout the year had served as "stepping stones" for many exotic lymnaeids and other non-native pulmonates on their way to Europe and other countries lying within the temperate latitudes (see, among others, Meeuse and Hubert 1949; Eichler 1952; Leiß and Reischütz 1995; Horsák et al. 2004; Pointier 2015; Vinarski 2017; Vinarski et al. 2022a). Using such "thermal refugia", some exotic pulmonates have managed even to colonize waterbodies situated in the Arctic zone (Nekhaev and Palatov 2016; Vinarski et al. 2021).

Evidently, the only freshwater environment not conquered by the lymnaeids is the subterraneous waterbodies, including the cave streams and lakes.

The most widespread lymnaeid species are, as a rule, the most ecologically tolerant and eurybiontic, being found in very different types of waterbodies. The typical examples of such species are the Holarctic lymnaeids *Lymnaea stagnalis* (Linnaeus, 1758) s. lato, *Radix auricularia* (Linnaeus, 1758), and *Ladislavella catascopium* (Say, 1817). Another instance is the native Nearctic *Pseudosuccinea columella* (Say, 1817), which has become an effective sub-global invader, being now found in most continents beyond the Americas. The African *Radix natalensis* (Krauss, 1848) was observed by us in Uganda to live in a variety of habitats, ranging from the inshore zone of Lake Victoria to small pools on the shores of this lake (Vinarski and Bolotov, pers. observations; see also van Someren 1946; Brown 1994, el-Dafrawi 2002, and Dar et al. 2010 on the ecology of this species).

Examples of lymnaeid species having a relatively narrow ecological niche and occurring in a few (or in a single) habitat types can also be cited. *Omphiscola glabra* (O.F. Müller, 1774), the mud pond snail, is endemic to Europe, a rare species known to spend life almost amphibiotically in small, sometimes ephemeral, ditches, mires, swampy meadows, and pools rich in vegetation; the periodical desiccation of a waterbody is a normal living condition for this snail which cannot be found in relatively large ponds and lakes (Økland 1990; Jackiewicz 1998; Vignoles et al. 2017; Glöer 2019). The members of the sister taxon of *Omphiscola*, the genus *Aenigmomphiscola*, which is endemic to West Russia and Kazakhstan, seem to lead approximately the same lifestyle though, due to their extreme rarity, the ecology of these snails remains poorly studied (Vinarski and Grebennikov 2012). In North America, an ecological analog is *Hinkleyia caperata* (Say, 1829)—a lymnaeid snail found most often in seasonal waterbodies. This species is characteristic of habitats like irrigation ditches, sloughs, and shallow ponds (Taylor et al. 1963; Clarke 1973).

An interesting case of convergence in shell shape caused, probably, by the need to adapt to life under similar environmental conditions, represent two lymnaeid species whose ranges are separated by thousands of kilometers. The first, *Erinna newcombi*

Fig. 9.3 (continued) Kabardinka settlement, Old Park, a bowl with water lilies (*Radix euphratica*). (**g**) Russia, Kamchatka Peninsula, Petropavlovsk-Kamchatskiy City, TPP channel (*Radix auricularia*). (**h**) Russia, Primorsky Territory, drainage ditch (*Radix auricularia*). Photos: Maxim V. Vinarski (A–D) & Olga V. Aksenova (E–H)

Fig. 9.4 Examples of hydrothermal habitats of various lymnaeid snails. (**a**) Iceland, thermal springs near Geysir (*Ampullaceana balthica*). (**b**) Russia, Kamchatka, Verkhe-Paratunskiye thermal springs (*Radix auricularia*). (**c**) USA, Yellowstone National Park, thermal lakelet near Mammoth Hot Springs (*Ladislavella catascopium*). (**d**) Russia, Kamchatka, thermal swamp near Karymshinskiye Hot Springs (*Kamtschaticana kamtschatica*). Photos: Olga V. Aksenova

Fig. 9.5 Examples of lymnaeid snails found on various surfaces and objects, and various substrates for their egg-laying. (**a**) Canada, Yukon Territory, Enger Lake, shrub branches (*Lymnaea stagnalis*). (**b**) Russia, Republic of Yakutia, the Lena River floodplain, leaves of macrophytes (*Ampullaceana lagotis*). (**c**) Canada, Yukon Territory, eutrophic lake, filamentous algae (*Ladislavella catascopium*). (**d**) Russia, Nenets Autonomous Okrug, Vashutkiny Lakes, silty sand (*Stagnicola palustris*). (**e**) Canada, Yukon Territory, Fox Lake, stones (*Ladislavella catascopium*). (**f**) Italy, Metauro River, stones (*Peregriana peregra*). (**g**) Russia, Republic of Yakutia, small eutrophic lake, old refrigerator on the shore (*Lymnaea stagnalis*). (**h**) Iceland, vicinity of Reykjavik, siding fragment from creek (*Ampullaceana balthica* and its egg masses). (**i**) Russia, Krasnoyarsk Territory, Krasnoyarsk City, the Yenisei River, an empty beer can submerged in water (*Radix auricularia* and its eggs). (**j**) Russia, Nenets Autonomous Okrug, Vashutkiny Lakes, eggs on the macrophytes (*Radix auricularia*). (**k**) Russia, Nenets Autonomous Okrug, Pymvashor thermal springs, egg-masses on the stones (*Ampullaceana lagotis*). (**l**) Russia, Kamchatka, Krasikovskoye Lake, an egg mass on the shell of mussels (*Radix auricularia*). (**m**) Russia, Chukotka Peninsula, Iultinsky District, Chirovoye Lake, an egg mass on the shell of pond snails (*Dallirhytis atkaensis*). Photos: Olga V. Aksenova

(H. & A. Adams, 1855) of the Hawaii Islands, according to Burch (1968, p. 28) "lives on rocks in swiftly flowing water, usually on the nearly vertical faces of waterfalls" (see also Hubendick 1952). The habitat of *Lantzia carinata* (Jousseaume, 1872), which is endemic to the Reunion Island in the Indian Ocean, is also waterfalls

(Brown 1994). Both species have independently evolved a similar neritiform shell, almost perfectly round, with a dramatically enlarged aperture and an almost invisible spire.

The existence of such independently evolving, not closely related, groups of lymnaeid species that demonstrate similar shell habitus and have roughly identical lifestyles has led some researchers to propose an ecomorphological classification of the family, with dividing it in a series of "life-forms" (Kruglov and Starobogatov 1987; Kruglov 2005; Nekhaev 2011). The term "life-form" has been taken from plant ecology where it has gained much more popularity than among zoologists (e.g., Du Rietz 1931; Humphreys et al. 2019). A lymnaeid "life-form" may be defined as a group of independently evolving genera and subgenera whose shells are characterized by a general resemblance in size and shape and that are adapted to similar living conditions. A good example is the "*Galba*-like" life-form which unites small-sized (typical shell length does not exceed 10–12 mm, usually smaller) amphibious or semi-amphibious lymnaeids living in temporary waterbodies of a different kind, including the smallest one (Kruglov and Starobogatov 1985; Kruglov 2005). The representatives of two genera, *Galba* (subfamily Lymnaeidae) and *Orientogalba* (subfamily Amphipepleinae) belong to this life-form (Kruglov 2005). Another example—the "*Myxas*-like" life-form comprising the European genus *Myxas* and the genera *Limnobulla* (the Falkland Islands), *Bullastra* (the Philippines) as well as *Austropeplea* of Australia and adjacent islands (Kruglov 2005).

Table 9.1 summarizes the vertical (altitudinal and bathymetric) distribution of the Lymnaeidae. The highest habitats of the pond snails are reported in the Tibetan Plateau, "the Roof of the World." The members of the endemic lymnaeid genus *Tibetoradix* are able to cope with very harsh abiotic conditions that exist at extreme altitudes above 4500 m a.s.l., which makes these snails one of the most high-altitude molluscs on Earth (see Bössneck 2012 for a review of high-altitudinal recordings of aquatic Mollusca). The highest lymnaeid habitat recorded in the literature (*Tibetoradix hookeri* collected at 18,000 feet = 5468 m a.s.l.; Reeve 1850) seems to be based on an error. No other records of *Tibetoradix*, or another freshwater snail species, from altitudes above 5000 m are known (Bössneck 2012; Aksenova et al. 2018; Gittenberger et al. 2021; Vinarski et al. 2022b).

The maximum depth which can be tolerated by the lymnaeid snails is, evidently, not exceeding 300 meters (see Table 9.1). The most deep-water populations of these molluscs were discovered as early as the late 1860s in the alpine lakes of Central Europe—in Switzerland and southern Germany (Forel 1869, 1874; Clessin 1877). These depth records are greatly overshadowed by some other representatives of aquatic pulmonates. For instance, in Lake Baikal, some endemic acroloxid snails were registered from depths ranging from 95 to 1000 meters (Shirokaya and Takhteev 2018). The populations of Lymnaeidae inhabiting large lakes are typically concentrated within the first meters of depth. Exceptions are relatively rare. Thus, *Radix auricularia* in Baikal can be found at depths between 0 and 30 meters (Schniebs et al. 2022), *R. auricularia* in Lake Issyk-Kul' in Kyrgyzstan occurs in water depths down to 65 meters (see Table 9.1). The same is true for the *Ampullaceana* species inhabiting lakes Ohrid and Prespa in southeast Europe.

Table 9.1 Vertical distribution of selected lymnaeid species

Species	Region, locality	Altitude, m a.s.l. / Depth, m	Reference(s)
High-altitude records			
Galba cousini (Jousseaume, 1887)	Venezuela	3594–4193	Pointier 2015
G. schirazensis (Küster, 1862)	Ecuador	2821–3158	Bargues et al. 2011
G. mweruensis (Connolly, 1929)	Uganda	1639–3939	Mahulu et al. 2019
G. truncatula (O.F. Müller, 1774)	Venezuela	1909–3988	Pointier 2015
G. viator (d'Orbigny, 1835)	Bolivia, the Bolivian Altiplano	3800–4200	Jabbour-Zahab et al. 1997;
Radix alticola (Izzatullaev et al., 1983)	Tajikistan, the Pamir Mts.	3360	Aksenova et al. 2018
R. auricularia (Linnaeus, 1758)	China, the Tibetan Plateau, the Lake Donggi Cona system	4090	Clewing et al. 2016; Vinarski et al. 2022b
R. narzikulovi (Izzatullaev et al., 1983[a]	Tajikistan, the Pamir Mts, Lakes Zorkul and Turumtaikul	4126–4213	Izzatullaev et al. 1983; Kruglov 2005
Tibetoradix spp. (six species)	China, the Tibetan Plateau	3544–4749	Vinarski et al. 2022b
T. cf. *hookeri* (Reeve, 1850)	Bhutan	4725	Gittenberger et al. 2021
Depth records from large lakes			
Ampullaceana balthica (Linnaeus, 1758)	Switzerland, Lake Geneva	30–50	Piaget 1912 (as *Limnaea foreli* and *L. limosa*)
A. balthica	Switzerland, Lake Geneva	100	Roszkowski 1914 (as *Limnaea ovata profunda*)
A. lagotis	Russia, Altai Republic, Lake Teletskoye	0–4	Our data
A. relicta (Poliński, 1929)	Albania/North Macedonia, lake Ohrid	< 30	Welter-Schultes 2012
"Limnaea" spp.	North America, Lake Superior	< 36.5	Cheatum 1934
"Limnaea" sp.	North America, Lake Michigan	25	Cheatum 1934
Lymnaea stagnalis (Linnaeus, 1758)	Switzerland, Lake Geneva	30–247	Piaget 1912 (as *Limnaea yungi*)

(continued)

Table 9.1 (continued)

Species	Region, locality	Altitude, m a.s.l. / Depth, m	Reference(s)
L. stagnalis	Switzerland, Lake Geneva	50	Clessin 1877 (as *Limnaea profunda*)
Radix auricularia	Germany, Lake Starnberger	15–25	Cheatum 1934
R. auricularia	Switzerland, Lake Geneva	20–25	Piaget 1913 (as *Limnaea limosa* var. *roszkowskiana* and *L. limosa* var. *sublittoralis*)
R. auricularia	Russia, Lake Baikal	0–30	Schniebs et al. 2022
R. obliquata (Martens, 1864)	Kyrgyzstan, Lake Issyk-Kul'	0–65	Pavlova 1983
Stagnicola palustris (O.F. Müller, 1774)	Switzerland, Lake Geneva	30–160	Piaget 1912 (as *Limnaea abyssicola*); Roszkowski 1914

[a]Most probably, a synonym of *Radix auricularia* (see Vinarski and Kantor 2016)

The lymnaeids inhabiting non-permanent habitats are adapted to aestivation that helps them survive when their waterbody is dried up. The snails are waiting out the dry period in moist soil or among dead leaves of aquatic plants (Cheatum 1934; McMahon 1983; Alyakrinskaya 2004). According to different researchers, the duration of stay in an anhydrous environment varies significantly among freshwater pulmonates, ranging from 1 month to 390 days (Alyakrinskaya 2004). Despite these adaptations, the survival rate of molluscs during the unfavorable season is very low (Fig. 9.2c). For instance, in two species of the genus *Stagnicola*, it was 19.5% after 3 months of drying up of the bog-type habitat (Stadnichenko 2006). Only 16.3% of *Peregriana peregra* individuals survived the 112-day drying up of a brook. According to Cheatum (1934), the survival rate in various lymnaeid snails subjected to a 62-day-long exposure to air (in soil) varied from 21 to 40%.

The survival rate of lymnaeids in shallow waterbodies that freeze to the bottom in winter varied in different species and different years from 1.3 to 23.7% (Stadnichenko 2006). The mortality rate is especially high among individuals infected with trematode larvae (Katkov 1971). The surviving individuals are severely dehydrated; thus, in the dwarf pond snail, water losses during the drying up of the reservoir reach 69% (Alyakrinskaya 2004).

9.2 Responses of the Lymnaeids to Abiotic Factors of Their Environment

The variety of abiotic factors the lymnaeid snails are faced with is enormous, and, quite usually, some of these factors are acting simultaneously, which creates a complex picture of many inorganic forces affecting snails and other aquatic animals in different directions. However, as a rule, the authors give an one-dimensional overview of the lymnaeid ecology analyzing, step by step, their response to particular abiotic factors (see, for example, Økland 1990; Dillon 2000). Among these factors, the ambient temperature, hydrogen ion (pH), and calcium concentration in water are usually considered the most important for the survival and reproduction of aquatic pulmonates, including the lymnaeids.

The following paragraphs provide a review of the responses of lymnaeids to selected abiotic factors, both of physical and chemical nature. An obvious shortcoming of our review is that it centered almost exclusively on the members of the family living in temperate regions of the North Hemisphere. This is explained both by the authors' research experience and the obvious fact that the ecology of the Lymnaeidae of the other world regions has attracted much less attention and thus remains much less known.

Temperature governs many aspects of the life of lymnaeid snails, including the duration of their embryonic development, postembryonic growth, longevity, the intensity of feeding, respiration as well as other metabolic processes (Vaughn 1953; Krkać 1979; Khmeleva et al. 1985; Pip 1993; Aziz and Raut 1996; Stadnichenko 2006; Leicht et al. 2017). The onset and end of their reproduction season (in a temperate climate) is also determined by this factor (Beriozkina and Starobogatov 1988). Most temperate species of this family have their temperature optima lying between 15 and 25 degrees of Celsius (Graevsky 1946; Pip 1993; Stadnichenko 2006; Johansson et al., 2016). The relatively cold-adapted snails, such as those living in cold springs, prefer temperatures between 10 and 13^0. The Lymnaeidae are relatively stenothermic animals, and the temperature below -0.4^0 C causes the cessation of their physical activity and a sharp decrease in the level of energy metabolism. Most lymnaeids, even those living under temperate climates, cannot survive the temperature drop below -2^0 C (Stadnichenko 2006). Lymnaeids hibernate buried in the ground, often among the roots of aquatic plants. Cases of wintering of young pond snails of different species in the leaf axils of hydrophytes were noted (Stadnichenko 2006). The activity of molluscs in winter is sharply reduced, they, as a rule, cease feeding, the intensity of gas exchange also decreases, and their lung is filled with water. Note, however, that at least some lymnaeid species, as well as other species of aquatic pulmonates, can overwinter being frozen into ice and recover to normal life after the ice is melted (Boldyreva 1930; Olsson 1981; Oswood et al. 1991). Such an ability is considered a prerequisite (i.e., preadaptation) for freshwater snails to live at low temperatures and colonize the high-latitude waterbodies (Vinarski et al. 2021).

The upper temperature threshold for the survival and growth in various species of temperate lymnaeids coincides roughly with 30^0 C (Vaughn, 1953; Axenov-Gribanov et al., 2015). No wonder, in tropical species it may be somewhat higher (i.e., Aziz and Raut 1996).

A case of special interest from the point of view of physiology, ecology, and microevolution is formed by those species of the Lymnaeidae (as well as of other freshwater gastropod families) that managed to infiltrate the geothermal ecosystems and to establish sustained populations in such extreme biotopes. The list of the "extremophilic" pond snails includes around 15 species, and the records of such thermal populations are known from all continents, apart from Africa (Vinarski et al. 2023; Vinarski and Pointier 2023) (Fig. 9.4). The springs, ponds, and swamps of hydrothermal sites are characterized by high temperature (> 25-30 ° C) that is maintained the year round, without drastic seasonal fluctuations. It makes them a unique environment for the high latitudes, inhabited by various taxa of invertebrates, including molluscs (Bespalaya et al. 2011; Bolotov et al. 2012).

The thermal populations of lymnaeids exhibit some degree of morphological and genetic specificity that makes them phenotypically and physiologically distant from their conspecifics living in "zonal" waterbodies (Bolotov et al. 2012; Aksenova et al. 2017a, b; Takhteev et al. 2019). Most probably, these changes reflect the first steps of the speciation process. For instance, Khmeleva et al. (1985) showed experimentally that individuals of *Radix* from geothermal sites of Kamchatka retain viability under low temperatures (5–7 ° C) but completely lose their ability to reproduce. In the temperature range, 12–20 ° C, the reproduction of these snails is inhibited (decreased frequency of egg-laying), while the duration of embryonic development increases drastically. These "hot" snails have their thermal optima shifted to 30–35^0 C, whereas the upper temperature threshold for their survival is near 42^0 C (Khmeleva et al., 1985). Their populations are characterized by drastically reduced life cycles, dwarf sizes, and untypical, "juvenilized" shell phenotypes (Vinarski et al. 2023). They have become able to reproduce all year round. However, most nominal species and varieties of the geothermal lymnaeids, which were described during the nineteenth and twentieth centuries (i.e., Kruglov and Starobogatov 1989), appeared to be junior synonyms of widely distributed species. Though the genetic methods have revealed unique haplotypes among these "hot" lymnaeids, their overall genetic distance is not substantial enough to regard them as anything more than the intraspecific ecological races (Bolotov et al. 2014; Aksenova et al. 2017a, b; Bolotov et al. 2017; Vinarski et al. 2023; but see Takhteev et al. 2019 for an alternative opinion).

The snails, living in such habitats, usually feed on the algal-bacterial mat and filamentous algae (Fig. 9.4). In most cases, the mollusc communities in geothermal springs and swamps consist of a single or two species, which makes the interspecific competition highly relaxed in such habitats. As a result, the lymnaeid populations can reach very high densities, up to several thousand individuals per square meter (i.e., Takhteev et al. 2006). Another ecological and physiological benefit of being adapted to live under extremely hot temperatures is that these snails do not need to cope with thermal fluctuations characteristic for the "seasonal" waterbodies and thus are not forced to develop special adaptations for overwintering, periodical food shortage, and so on (Bespalaya et al. 2011; Bolotov et al. 2012).

The factors of water velocity and wave action seem to be of lowered importance for the lymnaeid snails, as these molluscs prefer to dwell in lentic habitats (see above). Lymnaeid populations in brooks, streams, and rivers are not numerous and, typically, the density of snails in such environments is comparatively low. The lymnaeids that inhabit the offshore zones of large lakes, for example, the stony littoral of mountain lakes, are exposed to heavy wave action. The shell phenotypes of such lymnaeids, with their extremely inflated body whorl, diminutive spire, and thick walls (Fig. 9.5), represent a textbook example of a morphological modification caused by a need to cope with wave exposure (Kobelt 1871; Abel 1920). In many older works, it served as an illustration of the Lamarckian mechanism of the direct and heritable influence of the environment on animal morphology (i.e., Baker 1928). However, the correspondence between the ear-shaped shell and the habitat is not absolute. In other words, contrary to Kobelt's (1871) opinion, not in all cases when the appropriate conditions are present, the ear-shaped morphotype is formed. It was mentioned by Roszkowski as early as 1925 (Roszkowski 1925).

Light is another physical factor having some importance for the lymnaeids, most of which inhabits low depths where the sunlight is usually easily available. The abundance of light in such habitats determines the development of macrophytes, which serve as a primary food source for the lymnaeids and other phytophilous gastropods. The influence of photoperiod on physiological processes, growth, and reproduction of aquatic pulmonates (lymnaeids included) has been studied in many works (i.e., Kunigelis and Saleuddin 1978; Numata and Udaka 2009; Ter Maat et al. 2012; Kitai et al. 2021). The pond snails inhabiting mountain lakes at depths exceeding 250 m (see above) must spend their whole life in almost complete darkness, however, their physiological response to the lack of abundant light is virtually not studied.

It would be impossible to discuss here, even briefly, the impact of all chemical factors on the physiology, ecology, and life cycles of the lymnaeids. Therefore, we limit ourselves to a few factors which play the largest role in the well-being of the pond snails and the sustainability of their populations.

The calcium content in the water and food affects both distribution of the lymnaeids and their metabolic and morphogenetic processes. The importance of this element for snails' life has been discussed in many papers (Boycott 1936; Dussart 1979; McMahon 1983; Økland 1990; Piggott and Dussart 1995; Briers 2003). In short, the lymnaeids, as well as other freshwater snail species, can be divided into calciphiles, not able to survive in low calcium waters, and non-calciphiles, which can sustain the shortage of this element in their environment (to some extent, of course). According to Boycott (1936) and Briers' (2003) information, among the British lymnaeids, *Lymnaea stagnalis*, *Myxas glutinosa*, and *Radix auricularia* belong to the calcephile group, whereas *Galba truncatula*, *Omphiscola glabra*, and *Stagnicola palustris* can dwell in relatively soft-water habitats. In these soft waters, the snails receive much more calcium from their food than from water (Stadnichenko 2006). One of the results of the calcium deficiency is that snails exposed to it build thin-walled, fragile, and small-sized shells. No species, however, are unable to develop and reproduce successfully under conditions of strong calcium shortage. It has been experimentally established that a decrease in the concentration of calcium ions to 1.04 mg/l leads to its negative balance in the body and causes the death of lymnaeids (Kruglov 2005). According to Stadnichenko (2006), the optimum concentrations of calcium ions for most East European species of the family lie between 20 and 100 mg/L.

Another chemical parameter, vitally important for the Lymnaeidae, is the hydrogen-ion concentration (pH). The pH level determines some basic chemical processes like solubility of various plant nutrients as well as toxic ions and elements, which may have a direct impact on snails' physiological processes.

As Table 9.2 shows, most of the lymnaeid species studied in this respect occur in waters with pH ranging from 5.0 to 10.0. The optimum values of pH lie between 7.0 and 8.0. When this variable drops to 6.0–6.5, it causes in vivo destruction of the shell material and high embryonic mortality (Kruglov 2005). Under pH = 4.7–4.9, significantly lower adult growth rates and reduced gross fecundity are observed (Hunter 1990). Freshwater molluscs are totally absent from waterbodies with pH < 4.0 (Salazkin 1969; but see Stadnichenko 2006, who argues that some snails can live under pH = 3.4); however, the reactions of molluscs to the effect of this factor are often species- and genus-specific (Salazkin 1969; Økland 1990; Kruglov 2005). For example, the gelatinous pond snail, *Myxas glutinosa*, exhibits low resistance to water acidification which was revealed under experimental conditions (Berezina 2001). On the other hand, such species as *Lymnaea stagnalis* and *Galba truncatula* in Norway were found to dwell in a wide pH range, from 4.4 to 8.8. Both species demonstrated an increased frequency of occurrence in waterbodies with pH values = 7.4–8.8 (Økland 1990). It must be noted that the values of pH are not constant and can change within the same waterbody with depth and throughout the year, following the succession of seasons.

The concentration of oxygen dissolved in water is crucial for the lymnaeids as well as for other aquatic pulmonates despite that these animals can breathe atmospheric air. The presence of oxygen is needed for the normal embryonic development of the pond snails (Stadnichenko 2006), therefore, these molluscs cannot sustain anoxic conditions. The access to atmospheric air is limited if not entirely closed for the lymnaeids living at depths of several meters or more, especially if the wave action or ice cover prevents them from regularly rising to the water surface to breathe. In many such circumstances, the lung is filled with water and operates as the so-called adaptive gill.

Other chemical elements dissolved in water, which are important for the lymnaeid snails, are Mg, Fe, Cu, Cl, and some others. For example, molluscs need copper, as this element is part of their respiratory pigment (hemocyanin).

The overall mineralization of the water, which is the sum of the six most important ions, determines the osmotic properties of the aquatic environment and thus regulates both the water and salt metabolism of molluscs. In most lymnaeid habitats, the dissolved mineral compounds are represented by carbonates (up to 80%) followed by sulfates and chlorides (Stadnichenko 2006). According to

Table 9.2 pH-ranges reported for some species of the Lymnaeidae

Species	Region/ country	pH limits	Reference(s)
Acella haldemani (Binney, 1867)	Canada	7.4–8.5	Pip 1986
Ampullaceana balthica (Linnaeus, 1758)	Russia	6.0–9.0	Berezina 2001
Bulimnea megasoma (Say, 1824)	Canada	5.0–9.0	Pip 1986
Austropeplea tomentosa (L. Pfeiffer, 1855)	Australia	5.0–8.0	Boray 1964
Galba dalli (F.C. Baker, 1907)	Canada	7.5–8.3	Pip 1986 (as *Fossaria dalli*)
G. exigua (Lea, 1841)	Canada	6.6–9.8	Pip 1986 (as *Fossaria exigua*)
G. galbana (Say, 1825)	Canada	7.9–9,5	Pip 1986 (as *Fossaria decampi*)
G. modicella (Say, 1825)	Canada	6.2–10.5	Pip 1986 (as *Fossaria modicella*)
G. parva (Lea, 1841)	Canada	6.7–9.5	Pip 1986 (as *Fossaria parva*)
G. truncatula (O.F. Müller, 1774)	Norway	4.8–9.6	Økland 1990
Hinkleyia caperata (Say, 1829)	Canada	6.6–8.4	Pip 1986 (as *Stagnicola caperata*)
Ladislavella catascopium (Say, 1817)	Canada	7.3–9.5	Pip 1986 (as *Stagnicola catascopium*)
L. emarginata (Say, 1821)	Canada	6.2–10.5	Pip 1986 (as *Stagnicola palustris*)
Lymnaea stagnalis (Linnaeus, 1758)	Norway	6.0–9.5	Økland 1990
	Canada	6.2–10.0	Pip 1986
Myxas glutinosa (O.F. Müller, 1774)	Russia	6.0–9.0	Berezina 2001
Omphiscola glabra (O.F. Müller, 1774)	Norway	5.4–8.8	Økland 1990
Radix auricularia (Linnaeus, 1758)	Europe	7.0–9.9	Welter-Schultes 2012
R. natalensis (Krauss, 1848)	Egypt	5.0–9.0	el-Dafrawi 2002
	Kenya	6.0–9.5	Van Someren 1946 (as *Limnaea caillaudi*)
R. rubiginosa (Michelin, 1831)	Malaysia, Thailand	6.5–8.0	TROPMED Technical Group 1986
Stagnicola palustris (O.F. Müller, 1774)	Norway	5.4–9.5	Økland 1990

Stadnichenko (2006), the pond snails can survive in a broad range of waterbodies differing in their overall mineralization from <200 to >1000 mg/L (up to 3100 mg/L). However, the optimum lies between 200 and 1000 mg/L.

Regarding salinity, most representatives of the family are stenohaline organisms that reach the highest abundance, population density as well as species richness in the oligohaline habitats, with salinity varying from 0.5 to 5.0‰. The brackish water populations of a few lymnaeid snails are known from the desalinated parts of seas, for example, from the Finnish Gulf of the Baltic Sea under salinity around 2‰ or even higher (Dregol'skaya et al. 1988). In the Issyk-Kul Lake in Central Asia, the lymnaeid *Radix obliquata* lives under salinity 5.8–5.9‰ (Pavlova 1983). In Northern Kazakhstan, several representatives of the family were found in mineralized lakes with salinity equal to 4.95‰ (Andreeva et al. 2010). Moreover, some lymnaeids can survive in small steppe waterbodies of the South Urals, whose salinity fluctuates during the year and, at some moments, may reach 10–15‰ (Dulkin 1961).

The concentration of organic compounds of natural and artificial waterbodies varies greatly and depends on a plethora of factors, including the flow of organic matter from the surrounding lands, production of primary biomass by phytoplankton, algae, and macrophytes (remember the summer cyanobacterial blooms leading to the overproduction of organics, which, consequently, creates low dissolved oxygen conditions in the water column), and the tempo of its consumption and destruction by consumers and decomposers.

The lymnaeid snails are known to occur in waterbodies with different trophic statuses, dystrophic to eutrophic, with the highest species richness, abundance, and population density observed in oligotrophic and mesotrophic habitats (Fig. 9.2). The progressive eutrophication of a waterbody, which today is most often observed in habitats receiving large volumes of anthropogenic waste (fertilizers, industrial waste), leads to the degradation of the molluscan community and the decrease in snails' population characteristics (Stadnichenko 2006).

9.3 Biotic Interactions of the Lymnaeid Snails

The lymnaeid snails interact with a plethora of other organisms living in the same habitats, and the essence and importance of these interactions vary greatly depending on the species richness of a particular environment, the type of habitat, its geographic position, and so on (Fig. 9.6). One of the simplest schemes proposed to classify the biotic interactions of animal species is that of Beklemishev (1951; reprinted in Beklemishev 1970). According to this author, there are four basic kinds of biotic relationships (trophic, topic, fabric, and phoretic), whose meanings are explained in Box 9.1.

Trophic interactions, the lymnaeid snails are involved in, are multifarious. Tsikhon-Lukanina (1987), Tsikhon-Lukanina et al. (1998), and Dillon (2000) published overviews of feeding habits and diets of a plethora of aquatic mollusc taxa, including the Lymnaeidae. According to these overviews, the diet of pond snails consists typically of tissues of higher aquatic plants (macrophytes), macro- and microalgae, and plant and bacterial detritus. They prefer to consume dead parts, thus facilitating the process of natural decomposition of plant biomass in a waterbody. Larger species (like *Lymnaea stagnalis*) may occasionally be carnivorous, feeding on decaying corpses of various aquatic animals (frogs, fish, snails). The cases of cannibalism among lymnaeids, albeit rather rare, were observed, especially under conditions of laboratory rearing. Some authors even argue that animal food is

Fig. 9.6 A typical (left) and "large lake" (right) phenotypes of *Lymnaea stagnalis*. After Abel (1920), modified

necessary for the normal development of pond snails (Frömming 1956; Tsikhon-Lukanina 1987). Under laboratory conditions, the cultures of some lymnaeid species could grow and reproduce fed only filter paper or cellulose powder (Dillon 2000).

In their turn, the Lymnaeidae serve as a food item for a broad range of aquatic and semiaquatic animals, both vertebrates and invertebrates. These snails are consumed by many species of molluscivore fish, waterfowl, aquatic and semiaquatic mammals, predatory larvae or adult individuals of aquatic insects (beetles, bugs, flies), and leeches. There are anecdotal reports of terrestrial mammals feeding on lymnaeids (e.g., Lawson 1921), however, to the best of our knowledge, no representative of this family is used as a food item by humans. Amphibiotic lymnaeids, such as *Galba truncatula*, may be eaten by terrestrial birds (e.g., Patzer 1927; Belskii and Grebennikov 2014) and predatory land gastropods, e.g., *Zonitoides nitidus* (O.F. Müller, 1774). The latter circumstance may be used as a means of biological control of *G. truncatula*, which serves as an intermediate host for *Fasciola hepatica* (Rondelaud 1976; Rondelaud et al. 2006, 2023). On the other hand, predatory snail species, especially non-indigenous ones, can become a threat to vulnerable and endangered lymnaeid snails, as is the case with *Erinna newcombi*, endemic to Hawaii, which serves as prey for an invasive predatory snail *Euglandina rosea* (Férussac, 1821) of the family Spiraxidae H.B. Baker, 1939 (CABI 2021).

The dipteran family Sciomyzidae Fallén, 1820, or snail-killing flies, is distributed in all biogeographic regions. This group covers around 550 described species, most of which are natural enemies of various snails, including the lymnaeids, and also may be used in biological control of snails transmitting schistosomiasis and other trematode-borne diseases (see Maharaj et al. 2005; Knutson and Vala 2011; Marinoni and Murphy 2016 and the references therein).

The predatory and parasitic leeches feeding on aquatic Mollusca are classified within the genera *Helobdella* Blanchard, 1896 and *Glossiphonia* Johnson, 1816 of the family Glossiphoniidae Vaillant, 1890, which is also cosmopolitan. Different lymnaeid species serve as prey for malacophagous glossiphonid leeches (Klemm 1976; Lukin 1976; Brönmark 1992; Bielecki et al. 1999) (Fig. 9.7a).

The competitive interactions between the lymnaeid snails and other species of freshwater molluscs belonging to the same or other families have been studied in detail only in a few works (see Dillon 2000 for review). There are observations of probable replacing (or, perhaps, direct outcompeting) of one lymnaeid species by another (Adam and Lewis 1992). Rondelaud et al. (2016) reported a striking case of outcompeting of two species of pond snails (*Galba truncatula* and *Omphiscola glabra*) by an introduced physid (*Aplexa hypnorum*). The authors suggest that the latter species produces mucus which is toxic for the pond snails, therefore the competition for food may not explain this replacement. Brönmark et al. (1991) described an interesting case of interactions between two lymnaeid species and tadpoles of the common frog (*Rana temporaria*), in which snails indirectly facilitated tadpole growth, whereas tadpoles had a negative impact on snails through resource depression.

The parasitic relationships between snails and other organisms fall simultaneously to two Beklemishev's categories, i.e. trophic and *topic* (see Box 9.1). It is possible because parasitic animals, especially endoparasites, use their host(s) not only as a source of nutrients but also as their "home." In other words, the host's organism provides the first-order environment for its parasite, protecting it from weather fluctuations, changes in chemical content, humidity, as well as from predator attacks. Most interactions of a parasite with the outside world occur through the mediation of its host's body. As stated above, the details of host-parasitic relationships in which lymnaeids are involved are reviewed in other chapters of this volume (Rondelaud et al. 2023; Vázquez et al. 2023).

Close to parasitism, with which it has numerous border-line cases, is *commensalism*, i.e. an asymmetrical topic interaction in which species A provides species B with a living space without any apparent benefit to itself. In the case of commensalism, the strength of interaction between two species is generally thought to be much weaker and less obligatory than in parasitism. The most studied case of commensalism involving lymnaeid snails is, arguably, that of *Chaetogaster limnaei* (von Baer, 1827), the aquatic oligochaete belonging to the family Naididae Ehrenberg, 1831. As its Latin name hints, lymnaeid snails are among the most usual hosts of this cosmopolitan worm, which, however, interacts with a diversity of freshwater snails and bivalves (*Dreissena*; Sphaeriidae) as well as other aquatic animals ranging from sponges to acipenserid fish [see Buse 1974; Martins and Alves 2010; Smythe et al.

Fig. 9.7 Illustrations of biotic interactions between lymnaeid snails and other organisms. (**a**) Canada, British Columbia, Duck Lake. *Radix auricularia* as prey for a malacophagous leech. (**b**) Russia, Vologda Region, Sheksna Reservoir, shell of *Radix auricularia* as substrate for hydroids. (**c**) Russia, Vologda Region, Sheksna Reservoir, amphipod on the shell of *Lymnaea stagnalis*. (**d**) Russia, Chukotka, Chirovoye Lake, a planorbid snail on the shell of *Dallirhytis atkaensis*. (**e**) Russia, Chukotka, Chirovoye Lake, shell of *Dallirhytis atkaensis* as substrate for a range of algae. (**f**) Russia, Khabarovsk Territory, Tumnin River, eggs of unidentified invertebrates on the shell surface of *Radix auricularia*. Photos: Ilya V. Vikhrev (A) & Olga V. Aksenova (B–F)

2015]. Around 50 years ago, Gruffydd (1965a, b), based on an earlier observation made by Vaghin (1946), supposed that there may be two ecological subspecies of *Ch. limnaei*. The nominative one is a true commensal attaching itself to the outer or inner surface of snail shells, being often found in the mantle cavity. It can move freely and feed on minute aquatic animals (for example, rotifers) and algae. Another subspecies, *Ch. l. vaghini* Gruffydd, 1965, was thought to be a true parasite of molluscs that lives in the kidney of its host. The latter is unable to survive and feed when transferred to the external surface of the snail (Gruffydd 1965a; Smythe et al. 2015). A recent molecular study has revealed, although, that *Ch. l. vaghini* is not a distinct species or subspecies but rather a "part of a large, plastic, lineage of worms that are able to exploit both the internal kidney habitat and the external habitat in the host's pallial cavity or just under the shell" (Smythe et al. 2015, p. 323). However, the authors added a cautionary note that their results are based on a single-locus study, and the data from additional loci or infection experiments with offspring from both forms are needed.

It should be noted that the view of *Ch. limnaei* as a true commensal, let alone parasite, may be incorrect. There are data that this worm can defend their snail hosts against trematode larvae (Vaghin 1931; Khalil 1961; Hopkins et al. 2013). In this case, the characteristics of their interaction must be changed from commensalism to symbiotic mutualism (but see Stoll et al. 2013; Mitchell and Leung 2016 on the detrimental effects of the oligochaete on its snail hosts).

There are diversified *topic* interactions between the lymnaeid snails and aquatic macrophytes. The plants create a suitable environment for phytophilous species of snails providing them with living space, substrate, and, to some extent, shelter from predator attacks and certain abiotic factors (wave action, heavy current). Many lymnaeid species spend a significant part of their lifespan (if not whole life) crawling and feeding on the leaves and stems of macrophytes that serve also as a substrate for oviposition (Fig. 9.5). On the other hand, lymnaeid shells sometimes provide a suitable substrate for a range of filamentous algae, which form dense thickets that can cover the shell almost entirely. There are records of varied minute invertebrates (oligochaetes, aquatic mites, hydroids, rotifers) inhabiting these thickets (Stadnichenko 2006) (Fig. 9.7).

Dead shells of large lymnaeid species (*Lymnaea stagnalis*, *Radix auricularia*) are often used by some aquatic spiders as their homes, which can be classified as another example of topic relationships.

The Lymnaeidae are actively involved in direct *phoretic* interactions with a broad range of animal species, serving to these snails as vectors of their passive dispersal. Such dispersal can proceed in different ways, including the passage of alive snails through the gut of molluscivorous fish and waterfowl. Though the percentage of individuals able to survive after being swallowed by a predator should be very low, the reality of this mode of dispersal for some species of continental snails has been demonstrated in some empirical studies (Malone 1965; Brown 2007; Cadée 2011; van Leeuwen et al. 2012; Simonová et al. 2016). Since Darwin (1878, 1882) and Kew (1893), the naturalists believed in the effectiveness of long-distance dispersal of living freshwater molluscs (including lymnaeids) or their eggs on the extremities or plumage of birds, legs of domestic cattle, and even, probably, with large flying insects. A huge body of empirical observations of such dispersal has been accumulated (see reviews of Madsen and Frandsen 1989; Bilton et al. 2001; Kappes and Haase 2012; van Leeuwen et al. 2013). The human-mediated dispersal of the Lymnaeidae is also not uncommon. Some members of the family (e.g., *Pseudosuccinea columella, Radix rubiginosa*) are popular among aquarists

Fig. 9.8 A caddisfly (? *Lymnephilus flavicornis*) larval case taken from two angles. It is built of shells belonging to four mollusc families (Lymnaeidae, Planorbidae, Sphaeriidae, Viviparidae). The case was collected in Omsk Region (West Siberia, Russia). Photo: Maxim V. Vinarski

(Ng et al. 2016; Lounnas et al. 2017) that makes aquarium trade an important medium for their dispersal throughout the world (Patoka et al. 2017).

We frequently found living pond snails, chiefly juvenile ones, being attached to bottoms of riverboats, to fishermen's nets, and other artificial objects which are occasionally transferred from one aquatic habitat to another. The use of floating vehicles by varied species of freshwater gastropods and bivalves as the way of their transportation has been reported (De Ventura et al. 2016). Starobogatov (1970) described his findings of living lymnaeids of two species (*Galba truncatula*, *Peregriana peregra*) in the grooves of the tire treads of trucks. By traveling in this manner, the snails are potentially able to cover vast distances and find themselves in waterbodies situated far from their native habitat.

The *fabric* interactions, in which lymnaeids are engaged, are relatively not abundant and not diversified. The most evident example of such interactions is, perhaps, the use of empty shells of these gastropods (or even living snails) by some Trichoptera larvae as materials for their cases (Fig. 9.8). The choice of shells by these larvae is not species-specific, and insects can use various species of molluscs, both bivalves and gastropods, as the source of material for their larval cases. Notably enough that the fact of being incorporated into caddisfly cases does not have an apparent negative effect on the vital functions of snails, which can long stay alive and even reproduce (Anistratenko 1987; Martynov et al. 2018).

9.4 Life Cycles and Reproduction

Most available data on the life cycles, growth, and reproduction of the Lymnaeidae were obtained from studies on species inhabiting the temperate regions of the North Hemisphere, whereas far less research has focused on the lymnaeids of lands lying beyond this area (but see studies in exotic lymnaeid snails by Pagulayan et al. 1983; Lam 1994; Aziz and Raut 1996). According to the accessible data, the vast majority of lymnaeid snails are characterized by relatively short life cycles. In most studied cases, the life span of pond snails varied between 1 and 2 years (Russell-Hunter 1978; Beriozkina and Starobogatov 1988, 1992; Heller 1990). The large-bodied species (*Lymnaea stagnalis*, *Radix auricularia*, *Stagnicola corvus*, and like) tend to have 2-year life cycles, whereas 1-year life cycles are characteristic for species of smaller body size, like *Galba truncatula* and *Peregriana peregra* (Stadnichenko 2006). However, in older papers, alternative estimates of the duration of the lymnaeid life cycle can be found. For instance, Hazay (1881), Schodduyn (1925), and Levina (1973) reported that the lifespan of the great pond snail, *Lymnaea stagnalis*, in Western and Central Europe is 4–5 years, which is, most probably, an exaggeration (see Beriozkina and Starobogatov 1988).

Lymnaeid snails are typical *r*-strategists characterized by relatively short life spans, high fecundity, increased rate of food consumption, and fast generation turnover. The number of generations per year in temperate waterbodies highly depends on local conditions and the geographic position of the waterbody (Russell-Hunter 1964, 1978) (Fig. 9.9). The basic quantitative parameters of the lymnaeid life cycles (growth rate, duration of time from hatching to maturity, fecundity, time of onset of breeding, survivorship, etc.) are rather flexible, and numerous variations in their pattern are observed both between species and between conspecific populations inhabiting different localities within a species' range. Some authors (e.g., Beriozkina and Starobogatov 1988) consider the ambient temperature the crucial external factor determining the specificity of the life cycle of a given species/population, whereas other workers emphasize the day-length as an important environmental trigger (see Russell-Hunter 1978 for review). Evidently, at least some of the life-cycle parameters may be genetically determined (e.g., Wullschleger and Jokela 2002).

A simple univoltine cycle, with breeding in late spring or early summer and complete replacement of generations, seems to be the most common pattern for the temperate Hygrophila (Calow 1978; Russell-Hunter 1978), however, many variations within both 1- and 2-year life cycles have been revealed (see Calow 1978; Russell-Hunter 1978; Beriozkina and Starobogatov 1988 for details). The lymnaeids

Fig. 9.9 Lymnaeid reproduction: some illustrations. (**a**) Russia, Arkhangelsk Region, Mezen River basin, small pond, copulation between individuals of *Lymnaea stagnalis* in the end of June. (**b**) Russia, Krasnoyarsk Territory, vicinity of Norilsk City, copulation between individuals of *Ampullacaeana lagotis* in the end of July. (**c**) Russia, Chukotka, Chirovoye Lake, the egg-laying of *Dallirhytis atkaensis* in the end of July. (**d**) Russia, Vashutkiny Lakes, egg mass of *Radix auricularia* in August. (**e**) Russia, Krasnoyarsk Territory, vicinity of Dudinka settlement, egg mass of *Ampullaceana lagotis.* (**f**) Russia, Arkhangelsk Region, North Dvina River basin, egg mass of *Lymnaea stagnalis* on leaves of macrophytes. (**g**) Russia, Nenets Autonomous Okrug, Pymvashor

with annual life cycles can breed one to three times per vegetative season (i.e., when the water temperature is warm enough to enable their reproduction), and the generation replacement in such populations can be either complete or partial. The autumn generation overwinters and starts to reproduce next spring or early summer.

A biennial life cycle, typical for relatively large-bodied pond snails, can be illustrated by the life-cycle characteristic of *Lymnaea stagnalis* in Eastern Europe (Central Ukraine, European Russia) as it was described in works of Beriozkina and Starobogatov (1988) and Stadnichenko (2006).

The hatchlings appear in the autumn, grow fast, and by winter attain shell height of about 20–30 mm. The onset of their first reproduction takes place in late June–early July, at shell height 30–36 mm. The oviposition may last 1.5–2 months, and the egg masses are laid till the end of August. As a result, the population density reaches its peak by late August–early September. By the start of their second overwintering, the snails reach a shell height of 45 mm or more. Next spring, the generation is ready for their second (and last) reproduction. This time, the onset of oviposition occurs in the late April–early May and lasts till the late June–early July. The dying out of this generation is observed in the second half of the summer, but some individuals may be found alive even in October [Boycott (1936) cited an anecdotal observation of "*Limnaea peregra*" found ovipositing in November]. Virtually none of them, however, is able to overwinter one more time. The shell size of the oldest individuals may reach 62 mm. Such a type of life cycle does not assume a complete replacement of generations, and, at any moment, snails of different ages/generations co-occur in a waterbody.

The growth rate in lymnaeid snails, as in the vast majority of other ectothermic invertebrates, reaches its peak before maturation and slows down (if not virtually stops) after the start of reproduction.

The dioecy of the Lymnaeidae, as well as their ability to reproduce by self-fertilization, has often been considered as an adaptation to life in non-permanent waterbodies as well as to rapid colonization of novel habitats (Boss 1978; Beriozkina and Starobogatov 1992; Jarne et al. 1993; Jordaens et al. 2007; Vinarski et al. 2021). The adaptive advantage of hermaphroditism is that it "allow[s] mating with any members of the species where the probability of encountering a suitable mate is low" (Boss 1978, p. 421). This reproductive strategy allows a lymnaeid species to gain a foothold in the marginal parts of its range, and even a single individual may become a founder of a sustainable population. High fecundity that is characteristic of many r-strategists secures fast population recovery after catastrophic events of dying out.

Despite the simultaneous hermaphroditism of freshwater Pulmonata, self-fertilization (*autogamy*) is not obligatory for these animals (Jarne et al. 1993). In most cases, cross-fertilization (*allogamy*) prevails, in which one individual plays the role of a "male" and the other acts as a "female," but the possibility of autogamy

Fig. 9.9 (continued) thermal springs, the embryos in egg capsules of *Ampullaceana lagotis* in September. (**h**) Juvenile *A. lagotis* in September. Photos: Olga V. Aksenova

remains though its frequency varies among different species (Beriozkina and Starobogatov 1988; Jordaens et al. 2007). It is believed (Beriozkina and Starobogatov 1992; Heller 1993; Dillon 2000) that the combination of auto- and allogamy has an important adaptive value, facilitating the penetration of snails in new habitats and formation of stable populations there. Furthermore, it acts as a mechanism for maintaining a high level of phenotypic variability. High rates of self-fertilization, up to 80%, are observed in the dwarf pond snail, *Galba truncatula*, which was found to be a preferential selfer (Trouvé et al. 2003, 2005). The world-wide invader lymnaeid *Pseudosuccinea columella* gives another instance of preferential selfing which, presumably, facilitates its march throughout the world. The selfing rate in this snail may reach 93% (Nicot et al. 2008; Lounnas et al. 2017).

The ability to self-fertilize also determines the phenomenon of aphally—the absence of the copulatory apparatus, which is known in some representatives of Hygrophila (Harutyunova 1977; Starobogatov et al. 1989). In this case, the molluscs reproduce exclusively autogamously. In some species, the aphally arises as a rare aberration (Harutyunova 1977), in others, it is obligate, like in the Far East physid snail *Aplexa aphallica* Starobogatov & Zatravkin in Starobogatov, Prozorova & Zatravkin, 1989 (see Starobogatov et al. 1989). Completely aphallic species are unknown among the Lymnaeidae, however.

The age of maturity in lymnaeids is flexible and depends on many conditions, both external and internal. These molluscs are able to suspend development for a long time in case of unfavorable circumstances (Beriozkina and Starobogatov 1988). It is reasonable to think that, in general, the warmer their environment is the faster snails grow and the earlier they attain maturity. However, some deviations from this simple relationship have been observed. For example, in a laboratory culture of *Stagnicola palustris* studied by Beriozkina and Starobogatov (1988), some individuals started oviposition being 60 days old, whereas some others did not reproduce until 123-day age. The mean age of the first oviposition in this culture was between 80 and 90 days. The period of oviposition is rather long and can last up to 1.5–2 months; the maximum number of egg masses produced per day by a single individual of *S. palustris* could reach 92. The absolute fecundity (i.e., the sum of egg masses laid by a single animal during the season of reproduction) in some lymnaeid species may reach 5500 though typically it is significantly lower (Beriozkina and Starobogatov 1988).

The fecundity of lymnaeids depends on several factors, including the animal size (the larger the snail is, the more eggs it can lay), ambient conditions, and, in some circumstances, the availability of a sexual partner. The latter is essential for the lymnaeid species, demonstrating a low rate of self-fertilization. Individuals of such species, lacking suitable partners, demonstrate depressed fecundity and low oviposition intensity (Beriozkina and Starobogatov 1988; Golubev et al. 2013).

The egg masses are deposited on a variety of substrates such as submerged stones and logs, stems and leaves of macrophytes, hard pieces of anthropogenic rubbish, and, when the population density is relatively high, on shells of conspecific snails (i.e., Dussart 1979) (Fig. 9.5). The number of egg capsules in an egg mass varies considerably and, as a rule, depends on the body size of the molluscs (Fig. 9.9). In large-sized species of Lymnaeidae, such as *Lymnaea stagnalis*, the number of egg capsules per egg mass can exceed 100 (Beriozkina and Starobogatov 1988; Kruglov and Starobogatov 1992), while in some members of the genus *Radix* it reaches 250 (Prozorova 1992). On the other hand, species of the genus *Galba* are characterized by the small size of egg masses and, as a consequence, a few egg capsules (Kruglov and Starobogatov 1992).

The duration of the embryogenesis is ruled mainly by the factor of aquatic temperature, and, according to data reviewed by Stadnichenko (2004), varies from 11 to 24 days at 18–20^0 C. Even a slight increase in ambient temperature results in a significant increase of the development rate. Thus, Kruglov (1980) and Stadnichenko (2004) reported that the embryogenesis of *Radix auricularia* at 18^0 C lasts 18 days, at 22–23^0 C – 14 days, and takes only 11 days at 24^0 C. If the temperature rises above or drops below some thresholds, embryogenesis becomes impossible. For example, in *Lymnaea stagnalis*, no embryos hatched when reared under temperature above 30^0 C (Richards 1965). The lower temperature threshold for the successful development of this snail lies approximately at 9^0 C (Meshcheryakov 1990).

Apart from the temperature, other external factors are known to regulate the rate of embryogenesis in the Lymnaeidae and other aquatic pulmonates. For example, as was shown by Voronezhskaya et al. (2004), the developing embryos of *L. stagnalis* are able to receive chemical signals from newly hatched juvenile conspecifics. These signals bring information about unfavorable conditions, i.e. food shortage and/or crowding, which results in twofold prolongation of the embryonic development (see Dillon 2000 for details on the detrimental effects of crowding on aquatic snails). Interestingly, these signals are not species-specific and affect the developing embryos of other snail species (Voronezhskaya et al. 2007).

The population density in the Lymnaeidae can vary numerically over a broad range, changing from a few to several thousand specimens per square meter. The highest densities are more often observed in small-bodied pond snails (like *Galba truncatula* or juvenile generations of larger lymnaeids). The factors controlling this parameter are numerous—food availability, depth, wave action and water turbidity, substrate suitability, predator and parasite pressure, interspecific competition, human activity, etc. The lymnaeid snail density in a waterbody also fluctuates following the population cycles of reproduction and dying out. The short lifespan and fast generation turnover in most lymnaeids make their density a very flexible parameter that is not constant even for a particular habitat over a series of successive seasons. The knowledge of patterns of these fluctuations is important for applied (epidemiological, fisheries research) studies as the lymnaeids often constitute a substantial portion of benthic biomass in a waterbody, whereas the high abundance of these snails is of interest for parasitologists as it can be related to the abundance of parasitic larvae.

Some authors reported certain geographical trends in lymnaeid abundance and density. For example, Nekhaev (2021) discusses the depressed abundance of aquatic snails (including the lymnaeids) in the tundra waterbodies of the north of European Russia. According to this author, this low abundance is explained by the shortage of aquatic vegetation in the Arctic waterbodies accompanied by the lowered flow of organics from terrestrial ecosystems.

Acknowledgment The authors are grateful to the Russian Science Foundation (projects No. 19-14-00066/P and and 21-74-10155) for the financial support of their lymnaeid studies.

Box 9.1 A Classification of Biotic Interactions Between Species (After Beklemishev 1970).

Biotic Interactions

Fabric [from Latin "fabric"—a workshop, art, trade, etc.]: **species A uses the bodies of individuals of species B, or parts of these bodies, or their remains, as material for its buildings.**

Phoretic [from Greek "φόρεσις"—an act of bearing]: **species A uses individuals of species B as the means of its dispersal.**

Topic [from Greek "τοπίκός"—a place]: **species A changes the living conditions of species B for the better or worse.**

Trophic [from Greek "τροφίκός"—a place]: **species A feeds on individuals of species B, or their corpses, excrements, or discarded body parts.**

References

Abel O (1920) Lehrbuch der Paläozoologie. G. Fischer Verlag, Jena

Adam M, Lewis J (1992) The lack of co-existence between *Lymnaea peregra* and *Lymnaea auricularia* (Gastropoda: Pulmonata). J Mollus Stud 58:227–228. https://doi.org/10.1093/mollus/58.2.227

Aksenova O, Vinarski M, Bolotov I et al (2017a) An overview of recent studies of Lymnaeid snails from geothermal habitats of Northern Palearctic. Book of abstracts, 8th European Congress of Malacological Societies, September 10–14, 2017. Kraków, Poland. P. 139

Aksenova O, Vinarski M, Bolotov I et al (2017b) Two *Radix* spp. (Gastropoda: Lymnaeidae) endemic to thermal springs around Lake Baikal represent ecotypes of the widespread *Radix auricularia*. J Zool Syst Evol Res 55(4):298–309. https://doi.org/10.1111/jzs.12174

Aksenova OV, Bolotov IN, Gofarov MY et al (2018) Species richness, molecular taxonomy and biogeography of the radicine pond snails (Gastropoda: Lymnaeidae) in the Old World. Sci Rep-UK 8:11199. https://doi.org/10.1038/s41598-018-29451-1

Alyakrinskaya IO (2004) Drying resistance of aquatic molluscs. Izvestiya RAN, seriya biologicheskaya 3:367–374. [in Russian]

Andreeva SI, Andreev NI, Vinarski MV (2010) Key to freshwater gastropods of Western Siberia (Mollusca: Gastropoda). V. 1. Gastropoda: Pulmonata. Fasc. 1. Families Acroloxidae and Lymnaeidae. The authors, Omsk. [in Russian]

Anistratenko VV (1987) The use of empty shells and alive mollusks by Trichoptera larvae for construction of their cases. Vestnik zoologii 3:75–76. [in Russian]

Axenov-Gribanov DK, Vereshchagina Y, Lubyaga A (2015) Stress response at the cellular and biochemical levels indicates the limitation of the environmental temperature range for Eastern Siberian populations of the common gastropod *Lymnaea stagnalis*. Malacologia 59:33–44. https://doi.org/10.4002/040.059.0105

Aziz MDA, Raut SK (1996) Thermal effect on the life-cycle parameters of the medically important freshwater snail species *Lymnaea* (*Radix*) *luteola* (Lamarck). Mem Inst Oswaldo Cruz 97(1): 119–128. https://doi.org/10.1590/S0074-02761996000100022

Baker FC (1928) Influence of a changed environment in the formation of new species and varieties. Ecology 9:271–283. https://doi.org/10.2307/1932368

Bargues MD, Artigas P, Khoubbane M et al (2011) *Lymnaea schirazensis*, an overlooked snail distorting fascioliasis data: genotype, phenotype, ecology, worldwide spread, susceptibility, applicability. PLoS One 6:e24567. https://doi.org/10.1371/journal.pone.0024567

Beklemishev VN (1951) On classification of biotic (symphysiological) relationships. Byulleten Moskovskogo obshchestva ispytateley prirody. Otdel biologicheskiy 56:3–30. [in Russian]

Beklemishev VN (1970) The biocoenological fundamentals of comparative parasitology: a collection of essays. Nauka Publishers, Moscow. [in Russian]

Belskii E, Grebennikov M (2014) Snail consumption and breeding performance of pied flycatchers (*Ficedula hypoleuca*) along a pollution gradient in the Middle Urals, Russia. Sci Total Environ 490:114–120. https://doi.org/10.1016/j.scitotenv.2014.04.116

Berezina NA (2001) Influence of ambient pH on freshwater invertebrates under experimental conditions. Russ J Ecol 32:343–351. https://doi.org/10.1023/A:1011978311733

Beriozkina GV, Starobogatov YI (1988) Reproductive ecology and egg masses of freshwater Pulmonata. Trudy Zoologicheskogo Instituta AN SSSR 174:1–306. [in Russian]

Beriozkina GV, Starobogatov YI (1992 "1991"). The systems of reproduction of freshwater Pulmonata. Trudy Zoologicheskogo Instituta AN SSSR 228:130–139. [in Russian]

Bespalaya YV, Bolotov IN, Usacheva OV (2011) Mollusk population of a Subarctic hydrothermal ecosystem in winter. Zool Zh 90(11):1304–1322. [in Russian]

Bielecki A, Rybak J, Lukowiak-Bielecka M (1999) Glossiphoniidae Vaillant, 1850 [Hirudinea] of Poland – systematics and perspectives of studies. Ann Parasitol 45(1):29–61

Bilton DT, Freeland JR, Okamura B (2001) Dispersal in freshwater invertebrates. Annu Rev Ecol Syst 32:159–181. https://doi.org/10.1146/annurev.ecolsys.32.081501.114016

Boettger CR (1944) Basommatophora. Die Tierwelt der Nord- und Ostsee. Akademische Verlagsgesellschaft Becker and Erler Kom.-Ges., Leipzig 9(b2):241–478

Boldyreva NV (1930) On overwintering of aquatic organisms frozen into ice. Russkiy Gidrobiologicheskiy Zhurnal 9(1):45–84. [in Russian with German resume]

Bolotov IN, Bespalaya YV, Usacheva OV (2012) Ecology and evolution of hydrobionts in hot springs of the Subarctic and Arctic: formation of similar assemblages, adaptation of species, and microevolutionary processes. Biol Bull Reviews 2(4):340–348. https://doi.org/10.1134/S2079086412040020

Bolotov I, Bespalaya Y, Aksenova O et al (2014) A taxonomic revision of two local endemic *Radix* spp. (Gastropoda: Lymnaeidae) from Khodutka geothermal area, Kamchatka, Russian Far East. Zootaxa 3869(5):585–593. https://doi.org/10.11646/zootaxa.3869.5.9

Bolotov IN, Aksenova OV, Bespalaya YV et al (2017) Origin of a divergent mtDNA lineage of a freshwater snail species, *Radix balthica*, in Iceland: cryptic glacial refugia or a postglacial founder event? Hydrobiologia 787:73–98. https://doi.org/10.1007/s10750-016-2946-9

Boray JC (1964) Studies on the ecology of *Lymnaea tomentosa*, the intermediate host of *Fasciola hepatica*. 1. History, geographical distribution, and environment. Aust. J Zoology 12(2): 217–230. https://doi.org/10.1071/ZO9640217

Boss (1978) On the evolution of gastropods in ancient lakes. In: Fretter V, Peake J (eds) Pulmonates, vol 2A. Academic Press, London, pp 385–428

Bössneck U (2012) Leben am Limit: Besiedlung von Süßwasser-Habitaten extremer Hochlagen Asiens, Amerikas und Afrikas durch Mollusken (Mollusca: Bivalvia & Gastropoda). In: Hartmann M, Weipert J (eds) Biodiversität und Naturausstattung im Himalaya IV. Verein der Freunde und Förderer des Naturkundemuseums Erfurt e. V, Erfurt, pp 103–106

Boycott AE (1936) The habitats of fresh-water Mollusca in Britain. J Anim Ecol 5(1):116–186. https://doi.org/10.2307/1096

Briers RA (2003) Range size and environmental calcium requirements of British freshwater gastropods. Glob Ecol Biogeogr 12:47–51. https://doi.org/10.1046/j.1466-822X.2003.00316.x

Brönmark C (1992) Leech predation on juvenile freshwater snails: effects of size, species and substrate. Oecologia 91:526–529. https://doi.org/10.1007/BF00650326

Brönmark C, Rundle SD, Erlandsson A (1991) Interactions between freshwater snails and tadpoles: Competition and facilitation. Oecologia 87(1):8–18. https://doi.org/10.1007/BF00323774
Brown DS (1994) Freshwater snails of Africa and their medical importance. Taylor and Francis, London. https://doi.org/10.1201/9781482295184
Brown RJ (2007) Freshwater mollusks survive fish gut passage. Arctic 60(2):124–128. https://doi.org/10.14430/arctic237
Burch JB (1968) *Erinna newcombi* of Hawaii and *Limnaea onychia* of Japan. Malacol Rev 1:15–30
Buse A (1974) The relationship of *Chaetogaster limnaei* (Oligochaeta: Naididae) with a variety of gastropod species. J Anim Ecol 43(3):821–837. https://doi.org/10.2307/3538
CABI – Invasive species compendium (2021). Accessed at https://www.cabi.org/isc/datasheet/23113 on 1 November 2021
Cadée GC (2011) *Hydrobia* as "Jonah in the Whale": Shell repair after passing through the digestive tract of shellducks alive. Palaios 26:245–249. https://doi.org/10.2110/palo.2010.p10-095r
Calow P (1978) The evolution of life-cycle strategies in fresh-water gastropods. Malacologia 17: 351–364
Cheatum EP (1934) Limnological investigations on respiration, annual migratory cycle, and other related phenomena in fresh-water pulmonate snails. T Am Misrosc Soc 53(4):348–407. https://doi.org/10.2307/3222383
Clarke AH (1973) The freshwater molluscs of the Canadian Interior Basin. Malacologia 13:1–509
Clessin S (1877) Die Mollusken der Tiefenfauna unserer Alpenseen. Malakozoologische Blätter 24: 159–185
Clewing C, Albrecht C, Wilke T (2016) A complex system of glacial sub-refugia drives endemic freshwater biodiversity on the Tibetan Plateau. PLoS One 11:e0160286. https://doi.org/10.1371/journal.pone.0160286
Dar Y, Djuikwo-Teukeng FF, Vignoles P et al (2010) *Radix natalensis* (Gastropoda: Lymnaeidae), a potential intermediate host of *Fasciola hepatica* in Egypt. Parasite 17:251–256. https://doi.org/10.1051/parasite/2010173251
Darwin C (1878) Transplantation of shells. Nature 18:120
Darwin C (1882) On the dispersal of freshwater Bivalves. Nature 25:529–530
De Ventura L, Weissert N, Tobias R et al (2016) Overland transport of recreational boats as a spreading vector of zebra mussel *Dreissena polymorpha*. Biol Invasions 18:1451–1466. https://doi.org/10.1007/s10530-016-1094-5
Dillon RT (2000) The ecology of freshwater molluscs. Cambridge University Press, Cambridge etc. https://doi.org/10.1017/CBO9780511542008
Dregol'skaya IN, Korotneva NV, Starobogatov YI (1988) Molluscs of the eastern part of the Finnish Gulf and their cytoecological study. Zool Zh 67(11):1631–1637. [in Russian]
Du Rietz GE (1931) Life-forms of terrestrial flowering plants. Acta Phytogeographica Suecica 3(1): 1–95
Dulkin AL (1961) Water chemistry of the lakes of the Southern Urals and Trans-Urals and [their] mollusks. Zool Zh 40(10):1461–1464. [in Russian]
Dussart GBJ (1979) Life cycles and distribution of the aquatic gastropod molluscs *Bithynia tentaculata* (L.), *Gyraulus albus* (Muller), *Planorbis planorbis* (L.) and *Lymnaea peregra* (Muller) in relation to water chemistry. Hydrobiologia 67(3):223–239. https://doi.org/10.1007/BF00023179
Eichler W (1952) Die Tierwelt der Gewächshäuser. Academische Verlagsgesellschaft, Gees & Portig, Leipzig
el-Dafrawi SM (2002) Morphological and ecological studies on *Lymnaea natalensis*, the snail vector of *Fasciola gigantica* in Egypt. J Egypt Soc Parasitol 32(2):447–456
Forel FA (1869) Introduction à l'étude de la faune profonde du lac Léman. Bulletin de la Société vaudoise des sciences naturelles 10:217–223
Forel FA (1874) Matériaux pour servir à l'étude de la faune profonde du lac Léman. 1″ série. Bulletin de la Société vaudoise des sciences naturelles 13:1–164

Frömming E (1956) Biologie der mitteleuropäischen Süßwasserschnecken. Dunker-Humblot, Berlin

Gittenberger E, Tobgay K, Gyeltshen C et al (2021) Two high altitude species of molluscs, new for Bhutan (Bivalvia, Sphaeriidae – Gastropoda, Lymnaeidae). Basteria 85(1):13–17

Glöer P (2019) The freshwater gastropods of the West-Palaearctis. Volume 1. Fresh- and brackish waters except spring and subterranean snails. Identification key, anatomy, ecology, distribution. The author, Hetlingen

Golubev AP, Bodilovskaya OA, Slesareva LE et al (2013) Population dynamics of the pulmonate mollusk *Lymnaea stagnalis* under the conditions of prolonged reproduction by self-fertilization. Dokl Biol Sci 452:300–304. https://doi.org/10.1134/s0012496613050165

Graevsky EA (1946) The temperature-preference and the temperature optimum of fresh-water molluscs and Arthropoda. Zh Obshch Biol 7:455–472. [in Russian]

Gruffydd LD (1965a) Evidence for the existence of a new subspecies of *Chaetogaster limnaei* (Oligochaeta) in Britain. J Zool 146:175–196. https://doi.org/10.1111/j.1469-7998.1965.tb05208.x

Gruffydd LD (1965b) The population biology of *Chaetogaster limnaei limnaei* and *Chaetogaster limnaei vaghini* (Oligochaeta). J Anim Ecol 34:667–690. https://doi.org/10.2307/2455

Harutyunova LD (1977) Aphally in a population of *Radix* (Gastropoda; Lymnaeidae) of Armenia. Biologicheskiy Zhurnal Armenii 30(3):89–90. [in Russian]

Hazay J (1881) Die Molluskenfauna von Budapest. III. Biologische Theil. Zur Entwicklungs- und Lebensgeschichte der Land- und Süsswasser-Mollusken. Malakozoologische Blatter, Neue Folge 4:43–224

Heller J (1990) Longevity in molluscs. Malacologia 31(2):259–295

Heller J (1993) Hermaphroditism in molluscs. Biol J Linn Soc 48:19–42. https://doi.org/10.1111/j.1095-8312.1993.tb00874.x

Hopkins SR, Wyderko JA, Sheehy RR et al (2013) Parasite predators exhibit a rapid numerical response to increased parasite abundance and reduce transmission to hosts. Eco Evol 3(13):4427–4438. https://doi.org/10.1002/ece3.634

Horsák M, Dvořák L, Juřičková L (2004) Greenhouse gastropods of the Czech Republic: current stage of research. Malakológiai Tájekoztató (Malacological Newsletter) 22:141–147

Hubendick B (1952) Hawaiian Lymnaeidae. Occasional Papers of Bernice P. Bishop Museum, Honolulu, Hawaii 20(19):307–328

Humphreys AM, Govaerts R, Ficinski SZ et al (2019) Global dataset shows geography and life form predict modern plant extinction and rediscovery. Nat Ecol Evol 3:1043–1047. https://doi.org/10.1038/s41559-019-0906-2

Hunter RD (1990) Effects of low pH and low calcium concentration on the pulmonate snail *Planorbella trivolvis*, a laboratory study. Can J Zool 68:1578–1583. https://doi.org/10.1139/z90-233

Izzatullaev Z, Kruglov ND, Starobogatov YI (1983) New and ill-known species of mollusks of the subgenus *Radix* of the genus *Lymnaea* of the USSR fauna from the Central Asia (Gastropoda, Pulmonata). Izvestiya Akademii Nauk Tajikskoy SSR, otdelenie biologicheskikh nauk 4(93):53–57. [in Russian]

Jabbour-Zahab R, Pointier JP, Jourdane J et al (1997) Phylogeography and genetic divergence of some lymnaeid snails, intermediate hosts of human and animal fascioliasis with special reference to lymnaeids from the Bolivian Altiplano. Acta Trop 64:191–203. https://doi.org/10.1016/S0001-706X(96)00631-6

Jackiewicz M (1998) European species of the family Lymnaeidae (Gastropoda, Pulmonata, Basommatophora). Genus 9(1):1–93

Jarne P, Vianey-Liaud M, Delay B (1993) Selfing and outcrossing in hermaphrodite freshwater gastropods (Basommatophora): where, when and why. Biol J Linn Soc 49:99–125. https://doi.org/10.1111/j.1095-8312.1993.tb00893.x

Johansson MP, Ermold F, Kristjánsson BK et al (2016) Divergence of gastropod life history in contrasting thermal environments in a geothermal lake. J Evol Biol 29:2043–2053. https://doi.org/10.1111/jeb.12928

Jordaens K, Dillen L, Backeljau T (2007) Effects of mating, breeding system and parasites on reproduction in hermaphrodites: pulmonate gastropods. Anim Biol 57(2):137–195. https://doi.org/10.1163/157075607780377965

Kappes H, Haase P (2012) Slow, but steady: dispersal of freshwater molluscs. Aquat Sci 74:1–14. https://doi.org/10.1007/s00027-011-0187-6

Katkov MV (1971) On the survival of planorbid molluscs (Pulmonata) in non-permanent waterbodies and the persistence of trematode larvae in them. Zool Zh 50(4):584–586. [in Russian]

Kew HW (1893) The dispersal of shells, an inquiry into the means of dispersal possessed by freshwater and land Mollusca. The International Scientific Series 75:xi+1–291. https://doi.org/10.5962/bhl.title.10376

Khalil LF (1961) On the capture and destruction of miracidia by *Chaetogaster limnaei* (Oligochaeta). J Helminthol 35(3–4):269–274. https://doi.org/10.1017/S0022149X00004648

Khmeleva NN, Golubev AP, Laenko TN (1985) The ecology of gastropod molluscs from geothermal springs in Kamchatka. Zh Obshch Biol 46(2):230–240. [in Russian]

Kitai H, Kakuda U, Goto S et al (2021) Photoperiod controls egg laying and caudodorsal cell hormone expression but not gonadal development in the pond snail *Lymnaea stagnalis*. J Comp Physiol A 207:523–532. https://doi.org/10.1007/s00359-021-01494-2

Klemm DJ (1976) Leeches (Annelida: Hirudinea) found in North American mollusks. Malac Review 9:62–76

Knutson LV, Vala JC (2011) Biology of snail-killing Sciomyzidae flies. Cambridge University Press, Cambridge etc

Kobelt W (1871) Zur Kenntniss der europäischen Limnäen. II. *L. stagnalis* Linné. Malakozoologische Blätter 18:108–119

Krkač N (1979) Temperature and reproductive cycle relations in *Radix peregra* O.F. Müller. Malacologia 18:227–232

Kruglov ND (1980) The reproductive biology of the freshwater pulmonate molluscs (Pulmonata). Zool Zh 59(7):986–995. [in Russian]

Kruglov ND (2005) Lymnaeid snails of Europe and Northern Asia. Smolensk State Pedagogical University Press, Smolensk. [in Russian]

Kruglov ND, Starobogatov YI (1985) The volume of the subgenus *Galba* and of other similar subgenera of the genus *Lymnaea* (Gastropoda, Pulmonata). Zool Zh 64(1):24–35. [in Russian]

Kruglov ND, Starobogatov YI (1987) Life forms in lymnaeids and some problems of their systematics. In: Starobogatov YI, Golikov AN, Likharev IM (eds) Molluyski: rezultaty i perspektivy ikh issledovaniy. Vsesoyuznoye soveshchaniye po izucheniyu mollyuskov. Avtoreferaty dokladov, vol 8. Nauka Publishers, Leningrad, pp 68–70. [in Russian]

Kruglov ND, Starobogatov YI (1989) Morphology and taxonomy of the molluscs from the subgenus *Radix* of the genus *Lymnaea* (Gastropoda, Pulmonata, Lymnaeidae) from Siberia and far east of the USSR. Zool Zh 68(5):17–30. [in Russian]

Kruglov ND & Starobogatov YI (1992 ["1991"]) The unknown egg masses of molluscs of the genus *Lymnaea* (Gastropoda Pulmonata Lymnaeidae). Trudy Zoologicheskogo Instituta Akademii Nauk SSSR 228:111–129. [in Russian]

Kunigelis SC, Saleuddin ASM (1978) Regulation of growth in the pulmonate gastropod *Helisoma duryi*. Can J Zool 56:1975–1980. https://doi.org/10.1139/z78-265

Lam PKS (1994) Intraspecific life-history variation in *Radix plicatulus* (Gastropoda, Pulmonata, Lymnaeidae). J Zool 232:435–446. https://doi.org/10.1111/j.1469-7998.1994.tb01584.x

Lawson AK (1921) *Limnaea stagnalis* destroyed by rats. J Conchol 16(5):144

Leicht K, Seppälä K, Seppälä O (2017) Potential for adaptation to climate change: family-level variation in fitness-related traits and their responses to heat waves in a snail population. BMC Evol Biol 17:140. https://doi.org/10.1186/s12862-017-0988-x

Leiß A, Reischütz PL (1995) Ein Beitrag zur Kenntnis der Molluskenfauna der Gewächshäuser in Wien und Niederösterreich. Bufus-Info – Mitteilungsblatt der Biologischen Unterwasserforschungsgruppe der Universität Salzburg, vol 16, pp 23–30

Levina OV (1973) The fecundity of the freshwater molluscs *Limnaea stagnalis* and *Radix ovata*. Zool Zh 52(5):676–684. [in Russian]

Lounnas M, Correa AC, Vázquez AA et al (2017) Self-fertilization, long-distance flash invasion and biogeography shape the population structure of *Pseudosuccinea columella* at the worldwide scale. Mol Ecol 26:887–903. https://doi.org/10.1111/mec.13984

Lukin EI (1976) The leeches of fresh- and brackish waterbodies. Fauna SSSR. Piyavki 1:1–484. Nauka, Leningrad. [in Russian]

Madsen H, Frandsen F (1989) The spread of freshwater snails including those of medical and veterinary importance. Acta Trop 46:139–146. https://doi.org/10.1016/0001-706x(89)90030-2

Maharaj R, Naidoo I, Appleton C (2005) Susceptibility of schistosome host snails to predation by sciomyzid flies (Diptera: Sciomyzidae) in South Africa. Afr J Aquat Sci 30(2):175–178. https://doi.org/10.2989/16085910509503853

Mahulu A, Clewing C, Stelbrink B et al (2019) Cryptic intermediate snail host of the liver fluke *Fasciola hepatica* in Africa. Parasite Vectors 12:573. https://doi.org/10.1186/s13071-019-3825-9

Malone CR (1965) Dispersal of aquatic gastropods via the intestinal tract of water birds. The Nautilus 78:135–139

Marinoni L, Murphy WL (2016) Family Sciomyzidae. Zootaxa 4122(1):642–647. https://doi.org/10.11646/zootaxa.4122.1.53

Martins RT, Alves RG (2010) Occurrence of *Chaetogaster limnaei* K. von Baer, 1927 (Oligochaeta, Naididae) associated with Gastropoda mollusks in horticultural channels in Southeastern Brazil. Braz J Biol 70(4):1055–1057. https://doi.org/10.1590/S1519-69842010000500020

Martynov AV, Anistratenko VV, Furyk YI (2018) Endemic spring snails *Terrestribythinella* (Mollusca) as unusual material for larval case of *Crunoecia irrorata* (Trichoptera: Lepidostomatidae) in Transcarpathian Ukraine. Ecologica Montenegrina 18:120–125. https://doi.org/10.37828/em.2018.18.11

McMahon RF (1983) Physiological ecology of freshwater pulmonates. In: Russell-Hunter WD (ed) The Mollusca, vol 6. Academic Press, Orlando etc, pp 360–430

Meeuse ADJ, Hubert B (1949) The mollusc fauna of glasshouses in The Netherlands. Basteria 13(1/3):1–44

Meshcheryakov VN (1990) The common pond snail, *Lymnaea stagnalis*. In: Dettlaff TA, Wassetsky SG (eds) Animal species for developmental studies: Invertebrates. Plenum Publishing Corp, New York, pp 69–132

Mitchell DR, Leung TLF (2016) Sharing the load: a survey of parasitism in the invasive freshwater pulmonate, *Physa acuta* (Hygrophila: Physidae) and sympatric native snail populations. Hydrobiologia 766:165–172. https://doi.org/10.1007/s10750-015-2452-5

Nekhaev IO (2011) Geographic distribution features of subgenus and life forms of Lymnaeidae (Gastropoda, Pulmonata) in eastern Europe. Biol Bull 38(9):406–411. https://doi.org/10.1134/S1062359011040121

Nekhaev IO (2021) Freshwater gastropods of the western part of the Kola Peninsula and northern Karelia (northern Europe). Ruthenica. Russian Malacol J 31(4):147–175. https://doi.org/10.35885/ruthenica.2021.31(4).1

Nekhaev IO, Palatov DM (2016) From the Black Sea to the White Sea: The first record of the invasive mollusc *Physella acuta* in the extreme North of the Europe. Russ J Biol Invasions 7(4): 351–354. https://doi.org/10.1134/S2075111716040056

Ng TH, Tan SK, Wong WH et al (2016) Molluscs for sale: Assessment of freshwater gastropods and bivalves in the ornamental pet trade. PLoS One 11(8):e0161130. https://doi.org/10.1371/journal.pone.0161130

Nicot A, Dubois M-P, Debain C et al (2008) Characterization of 15 microsatellite loci in the pulmonate snail *Pseudosuccinea columella* (Mollusca, Gastropoda). Mol Ecol Resour 8:1281–1284. https://doi.org/10.1111/j.1755-0998.2007.02065.x

Numata H, Udaka H (2009) In: Nelson RJ, Denlinger DL, Somers DE (eds) Photoperiodism in mollusks. Oxford University Press, Oxford, pp 173–192. https://doi.org/10.1093/acprof:oso/9780195335903.003.0008

Økland J (1990) Lakes and snails. Environment and Gastropoda in 1,500 Norwegian lakes, ponds and rivers. Universal Book Services / Dr. W. Backhuys Publ, Oegstgeest

Olsson TI (1981) Overwintering of benthic macroinvertebrates in ice and frozen sediment in a North Swedish river. Ecography 4:161–166. https://doi.org/10.1111/j.1600-0587.1981.tb00993.x

Oswood MW, Miller LK, Irons JG (1991) Overwintering of freshwater benthic macroinvertebrates. In: Lee RE, Denlinger DL (eds) Insects at low temperature. Springer, Boston MA, pp 360–375. https://doi.org/10.1007/978-1-4757-0190-6_15

Pagulayan IF, Umaly RC, Enriquez GL (1983) Studies on the biology of *Radix quadrasi*: shell color, mantle pigmentation, and chromosome numbers. Nat Appl Sci Bull 35(2):147–155

Patoka J, Kopecký O, Vrabec V et al (2017) Aquarium molluscs as a case study in risk assessment of incidental freshwater fauna. Biol Invasions 19:2039–2046. https://doi.org/10.1007/s10530-017-1412-6

Patzer HE (1927) Beiträge zur Biologie der Leberegelschnecke, *Galba* (*Limnaea*) *truncatula* Müll. Zool Jahrb Abt Syst Ökol Geogr Tiere 53(4/5):321–372

Pavlova MV (1983) Biology and productivity of *Limnaea* (*Radix*) *auricularia* (L.) var. *obliquata* Martens in the Lake Issyk-Kul. In: Konurbaev AO (ed) Limnologicheskiye issledovaniya v Kirgizii. Frunze, Ilim, pp 34–46. [in Russian]

Piaget J (1912) Les récents dragages malacologiques de M. le Prof. Émile Yung dans le lac Léman. Journal de Conchyliologie 60:205–233

Piaget J (1913) Les mollusques sublittoraux du Léman recueillis par M. le Prof. Yung. Zool Anz 42: 615–624

Piggott H, Dussart G (1995) Egg-laying and associated behavioural responses of *Lymnaea peregra* (Müller) and *Lymnaea stagnalis* (L.) to calcium in their environment. Malacologia 37(1):13–21

Pip E (1986) The ecology of freshwater gastropods in the central Canadian region. The Nautilus 100:56–66

Pip E (1993) The distribution of freshwater gastropods in central North America in relation to water temperature. Heldia 2(1/2):21–27

Pointier JP (ed) (2015) Freshwater molluscs of Venezuela and their medical and veterinary importance. Conchbooks, Harxheim

Prozorova LA (1992 ["1991"]) Morphology of gastropod egg-clusters of the Primorye Region. Trudy Zoologicheskogo Instituta Akademii Nauk SSSR 228:74–110. [in Russian]

Reeve LA (1850) On a new species of Lymnaea from Thibet. Proc Zool Soc Lond 18:49. https://www.biodiversitylibrary.org/page/12858725

Richards AG (1965) The developmental rate and oxygen consumption of snail eggs at various temperatures. Z Naturforsch B 20:347–349. https://doi.org/10.1515/znb-1965-0417

Rondelaud D (1976) L'interet d'une association de mollusques Zonitidae dans le contróle biologique de *Lymnaea* (*Galba*) *truncatula* Müller sur terrains siliceux et calcaires. Ann Parasitologie 51:41–49

Rondelaud D, Vignoles P, Dreyfuss G et al (2006) The control of *Galba truncatula* (Gastropoda: Lymnaeidae) by the terrestrial snail *Zonitoides nitidus* on acid soils. Biol Control 39:290–299. https://doi.org/10.1016/j.biocontrol.2006.07.015

Rondelaud D, Vignoles P, Dreyfuss G (2016) *Aplexa hypnorum* (Gastropoda: Physidae) exerts competition on two lymnaeid species in periodically dried ditches. Ann Limnol 52:379–386. https://doi.org/10.1051/limn/2016022

Rondelaud D, Vignoles G, Dreyfuss G et al (2023, this volume) Control of fasciolosis-transmitting lymnaeids in the field. In: Vinarski MV, Vàzquez AA (eds) The Lymnaeidae: a handbook. Springer, Cham, pp 415–445

Roszkowski W (1914) Contribution a l'étude des limnées du Lac Léman. Rev Suisse Zool 22:457–539

Roszkowski W (1925) Contributions to the knowledge of the family Lymnaeidae. II. Some new data to the knowledge of the Altai *Lymnaea*. Annales Zoologici Musei Polonici Historiae Naturalis 4(4):277–281

Russell-Hunter WD (1964) Physiological aspects of ecology in non-marine mollusks. In: Wilbur KM, Yonge CM (eds) Physiology of Mollusca. Academic Press, London, pp 83–125

Russell-Hunter WD (1978) Ecology of freshwater pulmonates. In: Fretter V, Peake J (eds) Pulmonates, vol 2A. Academic Press, London, pp 335–384

Salazkin AA (1969) On some pecularities of distribution of gastropods in lakes of different type in the humid zone of Western Siberia. In: Johanzen BG (ed) Voprosy malakologii Sibiri. Tomsk State University, Tomsk, pp 57–60. [in Russian]

Schniebs K, Sitnikova TY, Vinarski MV et al (2022) Morphological and Genetic Variability in *Radix auricularia* (Mollusca: Gastropoda: Lymnaeidae) of Lake Baikal Siberia: The Story of an Unfinished Invasion into the Ancient Deepest Lake Diversity 14(7):527. https://doi.org/10.3390/d14070527

Schodduyn R (1925) Observations biologiques en microaquarium. Ann Biol Lacustre 14:39–47

Shirokaya AA, Takhteev VV (2018) The key for gastropods of the family Acroloxidae Thiele, 1931 (Pulmonata: Hygrophila) of Lake Baikal, with remarks on the species' ecology and distribution. V.B. Sochava Institute of Geography SB RAS, Irkutsk. [in Russian]

Simonová J, Simon OP, Kapic Š et al (2016) Medium-sized forest snails survive passage through birds' digestive tract and adhere strongly to birds' legs: more evidence for passive dispersal mechanisms. J Mollus Stud 82(3):422–426. https://doi.org/10.1093/mollus/eyw005

Smythe AB, Forgrave K, Patti A et al (2015) Untangling the ecology, taxonomy, and evolution of *Chaetogaster limnaei* (Oligochaeta: Naididae) species complex. J Parasitol 101(3):320–326. https://doi.org/10.1645/13-465.1

Stadnichenko AP (2004) Pond snails and limpet snails (Lymnaeidae and Acroloxidae) of Ukraine. Tsentr uchebnoj literatury, Kiev. [in Russian]

Stadnichenko AP (2006) Lymnaeidae and Acroloxidae of Ukraine: sampling and studying methods, biology, ecology, their practical importance. Ruta Press, Zhytomyr. [in Russian]

Starobogatov YI (1970) Molluscan fauna and zoogeographic zonation of continental freshwater bodies of the world. Nauka, Leningrad. [in Russian]

Starobogatov YI, Prozorova LA, Zatravkin MN (1989) Species composition of the family Physidae (Gastropoda Pulmonata Lymnaeiformes) of Siberia and Far East of the USSR (with notes on European physids). Byulleten' Moskovskogo obshchestva ispytateley pripody. Otdel biologicheskiy 94(1):62–78. [in Russian]

Stoll SD, Frueh B, Westerwald N et al (2013) Density-dependent relationship between *Chaetogaster limnaei limnaei* (Oligochaeta) and the freshwater snail *Physa acuta* (Pulmonata). Freshw Sci 32:642–649. https://doi.org/10.1899/12-072.1

Takhteev VV, Sudakova EA, Egorova IN et al (2006) To the characteristics of aquatic and land biocoenoses in the sites of effusion of thermal springs in east Prybaikaliye, vol 6. Trudy Biologo-Pochvennogo Fakulteta IGU, Irkutsk, pp 111–136. [in Russian]

Takhteev VV, Eropova IO, Egorova IN et al (2019) Structure of hydrobiocenoses in mineral and thermal springs of the Lake Baikal region: a review. Contemp Probl Ecol 12(2):126–142. https://doi.org/10.1134/S1995425519020082

Taylor DW, Walter HJ, Burch JB (1963) Freshwater snails of the subgenus *Hinkleyia* (Lymnaeidae: *Stagnicola*) from the Western United States. Malacologia 1(2):237–281

Ter Maat A, Pieneman AW, Koene JM (2012) The effect of light on induced egg laying in the simultaneous hermaphrodite *Lymnaea stagnalis*. J Mollus Stud 78(3):262–267. https://doi.org/10.1093/mollus/eys008

TROPMED Technical Group (1986) Snails of medical importance in Southeast Asia. Southeast Asian J Trop Med Public Health 17(2):282–322

Trouvé S, Degen L, Renaud F et al (2003) Evolutionary implications of a high selfing rate in the freshwater snail *Galba truncatula*. Evolution 57(10):2302–2314. https://doi.org/10.1554/02-452

Trouvé S, Degen L, Goudet J (2005) Ecological components and evolution of selfing in the freshwater snail *Galba truncatula*. J Evol Biol 18:358–370. https://doi.org/10.1111/j.1420-9101.2004.00831.x

Tsikhon-Lukanina EA (1987) Trophology of aquatic molluscs. Nauka, Moscow. [in Russian]

Tsikhon-Lukanina EA, Reznichenko OG, Lukasheva TA (1998) Diet composition and food diversity in marine and freshwater gastropods. Zool Zh 77(3):270–277. [in Russian]

Vaghin [=W]V (1931) *Chaetogaster limnaei* K. Baer als Cercarienvertilger. Zool Anz 95:55–59

Vaghin [=W]V (1946) On the biological species of *Chaetogaster limnaei*, k. Baer. Doklady Akademii Nauk SSSR 51(6):481–484. [in Russian]

van Leeuwen CHA, van der Velde G, van Lith B et al (2012) Experimental quantification of long distance dispersal potential of aquatic snails in the gut of migratory birds. PLoS One 7(3):32292. https://doi.org/10.1371/journal.pone.0032292

van Leeuwen CHA, Huig N, van der Velde G et al (2013) How did this snail get here? Several dispersal vectors inferred for an aquatic invasive species. Freshw Biol 58:88–99. https://doi.org/10.1111/fwb.12041

Van Someren VD (1946) The habitats and tolerance ranges of *Lymnaea* (*Radix*) *caillaudi*, the intermediate snail host of liver fluke in East Africa. J Anim Ecol 15(2):170–197. https://doi.org/10.2307/1557

Vaughn CM (1953) Effects of temperature on hatching and growth of *Lymnaea stagnalis appressa* Say. Am Midl Nat 49:214–228. https://doi.org/10.2307/2422289

Vàzquez AA, Alba A, Alda P et al (2023, this volume) Lymnaeid snails and the transmission of fasciolosis: understanding the differential risks from local to global scale. In: Vinarski MV, Vàzquez AA (eds) The Lymnaeidae: a handbook. Springer, Cham, pp 359–394

Vignoles P, Dreyfuss G, Rondelaud D (2017) Ecologie et parasitisme de la Limnée e?troite (*Omphiscola glabra*). PULIM, Limoges

Vinarski MV (2017) The history of an invasion: phases of the explosive spread of the physid snail *Physella acuta* through Europe, Transcaucasia and Central Asia. Biol Invasions 19(4):1299–1314. https://doi.org/10.1007/s10530-016-1339-3

Vinarski MV, Grebennikov ME (2012) An overview of species of the genus *Aenigmomphiscola* Kruglov et Starobogatov, 1981 (Gastropoda: Pulmonata: Lymnaeidae). Ruthenica Russian Malacol J 22(2):159–170. [in Russian]

Vinarski MV, Kantor YI (2016) Analytical catalogue of fresh and brackish water molluscs of Russia and adjacent countries. A.N. Severtsov Institute of Ecology and Evolution of RAS, Moscow

Vinarski MV, Palatov DM (2018) *Ferrissia californica*: the first record of a global invader in a cave habitat. J Nat Hist 52(17–18):1147–1155. https://doi.org/10.1080/00222933.2018.1450904

Vinarski MV, Pointier JP (2023, this volume) General characteristics of the family Lymnaeidae. In: Vinarski MV, Vàzquez AA (eds) The Lymnaeidae: a handbook. Springer, Cham, pp 25–66

Vinarski MV, Bolotov IN, Aksenova OV et al (2021) Freshwater Mollusca of the circumpolar Arctic: a review on their taxonomy, diversity and biogeography. Hydrobiologia 848:2891–2918. https://doi.org/10.1007/s10750-020-04270-6

Vinarski MV, Aksenova OV, Bolotov IN et al (2022a) A new alien snail *Ampullaceana balthica* for the Canadian fauna, with an overview of Transatlantic malacofaunal exchange in the Anthropocene. Aquat Invasions 17(1):21–35. https://doi.org/10.3391/ai.2022.17.1.02

Vinarski MV, von Oheimb PV, Aksenova OV et al (2022b) Trapped on the Roof of the World: Taxonomic diversity and evolutionary patterns of Tibetan Plateau endemic freshwater snails (Gastropoda: Lymnaeidae: *Tibetoradix*). Integr Zool 17:825–848. https://doi.org/10.1111/1749-4877.12600

Vinarski MV, Aksenova OV, Bespalaya YV et al (2023) How an Ecological Race Is Forming: Morphological and Genetic Disparity among Thermal and Non-Thermal Populations of Aquatic

Lymnaeid Snails (Gastropoda: Lymnaeidae). Diversity 15(4):548. https://doi.org/10.3390/d15040548

Voronezhskaya EE, Khabarova MY, Nezlin LP (2004) Apical sensory neurones mediate developmental retardation induced by conspecific environmental stimuli in freshwater pulmonate snails. Development 131:3671–3680. https://doi.org/10.1242/dev.01237

Voronezhskaya EE, Khabarova MY, Chaban AM et al (2007) Role of chemical signalling in release of motor programs during embryogenesis of freshwater snails *Lymnaea stagnalis* and *Helisoma trivolvis*. Ontogenez 38(2):94–104. [in Russian]

Welter-Schultes F (2012) European non-marine molluscs: a guide for species identification. Planet Poster Editions, Göttingen

Wesenberg-Lund C (1939) Biologie der Süsswassertiere. Wirbellose Tiere. Julius Springer Verlag, Wien

Wullschleger EB, Jokela J (2002) Morphological plasticity and divergence in life-history traits between two closely related freshwater snails, *Lymnaea ovata* and *Lymnaea peregra*. J Mollus Stud 68:1–5. https://doi.org/10.1093/mollus/68.1.1

Chapter 10
Reproductive Strategies, Genetic Diversity, and Invasive Ability in Lymnaeidae

Nicolás Bonel, Yumi Nakadera, Julia Pizá, Antonio A. Vázquez, Joris M. Koene, Patrice David, Philippe Jarne, and Pilar Alda

Abstract Lymnaeid snails are simultaneous hermaphrodites that have a worldwide distribution, inhabiting freshwater areas from almost all continents ranging from tropical to arctic regions and from sea level to very high altitudes. In this chapter, we review the reproductive anatomy, behavioral and physiological traits, and mating strategies associated with increased survival and invasiveness of lymnaeids across different ecosystems around the globe. We also discuss the biotic and abiotic factors that can affect mating systems in this family, and how they have expanded their geographical range by natural, as well as human-mediated ways, likely promoting the spread of infectious diseases. Finally, we discuss why we believe that lymnaeids are suitable model organisms for studying mechanisms and processes involved in the ecology and evolution of mating systems and biological invasions.

N. Bonel (✉) · J. Pizá
Centro de Recursos Naturales Renovables de la Zona Semiárida (CERZOS, CCT, UNS-CONICET Bahía Blanca), Bahía Blanca, Argentina

Y. Nakadera · J. M. Koene
A-LIFE, Faculty of Science, Vrije Universiteit, Amsterdam, The Netherlands
e-mail: y.nakadera@vu.nl; joris.koene@vu.nl

A. A. Vázquez
Laboratory of Malacology, Institute of Tropical Medicine "Pedro Kourí,", Havana, Cuba

MIVEGEC, University of Montpellier, CNRS, IRD, Montpellier, France

P. David · P. Jarne
CEFE, University of Montpellier, CNRS, EPHE, IRD, Montpellier, France
e-mail: Patrice.DAVID@cefe.cnrs.fr; philippe.jarne@cefe.cnrs.fr

P. Alda
Genética y Ecología Evolutiva (CERZOS–CCT–CONICET Bahía Blanca), Bahía Blanca, Argentina

MIVEGEC, University of Montpellier, CNRS, IRD, Montpellier, France

M. V. Vinarski, A. A. Vázquez (eds.), *The Lymnaeidae*, Zoological Monographs 7,
https://doi.org/10.1007/978-3-031-30292-3_10

10.1 Introduction

The family Lymnaeidae is a worldwide-distributed group that originated and diversified 200 Myr ago, which now inhabits freshwater areas from all continents (except for Antarctica) ranging from tropical to arctic regions and from sea level to very high altitudes. Researchers have studied genetic diversity, life-history traits, and reproductive strategies to better understand their ecological and evolutionary history, and how they are related to the current distribution of the family. Yet, compared to other zoological groups (e.g., insects, vertebrates), relatively few studies have been conducted on this group, and they have focused mainly on a handful of species.

In this chapter, we review the life-history traits associated with lymnaeid survival and invasion by focusing on evolutionary and ecological studies performed in this family. We also call upon studies performed in other animal groups, and sometimes in plants, in an attempt to understand lymnaeid ecological and evolutionary success. We review the morphological, behavioral, and physiological traits involved in lymnaeid reproduction. We also discuss the factors that can affect lymnaeid mating systems, and how lymnaeids have expanded their geographical range by natural as well as human-mediated ways. Finally, we explain why we believe that lymnaeids are suitable model organisms for studying mechanisms and processes involved in the ecology and evolution of mating systems and biological invasions. In the whole chapter, we draw heavily on results obtained in the most studied species—that is, the great pond snail *Lymnaea stagnalis* and the *Galba* genus, putting emphasis on the mating system (selfing vs. outcrossing).

10.2 Reproductive Anatomy and Traits Potentially Enhancing Invasion in Lymnaeids

In this section, we first provide an overview of the reproductive anatomy of lymnaeids. Next, we move on to describing the morphological, behavioral, and physiological features that could enhance their invasiveness. Most of the following insights are from studies of *Lymnaea stagnalis*, as this species is extensively studied in multiple research fields (e.g., neurophysiology, ecotoxicology). Previous reviews in the Hygrophila, i.e. the superorder including the Lymnaeidae family, have developed this matter in greater detail (Geraerts and Joosse 1984; Jarne et al. 2010, 1993).

10.2.1 Reproductive Anatomy of Lymnaeids

As in other Hygrophila species, the reproductive morphology of Lymnaeids is rather complex with enough variation among species to serve as a rather accurate tool in systematics (Geraerts and Joosse 1984; Jarne et al. 2010; Fig. 10.1). Lymnaeids, as

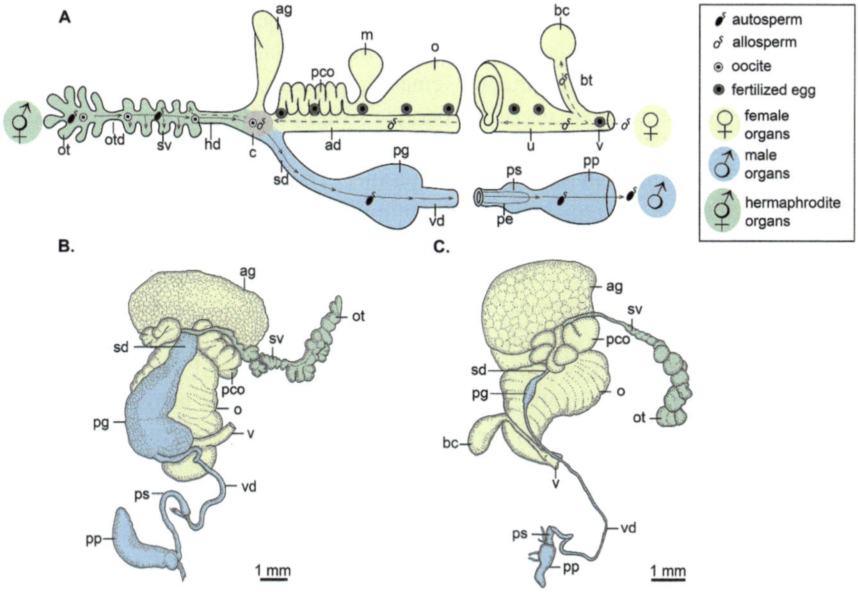

Fig. 10.1 (**a**) Schematic representation of the reproductive apparatus of Lymnaeidae (modified from Jarne et al. 2010). Reproductive apparatus drew using a camera lucida attachment. (**b**) from the outcrossing species *Ampullaceana balthica* and (**c**) from the selfing species *Pseudosuccinea columella*. Hermaphrodite, female and male organs are colored in green, yellow and blue, respectively. The route of oocytes, fertilized eggs, allo- and autosperm is also indicated. Abbreviations: ad, allosperm duct; ag, albumen gland; bc, bursa copulatrix; bt, bursa copulatrix tract; c, carrefour (= fertilization pouch); hd, hermaphroditic duct; m, muciparous gland; o, oothecal or nidamental gland; ot, ovotestis; otd, ovotestis duct; pco, pars contorta (= oviduct); pe, penis; pg, prostate gland; pp preputium; ps, penis sheath; sd, sperm duct; sv, seminal vesicles; u, uterus; v, vagina; vd, vas deferens

all Hygrophila and a significant fraction of gastropods, are simultaneous hermaphroditic organisms, having male and female functions within a single body at the same time (Jarne and Auld 2006). Thus, they have two choices of sperm for fertilization: autosperm (own sperm) for self-fertilization (or selfing), and allosperm (partners' sperm) for cross-fertilization (or outcrossing). Sperm and eggs are produced in a single hermaphroditic organ, called the ovotestis. After spermatogenesis, mature spermatozoa are stored in the seminal vesicles after descending through the hermaphroditic duct. The latter separates into female and male tracts at the level of the carrefour where the fertilization pouch is positioned. Mature oocytes also exit the ovotestis and go down the hermaphroditic duct. The exact sites for both cross- and self-fertilization remain elusive, but should occur upstream of the carrefour. Then, fertilized eggs move down through several glands (albumen gland, pars contorta [oviduct], muciparous gland [oviduct pouch], oothecal gland) packaging eggs into egg capsules and an egg mass, which can contain from a few to a few hundred eggs. The egg mass is then laid through the female gonopore and deposited onto some surface in the environment to which they adhere.

When an individual plays the male role, it everts the preputium. The penis is usually invaginated within the preputium. During copulation, the male-acting snail sticks the preputium into the partner's female gonopore, then its penis devaginates from within the preputium to inside the partner's vagina. Mature spermatozoa stored in the seminal vesicles are transferred, furnished with seminal fluids in the prostate gland, into the vagina of the female-acting partner. The ejaculate is transferred to the partner via the vas deferens and the preputium (that stays outside) and the penis (inside the partner) toward the female part of the partner's apparatus. There, most of the ejaculate is directed toward the bursa copulatrix where it supposedly gets digested (data are from a species from another gastropod family, the garden snail *Cornu aspersum* (Rogers and Chase 2001)), and a small fraction of spermatozoa travels up the female tracts toward the carrefour to get stored and used for later fertilization (Koene et al. 2009). In contrast, the trajectory of the autosperm used for self-fertilization is different. It is highly unlikely that the autosperm travel through the male ducts of the donor and then female ducts of the recipient up to the carrefour for fertilization, meaning that autosperm does not come with seminal fluid. Thus, autosperm and allosperm may be in different states, allowing for differential success in fertilization, and leaving room for a potential control over selfing rate by the mother.

10.2.2 Mating Strategies and Behavioral and Physiological Traits

10.2.2.1 Selfing

In common with all Hygrophila, the species of the Lymnaeid family can self-fertilize, meaning that a single adult snail can initiate a population without any mates ("reproductive assurance"; Baker's law: Baker 1955; Cheptou 2012). In freshwater snails, selfing is associated with the modification of traits associated with reproduction, including morphological ones, which has been referred to as the selfing syndrome (Doums et al. 1996). This syndrome encompasses the wide range of traits associated with preferential selfing, such as decreased inbreeding depression, limited male copulatory behavior, and reduced heterozygosity. Further, the selfing syndrome is supported by the predictions from the sex allocation theory: selection would drive the adaptive optima toward allocating fewer resources to the male function because of reduced sexual selection in selfing species (Charnov 1982; Schärer 2009). In species prone to selfing, it is therefore expected a lower investment into the prostate gland and prostatic fluids relative to outcrossing ones. Such variation in sex allocation is exemplified in Fig. 10.1, which compares the reproductive morphology of *Pseudosuccinea columella* (a selfer) and *Ampullaceana (Radix) balthica* (an outcrosser) where the prostate gland exhibits a simpler internal structure in selfing relative to an outcrossing species (Jarne et al. 2010; Swart et al. 2020). However, the analysis of the relationship between selfing rates and reproductive morphology is limited to few species in lymnaeids (Escobar et al. 2011). Thus, a comparative study including a sufficient number of species, associated with phylogenetic analysis to account for phylogenetic correlations, would be needed to confirm if selfing syndrome is indeed affecting the evolution of reproductive morphology.

As Hygrophila in general, the Lymnaeid family displays a diversity of reproductive strategies with respect to selfing rate, which generally falls into two well-separated groups: preferential outcrossers and preferential selfers (strongly U-shaped distribution, based on molecular estimates; Escobar et al. 2011). On the one hand, preferential outcrossers (including, as far as we know, *Lymnaea stagnalis* and the *Radix* clade) show selfing rates below 20% in natural populations, and self-fertilization is generally associated with inbreeding depression (Fig. 10.2). Hence, these species outcross their eggs whenever possible, and they self-fertilize only when they are unable to find a mate, after waiting some time for potential mates. The latter strategy is likely because it is better paying the price of depressed fitness with selfing than not reproducing at all (see review in Escobar et al. 2011). Note that the pond snail *L. stagnalis* is an exception (and a puzzle): this species is an outcrosser, they do wait for a potential mate, exhibiting little inbreeding depression, though initiating selfing at an older age than outcrossing (Coutellec and Caquet 2011; Koene et al. 2008; Puurtinen et al. 2007). On the other hand, preferential selfers, including *Pseudosuccinea columella* (Lounnas et al. 2017a; Nicot et al. 2008) and the *Galba* spp. clade (Chapuis et al. 2007; Hurtrez-Boussès et al. 2010; Meunier et al. 2004; Trouvé et al. 2003; Lounnas et al. 2018, 2017b), show natural selfing rates in excess of 80% and there is no evidence for the existence of any significant delay in egg production upon the lack of a partner, or of inbreeding depression (see Fig. 10.2). In both preferential outcrossers and preferential selfers, selfed progeny is readily obtained and raised in laboratory conditions (Alba et al. 2019; Nakadera et al. 2019). Accordingly, selfing could serve, for instance, as a (temporary or permanent) strategy to successfully invade new areas and overcome difficulties associated with low densities and low population increase. Regular selfers, or species like *L. stagnalis*, that lack inbreeding depression and/or reproductive delays associated with self-fertilization, may pay even lower costs during founding events, which occur repeatedly when an invasive species spreads through fragmented freshwater habitats.

10.2.2.2 Multiple Mating

Among outcrossing lymnaeids, multiple matings with different male and female partners are frequent. This behavior has largely been documented in *L. stagnalis* under laboratory conditions with up to two inseminations per day in the male role for some individuals (Koene and Ter Maat 2007). The male mating drive is determined by mating history, and virgin individuals or individuals that have been isolated for several days are eager to copulate as male (Van Duivenboden and Ter Maat 1985). This motivation is associated with the fullness of the prostate gland (De Boer et al.

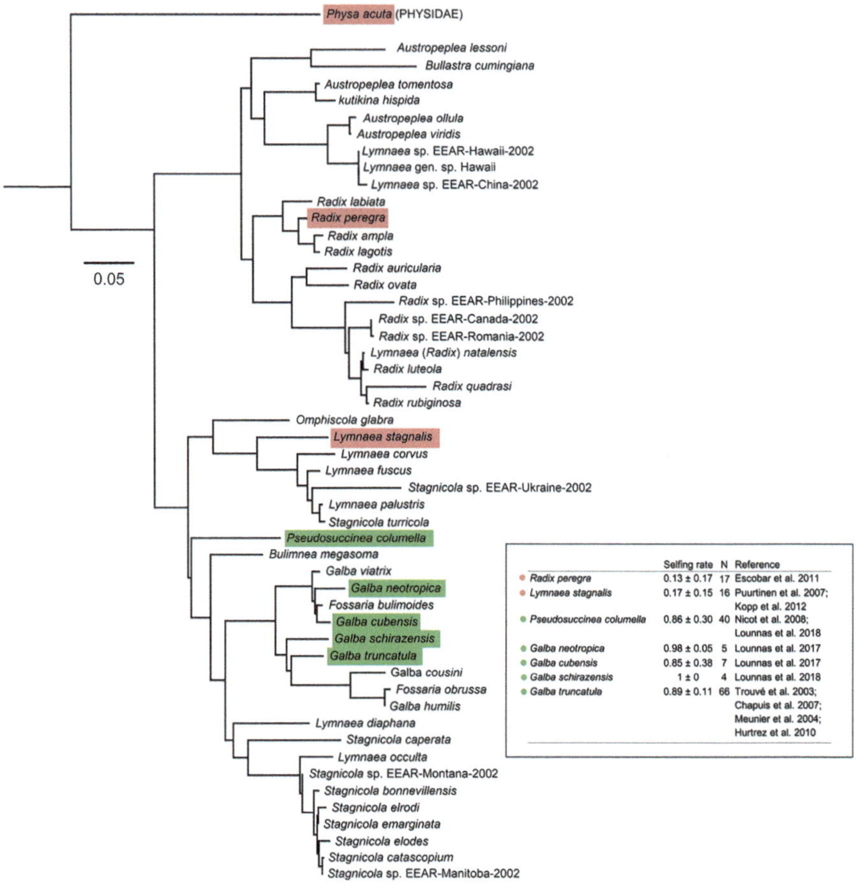

	Selfing rate	N	Reference
Radix peregra	0.13 ± 0.17	17	Escobar et al. 2011
Lymnaea stagnalis	0.17 ± 0.15	16	Puurtinen et al. 2007; Kopp et al. 2012
Pseudosuccinea columella	0.86 ± 0.30	40	Nicot et al. 2008; Lounnas et al. 2018
Galba neotropica	0.98 ± 0.05	5	Lounnas et al. 2017
Galba cubensis	0.85 ± 0.38	7	Lounnas et al. 2017
Galba schirazensis	1 ± 0	4	Lounnas et al. 2018
Galba truncatula	0.89 ± 0.11	66	Trouvé et al. 2003; Chapuis et al. 2007; Meunier et al. 2004; Hurtrez et al. 2010

Fig. 10.2 Phylogeny of the Lymnaeidae indicating the preferentially outcrossing (pink) and preferentially selfing species (green). The tree was obtained by concatenating the 16S, ITS1, and ITS2 sequences and includes 50 species and *Physa acuta* as an outgroup (modified from Correa et al. 2010). The embedded table shows the mean and the standard deviation of the selfing rate, the number of studied populations (N), and the references that have been used to estimate the selfing rate for each lymnaeid species

1997). Field investigations using genetic markers support that multiple mating is indeed prevalent in natural populations of outcrossing lymnaieds, although this approach would underestimate the degree of multiple mating due to the potential matings did not yield any offspring (*L. stagnalis*, Nakadera et al. 2017; *A. balthica*, Coutellec-Vreto et al. 1997; Bürkli and Jokela 2017). One of the consequences of promiscuity and multiple paternity is increased genetic variance among offspring, which could be beneficial to cope with environmental challenges. Also, promiscuity is likely associated with strong male–male competition, both precopulatory (access to mating partners) and postcopulatory (sperm competition), and, consequently, sexual selection on the male function (e.g., Hoffer et al. 2017 for study of sexual selection in *L. stagnalis*). Such selective pressures on the male function in outcrossing lymnaeids would promote purging deleterious mutations and limiting genetic load in populations, as shown in other freshwater snail species (Bonel et al. 2018; Noël et al. 2019).

10.2.2.3 Mate Preference and Mating Behavior

Contrasting with separate-sex species, in which individuals need the opposite sex to reproduce, simultaneous hermaphrodites can reproduce with any conspecific, thus increasing the potential to initiate a population and overcoming the Allee effect in low-density populations (see above for the further advantage of selfing). *Lymnaea stagnalis* prefers to mate with a new partner than a previously mated one, which is called the Coolidge effect (Koene and Ter Maat 2007), which would increase the genetic diversity of offspring by getting access to new partners. Moreover, other types of mate choice, depending on the relatedness or infectious status, would influence their capability of invading a new area, but it remains to be evaluated in lymnaeids.

Sex role preference is associated with unilateral mating, meaning that in a given mating event, one snail plays the male role while the other plays the female role (the most common mating pattern in Hygrophila; Jarne et al. 2010). We already mentioned that individuals without access to partners for several days are very eager to mate in outcrossing species (Coolidge effect), as shown experimentally in *L. stagnalis* (van Duivenboden 1983). Such a role preference and mating eagerness may generate a conflict in mating encounters because smaller individuals are more likely to play the male role but bigger ones are also motivated to copulate as males (Nakadera et al. 2014). After the first copulation, individuals may exchange roles where the female-acting snail often extends its body to grasp the male-acting partner to exchange role, leading to bilateral sperm exchange in a unilaterally-copulating species (Koene and Ter Maat 2005). The male mating behavior has been extensively examined in *L. stagnalis*, but recent studies on the female behavior of this species elucidated the unexplored dimension of mating interaction to examine the sex role preferences (Daupagne and Koene 2020; Moussaoui et al. 2018).

10.2.2.4 Sperm Storage

Gastropods have developed efficient sperm storage capacity (Backeljau et al. 2007), and lymnaeids are no exception, even if this capacity has been studied only in *L. stagnalis*. From a single mating event, individuals can store allosperm up to 2.5 months, a rather long period given their reproduction is initiated at 3 months after hatching in this species (Nakadera et al. 2014). This prolonged sperm storage ability would be beneficial to start a new population. We still ignore, however, where and how freshwater snails store allosperm, how sperm choice occurs, and how allosperm and autosperm are interacting (Koene et al. 2009). Sperm choice might indeed be critical in the context of range shift, provided that the partner's quality can be evaluated based on sperm characteristics (e.g., motility).

10.2.2.5 Egg Production

It is crucial to decide when to produce eggs and egg capsules, and how much to invest in them. Mature individuals in lymnaeid species continuously produce eggs, packaged in egg mass, and deposited them on various types of surfaces in the environment (e.g., branches or stones in the wild, box wall in the laboratory). The number of eggs and egg masses depend on the species. For instance, adult *L. stagnalis* lay, on average, 1–3 egg masses per week, and each egg mass can contain from 50 to 150 eggs, whereas this figure has been reported to be much lower in *Galba* species (2 eggs per egg capsule; e.g., Chapuis et al. 2007). Lymnaeid species show a stereotyped egg-laying behavior, including extensive cleaning on the surface to lay egg mass and inspection after laying (Jarne et al. 2010). Moreover, it has been reported in the same species that the exposure to clean water and surface immediately triggers egg-laying within 2 h (clean water stimulus; Ter Maat et al. 1983), indicating that environmental conditions play a role in egg-laying strategy. That is, lymnaeid species do not provide any parental care after depositing egg masses. Hence, it is a critical choice where and when eggs masses should be laid for successful juvenile development before and after hatching. Note that high fecundity also allows "not to put all eggs in the same basket," which can be interpreted as a some sort of reproductive assurance.

Mating does not result only in the transfer of spermatozoa, but also of seminal fluids, including seminal fluid protein (SFP). Such proteins have been shown to strongly affect the behavior and fecundity of receiving individuals in a wide range of animal species (e.g., Avila et al. 2011; Herndon and Wolfner 1995; Patlar et al. 2020). Their effects on egg production have been primarily studied in *L. stagnalis*. For instance, receiving Ovipostatin (an SFP also called LyAcp10) during copulation triggers a significant delay of egg laying (Koene et al. 2010) and a reduction in the number of eggs laid associated with a higher investment per egg (Swart et al. 2020). These results need to be set in a broader fitness analysis to assess the extent to which transferring ovipostatin is adaptive and how receivers have developed a counter-adaptation. Whether the production of SFP is reduced in selfing species, as expected from the theory of sex allocation, remains unknown.

In addition, the role of the gelatinous substance in egg mass has been well speculated but has yet not been systematically evaluated. Eggs indeed do not hatch synchronously with a hatching range of 34 days for an incubation time of ca. 1 week as in *Lymnaea peregra*, for instance (Jarne and Delay 1990). However, Marois and Croll (1991) reported for *L. stagnalis* that removing the gelatinous substance largely increases the synchronicity of hatching timing. It is therefore likely that the access to oxygen and other essential substances from the outer environment that are required for egg development is heterogeneous within a capsule. This gelatinous substance might also protect against bacterial or fungal infection, and slow down desiccation, a useful characteristic both in case of drought and of transportation by birds or other larger animals; this certainly opens the possibility of long-distance migration to reach new habitats (Van Leeuwen et al. 2013).

10.2.2.6 Sperm Digestion

Gastropod reproductive apparatus includes an organ called bursa copulatrix (Fig. 10.1), and it has been shown in the land snail *Cornu aspersum* that the major part of the received ejaculate ends up in this organ where it is digested (Rogers and Chase 2001). This suggests that the ejaculate could be interpreted as a "nuptial gift," since mating partners can extract energy from received ejaculates. This seems rather unlikely, at least in *L. stagnalis*, since the energy brought in by ejaculates is very small compared to the energy required for egg production (Lodi et al. 2017). However, it might play a role in limiting the effect of SFPs received during copulation, or even of sexually-transmitted diseases. These ideas are purely speculative and need to be tested based on experiments in the great pond snail.

10.3 Is Selfing an Evolutionary Dead End?

Biologists have studied transitions between predominantly outcrossing and predominantly selfing systems for a long time in plants, a literature dominated by the concept of predominant selfing as an evolutionary dead end (Burgarella and Glémin 2017; Igic and Busch 2013). As illustrated by the case of *Arabidopsis thaliana* (Durvasula et al. 2017; Tang et al. 2007), predominantly selfing species sometimes rapidly evolve from outcrossing ancestors. Indeed, as mentioned above, keeping ovules for self-fertilization has some short-term advantages, including higher fidelity of gene transmission and reproductive assurance, which may be especially relevant during the establishment of new populations or demographic bottlenecks. If not opposed by high inbreeding depression, selfing variants may therefore rapidly spread through populations (Burgarella and Glémin 2017; Charlesworth and Willis 2009). Increased selfing may also promote purging of some deleterious mutations (e.g., recessive semi-lethals), reducing inbreeding depression (Noël et al. 2019). This, in turn, would increase the advantage of selfing in a self-sustained loop, leading to very high selfing rates (Burgarella and Glémin 2017; Charlesworth and Willis 2009). This evolutionary loop between mutation load and selfing rate may reinforce the stability of extreme states—predominant outcrossers with high inbreeding depression, and predominant selfers with low inbreeding depression—and explain the rarity of transitions.

The asymmetry in transitions (more transitions from outcrosser to selfer than the reverse) may be explained by both short-term and long-term processes. On the short term, reproductive assurance may favor selfing in outcrossing populations despite high inbreeding depression under particular conditions such as lack of pollinators.

By contrast, there are no obvious conditions that would give an advantage to an outcrossing mutant in a selfing population with low inbreeding depression. On the long term, highly selfing taxa get extinct with higher probability than outcrossing ones, because high selfing rates reduce effective recombination, decreasing evolutionary potential and facilitating the fixation of deleterious mutations in genomes over time scales in the order of 1–2 Myr (Goldberg et al. 2010; Slotte et al. 2013).

Lymnaeids have been much less studied than plants, and we know relatively little about the frequency of transitions from predominant outcrosser to predominant selfer in their evolutionary history. They offer though an interesting situation that is not found in plants: the *Galba* genus, which age is estimated at 20 Mya (Burgarella et al. 2015), is essentially made of highly selfing species (see Fig. 10.2), suggesting that selfing has been the main mating system for a long time (Alda et al. 2021). Although the genomes of selfing *Galba* tend to accumulate relatively high numbers of mutations, as predicted from the long-term effects of selfing (Burgarella et al. 2015), the long-term persistence of these species contradicts the "dead-end claim" for selfing, but the circumstances under which this is possible remain unclear. Burgarella et al. (2015) suggested that this might be due to the fact that *Galba* species have colonized a new adaptive zone. They are indeed more amphibious than aquatic, being able to survive on mud banks or swamped meadows, when competition with other Hygrophila is limited. This idea, however, remains to be evaluated adequately.

Although we have currently no example of recent or ongoing transitions between mating systems in Lymnaeids (and Hygrophila in genera), the forces involved in such transitions could be experimentally approached, as has been done in *Physa acuta* (belonging to the Hygrophila) with continuous or recurrent selfing (Noël et al. 2019, 2017). This study has confirmed a loss of evolutionary potential after increasing selfing rates (Noël et al. 2017), with potential consequences on the capacity to adapt to changing environmental conditions.

Lymnaea stagnalis is a good candidate for similar investigations because this outcrossing species has a low inbreeding depression (a characteristic that may favor selfing variants) and can be inbred for tens of generations without lethality (Colton and Pennypacker 1934; J. Koene, unpublished data). In New Zealand, invasive populations of *L. stagnalis* have practically no polymorphism (Kopp et al. 2012), as would be the case after a few recurrent generations of selfing. The lack of variation precludes the estimation of spontaneous selfing rates in extant New Zealand populations. In this case, however, self-fertilization (compared to outcrossing) is expected to have no particular effect on heterozygosity and fitness. This creates a favorable context to invasion by predominantly selfing strategies and should be studied in more detail.

10.4 Mating Systems in a Changing World

Human-induced environmental changes are altering ecosystems by triggering a wide array of irreversible consequences. These environmental changes are predicted to affect ecological processes across several levels of organization, likely over the coming decades, by altering the performance of individual organisms, the dynamics of populations, and the distribution of species, ultimately modifying ecosystem properties and functioning (Lonhart et al. 2019). If organisms have difficulties in adapting and fail to track projected environmental changes, populations become vulnerable to decline and extinction (Hill et al. 2011). Research has documented several mechanisms allowing rapid adaptation, which is crucial for the long-term persistence of populations facing new and stressful environment conditions brought about by global change and range shifting (Hill et al. 2011).

As mentioned above, mating systems have a deep impact on individual fitness and may therefore contribute to understanding how species respond to stressful environmental conditions (Aanen et al. 2016; Bijlsma and Loeschcke 2005). Some lymnaeid species, for instance, can survive drought conditions for several weeks and even months by adhering their aperture to the soil surface or burying themselves into deeper ground where humidity is conserved (Chapuis et al. 2007). When moist conditions return, one or a few individuals can self-fertilize and recolonize the site. Individuals can also colonize a new habitat if they have been transported by flooding, birds, mammals, or humans (Kappes and Haase 2012; Van Leeuwen et al. 2013, 2012). In this sense, self-fertilization facilitates and ensures colonization success, which may explain how two lymnaeid selfing species such as *P. columella* and *G. schirazensis* have successfully established across different ecosystems worldwide (Lounnas et al. 2017a, 2018). Indeed, human activities drastically accelerated lymnaeid invasion worldwide. The aquarium trade has a long history of transporting and introducing snails, among plants and fish, into regions where they are not native (Duggan 2010). This is particularly interesting in long-range invasions because it may intensify selection for selfing species. In other words, aquarium populations are typically recurrently bottlenecked and re-seeded with a few eggs or individuals on a plant. Such populations may selectively propagate selfing species such as *P. columella* or may also force preferential outcrossers through several cycles of selfing, which would be, for instance, the case of *L. stagnalis* introduced in New Zealand (Kopp et al. 2012).

A drastic loss of genetic diversity was observed in the invaded regions in the selfing species *P. columella* (Lounnas et al. 2017a), *G. schirazensis* (Lounnas et al. 2018), and *G. truncatula* (Meunier et al. 2001). Populations in invaded areas analyzed in these studies are characterized by an extremely low genetic diversity, with sometimes a single (or a few) genotype. Selfing explains much of this depletion because species are indeed made of little-recombining genotypes and have lower variation. This could result from bottlenecks (associated with a low propagule pressure), which efficiently sieve out a few genotypes and then proliferate through selfing. All of these studies suggest that multiple ecological and evolutionary factors shape the distribution of genetic variation throughout species ranges. Yet, the only aspect that has been studied is neutral genetic variation, and we lack data on many aspects related to the invasion. The question that inevitably arises is how lymnaeids succeed in their invasion process with genetically depleted populations that originated by selfing. This dilemma is known as the genetic paradox of invasion (Allendorf and Lundquist 2003).

A growing number of studies propose different mechanisms that allow invasive populations to increase population fitness and compensate for the loss in genetic diversity during introduction events and at the invasion front (see Estoup et al. 2016 for further details). Purging of the genetic load is one possibility (Crnokrak and Barrett 2002). Recurrent introductions of different genotypes can also increase the genetic variance and adaptability (Facon et al. 2008). However, the same single genotype can repeatedly invade new regions, resulting in low genetic variability in those regions. This would be the case of *P. columella* because an extensive sampling coupled with population genetic analysis showed that the depletion of genetic variation is extremely strong and that this species has a unique "worldwide invader" genotype (Lounnas et al. 2017a). Adaptive phenotypic plasticity is another possible mechanism that could compensate for genetic diversity loss and explain the high invasiveness of populations (Estoup et al. 2016; Ghalambor et al. 2007). Future work should focus on investigating which possible mechanisms, processes, or strategies that facilitate genetically depleted populations of lymnaeids to rapidly respond to novel stressful environmental conditions when shifting their distribution and colonizing new areas.

10.5 Can Parasites Shift Host Mating Systems?

The Red Queen Hypothesis (RQH) suggests that hosts continuously evolve to evade strategies developed by their parasites and, in turn, parasites must counter-adapt to host changes. Parasites therefore have a profound impact on their hosts as they can impose selection against common host genotypes, resulting in an advantage to producing outcrossed, genetically variable progeny (Hamilton 1980). The original version of the RQH was developed for the evolution of sex in plants and animals (Lively 1987; Verhoeven and Biere 2013) but has been extended for the evolution of selfing. Coevolution, in this case, leads to selection for outcrossing of both parasites and hosts (Agrawal and Lively 2001; Hurtrez-Boussès et al. 2001). Recent laboratory experiments have shown that coevolving parasites can favor increased rates of outcrossing in mixed-mating (outcrossing and selfing) host populations of the nematode *Caenorhabditis elegans* (Slowinski et al. 2016). Computer simulations have also predicted that if selective pressure imposed by parasites is high enough to favor outcrossing in hosts, it could promote the occurrence of resistant mutants among the host population (e.g., Agrawal and Lively 2001).

Lymnaeids and the parasites that infect them, such as trematodes like *Fasciola hepatica*, are excellent models to study the evolution of mating systems because both are usually simultaneous hermaphrodites that reproduce by selfing or outcrossing (Chapuis et al. 2007; Hurtrez-Boussès et al. 2004; Trouvé et al. 1996). Yet, there is limited or no evidence that the mating system evolved in hosts and parasites as a response to their interaction. Further research in simultaneously hermaphroditic animals is necessary to evaluate which and how different stressors affect selfing and outcrossing rates, considering the underlying genetic and phenotypic basis of both mating systems and the basis for infection and virulence in parasites.

The mating strategy resulting from environment or parasite pressure is likely to have an impact on the transmission of infectious diseases. Theoretical models indeed show that low host genetic diversity—as a result of the type of mating system and of the rate of migration—can promote disease spread (e.g., Lively 2010). It is widely accepted that genetically homogenous host populations, resulting from selfing or low propagule pressure, are more vulnerable to infection than genetically diverse ones (King and Lively 2012). This has been observed in plants (Mundt, 2002; Zhu et al. 2000) and in animals (Altermatt and Ebert 2008; Ellison et al. 2011). For instance, the rapid expansion of the selfing freshwater snail *Biomphalaria pfeifferi*, an intermediate host for human schistosomiasis, might result in the establishment of genetically homogenous populations in pristine water bodies in Zimbabwe that were susceptible to the parasite thus promoting the resurgence of the disease in this country (Campbell et al. 2010). As most invasive host populations have no genetic variability to evolve resistance mechanisms, the parasite could therefore freely expand without opposition, representing a health threat (Lounnas et al. 2017a).

10.6 What's Next?

Lymnaeids are suitable model organisms for tackling issues that are of general interest in ecology and evolution of mating systems and invasion. Up to now, only three lymnaeid species groups have monopolized studies regarding these issues: *L. stagnalis*, *Galba* species and the sister species *P. columella*, and to a lesser extent some *Radix* species. There are tens of other lymnaeid species that have never or very little been studied that could bring light in our current understanding about the ecology and evolution of mating systems and invasion in lymnaeids.

Experimental evolution approaches and genomic and phylogenetic analyses that explore the genetic architecture of lymnaeid reproductive traits should be performed to better understand the evolution of mating systems. A good starting point for digging into lymnaeid mating systems would be a thorough phylogenetic comparison of selfing rates in lymnaeid species. Our current knowledge about the evolution of mating systems in lymnaeids (see Fig. 10.2) is far from being sufficient because, in some cases, genetic evidence for self-fertilization in natural population is hard to be obtained. This is because it is difficult to estimate selfing rates in highly selfing populations showing very low variation (Lounnas et al. 2018). In these cases, genetic approaches such as RAD seq that use hundreds of genetic markers must be applied (Miller et al. 2007). This phylogenetic approach could also include a comparison of morphological reproductive traits at a family level. The hypothesis that outcrossing species show a complex reproductive anatomy and that selfing species a simplified one needs to be tested in lymnaeids. In addition, it would be interesting to investigate whether there are genetic differences in SFPs between selfing and outcrossing species.

Lymnaea stagnalis and *A. balthica* are good lymnaeid species for conducting experimental approaches since they are easy to breed under experimental conditions, unlike other lymnaeid species. Yet, the distribution of selfing rates among populations of both species is still unclear, and it should be studied at the species scale. *Galba* species are difficult to breed under experimental conditions, but they are a very promising species group for studying the evolution of selfing. Recently, Alda et al. (2021) suggested that a group of *Galba* snails would have reverted its mating system from selfing to outcrossing—something that would contradict Stebbins' law that claims: "the transition from outcrossing to selfing is unidirectional." This seemingly unexpected reversion in *Galba cousini*, a species that inhabits high altitudes in the north-western Andes, needs to be properly tested. If true, it would be the first evidence in the history of the study of mating systems of hermaphrodites that would have evolved from selfing to outcrossing. Further studies should explore which are the genetic architecture and morphological traits that accompanied this reversion, and which are the ecological factors that could prompt it.

Lymnaeid selfing species could be used as model organisms in systematics to test species concepts in selfing species. Most species-delimitation models currently used in systematics are based on Multispecies Coalescence that assumes that gene flow occurs within species though not between species (Sukumaran and Knowles 2017). Thus, the use of these models would be inappropriate in a fully selfing species groups. The selfing rate in lymnaeid species and populations is high (often ~0.9), but outcrossing does occur (Chapuis et al. 2007; Lounnas et al. 2017a, 2017b). Further research could explore how selfing may affect speciation (Cutter 2019). Lymnaeid species would be excellent models to investigate such questions, since selfing rates vary among species and populations.

Studies of population genetics, demography, and phylogeography of both source and invading populations will be useful for understanding the underlying mechanisms driving the success of lymnaeid invasive species, their geographic patterns of invasion and range expansion and potential for evolutionary responses to new environments and their parasites and other ecological challenges. Compiling the existing data about the effect of pesticides, herbicides, fertilizers, pharmaceuticals, and other pollutants in many lymnaeids species could help us to predict which species could potentially become urban invaders.

Finally, it is imperative to conduct studies investigating the consequences of lymnaeid invasions in the spread of infectious diseases. For that purpose, an investigation of the interaction between invader genotypes and local populations of parasites is needed. Such population genetic studies in snail hosts and parasites would bridge the gap between biological invasions and epidemiology; research areas that are interconnected but rarely studied together. The snail-parasite system constituted by *P. columella*, the species from the genus *Galba*, and the trematode *F. hepatica* would be an appropriate model to investigate what are the mechanisms and processes behind the interactions between hosts, parasites, and environment that enhance the spread of infection diseases in wildlife, domestic animals, and humans. This goal would completely match the scope of the One Health Initiative—an approach that seeks to understand infectious diseases to then apply programs, policies, and legislation in which multiple sectors work together to achieve better public health outcomes (e.g., Destoumieux-Garzón et al. 2018).

Acknowledgments We thank Jean-Pierre Pointier for providing drawings, references, and insights to this chapter.

References

Aanen D, Beekman M, Kokko H (2016) Weird sex: the underappreciated diversity of sexual reproduction. Phil Trans R Soc B 371:20160262. https://doi.org/10.1098/rstb.2016.0262
Agrawal AF, Lively CM (2001) Parasites and the evolution of self-fertilization. Evolution 55:869–879. https://doi.org/10.1554/0014-3820(2001)055[0869:PATEOS]2.0.CO;2
Alba A, Vázquez AA, Sánchez J et al (2019) Patterns of distribution, population genetics and ecological requirements of field-occurring resistant and susceptible *Pseudosuccinea columella* snails to *Fasciola hepatica* in Cuba. Sci Rep-UK 9:14359. https://doi.org/10.1038/s41598-019-50894-7
Alda P, Lounnas M, Vázquez AA et al (2021) Systematics and geographical distribution of *Galba* species, a group of cryptic and worldwide freshwater snails. Molec Phyl Evol 157:107035. https://doi.org/10.1016/j.ympev.2020.107035
Allendorf FW, Lundquist LL (2003) Introduction: population biology, evolution, and control of invasive species. Conserv Biol 17:24–30. https://doi.org/10.1046/j.1523-1739.2003.02365.x
Altermatt F, Ebert D (2008) Genetic diversity of *Daphnia magna* populations enhances resistance to parasites. Ecol Lett 11:918–928. https://doi.org/10.1111/j.1461-0248.2008.01203.x
Avila FW, Sirot LK, LaFlamme BA et al (2011) Insect seminal fluid proteins: identification and function. Annu Rev Entomol 56:21–40. https://doi.org/10.1146/annurev-ento-120709-144823
Backeljau T, Jordaens K, Dillen L (2007) Effects of mating, breeding system and parasites on reproduction in hermaphrodites: pulmonate gastropods (Mollusca). Anim Biol 57:137–195. https://doi.org/10.1163/157075607780377965
Baker HG (1955) Self-compatibility and establishment after "long-distance" dispersal. Evolution 9: 347–349. https://doi.org/10.2307/2405656
Bijlsma R, Loeschcke V (2005) Environmental stress, adaptation and evolution: an overview. J Evol Biol 18:744–749. https://doi.org/10.1111/j.1420-9101.2005.00962.x
Boer P, Jansen R, Koene J, Maat M (1997) Nervous control of male sexual drive in the hermaphroditic snail. J Exp Biol 200:941–951
Bonel N, Noël E, Janicke T et al (2018) Asymmetric evolutionary responses to sex-specific selection in a hermaphrodite. Evolution 72:2181–2201. https://doi.org/10.1111/evo.13565
Burgarella C, Glémin S (2017) Population genetics and genome evolution of selfing species. In: John Wiley & Sons Ltd (ed) eLS. John Wiley & Sons Ltd, Chichester, pp 1–8. https://doi.org/10.1002/9780470015902.a0026804
Burgarella C, Gayral P, Ballenghien M et al (2015) Molecular evolution of freshwater snails with contrasting mating systems. Mol Biol Evol 32:2403–2416. https://doi.org/10.1093/molbev/msv121

Bürkli A, Jokela J (2017) Increase in multiple paternity across the reproductive lifespan in a sperm-storing, hermaphroditic freshwater snail. Mol Ecol 26:5264–5278. https://doi.org/10.1111/mec.14200

Campbell G, Noble LR, Rollinson D et al (2010) Low genetic diversity in a snail intermediate host (*Biomphalaria pfeifferi* Krauss, 1848) and schistosomiasis transmission in the Senegal River basin. Mol Ecol 19:241–256. https://doi.org/10.1111/j.1365-294X.2009.04463.x

Chapuis E, Trouve S, Facon B et al (2007) High quantitative and no molecular differentiation of a freshwater snail (*Galba truncatula*) between temporary and permanent water habitats. Mol Ecol 16:3484–3496. https://doi.org/10.1111/j.1365-294X.2007.03386.x

Charlesworth D, Willis JH (2009) The genetics of inbreeding depression. Nat Rev Genet 10:783–796. https://doi.org/10.1038/nrg2664

Charnov EL (1982) The theory of sex allocation. Monogr Popul Biol 18:1–355

Cheptou PO (2012) Clarifying Baker's Law. Ann Bot-London 109:633–641. https://doi.org/10.1093/aob/mcr127

Colton HS, Pennypacker M (1934) The results of twenty years of self-fertilization in the pond snail *Lymnaea columella* say. Am Nat 68:129–136. https://doi.org/10.1086/280532

Correa AC, Escobar JC, Durand P et al (2010) Bridging gaps in the molecular phylogeny of the Lymnaeidae (Gastropoda: Pulmonata), vectors of fascioliasis. BMC Evol Biol 10:381. https://doi.org/10.1186/1471-2148-10-381

Coutellec MA, Caquet T (2011) Heterosis and inbreeding depression in bottlenecked populations: a test in the hermaphroditic freshwater snail *Lymnaea stagnalis*: genetic load in a freshwater snail. J Evol Biol 24:2248–2257. https://doi.org/10.1111/j.1420-9101.2011.02355.x

Coutellec-Vreto MA, Madec L, Guiller A (1997) Selfing and biparental inbreeding: a mating system analysis in *Lymnaea peregra* (Gastropoda: Lymnaeidae). Heredity 79:277–285. https://doi.org/10.1038/hdy.1997.155

Crnokrak P, Barrett SCH (2002) Perspective: purging the genetic load: a review of the experimental evidence. Evolution 56:2347–2358. https://doi.org/10.1111/j.0014-3820.2002.tb00160.x

Cutter AD (2019) Reproductive transitions in plants and animals: selfing syndrome, sexual selection and speciation. New Phytol 224(3):1080–1094

Daupagne L, Koene JM (2020) Disentangling female postmating responses induced by semen transfer components in a simultaneous hermaphrodite. Anim Behav 166:147–152. https://doi.org/10.1016/j.anbehav.2020.06.009

Destoumieux-Garzón D, Mavingui P, Boetsch G et al (2018) The one health concept: 10 years old and a long road ahead. Front Vet Sci 14

Doums C, Viard F, Pernot AF et al (1996) Inbreeding depression, neutral polymorphism, and copulatory behavior in freshwater snails: a self-fertilization syndrome. Evolution 50:1908–1918. https://doi.org/10.1111/j.1558-5646.1996.tb03578.x

Duggan IC (2010) The freshwater aquarium trade as a vector for incidental invertebrate fauna. Biol Invasions 12:3757–3770. https://doi.org/10.1007/s10530-010-9768-x

Durvasula A, Fulgione A, Gutaker RM et al (2017) African genomes illuminate the early history and transition to selfing in *Arabidopsis thaliana*. Proc Natl Acad Sci U S A 114:5213–5218. https://doi.org/10.1073/pnas.1616736114

Ellison A, Cable J, Consuegra S (2011) Best of both worlds? Association between outcrossing and parasite loads in a selfing fish. Evolution 65:3021–3026. https://doi.org/10.1111/j.1558-5646.2011.01354.x

Escobar JS, Auld JR, Correa AC et al (2011) Patterns of mating-system evolution in hermaphroditic animals: correlations among selfing rate, inbreeding depression, and the timing of reproduction: mating-system evolution in hermaphroditic animals. Evolution 65:1233–1253. https://doi.org/10.1111/j.1558-5646.2011.01218.x

Estoup A, Ravigné V, Hufbauer R et al (2016) Is there a genetic paradox of biological invasion? Annu Rev Ecol Evol Syst 47:51–72. https://doi.org/10.1146/annurev-ecolsys-121415-032116

Facon B, Pointier JP, Jarne P (2008) High genetic variance in life-history strategies within invasive populations by way of multiple introductions. Curr Biol 18:363–367. https://doi.org/10.1016/j.cub.2008.01.063

Geraerts WPM, Joosse J (1984) Freshwater snails (Basommatophora). In: Wilbur KM (ed) The Mollusca, Reproduction, vol Vol. 7. Academic Press, New York, pp 141–207

Ghalambor CK, McKay JK, Carroll SP et al (2007) Adaptive versus non-adaptive phenotypic plasticity and the potential for contemporary adaptation in new environments. Funct Ecol 21: 394–407. https://doi.org/10.1111/j.1365-2435.2007.01283.x

Goldberg EE, Kohn JR, Lande R et al (2010) Species selection maintains self-incompatibility. Science 330:493–495. https://doi.org/10.1126/science.1194513

Hamilton WD (1980) Sex versus non-sex versus parasite. Oikos 35:282–290. https://doi.org/10.2307/3544435

Herndon LA, Wolfner MF (1995) A *drosophila* seminal fluid protein, Acp26Aa, stimulates egg laying in females for 1 day after mating. Proc Natl Acad Sci U S A 92:10114–10118. https://doi.org/10.1073/pnas.92.22.10114

Hill JK, Griffiths HM, Thomas CD (2011) Climate change and evolutionary adaptations at species' range margins. Annu Rev Entomol 56:143–159. https://doi.org/10.1146/annurev-ento-120709-144746

Hoffer JN, Mariën J, Ellers J et al (2017) Sexual selection gradients change over time in a simultaneous hermaphrodite. eLife 6:e25139. https://doi.org/10.7554/eLife.25139

Hurtrez-Boussès S, Meunier C, Durand P et al (2001) Dynamics of host–parasite interactions: the example of population biology of the liver fluke (*Fasciola hepatica*). Microbes Infect 3:841–849. https://doi.org/10.1016/S1286-4579(01)01442-3

Hurtrez-Bousses S, Durand P, Jabbour-Zahab R (2004) Isolation and characterization of microsatellite markers in the liver fluke (*Fasciola hepatica*). Molec Ecol Notes 4:689–690. https://doi.org/10.1111/j.1471-8286.2004.00786.x

Hurtrez-Boussès S, Hurtrez JE, Turpin H et al (2010) Hydrographic network structure and population genetic differentiation in a vector of fasciolosis, *Galba truncatula*. Infect Genet Evol 10: 178–183. https://doi.org/10.1016/j.meegid.2010.01.005

Igic B, Busch JW (2013) Is self-fertilization an evolutionary dead end? New Phytol 198:386–397. https://doi.org/10.1111/nph.12182

Jarne P, Auld JR (2006) Animals mix it up too: the distribution of self-fertilization among hermaphroditic animals. Evolution 60:1816–1824. https://doi.org/10.1111/j.0014-3820.2006.tb00525.x

Jarne P, Delay B (1990) Inbreeding depression and self-fertilization in *Lymnaea peregra* (Gastropoda: Pulmonata). Heredity 64:169–175. https://doi.org/10.1038/hdy.1990.21

Jarne P, Vianey-Liaud M, Delay B (1993) Selfing and outcrossing in hermaphrodite freshwater gastropods (Basommatophora): where, when and why. Biol J Linn Soc 49:99–125. https://doi.org/10.1111/j.1095-8312.1993.tb00893.x

Jarne P, Pointier JP, David P et al (2010) Basommatophoran gastropods. In: Leonard J, Cordoba-Aguilar A (eds) The evolution of primary sexual characters in animals. Oxford University Press, Oxford and New York, pp 173–196

Kappes H, Haase P (2012) Slow, but steady: dispersal of freshwater molluscs. Aquat Sci 74:1–14. https://doi.org/10.1007/s00027-011-0187-6

King KC, Lively CM (2012) Does genetic diversity limit disease spread in natural host populations? Heredity 109:199–203. https://doi.org/10.1038/hdy.2012.33

Koene JM, Ter Maat A (2005) Sex role alternation in the simultaneously hermaphroditic pond snail *Lymnaea stagnalis* is determined by the availability of seminal fluid. Anim Behav 69:845–850. https://doi.org/10.1016/j.anbehav.2004.07.012

Koene JM, Ter Maat A (2007) Coolidge effect in pond snails: male motivation in a simultaneous hermaphrodite. BMC Evol Biol 7:212. https://doi.org/10.1186/1471-2148-7-212

Koene JM, Loose MJ, Wolters L (2008) Mate choice is not affected by mating history in the simultaneously hermaphroditic snail *Lymnaea stagnalis*. J Mollus Stud 74:331–335. https://doi.org/10.1093/mollus/eyn020

Koene JM, Montagne-Wajer K, Roelofs D et al (2009) The fate of received sperm in the reproductive tract of a hermaphroditic snail and its implications for fertilisation. Evol Ecol 23:533–543. https://doi.org/10.1007/s10682-008-9253-5

Koene JM, Sloot W, Montagne-Wajer K et al (2010) Male accessory gland protein reduces egg laying in a simultaneous hermaphrodite. PLoS One 5:e10117. https://doi.org/10.1371/journal.pone.0010117

Kopp KC, Wolff K, Jokela J (2012) Natural range expansion and human-assisted introduction leave different genetic signatures in a hermaphroditic freshwater snail. Evol Ecol 26:483–498. https://doi.org/10.1007/s10682-011-9504-8

Lively CM (1987) Evidence from a New Zealand snail for the maintenance of sex by parasitism. Nature 328:519–521. https://doi.org/10.1038/328519a0

Lively CM (2010) The effect of host genetic diversity on disease spread. Am Nat 175:E149–E152. https://doi.org/10.1086/652430

Lodi M, Meijer FW, Koene JM (2017) Ejaculates are not used as nuptial gifts in simultaneously hermaphroditic snails. Zoology 123:30–36. https://doi.org/10.1016/j.zool.2017.05.007

Lonhart SI, Jeppesen R, Beas-Luna R (2019) Shifts in the distribution and abundance of coastal marine species along the eastern Pacific Ocean during marine heatwaves from 2013 to 2018. Mar Biodivers Rec 12:13. https://doi.org/10.1186/s41200-019-0171-8

Lounnas M, Correa AC, Vázquez AA et al (2017a) Self-fertilization, long-distance flash invasion and biogeography shape the population structure of *Pseudosuccinea columella* at the worldwide scale. Mol Ecol 26:887–903. https://doi.org/10.1111/mec.13984

Lounnas M, Vázquez AA, Alda P (2017b) Isolation, characterization and population-genetic analysis of microsatellite loci in the freshwater snail *Galba cubensis* (Lymnaeidae). J Mollus Stud 83:63–68. https://doi.org/10.1093/mollus/eyw041

Lounnas M, Correa AC, Alda P et al (2018) Population structure and genetic diversity in the invasive freshwater snail *Galba schirazensis*. Can J Zool 96(5):425–435. https://doi.org/10.1139/cjz-2016-0319

Marois R, Croll RP (1991) Hatching asynchrony within the egg mass of the pond snail, *Lymnaea stagnalis*. Invertebr Reprod Dev 19:139–146. https://doi.org/10.1080/07924259.1991.9672167

Meunier C, Tirard C, Hurtrez-Boussès S et al (2001) Lack of molluscan host diversity and the transmission of an emerging parasitic disease in Bolivia. Mol Ecol 10:1333–1340. https://doi.org/10.1046/j.1365-294x.2001.01284.x

Meunier C, Hurtrez-Boussès S, Durand P et al (2004) Small effective population sizes in a widespread selfing species, *Lymnaea truncatula* (Gastropoda: Pulmonata). Mol Ecol 13:2535–2543. https://doi.org/10.1111/j.1365-294X.2004.02242.x

Miller MR, Dunham JP, Amores A, Cresko WA, Johnson EA (2007) Rapid and cost-effective polymorphism identification and genotyping using restriction site associated DNA (RAD) markers. Genome Res 17(2):240–248. https://doi.org/10.1101/gr.5681207. Epub 2006 Dec 22. PMID: 17189378; PMCID: PMC1781356

Moussaoui R, Verdel K, Benbellil-Tafoughalt S et al (2018) Female behaviour prior to additional sperm receipt in the hermaphroditic pond snail *Lymnaea stagnalis*. Invertebr Reprod Dev 62:82–91. https://doi.org/10.1080/07924259.2017.1415988

Mundt CC (2002) Use of multiline cultivars and cultivar mixtures for disease management. Annu Rev Phytopathol 40:381–410. https://doi.org/10.1146/annurev.phyto.40.011402.113723

Nakadera Y, Blom C, Koene JM (2014) Duration of sperm storage in the simultaneous hermaphrodite *Lymnaea stagnalis*. J Mollus Stud 80:1–7. https://doi.org/10.1093/mollus/eyt049

Nakadera Y, Mariën J, Van Straalen NM et al (2017) Multiple mating in natural populations of a simultaneous hermaphrodite, *Lymnaea stagnalis*. J Mollus Stud 83:56–62. https://doi.org/10.1093/mollus/eyw043

Nakadera Y, Giannakara A, Ramm SA (2019) Plastic expression of seminal fluid protein genes in a simultaneously hermaphroditic snail. Behav Ecol 30(4):904–913. https://doi.org/10.1093/beheco/arz027

Nicot A, Dubois MP, Debain C (2008) Characterization of 15 microsatellite loci in the pulmonate snail *Pseudosuccinea columella* (Mollusca, Gastropoda). Mol Ecol Resour 8:1281–1284. https://doi.org/10.1111/j.1755-0998.2007.02065.x

Noël E, Jarne P, Glémin S et al (2017) Experimental evidence for the negative effects of self-fertilization on the adaptive potential of populations. Curr Biol 27:237–242. https://doi.org/10.1016/j.cub.2016.11.015

Noël E, Fruitet E, Lelaurin D et al (2019) Sexual selection and inbreeding: two efficient ways to limit the accumulation of deleterious mutations. Evolution Letters 3:80–92. https://doi.org/10.1002/evl3.93

Patlar B, Weber M, Temizyürek T et al (2020) Seminal fluid-mediated manipulation of post-mating behavior in a simultaneous hermaphrodite. Curr Biol 30:143–149.e4. https://doi.org/10.1016/j.cub.2019.11.018

Puurtinen M, Emily Knott K, Suonpää S et al (2007) Predominance of outcrossing in *Lymnaea stagnalis* despite low apparent fitness costs of self-fertilization. J Evol Biol 20:901–912. https://doi.org/10.1111/j.1420-9101.2007.01312.x

Rogers D, Chase R (2001) Dart receipt promotes sperm storage in the garden snail *Helix aspersa*. Behav Ecol Sociobiol 50:122–127. https://doi.org/10.1007/s002650100345

Schärer L (2009) Tests of sex allocation theory in simultaneously hermaphroditic animals. Evolution 63:1377–1405. https://doi.org/10.1111/j.1558-5646.2009.00669.x

Slotte T, Hazzouri KM, Ågren JA et al (2013) The *Capsella rubella* genome and the genomic consequences of rapid mating system evolution. Nat Genet 45:831–835. https://doi.org/10.1038/ng.2669

Slowinski SP, Morran LT, Parrish RC (2016) Coevolutionary interactions with parasites constrain the spread of self-fertilization into outcrossing host populations. Evolution 70:2632–2639. https://doi.org/10.1111/evo.13048

Sukumaran J, Knowles LL (2017) Multispecies coalescent delimits structure, not species. Proc Natl Acad Sci U S A 114:1607–1612. https://doi.org/10.1073/pnas.1607921114

Swart EM, Starkloff NC, Ypenburg S et al (2020) The effect of mating on female reproduction across hermaphroditic freshwater snails. Invertebr Biol 139(1):e12275. https://doi.org/10.1111/ivb.12275

Tang C, Toomajian C, Sherman-Broyles et al (2007) The evolution of Selfing in *Arabidopsis thaliana*. Science 317:1070–1072. https://doi.org/10.1126/science.1143153

Ter Maat A, Lodder JC, Wilbrink M (1983) Induction of egg-laying in the pond snail *Lymnaea stagnalis* by environmental stimulation of the release of ovulation hormone from the caudo-dorsal cells. Int J Invert Repr 6:239–247. https://doi.org/10.1080/01651269.1983.10510048

Trouvé S, Renaud F, Durand P (1996) Selfing and outcrossing in a parasitic hermaphrodite helminth (Trematoda, Echinostomatidae). Heredity 77:1–8. https://doi.org/10.1038/hdy.1996.101

Trouvé S, Degen L, Renaud F et al (2003) Evolutionary implications of a high selfing rate in the freshwater snail *Lymnaea truncatula*. Evolution 57:2303–2314. https://doi.org/10.1554/02-452

van Duivenboden YA (1983) Transfer of semen accelerates the onset of egg-laying in female copulants of the hermaphrodite freshwater snail, *Lymnaea stagnalis*. Int J Invert Repr 6:249–257. https://doi.org/10.1080/01651269.1983.10510049

Van Duivenboden YA, Ter Maat A (1985) Masculinity and receptivity in the hermaphrodite pond snail, *Lymnaea stagnalis*. Anim Behav 33:885–891. https://doi.org/10.1016/S0003-3472(85)80022-1

Van Leeuwen CHA, Van der Velde G, Van Groenendael JM et al (2012) Gut travellers: internal dispersal of aquatic organisms by waterfowl. J Biogeogr 39:2031–2040. https://doi.org/10.1111/jbi.12004

Van Leeuwen CHA, Huig N, Van Der Velde G et al (2013) How did this snail get here? Several dispersal vectors inferred for an aquatic invasive species. Freshw Biol 58:88–99. https://doi.org/10.1111/fwb.12041

Verhoeven KJ, Biere A (2013) Geographic parthenogenesis and plant-enemy interactions in the common dandelion. BMC Evol Biol 13:23. https://doi.org/10.1186/1471-2148-13-23

Zhu Y, Chen H, Fan J (2000) Genetic diversity and disease control in rice. Nature 406:718–722. https://doi.org/10.1038/35021046

Chapter 11
Biology of the Lymnaeidae–Parasite Interaction

Annia Alba and Benjamin Gourbal

Abstract Herein, general aspects of the relationship between lymnaeid snails and their parasites are reviewed from a mechanistic point of view. It includes an outlook to encounter filters whilst emphasizing on compatibility filters driven by the different components of the internal defense system within Lymnaeidae, and to the immunobiological interactions that determined the recognition and elimination of the pathogen upon contact, or the establishment of the dynamic equilibrium characteristic of a successful infection. The negative effects caused by parasitism on the physiology and life history traits of the lymnaeid hosts are briefly reviewed, as well as the role of ecological factors on modifying encounter and compatibility filters and the ultimate infection outcome. Finally, insights into the significance of building up knowledge on the biology of host–parasite interaction will be discussed to introduce the opportunities and challenges that lies ahead concerning the family Lymnaeidae and their most significant pathogens. Most of the published works on the subject deals with Lymnaeidae–trematode interaction and consequently, this represents the main topic reviewed in this chapter.

11.1 Introduction

Lymnaeid snails are known hosts of a plethora of pathogens, mainly within the class Trematoda, among which there are some veterinary and/or medically-relevant parasites, e.g. *Fasciola* spp., *Diplostomum* spp., *Trichobilharzia* spp., *Echinostoma* spp.. For instance, in a family of around 100 species described, the species *Radix natalensis* is involved in the life cycles of at least 71 trematodes belonging to 13 different families that affect birds and both domestic and wild mammals as

A. Alba (✉)
Laboratory of Malacology, Institute of Tropical Medicine "Pedro Kourí", Havana, Cuba

B. Gourbal
Interactions Hosts Pathogens Environments, University of Montpellier, CNRS, Ifremer, University of Perpignan Via Domitia, Perpignan, France
e-mail: benjamin.gourbal@univ-perp.fr

M. V. Vinarski, A. A. Vázquez (eds.), *The Lymnaeidae*, Zoological Monographs 7,
https://doi.org/10.1007/978-3-031-30292-3_11

definitive hosts (Brown 1978), and this without counting digeneans of lower vertebrates such as amphibians. Therefore, most of the works published deals with trematode–Lymnaeidae interactions and consequently, this represents the main topic reviewed within the present chapter.

In the next sections, we will review some general aspects of the relationship between lymnaeid snails and their parasites from a mechanistic point of view. In this sense, several features, mechanisms, and conditions which involve not only the parasite and the lymnaeid snail but also the environment relevantly contribute in shaping the landscape of snail host specificity and infection success; from the pre-penetration step to the final outcome of each interaction. Two "filters" have been proposed by Combes (2001) aiming to distinguish the main components encompassing parasite success: the encounter filter and the compatibility filter. The first one comprises the elements affecting the probability of contact between the parasite and the host and includes appropriate spatial and temporal overlap as well as the ability to enter the potential host. The compatibility filter determines whether the parasite succeeds in infecting and stablishing itself in the host. Among the determining elements of the compatibility filter, the morphological, biochemical, immunological, and physiological interplays between both protagonists can be referred as they permit or impede the development and transmission of the parasite (Combes 2001). Selection will act on these filters to increase or to decrease their permeability and specificity, changing the diameter of the encounter filter or the angle of the compatibility filter (Poulin 2007). Furthermore, the development of human civilization imposes new challenges to our understanding of the host–parasite interplay as factors such as the global effects of urbanization, climate change, or increased pollution heavily influenced environmental dynamics which are known to affect both encounter and compatibility filters, and to ultimately influence infection outcome (Fig. 11.1).

Herein, emphasis will be given to compatibility and immunobiological interactions, and to ecological features affecting infection outcome and transmission. Insights into the significance of building up knowledge on the biology of host–parasite–environment interactions will be also discussed to introduce the opportunities and challenges that lies ahead concerning the family Lymnaeidae and their most significant pathogens.

11.2 How Do Parasites Meet Their Lymnaeid Hosts?

Whilst the nature of the compatibility filter is often amenable to experimentation as infective stages of parasites are provided access to potential hosts, direct assessment of the encounter filter is difficult to quantify, particularly in the field where the use of sentinel snails may be required (see Allan et al. 2013, for example). However, no successful parasite infection can occur without a successful contact of the parasite with the host. In this sense, the encounter filter includes the large- and small-scale spatial distribution of the host and parasite (i.e., geography and habitat use), and their

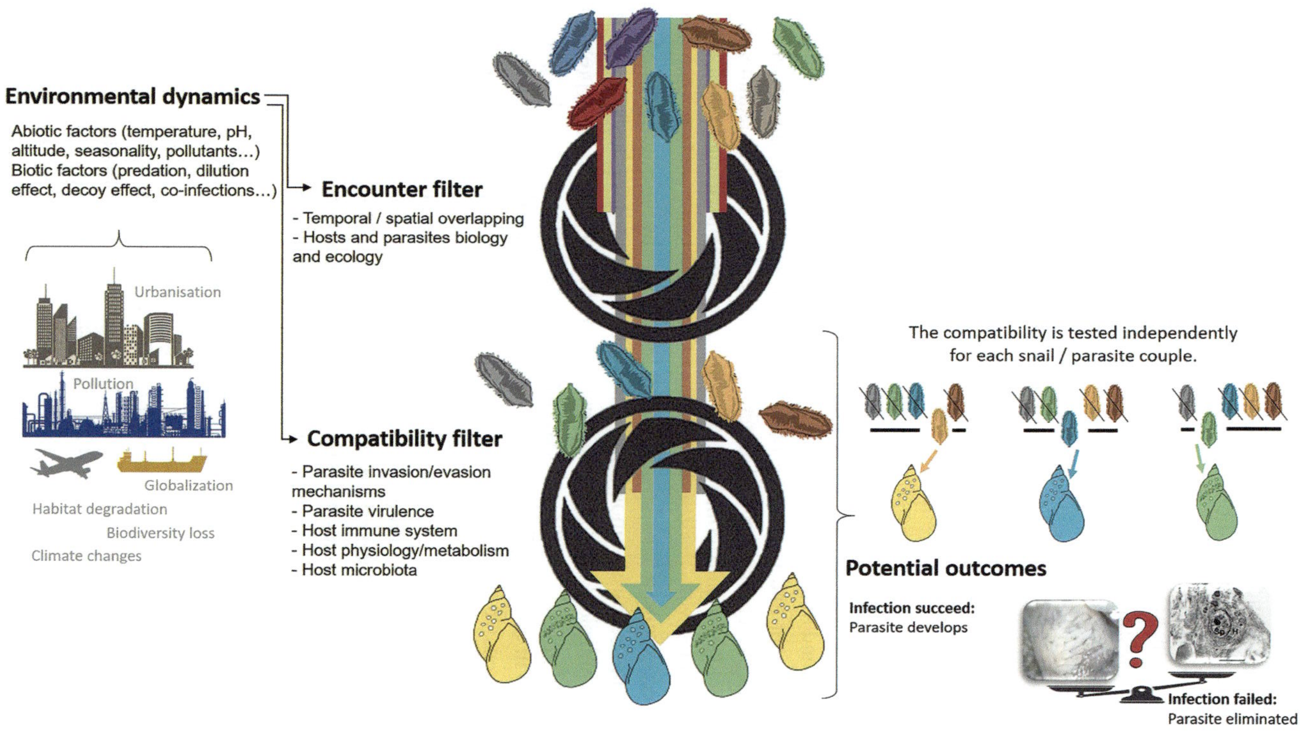

Fig. 11.1 Integrative scheme representing the encounter and compatibility filters and some extrinsic biotic processes and environmental factors all of which primarily and ultimately govern the probability of infection of a Lymnaeidae snail by any given parasite (exemplified by a trematode parasite), and the balance of the interaction and its outcome

behavior (e.g., circadian rhythms/chronobiology) and biology (e.g., for trophically-transmitted parasites, whether the host ever eats the next host of the transmission chain; see Combes 2001).

The family Lymnaeidae is present in all inhabited continents with species occurring in different climatic conditions, from tropical to temperate to artic conditions (Vázquez et al. 2018). Certain species such as *Galba cousini* and *G. truncatula* are known to occur at high altitudes (Celi-Erazo et al. 2020; Roldán et al. 2020). Representatives within this family are also commonly found in a wide variety of freshwater environments; from rivers, lakes and streams to ponds, ditches, puddles, and even, on man-made water systems, and they present a tendency to occur above or near the shore and to inhabit ecosystems with slow-moving currents that could be even temporal (Utzinger and Tanner 2000; Kock et al. 2003; Schweizer et al. 2007; Vázquez et al. 2009; Dida et al. 2014). These ecosystems usually present high productivity and biomass content, and a diversity of other species within or associated to them, including parasites (Soldánová et al. 2011; Preston et al. 2013, 2021; Schmid-Araya et al. 2020). In such habitats lymnaeid snails also become important members of the biotic community due to their propensity for feeding on numerous species of algae, and macrophytes, foraging behavior when food is scarce and certain resilience to drought and to playing roles in structuring the plant and macrophyte communities in the environment (Hunter 1980). These ecological features of the family Lymnaeidae should contribute to the establishment of ecological equivalences for a number of parasites. For instance, snail feeding ecology and habitat preference were suggested to be important factors in the infection dynamics of *Echinostoma* spp. cercariae among the snail second intermediate hosts, with *Physa* spp. and the lymnaeid *Pseudosuccinea columella* showing the highest prevalence and intensity of infection, respectively, when compared to compatible sympatric planorbid snails in the field. The pattern of infection displayed by *P. columella* was related to its extensive vagility while foraging, in order to acquire sustainable food resources, and to an increased probability of autoinfection as they are more likely to harbor *Echinostoma* spp. as first intermediate hosts. Furthermore, its amphibious tendency was speculated to aid transmission of the parasite to the definitive host as they can locate closer to the surface (Zimmermann et al. 2014).

Besides ecological affinities between host and parasites, both entities can "try" to open or close the encounter filters by modifying their behavior to chase or to avoid each other. Several parasites like water-associated digeneans present free-living stages that seek hosts in the environment primarily moving toward suitable micro-habitats based on cues like gravity, temperature, or light intensity when the snail host is not present within their vicinity, all of which have been linked to a selection driven by the ecology of the host and by the probability of encounter (Christensen 1980; Combes 2001). Once in the proximity of a host, molecular sensors that are believed to exist in the parasitic larva facilitate a discriminatory chemotaxis toward the suitable snail through the gradient of biochemical cues segregated by the host (Wilson and Denison 1970a, b). Chemical attraction of *Fasciola hepatica* and *Trichobilharzia szidati* [European isolates have been erroneously classified as *Trichobilharzia ocellata*; see (Rudolfova et al. 2005; Aldhoun et al. 2016)] miracidia by species-specific mucin-type glycoproteins released by *Galba truncatula* and *Lymnaea stagnalis* is supposed to influence host finding in these snail–parasite systems (Kalbe et al. 1997, 2000).

There are different degrees of "openness" of encounter (and of compatibility) filters irrespective of the biological level (species, population, individuals; see Combes, 2001), and even known host snails vary their degrees of attractiveness through parasite chemo-orientation behavior at the species or individual levels (Christensen et al. 1976; Langeloh and Seppälä 2018). In this sense, an "open" encounter filter does not determine that the contact will be followed by infection success as physiological, biochemical, and immunological interactions with the host (determined by compatibility filters), as well as environmental factors will ultimately dictate the overall success of the parasite in infecting the snail host (Combes 2001; Langeloh and Seppälä 2018).

11.3 How Does the Immunobiological Interaction Between the Lymnaeid Host and the Parasite Take Place?

i. **The immune system of Lymnaeidae snails**

Lymnaeid snails, as all invertebrates, rely on an internal defense system (IDS) to protect themselves against pathogens. Even though lymphocytes, antibodies, and adaptive responses to specific antigens are lacking, their defense strategies are far from being simple or non-effective (Loker et al. 2004). Homologs of molecules and immune mechanisms between vertebrate and snail (invertebrate) immune defenses exist although, there are unique immune features that are specifics to their lineage and are the result of a long independent evolutionary process (see Schultz and Adema 2017 for details). In general, three basic components from the snail host's IDS are included in the multistep infection processes determining the outcome of any lymnaeid–parasite interaction: (1) physical, chemical, and biological barriers (e.g., shell, mucus, microbiota), which impose the first challenge to overcome for any given pathogen, (2) elements to discriminate self from dangerous non-self (e.g., pattern recognition receptors, PRR), and (3) an arsenal of cellular and humoral effectors (e.g., hemocytes, reactive oxygen and nitrogen species (ROS, RNS), opsonins, proteases) and defense mechanisms which are triggered to actively control the injury/infection. Cellular and humoral elements co-operate in the process of self-non-self-recognition as well as in the killing of the non-self, whereas the communication and integration between the different immune pathways allow to orchestrating a multifactorial defense response of the snail host toward pathogens (Pila et al. 2017; Alba et al. 2020). Evidences of specificity and innate immune memory within snail's immune system also exist (Sire et al. 1998; Portela et al. 2013; Pinaud et al. 2016; Seppälä et al. 2021). Significantly, comparative studies between juveniles and adult lymnaeid snails have brought forward some functional differences of the IDS between both age groups as the former showed aspects related to immunological immaturity, e.g. lower and/or delayed capacity concerning phagocytosis, humoral defenses (namely opsonization and agglutination), and elimination of pathogens (Rondelaud and Barthe 1980a; McReath et al. 1982; Dikkeboom et al. 1985).

Barriers Mucus itself is an excellent physical barrier to cells, microorganisms, and pathogens; its visco-elastic properties may act as an adhesive providing an initial physical trap created by cross-linked glycoproteins (mucoprotein matrices) and preventing colonization by microorganism and pathogens (Vieira et al. 2004; Allam and Espinosa 2015). In addition to representing an efficient physical barrier, mucus matrices contain various cells and bioactive molecules and have gained prominence in the last few decades as main components of the innate and acquired immune systems (Allam and Espinosa 2015). The biochemical nature of the mucus may have anti-pathogenic activities, whether by the excretion-secretion of host-produced factors or by the exogenous products secreted by the microbiota (Ehara et al. 2002; Guo et al. 2009; Allam and Espinosa 2015). Particularly, the mucus has been pointed as a mediator of snail–digenean compatibility when contact is established. The in vitro effect of snail mucus of two different species on *Fascioloides magna* miracidia was assayed (Coyne et al. 2015). While no damage was observed when parasite larvae were exposed to the mucus of the lymnaeid species *Ladislavella elodes* (compatible host), significant tegument damage and/or larval death were recorded in the miracidia when tested against the mucus of an incompatible snail (*Helisoma trivolvis*) and related to the presence of a cytotoxic protein-like factor (Coyne et al. 2015).

In addition to immunological and enzymatic factors such as PRRs, humoral effectors, heat shock proteins, etc., the diverse mucus matrices in mollusks are also enriched in specialized microbiota that are recruited and maintained by the mucus layer and kept in check by the IDS. However, microbiota is believed to also influence and be influenced by snail–pathogen interactions (Allam and Espinosa 2015) and for instance, differences regarding the diversity and composition of bacterial microbiota exist between resistant and susceptible strains of *Biomphalaria glabrata* to the trematode *Schistosoma mansoni* (Allan et al. 2018; Huot et al. 2020). Furthermore, a shift in the composition of *B. glabrata* microbiota during early infection by *S. mansoni* (Portet et al. 2021) or by the nematode *Angiostrongylus cantonensis* (Osorio et al. 2020) has been observed. Changes in the expression of some antimicrobial peptides were recorded in accordance to the changes in the structure of the microbial community in *B. glabrata–S. mansoni* system (Portet et al. 2021). While this topic will deserve further investigations in lymnaeid snails, evidences exist of modulation of the bactericidal activity from in vitro bacterial clearance assays following infection of *L. stagnalis* with *T. szidati* (van der Knaap et al. 1987; Núñez et al. 1994).

Recognition Elements A prerequisite for the functioning of any internal defense system is the recognition of foreignness and therefore, recognition molecules are pivotal in orchestrating the immune response of the host that follows parasite's penetration. In molluscs, recognition of dangerous non-self and subsequent immune activation is significantly mediated through lectins and, as other PRRs, they can occur freely in the plasma or on the surface of immune phagocytic cells named hemocytes (Schultz and Adema 2017). These molecules are non-enzymatic, non-antibody proteins that function as PRRs by binding to repetitive carbohydrate surface determinants that characterize groups of pathogens and by activating immune responses (Horák and van der Knaap 1997). When soluble, lectins can function as agglutinins/opsonins; directly neutralizing the target element or creating carbohydrates bridges between the hemocyte receptors and the target element (Horak and Deme 1998).

Little is still known about the molecular arsenal of PRRs of lymnaeid snails and thorough functional and molecular characterizations are surely needed to elucidate the level of diversity within such family of molecules. However, it has been proved that recognition of and defense response against several degrees of foreignness is possible in lymnaeid snails; from abiotic materials to xenografts, septic injuries and parasitic infections, (e.g., Sminia et al. 1974; van der Knaap et al. 1981; Gutiérrez et al. 2003). In this sense, several lectins have been found in *L. stagnalis* and *P. columella* and some of their properties and/or binding abilities have been identified (e.g., van der Knaap et al. 1983b; Horak and Deme 1998; Alba et al. 2019a, 2020; Seppälä et al. 2021). The presence of most on the hemocytes of lymnaeid snails has been recognized by means of saccharide-dependent inhibition of phagocytosis/encapsulation, as saturation of lectins with sugars reduces lectin recognition (Horak and Deme 1998; Horák et al. 1998; Alba et al. 2020), whereas evidence of reversed bound to hemocytes of cell-free lectins has been also found (van der Knaap et al. 1983b). Recently, the transcriptome of *L. stagnalis* under different biotic and abiotic stressors was elucidated and several transcripts within different types of lectins were identified in the reference transcriptome; i.e. galectin, Chi-lectin, L-type lectin, M-type lectin (Seppälä et al. 2021). Specifically, non-self-recognition lectin-like soluble factors belonging to the immunoglobulin superfamily, i.e. fibrinogen-related proteins of the variable immunoglobulin and lectin domain-containing molecules (FREP; a C-terminal fibrinogen-related domain and one or two Ig domains, see Seppälä et al. 2021), and a molluscan defense molecule (MDM; five tandemly arranged Ig domains; see Hoek et al. 1996) have been described in *L. stagnalis*. In *P. columella*, the characterization of the differential transcriptome between susceptible and resistant snails to *F. hepatica* infection resulted in the identification of several lectins; including variants of C-type lectins, different mannose and galactose binding molecules, galectin, and CD109, which were found to be overrepresented in the resistant phenotype (Alba et al. 2019a). Although no specific molecules have been related to *P. columella–F. hepatica* recognition, the role of lectins in this interaction was brought forward as the in vitro encapsulation of the parasite larvae by the snail's hemocytes was significantly diminished by adding sugars; with D-mannose exerting the strongest inhibition (Alba et al. 2020).

Furthermore, stimulation of *L. stagnalis* hemocytes with pathogen-associated molecular patterns such as laminarin (a β-glucan of brown algae), zymosan (-β-glucan of fungi), and bacteria lipopolysaccharide (LPS) leads to the activation of the immune cells (Plows et al. 2004; Lacchini et al. 2006; Wright et al. 2006) suggesting the presence of specific β-1, 3-glucan-binding proteins (BGBP) and

LPS receptors on the surface of the immune cells that has yet to be characterized. Furthermore, several transcripts of gram-negative binding protein have been identified in the reference transcriptome of *L. stagnalis* (Seppälä et al. 2021) and proteins containing peptidoglycan binding domains and Toll-like receptors have been found in both *L. stagnalis* and *P. columella* snails (Alba et al. 2019a; Seppälä et al. 2021).

Cellular Effectors At least four types of cells participate in snail defenses; three are "fixed" cells that have been poorly studied concerning snail–pathogen interactions; i.e. antigen-trapping endothelial cells, reticulum cells, and pore cells, whereas the fourth type comprises macrophage-like cells that dwell in the tissues or circulate with the blood fluid (hemolymph) of mollusks and are called hemocytes or amebocytes (van der Knaap and Loker 1990). Hemocytes move freely to and from the tissues continuously sensing the internal environment, and prompt to cellular immune responses and cytotoxic reactions involving lysosomal enzymes and the production of ROS and RNS when stimulated by immune elicitors (van der Knaap and Loker 1990; Adema et al. 1991; Lacchini et al. 2006; Wright et al. 2006; Zelck et al. 2007). These cells originate from connective tissues or specialized organs (e.g., the amebocyte producing organ (APO) in some gastropods, or the white body organ in cephalopods) and their continuous and regulated hematopoiesis is important not only for immune functions, but also for the survival of the animal as a whole as it replenish the number of circulating hemocytes that are depleted from defense reactions, migration to tissues or cellular turn-over or senescence (Pila et al. 2016b).

The localization and structure of the APO varies among snail species and some lymnaeid snails may exhibit a diffuse hematopoietic site as it is believed to be the case in *L. stagnalis*; the hemocytes are possibly formed throughout the connective tissue and the vascular system of this snail species (Sminia 1974). Contrastingly, a lymphoid organ located beneath the mantle, between the pericardial cavity and the saccular kidney, and presenting a central lumen that is continuous with the kidney lumen has been described in *G. truncatula* (Rondelaud and Barthe 1981, 1982). In this species, the APO involves the proximal part of the kidney sac and presents a dual structure: the main part, containing an epithelial layer of podocytes that separates the broad hemal spaces adapted for hemolymph filtration from the urinary space, and the apical portion, adhering to the mantle and pericardial epithelia, and consisting in thicker connective-tissue frame in which hemocyte nodules may develop (Monteil and Matricon-Gondran 1991a). An ill-defined tissue located in the region of the lung roof close to the heart of *Ladislavella catascopium* contained massive numbers of hemocytes but whether it is the site for hemocyte production remains to be clearly demonstrated (Loker 1979).

During infection, the APO can increase its size compared to non-infected snails (Rondelaud and Barthe 1981). However, in lymnaeid snails, proliferation of hemocytes can occur outside of the hematopoietic organ such as in the connective tissue (Sminia 1974), in various scattered nodular regions (Monteil and Matricon-Gondran 1991a), and even in the hemolymph (Sminia 1974; Alba et al. 2020).

Another feature of hemocytes within Mollusca is their biochemical, functional, and morphological heterogeneity, whilst the variety of approaches used for their characterization (as there is a lack of specific universal molecular markers) has resulted in a non-unified classification system (Pila et al. 2016b). From a rough morphological perspective, based on their spreading capabilities, two main types of hemocytes can be defined: cells that typically spread, forming filopodia or lamellipodia in the process, are believed to be the mature effectors in phagocytosis and encapsulation processes and constitute the majority (if not all) of hemocyte populations, and smaller cells that do not spread or spread minimally when placed on artificial surfaces and present high nucleus/cytoplasm ratio, which are often called round cells or blast-like cells (Loker 2010). Spreading cells have been further segregated, mainly based on their granular content, into two groups: (i) granulocytes, which present conspicuous granules located mainly in the endoplasm and produce extensive pseudopodia in all directions, and the (ii) hyalinocytes which are smaller in size, agranular or less granular compared to the granulocytes, generally spherical or slightly oval with lobose and shorted pseudopodia (Pila et al. 2016b). Granulocytes are believed to be more active (e.g., phagocytic and enzymatic activities) than hyalinocytes (Mahilini and Rajendran 2008; Travers et al. 2008). Two hypothesis exist behind the morphological and functional heterogeneity of hemocytes; whereas some researchers consider that the different morphotypes represent different lineages of the cells, others state that they reflect different developmental stages of a single cell type (Pila et al. 2016b).

In *G. truncatula*, the ultrastructure of circulating hemocytes revealed a preponderance of spreading cells at different stages of differentiation, whereas a few round cells were also observed and suggested as a distinct population. Endogenous peroxidase activity was localized in secretory granules as well as in multivesicular bodies of the spreading hemocytes (Monteil and Matricon-Gondran 1993).

Fluorescence staining of cytoplasm and nucleus, and the estimation of size/cytoplasmic complexity by flow cytometry allowed the description of two main morphotypes in *P. columella*: (i) large adherent cells and (ii) non-adherent small, roundly-shaped hemocytes with low cytoplasmic complexity and high nucleus-to-cytoplasm ratio; i.e. blast-like cells (Alba et al. 2020). In *P. columella*, the blast-like cells were related to proliferation reactions in the hemolymph, particularly in resistant snails exposed to *F. hepatica* (Alba et al. 2020), endorsing the hypothesis of the pluripotency of this morphotype related with certain stem-cell capacity (Wang et al. 2012; Rebelo et al. 2013).

In *S. palustris*, three populations of circulating hemocytes, morphologically and cytochemically distinct, were observed; (i) round non-adherent cells showing a central spherical nucleus and a low nucleus-to-cytoplasm ratio, (ii) adherent irregular hyalinocytes showing an eccentric kidney-shaped nucleus surrounded by a thin ring of cytoplasm, a light basophilic cytoplasm containing small vacuoles and few basophilic, rarely neutrophilic, granules, and (iii) large granulocytes producing long, filamentous pseudopodia and characterized by an acidophilic cytoplasm containing many acidophilic granules (Russo and Lagadic 2000).

In *L. stagnalis*, size differentiation following a density gradient centrifugation segregated the hemocytes in: small cells (3.6–4.8 μm; 4.4 ± 0.6) with high nucleus/cytoplasm ratio, and two types of spreading cells; medium-size cells (6–12 μm;

9.04 ± 2.00) displaying some pseudopodia, and large hemocytes (13–17 μm; 14.22 ± 1.00). Although it is worth mentioning, that no complete separation among cellular fractions was achieved. Moreover, cytochemical assays for lysosomal enzymes revealed α-naphthyl acetate esterase activity in all three size groups, whereas acid phosphatase activity was weak or absent in most of the large-size cells (Adema et al. 1994a). The previous size/appearance based-segregation was more recently observed by (Boisseaux et al. 2016) in side-scattered signal outputs as three clusters, differing in size and granulometry, were also resolved; a small cells cluster (clearly segregated from the others), and intermediate and large cells clusters that sometimes overlapped. However, differences on intermediate and large cells clusters observed in the cytograms were believed to be caused by the variety of nucleus lobulation (hemocytes with one up to four nuclear lobules were observed) rather than to a cytoplasm granulation gradient as light microscopy made difficult to discriminate granulocytes from hyalinocytes (Boisseaux et al. 2016). Further ultrastructural analysis showed only one cell population, evidences of high cellular activity and the presence of two interesting organelles, so far, only described from *L. stagnalis*; (i) a large organelle with a spectacular fibrillary arrangement called lysosomal-like structures and (ii) a massive structure, usually located within the perinuclear regions, with a dense aspect and vesicles sprouting out from the internal nuclear membrane was also depicted filling intracellular compartments and occupying a vast cytosolic space in the vast majority of hemocytes (Boisseaux et al. 2016). The hypothesis that hemocytes from *L. stagnalis* originate from a single cell lineage that mature progressively as a differentiation/maturation continuum reflected in the cytograms clusters and in the small differences observed in microscopy, is strongly suggested (Boisseaux et al. 2016).

Phagocytosis (Plows et al. 2006b), encapsulation (Loker et al. 1982), and production of cytotoxic molecules (Lacchini et al. 2006; Wright et al. 2006) involved in pathogen killing and elimination are among the significant roles of hemocytes in mollusks. However, they also participate in other vital processes such as wound healing (Franchini and Ottaviani 2000) and nerve repair (Hermann et al. 2005), as well as in the production of a plethora of molecules involved in signal transduction, defense regulation, or effector capacities (Baeza-Garcia et al. 2010; Galinier et al. 2013; Pila et al. 2016a). All these functions ensure the homeostasis of the organism and require the activation of the cells through different signaling pathways and regulatory molecules, many of which occurs in concert with humoral factors.

Signaling/Regulatory Molecules and Pathways, and Defense Reactions Hemocytes can phagocytize small particles and cells, whereas encapsulation is reserved for larger elements or organism (such as metazoan parasites like digenean larvae). Both processes involve recognition and adherence to the surface of non-self-particles through cellular receptors and, in the case of encapsulation, recognition and adherence must be extended to other hemocytes that must significantly spread to form layers of overlapping cells that will eventually surround and isolate the element (Loker et al. 1982). The involvement of the mitogen-activated protein kinase (MAPK) pathways has been particularly described in several snail species in association to adhesion, motility and spreading cellular events required for phagocytosis, encapsulation, and regulation of cytotoxic molecules/reactions (e.g., Humphries and Yoshino 2003, 2008; Iakovleva et al. 2006). In *L. stagnalis*, the pivotal role of the MAPK and phosphatidylinositol 3-kinase (PI3-K) signal cascades in the immune responses of the snail's hemocytes was elucidated using different specific inhibitors (Zelck et al. 2007). To note, the extracellular signal-regulated kinase (ERK) inhibitor, PD098059, substantially impaired hemocyte spreading and aggregation, phagocytosis, encapsulation and H_2O_2 production, making patent the involvement of the ERK cascade in all phases of the cellular processes. The MEK/ERK inhibitor showed the strongest effect when added before hemocyte attachment, supporting the significant role for this signaling pathway in the initial processes of cell migration and cell defense responses (Zelck et al. 2007).

Upstream events involved in the defense activity of *L. stagnalis* hemocytes have been related to the activation of cell-surface bound αvβ3 and β1-like integrins mediating Ca^{2+}/Mg^{2+}-dependent cell adhesion; their inhibition with blocking peptides resulted in a significant diminished of the spreading capacity of the cells and in a significant inhibition of the phagocytosis of *Escherichia coli* bioparticles. It has been proposed that integrin engagement promotes outside-in signaling events in *L. stagnalis* hemocytes (Plows et al. 2006a; Walker et al. 2010) as it mediates association of protein kinase C (PKC) with focal adhesion kinase (FAK)/tyrosine kinase Src complexes to promote focal adhesion assembly, and to possibly modulate hemocyte early adhesion to pathogens (Walker et al. 2010). This model has been endorsed by the findings that phosphorylated (activated) PKC colocalized with focal adhesion sites (where a FAK-like protein was also found phosphorylated; (Plows et al. 2006a)), whereas specific inhibitors of PKC and of Src (which phosphorylates and associates with FAK) attenuated hemocyte spreading (Walker et al. 2010). Moreover, phosphatidyl inositol-3 phosphate kinase (PI3-K) and ERK have been also pointed out to strongly regulate cell motility and spreading of *L. stagnalis* hemocytes as the addition, prior to hemocyte attachment, of the MEK/ERK inhibitor, PD98059, or the PI3-K inhibitor, wortmannin, severely suppressed hemocyte spreading and aggregate formation in a concentration-dependent manner. No evidences of a significant participation of c-Jun N-terminal kinase (JNK) inhibitor, SP600125, and the p38 kinase inhibitor, SB203580, on hemocyte spreading were observed (Zelck et al. 2007).

A variety of targets such as zymosan (Dikkeboom et al. 1987), gram-positive and gram-negative bacteria (Dikkeboom et al. 1987; Plows et al. 2006b), red blood cells (Zelck et al. 2007), or latex particles (Dikkeboom et al. 1987; Russo and Lagadic 2000) have been used to elicit phagocytosis by lymnaeid snails' hemocytes. In vitro phagocytosis assays of *E. coli* by *L. stagnalis* hemocytes in the presence of inhibitors of PI3-K and of protein kinase A (PKA) demonstrated that this reaction is dependent on the activation of the first enzyme (Plows et al. 2006b). In another study, the stimulation of *L. stagnalis* hemocytes with bacterial LPS resulted in a transient activation of ERK, whereas inhibition of MEK activity impaired phosphorylation of ERK, its colocalization in the nuclear and perinuclear regions after LPS challenge, as well as the phagocytosis of *E. coli* bioparticles (Plows et al. 2004). Furthermore, addition of GF109203X, a PKC inhibitor, or of a FTase inhibitor I, which blocks farnesylation of the upstream MAPK component Ras, resulted in a reduced phagocytosis suggesting that activation of ERK can be driven by both Ras-dependent (with MEK as upstream kinase of ERK) and Ras-independent mechanisms (which may involve PKC, PKA or PI-3-kinase; Plows et al. 2004). ERK, and JNK, and p38 Map kinase in a lesser extent, were also found to contribute in signal transduction mediating the phagocytosis of sheep red blood cells by *L. stagnalis* hemocytes, indicating that regulation of this process may depend on the activation of and the cross-talk among all three MAPK pathways. Moreover, trematode encapsulation by *L. stagnalis* was also impaired by inhibitors for MEK/ERK and PI3-K (Zelck et al. 2007).

Activated hemocytes trigger cytotoxic reactions toward phagocytized or encapsulated elements. Contact between the immune elicitor and the hemocyte's surface stimulates a membrane-bound system to generate and to release oxygen radicals that can be sensed in the extracellular milieu (Adema et al. 1991). The encapsulation reaction allows the hemocytes to isolate the menace and to neutralize it within the formed capsule without seriously endanger the integrity of the surrounded tissues (Loker et al. 1982; McReath et al. 1982). When internalization of the immune elicitor occurs through phagocytosis, the production of oxygen radicals localized inside the phagolysosome (Dikkeboom et al. 1987; Adema et al. 1991).

The mechanism for production of toxic ROS in *L. stagnalis* (as well as in other mollusk species) has been proposed to be homologous to that of the respiratory burst of vertebrate phagocytes as both are sensitive to catechol-like phenol inhibitors of the vertebrate NADPH-oxidase enzyme complex that generates superoxide (Adema et al. 1993; Lacchini et al. 2006). It is postulated that, in *L. stagnalis* hemocytes, an active NADPH-oxidase enzyme complex has to be assembled from putative cytosolic and membrane-associated components (Adema et al. 1993). At the molecular level, ERK appears to play a major role in the activation of the oxidative burst of *L. stagnalis* hemocytes following phagocytosis or encapsulation reactions (Zelck et al. 2007). Challenge of freshly collected hemocytes with 1,3-glucan laminarin promoted PKC-mediated signaling and downstream MEK/ERK activation and H_2O_2 release (Lacchini et al. 2006). Moreover, ERK1/2 and PKC (to a lesser extend) signaling seems to play a pivotal role in mediating nitric oxide synthase (NOS) activation and the subsequent NO production following stimulation of *L. stagnalis* hemocytes by laminarin and the phorbol ester, phorbol myristate acetate (Wright et al. 2006). However, as with phagocytosis, observations that PD098059, a specific inhibitor of ERK/MEK, did not completely inhibit H_2O_2 generation suggest that multiple signaling cascades, possibly other MAPK pathways, may contribute to the orchestration of cellular defense responses in lymnaeid snails (Zelck et al. 2007).

Another significant pathway mediating immune reactions in different phyla is the Toll-like receptor (TLR) pathway. In mollusks, TLRs can be activated and their expression can be upregulated after septic injury to promote hemocyte activation (Nie et al. 2018), whereas the positive influence of TLRs and their activation in the protective response toward pathogens have been experimentally stated in *B. glabrata–S. mansoni* system (see Pila et al. 2016c). One Toll-like receptor

(TLR) with transmembrane domain and several transcripts belonging to downstream mediators within the TLR pathway (e.g., MyD88, IkB, and NF-kBp65) have been recorded in the reference transcriptome of *L. stagnalis*. It was also observed that IκB, a regulator of TLR signaling, showed increased transcription to different immune elicitors (gram-negative and gram-positive bacteria, trematode, injury) except wounding and even though this effect was not seen in all individuals, it suggests that the TLR pathway generally participates in the control of the defense responses in *L. stagnalis* (Seppälä et al. 2021). In *P. columella*, PKC, leucine-rich repeat-containing molecules, and several Toll-like receptors (TLR) and some of their downstream molecules, were found to be overrepresented in *F. hepatica*-resistant snails (Alba et al. 2019a).

Other transcripts involved in regulation/signaling such as granulocyte colony stimulatory factor receptor (G-CSFR), transforming growth factor 1-beta (TGF1β), members of the superfamily of tumoral necrosis factor (TNF) receptors and interferon regulatory factors (IRF) have been recorded in *P. columella* transcriptome (Alba et al. 2019a). Similarly, several variants of the cytokine macrophage migration inhibitory factor (MIF) have been found in *L. stagnalis* and *P. columella* (Alba et al. 2019a; Seppälä et al. 2021), whereas three variants of the TNF and 23 variants of interleukin 17 families were depicted in the reference transcriptome of *L. stagnalis* (Seppälä et al. 2021). Furthermore, expression of granulin, a cytokine with pleiotropic functions in mammals (Jian et al. 2013) that also promotes proliferation, differentiation, and activation of *B. glabrata* hemocytes in response to *S. mansoni* (Pila et al. 2016a; Hambrook et al. 2019), was specifically studied in *P. columella* by qRT-PCR following exposure to *F. hepatica* (Alba et al. 2020). In this study, the granulin gene was found to be significantly upregulated in resistant *P. columella* snails during infection in consistency with the increase of hemocyte proliferation and spreading that was also associated to this host phenotype (Alba et al. 2020).

Apart from the mentioned cellular reactions, the phenoloxidase (PO) system is another immune trait with a major role in invertebrates and their defenses against pathogens and damaged tissues that eventually leads to melanization (Cerenius and Söderhäll 2004). In *L. stagnalis*, PO-like activity, measured as an increase in optical density of a solution in which PO enzymes from snail hemolymph oxidize the substrate L-dopa, has been assessed after experimental challenges with different immune elicitors or environmental conditions (e.g., Seppälä and Jokela 2010; Seppälä and Leicht 2013). Although, this assay may also measure the combined activity of different factors (see Le Clec'h et al. 2016), the finding of several variants of laccase and tyrosinase, enzymes involved in phenoloxidase/melanization type-reactions, in the reference transcriptome of *L. stagnalis* warrants further investigations about their significance in protective immune defenses (Seppälä et al. 2021).

In addition, hemocytes of the lymnaeid snails *L. stagnalis* and *Radix lagotis,* and of the planorbid *Planorbarius corneus* have recently been shown to generate extracellular trap-like (ET-like) fibers in vitro, although their induction following stimulation of hemocytes with immune elicitors was rare (Skála et al. 2018). This mechanism was firstly described in mammalian neutrophils, which generate extracellular chromatin fibers that facilitates immobilization and elimination of unicellular pathogens (Kaplan and Radic 2012). In terrestrial slugs and snails, ET-like fibers were observed to be extruded from hemocytes following exposure to metastrongyloid parasites and to participate in the contact, entrapment, and immobilization of the larva (Lange et al. 2017). Further characterization of this phenomenon in the family Lymnaeidae is needed to elucidate its significance in the IDS.

Humoral Effectors and Antioxidant System The role of plasma (cell-free hemolymph) in digenean–snail compatibility has proved crucial since significant damage to sporocyst (digenean intramolluscan larva) tegument and diminishing of its survival can be seen after parasite's exposure to plasma from incompatible hosts. In 9 of 10 combinations tested, lymnaeid plasma was toxic to the parasites of planorbid snails and in 2 of 4 combinations, planorbid plasma was toxic to the parasites of lymnaeid snails. The effect was not associated to the physic-chemical conditions of the plasma and thus, the existence of particular humoral factors dictating patterns of snail specificity was speculated (Sapp and Loker 2000).

Further studies have described the presence of humoral factors with potential defense activity against infection in lymnaeid snails, such as different variants within families of antimicrobial peptides, i.e. hydramacin in *P. columella* (Alba et al. 2019a) and macin in *L. stagnalis* (Seppälä et al. 2021). Soluble lectins in the plasma can function as opsonins or agglutinins neutralizing pathogens, or promoting recognition and effector capacity of cellular defenses (Sminia et al. 1979; van der Knaap et al. 1981, 1983b; Horak and Deme 1998). Moreover, an opsonin named granularin, a cysteine-rich protein that shares amino acid sequence identity with domains present in Von Willebrand factor and thrombospondin, has been described in *L. stagnalis* to stimulate phagocytosis of foreign particles by hemocytes (Smit et al. 2004). Several proteins with potential activity in antimicrobial defenses were identified in the reference transcriptome of *L. stagnalis*; i.e. lipopolysaccharide-binding protein/bactericidal permeability-increasing proteins (LBP/BPI), lysozymes and L-amino acid oxidases. In addition, different transcripts encoding for cytolytic β pore-forming toxins, called as *Lymnaea*-lysin and stagnalysin, were also found (Seppälä et al. 2021). In *P. columella*, a G-type lysozyme and a LBP/BPI were only observed in the differential proteome of resistant *P. columella* to *F. hepatica* (Alba et al. 2019a). It is worth mentioning that, whereas hemocytes are believed to play a crucial role in the synthesis and secretion of humoral factors (Baeza-Garcia et al. 2010; Galinier et al. 2013), other cellular types and organs have been proved to be involved in this activity in lymnaeid snails, such as the granular cells (Hoek et al. 1996; Smit et al. 2004) and the albumen gland (Alba et al. 2019a).

Enzymes and proteins with roles in regulating the oxidative damage, i.e. superoxide dismutases (Cu/Zn and Fe/Mn families; SOD), glutathione peroxidases, glutathione reductase, peroxiredoxins, glutathione S-transferases (GST) and catalase, were represented in a transcriptomic study on *L. stagnalis* focalized on responses to environmental stress (Bouétard et al. 2012). In naïve resistant *P. columella,* transcripts identified as catalase, superoxide dismutase and probable deferrochelatase peroxidase have been identified as overrepresented, as well as molecules involved in acute phase response like ferritin and NOS (Alba et al. 2019a). Such antioxidant potential is essential to maintain the redox balance of the host prior and during infection and has proved significant to resist bacterial infection in the clam *Meretrix meretrix* (Wang et al. 2013) or to be positively associated to anti-parasite resistance in *B. glabrata* snails as higher constitutive levels of Cu/Zn SOD make killing of the trematode larva more effective (Goodall et al. 2004; Lockyer et al. 2012).

To summarize the main content of the present section, a representation of some of the elements, signaling pathways, and immune reactions described to be involved in the IDS of Lymnaeidae is depicted in Fig. 11.2.

ii. **The interplay between the lymnaeid host and the parasite: defenses vs. counter defenses**

The penetration of a pathogen into the lymnaeid snail and its colonization stimulate the IDS of the host through the recognition of the somatic components or the excretion-secretion products (ESP) of the parasite (Núñez et al. 1994; Skála et al. 2014). However, whether these events will result in the destruction of the pathogen or in the dynamic equilibrium that characterized an established infection will depend on the interplay between host's defenses and pathogen's counter defenses at each step of the interaction (van der Knaap and Loker 1990). The interplay is, therefore, dependent on (epi)genetic factors that control the infectivity and pathogenicity of the parasite and the level of resistance of the host (Mitta et al. 2017).

Destruction of the Pathogen The sensing of a trematode occurs early in the infection and it may lead to the activation of the IDS of the snail host as shown by the enhanced in vitro phagocytosis of zymosan particles or live bacteria by *L. stagnalis* hemocytes between 1.5 and 3 h post-exposure (p.e.) to *T. szidati* (van der Knaap et al. 1987; Núñez et al. 1994). It can be featured by the recruitment of hemocytes to the site of infection as it was proposed when a decrease of hemocytes counts in the hemolymph of *P. columella* was observed as early as 30 minutes p.e. to *F. hepatica* (Ribeiro et al. 2017). In *R. lagotis* infected by *Trichobilharzia regenti*, histological sections showed that circulating hemocytes migrate and accumulate nearby the developing parasite between 2 and 36 h p.e. (Skála et al. 2014). In *S. palustris,* hemocytes gathered around *F. hepatica* sporocyst between 24 and 72 h p.e., although the length of the mobilization period and of the defense reactions decreases as the age of the snail increases (McReath et al. 1982).

An increase of circulating hemocytes (Ribeiro et al. 2017; Alba et al. 2020), changes in the dynamic of the different morphotypes (Russo and Lagadic 2000) and even qualitative alterations of their morphology and functionality can also result from the contact with trematodes (Alba et al. 2020). In resistant *P. columella* challenged with *F. hepatica*, an increase of the proliferative activity of blast-like cells and of the spreading of large hemocytes, which tended to aggregate into a sheet-like spreading pattern and to display an increased length of broad-based cytoplasmic projections, was observed at 24 h p.e. (Alba et al. 2020). In *G. truncatula,* the APO is

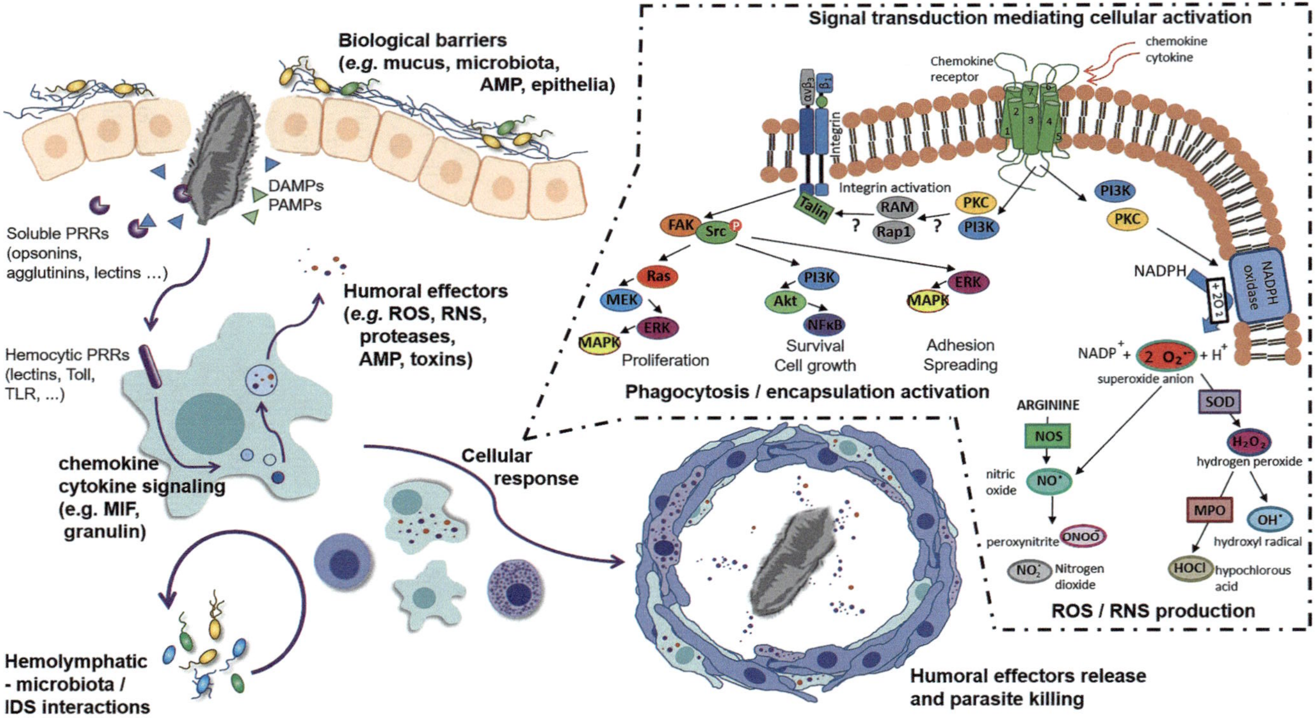

Fig. 11.2 Overview on *Lymnaea* internal defense system. Snail defense system could be separated into 4 steps, (1) physical, chemical, and biological barriers dedicated to block pathogens and avoid their penetration in snail tissues; (2) immune recognition to discriminate self from non-self based on Pattern Recognition Receptors (PRR) able to recognize Pathogen Associated Molecular Patterns (PAMPs) or Danger Associated Molecular Patterns (DAMPs) associated with pathogen-dependent wounding or tissue lysis; (3) signaling pathways regulate immune and stress response (cytokine, TLR, MAP kinase, . . .), and (4) an arsenal of cellular and humoral effectors (e.g., hemocytes, reactive oxygen and nitrogen species (ROS, RNS), opsonins, proteases) are produced as humoral hemolymphatic factors or degranulated inside hemocytic capsule (encapsulation process) to kill and eliminate pathogens. Endosymbiotic microbiota

Fig. 11.2 (continued) pathogen interactions, this resulting in changes in host fitness and defense capabilities. *Abbreviations*: ROS: reactive oxygen species, RNS: reactive nitrogen species, AMP: antimicrobial peptide, NOS: nitric oxide synthase, SOD: super oxide dismutase, MOP: myeloperoxidase, MIF: macrophage migration inhibitory factor, PKC: protein kinase C, MAPK: mitogen-activated protein kinase, ERK: extracellular-signaling regulated kinases, MEK: mitogen-activated protein kinase kinase, PI3K: phosphatidylinositol kinase, NF-kB: nuclear factor-kappa B, Rap1: Ras family GTPase, Ras: Rat sarcoma virus small GTPase, FAK: focal adhesion kinase, Src: steroid receptor coactivator, RAM: regulation of Ace2p activity and cellular morphogenesis, Akt: protein kinase B

significantly enlarged upon exposure to *F. hepatica* and showed an increase of hemocyte proliferation and density during early infection compared to uninfected snails (Rondelaud and Barthe 1981).

The regulation of the expression of several immune molecules also occurs following IDS activation, which results in mounting and controlling the immune response during all the infection process. In resistant *P. columella* challenged with *F. hepatica,* the transcription of the cytokine granulin is significantly upregulated at 6 h p.e. compared to naïve snails, whilst at 96 h p.e. a higher increase was recorded in exposed snails irrespective of the phenotype (Alba et al. 2020). In *L. stagnalis* snails infected with *T. szidati*, the gene encoding the opsonin granularin is upregulated from 1.5 h p.e. onwards (Smit et al. 2004). Upregulation of laccase transcription following injection of soluble extracts from trematode infected snails has been also reported in *L. stagnalis* (Seppälä et al. 2021).

All the reactions elicited in the snail by the infective trematodes might eventually lead to an increase of the phagocytosis/encapsulation activity specifically directed toward the parasite and to the activation of the respiratory burst. Highly antigenic miracidial ciliated plates shed by the parasite upon penetration have been observed within phagosomes of hemocytes of *R. lagotis* after infection with *T. regenti* (Skála et al. 2014). On the other hand, stimulation of *L. stagnalis* with zymosan or trematode larvae triggered the generation of elevated amounts of H_2O_2 in the snail hemocytes 2 to 48 h p.e.; maximal inductions were observed at 45 min during phagocytosis and at 12 h during encapsulation reactions (Zelck et al. 2005). The respiratory burst and the concomitant production of ROS and RNS should follow the proper activation of hemocytes (Adema et al. 1991; Lacchini et al. 2006; Zelck et al. 2007) and the cytotoxic effect exerted by reactive oxygen intermediates like superoxide (O_2-) and hydrogen peroxide (H_2O_2) has been proved effective against schistosomes (Adema et al. 1994b).

In particular, the encapsulation and further killing of the encapsulated parasite are the effective snail defense that is generally associated to protection against trematodes (Loker 1979; McReath et al. 1982; Gutiérrez et al. 2003). The velocity and effectiveness of mounting such a response significantly increases with the age of the snails; in 6 to 9 days old *S. palustris*, encapsulation of 50% of *F. hepatica* sporocysts was observed at 96 h p.e and the infection was resolved within 7 days p.e., whereas in 18–21 days old snails, 50% of parasite larva were already encapsulated within the first 24 h and total clearance was attained at day 5 p.e. (McReath et al. 1982).

On the other hand, even though most studies have encompassed the immune response toward trematodes, defense reactions against other types of pathogens have been explored in lymnaeid snails through experimental challenges using model bacteria species. As with digenean trematodes, evidences of certain specificity of the immune response can be observed as distinctive molecular patterns have emerged following exposure. In this sense, experimental challenge of *L. stagnalis* with the gram-negative *E. coli* bacteria results in the differential expression of the TLR and of several molecules of the TLR signaling pathway (IκB, NF-κB) as well as of the antibacterial cytotoxins *Lymnaea*-lysins. In contrast, exposure to the gram-positive bacteria *Micrococcus lysodeikticus* leads to the overexpression of the lectin

FREP and the cytokine IL-17. The enzyme laccase and the component of the TLR pathway IκB were found overexpressed 6 h after both bacterial challenges (Seppälä et al. 2021). At the phenotypic level, injection of *M. lysodeikticus* significantly increases antibacterial activity of the hemolymph of *L. stagnalis* at 6 h p.e. (Seppälä and Leicht 2013). Exposure to *Staphylococcus saprophyticus* or *E. coli* results in early phagocytosis of the bacteria by hemocytes with the number of fusions of lysosomes and phagosomes as well as of the fusions among phagosomes increasing from 30 min p.e. onwards, and 99% of bacterial clearance attained as early as 2 h p.e. (van der Knaap et al. 1981).

Nevertheless, all the previous immunobiological events will result in the protective response only when resistant or incompatible snail–pathogen combinations take place (Núñez and De Jong-Brink 1997; Gutiérrez et al. 2003; Alba et al. 2020). In *P. columella*, two contrasting phenotypes concerning infection by the trematode *F. hepatica* exist; naturally resistant and susceptible snail populations (Gutiérrez et al. 2003; Alba et al. 2018, 2019a, b). Although defense reactivity seems to be constitutively enhanced in naïve resistant snails compared to susceptible ones (Alba et al. 2019a, 2020), the experimental challenge with *F. hepatica* ends up boosting the encapsulation capacity of the hemocytes toward the sporocysts particularly in resistant snails, both in vitro (Alba et al. 2020) and in vivo (Gutiérrez et al. 2003). Apparently, the defense response elicited by *F. hepatica* in resistant *P. columella* is potent enough to eliminate the infection regardless of the infective dose, infection scheme, or the geographic origin of the parasite (Calienes et al. 2004; Vázquez et al. 2014; Alba et al. 2018). However, when analyzing susceptible *P. columella* and other compatible lymnaeid snails high among-individual variation in terms of the immune response toward trematodes has been observed at the molecular (Seppälä et al. 2021) and phenotypical levels (Vázquez et al. 2014; Alba et al. 2020). Even within a single snail host individually infected by multiple miracidia, evidences of effective encapsulation of certain larva whilst other remain free to develop and multiply attest for this mosaic of phenotypic responses to infection within susceptible (= compatible) hosts (e.g., (Loker 1979; Thèron and Coustau 2005; Coustau et al. 2015). This is the result of a polymorphism of compatibility, which is driven by genotype/genotype interaction from each single host–parasite combination (Thèron and Coustau 2005).

Dynamic Equilibrium of a Successful Infection In snail–trematode systems, although all intramolluscan developmental stages are continuously confronted with a potentially lethal host response, compatible parasites are left unharmed as the countermeasures displayed by them are effective enough as to impair the orchestration of a protective response and to succeed in guarantying the infection and development within the host. Two main strategies; evasion of immunorecognition, and selective and direct interference of the antiparasitic activities elicited upon infection, are used by compatible parasites to overcome host defenses, and can act independently or concomitantly depending on the host–parasite system (van der Knaap and Loker 1990).

The avoidance of recognition by the host's IDS can be achieved through antigenic variation or molecular disguise (i.e., molecular masking and molecular mimicry; see van der Knaap and Loker 1990). The miracidium-to-sporocyst (-to rediae) transformation of digenean trematodes involves considerable changes in the molecular composition of the larvae, including the main targets for non-self-recognition via host's lectins, i.e. the carbohydrate moieties, as demonstrated on the surface of *T. szidati* and *F. hepatica* by the use of a variety of lectin probes (Gerhardus et al. 1991; Horák 1995; Georgieva et al. 2016). Moreover, the intramolluscan stages of *T. szidati* (miracidium, mother sporocyst, and cercaria) contain a polypeptide exhibiting N-acetyl-galactosaminyl-transferase activity that controls the synthesis of N'N-diacetyllactosediamino LacdiNAc-type glycans that are also common to *L. stagnalis* and might, therefore, contribute to molecular mimicry (Neeleman et al. 1994). In *F. hepatica–P. columella/Galba cubensis*, cross-reactivity of anti-rediae polyclonal antibodies raised in mice and rabbits with antigens of naïve snail is suggestive of the existence of some immunogenic determinants of the parasite similar to those of the snail hosts, mainly associated with low molecular weight proteins that appear to be parasite-specific (Alba et al. 2014). In addition, it has been observed that some glycan domains are shared between *F. hepatica* and some of the tissues of *G. truncatula* where the parasite establishes itself (Georgieva et al. 2016). Ultrastructural observations of *Haplometra cylindracea* sporocysts revealed that *G. truncatula* hemocytes, which are modified upon contact with the parasite, end up covering the sporocyst in a structure called "paletot," an intermediate state between typical free hemocytes and paletot cells derived from the transformation of the formers upon contact with the parasite larva. This results in hemocytes becoming fixed cells, losing most pseudopodia, endogenous peroxidase activity and secretory granules, acquiring a basal lamina and presenting a cytoplasm gradually invaded by glycogen and lipid inclusions and in a final structure that fails to encapsulate the larva but could aid in interfering with IDS recognition and/or defense while providing nutrition for the larvae (Monteil and Matricon-Gondran 1991b).

Direct interference from the parasite with defense activities of the host can be also cited from Lymnaeidae–trematode interactions. In compatible *L. stagnalis–T. szidati* combinations, snail hemocytes displayed an initial enhance of phagocytic activity that is followed by a reduced capacity to eliminate foreign particles from 12 to 96 h p.e. (e.g., Núñez et al. 1994). In this sense, glycoproteins present in the high molecular weight fraction of *T. szidati* ESPs and produced between 33 and 72 h of culture inhibited the phagocytosis of bacteria suggesting an immunosuppressive role (Núñez and De Jong-Brink 1997). Exposure to *Plagiorchis mutationis* causes a severe immunosuppression of the cellular immune response in *L. stagnalis*. In this study, hemocytes of both uninfected and infected snails did not form a capsule around live cercariae, whereas all dead *P. mutationis* larva were encapsulated by hemocytes in the uninfected snails but only 20% of all dead cercariae were completely surrounded in vitro by hemocytes from infected snails (Kryukova et al. 2014). Similarly, only few hemocytes of *L. catascopium* occasionally adhere to the body wall of compatible *Schistosomatium douthitti* mother sporocyst that typically neither flatten nor accumulate in any numbers and thus, no effective host responses are observed (Loker 1979). On the other hand, the early recruitment of hemocytes of *R. lagotis* infected by *T. regenti* does not result in the destruction of the parasite. From 20 to 36 h p.e., hemocytes did not accumulate in layers and appear scattered around the parasite without contacting parasite's tegument, whilst at 44–49 h p.e. no hemocytes were further observed in proximity to the sporocysts. In the same study, phagocytic activity and H_2O_2 production of these cells were found diminished during the patent period of infection compared to those from non-infected snails, showing also lower levels of phosphorylation of PKC and of ERK (Skála et al. 2014). In *G. truncatula* infected by *F. hepatica*, interference with hemocyte functions was suggested as in the few capsules that occurred in parasitized snails the hemocytes did not spread normally and their peroxidase granules were resorbed into multivesicular bodies (Monteil and Matricon-Gondran 1993). Evidences of a putative *F. hepatica* driven inhibition of host immunity in susceptible *P. columella* snails were recorded by Alba et al. (2020). No significant increase of total hemocyte counts, no patent proliferation of blast-like cells, and a shrinkage of large adherent cells which appeared more scattered, without a clear tendency to aggregate and showing a discrete spreading morphology with shorter pseudopodia when compared to naïve conditions, were recorded following experimental infection by the parasite. In addition, a significant decrease of production of the cytokine granulin at 12 and 24 h p.e. compared with the constitutive level in uninfected snails also features the dynamic of the interplay in *F. hepatica*-susceptible *P. columella* (Alba et al. 2020).

Even though compatibility filters are early gauged, countermeasures and immune regulation associated with parasite development are also present during mid and late-infections in compatible host–parasite systems. An increase of the density of *G. truncatula* hemocytes at the sites where *F. hepatica* occurs is significantly patent at day 7 p.e., reached a maximum between days 35 and 49 p.e, peaking simultaneously with the redial multiplication, to disappear almost completely at day 70 p.-e. (Rondelaud and Barthe 1980a). In the APO, the number of cells reached a maximum at day 21 p.e. (Rondelaud and Barthe 1981). However, although a certain hemocyte reaction can be observed, no encapsulation of the parasitic larva occurred (Rondelaud and Barthe 1980a). In this sense, an increase in the number of hemocytes has been detected in advanced infections in different Lymnaeidae–trematode models; e.g. *L. stagnalis–T. szidati* (van der Knaap et al. 1987), *S. palustris–Metaleptocephalus* sp. (Russo and Lagadic 2000), *R. lagotis–T. regenti* (Skála et al. 2014), *L. catascopium–S. douthitti* (Loker 1979) possibly to cope with the effects of parasite multiplication and/or cercarial shedding. In *L. stagnalis*, the expression of the MDM receptor is suppressed by the avian trematode parasite *T. szidati* with a significant downregulation to 76% of the non-parasitized level observed at 2 weeks p.e. that continues at least up to 8 week p.e. where a 21% decrease of expression was still observable (Hoek et al. 1996). Moreover, a decrease on the phagocytic activity of the hemocytes was recorded from 5 weeks p.e. onwards (van der Knaap et al. 1987), whereas Horak and Deme (1998) reported just minor differences between infected and uninfected snails in terms of hemocyte phagocytic activity at 10 weeks p.e. Hemocytes from infected *L. stagnalis* shedding *P. mutationis* cercaria showed a significant diminish of their phagocytic and encapsulation capacities against *E. coli* bioparticles and Sephadex beads, respectively, compared to uninfected snails (Kryukova et al. 2014). Similarly, patent infection of *T. regenti* in *R. lagotis* results in a significant reduction of hemocyte phagocytic activity and H_2O_2 production when compared to uninfected counterparts, an effect that was associated with a lower PKC and ERK activity in cell adhesion assays (Skála et al. 2014).

11.4 What Could Be the Effects of a Parasitic Infection in the Lymnaeid Host?

In compatible host–parasite combinations, the successful infection of the snail host by a pathogen, once established, concomitantly leads to the development and/or reproduction of the parasite and to the continuity of its transmission cycle. This causes profound effects in several aspects within the host; from the physical damage to organs and tissues where the pathogen settled to changes in the metabolism and physiology of the host that ultimately result in changes in life history traits. Examples involving infection-associated effects that somehow contributes to reproduction and transmission of the parasite, particularly within snail–digenean systems, e.g. modulation of energy reserves, energetic budget and neuroendocrine metabolism (de Jong-Brink et al. 1997), behavioral changes (Żbikowska and Żbikowski 2015), host castration (Schallig et al. 1991), gigantism (Sorensen and Minchella 1998), or growth stunting (Muñoz-Antoli et al. 2007), are reported.

i. **On the integrity of internal organs and tissues**

The development of an infection, particularly those characterized by high parasite burdens can result in a significant damage to the internal systems of the snail hosts. In the case of digenean trematodes, those involving the production of rediae within the host can inflict serious injuries to the snail's tissues partly associated with the voracious behavior of this actively-feeding larva (Esch et al. 2002). For instance, structural modifications, degeneration, and necrosis have been described in the digestive and reproductive tissues of *G. truncatula* infected by *F. hepatica*, in association to the mechanic trauma exerted by the establishment and multiplication of the parasite in these organs (Rondelaud and Barthe 1980b; Wilson and Denison 1980). The lesions relate to the intensity of the evolutive infection: necrosis of the digestive gland is found in moderate to low infections, whereas gonadal atrophy with epithelial necrosis is usually found in highly infected snails (Sindou et al. 1991). Mid and late-infections by *F. hepatica* in *G. truncatula* also cause the destruction of renal epithelial cells followed by epithelial hyperplasia (Rondelaud and Barthe 1983), whereas in *Omphiscola glabra, S. palustris,* and *Ampullaceana balthica*, epithelial necrosis of the kidney has been recorded (Sindou et al. 1991). In addition, in *O. glabra*, multifocal and generalized epithelial necrosis, and epithelial reconstitution with cell hyperplasia can be observed in the digestive gland of

F. hepatica-infected snails, with the latter event occurring more frequently and appearing sooner in individuals showing bigger shells (Sindou et al. 1991). In *Galba viator* infected with *F. hepatica*, severe epithelial damage and leakage of hemolymph into the vicinity of the digestive gland leading to a persistent degeneration and necrotic process of the organ, was observed and related to the presence the parasite rediae (Magalhães et al. 2008). Edema, cellular degeneration, and loss of cytoplasmic content of the hepatic cells and glycogen clumps in the luminal boundary of the hepatopancreatic cells were noted in *Radix auricularia* infected with *Fasciola gigantica* (Patnaik 1968).

It was observed that infection by *Echinostoma revolutum* in young *S. palustris* snails partially inhibited the development of the hepatopancreas, and the hepatopancreatic epithelium was reduced to a syncytium showing necrosis and loss of tissue. In infected adult snails, the hepatopancreas was crowded to the periphery and the epithelium commonly underwent vacuolar degeneration, whereas gonads were degenerated (Zischke 1972). In *L. elodes*, the growth of the rediae population of *E. revolutum* had a strong negative effect on the ovotestis and the digestive gland of the snail host; the size and integrity of the organs decreased with the progression of the infection due to their apparent consumption by the parasitic larvae (Sorensen and Minchella 1998).

In *L. catascopium*–*S. douthitti*, the migration and establishment of daughter sporocyst results in atrophy of the digestive gland and, in consequence, many smaller lobules of the digestive gland disappear, whilst the remaining lobules became more widely separated. In occasions, the pressure resulting from growth of mother sporocysts has proved sufficient to rupture the transverse membrane that normally separates the cephalopedal and visceral sinuses in lymnaeid snails (Loker 1979). In the kidney of *L. stagnalis* parasitized by *T. ocellata* (North American isolate), vacuolization was observed within the cells of the tubule epithelium, although no evidences of mechanical trauma related to the development and multiplication of the sporocysts were recorded. Significantly, the reproductive tract of infected snails failed to develop, remaining rudimentary with the albumen and muciparous glands often being indiscernible and the ovotestis regressed (McClelland and Bourns 1969).

Experimental infection of *P. columella* with the entomopathogenic nematode *Heterorhabditis baujardi* results in severe damage of the cephalopodal mass and the digestive gland due to the intense cell disorganization and the formation of granulomatous reactions caused by the inflammatory response occurring in the infected snails (Tunholi et al. 2017).

ii. **On the physiology and the metabolism**

Pathogens profit from host resources to support their development which causes significant physiological and metabolic alterations aiming at balancing the energy budget (Patnaik 1968; de Jong-Brink et al. 1997; Humiczewska and Rajski 2005). Furthermore, parasitism can also alter the biological rhythms, foraging behavior, and responses to abiotic stimuli of the host, usually benefiting parasite development and/or transmission (Voutilainen 2010; Żbikowska and Żbikowski 2015). For instance, changes in ionic balance and calcium deposition can result from infection. Significant hypercalcification of the shell has been observed in infected compared to uninfected *L. stagnalis* when inhabiting lakes with low calcium content (Żbikowska 2003), whereas the calcium content of the shells and soft parts of *Radix natalensis* shedding *F. gigantica* cercariae was lower compared to uninfected snails (Mostafa 2008). In the latter study, the concentration of lead and sodium ions in infected *R. natalensis* showed a significant decrease during the patent period whilst higher concentrations of zinc, potassium, and copper were found in infected compared to non-infected snails.

Infection of *R. auricularia* by *E. revolutum* results in depletory changes in glycogen reserves within host tissues which are shifted toward the development of or deposition in the larval trematode (Patnaik 1968). Lower rates of oxygen consumption have been reported in *E. revolutum*-infected *S. palustris* in association with the parasite burden (Zischke 1972), whereas the galactogen concentration in the albumen gland of *P. columella* exposed to the nematode *Heterorhabditis baujardi* declined significantly as infection progressed in relation to non-exposed control group (Tunholi et al. 2017). Infection of *P. columella* by *Echinostoma paraensei* significantly altered the contents of carbohydrates (60% reduction) and total proteins (76.4% reduction) in the digestive gland during the prepatent period, whereas total protein amounts were also 80% lower in the hemolymph of parasitized snails compared to uninfected individuals. Nitrogen catabolism was estimated by measuring uric acid and urea; the former was reduced in both organs after 10 days p.e, whereas urea increased in the hemolymph, peaking at day 10 to decrease afterwards, and it also increased in the albumen gland by 20 days p.e. when the larvae were well-established. A shift in infected host from using carbohydrates to proteins as main energy source was suggested possibly in response to the depletion of snail glycogen deposits caused by the parasites (Pinheiro et al. 2009).

In *G. truncatula*, infection by *F. hepatica* resulted in increased cellular vacuolization and lysosomal enzyme activity that apparently leads to an increase in intracellular digestive processes, including autolysis, most likely as a result of a host's attempt to maintain its nutritional requirements under parasitic stress (Moore and Halton 1973). Furthermore, it leads to a significant decrease of up to 80% of the lipids reserves in the digestive gland of the snail host from early infection onwards, which was associated to a mobilization of lipid energy reserves to compensate for the deficiency of carbohydrates as they are used by *F. hepatica*, and to the direct accumulation of the lipid content by the different parasitic larvae (Humiczewska and Rajski 2005). An increase of 38% in the free fatty acid content probably correlated to a decrease of 56% and 52% in the monoglyceride and the triglyceride fractions has been observed in the hepatopancreas of *F. hepatica*-infected *G. truncatula* compared to uninfected snails, which was linked to a direct increase of hydrolysis of host's triglycerides and diglycerides. Moreover, phospholipids were reduced by 13% compared to uninfected snails, with the most significant change being the 80% decrease of the phosphatidyl ethanolamine fraction (Southgate 1970).

On the other hand, thermo-behavioral study of *L. stagnalis* shedding *Notocotylus attenuatus* cercariae showed an induced anapyrexia that contrasted with the behavior of non-infected snails, which preferred the warmer end of the thermal gradient for longer periods of time. It was speculated that such behavioral change favored parasite transmission as snail hosts kept at a decreased temperature present increased life span and release significantly more cercariae (Żbikowska and Żbikowski 2015). In the case of *Diplostomum pseudospathaceum* infection, it has been suggested that, even though parasitized *L. stagnalis* snails display fewer feeding events per individual than their uninfected counterparts, they are more active in the presence of a fish predator (parasite's second intermediate host) than in the absence of other non-fish predators. As *D. pseudospathaceum* cercariae have low motility and a short infective span once release in the water, this change of the foraging pattern of the infected snail enhances the success of transmission of the parasite to its next host, whether it is a direct result of parasite's manipulation or a by-product of infection (Voutilainen 2010).

One of the most studied examples of parasite-mediated regulation of the neuroendocrine system of the host is that of *T. szidati* when infecting *L. stagnalis*; its effect resulting in the castration of infected hosts (Schallig et al. 1991). Schistosomin, a host neuropeptide elicited upon infection (Schallig et al. 1991), has been strongly associated to physiological and metabolic effects observed on parasitized snails. They are featured by the inhibition of the bioactivity of female gonadotropic hormones, i.e. the caudodorsal cell hormone, calfluxin, and dorsal body hormone, via the antagonizing effect of schistosomin on target organs, the changes in the electrophysiological properties and the release growth- and reproduction-controlling peptides by neuroendocrine centers, and the changes in the gene expression in the central nervous system of infected *L. stagnalis* snails (reviewed in de Jong-Brink et al. 1997).

iii. **On the life history traits**

Energy constrains during infection affect host's life history traits, as both, the host and the pathogen attempt to use the same limited resources to maximize their own growth, reproduction, and survival. At the light of host–parasite interactions, host's traits must be interpreted in terms of the life history strategies of both, the parasite and of the host (Sorensen and Minchella 2001).

In rediae-producing infections, the strongest negative effect on host survival usually occurs during the prepatent period, whilst infections characterized by the development of mother and daughter sporocysts generally decrease host survival during patency (Sorensen and Minchella 2001). Moreover, survival usually decreases as the infective dose or the number of successive exposures increases (Rondelaud and Barthe 1982; Alba et al. 2018). In any case, although survival of the snail host can be significantly affected by infection (see (Sorensen and Minchella 1998; Gutiérrez et al. 2000, 2002; Salazar et al. 2006; Muñoz-Antoli et al. 2007) for examples on parasitized lymnaeid snails), when analyzing this parameter it should be considered that the degree of damage caused by the parasite is kept in check by the requirement that the host survives long enough until the time comes for the pathogen to be transmitted (Sorensen and Minchella 1998; Żbikowska 2011). For instance, in one study, infection of *L. stagnalis* by *T. ocellata* resulted in an increase of the life span of infected snails; 90% of the exposed group survived at week 28 of age when all uninfected snails were already dead (McClelland and Bourns 1969). In addition, snail age and host density also appear to affect host survival as well as compatibility. Death observed during mid and latter infections presumably indicates excessive energy demands or tissue damage imposed by the parasite, whilst mortality of exposed snail, especially during early infection may reflect compatibility issues (Sorensen and Minchella 2001). Therefore, survival has been used as an additional element to depict gradients within compatible host–parasite systems (Alba et al. 2018; Vázquez et al. 2019). The dynamic balance allowing parasite's exploitation of host resources without severely endangered critical resources for host's survival, at least before transmission is possible, generally occurs at the expenses of other hosts traits such as reproduction (Sorensen and Minchella 2001).

Reduction in the reproductive parameters of the host can result from parasitism. The entomopathogenic nematode *H. baujardi* negatively affects oviposition, the number of eggs and eggs masses per snail, and the hatching rate of infected *P. columella*, significantly impairing reproduction (Tunholi et al. 2017). Some studies have shown that infection of *P. columella* by the digeneans trematodes *F. magna* and *F. hepatica* is accompanied by a reduction in egg mass production and by a decrease in the number of eggs per egg mass (Salazar et al. 2006; Pankrác et al. 2016). In *G. cubensis–F. hepatica*, infected snails showed lower fecundity parameters than non-exposed individuals (Gutiérrez et al. 2000) whereas, egg laying progressively declined during infection of *S. palustris* snails by *E. revolutum* until complete castration occurred (Zischke 1972). During experimental infection of *Lymnaea olulla* by *F. gigantica*, a consistently lower levels of egg and embryo production was observed throughout a 7 weeks observation period in parasitized snails (Kumkate et al. 2009). A fecundity compensation can occur with parasitic castrators in which large numbers of eggs are produced by infected individuals before a significant impairment of oviposition ensues (Schallig et al. 1991; Sorensen and Minchella 2001; Gutiérrez et al. 2002). In any case, the ultimate reduction in the reproductive outputs of parasitized snails may be directly linked to the redirection of limited host resources to better support infection, as in *L. stagnalis–T. ocellata/T. szidati* systems (McClelland and Bourns 1969; Schallig et al. 1991), or to a direct result of biochemical, histopathological, and/or mechanical damage caused by the development of the parasite in the snail's reproductive organs, as observed in *G. truncatula* when heavily infected with *F. hepatica* (Wilson and Denison 1980). Regarding the first cause, there is debate concerning whether parasite-mediated fecundity reduction reflects "strategy" directed by the host or the parasite. However, parasitic castrators that actively divert host resources away from reproduction can catalyze another phenomenon: enhanced growth of infected snails (Sorensen and Minchella 2001; Hall et al. 2007). Thus, it is not surprising that examples of both, a significant decrease of reproductive traits and an increase of growth rate and shell size in parasitized snails have been observed, e.g. *L. stagnalis–T. szidati/D. pseudospathaceum* (Schallig et al. 1991; Seppälä et al. 2013), *L. elodes–E. revolutum/Plagiorchis elegans* (Sorensen and Minchella 1998; Zakikhani and Rau 1999). The overall increase on the shell length on infected snails may vary with factors like food availability (see Sorensen and Minchella 2001), the infective dose or the age of the snail host (Zakikhani and Rau 1999). Conversely to the previous examples of gigantism in infected hosts, it has been reported that the infection of *Peregriana peregra* by *Echinostoma friedi* not only produces a total parasitic castration but also reduction of normal development (stunting) of infected snails (Muñoz-Antoli et al. 2007).

Independently of reproductive constrains, evidence of parasitism by digenean trematodes on the size and shape of lymnaeid snails has been observed in both, field and experimental studies (see Sorensen and Minchella 2001 for review). A decrease of the growth rate during early and mid-infection and a reduction of the rate of shell formation in young snails have been reported in *E. revolutum*-infected *S. palustris* in association with the parasite burden (Zischke 1972). In addition, it has been observed that field-occurring *L. stagnalis* snails infected with *Echinoparyphium aconiatum*, *E. revolutum*, *D. pseudospathaceum*, and *Opisthioglyphe ranae* differed in shell shape compared to uninfected individuals; a slenderer shell, with a more elongate spiral was recorded in the former group and suggested to be associated with infection (Żbikowska and Żbikowski 2005).

11.5 How Can the Environmental Factors Affect the Infection Outcome of Lymnaeid Snails: Parasite Interaction?

Host–parasite interactions result from the conjunction of the interplay between biological factors (encounter filter and compatibility filter), which are primarily determined by the genetic and epigenetic structure of each partner, and a plethora of environmental factors, as changes in environmental variables (e.g., temperature, droughts, pollution, co-infections) may affect the exposure to parasites and the outcome of the infection, even in highly compatible host–parasite systems (Sandland and Minchella 2003; Wolinska and King 2009). It is important to note, that the same factor could have positive and negative effects on the different encounter and compatibility filters and thus, it is the final balance between the multiple influences what will decide the faith or the extent of the contact/infection. In this sense, environmental factors can affect crucial features of parasite's infectivity and/or survival (e.g., Rondelaud and Barthe 1980b; Ford et al. 1998; Morley et al. 2007) or the immunological and/or life history traits in the host (e.g., Sandland and Minchella 2003; Leicht et al. 2013), all of which could modify encounter and compatibility filters and the ultimate phenotype resulting from exposure to the pathogen.

In addition, ecological features of the host population (e.g., density, abundance, aggregation, demographic structure, population genetics), by themselves, are also significant influencers of encounter and compatibility filters. For instance, although parasitism, in general, should have a negative effect on host population density, a positive correlation between *L. stagnalis* density and *Diplostomum* sp. prevalence was reported as the parasite was fairly common in *L. stagnalis* populations in small lakes but rare in large lakes (Voutilainen et al. 2009). The demographic structure of the host population can also influence contact and compatibility events with parasites; adult snails showing higher sizes could be better in attracting parasites as they may produce higher amounts of chemical clues (Langeloh and Seppälä 2018) although, contrastingly, susceptibility to infection usually decreases with age (e.g., Zakikhani and Rau 1999). Moreover, the high transmission of *F. hepatica* through *G. truncatula* snails in the Bolivian Altiplano has been related to a depleted genetic diversity within the host population featured by the occurrence of a single and unique genotype that is highly compatible with the parasite (Meunier et al. 2001). Similar as to what occurred at the individual level, the ecological features of host populations are also constantly and ultimately influenced and regulated by a plethora of environmental factors and their dynamics.

i. **Abiotic factors**

The degree of openness of the encounter filter ultimately depends on host availability which is determined by environmental factors that define the temporal and spatial occurrence of a given host–parasite system, as well as its meeting rates (Detwiler and Minchella 2009). Factors such as altitude (Walker et al. 2008), pH and water hardness (Alba et al. 2019b) have been reported as significant in determining the occurrence and habitat preference of different lymnaeid snails. Furthermore, certain seasonal patterns of transmission have been informed for different lymnaeid-borne trematodes species in correspondence to fluctuations of abiotic factors such as rainfall pattern, temperature, water flow/level, and soil evapotranspiration driving variations on parasite's infectivity and/or intermediate hosts populations (e.g., Luzón-Peña et al. 1995; Yurlova et al. 2006; Żbikowska et al. 2006). In this sense, although many lymnaeid species show resilience to desiccation, estivation of infected snails during drought can result in a significant degeneration of the intramolluscan parasitic larvae most likely related to an increased stress and to the negative effect on host nutrition, as observed in *G. truncatula–F. hepatica* (Rondelaud and Barthe 1980b). In addition, water with high turbidity levels negatively influence host-finding capacity of *F. hepatica* miracidium (Christensen 1980), whereas the oxygen content and the pH of the water were identified to be associated with the probability of infection by trematodes of field-occurring *Radix auricularia* and *L. stagnalis* (Soldánová et al. 2010). Negative and positive correlations between the mean abundance and the water level have been reported for *Cotylurus cornutus* and *Molinella anceps* trematodes in *L. stagnalis* inhabiting a Siberian lake (Yurlova et al. 2006). An increased risk of infection of *G. truncatula* by *F. hepatica* has been associated with the type of habitat; higher risk was related to snail populations originating from spring swamps, wells, and reeds compared to populations from streams (Schweizer et al. 2007).

The influence of temperature on hosts, parasites and their interplay has been studied from different points of view, particularly in the case of ectothermic hosts such as snails. Overall, an increase of temperature within the optimal range usually favors higher metabolic rates and physiological process such as the motility of free-living parasitic stages (Morley et al. 2007; Selbach and Poulin 2020), and the growth rate and fecundity of the snail host (Vaughn 1953; Leicht et al. 2013). However, a decrease or elevation of the temperature away from the biological range of tolerance for each partner will produce negative effects on host's life traits, as well as on parasite's infectivity and survival (e.g., Vaughn 1953; Luzón-Peña et al. 1995; McCarthy 1999). Furthermore, intramolluscan colonization, development and further transmission of the parasite are also influenced by temperature (Wilson and Draskau 1976; Abrous et al. 1999; Morley et al. 2007) in a species-specific manner (see Selbach and Poulin 2020) for examples among trematodes). In this sense, immune responsiveness of the host is another parameter heavily affected by changes in this environmental factor. For instance, an increased transcription of the cytokine MIF was recorded in *L. stagnalis* snails exposed to high temperature (Seppälä et al. 2021), whereas a long heat wave (i.e., for more than 7 days) was found to decrease hemocyte concentration and PO-like activity (Leicht et al. 2013). It has been referred that phagocytosis in this snail species is optimal between 20–25 ° C range (van der Knaap et al. 1983b). An increased susceptibility of *L. stagnalis* snails to *E. aconiatum* was observed when the host population was exposed to a heat wave for 7 days, in association to the impairment of host's immune reactiveness known to occur under such treatment (Leicht and Seppälä 2014).

Chemical pollution, particularly those associated to human practices such as agriculture, can also have potential effects on parasite transmission. Several studies have experimentally assessed the influence of toxicants on the biology of the lymnaeid snails. A myriad of effects on their life history traits and population genetic structure (see Coutellec et al. 2013; Amorim et al. 2019 for examples on *Lymnaea stagnalis*), on the IDS and snail immunoreactivity (e.g., Russo and Lagadic 2000; Russo et al. 2008), and on host–parasite interactions, in particular (e.g., in *R. natalensis–F. gigantica* model, see Soliman 2009); *P. peregra*/*L. stagnalis–Diplostomum spathaceum*, see Morley et al. 2005), have been recorded. Others discuss about a potential relation between the use of herbicides and chemical pollutants on food crops and the punctual decline of lymnaeid snails' density in the field (Nguyen et al. 2017). Moreover, it has been demonstrated that pollution by the herbicide Paraquat of the aquatic environment not only adversely affects *R. natalensis* metabolism but also resulted in a significant reduction of the infectivity of *F. gigantica* miracidia to the snail (Bakry et al. 2016).

ii. **Biotic factors**

Biotic factors add complexity to host–parasite interplays as other organisms, co-occurring with the system, can significantly influence encounter and compatibility filters and transmission success through a variety of interactions (predation, competence, dilution effect, decoy effect, invasive species, etc.). This implies that biodiversity, in general, should have significant effects on parasite transmission (Thieltges et al. 2008). In this sense, food resources have proved significant in host–parasite interactions as factors like the quantity and quality of the diet may significantly affect the encounter filter and can modify the physiological characteristics and conditions of the host. For instance, strong long-term, but not short-term food deprivation reduced *L. stagnalis* chemical attractiveness to *E. aconiatum* cercariae and thus, may negatively affect the probability of encounter (Seppälä and Leicht 2015; Langeloh and Seppälä 2018). Furthermore, starvation can directly modify the IDS and in consequence, the immune reactivity and the ultimate compatibility of the host. In this sense, food deprivation downregulates the transcription of an M-type lectin, and GST, and enhances the transcription of MIF, LBP/BPI and lysozyme in *L. stagnalis* (Seppälä et al. 2021). Concerning hemocytes, those derived from 14 days-starved *L. stagnalis* snails phagocytized less avidly than cells of individuals fed ad libitum and showed significantly higher titers of agglutinin in the hemolymph (van der Knaap et al. 1983b). Moreover, *E. revolutum*-infected *L. elodes* snails subject to a low-protein diet exhibited lower survival than infected snails fed on a high-protein diet and thus, influencing transmission to the next host (Sandland and Minchella 2003). Food deprivation constrained the overall parasite within-host reproduction in *L. stagnalis–D. pseudospathaceum* system, as the release of parasite transmission stages was reduced (Seppälä et al. 2015). In *G. truncatula*, the cercarial production of *F. hepatica*, *F. gigantica*, and *Calicophoron daubneyi* varied depending on the different food sources used (Rondelaud et al. 2002a).

In addition, an increased diversity and abundance of co-occurring non-host species could reduce the risk of contact and further infection of suitable snails, a phenomenon commonly known as "dilution effect" (Civitello et al. 2015) where non-host species can disrupt parasite–host encounter via several mechanisms like predation of hosts and of free-living stages of parasites (Johnson et al. 2010), or by acting as decoys attracting the pathogen (Marszewska et al. 2018). In particular, besides directly reducing the abundance of the host, it was observed through experimental infections of *G. truncatula* with *F. hepatica* that the stress caused by the presence of predators (sciomyzid larvae or zonitid snails) in the host during infection significantly influences survival of parasitized snails; the duration of the cercarial shedding and the number of emitted cercariae were reduced compared to controls (infected snails not subjected to predators; see Rondelaud et al. 2002b).

As each snail host presents limited nutrient and spatial resources, a structuration of the parasite community, driven by complex interactions among infecting pathogens (particularly trematodes) related to competition, facilitation, or predation, can occur within the host, during a multi-infection process (Lie et al. 1973; Abrous et al. 1996; Soldánová et al. 2012). For instance, infection synergism between *C. daubneyi* followed by *F. hepatica*, and between *Trichobilharzia brevis* followed by *Echinostoma hystricosum* has being reported when successive cross-exposures, in the given order, were performed in *G. truncatula* or *L. rubiginosa*, as higher infection rates were attained compared to each mono species-infection (Lie et al. 1973; Augot et al. 1996). Conversely, if *T. brevis* followed a primary infection by *E. hystricosum*, the predatory rediae of *E. hystricosum* consumed *T. brevis* sporocysts impairing the establishment of the second parasite (Lie et al. 1973). A field study on the trematode infracommunity of *L. stagnalis* reported two pairs of trematodes, i.e. *P. elegans/T. szidati* and *M. anceps/Notocotylus attenuatus*, that were more frequently found infecting in combination than would be expected by chance suggesting facilitation. Moreover, competitive relationships among trematode species were hypothesized as a dominance hierarchy in which seven top dominant species with putatively similar competitive abilities reduced the prevalence of the other trematode species found (which develop in sporocysts only; see Soldánová et al. 2012).

11.6 What Can the Study of Interaction Between Lymnaeid Snails and Their Parasite Bring to Evolutionary Biology and Disease Control, and Where Do We Go from Here?

Since some snail-borne parasites cause significant diseases in humans and domestic animals, the study of the host snails and of their specific interaction may have a great impact on veterinary and human health sciences. In particular, given that it is mandatory for digenean parasites to develop inside their snail hosts, controlling the populations of the latter is a suitable strategy for directly tackling snail-borne trematodiasis. However, the mechanisms and factors promoting the emergence or re-emergence of these diseases are still poorly understood making any predictions on their establishment and spreading, in a context of environmental changes, and successfully preventing their propagation, particularly challenging. In this sense, a thorough look into snail–parasite interaction is mandatory (Adema et al. 2012). Aspects such as the factors determining the population structure and geographical distribution of snails and thus, possible transmission foci, the effects of environmental perturbations on the distribution and transmission of parasites through their vector snails, and the extent of the influence of encounter and compatibility filters between hosts and parasites versus the ecological factors that limit infections, are among the significant questions that can be solved by investigating the biology of snail–parasite interactions. Moreover, the deep study of the immunobiological features of each particular digenean–mollusk system could allow to identify molecules with potential biomedical relevance, and could also serve as model to study the impact of pollutants or other stressors on immune function, to monitor climate change and to record its impact, or ultimately to predict their effects on parasite transmission and vectorial competence (Loker 2010). The knowledge on how immunity and virulence have evolved to influence the snail–parasite compatibility or which are the genetic markers determining such compatibility or the regulators of parasite transmission in natural snail populations can only be attained through more investigations encompassing snail–parasite dynamics (Adema et al. 2012). They will also bring light on evolutionary and environmental immunology.

Nevertheless, in spite of the crucial contribution of this field to move securely toward the control of human trematodiasis or a major understanding of invertebrate's biology and immunology, and the progresses achieved, significant knowledge gaps remain, specially within Lymnaeidae–parasites systems. Encounter filter and its association to susceptibility/virulence of host and parasites needs to be deeply characterized under experimental as well as natural conditions. The role of microbiota in lymnaeid snails–parasite dynamics has been largely overlooked, although the host genome and the microbiota meta-genome are a real unit of selection, one that is not only associated to the host genotype and/or physiological status but also to environmental variations, and that affects host plasticity and adaptive responses (Zilber-Rosenberg and Rosenberg 2008). Therefore, identifying microbial communities more or less prone to modify parasite transmission in lymnaeid host would have critical consequences in terms of snail control strategies for fighting Lymnaeidae-borne diseases in the field and to understand evolutionary dynamics. Furthermore, and despite the advances, our knowledge of PRRs, receptor–ligand interactions, downstream mechanisms of signaling transduction and molecular and mechanistic regulators, humoral effectors and defense reactions to convey an appropriate immune response is currently poor in snails in general, and in the family Lymnaeidae in particular. Therefore, it warrants further investigations by multi-"omics" methods and functional validations as it is crucial for depicting the (epi)genetic and molecular bases of the defense responses. In this sense, although RNA interference is a powerful transient gene-expression repression system widely employed in snails including the Lymnaeidae (e.g., Korneev et al. 2002; Lu and Feng 2011), the potential of the CRISPR-Cas9 gene editing technique offers a wider variety of applications (from complete gene knock-out to gene editing), appears to be a more determinant approach that it is already under development to generate transgenic mollusc (Perry and Henry 2014). A recent report demonstrates its successful application for the genome editing of *L. stagnalis* (Abe and Kuroda 2019) and opens up significant opportunities for functional genomics to investigate the role of specific genes in snail immunobiology. Moreover, the lack of molecular markers to properly classify hemocytes, its ontogeny and functional differences deserves further efforts in order to properly elucidate the different lineages or types, and their specific role during infection dynamics. In-depth characterization of hemocyte populations and their interaction with trematodes to discriminate transcriptional infra-populations, trajectories of differentiation and specific genes related to each cellular stage/lineage can now be performed using Single-Cell RNA sequencing technology (scRNA-seq; see Fu et al. 2020).

It is important to note, that most of the immunobiological information concerning lymnaeid snails comes primarily from the study of a single species, the pond snail *L. stagnalis*. Nonetheless, as the response of different species of snails to parasitic infections is greatly varied, the information about one snail host/parasite model cannot be strictly applied to other systems. Hence, it is crucial for future investigations to encompass other species in order to widen up the view of the IDS of the Lymnaeidae and its interaction with pathogens. In parallel, further characterization of virulence factors of pathogens affecting lymnaeid snails should be carried out in order to completely elucidate the evolutionary bases of each host–parasite interaction. Topics such as the specificity of the defense response (e.g., (van der Knaap et al. 1983a; Seppälä et al. 2021), immune memory (van der Knaap et al. 1983a), immune priming (Alba et al. 2018), and transgenerational priming (Vorontsova et al. 2019) have been only modestly embraced within the Lymnaeidae and further efforts should be directed toward them.

In addition, investigations of the effects of host–parasite interplay on lymnaeid snail physiology, metabolism and ultimately, on life history traits must be emphasized and approached by integrative studies. Lastly, as the Anthropocene imposes significant challenges for environmental dynamics, it is pivotal to develop a more mechanistic understanding of how anthropic disturbances and environmental degradation, habitat transformation, biodiversity loss and invasive species, and climate change will affect hosts and parasites communities and their interplay if we aimed at predicting changes in transmission dynamics and disease epidemiology. In this sense, field studies, although difficult to carried out, should be particularly stimulated not only with the finality of building up basic knowledge on host–parasite–environment interplay, but also of exploring ways to practically exploit snail-based control procedures of parasite transmission. Vector mediation by transgenesis strategies either using introgression of selected snail genotypes refractory to trematodiasis infection into, or direct genetic modification by CRISPR-Cas9 (transfection with a multi-resistant gene) of natural populations might be envisioned for increasing snail resistance, reducing their vectorial competence and tackling pathogen transmission. Profiting on snail's microbiota for potentially controlling snail-borne diseases, either through chemical modification or using paratransgenesis solutions (using transgenic bacteria expressing foreign gene products; see Coutinho-Abreu et al. 2010) to favor snail's microbial communities rendering the host more refractory or resistant to parasite infection, could be also projected.

The incorporation of additional research efforts, new competences, technology, and funding is as challenging as it is essential since global attention is traditionally diverted toward other host–parasite systems mostly driven by the notoriety of other vectors and diseases in global health agendas. The scarce resources and interest devoted to tackle snails and the diseases they transmit should be always considered, as it is a major limitation for exploring related scientific and practical opportunities. In addition, the list of scientists working on snails and the diseases they transmit continues to grow thinner (Adema et al. 2012). Nevertheless, it must be brought forward that the variety of models to study, the exciting investigations and discoveries that lay ahead, the applicability of such knowledge in different disciplines and their usefulness for science and society, present to scientist and decision-makers an attractive road deserving to be explored.

References

Abe M, Kuroda R (2019) The development of CRISPR for a mollusc establishes the formin *Lsdia1* as the long-sought gene for snail dextral/sinistral coiling. Development 146:175976. https://doi.org/10.1242/dev.175976

Abrous M, Rondelaud D, Dreyfuss G (1996) *Paramphistomum daubneyi* and *Fasciola hepatica*: the effect of dual infection on prevalence and cercarial shedding in preadult *Lymnaea glabra*. J Parasitol 82:1026–1029

Abrous M, Rondelaud D, Dreyfuss G (1999) Influence of low temperatures on the cercarial shedding of *Paramphistomum daubneyi* from the snail *Lymnaea truncatula*. Parasite 6:85–88. https://doi.org/10.1051/parasite/1999061085

Adema CM, van Deutekom-Mulder EC, van der Knaap WP et al (1991) Generation of oxygen radicals in hemocytes of the snail *Lymnaea stagnalis* in relation to the rate of phagocytosis. Dev Comp Immunol 15:17–26

Adema CM, van Deutekom-Mulder EC, van der Knaap WP et al (1993) NADPH-oxidase activity: the probable source of reactive oxygen intermediate generation in hemocytes of the gastropod *Lymnaea stagnalis*. J Leukoc Biol 54:379–383. https://doi.org/10.1002/jlb.54.5.379

Adema CM, Mohandas A, van der Knaap WP et al (1994a) Separation of *Lymnaea stagnalis* hemocytes by density gradient centrifugation. Dev Comp Immunol 18:25–31

Adema CM, van Deutekom-Mulder EC, Van der Knaap WP et al (1994b) Schistosomicidal activities of *Lymnaea stagnalis* hemocytes: the role of oxygen radicals. Parasitology 109:479–485

Adema CM, Bayne CJ, Bridger JM et al (2012) Will all scientists working on snails and the disease they transmit please stand up? PLoS Negl Trop Dis 6:e1835. https://doi.org/10.1371/journal.pntd.0001835

Alba A, Hernández H, Marcet R et al (2014) Exploring the antigenic features of *Fasciola hepatica* (Trematoda: Digenea) rediae through the evaluation of three different antigenic candidates for further monoclonal antibody development. Parasitol Res 113:3185–3193

Alba A, Vázquez AA, Sánchez J et al (2018) Fasciola hepatica - Pseudosuccinea columella interaction: effect of increasing parasite doses, successive exposures and geographic origin on the infection outcome of naturally-resistant and susceptible snails from Cuba. Parasite Vector 11:559. https://doi.org/10.1186/s13071-018-3155

Alba A, Tetrau G, Chaparro C et al (2019a) Field-occurring resistance to *Fasciola hepatica* (Trematoda) in *Pseudosuccinea columella* snails: a review from literature and insights from comparative "omic" analyses. Dev Comp Immunol 101:103463. https://doi.org/10.1016/j.dci.2019.103463

Alba A, Vázquez AA, Sánchez J et al (2019b) Patterns of distribution, population genetics and ecological requirements of field-occurring resistant and susceptible *Pseudosuccinea columella* snails to *Fasciola hepatica* in Cuba. Sci Rep 9:14359. https://doi.org/10.1038/s41598-019-50894-7

Alba A, Duval D, Sánchez J et al (2020) The immunobiological interplay between *Pseudosuccinea columella* resistant/susceptible snails with *Fasciola hepatica*: hemocytes in the spotlight. Dev Comp Immunol 102:103485. https://doi.org/10.1016/j.dci.2019.103485

Aldhoun J, Kment P, Horak P (2016) Historical analysis of the type species of the genus *Trichobilharzia* Skrjabin et Zakharov, 1920 (Platyhelminthes: Schistosomatidae). Zootaxa 4084:593–595. https://doi.org/10.11646/zootaxa.4084.4.10

Allam B, Espinosa E (2015) Mucosal immunity in mollusks. In: Beck BH, Peatman E (eds) Mucosal health in aquaculture. Academic Press, Amsterdam, pp 325–370. https://doi.org/10.1016/B978-0-12-417186-2.00012-1

Allan F, Dunn AM, Emery AM, Stothard JR et al (2013) Use of sentinel snails for the detection of *Schistosoma haematobium* transmission on Zanzibar and observations on transmission patterns. Acta Trop 128:234–240

Allan ER, Tennessen JA, Sharpton TJ et al (2018) Allelic variation in a single genomic region alters the microbiome of the snail *Biomphalaria glabrata*. J Hered 109:604–609. https://doi.org/10.1093/jhered/esy014

Amorim J, Abreu I, Rodrigues P et al (2019) *Lymnaea stagnalis* as a freshwater model invertebrate for ecotoxicological studies. Sci Total Environ 669:11–28

Augot D, Abrous M, Rondelaud D et al (1996) *Paramphistomum daubneyi* and *Fasciola hepatica*: the redial burden and cercarial shedding in *Lymnaea truncatula* subjected to successive unimiracidial cross-exposures. Parasitol Res 82:623–627

Baeza-Garcia A, Pierce RJ, Gourbal B et al (2010) Involvement of the cytokine MIF in the snail host immune response to the parasite *Schistosoma mansoni*. PLoS Pathog 6:e1001115. https://doi.org/10.1371/journal.ppat.1001115

Bakry FA, Eleiwa ME, Taha SA et al (2016) Comparative toxicity of Paraquat herbicide and some plant extracts in *Lymnaea natalensis* snails. Toxicol Ind Health 32:143–153. https://doi.org/10.1177/0748233713498457

Boisseaux P, Delignette-Muller ML, Abbaci K et al (2016) Analysis of hemocytes in *Lymnaea stagnalis*: characterization and effects of repeated hemolymph collections. Fish Shellfish Immunol 57:116–126

Bouétard A, Noirot C, Besnard AL et al (2012) Pyrosequencing-based transcriptomic resources in the pond snail *Lymnaea stagnalis*, with a focus on genes involved in molecular response to diquat-induced stress. Ecotoxicology 21:2222–2234. https://doi.org/10.1007/s10646-012-0977-1

Brown DS (1978) Pulmonate molluscs as intermediate hosts for digenetic trematodes. In: Fretter V, Peake J (eds) Pulmonates, vol 2A. Academic Press, London, pp 287–333

Calienes AF, Fraga J, Pointier JP et al (2004) Detection and genetic distance of resistant populations of *Pseudosuccinea columella* (Mollusca: Lymnaeidae) to *Fasciola hepatica* (Trematoda: Digenea) using RAPD markers. Acta Trop 92:83–87

Celi-Erazo M, Alda P, Montenegro-Franco M et al (2020) Prevalence of *Fasciola hepatica* infection in *Galba cousini* and *Galba schirazensis* from an Andean region of Ecuador. Vet Parasitol Reg Stud Rep 20:100390. https://doi.org/10.1016/j.vprsr.2020.100390

Cerenius L, Söderhäll K (2004) The prophenoloxidase-activating system in invertebrates. Immunol Rev 198:116–126. https://doi.org/10.1111/j.0105-2896.2004.00116.x

Christensen NO (1980) A review of the influence of host- and parasite-related factors and environmental conditions on the host-finding capacity of the trematode miracidium. Acta Trop 37:303–318

Christensen NO, Nansen P, Frandsen F (1976) Molluscs interfering with the capacity of *Fasciola hepatica* miracidia to infect *Lymnaea truncatula*. Parasitology 73:161–167. https://doi.org/10.1017/s0031182000046849

Civitello DJ, Cohen J, Fatima H et al (2015) Biodiversity inhibits parasites: broad evidence for the dilution effect. Proc Natl Acad Sci USA 28:8667–8671

Combes C (2001) Parasitism. The ecology and evolution of intimate interactions. University of Chicago Press, Chicago, IL

Coustau C, Gourbal B, Duval D et al (2015) Advances in gastropod immunity from the study of the interaction between the snail *Biomphalaria glabrata* and its parasites: a review of research progress over the last decade. Fish Shellfish Immunol 46:5–16

Coutellec MA, Besnard AL, Caquet T (2013) Population genetics of *Lymnaea stagnalis* experimentally exposed to cocktails of pesticides. Ecotoxicology 22:879–888

Coutinho-Abreu IV, Zhu KY, Ramalho-Ortigao M (2010) Transgenesis and paratransgenesis to control insect-borne diseases: current status and future challenges. Parasitol Int 59:1–8. https://doi.org/10.1016/j.parint.2009.10.002

Coyne K, Laursen JR, Yoshino TP (2015) In vitro effects of mucus from the mantle of compatible (*Lymnaea elodes*) and incompatible (*Helisoma trivolvis*) snail hosts on *Fascioloides magna* miracidia. J Parasitol 101:351–357

de Jong-Brink M, Bergamin-Sassen M, Solis Soto M (1997) Multiple strategies of schistosomes to meet their requirements in the intermediate snail host. Parasitology 123:S129–S141

Detwiler JT, Minchella DJ (2009) Intermediate host availability masks the strength of experimentally-derived colonisation patterns in echinostome trematodes. Int J Parasitol 39: 585–590

Dida GO, Gelder FB, Anyona DN et al (2014) Distribution and abundance of schistosomiasis and fascioliasis host snails along the Mara River in Kenya and Tanzania. Infect Ecol Epidemiol 4(1): 24281. https://doi.org/10.3402/iee.v4.24281

Dikkeboom R, Van der Knaap WP, Meuleman EA et al (1985) A comparative study on the internal defence system of juvenile and adult *Lymnaea stagnalis*. Immunology 55:547–553
Dikkeboom R, Tunagel JMGH, Mulder EC et al (1987) Hemocytes of the pond snail *Lymnaea stagnalis* generate reactive forms of oxygen. J Invert Pathol 49:321–331
Ehara T, Kitajima S, Kanzawa N et al (2002) Antimicrobial action of achacin is mediated by l-amino acid oxidase activity. FEBS Lett 531:509–512
Esch GW, Barger MA, Fellis J (2002) The transmission of digenetic trematodes: style, elegance, complexity. Integr Comp Biol 42:304–312. https://doi.org/10.1093/icb/42.2.304
Ford DM, Nollen PM, Romano MA (1998) The effects of salinity, pH and temperature on the half-life and longevity of *Echinostoma caproni* miracidia. J Helminthol 72:325–330
Franchini A, Ottaviani E (2000) Repair of molluscan tissue injury: role of PDGF and TGF-beta1. Tissue Cell 32:312–321. https://doi.org/10.1054/tice.2000.0118
Fu Y, Huang X, Zhang P et al (2020) Single-cell RNA sequencing identifies novel cell types in *Drosophila* blood. J Genet Genomics 47:175–186. https://doi.org/10.1016/j.jgg.2020.02.004
Galinier R, Portela J, Moné Y et al (2013) Biomphalysin, a new β pore-forming toxin involved in *Biomphalaria glabrata* immune defense against *Schistosoma mansoni*. PLoS Pathog 9: e1003216. https://doi.org/10.1371/journal.ppat.1003216
Georgieva K, Georgieva L, Mizinska-Boevska Y et al (2016) Study of surface carbohydrates in *Galba truncatula* tissues before and after infection with *Fasciola hepatica*. Mem Inst Oswaldo Cruz 111:475–483
Gerhardus MJT, Baggen JMC, van der Knaap WP et al (1991) Analysis of surface carbohydrates of *Trichobilharzia ocellata* miracidia and sporocyst using lectin binding techniques. Parasitology 103:51–59
Goodall CP, Bender RC, Broderick EJ et al (2004) Constitutive differences in Cu/Zn superoxide dismutase mRNA levels and activity in hemocytes of *Biomphalaria glabrata* (Mollusca) that are either susceptible or resistant to *Schistosoma mansoni* (Trematoda). Mol Biochem Parasitol 137: 321–328
Guo F, Huang ZB, Huang MQ et al (2009) Effects of small abalone, *Haliotis diversicolor*, pedal mucus on bacterial growth, attachment, biofilm formation and community structure. Aquaculture 293:35–41
Gutiérrez A, Perera G, Yong M et al (2000) Life-history traits of *Fossaria cubensis* (Gastropoda: Lymnaeidae) under experimental exposure to *Fasciola hepatica* (Trematoda: Digenea). Mem Inst Oswaldo Cruz 95:747–752
Gutiérrez A, Perera G, Yong M et al (2002) *Fasciola hepatica* (Trematoda: Digenea): its effect on the life history traits of *Pseudosuccinea columella* (Gastropoda: Lymnaeidae), an uncommon interaction. Parasitol Res 88:535–539
Gutiérrez A, Pointier JP, Yong M et al (2003) Evidence of phenotypic differences between resistant and susceptible isolates of *Pseudosuccinea columella* (Gastropoda: Lymnaeidae) to *Fasciola hepatica* (Trematoda: Digenea) in Cuba. Parasitol Res 90:129–134
Hall SP, Becker C, Cáceres CE (2007) Parasitic castration: a perspective from a model of dynamic energy budgets. Int Comp Biol 47:295–309
Hambrook JR, Gharamah AA, Pila EA et al (2019) *Biomphalaria glabrata* Granulin increases resistance to *Schistosoma mansoni* infection in several *Biomphalaria* species and induces the production of reactive oxygen species by Haemocytes. Genes 11:38. https://doi.org/10.3390/genes11010038
Hermann PM, Nicol JJ, Nagle GT et al (2005) Epidermal growth factor-dependent enhancement of axonal regeneration in the pond snail *Lymnaea stagnalis*: role of phagocyte survival. J Comp Neurol 492:383–400
Hoek RM, Smit AB, Frings H et al (1996) A new Ig-superfamily member, molluscan defence molecule (MDM) from *Lymnaea stagnalis*, is down-regulated during parasitosis. Eur J Immunol 26:939–944
Horák P (1995) Developmentally regulated expression of surface carbohydrate residues on larval stages of the avian schistosome Trichobilharzia szidati. Folia Parasitol 42:255–265

Horak P, Deme R (1998) Lectins and saccharides in *Lymnaea stagnalis* haemocyte recognition. Comp Haematol Int 8:210–218

Horák P, van der Knaap WP (1997) Lectins in snail-trematode immune interactions: a review. Folia Parasitol 44:161–172

Horák P, Neeleman AP, van der Knaap WP et al (1998) Effect of saccharides on plasma mediated haemagglutination and in vitro phagocytosis by haemocytes of the snail *Lymnaea stagnalis*. Comp Haematol Int 8:43–49

Humiczewska M, Rajski K (2005) Lipids in the host-parasite system: digestive gland of *Lymnaea truncatula* infected with the developmental stages of *Fasciola hepatica*. Acta Parasitol 50:235–239

Humphries JE, Yoshino TP (2003) Cellular receptors and signal transduction in molluscan hemocytes: connections with the innate immune system of vertebrates. Integr Comp Biol 43:305–312

Humphries JE, Yoshino TP (2008) Regulation of hydrogen peroxide release in circulating haemocytes of the planorbid snail *Biomphalaria glabrata*. Dev Comp Immunol 32:554–562

Hunter R (1980) Effects of grazing on the quantity and density of freshwater aufwuchs. Hydrobiologia 69:251–259. https://doi.org/10.1007/BF00046800

Huot C, Clerissi C, Gourbal B et al (2020) Schistosomiasis vector snails and their microbiota display a phylosymbiosis pattern. Front Microbiol 10:3092. https://doi.org/10.3389/fmicb.2019.03092

Iakovleva NV, Gorbushin AM, Storey KB (2006) Modulation of mitogen-activated protein kinases (MAPK) activity in response to different immune stimuli in haemocytes of the common periwinkle *Littorina littorea*. Fish Shellfish Immunol 21:315–324

Jian J, Konopka J, Liu C (2013) Insights into the role of progranulin in immunity, infection, and inflammation. J Leukoc Biol 93:199–208. https://doi.org/10.1189/jlb.0812429

Johnson PT, Dobson A, Lafferty KD et al (2010) When parasites become prey: ecological and epidemiological significance of eating parasites. Trends Ecol Evol 25:362–371

Kalbe M, Haberl B, Haas W (1997) Miracidial host-finding in *Fasciola hepatica* and *Trichobilharzia ocellata* is stimulated by species-specific glycoconjugates released from the host snails. Parasitol Res 83:806–812

Kalbe M, Bernhard H, Haas BJ (2000) Snail host finding by *Fasciola hepatica* and *Trichobilharzia ocellata*: compound analysis of "Miracidia-attracting glycoproteins". Exp Parasitol 96:231–242

Kaplan MJ, Radic M (2012) Neutrophil extracellular traps (NETs): double-edged swords of innate immunity. J Immunol 189:2689–2695. https://doi.org/10.4049/jimmunol.1201719

Kock KN, Wolmarans CT, Bornman M (2003) Distribution and habitats of the snail *Lymnaea truncatula*, intermediate host of the liver fluke *Fasciola hepatica*, in South Africa. J South Afr Vet Assoc 74:117–122

Korneev SA, Kemenes I, Straub V et al (2002) Suppression of nitric oxide (NO)-dependent behavior by double-stranded RNA-mediated silencing of a neuronal NO synthase gene. J Neurosci 22(11):RC227. https://doi.org/10.1523/JNEUROSCI.22-11-j0003.2002

Kryukova NA, Yurlova NI, Rastyagenko NM et al (2014) The influence of *Plagiorchis mutationis* larval infection on the cellular immune response of the snail host *Lymnaea stagnalis*. J Parasitol 100:284–287

Kumkate S, Onmek N, Boonburapong B et al (2009) Estrogen-related fecundity reduction of *Lymnaea ollula* following *Fasciola gigantica* infection. J Parasitol 95:1391–1396. https://doi.org/10.1645/GE-2080.1

Lacchini AH, Davies AJ, Mackintosh D et al (2006) beta-1, 3-glucan modulates PKC signalling in *Lymnaea stagnalis* defence cells: a role for PKC in H_2O_2 production and downstream ERK activation. J Exp Biol 209:4829–4840

Lange MK, Penagos-Tabares F, Muñoz-Caro T et al (2017) Gastropod-derived haemocyte extracellular traps entrap metastrongyloid larval stages of *Angiostrongylus vasorum*, *Aelurostrongylus abstrusus* and *Troglostrongylus brevior*. Parasites Vector 10:50. https://doi.org/10.1186/s13071-016-1961-z

Langeloh L, Seppälä O (2018) Relative importance of chemical attractiveness to parasites for susceptibility to trematode infection. Ecol Evol 8:8921–8929. https://doi.org/10.1002/ece3.4386

Le Clec'h W, Anderson TJC, Chevalier FD (2016) Characterisation of hemolymph phenoloxidase activity in two *Biomphalaria* snail species and impact of *Schistosoma mansoni* infection. Parasite Vector 9:32. https://doi.org/10.1186/s13071-016-1319-6

Leicht K, Seppälä O (2014) Infection success of *Echinoparyphium aconiatum* (Trematoda) in its snail host under high temperature: role of host resistance. Parasites Vector 7:192. https://doi.org/10.1186/1756-3305-7-192

Leicht K, Jokela J, Seppälä O (2013) An experimental heat wave changes immune defense and life history traits in a freshwater snail. Ecol Evol 3:4861–4871. https://doi.org/10.1002/ece3.874

Lie KJ, Lim HK, Ow-Yang CK (1973) Synergism and antagonism between two trematodes species in the snail *Lymnaea rubiginosa*. Int J Parasitol 3:729–733

Lockyer AE, Emery AM, Kane RA et al (2012) Early differential gene expression in haemocytes from resistant and susceptible *Biomphalaria glabrata* strains in response to *Schistosoma mansoni*. PLoS One 7:e51102. https://doi.org/10.1371/journal.pone.0051102

Loker ES (1979) Pathology and host responses induced by *Schistosomatium douthitti* in the freshwater snail *Lymnaea catascopium*. J Invert Pathol 33:265–273

Loker ES (2010) Gastropod immunobiology. Adv Exp Med Biol 708:17–43. https://doi.org/10.1007/978-1-4419-8059-5_2

Loker ES, Bayne CJ, Buckley PM et al (1982) Ultrastructure of encapsulation of *Schistosoma mansoni* mother sporocysts by hemocytes of juveniles of the 10-R2 strain of *Biomphalaria glabrata*. J Parasitol 68:84–94

Loker ES, Adema CM, Zhang SM et al (2004) Invertebrate immune systems – not homogeneous, not simple, not well understood. Immunol Rev 198:10–24

Lu TZ, Feng Z-P (2011) A sodium-leak current regulates pacemaker activity of adult central pattern generator neurons in *Lymnaea stagnalis*. PLoS One 6:e18745. https://doi.org/10.1371/journal.pone.0018745

Luzón-Peña M, Rojo-Vázquez FA, Gómez-Bautista M (1995) Seasonal availability of *Fasciola hepatica* metacercariae in a temperate Mediterranean area (Madrid, Spain). Zentralbl Veterinarmed B 42:577–585. https://doi.org/10.1111/j.1439-0450.1995.tb00751.x

Magalhães KC, Jannotti-Passos LK, Caldeira RL et al (2008) Isolation and detection of *Fasciola hepatica* DNA in *Lymnaea viatrix* from formalin-fixed and paraffin-embedded tissues through multiplex-PCR. Vet Parasitol 152:333–338

Mahilini H, Rajendran A (2008) Categorization of hemocytes of three gastropod species *Trachea vittata* (Muller), *Pila globosa* (Swainson) and *Indoplanorbis exustus* (Dehays). J Invertebr Pathol 97:e26. https://doi.org/10.1016/j.jip.2007.07.007

Marszewska A, Cichy A, Bulantová J et al (2018) *Potamopyrgus antipodarum* as a potential defender against swimmer's itch in European recreational water bodies-experimental study. PeerJ 6:e5045. https://doi.org/10.7717/peerj.5045

McCarthy AM (1999) The influence of temperature on the survival and infectivity of the cercariae of *Echinoparyphium recurvatum* (Digenea: Echinostomatidae). Parasitology 118:383–388. https://doi.org/10.1017/s003118209900400x

McClelland G, Bourns TKR (1969) Effects of *Trichobilharzia ocellata* on growth, reproduction, and survival of *Lymnaea stagnalis*. Exp Parasitol 24:137–146

McReath AM, Reader TA, Southgate VR (1982) The development of the host-response in juvenile *Lymnaea palustris* to invasion by *Fasciola hepatica*. Z Parasitenkd 67:175–184. https://doi.org/10.1007/BF00928113

Meunier C, Tirard C, Hurtrez-Boussès S et al (2001) Lack of molluscan host diversity and the transmission of an emerging parasitic disease in Bolivia. Mol Ecol 10:1333–1340

Mitta G, Gourbal B, Grunau C et al (2017) The compatibility between *Biomphalaria glabrata* snails and *Schistosoma mansoni*: an increasingly complex puzzle. Adv Parasitol 97:111–145. https://doi.org/10.1016/bs.apar.2016.08.006

Monteil JF, Matricon-Gondran M (1991a) Hemocyte production in trematode-infected *Lymnaea truncatula*. Parasitol Res 77:491–497

Monteil JF, Matricon-Gondran M (1991b) Interactions between the snail *Lymnaea truncatula* and the plagiorchiid trematode *Haplometra cylindracea*. J Invert Pathol 58:127–135

Monteil JF, Matricon-Gondran M (1993) Structural and cytochemical study of the hemocytes in normal and trematode-infected *Lymnaea truncatula*. Parasitol Res 79:675–682. https://doi.org/10.1007/BF00932510

Moore MN, Halton DW (1973) Histochemical changes in the digestive gland of *Lymnaea truncatula* infected with *Fasciola hepatica*. Z Parasitenk 43:1–16

Morley NJ, Crane M, Lewis JW (2005) Changes in survival characteristics of *Diplostomum spathaceum* cercariae emerged from cadmium-exposed *Lymnaea stagnalis*. J Helminthol 79: 55–59. https://doi.org/10.1079/joh2004271

Morley NJ, Adam ME, Lewis JW (2007) The effects of temperature on the distribution and establishment of *Echinoparyphium recurvatum* metacercariae in *Lymnaea peregra*. J Helminthol 81:311–315. https://doi.org/10.1017/S0022149X07818530

Mostafa OMS (2008) Effects of *Fasciola gigantica* experimental infection on some inorganic elements in the snail host *Lymnaea natalensis*. Exp Parasitol 118:468–471. https://doi.org/10.1016/j.exppara.2007.10.009

Muñoz-Antoli C, Marín A, Toledo R et al (2007) Effect of *Echinostoma friedi* (Trematoda: Echinostomatidae) experimental infection on longevity, growth and fecundity of juvenile *Radix peregra* (Gastropoda: Lymnaeidae) and *Biomphalaria glabrata* (Gastropoda: Planorbidae) snails. Parasitol Res 101:1663–1670

Neeleman AP, van der Knaap WP, van den Eijnden DH (1994) Identification and characterization of a UDP-GalNAc:GlcNAc beta-R beta-->4-N-acetylgalactosaminyl transferase from cercariae of the schistosome *Trichobilharzia ocellata*. Catalysis of a key step in the synthesis of N'N-diacetyllactosediamino (lacdiNAc)-type glycans. Glycobiology 4:641–651

Nguyen NT, Le TC, Vo MDC et al (2017) High prevalence of cattle fascioliasis in coastal areas of Thua Thien Hue province, Vietnam. J Vet Med Sci 79:1035–1042. https://doi.org/10.1292/jvms.16-0331

Nie L, Cai SY, Shao JZ et al (2018) Toll-like receptors, associated biological roles, and signaling networks in non-mammals. Front Immunol 9:1523–1542. https://doi.org/10.3389/fimmu.2018.01523

Núñez PE, De Jong-Brink M (1997) The suppressive excretory-secretory product of *Trichobilharzia ocellata*: a possible factor for determining compatibility in parasite-host interactions. Parasitology 115:193–203. https://doi.org/10.1017/s0031182097001169

Núñez PE, Adema CM, de Jong-Brink M (1994) Modulation of the bacterial clearance activity of phagocytes from the freshwater mollusc, *Lymnaea stagnalis*, by the avian schistosome, *Trichobilharzia ocellata*. Parasitology 109:299–310. https://doi.org/10.1017/s0031182000078331

Osorio JB, de Mattos PL, Giongo A et al (2020) Mollusk microbiota shift during *Angiostrongylus cantonensis* infection in the freshwater snail *Biomphalaria glabrata* and the terrestrial slug *Phillocaulis soleiformis*. Parasitol Res 119:2495–2503. https://doi.org/10.1007/s00436-020-06743-y

Pankrác J, Novobilský A, Rondelaud D et al (2016) Effect of *Fascioloides magna* (Digenea) on fecundity, shell height, and survival rate of *Pseudosuccinea columella* (Lymnaeidae). Parasitol Res 115:3119–3125. https://doi.org/10.1007/s00436-016-5068-4

Patnaik MM (1968) Notes on glycogen deposits in *Lymnaea auricularia* var. *rufescens* and the parasitic larval stages of *Fasciola gigantica* and *Echinostoma revolutum*. A histochemical study. Ann Parasitol Hum Comp 43:449–456. https://doi.org/10.1051/parasite/1968434449

Perry KJ, Henry JQ (2014) CRISPR/Cas9-mediated genome modification in the mollusc, *Crepidula fornicata*. Genesis 53:237–244

Pila EA, Gordy MA, Phillips VK et al (2016a) Endogenous growth factor stimulation of hemocyte proliferation induces resistance to *Schistosoma mansoni* challenge in the snail host. Proc Natl Acad Sci USA 113:5305–5310

Pila EA, Sullivan JT, Wu XZ et al (2016b) Haematopoiesis in molluscs: a review of haemocyte development and function in gastropods, cephalopods and bivalves. Dev Comp Immunol 58: 119–128

Pila EA, Tarrabain M, Kabore AL et al (2016c) A novel toll-like receptor (TLR) influences compatibility between the gastropod *Biomphalaria glabrata*, and the digenean trematode *Schistosoma mansoni*. PLoS Pathog 12:e1005513. https://doi.org/10.1371/journal.ppat.1005513

Pila EA, Li H, Hambrook JR et al (2017) Schistosomiasis from a snail's perspective: advances in snail immunity. Trends Parasitol 33:845–857

Pinaud S, Portela J, Duval D et al (2016) A shift from cellular to humoral responses contributes to innate immune memory in the vector snail *Biomphalaria glabrata*. PLoS Pathog 12:e1005361. https://doi.org/10.1371/journal.ppat.1005361

Pinheiro J, Maldonado Júnior A, Lanfredi RM (2009) Physiological changes in *Lymnaea columella* (Say, 1817) (Mollusca, Gastropoda) in response to *Echinostoma paraensei* Lie and Basch, 1967 (Trematoda: Echinostomatidae) infection. Parasitol Res 106:55–59. https://doi.org/10.1007/s00436-009-1630-7

Plows LD, Cook RT, Davies AJ et al (2004) Activation of extracellular-signal regulated kinase is required for phagocytosis by *Lymnaea stagnalis* haemocytes. Biochim Biophys Acta 1692(1): 25–33. https://doi.org/10.1016/j.bbamcr.2004.03.002

Plows LD, Cook RT, Davies AJ et al (2006a) Integrin engagement modulates the phosphorylation of focal adhesion kinase, phagocytosis, and cell spreading in molluscan defence cells. Biochim Biophys Acta 1763(8):779–786. https://doi.org/10.1016/j.bbamcr.2006.04.008

Plows LD, Cook RT, Davies AJ et al (2006b) Phagocytosis by *Lymnaea stagnalis* haemocytes: a potential role for phosphatidylinositol 3-kinase but not protein kinase A. J Invert Pathol 91:74–77

Portela J, Duval D, Rognon A et al (2013) Evidence for specific genotype-dependent immune priming in the lophotrochozoan *Biomphalaria glabrata* snail. J Innate Immunol 5:261–276

Portet A, Toulza E, Lokmer A et al (2021) Experimental infection of the *Biomphalaria glabrata* vector snail by *Schistosoma mansoni* parasites drives snail microbiota dysbiosis. Microorganisms 9:1084. https://doi.org/10.3390/microorganisms9051084

Poulin R (2007) Evolutionary ecology of parasites. Princeton University Press, Princeton, NJ

Preston DL, Orlofske SA, Lambden JP et al (2013) Biomass and productivity of trematode parasites in pond ecosystems. J Anim Ecol 82:509–517. https://doi.org/10.1111/1365-2656.12030

Preston DL, Layden TJ, Segui LM et al (2021) Trematode parasites exceed aquatic insect biomass in Oregon stream food webs. J Anim Ecol 90:766–775. https://doi.org/10.1111/1365-2656.13409

Rebelo MF, Figueiredo ES, Mariante RM et al (2013) New insights from the oyster *Crassostrea rhizophorae* on bivalve circulating hemocytes. PLoS One 8:e57384. https://doi.org/10.1371/journal.pone.0057384

Ribeiro VMA, Coaglio AL, Pereira Oliveira FL et al (2017) Dynamics of haemocytes from *Pseudosuccinea columella* circulating infected by *Fasciola hepatica*. Braz J Vet Parasitol 26: 411–418

Roldán C, Begovoeva M, López-Olvera J et al (2020) Endemic occurrence of *Fasciola hepatica* in an alpine ecosystem, Pyrenees Northeastern Spain. Transbound Emerg Dis 68(4):2589–2594. https://doi.org/10.1111/tbed.13865

Rondelaud D, Barthe D (1980a) Etude descriptive d'une réaction amibocytaire chez *Lymnaea truncatula* Müller infestée par *Fasciola hepatica* L. Z Parasitenkd 61:187–196. https://doi.org/10.1007/BF00925464

Rondelaud D, Barthe D (1980b) *Fasciola hepatica* L.: Les formes larvaires non évolutives ou en dégénérescence chez *Lymnaea truncatula* Müller. Z Parasitenk 62:95–104. https://doi.org/10.1007/bf00925370

Rondelaud D, Barthe D (1981) The development of the amoebocyte-producing organ in *Lymnaea truncatula* Müller infected by Fasciola hepatica L. Z Parasitenkd 65:331–341. https://doi.org/10.1007/bf00926728

Rondelaud D, Barthe D (1982) Relationship of the amoebocyte-producing organ with the generalized amoebocytic reaction in *Lymnaea truncatula* Muller infected by *Fasciola hepatica* L. J Parasitol 68:967–969

Rondelaud D, Barthe D (1983) Les modifications structurales du rein chez *Lymnaea truncatula* Muller infestée par *Fasciola hepatica*. Ann Parasitol Hum Comp 58:109–116

Rondelaud D, Abrous M, Dreyfuss G (2002a) The influence of food sources on cercarial production in *Lymnaea truncatula* experimentally infected with Digenea. Vet Res 33:95–100

Rondelaud D, Vignoles P, Dreyfuss G (2002b) The presence of predators modifies the larval development of *Fasciola hepatica* in surviving *Lymnaea truncatula*. J Helminthol 76:175–178

Rudolfova J, Hampl V, Bayssade-Dufour C et al (2005) Validity reassessment of *Trichobilharzia* species using *Lymnaea stagnalis* as the intermediate host. Parasitol Res 95:79–89

Russo J, Lagadic L (2000) Effects of parasitism and pesticide exposure on characteristics and functions of hemocyte populations in the freshwater snail *Lymnaea palustris* (Gastropoda, Pulmonata). Cell Biol Toxicol 16:15–30

Russo J, Madec L, Brehélin M (2008) Effect of a toxicant on phagocytosis pathways in the freshwater snail *Lymnaea stagnalis*. Cell Tissue Res 333:147–158. https://doi.org/10.1007/s00441-008-0578-8

Salazar L, Estrada VE, Velásquez LE (2006) Effect of the exposure to *Fasciola hepatica* (Trematoda: Digenea) on life history traits of *Lymnaea cousini* and *Lymnaea columella* (Gastropoda: Lymnaeidae). Exp Parasitol 114:77–83

Sandland GJ, Minchella DJ (2003) Effects of diet and *Echinostoma revolutum* infection on energy allocation patterns in juvenile *Lymnaea elodes* snails. Oecologia 134:479–486. https://doi.org/10.1007/s00442-002-1127-x

Sapp KK, Loker ES (2000) Mechanisms underlying digenean-snail specificity: role of miracidial attachment and host plasma factors. J Parasitol 86:1012–1019

Schallig HD, Sassen MJ, Hordijk PL et al (1991) *Trichobilharzia ocellata*: influence of infection on the fecundity of its intermediate snail host *Lymnaea stagnalis* and cercarial induction of the release of schistosomin, a snail neuropeptide antagonizing female gonadotropic hormones. Parasitology 102:85–91. https://doi.org/10.1017/s0031182000060376

Schmid-Araya JM, Schmid PE, Majdi N et al (2020) Biomass and production of freshwater meiofauna: a review and a new allometric model. Hydrobiologia 847:2681–2703. https://doi.org/10.1007/s10750-020-04261-7

Schultz JH, Adema CM (2017) Comparative immunogenomics of molluscs. Dev Comp Immunol 75:3–15

Schweizer G, Meli ML, Torgerson PR et al (2007) Prevalence of *Fasciola hepatica* in the intermediate host *Lymnaea truncatula* detected by real time TaqMan PCR in populations from 70 Swiss farms with cattle husbandry. Vet Parasitol 150:164–169. https://doi.org/10.1016/j.vetpar.2007.08.006

Selbach C, Poulin R (2020) Some like it hotter: trematode transmission under changing temperature conditions. Oecologia 194:745–755

Seppälä O, Jokela J (2010) Maintenance of genetic variation in immune defense of a freshwater snail: role of environmental heterogeneity. Evolution 64:2397–2407. https://doi.org/10.1111/j.1558-5646.2010.00995.x

Seppälä O, Leicht K (2013) Activation of the immune defence of the freshwater snail *Lymnaea stagnalis* by different immune elicitors. J Exp Biol 216:2902–2907. https://doi.org/10.1242/jeb.084947

Seppälä K, Leicht K (2015) Quality attracts parasites: host condition-dependent chemo-orientation of trematode larvae. Funct Ecol 29:791–795

Seppälä O, Karvonen A, Kuosa M et al (2013) Are sick individuals weak competitors? Competitive ability of snails parasitized by a gigantism-inducing trematode. PLoS One 8:e79366. https://doi.org/10.1371/journal.pone.0079366

Seppälä O, Louhi KR, Karvonen A et al (2015) Relative reproductive success of co-infecting parasite genotypes under intensified within-host competition. Infect Genet Evol 36:450–455. https://doi.org/10.1016/j.meegid.2015.08.015

Seppälä O, Walser JC, Cereghetti T et al (2021) Transcriptome profiling of *Lymnaea stagnalis* (Gastropoda) for ecoimmunological research. BMC Genomics 22:144. https://doi.org/10.1186/s12864-021-07428-1

Sindou P, Cabaret J, Rondelaud D (1991) Survival of snails and characteristic lesions of *Fasciola hepatica* infection in four European species of *Lymnaea*. Vet Parasitol 40:47–58. https://doi.org/10.1016/0304-4017(91)90082-7

Sire C, Rognon A, Theron A (1998) Failure of *Schistosoma mansoni* to reinfect *Biomphalaria glabrata* snails: acquired humoral resistance or intra-specific larval antagonism? Parasitology 117:117–122. https://doi.org/10.1017/s0031182098002923

Skála V, Cerníková A, Jindrová Z et al (2014) Influence of *Trichobilharzia regenti* (Digenea: Schistosomatidae) on the defence activity of *Radix lagotis* (Lymnaeidae) haemocytes. PLoS One 9:e111696. https://doi.org/10.1371/journal.pone.0111696

Skála V, Walker A, Horák P (2018) Extracellular trap-like fiber release may not be a prominent defence response in snails: evidence from three species of freshwater gastropod molluscs. Dev Comp Immunol 79:137–141. https://doi.org/10.1016/j.dci.2017.10.011

Sminia T (1974) Haematopoiesis in the freshwater snail *Lymnaea stagnalis* studied by electron microscopy and autoradiography. Cell Tissue Res 150:443–454

Sminia T, Borghart-Reinders E, van de Linde AW (1974) Encapsulation of foreign materials experimentally introduced into the freshwater snail *Lymnaea stagnalis*. An electron microscopic and autoradiographic study. Cell Tissue Res 153:307–326

Sminia T, van der Knaap WP, Edelenbosch P (1979) The role of serum factors in phagocytosis of foreign particles by blood cells of the freshwater snail *Lymnaea stagnalis*. Dev Comp Immunol 3:37–44

Smit AB, De Jong-Brink M, Li KW et al (2004) Granularin, a novel molluscan opsonin comprising a single vWF type C domain is up-regulated during parasitation. FASEB J 18:845–847. https://doi.org/10.1096/fj.03-0590fje

Soldánová M, Selbach C, Sures B et al (2010) Larval trematode communities in *Radix auricularia* and *Lymnaea stagnalis* in a reservoir system of the Ruhr River. Parasite Vector 3:56. https://doi.org/10.1186/1756-3305-3-56

Soldánová M, Faltýnková A, Scholz T et al (2011) Parasites in a man-made landscape: contrasting patterns of trematode flow in a fishpond area in Central Europe. Parasitology 138:789–807. https://doi.org/10.1017/s0031182011000291

Soldánová M, Kuris AM, Scholz T et al (2012) The role of spatial and temporal heterogeneity and competition in structuring trematode communities in the great pond snail, *Lymnaea stagnalis* (L.). J Parasitol 98:460–471. https://doi.org/10.1645/GE-2964.1

Soliman MF (2009) *Fasciola gigantica*: cercarial shedding pattern from *Lymnaea natalensis* after long-term exposure to cadmium at different temperatures. Exp Parasitol 121:307–311

Sorensen RE, Minchella DJ (1998) Parasite influences on host life history: *Echinostoma revolutum* parasitism of *Lymnaea elodes* snails. Oecologia 115:188–195. https://doi.org/10.1007/s004420050507

Sorensen RE, Minchella DJ (2001) Snail-trematode life history interactions: past trends and future directions. Parasitology 123:S3–S8. https://doi.org/10.1017/s0031182001007843

Southgate VR (1970) Observations on the effect of the rediae of *Fasciola hepatica* on the lipid composition of the hepatopancreas of *Lymnaea truncatula*. Parasitology 61:293–299. https://doi.org/10.1017/s0031182000041123

Thèron A, Coustau C (2005) Are *Biomphalaria* snails resistant to *Schistosoma mansoni*? J Helminthol 79:189–191

Thieltges DW, Jensen KT, Poulin R (2008) The role of biotic factors in the transmission of free-living endohelminth stages. Parasitology 135:407–426

Travers MA, Mirella da Silva P, Le Go N et al (2008) Morphologic, cytometric and functional characterisation of abalone (*Haliotis tuberculata*) haemocytes. Fish Shellfish Immunol 24:400–411

Tunholi VM, Lorenzoni PO, da Silva YH et al (2017) Molluscicidal potential of *Heterorhabditis baujardi* (Rhabditida: Heterorhabditidae), strain LPP7, on *Lymnaea columella* (Gastropoda: Pulmonata): an alternative for biological control of fasciolosis. Acta Trop 173:23–29

Utzinger J, Tanner M (2000) Microhabitat preferences of *Biomphalaria pfeifferi* and *Lymnaea natalensis* in a natural and a man-made habitat in southeastern Tanzania. Mem Inst Oswaldo Cruz 95:287–294

van der Knaap WP, Loker ES (1990) Immune mechanism in trematode – snail interactions. Parasitol Tod 6:175–182

van der Knaap WP, Sminia T, Kroese FG et al (1981) Elimination of bacteria from the circulation of the pond snail *Lymnaea stagnalis*. Dev Comp Immunol 5:21–32

van der Knaap WP, Boots AMH, van Asselt LA et al (1983a) Specificity and memory in increased defence reactions against bacteria in the pond snail *Lymnaea stagnalis*. Dev Comp Immunol 7: 435–443

van der Knaap WP, Sminia T, Schutte R et al (1983b) Cytophilic receptors for foreignness and some factors which influence phagocytosis by invertebrate leucocytes: in vitro phagocytosis by amoebocytes of the snail *Lymnaea stagnalis*. Immunology 48:377–383

van der Knaap WP, Meuleman EA, Sminia T (1987) Alterations in the internal defence system of the pond snail *Lymnaea stagnalis* induced by infection with the schistosome *Trichobilharzia ocellata*. Parasitol Res 73:57–65. https://doi.org/10.1007/BF00536337

Vaughn CM (1953) Effects of temperature on hatching and growth of *Lymnaea stagnalis appressa* Say. Am Midland Nat 49:214–228

Vázquez AA, Hevia Y, Sánchez J (2009) Distribución y preferencia de hábitats de moluscos hospederos intermediarios de *Fasciola hepatica* en Cuba. Rev Cubana Med Trop 61:248–253

Vázquez AA, Sánchez J, Pointier JP et al (2014) *Fasciola hepatica* in Cuba: compatibility of different isolates with two intermediate hosts, *Galba cubensis* and *Pseudosuccinea columella*. J Helminthol 88:434–440. https://doi.org/10.1017/s0022149x13000382

Vázquez AA, Pilar A, Lounnas M et al (2018) Lymnaeid snails hosts of *Fasciola hepatica* and *Fasciola gigantica* (Trematoda: Digenea): a worldwide review. CAB Rev 13:1–15. https://doi.org/10.1079/PAVSNNR201813062

Vázquez A, Vargas M, Alba A et al (2019) Reviewing *Fasciola hepatica* transmission in the West Indies and novel perceptions from experimental infections of sympatric vs. allopatric snail/fluke combinations. Vet Parasitol 275:108955. https://doi.org/10.1016/j.vetpar.2019.108955

Vieira TC, Costa-Filho A, Salgado NC et al (2004) Acharan sulfate, the new glycosaminoglycan from *Achatina fulica* Bowdich 1822. Structural heterogeneity, metabolic labeling and localization in the body, mucus and the organic shell matrix. Eur J Biochem 271:845–854

Vorontsova YL, Slepneva IA, Yurlova NI et al (2019) The effect of trematode infection on the markers of oxidative stress in the offspring of the freshwater snail *Lymnaea stagnalis*. Parasitol Res 118(12):3561–3564. https://doi.org/10.1007/s00436-019-06494-5

Voutilainen A (2010) Interactive effects of predation risk and parasitism on the circadian rhythm of foraging activity in the great pond snail *Lymnaea stagnalis* (Gastropoda: Lymnaeidae). Ann Limnol Int J Lim 46:217–223. https://doi.org/10.1051/limn/2010025

Voutilainen A, van Ooik T, Puurtinen M et al (2009) Relationship between prevalence of trematode parasite *Diplostomum* sp. and population density of its snail host *Lymnaea stagnalis* in lakes and ponds in Finland. Aquat Ecol 43:351–357. https://doi.org/10.1007/s10452-008-9203-x

Walker SM, Makundi AE, Namuba FV et al (2008) The distribution of *Fasciola hepatica* and *Fasciola gigantica* within southern Tanzania – constraints associated with the intermediate host. Parasitology 135:495–503

Walker AJ, Lacchini AH, Sealey KL et al (2010) Spreading by snail (*Lymnaea stagnalis*) defence cells is regulated through integrated PKC, FAK and Src signaling. Cell Tissue Res 341:131–145

Wang Y, Hu M, Chiang MW et al (2012) Characterization of subpopulations and immune-related parameters of hemocytes in the green-lipped mussel *Perna viridis*. Fish Shellfish Immunol 32: 381–390

Wang C, Yue X, Lu X et al (2013) The role of catalase in the immune response to oxidative stress and pathogen challenge in the clam *Meretrix meretrix*. Fish Shellfish Immunol 34:91–99

Wilson RA, Denison J (1970a) Short-chain fatty acids as stimulants of turning activity by the miracidium of *Fasciola hepatica*. Comp Biochem Physiol 32:511–517

Wilson RA, Denison J (1970b) Studies on the activity of the miracidium of the common liver fluke, *Fasciola hepatica*. Comp Biochem Physiol 32:301–313

Wilson RA, Denison J (1980) The parasitic castration and gigantism of *Lymnaea truncatula* infected with the larval stages of *Fasciola hepatica*. Z Parasitenk 61:109–119. https://doi.org/10.1007/bf00925458

Wilson RA, Draskau T (1976) The stimulation of daughter redia production during the larval development of *Fasciola hepatica*. Parasitology 72:245–257. https://doi.org/10.1017/s0031182000049465

Wolinska J, King KC (2009) Environment can alter selection in host-parasite interactions. Trends Parasitol 25:236–244

Wright B, Lacchini AH, Davies AJ et al (2006) Regulation of nitric oxide production in snail (*Lymnaea stagnalis*) defence cells: a role for PKC and ERK signalling pathways. Biol Cell 96: 265–278

Yurlova NI, Vodyanitskaya SN, Serbina EA et al (2006) Temporal variation in prevalence and abundance of metacercariae in the pulmonate snail Lymnaea stagnalis in Chany Lake, West Siberia, Russia: long-term patterns and environmental covariates. J Parasitol 92:249–259. https://doi.org/10.1645/ge-544r2.1

Zakikhani M, Rau ME (1999) *Plagiorchis elegans* (Digenea: Plagiorchiidae) infections in *Stagnicola elodes* (Pulmonata: Lymnaeidae): host susceptibility, growth, reproduction, mortality, and Cercarial production. J Parasitol 85:454–463

Żbikowska E (2003) The effect of Digenea larvae on calcium content in the shells of *Lymnaea stagnalis* (L.) individuals. J Parasitol 89:76–79

Żbikowska E (2011) One snail – three Digenea species, different strategies in host-parasite interaction. Anim Biol 61:1–19. https://doi.org/10.1163/157075511X554383

Żbikowska E, Żbikowski J (2005) Differences in shell shape of naturally infected *Lymnaea stagnalis* (l.) individuals as the effect of the activity of digenetic trematode larvae. J Parasitol 91:1046–1051

Żbikowska E, Żbikowski J (2015) Digenean larvae—the cause and beneficiaries of the changes in host snails' thermal behavior. Parasitol Res 114:1063–1070

Żbikowska E, Kobak J, Żbikowski J et al (2006) Infestation of *Lymnaea stagnalis* by digenean flukes in the Jeziorak Lake. Parasitol Res 99:434–439. https://doi.org/10.1007/s00436-006-0178-z

Zelck UE, Janje B, Schneider O (2005) Superoxide dismutase expression and H_2O_2 production by hemocytes of the trematode intermediate host *Lymnaea stagnalis* (Gastropoda). Dev Comp Immunol 29:305–314

Zelck UE, Gege BE, Schmid S (2007) Specific inhibitors of mitogen-activated protein kinase and PI3-K pathways impair immune responses by hemocytes of trematode intermediate host snails. Dev Comp Immunol 31:321–331. https://doi.org/10.1016/j.dci.2006.06.006

Zilber-Rosenberg I, Rosenberg E (2008) Role of microorganisms in the evolution of animals and plants: the hologenome theory of evolution. FEMS Microbiol Rev 32:723–735

Zimmermann MR, Luth KE, Esch GW (2014) Differences in snail ecology lead to infection pattern variation of *Echinostoma* spp. larval stages. Acta Parasitol 59:502–509

Zischke DP (1972) The effects of *Echinostoma revolutum* larval infection on the histology, growth and fecundity of the snail, *Stagnicola palustris*. PhD Thesis, Louisiana State University and Agricultural and Mechanical College, LSU Historical Dissertations and Theses

Chapter 12
Overview of Interactions Between Parasitic Digenea and Their Molluscan Hosts, with Special Emphasis on the Lymnaeidae

Sylvie Hurtrez-Boussès, Annia Alba, Pilar Alda, Elodie Chapuis, Dominique Faugère, Benjamin Gourbal, Jean-Pierre Pointier, Jorge Sánchez, Marion Vittecoq, and Antonio A. Vázquez

Abstract Digeneans (Platyhelminthes, Trematoda) are considered as a very ancient and as the most diversified group of internal metazoan parasites of animals. They can have important detrimental effects on human and animal health. They also affect populations dynamics of their hosts, are able to shape the structure of communities, and can act as determinants of ecosystems structure and functioning. Whereas life-cycles are diversified, the fundamental life-cycle of digeneans is characterized by a

S. Hurtrez-Boussès (✉)
MIVEGEC, University of Montpellier, CNRS, IRD, Montpellier, France

Département de Biologie-Écologie, Faculté des Sciences, Université de Montpellier, Montpellier, France
e-mail: sylvie.hurtrez@ird.fr

A. Alba
Laboratory of Malacology, Institute of Tropical Medicine "Pedro Kourí", Havana, Cuba

Interactions Hosts Pathogens Environments, University of Montpellier, CNRS, Ifremer, University of Perpignan Via Domitia, Perpignan, France

P. Alda
MIVEGEC, University of Montpellier, CNRS, IRD, Montpellier, France

Genética y Ecología Evolutiva (CERZOS–CCT–CONICET Bahía Blanca), Bahía Blanca, Argentina

E. Chapuis · D. Faugère
MIVEGEC, University of Montpellier, CNRS, IRD, Montpellier, France
e-mail: elodie.chapuis@ird.fr; dominique.faugere@ird.fr

B. Gourbal
Interactions Hosts Pathogens Environments, University of Montpellier, CNRS, Ifremer, University of Perpignan Via Domitia, Perpignan, France
e-mail: benjamin.gourbal@univ-perp.fr

J.-P. Pointier
PSL Research University, USR 3278 CNRS–EPHE, CRIOBE Université de Perpignan, Perpignan, France
e-mail: pointier@univ-perp.fr

M. V. Vinarski, A. A. Vázquez (eds.), *The Lymnaeidae*, Zoological Monographs 7,
https://doi.org/10.1007/978-3-031-30292-3_12

mollusc as the first intermediate host, where an important asexual multiplication occurs and by a vertebrate as definitive host. The interactions between digeneans and molluscs are complex; they start with the infection of the mollusc by the miracidial larval stage, which implies molecular mechanisms of attraction, recognition, and attachment. Once the penetration successful, an intimate association starts between the mollusc and the digenean, where the mollusc displays an intense immune response and, in turn, the digenean counter-attacks by different mechanisms. All these complex interactions can lead to high reciprocal specializations which can lead to host–parasite specificity. After its installation, the miracidium converts into a sporocyst which develops and starts an intense asexual multiplication, highly detrimental for the molluscan fitness. At the end of several cycles of asexual reproduction, the last larval stages, the cercariae, are emitted and will serve to infect the following host. Taking into account the complexity of mollusc–digenean interactions, their sensitivity to environmental perturbations and their importance in ecological and epidemiological risks, such interactions have to be particularly scrutinized. In this chapter, we review such host–parasite interactions with emphasis on the Lymnaeidae.

12.1 Introduction

Water- and food-borne infectious diseases are probably those having the stronger impact on human populations. Emerging countries suffer the largest burden and global changes might increase the load in the future (Jones 1990; World Health Organization 2015; Cissé 2019). Among these diseases, those caused by Trematodes are widely neglected (Petney et al. 2016). However, humans can be infected by more than 70 species of food-borne or water-borne trematodes (Johansen et al. 2010). Be they in blood, lungs, liver, or intestine, all these flukes have detrimental effects on their hosts. For instance, more than 140 million people are affected by schistosomiases (caused by the genus *Schistosoma*) and it is estimated that at least 700 million people, mainly living in the tropical belts, are at risk of infection. In an extensive overview on food-borne hazards, World Health Organization (2015) highlighted that four food-borne trematodoses (caused by *Clonorchis sinensis*, *Fasciola* spp., *Opisthorchis* spp., and *Paragonimus* spp.) are responsible for 200,000 illnesses,

J. Sánchez
Laboratory of Malacology, Institute of Tropical Medicine "Pedro Kourí", Havana, Cuba
e-mail: jsnoda@ipk.sld.cu

M. Vittecoq
Institut de Recherche de la Tour du Valat, Arles, France
e-mail: vittecoq@tourduvalat.org

A. A. Vázquez
MIVEGEC, University of Montpellier, CNRS, IRD, Montpellier, France

Laboratory of Malacology, Institute of Tropical Medicine "Pedro Kourí", Havana, Cuba

including hepatitis, pneumopathia, and cancers. The consequences are more than 7000 deaths/year and 2 million DALYs (Disability-adjusted life years). It is estimated that more than 100 million people are infected with a least one food-borne trematode species (Khurana and Malla 2014) and that more than 10% of the world population is potentially at risk of food-borne trematodoses (Rim et al. 1994; Keiser and Utzinger 2005, 2009). Besides their effects on human health, food- and water-borne trematodoses also have a strong economic impact. Indeed, several of them affect livestock or aquacultures causing important losses of productivity (meat, milk, wool), restrictions on commercial exchanges, and decrease of consumer demand (WHO 2015).

Trematodes not only cause problems to human societies; they can also shape populations or communities or modify ecosystems. For instance, by causing malformations and by increasing mortality, the trematode *Ribeiroia ondatrae* is considered as one of the factors responsible for the decline of populations of several North American amphibian species (see Blaustein et al. 2012). Thomas et al. (1998) have demonstrated that another trematode, *Curtuteria australis,* acts as an ecosystem engineer in shores in New Zealand. When infected by this parasite, the intermediate host, the cockle *Austrovenus stutchburyi,* becomes unable to burrow efficiently in the mud and lies at the surface of the sediment. The parasitized cockles offer therefore a new substrate for benthic invertebrates. Whereas sea anemones are still preferring non-infected cockles, limpets are more likely to install on parasitized ones. When *C. australis* is present, the competition for substrate between limpets and anemones can therefore be reduced and the coexistence between these two species might be favored. More recently, Friesen et al. (2020) showed experimentally that, by differentially affecting the population dynamics of four host species, the trematode *Maritrema poulini* is able to shape the community of freshwater crustaceans.

Despite their detrimental effects on human and veterinary health, their economic impacts and their ecological roles, trematodes are still poorly studied and are widely neglected in health or conservation programs. A better knowledge of their biology and their interactions with the hosts they infect is therefore needed.

12.2 Who Are These Enemies?

The Trematoda belong to the phylum of Platyhelminthes, in the clade of Lophotrochozoa (Eernisse and Peterson 2004). The monophyletic group of Platyhelminthes includes free-living organisms, the Turbellariae (paraphyletic group comprising at least seven clades; see Park et al. 2007), and the clade of Neodermates, all parasites, characterized by a syncytial tegument (Littlewood et al. 1999). Among the Neodermates, three distinct clades are delineated: Monogenes, Cestodes, and Trematodes (Olson and Tkach 2005), although their relationships are not clearly elucidated yet (Littlewood et al. 1999; Olson et al. 2001; Lockyer et al. 2003; Park et al. 2007). The class Trematoda includes two sister groups:

Aspidogastrea and Digenea. With only 80 known species (Rohde 2001), aspidogastreans are considered as a small group, without any documented economic impact. Their larval stages, which do not multiply, live into a mollusc and adults can infect aquatic vertebrates (fishes or turtles). These parasites are characterized by a large intermediate and definitive host spectrum. At the opposite, their closely-related group, the digeneans, comprises 18,000–25,000 nominal species (Cribb et al. 2001; Esch et al. 2002), clustered in more than 140 families (Gibson et al. 2002) and is considered as the most diversified group of internal metazoan parasites of animals (Olson and Tkach 2005). Their life-cycles are highly diversified, with one to four hosts (Galaktionov and Dobrovolskij 2003a); they can have three different larval generations, and the ways by which they infect their hosts are diverse (Poulin and Cribb 2002). Most of them have a life-cycle which involves (at least) two hosts: a mollusc as intermediate host, in which larvae intensively multiply, and a vertebrate as definitive host, in which sexual reproduction of adults occurs (see review in Galaktionov and Dobrovolskij 2003a). The digeneans infect a wide spectrum of vertebrates, in which they can occupy different locations: digestive tract or glands, gills, lung, blood, urinary bladder, body cavity, ovary, eye. Despite descriptions of numerous cycles (e.g., Stunkard 1963; Dawes 1968; Galaktionov and Dobrovolskij 2003a), attempts to propose evolutionary schemes were hampered until the knowledge of phylogenetic scenarios.

12.3 Interactions Between Digeneans and Molluscs: A Long-Term and Complex Story

Olson et al. (2003) provided a first extended phylogeny of Trematoda. These authors included in the analyses 170 digenean taxa belonging to 77 nominal families; seven aspidogastrean taxa, representing three families, were used as outgroup. The results are congruent with those obtained by more recent phylogenetic analyses with more taxa and updated databases of gene markers (Littlewood et al. 2015; Pérez-Ponce de León and Hernández-Mena 2019). They revealed the existence of two major clades in Digenea: the Diplostomida, which comprises eight families (among which are the Schistosomatidae), and the Plagiorchiida. This last group is highly diversified, with 69 families, clustered in 13 independent lineages.

Based on these results, Cribb et al. (2003) attempted to solve the major features of the origin and evolution of the digeneans. For this purpose, these authors mapped the cycle's characteristics on the phylogenetic tree and computed the number of evolutionary steps required to reach each studied state. According to Littlewood et al. (1999), they confirmed that the clade of Neodermata emerged as parasites of vertebrates. Littlewood et al. (1999) proposed that the common ancestor of Trematoda adopted molluscs as intermediate hosts. Digeneans diverge from aspidogasters by the fact that their larvae intensively multiply into the molluscan host, with a succession of larval generations. Concerning definitive hosts, Cribb et al.

(2003) suggested that digeneans were primarily intestinal parasites of teleosts. It is considered that this emergence occurred at least 200 million years ago (Ma) (Blair et al. 2001). The oldness of this group has been recently confirmed by Barrios-de Pedro et al. (2020) who identified eggs of Opisthorchiidae in 126–129 Ma aged coprolites, probably attributed to teleostean fishes. Several events of host-switchings have occurred, leading to the high diversity of actual vertebrate hosts.

12.4 Digenean Stages into the Mollusc: Long Is the Road

Several authors have described and summarized the steps of the infection of molluscs by digeneans [see, e.g., Esch et al. (2002) for a synthesis]. Briefly, the adults of the parasite produce eggs, which leave the definitive host, overflowed by urine, feces, or sputum. Although strongly variable among species, the fecundity of digeneans is often outstanding and can reach 25,000 eggs per day, as described by Boray (1969) for *Fasciola hepatica* [see comparisons in Whitfield and Evans (1983)]. Such an impressive egg production can be interpreted as an adaptation to low egg elimination by the definitive host and hatching successes: for instance, Warren (1978) considered that 50% of the eggs of human schistosomes fail in leaving their host, and it has been suggested that only 0.017% of *Fasciola hepatica* eggs emitted in a pasture are able to hatch (Morley and Lewis 2017). Depending on the digenean species, at laying, eggs can be unembryonated and accomplish their embryonation after their emission. In other species, they are partially embryonated and achieve their development outside the definitive host. The third category comprises fully embryonated eggs that hatch immediately after their emission.

In all cases, hatching gives rise to a ciliated larva, called *miracidium*. In most species, the miracidium is free-swimming and infects its molluscan (first) intermediate host by active penetration. This larva is unable to feed and its life span is about a few hours (reviewed in Prince 1954); its swimming ability and its infectivity decrease rapidly (Andrews 1999) since glycogen reserves of the larva decline in a few hours (Andrews et al. 2021). These constraints are offsetted by a panel of morphological, physiological, and behavioral adaptations allowing an active search, penetration, infection, and development in a suitable mollusc (Galaktionov and Dobrovolskij 2003b). Indeed, active miracidia are characterized by ciliated epithelial plates, which ensure locomotion. This locomotory system is connected to the nervous system and is completed by a sensory apparatus which allows detection and reaction of the miracidium to environmental cues like light, gravity, or chemical parameters (Andrews 1999; Andrews et al. 2021). The finding of a suitable host can be divided in three main steps. (1) First, the miracidium has to go to a "good place": newly emerged miracidia tend to migrate toward environments specific of their molluscan hosts. Eyespots are key determinants in photoreception that allow the orientation of miracidium along the water-column, toward regions where the chances to encounter a compatible host are higher (Born-Torrijo et al. 2017). For instance, it has been demonstrated that miracidia of *Fasciola hepatica*, which infect lymnaeid snails (mainly *Galba* spp.) living along the edge of small water-systems, are attracted by the light (Andrews et al. 2021). Similarly, Born-Torrijo et al. (2017) suggested that miracidia of the Strigeidae *Cardiocephaloides longicollis* move toward the light, where they can infect their nassariid snail hosts. Contrarily, miracidia of *Fasciola gigantica* which infect *Radix natalensis*, a snail occupying deeper habitats, are repelled by light (Taylor 1964). (2) The second step consists in random displacements in the water, until the presence of suitable hosts can be detected. (3) When a miracidium arrives in the vicinity of suitable hosts, a wide panel of molecules called "miracidia-attracting glycoproteins" can be detected by its molecular sensors, inducing a chemotaxic response; the ability to detect such cues plays a key role in mollusc–digenean specificity since miracidium can discriminate between suitable intermediate molluscan hosts and other unsuitable molluscan species (Kalbe et al. 2000). Sapp and Loker (2000) experimentally tested compatible and incompatible miracidia–molluscs associations with four digenean (*Echinostoma paraensei*, *E. trivolvis*, *Schistosoma mansoni*, and *Schistosomatium douthitti*) and five gastropod species (*Biomphalaria glabrata*, *Helisoma trivolvis, Helix aspersa* and two lymnaeids: *Lymnaea stagnalis* and *Stagnicola elodes*). They observed that miracidia are more likely to attach to or to penetrate compatible snails than incompatible ones. However, in half of the cases, miracidia of each digenean studied species were able to attach and to attempt penetration whatever the snail host, compatible or incompatible. This strongly suggests that mechanisms of recognition and attachment of the miracidium cannot explain alone the high mollusc–digenean specificity.

When the miracidium is in contact with its putative host, the mucus of the mollusc can act as an adhesive trap (Vieira et al. 2004). It also contains cytotoxic protein-like factors that can attack and eliminate the digenean larvae. This is another factor of the mollusc–trematode specificity. Indeed, on the parasite *Fascioloides magna*, Coyne et al. (2015) have experimentally demonstrated that the mucus of the incompatible planorbid snail *Helisoma trivolvis* had detrimental effects on the miracidia, whereas the mucus of the compatible lymnaeid snail *Stagnicola elodes* had no effect. The penetration of miracidial larva into the mollusc involves both mechanical action due to the apical papilla and chemical destruction of tissues due to the action of secreted proteolytic enzymes (review in Andrews et al. 2021).

In some other digenean species, the infection is passive: the egg (or rarely the miracidium; see Pandian 2020) is eaten by the mollusc and hatches after being ingested. This phenomenon has been described in species belonging to distinct superfamilies and families like Opisthorchioidea, Brachylaimoidae, Plagiorchioidea (LaRue, 1951), Hemiuroidea (Pandian 2020). For instance, in *Haplometra cylindracea*, eggs are fully embryonated when laid. They hatch only after ingestion by a lymnaeid snail, e.g. *Stagnicola palustris* or *Radix ovata* (Smyth and Smyth 1980). It has been observed that after ingestion, the miracidium becomes active into the egg, displaying slow contractions; hatching begins soon after (Galaktionov and Dobrovolskij 2003b). Galaktionov and Dobrovolskij (2003b) suggested that factors (probably small molecules) and bacterial flora from the digestive tract of the specific mollusc host activate locomotory and secretory systems of the miracidium inside its egg and, therefore, stimulate the hatching. Cross-feeding experiments with compatible versus incompatible digenean–mollusc pairs can give helpful information. Galaktionov and Dobrovolskij (2003b) reported that eggs of *Opisthioglyphe ranae*, whose specific host is *Lymnaea stagnalis*, hatch correctly in the digestive tract of lymnaeids (*L. stagnalis, Radix ovata*) whereas their hatching success is lowered in the digestive tract of planorbids *(Planorbarius corneus, Planorbis planorbis*) and they only transit in the digestive system of viviparids (*Viviparus contectus*). Similar results were obtained for *Haematoloechus similis* (parasite of *Planorbis planorbis*) for which better activation was obtained in the digestive tract of planorbids than that of lymnaeids.

In aquatic habitats, both active and passive strategies have been observed, whereas the passive infection has only been described in terrestrial cycles (Niewiadomska and Pojmanska 2011). Francisco et al. (2012) provided a study of passive miracidia of *Prosorhynchus crucibulum*, whose first intermediate host is the mussel (*Mytilus* spp.). They showed that as compared to active miracidium of *F. hepatica*, who has a larger size, several epithelial ciliated plates and eyespots, the miracidium of *P. crucibulum* is small, without eyespots and possesses only two epithelial plates with longer cilia (Francisco et al. 2012). The increased length of cilia can be interpreted as an adaptation to movements in viscose contains of the molluscan-hosts digestive tracts (Galaktionov and Dobrovolskij 2003b).

In all cases, the penetration is accompanied by shedding of ciliated plates, eliminating by this way major antigenic surface components that would have triggered intense immune response. Actually, for digeneans taking a few hours to penetrate (species found in Fasciolidae, Echinostomatidae, Psilostomidae), the shedding takes place after attachment to the host but before entry into its body; for other species (found in Paramphistomidae, Schistosomatidae, Heronimidae) plates are shedded once into the host (Galaktionov and Dobrovolskij 2003b).

If physical barriers (shell, mucus, tegument, microbiota) are successfully crossed, an intimate association between the parasitic trematode and its snail host starts. This leads in the mollusc to the activation of a complex immune response (Pila et al. 2017), including recognition of non-self (e.g., patterns recognition receptors) and mechanisms of defense involving humoral (a) and cellular (b) effectors. It is noteworthy that specific responses and immune memory have been described in molluscs (e.g., Portela et al. 2013; Pinaud et al. 2016). (a) Concerning the humoral defenses, Sapp and Loker (2000) experimentally tested the effects of plasma (i.e., hemolymph without cells) of five molluscan species (gastropods: two planorbids, one helicid and two lymnaeids) on larvae of four digeneans species (*Echinostoma paraensei, E. trivolvis, Schistosoma mansoni,* and *Schistosomatium douthitti*). They showed that, in compatible mollusc–digenean associations, the plasma of snails has no detrimental effect on parasite larvae. Conversely, in incompatible associations, snail plasma was harmful to 50% of first larval stages (sporocysts) and to 33% of more mature larvae (rediae). In most of the cases (9/10), lymnaeid plasma has toxic effects on digeneans of planorbids, and in half of the cases (2/4) planorbid plasma negatively affects digeneans of lymnaeids. The authors proved that differences in plasma osmolarity are not responsible for such effects and suggested that humoral factors shape the response of snails to infection. More recently, several humoral factors have been identified in gastropods like opsonins or agglutinins (Sanchez et al. 2006; Hanington et al. 2010), anti-microbial proteins (Guillou et al. 2007; López-Abarrategui et al. 2015), cytotoxins (Galinier et al. 2013), pattern-recognition receptors (Wu et al. 2017). (b) The encapsulation of larvae by mollusc hemocytes is able to block the development of digeneans [see, e.g., Alba et al. (2020) about the hemocytary response of *Pseudosuccinea columella* infection by *F. hepatica*]. This encapsulation permits a total isolation of the parasite and, thus, the privacy of nutrients from its host. At the same time, hemocytes deliver humoral effectors and cytotoxic factors that attack the digenean (Coustau et al. 2015). Production of toxic reactive oxygen and nitrogen species (ROS and RNS) is particularly important during encapsulation (Hahn et al. 2001a, b).

The efficiency of the immune system of molluscs against digeneans can be one of the factors explaining the relatively low prevalences of the parasites (often lower than 10%) observed (Fried and Graczyk 1997).

In front of the arsenal of defense displayed by its host, the digenean can counter-attack in different manners: (1) evading host recognition, (2) escaping or overcoming attacks from immune molecules or cells, or (3) controlling the immune system, by impairing the hemocyte recruitment, motility, or metabolism. (1) A striking example is that of the mucins (SmPoMuc) observed in *Schistosoma mansoni* (Roger et al. 2008): high genetic polymorphism in this multigenic family, combined with important variations in expression, leads to a wide antigenic diversity. Similarly, Alba et al. (2015) suggested that *F. hepatica* might escape to the recognition from its lymnaeid hosts by expressing different epitopes from one larval stage to another one. Such a variability can exert a pressure on the populations of hosts, in which balanced selection can therefore occur. This can, for instance, explain the high polymorphism observed in *Biomphalaria glabrata* among Fibrinogen-Related-Proteins (FREPs) (Adema 2015), a family of molecules involved in the immune recognition of digeneans, including *Schistosoma mansoni* (Zhang and Loker 2004). Another possible strategy to avoid recognition by the host immune system is molecular masking, consisting for the digenean to acquire epitopes of the molluscan host. Bayne et al. (1986) have suggested that *S. mansoni* might use this method to escape the immune recognition by its host, *B. glabrata*. Molecular mimicry may also act: in that case, the parasite is able to produce antigens with epitopes common with molecules expressed by the host. For instance, Yoshino et al. (2013) have revealed that larvae of *S. mansoni* can harbor glycan determinants similar to those produced by their host. In lymnaeids, it has been shown that polyclonal antibodies anti-*F. hepatica* larvae, experimentally produced in mammals, cross-react with snail antigens. Although this needs further confirmations, it suggests that the parasite would be able to produce antigenic determinants mimetic to those of its snail host (Alba et al. 2014). (2) In response to mollusc host defenses, digeneans exhibit various substances, like proteases and antioxidant molecules, able to counteract the molluscan molecules of immunity (e.g., Gourbal et al. 2008; Wu et al. 2009). (3) Digeneans can also produce factors of immunomodulation, like excretory-secretory products, able to lower hemocytes motility (Lodes and Yoshino 1990) and to reduce their production of

NO (Zahoor et al. 2009) and superoxide (Connors et al. 1991). In the interactions between digeneans and lymnaeids, it has been shown that immunomodulation can take many forms. For instance, Kryukova et al. (2014) have shown that the cellular immune response of *Lymnaea stagnalis* is strongly reduced after exposure to *Plagiorchis mutationis*. After penetration of *Trichobilharzia regenti* in *Radix lagotis*, numerous hemocytes of the snail are recruited around the parasite, but they fail in destroying it (Skála et al. 2014). This can be due to the fact that, as compared to those of non-infected snails, hemocytes of *T. regenti*-infected lymnaeids produce less hydrogen peroxide (H_2O_2) and have a lower phagocytic activity.

The package of complex interactions between molluscs and digeneans (Haas and Haberl 1997), including sophisticated mechanisms of recognition, attachment, and immune responses/counter-attacks, can result in high levels of reciprocal specializations (see extensive review in Galaktionov and Dobrovolskij 2003b). The consequence is a narrow mollusc-spectrum for the majority of digeneans: Adema and Loker (1997) underlined that it is scarce that a digenean species can be able to infect molluscs belonging to several families. Counter-examples exist, as for instance *Clonorchis sinensis,* able to parasitize molluscs belonging to different gastropod families (Bithyniidae, Thiaridae, Hydrobiidae, Viviparidae) or *Fasciola hepatica* typically infecting lymnaeids, which has been also recorded in planorbids (Abrous et al. 1998) or succineids (Relf et al. 2009), although the exact role of such groups in transmission is still debated (see Vázquez et al. 2023, this volume). At the specific level, most of the digeneans species are oioxenous (one mollusc species per parasite species) or stenoxenous (a few closely-related species of molluscs per parasite species). Again, *F. hepatica* can be considered as exceptional since it can infect a wide diversity of lymnaeid species (Correa et al. 2010; see Vázquez et al. 2023, this volume). As suggested by Vázquez et al. (2021a, b), this extraordinary diversity of intermediate hosts might be favored by the important genome size of *F. hepatica* and by the high polymorphism observed in its populations (e.g., for neutral polymorphism; see Hurtrez-Boussès et al. 2004; Vázquez et al. 2021b). Indeed, Wilensky et al. (2015) have revealed that *F. hepatica* has a genome larger (1.3 Gb) than other digeneans (e.g., 363–397 Mb for *Schistosoma* spp.) and is one of the largest genomes of pathogens known so far. Moreover, these authors found higher levels of non-synonymous polymorphism for genes exhibiting orthology with parasitic Platyhelminthes than for genes shared with free-living Platyhelminthes; such differences can explain high adaptability in *F. hepatica* genes involved in infection processes and in the interactions with the snail host.

Once entered into its host, the digenean larva migrates, mainly via hemolymph vessels, to its suitable location in the mollusc's body; depending on the digenean and mollusc species, the parasite can install, e.g. into the mantle, the digestive system (hepato-pancreas) or the gonad (LaRue 1951; Esch et al. 2002; Andrews et al. 2021). The miracidium regressively transforms into a sac-like called sporocyst. Thanks to immunomodulation mechanisms, the sporocyst will develop and will start an intense asexual multiplication. It contains masses of stems cells, in which each cell will in turn give daughter cells packed in germ balls. The product of this clonal reproduction can be daughter sporocysts which make in turn a new event of asexual reproduction. In other species, daughter larvae, called *rediae*, are more differentiated, with a mouth, a pharynx, and rough gut (Esch et al. 2002). The amplification of the number of larvae continues with asexual multiplication in rediae, following the same processes as described above for sporocysts. The digestive equipment of rediae makes them able to go through host tissues that they tear and swallow (Esch et al. 2002). They can also consume larvae of other parasites present in the same mollusc, but they are able to discriminate between foreign larvae and their relatives, the latter being spared the predation. This feeding behavior has been suspected to reduce intra-host competition and to shape the communities of parasites. For instance, Joe et al. (1965) experimentally infected *Radix rubiginosa* with combinations of five digeneans (two echinostomids, one strigeid, one schistosomid, and a xiphidiocercaria). They found that, whatever the timing order of infection, echinostome rediae eliminate the other trematode sporocysts (see below about possible ecological consequences of such predatory behavior).

Finally, the last intra-molluscan larval stage produced is the *cercaria*. The numerous cercariae produced after the cycles of clonal reproduction emerge by a birth pore located on the tegument of sporocysts or rediae. Roughly, cercariae look like small tadpoles, with a tail, conferring the motility, and a body comprising the oral sucker and the ventral sucker which location prefigures that of the adult (Bogitsh et al. 2019). There is a wide morphological variability among species (Dawes 1968). In most of the cases, cercariae leave the mollusc, and after a free-living-swimming period, contaminate the following host or encyst in the environment (see below for information on cycles).

12.5 Digenean 1: Mollusc 0

The consequences of infection and of the overdevelopment of digenean larvae are important and somewhat disastrous for the mollusc. The most obvious effect is the destruction of host tissues during the development of parasitic larvae (Graczyk and Fried 1999). However several other, sometimes unexpected, effects have been reported.

One of the most spectacular examples is that of phenotypic changes of the terrestrial succineids parasitized by *Leucochloridium* spp.: the tentacles of the host are invaded by sporocysts and look like funny caterpillars. This parasite exhibition facilitates the contamination of the following host, an insectivorous bird which is attracted by a so appetizing prey. Wesołowska and Wesołowski (2014) demonstrated that this facilitation is also accompanied by a manipulation of host behavior: as compared to their healthy conspecifics, *Succinea putris* infected by *Leucochloridium paradoxum* move to higher distances and expose themselves in more open and better illuminated places. This behavior increases the risk of being detected and eaten by a definitive host.

Different experiments have reported stunting in infected snails [e.g., Pan (1965) in the planorbid *B. glabrata* infected by *S. mansoni*; Krist and Lively (1998) in the hydrobiid *Potamopyrgus antipodarum* infected by *Microphallus* sp.; Muñoz-Antol et al. (2007) in the lymnaeid *Radix peregra* infected by *Echinostoma friedi*]. Such results may be explained by the deep modification of the resource allocation scheme into the host. Indeed, the mollusc expends energy to display its immune response that needs to increase the synthesis of additional molecules and to recruit amounts of hemocytes. Moreover, for its own development and protection against host defenses, the parasite drains resources that belong to its host. There is therefore competition between parasite and host for the same resources. Taking into account the velocity of parasite development, the host is probably strongly disadvantaged. The mechanisms involved in the changes of energy allocation can be complex. For instance, in *Lymnaea stagnalis* experimentally infected by *Trichobilharzia ocellata*, Hoek et al. (2005) have reported upregulation of neuropeptides (LFRFamide) inhibitors of neuroendocrine cells that control growth, metabolism, or reproduction.

The modifications of metabolism induced by parasitism can complicate the host responses to other stressors. For instance, although pollutants seem to have any noticeable effects on survival and fecundity of freshwater snails (López-Doval et al. 2014), they can negatively interact with digenean infection. There are examples showing how parasitism can affect the tolerance of snails to heavy metals: Guth et al. (1977) have shown that exposure to high zinc concentrations induces more rapid mortality in *L. stagnalis* infected with *Schistosomatium douthitti* or *Trichobilharzia* sp. than in non-infected snails (see also Allah et al. 1997). Similarly, Morley et al. (2003) showed that exposure to cadmium reduces the survival of *Radix peregra* and *L. stagnalis* exposed to diverse digeneans as compared to non-infected snails. Conversely, exposure to pesticides may also cause histological and physiological disruptions into the mollusc (Kanapala and Arasada 2013) that can lower immune defenses, leading to an increased susceptibility to digeneans infection and favoring their development.

Counterintuitively, several examples have shown that infection by digeneans can cause gigantism in their snail hosts (e.g., Sorensen and Minchella 1998). According to the review done by Sorensen and Minchella (2001), freshwater snails seem to be more prone to respond to digenean infected by an increase of their size than marine snails, whereas the opposite is observed for stunting. Among pulmonates, lymnaeids appeared to be more prone to display gigantism than planorbids (which exhibit more often stunting). However, Sturrock (1966) revealed accelerated shell growth in *Biomphalaria pfeifferi* infected by *S. mansoni*. Alternatively, Hodasi (1972) showed that *Galba truncatula* infected by *Fasciola hepatica* exhibits gigantism only when infection occurs for snails aged 2 or 4 weeks, whereas no effect is detected for 6 weeks old snails. Both experiment and observational studies suggest a tendency to digenean with rediae to induce more gigantism than those with only sporocysts. However, digenean with sporocysts might induce acceleration of growth more precociously than those with rediae (Sorensen and Minchella 2001). According to Gorbushin and Levakin (1999), it seems that shorter-lived snails are more prone to exhibit gigantism when infected than long-lived species. Although gigantism would seem a paradox, it results from reallocation of energy into growth and can be interpreted as a beneficial strategy for the digenean. Indeed, bigger snails offer more resources and more space to insure the multiplication and development of larval stages (Niewiadomska and Pojmanska 2011).

The most striking effect of digeneans on their first intermediate host is probably castration. As shown by Sorensen and Minchella (1998), in *Stagnicola elodes* infected by *Echinostoma revolutum*, the depletion in host reserves caused by the parasite development can alter the reproductive function: the fecundity of exposed snail decreases 3 weeks post-exposure and reproduction is totally interrupted 2–3 weeks later. Similarly, Muñoz-Antol et al. (2007) have obtained total castration in both *Radix peregra* and *Biomphalaria glabrata* after infection by *Echinostoma friedi*. As reviewed by Sorensen and Minchella (2001), the mechanisms involved can differ between digenean with sporocysts and those with rediae. In some cases, an increase in fecundity is observed before castration occurs (Sorensen and Minchella 2001). However, this seems to be limited to a few cases, like in infections by *Schistosoma* spp. whatever the host, planorbids or lymnaeids (review in Sorensen and Minchella 2001), or situations in which the host is placed under conditions of starvation (e.g., Keas and Esch 1997). Experimental infections performed with *F. hepatica* in the American lymnaeid species *Pseudosuccinea columella* and *Galba cubensis*, showed no cessation of host fecundity with increased egg laying after cercarial emission in some cases (Gutiérrez et al. 2000, 2002).

By constraining the chances of survival, reproduction and transmission of their parasites, molluscs exert selective pressures on digeneans. Since digeneans, by the way of movements of their definitive hosts, are expected to disperse more than their molluscan hosts, conditions required for local adaptation are satisfied (Gandon et al. 1996). In such a case, it is expected that infection is more severe (higher virulence and transmission) for sympatric hosts than for allopatric ones. With reciprocal cross-infection experiments, Lively (1989) has demonstrated that *Microphallus* sp. is more infective to *Potamopyrgus antipodarum* from its sympatric populations than for allopatric ones. In *Galba truncatula*, Dar et al. (2013) showed that allopatric *F. hepatica* strains induce lower mortality than sympatric ones. Similarly, Dreyfuss et al. (2012) found that *G. truncatula* dying without *F. hepatica* cercarial shedding is clearly predominant in the allopatric group than in the sympatric one.

Since digeneans strongly affect the fecundity, and sometimes the survival, of their first intermediate hosts, they are able to exert strong selective pressures on the molluscs. They can therefore act as drivers of the evolution of molluscan populations (e.g., Minchella and LoVerde 1981; Dybdahl and Lively 1995; Lively et al. 2004).

Such strong reciprocal pressures between molluscs and digeneans can lead to co-adaptations (Hurtrez-Boussès et al. 2001). Associated to a close specificity, they can lead to cospeciation events, like those described for digeneans belonging to the complex *Schistosoma japonicum* and prosobranchs of the family Pomatiopsidae (Adema and Loker 1997).

12.6 The Life-Cycles of Digeneans: How to Choose Its Own Way

In the typical cycle, cercariae leave the mollusc, most often stimulated by external factors like light or gravity, scarcely by chemical cues emitted by the following host (Combes et al. 1994). Generally, cercariae are non-feeding and short-lived. Depending on the parasite species, the life-cycle of digeneans can be more or less complex with one to four hosts (see review in Galaktionov and Dobrovolskij 2003a). Most of the digenean species have two or three hosts. According to Poulin and Cribb (2002), cycles with one or two hosts are observed in more than 32 independent families of digeneans. Among the three main types of cycles (hereafter (1) to (3)), the most complex involves three hosts: (1) after leaving the mollusc, the cercaria penetrates, actively or passively, a second intermediate host, usually an arthropod but sometimes another invertebrate or a vertebrate (most often a fish or an amphibian) (Esch et al. 2002). After their installation, cercariae shed their tail and transform into metacercariae which can encyst or not. Contamination of the definitive host occurs by ingestion of the second intermediate host. Once into their definitive host, the parasites migrate toward the target organ and undergo their sexual maturation. This cycle has been for example described by Grobbelaar et al. (2014) for *Diplostomum* sp. in southern Africa: lymnaeids are the most probable first intermediate hosts, cercariae infect a teleost fish and the definitive host is a piscivorous bird. Interestingly, after shedding from the first intermediate host (a lymnaeid or a physid snail), the cercariae of the diplostomid *Cotylurus flabelliformis* penetrate their second intermediate host which is again a lymnaeid (or sometimes a leech), like *Lymnaea stagnalis* (Yoder and Coggins 1998). In this second host, they evolve in metacercariae under a form called tetracotyle (Campbell 1973). If their molluscan host is already parasitized by another digenean, these tetracotyles are able to become hyperparasites of sporocysts or rediae they encounter (Goater et al. 2014). It is noticeable that snails previously infected by sporocysts of *C. flabelliformis* develop immunity against their own metacercariae (Campbell 1973).

In other cases, (2) cercariae do not penetrate a second intermediate host but encyst as metacercariae on vegetation or in the water. The definitive hosts acquire the parasite by consuming contaminated plants or water. This is the well-known case of *Fasciola* spp., which start their cycle in lymnaeids, encyst in aquatic vegetation or water, and infect vertebrates by ingestion (Hurtrez-Boussès et al. 2001; Alba et al. 2021; Andrews et al. 2021).

(3) A shorter cycle is adopted in blood flukes belonging to three independent families (Sanguinocolidae, Schistosomatidae, and Spirorchidae): after leaving their snail-first intermediate host, cercariae actively penetrate the definitive host (e.g., Nelwan 2019). The most known example is that of schistosomes, but it has also been described in several other species, like, for instance, *Sanguinicola inermis*: two generations of sporocysts develop into *Radix peregra* and produce free-swimming cercariae which directly infect the teleost definitive host (Kirk and Lewis 1993).

(4) Sometimes, the vertebrate host is deleted from the life-cycle: eggs are directly produced within the selfer progenetic metacercarial cyst. This is the case of *Alloglossidium* spp. (family Macroderoididae) for which, after a freshwater snail as first intermediate host, the second intermediate host, a leech or a crustacean, is also the definitive one (Font 1980). As proposed by Carney and Brooks (1991) and Smythe and Font (2001), based on phylogenetic analyses, the ancestor cycle in this group had three hosts (snail-crustacean-fish). Simplification occurred by the loss of host-fish, followed in a more recent clade by a transfer from crustacean to leech.

(5) Life-cycle truncation can also occur when the mollusc acts at the same time as the first and second intermediate host, encystment of metacercariae happening into the mollusc body. Diverse digenean families can present such a way of life. As described by Taft (1972), the miracidium of *Cyclocoelum oculeum* infects a freshwater snail, lymnaeid (e.g., *Radix ovata*, *Lymnaea stagnalis*, *Stagnicola reflexa*) or planorbid. The redial development gives birth to cercariae which encyst directly in the snail tissues. The bird-definitive host is contaminated by ingesting snails harboring metacercariae.

(6) In other species, the same host can act simultaneously as second intermediate and definitive host. An example is given by *Haplometra cylindracea*: after sporocysts multiplication in a lymnaeid snail (e.g., *Galba truncatula*; see Vignoles et al. 2011), cercariae are emitted and infect a frog (e.g., *Rana temporaria*). They encyst as metacercariae on the buccal mucosa of this host (Pandian 2020). After bursting, the larvae migrate to the frog's lungs, where they develop as adults. It can be noted that the metacercarial stage can be shunted, since cercariae are also able to join directly to the lung and to mature as adults.

(7) Total simplification is observed in some digeneans with monogenetic cycles, the whole cycle being accomplished into the same mollusc. Two ways exist: with or without sexual reproduction. In the first case (e.g., *Plagioporus sinitsini* in prosobranch snails; see Barger and Esch 2022), cercariae are sequestered in daughter sporocysts and maturate to become adults, passing through the metacercarial stage. In some other rare cases (e.g., *Mesostephanus haliasturis* in the prosobranch *Melanoides tuberculata*; see Barker and Cribb 1993) infective miracidia are directly produced from sporocyst inside the mollusc.

(8) The cycle of some digeneans can involve up to four hosts, as described for *Halipegus ovocaudatus* (Kechemir 1978) which infects sequentially a planorbid, a copepod, an odonate and finally a frog as definitive host. It is also suspected in the genus *Platynosomum*, which larvae first multiplicate into a terrestrial snail, then encyst as metacercariae into a terrestrial isopod. Excysted metacercariae found in the liver of lizards suggest that these vertebrates can act as a third intermediate hosts, before being ingested by birds or mammals definitive hosts (see review and discussion in Pinto et al. 2014).

To understand the diversity and the complexity of cycles, Poulin and Cribb (2002) defined critical events needed to ensure transmission. First, the miracidium should find a compatible mollusc as first intermediate host. Second, the free short-living cercaria should find a suitable second intermediate host or a good location to encyst. Third, the parasite should encounter its definitive host. Additionally, crossing filters of encounter and compatibility and maintaining host–parasite interactions render extremely complex and hazardous the success of infection. High adult fecundity and intense asexual multiplication, as well as active detection of the host by the miracidium and by the cercaria, facilitation and manipulation of intermediate hosts, resistance of eggs and metacercariae in ambient environment can therefore be interpreted as adaptive strategies that counteract the hazards of transmission and thus that favor the completion of the cycle (see reviews in Poulin and Cribb 2002; Galaktionov and Dobrovolskij 2003b).

12.7 A Success Story

From the evolutionary point of view, the diversity of life-cycles observed in digeneans offers a good support to assess adaptive and contingent events that shape their evolution. Poulin and Cribb (2002) suggested that the ancestral life-cycle of digeneans involved three hosts and that simplifications (i.e., losses of hosts) have occurred independently at least 20 times. However, according to the phylogeny provided by Cribb et al. (2003) [see also Littlewood et al. 2015; Pérez-Ponce de León and Hernández-Mena 2019], it is most likely that the ancestral cycle comprised only two hosts, a mollusc as intermediate and a vertebrate as definitive hosts. Most probably, the common ancestor to digeneans had eggs that produced miracidia able to penetrate actively the mollusc. In agreement with such hypothesis, Galaktionov and Dobrovolskij (2003a) present passive accumulation of miracidia as a secondary acquisition. Results obtained by Cribb et al. (2003) also suggest that, in the ancestral cycle, asexual multiplication sporocysts gave birth to rediae that produced free cercariae which contaminated by ingestion the definitive host, a teleost fish. Among the 23 superfamilies represented in the Digenea phylogeny, only eight contain species that have a bivalve as first intermediate host. By contrast, each one of these 23 superfamilies includes at least one species with a gastropod as first intermediate host (Cribb et al. 2003). The most parsimonious hypothesis is therefore that the first intermediate host of the common ancestor of digeneans was a (aquatic) gastropod, and that switches toward bivalves and also scaphopods occurred later. The wide distribution of gastropods and the diversity of ecological niches they occupy has probably helped the digeneans to diversify and to conquer new areas. Among gastropods, pulmonate snails play an important role in the transmission of parasites in semiaquatic and terrestrial ecosystems. Within Hygrophila, the group (superfamily) of Lymnaeoidea (Ayyagari and Sreerama 2020) has received particular attention since it comprises planorbids and lymnaeids, groups which include vectors of diseases with strong impact on human and veterinary health, like, respectively, schistosomiasis and fasciolosis. Lymnaeids come out on the top by the diversity of digeneans they can host. Among the digeneans described by Dawes (1968), at least 71 species, belonging to 13 superfamilies (Dawes 1968) have a lymnaeid snail as intermediate host. Moreover, some lymnaeid species are able to host a wide diversity of digeneans. For instance, Gordy and Hanington (2019) reported 40 different digenean species hosted by *Stagnicola elodes*. In the concerned digenean lineages, the use of lymnaeids as first intermediate hosts has probably amplified the widening of definitive host spectrum, since the infection of amphibious snails has allowed parasites to conquer more terrestrial habitats and to infect the hosts which inhabit these environments. The scenario proposed by Vázquez et al. (2021a) to explain *Fasciola* spp. expansion and radiation offers an interesting overview to envisage how the combined effects of definitive and intermediate host spectra, together with ecological features of such vertebrates and molluscs, might have favored the success story of digeneans. Initially transmitted by planorbids in their area of origin, some fasciolids have probably switched ca. 65–55.9 Ma ago to lymnaeid snails with aquatic way of life (like actual *Radix* spp.). At their arrival in Asia (ca. 18.5 Ma ago), the initial wide range of compatible definitive hosts of fasciolids has allowed them to conquer new hosts from the diversified bovid tribe. The contamination of bovids, that do not spend significant time into the water but prefer to graze in humid meadows, might have favored the encounter with semi-amphibious lymnaeids like *Austropeplea* spp. Gradually, the flukes may have conquered more terrestrial definitive hosts and more amphibious snails, which would have facilitated the spread of *F. hepatica* in Americas where the true amphibious *Galba* spp. are omnipresent and highly diversified (Alda et al. 2021). This geographical and ecological expansion has probably allowed the encounter of new definitive hosts like camelids.

As a result of the ancient and complex interactions between digeneans and their molluscan hosts, digeneans have conquered most of the ecosystems and are therefore expected to be important actors in the interaction networks.

12.8 Dynamics of Molluscs–Digeneans Interactions: From Populations to Communities

The ecological interactions between digeneans and their molluscan hosts can be studied at three main levels: populations (1), infra-host communities (digenean diversity within the snail) (2) and, at local or larger scale, component communities (digenean diversity within and among snail populations) (3).

(1) Taking into account the high specificity of digenean–mollusc systems and the fitness effects both partners exert each on the other, reciprocal effects on population dynamics of the host and parasite are expected. Considering the exceptionally high levels of asexual multiplication observed in digeneans, compatible molluscs would have poor chances to impair the total headcount of their parasites. However, by selecting the most compatible genotypes and/or by arbitrating competition among strains, molluscs can regulate the intraspecific diversity of their parasites. Karvonen et al. (2012) tested the consequences of infection with one versus two different genotypes of *Diplostomum pseudospathaceum* in *Lymnaea stagnalis*. They found no differences in cercarial output between the two groups, suggesting that, in case of

multiple infection, intraspecific competition lowers the per-genotype reproduction. If differences in compatibility and/or in infectivity exist between strains, such competition events can lead to the selection of the most compatible/infective genotype. This most locally adapted parasitic genotype can in turn counter-select its most compatible/susceptible host genotype; the frequency of the latter will decrease, leading to the counter-selection of the associated parasite... and so on, resulting in frequency-dependent dynamics (Thomas et al. 2010).

Although it is quite difficult to test parasitism effects on snail population dynamics *in natura*, some studies gave interesting insights. Results obtained by Brown et al. (1988) in a demographic study of three populations of *Stagnicola elodes*, combined with a parasite survey, suggested that digeneans can lower by 10–20% the host-population growth. Similarly, Fredensborg et al. (2005) have followed populations of the snail *Zeacumantus subcarinatus* infected by the digenean *Maritrema novaezelandensis*. They found a strong variation in prevalences among populations and a negative correlation between local digeneans prevalences and both density and biomass of molluscs.

(2) Without any interaction between digenean species, we should expect that the probability of multiple infections directly depends on prevalences of each single species. However, several examples of multiple infections over- or underrepresented as expected at random have been reported (review in Esch et al. 2001). For instance, Bourns (1963) reported in *L. stagnalis* multiple infections more frequent than expected at random. Ewers (1960) proposed that some digeneans might facilitate other species by increasing the susceptibility of their hosts to further infections. On the opposite, Cort et al. (1937) found that some infections occur lesser than expected at random, suggesting the existence of mechanisms limiting co-infections. Kuris (1990) and Lafferty (1993) proposed that digeneans with large rediae are able to predate, inside their molluscan host, potential competitors with smaller rediae or only with sporocyst stage. However, their model lacks experimental or observational evidence. On the opposite, Curtis (1996) and Curtis and Tanner (1999) suggested that parasite recruitment in *Ilyanassa obsoleta* may be too low to induce interspecific competitive pressures. Sousa (1990) showed that in digenean communities of *Cerithidea californica*, co-occurrence of species with large rediae is scarce. However, he concluded that the predation they exert cannot be the only factor explaining the poverty of digenean infra-communities. Other studies in marine prosobranchs (Curtis 1996; Curtis and Tanner 1999) and in freshwater physids (Snyder and Esch 1993) and planorbids (Fernandez and Esch 1991) suggested that spatial and temporal heterogeneities are most efficient in driving the composition of digenean infra-communities than predation/competition among rediae. In order to understand determinants of trematode infections in communities of intertidal gastropods, Poulin and Mouritsen (2003) analyzed data gathered from a literature survey. They detected a positive relationship between digenean richness per snail species and prevalence, which is congruent with the additive nature of digenean prevalences in snail-hosts populations. They suggested that this could be explained by negative interspecific interactions between different digenean species sharing the same intermediate host species. Indeed, intra-molluscan competitive exclusion among digeneans is expected

to lower the probability of multiple co-infections. Prevalences are therefore expected to be additive in the snail populations. This can have important consequences in dynamics and evolution of molluscs populations, since each new digenean species colonizing a habitat would impose a new additional cost to its host.

(3) Although intraspecific interactions between digeneans can contribute to explain, at least in some cases, the infra-communities diversity, Esch et al. (2001) suggested that, due to the patchiness of the habitat in which competition occurs, they poorly predict component communities. Indeed, dominant species at intra-host level are not necessarily the most abundant in component communities. In other words, independently of its intra-molluscan competitive abilities, a digenean species can be abundant in one location and rare in another one (Esch et al. 2001).

Since molluscs can be considered as the resource for digeneans, correlations between host availability and diversity (including species richness and abundance) of digeneans would be expected. In freshwater ecosystems, Gérard (2001) revealed a positive relationship between the frequency of digenean species in the community and the number of freshwater gastropod species acting as first intermediate hosts. Taking into account the fact that digeneans are mainly specialists, one can ask if composition of mollusc communities are good predictors of composition of digenean component communities. A prerequisite is that each mollusc species has only one, or a very limited number of, digenean species able to exploit it.

In Denmark, Duan et al. (2021) inventoried 10 freshwater snail species harboring a total of 22 digenean species. Overall prevalence was 12.6%, with large variability among snail species. The highest prevalence (18.2%) and digenean diversity (10 species) were found in *Lymnaea stagnalis* which contributes to 47.6% of the parasite diversity. In an extensive inventory made in Germany, Schwelm et al. (2021) examined freshwater snails from 15 species, belonging to seven families. Among them, 9 species harbored digeneans. The lymnaeid *Ampullaceana balthica*, which was also the most abundant snail, showed the highest digenean richness, with 13 species among the 40 reported in the total sample. *Bithynia tentaculata, Planorbarius corneus,* and *L. stagnalis* harbored, respectively, 11, 10, and 9 species. The other 11 snail species harbored only a few or no digenean species. Similarly, in a large-scale study made in lakes in Alberta (Canada), Gordy and Hanington (2019) used molecular tools to characterize the digenean diversity. They found 79 species brought by only five snail species. Among them, two lymnaeids harbor the highest digenean richness: *Stagnicola elodes* (40 species) and *L. stagnalis* (10 species). These results are consistent with those of Selbach et al. (2020) who observed, in German freshwater systems, that among the four lymnaeid and two planorbid studied the two most abundant taxa (*Radix auricularia* and *Gyraulus albus*) host 86% of the total digenean diversity. Particularly, *R. auricularia* harbors 23 of the 36 digenean species detected. It seems therefore that the parasite diversity is not determined by the snail diversity per se, but better by the presence/absence of some keystone species. According to the studies described above, some lymnaeids and, in a lesser extent, planorbids, can therefore be considered as key host species in freshwaters systems (Schwelm et al. 2021). This might have important implications in the control programs of digeneans of medical or veterinary importance, since they would concentrate on a few mollusc target species.

The stability of component communities can also be addressed. In freshwater systems in France, Gérard (2001) found that prevalences strongly vary in time, due to important variations in snails populations dynamics, combined with fluctuations in ecological parameters. This author also mentioned a negative correlation between the variable abundances of gastropods and digeneans prevalences. Stability of component communities of digeneans is also influenced by life-history traits of their hosts: according to Esch et al. (2001), it strongly depends on the life-span of molluscs. Based on an extensive review, these authors suggested that component communities of (mainly marine) long-lived gastropods may accumulate multiple infection events whereas in the (mainly freshwater) short-lived snails, community compositions can more rapidly change with cohort turnover. This can have important epidemiological importance since in systems with long-lived snails, ephemeral visits of contaminated definitive hosts can be sufficient to maintain the parasite over decades. Contrarily, in systems with only short-lived snails, the maintenance of digenean, and therefore of associated epidemiological risks, depends on the frequency of definitive hosts visits and on the seasonality of encounters between digenean and compatible snails.

Finally, the digenean communities can be modified in response to environmental changes. Indeed, as underlined by Sabourin et al. (2018) for fasciolosis, man-made managements (e.g., irrigation systems) and modifications of human practices (e.g., intensive livestock husbandry) can favor compatible snail and can increase probability of snail-miracidium encounter. In other models, Poulin and Mouritsen (2006) have shown that cercarial production is highly susceptible to increases in temperature. Both local and global changes can therefore modify mollusc–digeneans communities, which would be used as sentinels of changes in the ecosystems and would serve to evaluate ecological and epidemiological risks.

For this purpose, we therefore need to reinforce our understanding on digenean–mollusc systems. A special attention has to be paid to freshwater systems, which are the most threatened by anthropic perturbations. Particularly, lymnaeids which host most of the digenean diversity have to be scrutinized. In complement to traditional methods of inventory, molecular tools can help to improve the characterization of diversity in both lymnaeids and digeneans. For instance, barcoding should be generalized to confidently determine hosts and parasites, by avoiding taxonomic confusions and by solving the problem of cryptic species. Moreover, methods of environmental-DNA should be applied to better detect snails and to inventory free-living stages of digeneans.

Acknowledgments Special thanks are due to François Renaud who encouraged us to work on the fascinating world of digeneans–lymnaeids interactions. The authors are also most grateful to Daniel Rondelaud, Gilles Dreyfuss, and Patrick Durand for fruitful collaborations. The authors acknowledge the numerous colleagues and students for their contributions and helpful discussions. Finally, we are indebted to our families for their everyday life support and patience.

References

Abrous M, Rondelaud D, Dreyfuss G et al (1998) Unusual transmission of the liver fluke, *Fasciola hepatica*, by *Lymnaea glabra* or *Planorbis leucostoma* in France. J Parasitol 84:1257–1259

Adema CM (2015) Fibrinogen-related proteins (FREPs) in mollusks. In: Hsu E, Du Pasquier L (eds) Pathogen-host interactions: antigenic variation vs somatic adaptations. Springer, Cham, pp 111–129. https://doi.org/10.1007/978-3-319-20819-0_5

Adema CM, Loker ES (1997) Specificity and immunobiology of larval digenean-snail associations. In: Advances in trematode biology, vol 8. Routledge, New York, pp 229–263

Alba A, Hernández HM, Marcet R et al (2014) Exploring the antigenic features of *Fasciola hepatica* rediae (Trematoda: Digenea) through the evaluation of different antigenic candidates for further monoclonal antibody generation. Parasitol Res 113:3185–3193. https://doi.org/10.1007/s00436-014-3981-y

Alba A, Hernández HM, Marcet R et al (2015) A novel monoclonal antibody-based immunoenzymatic assay for epidemiological surveillance of the vector snails of *Fasciola hepatica* (Trematoda: Digenea). Int J Parasitol 45:113–119. https://doi.org/10.1016/j.ijpara.2014.09.007

Alba A, Duval D, Sánchez J et al (2020) The immunobiological interplay between *Pseudosuccinea columella* resistant/susceptible snails with *Fasciola hepatica*: Hemocytes in the spotlight. Dev Comp Immunol 102:103485. https://doi.org/10.1016/j.dci.2019.103485

Alba A, Vázquez AA, Hurtrez-Boussès S (2021) Towards the comprehension of fasciolosis (re-) emergence: an integrative overview. Parasitology 148:385–407. https://doi.org/10.1017/s0031182020002255

Alda P, Lounnas M, Vázquez AA et al (2021) Systematics and geographical distribution of *Galba* species, a group of cryptic and worldwide freshwater snails. Mol Phyl Evol 157:107035. https://doi.org/10.1016/j.ympev.2020.107035

Allah ATA, Wanas MQS, Thompson SN (1997) Effects of heavy metals on survival and growth of *Biomphalaria glabrata* Say (Gastropoda: Pulmonata) and interaction with schistosome infection. J Mollus Stud 63:79–86

Andrews SJ (1999) The life cycle of *Fasciola hepatica*. In: Dalton JP (ed) Fasciolosis. CAB International Publishing, Oxford, UK, pp 1–30

Andrews SJ, Cwiklinski K, Dalton JP (2021) The discovery of *Fasciola hepatica* and its life cycle. In: Dalton JP (ed) Fasciolosis. CABI, Wallingford, pp 1–22. https://doi.org/10.1079/9781789246162.0000

Ayyagari VS, Sreerama K (2020) Molecular phylogeny and evolution of Pulmonata (Mollusca: Gastropoda) on the basis of mitochondrial (16S, COI) and nuclear markers (18S, 28S): an overview. J Genet 99:17. https://doi.org/10.1007/s12041-020-1177-z

Barger MA, Esch GW (2022) *Plagioporus sinitsini* (Digenea: Opecoelidae): a one-host life cycle. J Parasitol 86:150–153. https://doi.org/10.1645/0022-3395(2000)086[0150:psdoao]2.0.co;2

Barker SC, Cribb TH (1993) Sporocysts of *Mesostephanus haliasturis* (Digenea) produce miracidia. Int J Parasitol 23:137–139

Barrios-de Pedro S, Osuna A, Buscalioni ÁD (2020) Helminth eggs from early cretaceous faeces. Sci Rep 10:18747. https://doi.org/10.1038/s41598-020-75757-4

Bayne CJ, Loker ES, Yui MA (1986) Interactions between the plasma proteins of *Biomphalaria glabrata* (Gastropoda) and the sporocyst tegument of *Schistosoma mansoni* (Trematoda). Parasitology 92:653–664

Blair D, Davis GM, Wu B (2001) Evolutionary relationships between trematodes and snails emphasizing schistosomes and paragonimids. Parasitology 123:229–243. https://doi.org/10.1017/s003118200100837x

Blaustein AR, Gervasi SS, Johnson PTJ et al (2012) Ecophysiology meets conservation: understanding the role of disease in amphibian population declines. Philos Trans R Soc B 367:1688–1707. https://doi.org/10.1098/rstb.2012.0011

Bogitsh BJ, Carter CE, Oeltmann TN (2019) Human parasitology. Academic Press, London, UK

Boray JC (1969) Experimental Fascioliasis in Australia. Adv Parasitol 7:95–210

Born-Torrijo A, Holze AS, Raga JA et al (2017) Description of embryonic development and ultrastructure in miracidia of *Cardiocephaloides longicollis* (Digenea, Strigeidae) in relation to active host finding strategy in a marine environment. J Morphol 278:1137–1148. https://doi.org/10.1002/jmor.20700

Bourns TKR (1963) Larval trematodes parasitizing *Lymnaea stagnalis appressa* Say in Ontario, with emphasis on multiple infections. Can J Zool 41:937–941

Brown KM, Leathers BK, Minchella DJ (1988) Trematode prevalence and the population dynamics of freshwater pond snails. Am Midl Nat 120:289–301. https://doi.org/10.2307/2426001

Campbell RA (1973) Studies on the biology of the life cycle of *Cotylurus flabelliformis* (Trematoda: Strigeidae). Trans Am Microsc Soc 92:629–640

Carney JP, Brooks DR (1991) Phylogenetic analysis of *Alloglossidium* Simer, 1929 (Digenea: Plagiorchiiformes: Macroderoididae) with discussion of the origin of truncated life cycle patterns in the genus. J Parasitol 77:890–900

Cissé G (2019) Food-borne and water-borne diseases under climate change in low- and middle-income countries: further efforts needed for reducing environmental health exposure risks. Acta Trop 194:181–188. https://doi.org/10.1016/j.actatropica.2019.03.012

Combes C, Fournier A, Moné H, Théron A (1994) Behaviours in trematode cercariae that enhance parasite transmission: patterns and processes. Parasitology 109:S3–S13

Connors VA, Lodes MJ, Yoshino TP (1991) Identification of a *Schistosoma mansoni* sporocyst excretory-secretory antioxidant molecule and its effect on superoxide production by *Biomphalaria glabrata* hemocytes. J Invertebr Pathol 58:387–395. https://doi.org/10.1016/0022-2011(91)90185-s

Correa A, Escobar J, Durand P et al (2010) Bridging gaps in the molecular phylogeny of the Lymnaeidae (Gastropoda: Pulmonata), vectors of Fascioliasis. BMC Evol Biol 10:381. https://doi.org/10.1186/1471-2148-10-381

Cort WW, McMullen D, Brackett S (1937) Ecological studies on the cercariae in *Stagnicola emarginata angulata* (Sowerby) in the Douglas Lake region, Michigan. J Parasitol 23:504–532

Coustau C, Gourbal B, Duval D et al (2015) Advances in gastropod immunity from the study of the interaction between the snail Biomphalaria glabrata and its parasites: a review of research progress over the last decade. Fish Shellfish Immunol 46:5–16. https://doi.org/10.1016/j.fsi.2015.01.036

Coyne K, Laursen JR, Yoshino TP (2015) In vitro effects of mucus from the mantle of compatible (*Lymnaea elodes*) and incompatible (*Helisoma trivolvis*) snail hosts on *Fascioloides magna* miracidia. J Parasitol 101:351–357. https://doi.org/10.1645/14-606.1

Cribb TH, Bray RA, Littlewood D et al (2001) The Digenea. In: Littlewood TDJ, Bray RA (eds) Interrelationships of the Platyhelminthes. Taylor & Francis, London, UK, pp 168–185

Cribb T, Bray R, Olson P et al (2003) Life cycle evolution in the Digenea: a new perspective from phylogeny. Adv Parasitol 54:197–254. https://doi.org/10.1016/s0065-308x(03)54004-0

Curtis LA (1996) The probability of a marine gastropod being infected by a trematode. J Parasitol 82:830–833

Curtis LA, Tanner NL (1999) Trematode accumulation by the estuarine gastropod *Ilyanassa obsoleta*. J Parasitol 85:419–425

Cwiklinski K, Dalton JP, Dufresne PJ et al (2015) The *Fasciola hepatica* genome: gene duplication and polymorphism reveals adaptation to the host environment and the capacity for rapid evolution. Genome Biol 16:71. https://doi.org/10.1186/s13059-015-0632-2

Dar Y, Lounnas M, Djuikwo Teukeng FF et al (2013) Variations in local adaptation of allopatric *Fasciola hepatica* to French *Galba truncatula* in relation to parasite origin. Parasitol Res 112: 2543–2549. https://doi.org/10.1007/s00436-013-3421-4

Dawes B (1968) The Trematoda: with special reference to British and other European forms. Cambridge University Press, London

Dreyfuss G, Vignoles P, Rondelaud D (2012) Local adaptation of the trematode *Fasciola hepatica* to the snail *Galba truncatula*. Parasite 19:271–275. https://doi.org/10.1051/parasite/2012193271

Duan Y, Al-Jubury A, Kania PW et al (2021) Trematode diversity reflecting the community structure of Danish freshwater systems: molecular clues. Parasit Vectors 14:43. https://doi.org/10.1186/s13071-020-04536-x

Dybdahl MF, Lively CM (1995) Host-parasite interactions: infection of common clones in natural populations of a freshwater snail (*Potamopyrgus antipodarum*). Proc R Soc B 260:99–103. https://doi.org/10.1098/rspb.1995.0065

Eernisse DJ, Peterson KJ (2004) The history of animals. In: Cracraft J, Donoghue MJ (eds) Assembling the tree of life. Oxford University Press, Oxford, UK, pp 197–208

Esch GW, Curtis LA, Barger MA (2001) A perspective on the ecology of trematode communities in snails. Parasitology 123:57–75. https://doi.org/10.1017/s0031182001007697

Esch GW, Barger MA, Fellis KJ (2002) The transmission of digenetic trematodes: Style, elegance, complexity. Integr Comp Biol 42:304–312. https://doi.org/10.1093/icb/42.2.304

Ewers WMH (1960) Multiple infections of trematodes in a snail. Nature 186:990. https://doi.org/10.1038/186990a0

Fernandez J, Esch GW (1991) Guild structure of larval trematodes in the snail *Helisoma anceps*: patterns and processes at the individual host level. J Parasitol 77:528–539

Font WF (1980) The effect of progenesis on the evolution of *Alloglossidium* (Trematoda, Plagiorchiida, Macroderoididae). Acta Parasitol Pol 27:15–28

Francisco CJ, Hermida MA, Santos MJ (2012) *Prosorhynchus crucibulum* (Digenea: Bucephalidae) miracidium morphology and its passive transmission pattern. Parasite 19:277–280. https://doi.org/10.1051/parasite/2012193277

Fredensborg B, Mouritsen K, Poulin R (2005) Impact of trematodes on host survival and population density in the intertidal gastropod *Zeacumantus subcarinatus*. Mar Ecol Prog Ser 290:109–117

Fried B, Graczyk T (1997) Advances in trematode biology. CRC Press, Boca Raton, FL

Friesen OC, Goellner S, Poulin R, Lagrue C (2020) Parasites shape community structure and dynamics in freshwater crustaceans. Parasitology 147:182–193. https://doi.org/10.1017/s0031182019001483

Galaktionov KV, Dobrovolskij AA (2003a) The main types of trematode life cycles. In: Fried B, Graczyk TK (eds) The biology and evolution of trematodes. Springer, Dordrecht, The Netherlands, pp 319–348

Galaktionov KV, Dobrovolskij AA (2003b) The trematode life cycle as a system of adaptations. In: Fried B, Graczyk TK (eds) The biology and evolution of trematodes. Springer, Dordrecht, The Netherlands, pp 215–317

Galinier R, Portela J, Moné Y et al (2013) Biomphalysin, a new β pore-forming toxin involved in *Biomphalaria glabrata* immune defense against *Schistosoma mansoni*. PLoS Pathog 9: e1003216. https://doi.org/10.1371/journal.ppat.1003216

Gandon S, Capowiez Y, Dubois Y et al (1996) Local adaptation and gene-for-gene coevolution in a metapopulation model. Proc R Soc Lond B 263:1003–1009. https://doi.org/10.1098/rspb.1996.0148

Gérard C (2001) Structure and temporal variation of trematode and gastropod communities in a freshwater ecosystem. Parasite 8:275–287. https://doi.org/10.1051/parasite/2001084275

Gibson AK, Drown DM, Lively CM (2002) Keys to the Trematoda. CABI Publishing and the Natural History Museum, Wallingford, UK

Goater TM, Goater CP, Esch GW (2014) Parasitism: the diversity and ecology of animal parasites. Cambridge University Press, Cambridge, UK

Gorbushin AM, Levakin IA (1999) The effect of trematode parthenitae on the growth of *Onoba aculeus*, *Littorina saxatilis* and *L obtusata* (Gastropoda: Prosobranchia). J Mar Biol Assoc 79: 273–280. https://doi.org/10.1017/S0025315498000307

Gordy MA, Hanington PC (2019) A fine-scale phylogenetic assessment of digenean trematodes in Central Alberta reveals we have yet to uncover their total diversity. Ecol Evol 9:3153–3238. https://doi.org/10.1002/ece3.4939

Gourbal B, Guillou F, Mitta G et al (2008) Excretory-secretory products of larval *Fasciola hepatica* investigated using a two-dimensional proteomic approach. Mol Biochem Parasitol 161:63–66. https://doi.org/10.1016/j.molbiopara.2008.05.002

Graczyk TK, Fried B (1999) Development of *Fasciola hepatica* in the intermediate host. In: Dalton JP (ed) Fasciolosis. CAB International Publishing, Oxford, UK, pp 31–46

Grobbelaar A, Van As LL, Butler HJB et al (2014) Ecology of diplostomid (Trematoda: Digenea) infection in freshwater fish in Southern Africa. Afr Zool 49:222–232. https://doi.org/10.1080/15627020.2014.11407638

Guillou F, Mitta G, Galinier R, Coustau C (2007) Identification and expression of gene transcripts generated during an anti-parasitic response in *Biomphalaria glabrata*. Dev Comp Immunol 31: 657–671. https://doi.org/10.1016/j.dci.2006.10.001

Guth DJ, Blankespoor HD, Cairns J (1977) Potentiation of zinc stress caused by parasitic infection of snails. Hydrobiologia 55:225–229. https://doi.org/10.1007/BF00017554

Gutiérrez A, Perera G, Yong M et al (2000) Life-history traits of *Fossaria cubensis* (Gastropoda: Lymnaeidae) under experimental exposure to *Fasciola hepatica* (Trematoda: Digenea). Mém Inst Oswaldo Cruz 95:747–752. https://doi.org/10.1590/s0074-02762000000500025

Gutiérrez A, Yong M, Perera G et al (2002) *Fasciola hepatica* (Trematoda: Digenea): its effect on the life history traits of *Pseudosuccinea columella* (Gasteropoda: Lymnaeidae), an uncommon interaction. Parasitol Res 88:535–539. https://doi.org/10.1007/s00436-002-0625-4

Haas W, Haberl B (1997) Host recognition by trematode miracidia in Fried B, Graczyk T (eds) advances in trematode biology. CRC Press, Boca Raton, FL, pp 197–227

Hahn UK, Bender RC, Bayne CJ (2001a) Involvement of nitric oxide in killing of *Schistosoma mansoni* sporocysts by hemocytes from resistant *Biomphalaria glabrata*. J Parasitol 87:778–785. https://doi.org/10.1645/0022-3395(2001)087[0778:ionoik]2.0.co;2

Hahn UK, Bender RC, Bayne CJ (2001b) Killing of *Schistosoma mansoni* sporocysts by hemocytes from resistant *Biomphalaria glabrata*: role of reactive oxygen species. J Parasitol 87:292–299. https://doi.org/10.1645/0022-3395(2001)087[0292:kosmsb]2.0.co;2

Hanington PC, Forys MA, Dragoo JW et al (2010) Role for a somatically diversified lectin in resistance of an invertebrate to parasite infection. Proc Natl Acad Sci U S A 107:21087–21092. https://doi.org/10.1073/pnas.1011242107

Hodasi JKM (1972) The effects of *Fasciola hepatica* on *Lymnaea truncatula*. Parasitology 65:359–369. https://doi.org/10.1017/s0031182000045145

Hoek RM, Li KW, Van Minnen J et al (2005) LFRFamides: a novel family of parasitation-induced -RFamide neuropeptides that inhibit the activity of neuroendocrine cells in *Lymnaea stagnalis*: characterization of a novel LFRFamide neuropeptide precursor. J Neurochem 92:1073–1080. https://doi.org/10.1111/j.1471-4159.2004.02927.x

Hurtrez-Boussès S, Meunier C, Durand P et al (2001) Dynamics of host–parasite interactions: the example of population biology of the liver fluke (*Fasciola hepatica*). Microbes Infect 3:841–849. https://doi.org/10.1016/S1286-4579(01)01442-3

Hurtrez-Boussès S, Durand P, Jabbour-Zahab R et al (2004) Isolation and characterization of microsatellite markers in the liver fluke (*Fasciola hepatica*). Mol Ecol Notes 4:689–690. https://doi.org/10.1111/j.1471-8286.2004.00786.x

Joe LK, Basch PF, Umathevy T (1965) Antagonism between two species of larval trematodes in the same snail. Nature 206:422–423. https://doi.org/10.1038/206422a0

Johansen MV, Sithithawor P, Bergquist R et al (2010) Towards improved diagnosis of zoonotic trematode infections in Southeast Asia. Adv Parasitol 73:171–195. https://doi.org/10.1016/s0065-308x(10)73007-4

Jones HR (1990) Population geography. Chapman, London, UK

Kalbe M, Haberl B, Haas W (2000) Snail host finding by *Fasciola hepatica* and *Trichobilharzia ocellata*: compound analysis of "miracidia-attracting glycoproteins". Exp Parasitol 96:231–242. https://doi.org/10.1006/expr.2000.4579

Kanapala VK, Arasada SP (2013) Histopathological effect of paraquat (Gramoxone) on the digestive gland of freshwater snail *Lymnaea luteola* (Lamarck, 1799) (Mollusca: Gastropoda). Int J Sci Res Environ Sci 1:224–230. https://doi.org/10.12983/ijsres-2013-p224-230

Karvonen A, Rellsta C, Louhi K-R, Jokela J (2012) Synchronous attack is advantageous: mixed genotype infections lead to higher infection success in trematode parasites. Proc R Soc B 279: 171–176. https://doi.org/10.1098/rspb.2011.0879

Keas BE, Esch GW (1997) The effect of diet and reproductive maturity on the growth and reproduction of *Helisoma anceps* (Pulmonata) infected by *Halipegus occidualis* (Trematoda). J Parasitol 83:96–104

Kechemir N (1978) Démonstration expérimentale d'un cycle biologique à quatre hôtes obligatoires chez les Trématodes Hémiurides. Ann Parasitol Hum Comp 53:75–92

Keiser J, Utzinger J (2005) Emerging foodborne trematodiasis. Emerg Infect Dis 11:1507–1514. https://doi.org/10.3201/eid1110.050614

Keiser J, Utzinger J (2009) Food-borne trematodiases. Clin Microbiol Rev 22:466–483. https://doi.org/10.1128/cmr.00012-09

Khurana S, Malla N (2014) Water- and food-borne trematodiases. In: Singh PP, Sharma V (eds) Humans in water and health. Springer India, New Delhi, pp 219–227. https://doi.org/10.1007/978-81-322-1029-0_15

Kirk RS, Lewis JW (1993) The life-cycle and morphology of *Sanguinicola inermis Pl*ehn, 1905 (Digenea: Sanguinicolidae). Syst Parasitol 25:125–133

Krist AC, Lively CM (1998) Experimental exposure of juvenile snails (*Potamopyrgus antipodarum*) to infection by trematode larvae (*Microphallus* sp): infectivity, fecundity compensation and growth. Oecologia 116:467–474

Kryukova NA, Yurlova NI, Rastyagenko NM et al (2014) The influence of *Plagiorchis mutationis* larval infection on the cellular immune response of the snail host *Lymnaea stagnalis*. J Parasitol 100:284–287. https://doi.org/10.1645/13-214.1

Kuris A (1990) Guild structure of larval trematodes in molluscan hosts: prevalence, dominance and significance of competition. In: Esch GW, Bush AO, Aho JM (eds) Parasite communities: patterns and processes. Springer, Dordrecht, The Netherlands, pp 69–100

Lafferty KD (1993) Effects of parasitic castration on growth, reproduction and population dynamics of the marine snail *Cerithidea californica*. Mar Ecol Prog Ser 96:229–229. https://doi.org/10.3354/meps096229

LaRue GR (1951) Host-parasite relations among the digenetic trematodes. J Parasitol 37:333–342

Littlewood DTJ, Rohde K, Bray RA et al (1999) Phylogeny of the Platyhelminthes and the evolution of parasitism. Biol J Linn Soc 68:257–287. https://doi.org/10.1006/bijl.1999.0341

Littlewood DTJ, Bray RA, Waeschenbach A (2015) Phylogenetic patterns of diversity in cestodes and trematodes. In: Morand S, Krasnov BR, Littlewood DTJ (eds) Parasite diversity and diversification. Cambridge University Press, Cambridge, UK, pp 304–319

Lively CM (1989) Adaptation by a parasitic trematode to local populations of its snail host. Evolution 43:1663–1671. https://doi.org/10.1111/j.1558-5646.1989.tb02616.x

Lively CM, Dybdahl MF, Jokela J et al (2004) Host sex and local adaptation by parasites in a snail-trematode interaction. Am Nat 164:S6–S18. https://doi.org/10.1086/424605

Lockyer AE, Olson PD, Littlewood DTJ (2003) Utility of complete large and small subunit rRNA genes in resolving the phylogeny of the Neodermata (Platyhelminthes): implications and a review of the cercomer theory. Biol J Linn Soc 78:155–171. https://doi.org/10.1046/j.1095-8312.2003.00141.x

Lodes MJ, Yoshino TP (1990) The effect of schistosome excretory-secretory products on *Biomphalaria glabrata* hemocyte motility. J Invertebr Pathol 56:75–85. https://doi.org/10.1016/0022-2011(90)90147-x

López-Abarrategui C, McBeth C, Mandal SM et al (2015) Cm-p5: an antifungal hydrophilic peptide derived from the coastal mollusk *Cenchritis muricatus* (Gastropoda: Littorinidae). FASEB J 29: 3315–3325. https://doi.org/10.1096/fj.14-269860

López-Doval JC, Poquet M, Muñoz I (2014) Sublethal effects of the herbicide diuron on the freshwater snail *Physella acuta*. Limnetica 33:205–215

Minchella DJ, Loverde PT (1981) A cost of increased early reproductive effort in the snail *Biomphalaria glabrata*. Am Nat 118:876–881. https://doi.org/10.1086/283879

Morley NJ, Lewis JW (2017) Thermodynamics of egg production, development and hatching in trematodes. J Helminthol 91:284–294. https://doi.org/10.1017/s0022149x16000249

Morley NJ, Crane M, Lewis JW (2003) Cadmium toxicity and snail-digenean interactions in a population of *Lymnaea* spp. J Helminthol 77:49–55. https://doi.org/10.1079/joh2002148

Muñoz-Antol C, Marín A, Toledo R et al (2007) Effect of *Echinostoma friedi* (Trematoda: Echinostomatidae) experimental infection on longevity, growth and fecundity of juvenile *Radix peregra* (Gastropoda: Lymnaeidae) and *Biomphalaria glabrata* (Gastropoda: Planorbidae) snails. Parasitol Res 101:1663–1670. https://doi.org/10.1007/s00436-007-0710-9

Nelwan ML (2019) Schistosomiasis: life cycle, diagnosis, and control. Curr Ther Res Clin Exp 91: 5–9. https://doi.org/10.1016/j.curtheres.2019.06.001

Niewiadomska K, Pojmanska T (2011) Multiple strategies of digenean trematodes to complete their life cycles. Wiad Parazytol 57:233–241

Olson PD, Tkach VV (2005) Advances and trends in the molecular systematics of the parasitic platyhelminthes. Adv Parasitol 60:165–243. https://doi.org/10.1016/s0065-308x(05)60003-6

Olson PD, Littlewood DTJ, Bray RA et al (2001) Interrelationships and evolution of the tapeworms (Platyhelminthes: Cestoda). Mol Phylogenet Evol 19:443–467. https://doi.org/10.1006/mpev.2001.0930

Olson PD, Cribb TH, Tkach VV et al (2003) Phylogeny and classification of the Digenea (Platyhelminthes: Trematoda). Int J Parasitol 33:733–755. https://doi.org/10.1016/s0020-7519(03)00049-3

Pan C (1965) Studies on the host-parasite relationship between *Schistosoma mansoni* and the snail *Australorbis glabratus*. Am J Trop Med Hyg 14:931–976. https://doi.org/10.4269/ajtmh.1965.14.931

Pandian TJ (2020) Reproduction and development in Platyhelminthes, Reproduction and development in aquatic invertebrates 5. CRC Press, Boca Raton, FL

Park J-K, Kim K-H, Kang S et al (2007) A common origin of complex life cycles in parasitic flatworms: evidence from the complete mitochondrial genome of *Microcotyle sebastis* (Monogenea: Platyhelminthes). BMC Evol Biol 7:11. https://doi.org/10.1186/1471-2148-7-11

Pérez-Ponce de León G, Hernández-Mena DI (2019) Testing the higher-level phylogenetic classification of Digenea (Platyhelminthes, Trematoda) based on nuclear rDNA sequences before entering the age of the 'next-generation' tree of life. J Helminthol 93:260–276. https://doi.org/10.1017/s0022149x19000191

Petney TN, Sithithaworn P, Andrews RH et al (2016) Foodborne trematodes: a diverse and challenging group of neglected parasites. Trans R Soc Trop Med Hyg 110:1–3. https://doi.org/10.1093/trstmh/trv112

Pila EA, Li H, Hambrook JR et al (2017) Schistosomiasis from a snail's perspective: advances in snail immunity. Trends Parasitol 33:845–857. https://doi.org/10.1016/j.pt.2017.07.006

Pinaud S, Portela J, Duval D et al (2016) A shift from cellular to humoral responses contributes to innate immune memory in the vector snail *Biomphalaria glabrata*. PLoS Pathog 12:e1005361. https://doi.org/10.1371/journal.ppat.1005361

Pinto HA, Mati VLT, de Melo AL (2014) New insights into the life cycle of *Platynosomum* (Trematoda: Dicrocoeliidae). Parasitol Res 113:2701–2707. https://doi.org/10.1007/s00436-014-3926-5

Portela J, Duval D, Rognon A et al (2013) Evidence for specific genotype-dependent immune priming in the lophotrochozoan *Biomphalaria glabrata* snail. J Innate Immun 5:261–276. https://doi.org/10.1159/000345909

Poulin R, Cribb TH (2002) Trematode life cycles: short is sweet? Trends Parasitol 18:176–183. https://doi.org/10.1016/s1471-4922(02)02262-6
Poulin R, Mouritsen K (2003) Large-scale determinants of trematode infections in intertidal gastropods. Mar Ecol Prog Ser 254:187–198. https://doi.org/10.3354/meps254187
Poulin R, Mouritsen KN (2006) Climate change, parasitism and the structure of intertidal ecosystems. J Helminthol 80:183–191. https://doi.org/10.1079/joh2006341
Prince BE (1954) Biological studies on the miracidia of *Paramphistomum microbothrioides*, Price and McIntosh, and its intermediate snail host, S*tagnicola* (*Nasonia*) *cubensis*, Pfeiffer LSU Historical Dissertations and Theses, 8095
Relf V, Good B, McCarthy E et al (2009) Evidence of *Fasciola hepatica* infection in *Radix peregra* and a mollusc of the family Succineidae in Ireland. Vet Parasitol 163:152–155. https://doi.org/10.1016/j.vetpar.2009.04.003
Rim H-J, Farag HF, Sornmani S et al (1994) Food-borne trematodes: ignored or emerging? Parasitol Tod 10:207–209
Roger E, Grunau C, Pierce RJ et al (2008) Controlled chaos of polymorphic mucins in a metazoan parasite (*Schistosoma mansoni*) interacting with its invertebrate host (*Biomphalaria glabrata*). PLoS Negl Trop Dis 2:e330. https://doi.org/10.1371/journal.pntd.0000330
Rohde K (2001) The Aspidogastrea, an archaic group of Platyhelminthes. In: Littlewood DTJ, Bray RA (eds) Interrelationships of the Platyhelminthes. London, UK, USA, pp 159–167
Sabourin E, Alda P, Vázquez A et al (2018) Impact of human activities on fasciolosis transmission. Trends Parasitol 34(10):891–903. https://doi.org/10.1016/j.pt.2018.08.004
Sanchez J-F, Lescar J, Chazalet V et al (2006) Biochemical and structural analysis of *Helix pomatia* agglutinin. J Biol Chem 281:20171–20180. https://doi.org/10.1074/jbc.m603452200
Sapp KK, Loker ES (2000) Mechanisms underlying digenean-snail specificity: role of miracidial attachment and host plasma factors. J Parasitol 86:1012–1019. https://doi.org/10.1645/0022-3395(2000)086[1012:mudssr]2.0.co;2
Schwelm J, Selbach C, Kremers J (2021) Rare inventory of trematode diversity in a protected natural reserve. Sci Rep 11:22066. https://doi.org/10.1038/s41598-021-01457-2
Selbach C, Soldánová M, Feld CK, Kostadinova A, Sures B (2020) Hidden parasite diversity in a european freshwater system. Sci Rep 10:2694. https://doi.org/10.1038/s41598-020-59548-5
Skála V, Černíková A, Jindrová Z et al (2014) Influence of *Trichobilharzia regenti* (Digenea: Schistosomatidae) on the defence activity of *Radix lagotis* (Lymnaeidae) haemocytes. PLoS One 9:e111696. https://doi.org/10.1371/journal.pone.0111696
Smyth JD, Smyth MM (1980) Frogs as host-parasite systems I. Macmillan Education, London, UK
Smythe AB, Font WF (2001) Phylogenetic analysis of *Alloglossidium* (Digenea: Macroderoididae) and related genera: life-cycle evolution and taxonomic revision. J Parasitol 87:386–391. https://doi.org/10.1645/0022-3395(2001)087[0386:paoadm]2.0.co;2
Snyder SD, Esch GW (1993) Trematode community structure in the pulmonate snail *Physa gyrina*. J Parasitol 79:205–215
Sorensen RE, Minchella DJ (1998) Parasite influences on host life history: *Echinostoma revolutum* parasitism of *Lymnaea elodes* snails. Oecologia 115:188–195
Sorensen RE, Minchella DJ (2001) Snail–trematode life history interactions: past trends and future directions. Parasitology 123:S3–S18. https://doi.org/10.1017/s0031182001007843
Sousa WP (1990) Spatial scale and the processes structuring a guild of larval trematode parasites. In: Esch GW, Bush AO, Aho JM (eds) Parasite communities: patterns and processes. Springer, Dordrecht, The Netherlands, pp 41–67
Stunkard HW (1963) Systematics, taxonomy, and nomenclature of the Trematoda. Q Rev Biol 38: 221–233. https://doi.org/10.1086/403858
Sturrock B (1966) The influence of infection with *Schistosoma mansoni* on the growth rate and reproduction of *Biomphalaria pfeifferi*. Ann Trop Med Parasitol 60:187–197. https://doi.org/10.1080/00034983.1966.11686405
Taft SJ (1972) Aspects of the life history of *Cyclocoelum oculeum* (Trematoda: Cyclocoelidae). J Parasitol 58:882–884

Taylor EL (1964) Fascioliasis and the liver fluke. Food and Agriculture Organization, Rome
Thomas F, Renaud F, De Meeûs T et al (1998) Manipulation of host behaviour by parasites: ecosystem engineering in the intertidal zone? Proc R Soc Lond B 265:1091–1096. https://doi.org/10.1098/rspb.1998.0403
Thomas F, Lefevre T, Raymond M (2010) Biologie évolutive. De Boeck-CNRS, Bruxelles-Paris
Vázquez AA, Alba A, Alda P et al (2021a) On the arrival of fasciolosis in the Americas. Trend Parasitol 38:195. https://doi.org/10.1016/j.pt.2021.12.001
Vázquez A, Sabourin E, Alda P et al (2021b) Genetic diversity and relationships of the liver fluke *Fasciola hepatica* (Trematoda) with native and introduced definitive and intermediate hosts. Transbound Emerg Dis 68(4):2274–2286. https://doi.org/10.1111/tbed.13882
Vázquez AA, Alba A, Alda P et al (2023, this volume) Lymnaeid snails and the transmission of fasciolosis: understanding the differential risks from local to global scale. In: Vinarski MV, Vázquez A (eds) The Lymnaeidae. A handbook on their natural history and parasitological significance. Springer, Cham, pp 359–394
Vieira TCRG, Costa-Filho A, Salgado NC et al (2004) Acharan sulfate, the new glycosaminoglycan from *Achatina fulica* Bowdich 1822: structural heterogeneity, metabolic labeling and localization in the body, mucus and the organic shell matrix. Eur J Biochem 271:845–854. https://doi.org/10.1111/j.1432-1033.2004.03989.x
Vignoles P, Dreyfuss G, Rondelaud D (2011) *Haplometra cylindracea* (Zeder, 1800) (Trematoda: Plagiorchiidae): variation in the dates of cercarial shedding for overwintering *Galba truncatula*. Parasite 18:181–184. https://doi.org/10.1051/parasite/2011182181
Warren KS (1978) The pathology, pathobiology and pathogenesis of schistosomiasis. Nature 273: 609–612
Wesołowska W, Wesołowski T (2014) Do *Leucochloridium* sporocysts manipulate the behaviour of their snail hosts? J Zool 292:151–155. https://doi.org/10.1111/jzo.12094
Whitfield PJ, Evans NA (1983) Parthenogenesis and asexual multiplication among parasitic platyhelminths. Parasitology 86:121–160
World Health Organization (2015) Health in 2015: from MDGs, millennium development goals to SDGs, Sustainable Development Goals. World Health Organization, Geneva
Wu X-J, Sabat G, Brown JF et al (2009) Proteomic analysis of *Schistosoma mansoni* proteins released during in vitro miracidium-to-sporocyst transformation. Mol Biochem Parasitol 164: 32–44. https://doi.org/10.1016/j.molbiopara.2008.11.005
Wu X-J, Dinguirard N, Sabat G et al (2017) Proteomic analysis of *Biomphalaria glabrata* plasma proteins with binding affinity to those expressed by early developing larval *Schistosoma mansoni*. PLoS Pathog 13:e1006081. https://doi.org/10.1371/journal.ppat.1006081
Yoder HR, Coggins JR (1998) Larval trematode assemblages in the snail *Lymnaea stagnalis* from Southeastern Wisconsin. J Parasitol 84:259–268
Yoshino TP, Wu X-J, Gonzalez LA et al (2013) Circulating *Biomphalaria glabrata* hemocyte subpopulations possess shared schistosome glycans and receptors capable of binding larval glycoconjugates. Exp Parasitol 133:28–36. https://doi.org/10.1016/j.exppara.2012.10.002
Zahoor Z, Davies AJ, Kirk RS et al (2009) Nitric oxide production by *Biomphalaria glabrata* haemocytes: effects of *Schistosoma mansoni* ESPs and regulation through the extracellular signal-regulated kinase pathway. Parasites Vectors 2:18. https://doi.org/10.1186/1756-3305-2-18
Zhang S-M, Loker ES (2004) Representation of an immune responsive gene family encoding fibrinogen-related proteins in the freshwater mollusc *Biomphalaria glabrata*, an intermediate host for *Schistosoma mansoni*. Gene 341:255–266. https://doi.org/10.1016/j.gene.2004.07.003

Chapter 13
Lymnaeid Snails and the Transmission of Fasciolosis: Understanding the Differential Risks from Local to Global Scale

Antonio A. Vázquez, Annia Alba, Pilar Alda, Marion Vittecoq, Elodie Chapuis, Dominique Faugère, Jean-Pierre Pointier, and Sylvie Hurtrez-Boussès

Abstract The family Lymnaeidae is globally recognized for its role in the transmission of fasciolosis. This chapter thoroughly reviews the species of lymnaeid snails that have a proven or suspected role in the transmission of the liver flukes *Fasciola hepatica* and *Fasciola gigantica*. An outline of the differential roles of each

A. A. Vázquez (✉)
Laboratory of Malacology, Institute of Tropical Medicine "Pedro Kourí", Havana, Cuba

MIVEGEC, University of Montpellier, CNRS, IRD, Montpellier, France

A. Alba
Laboratory of Malacology, Institute of Tropical Medicine "Pedro Kourí", Havana, Cuba

P. Alda
MIVEGEC, University of Montpellier, CNRS, IRD, Montpellier, France

Genética y Ecología Evolutiva (CERZOS–CCT–CONICET Bahía Blanca), Bahía Blanca, Argentina

M. Vittecoq
Institut de Recherche de la Tour du Valat, Arles, France
e-mail: vittecoq@tourduvalat.org

E. Chapuis · D. Faugère
MIVEGEC, University of Montpellier, CNRS, IRD, Montpellier, France
e-mail: elodie.chapuis@ird.fr; dominique.faugere@ird.fr

J.-P. Pointier
PSL Research University, USR 3278 CNRS–EPHE, CRIOBE Université de Perpignan, Perpignan, France
e-mail: pointier@univ-perp.fr

S. Hurtrez-Boussès
MIVEGEC, University of Montpellier, CNRS, IRD, Montpellier, France

Département de Biologie-Écologie, Faculté des Sciences, Université de Montpellier, Montpellier, France
e-mail: sylvie.hurtrez@ird.fr

M. V. Vinarski, A. A. Vázquez (eds.), *The Lymnaeidae*, Zoological Monographs 7,
https://doi.org/10.1007/978-3-031-30292-3_13

major clade is presented with relation to the ecology of the species that helps to understand the epidemiological risks at different geographic levels. The chapter is accompanied by a detailed list of recorded prevalences either natural or experimental for 31 species of Lymnaeidae reviewed by country.

13.1 Introduction

The species belonging to the family Lymnaeidae are especially notorious for their role in the transmission of several species of 'liver fluke' parasites within the family Fasciolidae (Trematoda: Digenea). The disease provoked by those parasites, usually referred to as fasciolosis, has received considerable attention given their impacts in human/veterinary health and economy worldwide (Alba et al. 2021). Within this family of trematodes, the genus *Fasciola* encompasses three extant species: *Fasciola hepatica* Linnaeus, 1758, *Fasciola gigantica* Cobbold, 1855, and *Fasciola nyanzae* Leiper, 1910 (Heneberg 2013; Lotfy et al. 2008). The latter, *F. nyanzae*, is the less studied species, perhaps because it only affects hippos and thus is restricted to some African regions today. There, it is transmitted by the largely distributed lymnaeid *Radix natalensis* (Krauss, 1848) (Dinnik and Dinnik 1961). The other two *Fasciola* spp. are, however, much largely distributed, affecting a wide range of wild and domestic mammals as well as humans (Alba et al. 2021) and may rely on a larger number of lymnaeid species to be transmitted (Vázquez et al. 2018). Trematodes are frequently highly specific towards their intermediate snail host, meaning that just one or a few close-related snail species are able to transmit the same trematode (Adema and Loker 2015). However, in the *F. hepatica*/*F. gigantica*–lymnaeid species interaction, both trematodes are able to use species of Lymnaeidae as intermediate hosts from virtually across its whole phylogeny (Correa et al. 2010) and thus considered as highly generalists (Fig. 13.1). This chapter will be dealing with the species of the Lymnaeidae that transmit *Fasciola* spp. in order to understand the differential risks at local, regional, or global scales.

13.2 Fasciolosis: A Brief Outline

Among trematode's diseases, fasciolosis stands out as the one with the largest distribution worldwide from tropical to temperate regions, from sea level to very high altitudes, and from wild to domestic environments (Sabourin et al. 2018; Torgerson and Claxton 1999). Such success would be due to the high diversity of suitable intermediate and definitive hosts occurring almost everywhere the parasites are introduced, whether through natural or anthropic events. In relation to the definitive hosts, there are many species of mammals reported to be susceptible to *Fasciola* spp. (for review, see Alba et al. 2021), some of worldwide distribution and usually key in the dissemination of the parasites (e.g. cows, goats, sheep, buffaloes,

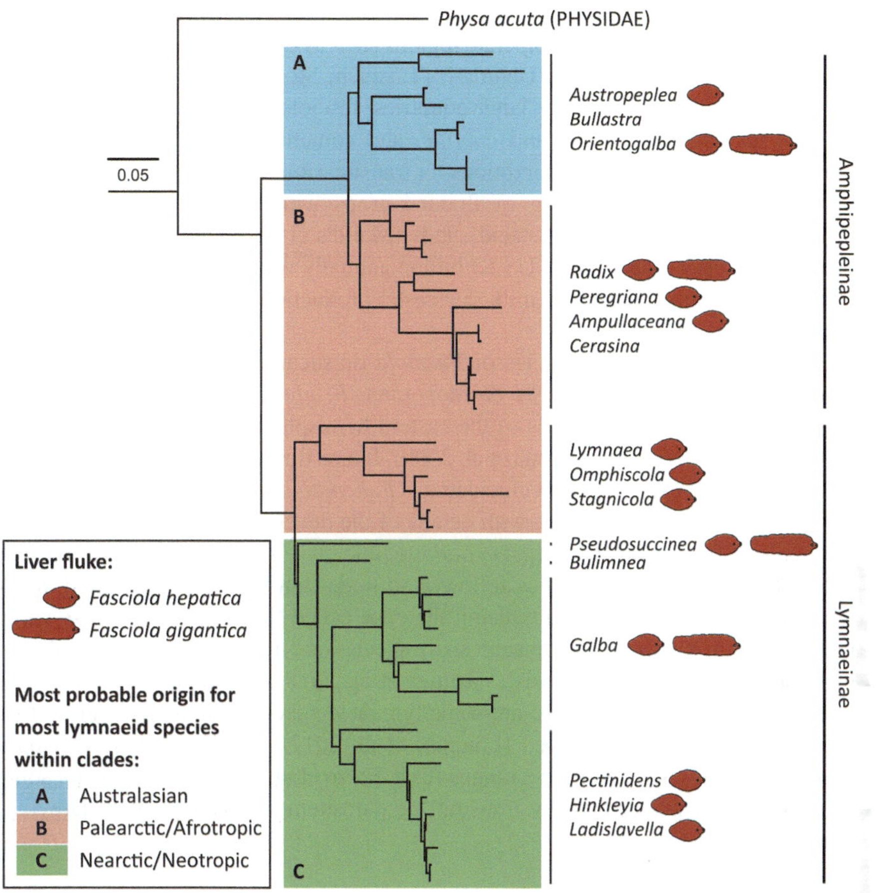

Fig. 13.1 Representation of the occurrence of intermediate lymnaeid snail hosts for each *Fasciola* species across the phylogeny of the Lymnaeidae with the probable most common native range for most species within each clade (adapted from the phylogeny of Correa et al. 2010 obtained through the concatenation of 16S, ITS-1 and ITS-2 molecular markers)

pigs, rats, hares, nutrias) and other endemic to restricted regions (e.g. kangaroos, lamas, guanacos). Moreover, the higher notoriety of fasciolosis probably lies in the fact that it also affects humans with the latest reports of about 35 million people infected and over 180 million at risk worldwide, particularly due to *F. hepatica* (Nyindo and Lukambagire 2015). These figures however are not equally represented in every region where *Fasciola* spp. are present. For instance, there are just a few hyperendemic human fasciolosis foci in the Bolivian Altiplano, Peru, Mexico, Ecuador, Egypt, and Vietnam (Esteban et al. 1997; Nguyen et al. 2011; Zumaquero et al. 2013). There are other meso/hypo-endemic foci from particular regions, such as Iran or countries from South-east Asia (Ashrafi et al. 2015; Le et al. 2008). These important foci from (sub)tropical regions have served for fasciolosis to be wrongly tagged as a 'Neglected Tropical Disease'. The disease is, however, present in many high-income countries from temperate regions such as France (Dreyfuss et al. 2005), Spain (Martínez et al. 2013), Denmark (Takeuchi-Storm et al. 2018), or the UK (McCann et al. 2010). In these latter countries, fasciolosis largely affects domestic animals, particularly bovine and caprine, and sometimes produces outbreaks in human populations when the conditions for transmission are enhanced: large amount of cattle infected, overlapping of flooded crops with infected mammals, large populations of compatible lymnaeid snails (Mailles et al. 2006). Economic losses related to fasciolosis ascend to US $3 billion annually worldwide, usually related to liver condemnation, meat and milk decreased production, and anthelminthic treatments (Toet et al. 2014).

The distribution of the species of *Fasciola* drastically differs at a global scale. While *F. hepatica* is considered cosmopolitan, *F. gigantica* appears restricted to tropical and subtropical regions of the eastern hemisphere, with just a few records from temperate areas (Mas-Coma et al. 2009). Interestingly, the trematode for which the complete life cycle was first elucidated is *F. hepatica* (see Fig. 13.2 for details on the life cycle of *Fasciola* spp.), with details of the development of the larvae inside the snails of the species *Galba truncatula* (Müller, 1774) (Leuckart 1882; Thomas 1882). Since then, many studies followed to understand which species and under which conditions were able to transmit the once called 'liver-rot'. Today, we know that a certain number of lymnaeid snail species are responsible for the above-mentioned success of fasciolosis (Vázquez et al. 2018), but their role is not quite the same for all of them. There are some lymnaeid species potentially transmitting everywhere they are introduced (Lounnas et al. 2017), there are others that only transmit in their restricted native ranges (Celi-Erazo et al. 2020), and there are some that poorly transmit or that not transmit the parasite at all (Boray 1978; Dreyfuss et al. 2015).

13.3 An Overview of Lymnaeid Snails Transmitting *Fasciola* spp.

Overall, the studies aimed at presenting the natural or experimental prevalences of lymnaeid snails infected with *F. hepatica* and *F. gigantica* are much biased towards the former, probably due to its larger distribution and human health relevance. In the literature, there are many old reports of field or experimental infections with snails from different regions. We advise that such reports must be considered cautiously, since the identification of the lymnaeid snail species is usually only supported by morphological traits. Even some species that are relatively easy to identify may have been mistaken in the past due to lack of experience. In addition, data on prevalence not always necessarily means that all reported individuals with larval forms of the liver flukes may actually shed viable cercariae. For instance, particularly in the case of molecular detection, these highly sensible and specific methods, usually able to

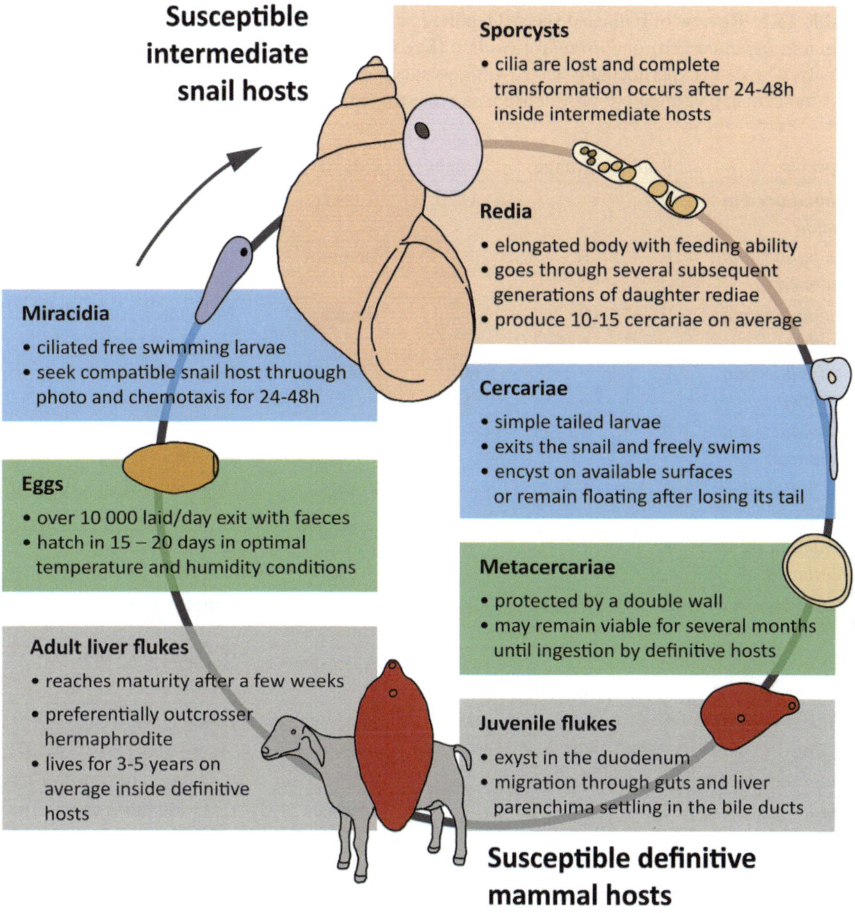

Fig. 13.2 Life cycle of *Fasciola* spp. with details on each developmental stage

detect the mere penetration of miracidia, may pose two kinds of problems when interpreting the transmission: (1) they ascertain the presence of *Fasciola* spp. in molluscs that are usually not compatible hosts, but because they occur in transmission foci, miracidia may eventually penetrate the snail and thus render samples positives (Relf et al. 2009); and (2) they detect very high field prevalence in lymnaeid snails which does not necessarily means that all snails will develop infection. Notwithstanding, Table 13.1 summarizes the largest compiled dataset with data of natural and experimental prevalence of each liver fluke by lymnaeid species and country.

Table 13.1 Review of lymnaeid species reported naturally (N) or experimentally (E) infected with *Fasciola hepatica* (Fh) or *Fasciola gigantica* (Fg) with their mean observed prevalences (%) by country ([a] means experimentally tested with no successful infection; NA means that the infection was molecularly verified but prevalence was not assessed)

Species	Report	Fh (range)	Fg (range)	Country	References
Ampullaceana ampla	E		1–25	Russia	Kruglov (1986)
Ampullaceana balthica	E	6		Belgium	Caron et al. (2007)
Ampullaceana balthica	E	9 (0.6–27)		France	Degueurce et al. (1999) Vázquez et al. (2020)
Ampullaceana balthica	E		60 (16–83)	Germany	Boray (1966)
Ampullaceana balthica	E	12–35	9–90	Russia	Kruglov (1986) Sazanov (1979)
Ampullaceana balthica	N	8.5		France	Degueurce et al. (1999)
Ampullaceana fontinalis	E	6	39–41	Russia	Kruglov (1986)
Ampullaceana lagotis	E	19–20	0[a]	Russia	Kruglov (1986)
Austropeplea tomentosa	E	77 (14–100)	70 (25–95)	Australia	Boray (1966, 1978)
Bullastra lessoni	E	0[a]	0[a]	Australia	Boray (1966)
Galba bulimoides	E	20		USA	Foreyt and Todd (1978)
Galba bulimoides	N	77		Mexico	Cruz-Mendoza et al. (2004)
Galba bulimoides	N	3 (0.3–6)		USA	Olsen (1944)
Galba cousini	E	34		Colombia	Salazar et al. (2006)
Galba cousini	N	13 (1–63)		Colombia	Brumpt et al. (1940) Pereira et al. (2020)
Galba cousini	N	46 (31–61)		Ecuador	Celi-Erazo et al. (2020) Villavicencio and de Vasconcellos (2005)
Galba cubensis	E	41 (21–61)		Argentina	Sanabria et al. (2012)
Galba cubensis	E	70 (20–100)		Cuba	Vázquez et al. (2014, 2019)
Galba cubensis	E	77 (63–90)		Dominican Republic	Vázquez et al. (2019)
Galba cubensis	E	75		Uruguay	Bargues et al. (2017)
Galba cubensis	E	21 (15–26)		USA	Cruz-Reyes and Malek (1987)
Galba cubensis	N	9 (1–34)		Cuba	Alba et al. (2016) Vázquez et al. (2015)

(continued)

Table 13.1 (continued)

Species	Report	Fh (range)	Fg (range)	Country	References
Galba cubensis	N	1 (0.1–3)		USA	Kaplan et al. (1997)
Galba cubensis	N	23		Venezuela	Morales and Pino (1982)
Galba humilis	N	33 (3–75)		Mexico	Cruz-Mendoza et al. (2004, 2005)
Galba modicella	N	4		USA	Rognlie et al. (1996)
Galba mweruensis	N	7		Ethiopia	Goll and Scott (1978)
Galba schirazensis	E	2		Dominican Republic	Dreyfuss et al. (2015)
Galba schirazensis	N	3		Ecuador	Caron et al. (2017) Celi-Erazo et al. (2020)
Galba truncatula	E	62 (47–77)		Algeria	Mekroud et al. (2006)
Galba truncatula	E	78 (39–69)		Belgium	Caron et al. (2007)
Galba truncatula	E	51		Bolivia	Bargues et al. (2017)
Galba truncatula	E	58 (34–89)		Egypt	Dar et al. (2013, 2014a)
Galba truncatula	E	53 (0.5–100)	37 (22–80)	France	Abrous et al. (2000, 2001) Busson et al. (1982) Dar et al. (2002, 2003) Degueurce et al. (1999) Dreyfuss and Rondelaud (1997) Dreyfuss et al. (2007) Gasnier et al. (2000) Goumghar et al. (2001) Lacourarie (1996) Rondelaud (1993) Rondelaud et al. (2013) Sanabria et al. (2012) Sindou et al. (1991) Vareille-Morel et al. (2002) Vázquez et al. (2019) Vignoles et al. (2001)
Galba truncatula	E	100		Germany	Boray (1966)
Galba truncatula	E	44		Morocco	Belfaiza et al. (2005)
Galba truncatula	E	60–85		Russia	Kruglov (1986)
Galba truncatula	E	7		Spain	Gasnier et al. (2000)
Galba truncatula	E	90		Sweden	Novobilský et al. (2013)
Galba truncatula	E	63 (25–90)		UK	Hodasi (1972)
Galba truncatula	N	5		Algeria	Mekroud et al. (2004)
Galba truncatula	N	16 (1–88)		Bolivia	Meunier (2002)
Galba truncatula	N	1		Colombia	Pereira et al. (2020)

(continued)

Table 13.1 (continued)

Species	Report	Fh (range)	Fg (range)	Country	References
Galba truncatula	N	5 (0.1–28)		France	Degueurce et al. (1999) Dreyfuss and Rondelaud (1997) Leimbacher (1973) Mage et al. (2002) Rondelaud et al. (2001) Vignoles et al. (2017b)
Galba truncatula	N	9 (1–17)		Iran	Ashrafi et al. (2007) Yakhchali et al. (2015)
Galba truncatula	N	14		Ireland	Relf et al. (2011)
Galba truncatula	N	3		Morocco	Khallaayoune and Hari (1991) Khallaayoune et al. (1991) Chlyeh (2002)
Galba truncatula	N	48		Peru	Londoñe et al. (2009)
Galba truncatula	N	27		Poland	Kozac and Wędrychowicz (2009)
Galba truncatula	N	1		Russia	Villavicencio et al. (2006)
Galba truncatula	N	0.1		Spain	Roldan et al. (2020)
Galba truncatula	N	11		Switzerland	Chapuis (2009) Schweizer et al. (2007)
Galba truncatula	N	19		Tunisia	Hammami et al. (2007)
Galba truncatula	N	6		UK	Jones et al. (2017)
Galba truncatula	N	71		Ukraine	Stadnichenko (2006)
Galba viator	E	67 (46–87)		Uruguay	Sanabria et al. (2012)
Galba viator	N	5 (0.9–14)		Argentina	Cucher et al. (2006) Kleiman et al. (2004)
Ladislavella occulta	E	44		Poland	Czapski (1962)
Lymnaea stagnalis	E	20		Belgium	Berghen (1964)
Lymnaea stagnalis	E	21 (4–50)		France	Busson et al. (1982) Dreyfuss et al. (2002) Vázquez et al. (2020)
Lymnaea stagnalis	E	1	14	Germany	Boray (1966)
Lymnaea stagnalis	E	1–5		Russia	Kruglov (1986)
Lymnaea stagnalis	N	1		Iran	Yakhchali et al. (2015)
Myxas glutinosa	E	23		France	Monny (1995)
Omphiscola glabra	E	23 (16–31)		France	Busson et al. (1982) Sindou et al. (1991)
Omphiscola glabra	N	0.3		France	Rondelaud et al. (2001)

(continued)

Table 13.1 (continued)

Species	Report	Fh (range)	Fg (range)	Country	References
Orientogalba ollula	E	83		Japan	Itagaki et al. (1988)
Orientogalba viridis	E	56		Australia	Boray (1978)
Orientogalba viridis	E	49		Nepal	Boray (1978)
Orientogalba viridis	E	64		Papua New Guinea	Boray (1978)
Orientogalba viridis	E	55		South Korea	Lee et al. (1995)
Orientogalba viridis	E		0.84 (0.84–2)	Vietnam	Dung et al. (2013)
Peregriana peregra	E	45		Belgium	Caron et al. (2007)
Peregriana peregra	E	43 (2–77)		Austria	Boray (1966)
Peregriana peregra	E	40		France	Busson et al. (1982)
Peregriana peregra	E	3		Germany	Boray (1966)
Peregriana peregra	E		73	Iran	Massoud and Sadjadi (1980)
Peregriana peregra	E		0[a]	Russia	Sazanov (1979)
Peregriana peregra	N	37		Ireland	Relf et al. (2009)
Peregriana peregra	E	2–21		Russia	Kruglov (1986)
Pseudosuccinea columella	E	68		Australia	Boray (1978)
Pseudosuccinea columella	E	24 (0–47)		Brazil	Pereira et al. (2020)
Pseudosuccinea columella	E	82		Colombia	Salazar et al. (2006)
Pseudosuccinea columella	E	61 (0–100)		Cuba	Alba et al. (2018) Vázquez et al. (2014, 2019)
Pseudosuccinea columella	E	53 (33–70)		Dominican Republic	Vázquez et al. (2019)
Pseudosuccinea columella	E	51 (38–64)	3 (2–3)	Egypt	Ahmed and Ramzy (1999) Dar et al. (2014a) Vignoles et al. (2015)
Pseudosuccinea columella	E	45 (33–57)		France	Vázquez et al. (2020) Vignoles et al. (2015)
Pseudosuccinea columella	E	64		New Zealand	Boray (1978)

(continued)

Table 13.1 (continued)

Species	Report	Fh (range)	Fg (range)	Country	References
Pseudosuccinea columella	E	51 (50–51)		USA	Cruz-Reyes and Malek (1987)
Pseudosuccinea columella	N	11 (7–18)		Argentina	Cucher et al. (2006) Moriena et al. (2008) Prepelitchi et al. (2003)
Pseudosuccinea columella	N	4 (1–5)		Brazil	Coelho and Lima (2003) Oliveira et al. (2002)
Pseudosuccinea columella	N	6 (1–11)		Cuba	Alba et al. (2019) Gutiérrez et al. (2011)
Pseudosuccinea columella	N		3	Egypt	Grabner et al. (2014)
Pseudosuccinea columella	N	41 (36–46)		Peru	Londoñe et al. (2009)
Pseudosuccinea columella	N	NA		South Africa	Malatji and Mukaratirwa (2019)
Radix auricularia	E		51	Iran	Ashrafi and Mas-Coma (2014)
Radix auricularia	E	39		Japan	Itagaki et al. (1988)
Radix auricularia	E	0[a]	85	Russia	Kruglov (1986)
Radix auricularia	N		0.3	Bangladesh	Chowdhury et al. (1994)
Radix cucunorica	N	93		China	Huang et al. (2019)
Radix euphratica	E	33	45	Iran	Cruz-Reyes and Malek (1987) Massoud and Sadjadi (1980)
Radix euphratica	N		2 (0.2–3)	Iran	Imani-Baran et al. (2012) Ashrafi et al. (2004)
Radix natalensis	E	12		Egypt	Dar et al. (2010)
Radix natalensis	E	0[a]		Kenya	Boray (1966)
Radix natalensis	E		96	West Africa	Rao (1966)
Radix natalensis	N		0.6	Egypt	Ahmed and Ramzy (1999)
Radix natalensis	N		25	Nigeria	Schillhorn van Veen (1980)
Radix rubiginosa	N		0.6	Indonesia	Estuningsih and Copeman (1996)
Radix rufescens	E		34	Pakistan	Rao (1966)
Stagnicola corvus	E	3–10		Russia	Kruglov (1986)
Stagnicola fuscus	E	33		France	Dreyfuss et al. (2000)

(continued)

Table 13.1 (continued)

Species	Report	Fh (range)	Fg (range)	Country	References
Stagnicola fuscus	E	13		Sweden	Novobilský et al. (2013)
Stagnicola palustris	E	50		Belgium	Berghen (1964)
Stagnicola palustris	E	12 (7–18)		France	Busson et al. (1982) Degueurce et al. (1999) Sindou et al. (1991)
Stagnicola palustris	E	0.2	0[a]	Germany	Boray (1966)
Stagnicola palustris	E	51		Sweden	Novobilský et al. (2013)
Stagnicola palustris	E	0–11		Russia	Kruglov (1986)
Stagnicola palustris	N	0.15		Sweden	Novobilský et al. (2013)

13.3.1 The Lymnaeinae

13.3.1.1 *Galba* Species: Amphibious, Cryptic, and Hazardous

Among lymnaeid snails, those within the genus *Galba* are perhaps the best known in relation to fasciolosis transmission, at least to the fasciolosis caused by *F. hepatica* (Bargues and Mas-Coma 1997). Most *Galba* species, but particularly *Galba truncatula* (O. F. Müller, 1774), remain tagged as the responsible for local transmission. In fact, the higher values of prevalence, whether natural or experimental, usually correspond to *Galba* spp. (Vázquez et al. 2018). Although there is still debate on the origin of this group (see Alda et al. (2021)), the genus *Galba* appears very well represented in the Americas where it reaches its highest species diversity (Alda et al. 2018). This is particularly interesting from an evolutionary point of view given the Eurasian assumed origin of the liver flukes 5–7 Ma (Choi et al. 2020). Today, there are studies pointing to a more ancient relationship between *F. hepatica* and its hosts in the Americas (Beltrame et al. 2020) than the one initially thought from the 1500s onward (Mas-Coma 2005). Whether that *F. hepatica* existed in the Americas since long ago or that ancient forms of *Galba* occurring in Eurasia encountered the liver fluke, both entities appear to have evolved a tight relationship allowing the transmission by almost every species in the genus (Vázquez et al. 2022).

Almost all species within the genus *Galba* are amphibious mud-dwelling snails, remaining for long periods above the water–land interface. One would think that this could be an issue for snail–parasite encounter probability. However, these species normally wander next to the water in very humid conditions except for extreme dry/cold periods in which they bury in estivation or cease activity (Belfaiza et al.

2009; Rognlie et al. 1996). But then, transmission is halted and usually marked by seasonality, at least in the direction of the intermediate host, though parasite development may continue during aestivation in infected snails (Belfaiza et al. 2009). Otherwise, snails would remain in permanent contact with mammals that graze in humid meadows, or that approach the waterbodies for drinking. Mammals that commonly engage in this behaviour are usually rodents, lagomorphs, cows, sheep, and goats; matching the typical definitive hosts of *F. hepatica*. These ecological features may have facilitated the tight relationships between amphibious lymnaeids and *F. hepatica*, supporting an overall susceptible phenotype in *Galba* spp. In the case of *F. hepatica*, miracidia commonly engage in a positive phototactic and negative geotactic behaviour bringing them to the surface where their common amphibious snail hosts occur (Yasuraoka 1954). However, *Galba* species may still go into the water and infect there as well with either *F. hepatica* or *F. gigantica*.

Within *Galba*, perhaps the pluri-continental dispersion observed in *G. truncatula* is the cause of considering this species as the main intermediate host of *F. hepatica* worldwide coupled with the fact that it is the main host in Europe (Correa et al. 2017), with also some reports of transmission of *F. gigantica* (Dar et al. 2003). This species is also the only present so far in the high-altitude hyperendemic foci of the Bolivian Altiplano (Meunier et al. 2001) with prevalences reaching up to 88% in Bolivia (Meunier 2002). It also occurs in other important transmission foci in Peru (48% prevalence as *G. viator* was recorded; see Bargues et al. 2012; Londoñe et al. 2009), Egypt (10% prevalence; Arafa et al. 2018), and Iran (17% prevalence; Yakhchali et al. 2015). Moreover, *G. truncatula* is largely accepted to be responsible for transmission in large grazing areas from several countries of Europe and Africa. For instance, in Europe, there are several reports of natural prevalences in France ranging from 0.8 to 10% (Dreyfuss and Rondelaud 1997), 0.1% above 2000 m altitude in the Spanish Pyrenees (Roldan et al. 2020), 26% in Poland (Kozac and Wędrychowicz 2009), 1% in European Russia (Villavicencio et al. 2006), and 6% in the UK (Jones et al. 2017). In Africa, field studies observed 19% prevalence from Tunisia (Hammami et al. 2007), 5% from Morocco (Chlyeh 2002), and 5% from Algeria (Mekroud et al. 2004). This species is also linked to the transmission of *F. gigantica* and has been tested experimentally several times (Dar et al. 2003; Dreyfuss and Rondelaud 1997; Rakotondravao et al. 1992) though its actual role in field transmission needs more studies.

However, fasciolosis is extensively present in the Americas, particularly from southern USA to Patagonia and throughout the Caribbean basin (Carmona and Tort 2017; Kaplan et al. 1997; Knapp et al. 1992; Vázquez et al. 2019). Here, transmission is mostly due to *Galba cubensis* (L. Pfaiffer, 1839) and *Galba viator* (d'Orbigny, 1835). For instance, *G. cubensis* is the most common intermediate host all over the insular Caribbean with many reports, particularly from Cuba with an overall 4% prevalence and a maximum of 34% recorded (Vázquez et al. 2015; Alba et al. 2016) and the Dominican Republic (Vázquez et al. 2019). From North America, *G. cubensis* is reported infected in the USA (4%; Kaplan et al. 1997) and in South America there are reports of 23% field infection in Venezuela (Morales and Pino 1982). The only record of *G. cubensis* out of the Americas appeared recently from rice paddies in the Delta del Ebro, Spain (Schniebs et al. 2018). Ecological and parasitological studies are much needed to assess the possible impacts in the overall transmission of fasciolosis in the region. In southern South America, snails identified as *G. viator* are found infected with prevalences of 0.9–14% in Argentina (Kleiman et al. 2007). This species is commonly marked as the main host in subtropical and temperate South America, particularly the Andean-Patagonian valleys (Kleiman et al. 2004).

However, other sibling species appear to maintain the transmission in more restricted areas according to their range. For instance, populations previously recorded as *Galba bulimoides* (I. Lea, 1841) were commonly found infected in some regions from the USA (0.2–7% prevalence; Olsen 1944) and Mexico (77%; Cruz-Mendoza et al. 2004). In Montana, USA, *Galba modicella* (Say, 1825) was found responsible for infection in a goat farm (Rognlie et al. 1996) and *Galba humilis* (Say, 1822) has been eventually observed transmitting *F. hepatica* in the field from Mexico (3–75% prevalence; Cruz-Mendoza et al. 2004). Notwithstanding, the reports pointing such particular species should be taken carefully as identification was solely made on morphological traits. In South America, the complex *G. cubensis/viator* drives most of the transmission (Vázquez et al. 2018), except in high altitude regions from Ecuador and Colombia. There, *Galba cousini* (Jousseaume, 1887) is adapted to such high altitudes maintaining the local transmission with several reports evidencing natural infections. For instance, *G. cousini* was found infected with prevalences ranging from 31–61% in Ecuador (Celi-Erazo et al. 2020; Villavicencio and de Vasconcellos 2005) and 1–63% in Colombia (Brumpt et al. 1940; Pereira et al. 2020). Yet another species, *Galba schirazensis* (Küster, 1862) that is considered native to the Americas (Lounnas et al. 2018), has received contradicting reports concerning *F. hepatica* transmission. Initially, some authors referred that this species has no role in fasciolosis transmission because some experimental infections failed to show a susceptible phenotype (Bargues et al. 2011). However, natural infections of *G. schirazensis* were observed in Ecuador (3% prevalence; Caron et al. 2017; Celi-Erazo et al. 2020). Some studies have included *G. schirazensis* when experimentally testing the compatibility of *F. hepatica* with several lymnaeid species. While some have failed to detect infection (Vázquez et al. 2019), there are others in which free cercariae were obtained after exposing several generations reared in the laboratory to the parasite (Dreyfuss et al. 2015). Thus, according to Vázquez et al. (2019), *G. schirazensis* cannot be completely ruled out because, though it seems generally a bad-performing host, it can still be a key driver of fasciolosis in certain areas (Pointier 2015).

These different levels of compatibility evidenced within the *Galba* become a problem when dealing with cryptic species occurring in sympatry. Crypticity is indeed a peculiar feature among *Galba* spp. and it is virtually impossible to distinguish species solely by shell morphology, with the only exception being *G. cousini* (Alda et al. 2018). Several records of small-bodied *Galba* snails identified in the past should be treated then with caution, particularly if they refer to parasite infection (see, for example, the case of *G. schirazensis* discussed in Lounnas et al. 2018). In a recent study, Mahulu et al. (2019) showed that several analysed sub-Saharan populations of *Galba* were different from *G. truncatula* and proposed the use of the formerly described *Galba mweruensis* (Connolly, 1929), a species that was already acknowledged as intermediate host of both *F. hepatica* and *F. gigantica* (Dinnik and Dinnik 1957). Thus, eventual transmission in countries such as Ethiopia, Tanzania, Lesotho, and Uganda that were thought to be due to *G. truncatula* might correspond to *G. mweruensis*. In fact, relatively high natural prevalences (7–8%) were observed in *G. mweruensis* from Ethiopian Central Highlands (recorded as *Lymnaea truncatula* in Goll and Scott 1978). Cryptic species co-occur in several high-transmission areas. For instance, *G. schirazensis* and *G. cubensis* coexist in some localities in the Dominican Republic (Vázquez et al. 2019), and in many parts of North and South America there are usually more than one species in the same region (e.g. *G. cubensis*/*G. bulimoides* in Mexico, *G. cubensis*/*G. humilis* in the USA, *G. cubensis*/*G. truncatula*/*G. schirazensis* in Peru, etc.). Therefore, pointing transmission to a given species could be challenging if molecular tools are not used (Alda et al. 2018) especially if the overall understanding of fasciolosis may rely on one particular species driving the epidemiology of the disease.

13.3.1.2 *Pseudosuccinea columella*: Invasive Facilitator for Transmission

The case of *Pseudosuccinea columella* (Say, 1817) is to some extent different to the one of *Galba* species. This lymnaeid species, as probably most of *Galba* spp., is native to North America (Lounnas et al. 2017) but its sounding reputation for transmission comes from abroad. In its native range, only few works have dealt with its role in transmission, mainly assessing its compatibility through experimental infections trials (Cruz-Reyes and Malek 1987). However, we should note that studies concerning the field transmission of fasciolosis in the USA or Canada are quite scarcer than in other regions, making that its actual role in its native range probably remains underestimated. The species *P. columella*, the only from its genus, is perhaps one of the most notorious introduced lymnaeids in tropical, subtropical, and temperate regions starting back nearly 200 years ago, likely facilitated by human activities related to the aquarium plants trade (Lounnas et al. 2017). However, prior to these events, it may have already covered a vast range down to the southern USA, eventually reaching out to Cuba (described as *Lymnaea francisca* Poey, 1858) and Brazil (described as *Lymnaea peregrina* Clessin, 1882). However, both species have been for long placed under synonymy with *P. columella* (Paraense 1994; Pointier et al. 2005) and most Cuban populations and all those explored in Brazil show little to none genetic diversity suggesting recent introductions (Alba et al. 2019; Marques et al. 2006).

Throughout the Caribbean, *P. columella* is presumed as an active host of *F. hepatica* (Vázquez et al. 2019). However, it has only been found naturally infected in Cuba (prevalence range from 1 to 10%) where its role in transmission seems secondarily to that of *G. cubensis* (Alba et al. 2019; Gutiérrez et al. 2011). While it is present in most of the insular Caribbean, the species has been tested for compatibility only from Cuba and the Dominican Republic advising of high risks of transmission (Alba et al. 2018; Vázquez et al. 2019). In South America, *P. columella* occurs throughout the continent down to the southern Pampas in Argentina (Martín et al. 2016). It is particularly important in the epidemiology of fasciolosis in Brazil (Coelho et al. 2009) and northern Argentina (Moriena et al. 2008), but may also play important roles in countries like Colombia and Venezuela, although studies are needed to validate its role (Pereira et al. 2020). However, throughout South America, it is perhaps only in Brazil where this species is marked as the main intermediate host of *F. hepatica* (Medeiros et al. 2014; Coelho and Lima 2003). While some parasitological studies attest that *G. viator* seems more efficient for transmission, the larger distribution of *P. columella* particularly in the South and Southeast has served to tag this species as the primary host transmitting fasciolosis (Medeiros et al. 2014).

Out of the Americas, *P. columella* is marked as a facilitator for fasciolosis mostly in Western Australia and New Zealand (Boray 1978; Molloy and Anderson 2006) but also in Egypt (Dar et al. 2014b). In Africa, it appears established in South Africa (Kock et al. 1989) where natural infection with *F. gigantica* was molecularly verified in KwaZulu-Natal and Eastern Cape (Malatji and Mukaratirwa 2019). Other studies report the presence of *P. columella* from Mozambique, Zimbabwe, Zambia, Madagascar, and La Réunion (Brown 1994; Lounnas et al. 2017). In Egypt, however, a case of parasite spillback was evidenced through a 3% prevalence of *F. gigantica* in *P. columella* in the Fayoum irrigation system (Grabner et al. 2014) after this species engaged in the transmission commonly driven by local *R. natalensis*. Throughout the Pacific islands, *P. columella* is particularly present in Hawaii (Cowie 2001) and also in Tahiti, Rapa, Tubuai, and Rurutu, but its role in fasciolosis is unknown (Pointier and Marquet 1990).

Passive introduction, associated with plants trade market (especially for aquariums), has been the source for its presence in many city-controlled environments throughout several European countries (Glöer 2019). However, several populations that may have escaped such controlled environments have now settled in the wild in Portugal (Martínez-Orti 2013) and southern France (Pointier et al. 2007). Their eventual insertion in the local transmission of fasciolosis is not ruled out, particularly if conditions are not too harsh for increasing their population effectives. In the ongoing world-changing climate, the risks of *P. columella* expansion towards higher latitudes may complicate the overall epidemiological scenario. Strong genetic bottlenecks following introductions in this species have been reported worldwide (Lounnas et al. 2017), with monomorphic populations matching highly susceptible phenotypes increasing the risks of transmission (Alba et al. 2019).

One of the most interesting facts with *P. columella*, however, is that it is the only species of Lymnaeidae that has shown an active true resistance to *F. hepatica* infection in some natural populations (Gutiérrez et al. 2003a). Individuals resistant to *F. hepatica* show a differential pigmentation of a belt-like shape pattern of very little and sharp white spots in the middle of the mantle (Gutiérrez et al. 2003b). All populations with this resistant phenotype are so far only reported from Cuba (Alba et al. 2019) and have been tested with several sympatric (i.e. Cuban) and allopatric (Dominican and French) isolates of *F. hepatica* (Alba et al. 2018; Vázquez et al.

2014, 2019). Some of these resistant populations were included in a world-scale study and clustered separated from the rest of susceptible populations at both mitochondrial and microsatellite markers (Lounnas et al. 2017) but their genetic distances are not sufficient for species separation (Gutiérrez et al. 2003a). Such resistant phenotype must have derived from susceptible populations long ago, as suggested by the divergence shown at mitochondrial markers of the populations spread in Cuba (Lounnas et al. 2017). Anyhow, the study of such resistance to *F. hepatica* infection may prove a key step in the control of fasciolosis. The introduction of snails from a species that commonly engage in transmission (i.e. miracidia will tend to enter the snails) but that do not develop infection in areas with risks of fasciolosis would be an effective strategy to tackle the transmission. Rational application after an ecological assessment of the target foci is however mandatory, as recent results advice of some ecological costs of resistance in *P. columella* in both field and laboratory studies (Alba et al. 2019).

13.3.1.3 *Hinkleyia*, *Ladislavella*, and *Pectinidens*: American Higher Latitude Hosts

The transmission in the Americas, as we have above-mentioned, is mainly due to the species from genera *Galba* and *Pseudosuccinea*. However, there are other species related to *F. hepatica*. In North America, *Hinkleyia caperata* (Say, 1829) has been linked to *F. hepatica* transmission from Montana, USA (Knapp et al. 1992) and *Ladislavella elodes* (syn. *Lymnaea proxima proxima*) was found naturally infected from eastern Washington by Lang (1977). We should note that the genus *Ladislavella* occurs also in cold latitudes from Eurasia, where the species *Ladislavella occulta* (Jackiewicz, 1959) has been found to serve as an important secondary host for *F. hepatica* in Poland (see (Czapski 1965, 1977; Jackiewicz 1998). According to Jackiewicz (1998), in Poland '*L. occulta* may play a great role as an intermediate host for the liver fluke (*Fasciola hepatica* L.), being in this respect even more important than *L. truncatula* (= *G. truncatula*)'.

In southernmost South America, the species *Pectinidens diaphanus* occurs in the Argentinean Patagonia and Chile (Paraense 1984). With virtually no other lymnaeid species occurring in the region with the southernmost report of *G. viator* being at 42° S from the north of the Chubut province in Argentina (Kleiman et al. 2004), infection in local cattle and wild mammals (guanacos) strongly suggest that transmission can only be due *P. diaphanus*. All these evidences imply that future studies are mandatory to understand the epidemiology of fasciolosis in such harsh environments and to clarify the role of these three species. Particularly in North America, a careful molecular revision is much needed as a plethora of published nominal lymnaeid species occur and thus, knowing the actual intermediate hosts of *F. hepatica* would bring lights into its epidemiology.

13.3.1.4 *Lymnaea*, *Stagnicola*, and *Omphiscola*: European Back Up Hosts?

In the European scenario of liver fluke transmission, *G. truncatula* is accepted as the primary intermediate host of *F. hepatica* driving the transmission of fasciolosis (Correa et al. 2017). However, the existence of several other species may allow the transmission in areas where *G. truncatula* does not occur, or complement its role in fully aquatic habitats not commonly exploited by this amphibious lymnaeid (Vignoles et al. 2017a). For instance, *Lymnaea stagnalis* (Linnaeus, 1758) has always been thought as a potential host from Europe, though reports on natural infections are still lacking. Since the works by Kendall (1949), its capacity for transmitting *F. hepatica* is accepted. Experimental infections that show 17% and 38% prevalence reinforce its presumptive role in field transmission (Busson et al. 1982; Dreyfuss et al. 2002). More recently, an experimental prevalence reaching 50% was found in *L. stagnalis* when several lymnaeid species were experimentally exposed to the current *F. hepatica* circulating strain from Camargue, southern France (Vázquez et al. 2020). However, it is only from out of Europe that *L. stagnalis* was reported naturally infected in Azerbaijan through molecular detection (1% prevalence; Yakhchali et al. 2015). In addition, introduced populations of *L. stagnalis* were found infected from eastern Washington, USA, in the field and later tested through experimental exposure (Lang 1977).

Similar to *L. stagnalis*, there are several species of stagnicolines that have been related to *F. hepatica* transmission, but with only a few natural infections recorded. Both *Stagnicola fuscus* (C. Pfeiffer, 1821) and *Stagnicola palustris* (O. F. Müller, 1774) have been experimentally infected with *F. hepatica* (13 and 51%, respectively) with the latter found infected in the field in Sweden (0.15% prevalence; Novobilský et al. 2013). Other works have also experimentally tested the compatibility of *S. fuscus* (33%), *S. palustris* (12%), and *Stagnicola corvus* (Gmelin, 1791) (3–10%) to *F. hepatica* (Degueurce et al. 1999; Dreyfuss et al. 2000; Kruglov 1986). Moreover, there are a few reports of natural infections of *S. palustris* introduced in North America, specifically from eastern Washington (Lang 1977). However, it is worth mentioning a field experimental trial in France where *F. hepatica* miracidia were introduced in ponds colonized by *S. palustris* for six consecutive years in which prevalence increased from 0.4 to 18% (Dreyfuss et al. 1994).

The case of *Omphiscola glabra* (O. F. Müller, 1774) is similar to the other European lymnaeids. This species has been found infected in the field from central France with prevalence ranging from 0.14–2% (Dreyfuss et al. 2003) while some experimental infections have shown prevalence of 16–31% (Busson et al. 1982; Sindou et al. 1991). Moreover, infected individuals of *O. glabra* seem able to overwinter, surviving until spring (Rondelaud et al. 2000), and some reports evidence their implication in *F. hepatica*-contaminated watercress beds from France (Rondelaud et al. 2005).

13.3.2 The Amphipepleinae

13.3.2.1 Radicine Relatives: Aquatic Old World Rulers of Fasciolosis

Radicine snails were not long ago all compiled within the genus *Radix*, now divided into several genera (Aksenova et al. 2018). They are largely known for their role in transmitting both *F. hepatica* and *F. gigantica* in Eurasia and Africa (Caron et al. 2007; Dar et al. 2010; Vázquez et al. 2018). Species of *Radix* s.l. (subgenera *Radix* and *Exsertiana*, see Vinarski et al. 2020) are overall more related to *F. gigantica* transmission in tropical and subtropical regions of Asia and Africa (Brown 1994; Kaset et al. 2010). As in the case of amphibious *Galba* snails and *F. hepatica*, this affinity might be the result of an overlapping ecology of the intervening hosts (*i.e.* more aquatic) that may have evolved different host-finding mechanisms. Opposite to those of *F. hepatica*, miracidia of *F. gigantica* show a negative phototactic response guiding the larvae towards its *Radix* typical benthic environments (Goater et al. 2014). For instance, in a study carried out in Iran, experimental infection trials showed that *Radix euphratica* (Mousson, 1874) was susceptible to *F. gigantica* with 45% prevalence (Massoud and Sadjadi 1980).

Specifically, *Radix natalensis* is largely responsible for fasciolosis due to *F. gigantica* in Africa (Brown 1994) where natural infections reports exist from Benin (Assogba et al. 2011), Nigeria (Adediran and Uwalaka 2013), and Ghana (Brown 1994) and its role as main host is presumed in Senegal (Vassiliades 1978), Ethiopia (Fromsa et al. 2011), and Uganda (Howell et al. 2012). Out of continental Africa, *R. natalensis* is suspected of *F. gigantica* transmission also from Madagascar and La Réunion (Barré et al. 1982; Daynes 1967). The species has also been suggested for *F. hepatica* transmission in Egypt (Dar et al. 2010) and Saudi Arabia (as *Lymnaea arabica* in Dajem 2012). Moreover, *R. natalensis* is the only species implicated so far in the transmission of *F. nyanzae*, which parasitizes hippos in Africa (Dinnik and Dinnik 1961).

There are several other species of *Radix* implicated in *F. gigantica* transmission. Out of Africa, *Radix auricularia* (Linnaeus, 1758) is perhaps the most sounded intermediate host of *F. gigantica* with natural prevalences of 0.3% recorded from Bangladesh (Chowdhury et al. 1994). In addition, *R. auricularia* is one of the main snail hosts of *F. gigantica* in Iran (Ashrafi and Mas-Coma 2014), India (Gupta and Yadav 1994), and Thailand (Charoenchai et al. 1997); and has been experimentally infected in Japan with Australian strains of *F. hepatica* (Itagaki et al. 1988). Another species, *Radix rubiginosa* (Michelin, 1831) was found infected with *F. gigantica* in the field from Java (0.6% prevalence; Estuningsih and Copeman 1996) and Thailand (Kaset et al. 2010). A third species from Iran, *R. euphratica*, showed a 0.15% natural prevalence with *F. gigantica* (= *R. gedrosiana*; Ashrafi et al. 2004). From Iraq, a published record under *Radix lagotis euphratica* (Kadhim and Altaif 1970) attests the susceptibility to *F. gigantica*, but it probably should be treated as *R. euphratica*. In Pakistan, a record of *Radix rufescens* (J. E. Gray, 1822) experimentally infected with African strains of *F. gigantica* was published by Rao (1966). Yet another lymnaeid in the Philippines, the species *R. swinhoei* (probably *R. plicatula*; see Vinarski et al. 2020) was found naturally infected with *F. gigantica* (Kaset et al. 2010). At last, a recent study from the Gannan Tibetan Autonomous Prefecture in China showed very high prevalences of *F. hepatica* (93%) through a nest-PCR in field collected *Radix cucunorica* (Möllendorff, 1902) from a small pond (Huang et al. 2019).

Other radicines now grouped under genera *Ampullaceana* and *Peregriana* (Aksenova et al. 2018) are more commonly related with *F. hepatica* transmission, particularly in Europe. For instance, *Ampullaceana balthica* (Linnaeus, 1758) has been found infected in the field from Belgium (0.16% prevalence; Caron et al. 2014) and France (9% prevalence recorded as *Lymnaea ovata*; Degueurce et al. 1999). In Russia, the species *Ampullaceana ampla* (W. Hartmann, 1821), *Ampullaceana fontinalis* (S. Studer, 1820), and *Ampullaceana lagotis* (Schrank, 1803) were successfully infected in the laboratory (Kruglov 1986). Moreover, both *A. balthica* and *Peregriana labiata* (Rossmässler, 1835) [= *Peregriana peregra* (O. F. Müller, 1774) sensu Vinarski et al. 2020] were experimentally exposed to *F. hepatica* miracidia showing 6 and 45% prevalence, respectively (Caron et al. 2007). Other studies have also tested the infection of *A. balthica* with *F. hepatica* suggesting that environmental pollution may disrupt the innate immunity of the snails and facilitate the larval development (as *L. ovata* in Dreyfuss et al. 2000). In addition, a study by Boray (1966) showed that *P. peregra*, usually considered an unsuitable host, might gradually adapt to *F. hepatica* infection after several laboratory reared generations showed experimental prevalence within 2–77% range. However, a recent molecular study showed a natural prevalence of 37% from the fields of Ireland (Relf et al. 2009). Finally, the species *Myxas glutinosa* (O. F. Müller, 1774), considered rare in Europe and also with only a few recorded localities from the Urals and Siberian regions (Vinarski et al. 2013), has proved susceptible under experimental conditions to *F. hepatica*, though not considered an important host (Monny 1995).

13.3.2.2 *Orientogalba* and *Austropeplea*: Transmitters of the Far East

The last clade of lymnaeid snail host of *Fasciola* spp. group all those occurring mainly in eastern Asia, Australia, and some Pacific islands. There, *Austropeplea tomentosa* (L. Pfeiffer, 1855) is strongly related to the transmission of *F. hepatica* in Australia, New Zealand, and Tasmania (Boray 1978; Molloy and Anderson 2006). Although studies reflecting the natural prevalence of *A. tomentosa* are much needed, the species has been experimentally tested with prevalences between 13 and 100% (Boray 1966, 1978). The species *Orientogalba viridis* (Quoy & Gaimard, 1833) was experimentally exposed to *F. hepatica* in South Korea with up to 55% prevalence observed and a production of metacercariae similar to that previously observed in *G. truncatula* or *P. columella* (Lee et al. 1995). Boray (1978) aimed at experimentally assessing the transmission capacity of *O. viridis* from different origins and showed prevalences of 56% (Australia), 49% (Nepal), and 64% (Papua New Guinea). Recently, *O. viridis* was found in the field from Delta del Ebro in Spain

(Schniebs et al. 2017), an introduction that may eventually complicate the transmission scenario if an expansion of this lymnaeid species occurs in the region. Concerning *F. gigantica* transmission, *O. viridis* was found infected in the field from Vietnam (0.11–2% prevalence; Dung et al. 2013). Moreover, an experimental infection trial with several species in Japan using an Australian strain of *F. hepatica* showed a prevalence of 83.3% in *Orientogalba ollula* (Gould, 1859) (Itagaki et al. 1988), a lymnaeid presumed responsible also for *F. gigantica* transmission in Hawaii (Alicata 1938). Another species, *Bullastra lessoni* (Deshayes, 1831), did not show larval development after being experimentally exposed with *F. hepatica* miracidia (Boray 1966).

13.3.2.3 Understanding the Risks of Fasciolosis Transmission by Lymnaeid Snails

As presented above, there is a quite fair number of lymnaeid species acting as intermediate hosts of either *F. hepatica* or *F. gigantica*, or both. In this chapter, we have compiled 31 species worldwide for which data on infection prevalence is available. We have also discussed other species that are suspected of transmitting in some regions (e.g. *H. caperata* and *P. diaphanus*) but for which more studies are needed to ascertain their true role. Such species occur practically throughout the complete phylogeny of the Lymnaeidae and are well distributed worldwide, although some clades are more represented in certain regions (Fig. 13.1). This means that fasciolosis is potentially able to occur virtually everywhere if conditions are not exceptionally harsh (e.g. desert or arctic environments) and both the hosts (intermediate and definitive) and the parasite co-occur. The risks posed by each snail species can be understood as the sum of several independent factors that may or may not concur in a particular scenario. For instance, if the snail species is widely distributed and adapted to the circulating strains of the liver fluke (higher compatibility in terms of surviving infection, redial burden, and cercarial shedding), and/or the snail species is capable of establishing itself in areas of cattle-related human activities (even very anthropized habitats) with ecological preferences that match those of the definitive hosts, and/or the species has a high invasive ability and the lack of genetic diversity of introduced individuals facilitate the transmission, then the risks increase considerably.

Perhaps worldwide illustrating examples of the above-mentioned combination of risks in the case of *F. hepatica* transmission are those of *G. truncatula* and *P. columella*. Both species have been proved susceptible to infection almost everywhere they are introduced, with some allopatric combinations tested experimentally displaying high compatibility (Alba et al. 2018; Vázquez et al. 2019). Their enormous risks of introduction have been demonstrated by their recorded presence in new temperate (*G. truncatula*) and tropical environments (*P. columella*) (Lounnas et al. 2017; Meunier et al. 2004). The lack of genetic diversity, as a direct result of founder effects after introductions, has been observed in both species from sites with known high fasciolosis transmission either in human populations (*G. truncatula* in Bolivia; Meunier et al. 2001) or cattle (*P. columella* in Cuba; Alba et al. 2019).

Unfortunately, a plethora of names within the Lymnaeidae (at both species and genus levels) in both ancient and recent literature usually confuses and entangles the comprehension of fasciolosis in a given area. Fortunately, however, with many now being reviewed (see, for example, Vinarski et al. 2020 and Alda et al. 2021) the question of who actually transmits whom is becoming clearer. This information is crucial for a full understanding of transmission patterns and overall epidemiology of fasciolosis. For instance, *G. cubensis* (also under *Lymnaea, Fossaria,* or *Bakerilymnaea* genera) is the recognized main snail host of *F. hepatica* in southern USA and the Caribbean (Vázquez et al. 2019). Now, this species is presumed a senior synonym over North American *G. bulimoides* transmitting largely in Mexico and some regions of the USA, or even perhaps a junior synonym of the South American *G. viator* responsible for transmission in Argentina and Peru (see the discussion from Alda et al. 2018, 2021). If that confirms, it would mean that a single species (i.e. *G. viator*) is actually responsible for transmission in a vast continental area such as the Americas. Something similar occurred with *R. natalensis*, host of *F. gigantica* especially in Africa (Brown 1994). This species was (re)described several times under different names from different countries of Africa (e.g. *L. africana*, Ethiopia; *L. electa*, Madagascar; *L. nyanzae*, Lake Victoria, Uganda; etc.) and Asia (e.g. *L. arabica*, Oman). Only studies that combine morphological and genetic traits coupled with reproductive compatibility in non-preferential selfer species would ultimately resolve the status of several epidemiologically relevant lymnaeids. Such studies have greatly helped in the acknowledgment of the most important actors in fasciolosis transmission. Although a given region may have several species able to transmit the parasite, usually only one or a few are actually responsible for the overall burden of the disease. These lymnaeid species are illustrated in Fig. 13.3 for each *Fasciola* sp. at a global scale, based on the reports at both field and experimental levels. Here, we see that, according to different regions, in the case of *F. hepatica*, the species *G. cubensis*/*G. viator*, *G. truncatula*, *O. viridis*, and *A. tomentosa* are actually the overall drivers of parasite transmission. Similarly, *F. gigantica* is overall mainly transmitted by *R. natalensis*, *R. auricularia*, *R. rufescens*, and *R. rubiginosa*.

Finally, the risks of fasciolosis transmission necessarily link directly to the ecology of their hosts. In the case of lymnaeid snails, it may differ for different genera and species, and such understanding might be the shorter way to prevent the infection or effectively cut the transmission, particularly in cattle populations. Furthermore, it appears that some genera have tighter relations with each of the *Fasciola* species. As such, authors have referred that the *Galba* group are best suited for hosting *F. hepatica* whereas the *Radix* group commonly relates to *F. gigantica* (Bargues and Mas-Coma 2005). Such differential role in the transmission is in fact tightly related to the compatibility established between the snail and the parasite, even at the population level. This 'affinity' usually has its origin in a linked evolutionary history facilitated by the ecology of the hosts (intermediate and definitive) in a particular group rather than within larger phylogenetic clades. This

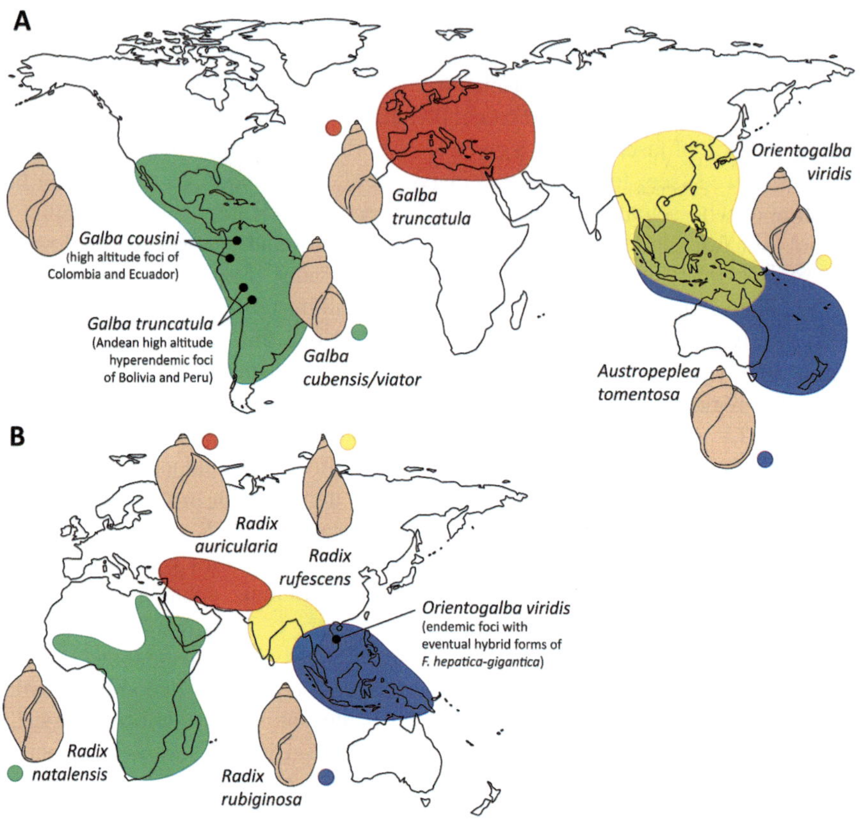

Fig. 13.3 Key lymnaeid species in the transmission of fasciolosis by region. (**a**) *Fasciola hepatica*; (**b**) *Fasciola gigantica*

preferential transmission is schematically illustrated in Fig. 13.4 using the reported prevalences after reviewing worldwide experimental infection trials. Here, we see that the overall probability of infection with either liver fluke for both the definitive and intermediate hosts differs in relation to the habitat. However, some overlapping occurs though more aquatic hosts match those commonly reported infected with *F. gigantica* at least in tropical regions. The reported prevalences in the case of *F. hepatica* suggest, however, a tight evolutionary history between amphibious lymnaeids and small mammals less related with aquatic environments but still with a larger range of hosts able to transmit the parasite.

Lymnaeid species may occur in populations with high densities that may occupy small or large effective areas (Rondelaud et al. 2009). In the case of amphibious or semi-amphibious lymnaeids, a population may settle over large humid pastures or mud not necessarily flooded as it has been shown for *G. cubensis* in Cuba (Vázquez and Gutiérrez 2007), *G. truncatula* in France (Rondelaud et al. 2009), or *A. tomentosa* and *G. truncatula* in New Zealand (Pullan et al. 1972). If conditions

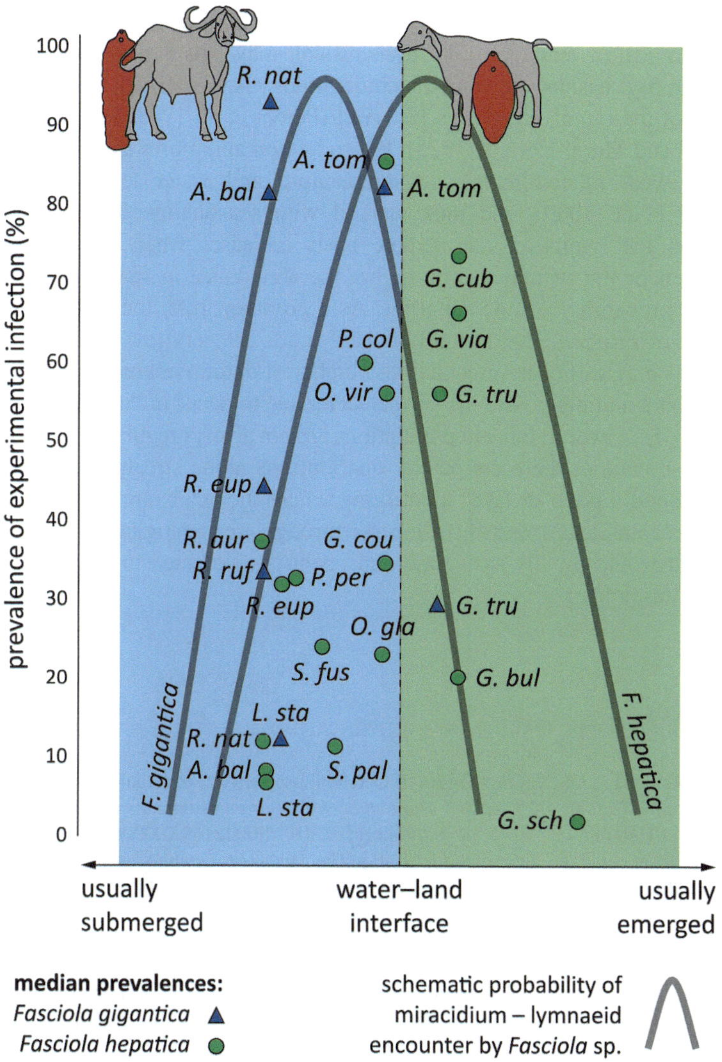

Fig. 13.4 Schematic representation of the overall transmission probability of each liver fluke in relation to the most common habitats of their most important intermediate lymnaeid hosts supported by the reported data of experimental infections from Table 13.1 (*A.bal, Ampullaceana balthica*; *A. tom, Austropeplea tomentosa*; *G.bul, Galba bulimoides*; *G.cou, Galba cousini*; *G.cub, Galba cubensis*; *Gsch, Galba schirazensis*; *G.tru, Galba truncatula*; *G.via, Galba viator*; *L.sta, Lymnaea stagnalis*; *O.gla, Omphiscola glabra*; *O.vir, Orientogalba viridis*; *P.col, Pseudosuccinea columella*; *Pper, Peregriana peregra*; *R.aur, Radix auricularia*; *R.eup, Radix euphratica*; *R.nat, Radix natalensis*; *R.ruf, Radix rufescens*; *Sfus, Satgnicola fuscus*; *S.pal, Stagnicola palustris*)

are favourable in some transmission foci, those populations may remain stable all year round and the life cycle of *Fasciola* spp. may remain uninterrupted. This is typical of some tropical regions and has been particularly studied in fasciolosis active foci from Cuba (Cañete et al. 2004; Gutiérrez et al. 2005). However, in some regions (tropical or temperate) the transmission may halt during harsh conditions (e.g. winter, dry seasons, etc.) and the epidemiology is then driven by the population dynamics of the lymnaeid species involved (Perera et al. 1995; Perera et al. 1986; Rondelaud and Morel-Veraille 1975). Reproduction and mortality peaks may occur across the year related mainly to temperature and water level (Perera 1996; Rondelaud et al. 2009), and thus marked with seasonality. The probability of infection in the lymnaeid intermediate hosts increases when the rate of snail–miracidia encounter increases. The higher the abundance in snail populations, the higher the probability of encountering. As a common rule, the reports of natural prevalence of *Fasciola* spp. in the lymnaeid snails are very low (less than 3%) but such figures may suffice to infect a high number of definitive hosts in relation to the type of habitat and the possibility of metacercariae to occur in large areas (Vázquez et al. 2015). However, lymnaeid populations are not always regularly scattered in the field, and sometimes there are patchy distributions along streams, irrigation channels, or around a pond or lake. Identifying where the snails are and which factors affect their abundances and distributions through malacological surveys are key actions that would greatly save time and resources allocated to prevent the transmission of fasciolosis.

References

Abrous M, Rondelaud D, Dreyfuss G (2000) Cercarial productivity of redial generations in single-miracidium infections of *Lymnaea truncatula* with *Paramphistomum daubneyi* or *Fasciola hepatica*. J Helminthol 74:1–5. https://doi.org/10.1017/S0022149X00000019

Abrous M, Rondelaud D, Dreyfuss G (2001) The stress of *Lymnaea truncatula* just before miracidial exposure with *Fasciola hepatica* increased the prevalence of infection. Exp Parasitol 99:49–51. https://doi.org/10.1006/expr.2001.4609

Adediran O, Uwalaka E (2013) Laboratory Breeding of *Lymnaea natalensis* (Krauss, 1848), Intermediate Host of *Fasciola gigantica* (Cobbold, 1856). N Y Sci J 6(8):55–57

Adema CM, Loker ES (2015) Digenean-gastropod host associations inform on aspects of specific immunity in snails. Dev Comp Immunol 48(2):275–283. https://doi.org/10.1016/j.dci.2014.06.014

Ahmed AH, Ramzy RM (1999) Infection of two lymnaeid snails with *Fasciola gigantica* in Giza, a field study. J Egypt Soc Parasitol 29(3):687–696

Aksenova OV, Bolotov IN, Gofarov MY et al (2018) Species richness, molecular taxonomy and biogeography of the radicine pond snails (Gastropoda: Lymnaeidae) in the Old World. Sci Rep 8:11199. https://doi.org/10.1038/s41598-018-29451-1

Alba A, Vázquez A, Sánchez J et al (2016) Assessment of the FasciMol-ELISA in the detection of the trematode *Fasciola hepatica* in field-collected *Galba cubensis*: a novel tool for the malacological survey of fasciolosis transmission. Parasit Vectors 9(22). https://doi.org/10.1186/s13071-016-1303-1

Alba A, Vázquez A, Sánchez J et al (2018) Fasciola hepatica-Pseudosuccinea columella interaction: effect of increasing parasite doses, successive exposures and geographical origin on the infection outcome of susceptible and naturally resistant snails from Cuba. Parasit Vectors 11:559. https://doi.org/10.1186/s13071-018-3155-3

Alba A, Vázquez AA, Sanchez J et al (2019) Patterns of distribution, population genetics and ecological requirements of field-occurring resistant and susceptible *Pseudosuccinea columella* snails to *Fasciola hepatica* in Cuba. Sci Rep 9(1):14359. https://doi.org/10.1038/s41598-019-50894-7

Alba A, Vázquez A, Hurtrez-Boussès S (2021) Towards the comprehension of fasciolosis (re-) emergence: an integrative overview. Parasitology 148(4):385–407. https://doi.org/10.1017/s0031182020002255

Alda P, Lounnas M, Vázquez A et al (2018) A new multiplex PCR assay to distinguish among three cryptic *Galba* species, intermediate hosts of *Fasciola hepatica*. Vet Parasitol 251:101–105. https://doi.org/10.1016/j.vetpar.2018.01.006

Alda P, Lounnas M, Vázquez AA et al (2021) Systematics and geographical distribution of *Galba* species, a group of cryptic and worldwide freshwater snails. Molec Phyl Evol 157:107035. https://doi.org/10.1016/j.ympev.2020.107035

Alicata J (1938) Observations on the life history of *Fasciola gigantica*, the common liver fluke of cattle in Hawaii, and the intermediate host, *Fossaria ollula*. Hawaii Agric Exp Stn Bull 80:1–22. https://doi.org/10.5962/bhl.title.87642

Arafa WM, Hassan AI, Snousi SAM et al (2018) *Fasciola hepatica* infections in cattle and the freshwater snail Galba truncatula from Dakhla Oasis, Egypt. J Helminthol 92(1):56–63. https://doi.org/10.1017/S0022149X17000086

Ashrafi K, Mas-Coma S (2014) *Fasciola gigantica* transmission in the zoonotic fascioliasis endemic lowlands of Guilan, Iran: experimental assessment. Vet Parasitol 205(1–2):96–106. https://doi.org/10.1016/j.vetpar.2014.07.017

Ashrafi K, Massoud J, Holakouei K et al (2004) Evidence Suggesting that *Fasciola gigantica* might be the most prevalent causal agent of Fascioliasis in Northern Iran. Iran J Publ Health 33(4): 31–37

Ashrafi K, Massoud J, Holakouei K et al (2007) Nuclear ribosomal DNA ITS-2 sequence characterization of *Fasciola hepatica* and *Galba truncatula*. Iran J Publ Health 36(4):42–49

Ashrafi K, Saadat F, O'Neill S et al (2015) The endemicity of human Fascioliasis in Guilan Province, Northern Iran: the baseline for implementation of control strategies. Iran J Publ Health 44(4):501–511

Assogba M, Dougnon J, Alkoiret I (2011) Epidémiologie de la fasciolose bovine au Bénin et essai de lutte biologique. Livest Res Rural Dev 23(12):248

Bargues MD, Mas-Coma S (1997) Phylogenetic analysis of lymnaeid snails based on 18S rDNA sequences. Mol Biol Evol 14(5):569–577. https://doi.org/10.1093/oxfordjournals.molbev.a025794

Bargues M, Mas-Coma S (2005) Reviewing lymnaeid vectors of fascioliasis by ribosomal DNA sequence analyses. J Helminthol 79:257–267. https://doi.org/10.1079/JOH2005297

Bargues MD, Artigas P, Khoubbane M et al (2011) *Lymnaea schirazensis*, an overlooked snail distorting fascioliasis data: Genotype, phenotype, ecology, worldwide spread, susceptibility, applicability. PLoS One 6:e24567. https://doi.org/10.1371/journal.pone.0024567

Bargues M, Artigas P, Khoubbane M et al (2012) Molecular characterisation of *Galba truncatula*, *Lymnaea neotropica* and *L. schirazensis* from Cajamarca, Peru and their potential role in transmission of human and animal fascioliasis. Parasites Vectors 5:174. https://doi.org/10.1186/1756-3305-5-174

Bargues M, Gayo V, Sanches J et al (2017) DNA multigene characterization of *Fasciola hepatica* and *Lymnaea neotropica* and its fascioliasis transmission capacity in Uruguay, with historical correlation, human report review and infection risk analysis. PLoS Negl Trop Dis 11(2): e0005352. https://doi.org/10.1371/journal.pntd.0005352

Barré N, Isautier H, Frandsen F et al (1982) Inventaire des mollusques d'eau douce de la Réunion. Revue d'Elevage et de Médecine Vétérinaires des Pays Tropicaux 35:35–41

Belfaiza M, Moncef M, Rondelaud D (2005) *Fasciola hepatica*: an unusual case of adaptation to a Moroccan population of *Galba truncatula*. Parasitol Res 95:374–378. https://doi.org/10.1007/s00436-004-1281-7

Belfaiza M, Moncef M, Vignoles P et al (2009) *Galba truncatula* (Mollusca Gastropoda, Lymnaeidae), le mollusque hôte de *Fasciola hepatica*: son aptitude à supporter un assèchement périodique en zone irriguée sous climat semi-aride. Rev Med Vet 160(2):103–107

Beltrame MO, Pruzzo C, Sanabria R et al (2020) First report of pre-Hispanic *Fasciola hepatica* from South America revealed by ancient DNA. Parasitology 147(3):371–375. https://doi.org/10.1017/S0031182019001719

Berghen P (1964) Some Lymnaeidae as intermediate hosts of *Fasciola hepatica* in Belgium. Exp Parasitol 15(2):118–124. https://doi.org/10.1016/0014-4894(64)90012-8

Boray JC (1966) Studies on the relative susceptibility of some lymnaeids to infection with *Fasciola hepatica* and *F. gigantica* and on the adaptation of *Fasciola* spp. Ann Trop Med Parasitol 60(1): 114–124. https://doi.org/10.1080/00034983.1966.11686394

Boray JC (1978) The potential impact of exotic *Lymnaea* spp. on fascioliasis in Australasia. Vet Parasitol 4:127–141. https://doi.org/10.1016/0304-4017(78)90004-3

Brown DS (1994) Freshwater snails of Africa and their medical importance. Taylor & Francis, London. https://doi.org/10.1201/9781482295184

Brumpt E, Velazquez J, Ucros H et al (1940) Mission E. Brumpt et L.-Ch. Brumpt en Colombie et au Venezuela. 1.- Découverte de l'hôte intermédiaire, *Limnaea bogotensis* Pilsbry, de la grande douve, *Fasciola hepatica*, en Colombie. Annales de Parasitologie Humaine et Comparee 17: 563–579

Busson P, Busson D, Rondelaud D et al (1982) Données expérimentales sur l'infestation des jeunes de cinq espèces de limnées par *Fasciola hepatica* L. Annales de Parasitologie 57(6):555–563

Cañete R, Yong M, Sánchez J et al (2004) Population dynamics of intermediate snail hosts of *Fasciola hepatica* and some environmental factors in San Juan y Martinez municipality, Cuba. Mem Inst Oswaldo Cruz 99(3):257–262. https://doi.org/10.1590/s0074-02762004000300003

Carmona C, Tort JF (2017) Fasciolosis in South America: epidemiology and control challenges. J Helminthol 91(2):99–109. https://doi.org/10.1017/S0022149X16000560

Caron Y, Lasri S, Losson B (2007) *Fasciola hepatica*: an assessment on the vectorial capacity of *Radix labiata* and *R. balthica* commonly found in Belgium. Vet Parasitol 149:95–103. https://doi.org/10.1016/j.vetpar.2007.07.012

Caron Y, Martens K, Lempereur L et al (2014) New insight in lymnaeid snails (Mollusca, Gastropoda) as intermediate hosts of *Fasciola hepatica* (Trematoda, Digenea) in Belgium and Luxembourg. Parasites Vectors 7:66. https://doi.org/10.1186/1756-3305-7-66

Caron Y, Celi-Erazo M, Hurtrez-Boussès S et al (2017) Is *Galba schirazensis* (Mollusca, Gastropoda) an intermediate host of *Fasciola hepatica* (Trematoda, Digenea) in Ecuador? Parasite 24(24):1–9. https://doi.org/10.1051/parasite/2017026

Celi-Erazo M, Alda P, Montenegro-Franco M et al (2020) Prevalence of *Fasciola hepatica* infection in *Galba cousini* and *Galba schirazensis* from an Andean region of Ecuador. Vet Parasitol Reg Stud Rep 20:100390. https://doi.org/10.1016/j.vprsr.2020.100390

Chapuis E (2009) Correlation between parasite prevalence and adult size in a trematode-mollusc system: evidence for evolutionary gigantism in the freshwater snail *Galba truncatula*? J Mollus Stud 75:391–396. https://doi.org/10.1093/mollus/eyp035

Charoenchai A, Tesana S, Pholpark M (1997) Natural infection of trematodes in *Lymnaea* (*Radix*) *auricularia rubiginosa* in water reservoirs in Amphoe Muang, Khon Kaen Province. Southeast Asian J Trop Med Public Health 28(Suppl 1):209–212

Chlyeh G (2002) Dynamique et génétique des populations de mollusques hôtes intermediaires des parasites eau-dépendantes (bilharzioses et fascioloses). PhD, Institut Agronomique et Vétérinaire Hassan II, Maroc

Choi YJ, Fontenla S, Fischer PU et al (2020) Adaptive radiation of the flukes of the family Fasciolidae inferred from genome-wide comparisons of key species. Mol Biol Evol 37(1): 84–99. https://doi.org/10.1093/molbev/msz204

Chowdhury S, Mondal M, Rahman M (1994) Prevalence of *Fasciola hepatica* cercariae in lymnaeid snails in Bangladesh. AJAS 7(3):401–403

Coelho L, Lima W (2003) Population dynamics of *Lymnaea columella* and its natural infection by *Fasciola hepatica* in the State of Minas Gerais, Brazil. J Helminthol 77:7–10. https://doi.org/10.1079/JOH2002138

Coelho L, Lima W, Guimarães M (2009) Sympatric and allopatric combinations of *Lymnaea columella* and *Fasciola hepatica* from southern and south-eastern Brazil. J Helminthol 83: 285–288. https://doi.org/10.1017/S0022149X09222930

Correa A, Escobar J, Durand P et al (2010) Bridging gaps in the molecular phylogeny of the Lymnaeidae (Gastropoda: Pulmonata), vectors of Fascioliasis. BMC Evol Biol 10:381. https://doi.org/10.1186/1471-2148-10-381

Correa A, De Meeûs T, Dreyfuss G et al (2017) *Galba truncatula* and *Fasciola hepatica*: genetic costructures and interactions with intermediate host dispersal. Infect Genet Evol 55:186–194. https://doi.org/10.1016/j.meegid.2017.09.012

Cowie R (2001) Invertebrate invasions on Pacific Islands and the replacement of unique native faunas: a synthesis of the land and freshwater snails. Biol Invasions 3:119–136. https://doi.org/10.1023/A:1014529019000

Cruz-Mendoza I, Figueroa J, Correa D et al (2004) Dynamics of *Fasciola hepatica* infection in two species of snails in a rural locality of Mexico. Vet Parasitol 121:87–93. https://doi.org/10.1016/j.vetpar.2004.02.006

Cruz-Mendoza I, Ibarra F, Quintero M (2005) Seasonal transmission of *Fasciola hepatica* in cattle and *Lymnaea* (*Fossaria*) *humilis* snails in central Mexico. Parasitol Res 95:283–286. https://doi.org/10.1007/s00436-004-1287-1

Cruz-Reyes A, Malek E (1987) Suitability of six lymnaeid snails for infection with *Fasciola hepatica*. Vet Parasitol 24(3–4):203–210. https://doi.org/10.1016/0304-4017(87)90041-0

Cucher M, Carnevale S, Prepelitchi L et al (2006) PCR diagnosis of *Fasciola hepatica* in field-collected *Lymnaea columella* and *Lymnaea viatrix* snails. Vet Parasitol 137:74–82. https://doi.org/10.1016/j.vetpar.2005.12.013

Czapski Z (1962) The snail *Galba occulta* Jackiewicz, 1959 – another intermediate host of *Fasciola hepatica* L. Z Tropenmed Parasitol 13:332–337

Czapski Z (1965) Z badań nad biologią *Galba occulta* Jack., 1959 – nowego żywiciela pośredniego *Fasciola hepatica* L. Wiadomosci parazytołogyczne 11:273–277

Czapski Z (1977) Biologicznc aspckty epidemiologii fasciolozy. Monografie Akademii Wychowania Fizycznego w Poznaniu, Poznań, Poland

Dajem S (2012) Molecular typing of the fresh water snail *Lymnaea arabica*, the possible intermediate host of *Fasciola hepatica*, collected from Saudi Arabia, by RAPD-PCR. Egypt Acad J Biol Sci 4(1):173–181

Dar Y, Vignoles P, Rondelaud D et al (2002) *Fasciola gigantica*: the growth and larval productivity of redial generations in the snail *Lymnaea truncatula*. Parasitol Res 88:364–367. https://doi.org/10.1007/s00436-001-0552-9

Dar Y, Rondelaud D, Dreyfuss G (2003) Cercarial shedding from *Galba truncatula* infected with *Fasciola gigantica* of distinct geographic origins. Parasitol Res 89:185–187. https://doi.org/10.1007/s00436-002-0732-2

Dar Y, Djuikwo F, Vignoles P et al (2010) *Radix natalensis* (Gastropoda: Lymnaeidae), a potential intermediate host of *Fasciola hepatica* in Egypt. Parasite 17:251–256. https://doi.org/10.1051/parasite/2010173251

Dar Y, Lounnas M, Djuikwo F et al (2013) Variations in local adaptation of allopatric *Fasciola hepatica* to French *Galba truncatula* in relation to parasite origin. Parasitol Res 112:2543–2549. https://doi.org/10.1007/s00436-013-3421-4

Dar Y, Rondelaud D, Vignoles P et al (2014a) *Fasciola hepatica*: development of redial generations in experimental infections of *Pseudosuccinea columella*. Parasitol Res 113:2467–2473. https://doi.org/10.1007/s00436-014-3893-x

Dar Y, Vignoles P, Rondelaud D et al (2014b) Role of the lymnaeid snail *Pseudosuccinea columella* in the transmission of the liver fluke *Fasciola hepatica* in Egypt. J Helminthol 89(6):699–706. https://doi.org/10.1017/S0022149X14000406

Daynes P (1967) La distomatose à Madagascar. Rapport sur des travaux réalisés de *Fasciola gigantica*. Revue d'Elevage et de Médecine Vétérinaire des Pays Tropicaux 20(4):557–562
Degueurce F, Abrous M, Dreyfuss G et al (1999) *Paramphistomum daubneyi* and *Fasciola hepatica*: the prevalence of natural or experimental infections in four species of freshwater snails in eastern France. J Helminthol 73:197–202. https://doi.org/10.1007/s004360050628
Dinnik J, Dinnik N (1957) A mud snail, *Lymnaea mweruensis* Connolly as an intermediate host of both liver flukes *Fasciola hepatica* L. and *Fasciola gigantica* Cobbold. Rep E Afr Vet Res Org 1:50–52
Dinnik J, Dinnik N (1961) On the morphology, and life history of *Fasciola nyanzae* Leiper, 1910 from the hippopotamus. J Helminthol 65(S1):53–62. https://doi.org/10.1017/S0022149X00017570
Dreyfuss G, Rondelaud D (1997) *Fasciola gigantica* and *F. hepatica*: a comparative study of some characteristics of *Fasciola* infection in *Lymnaea truncatula* infected by either of the two trematodes. Vet Res 28:123–130
Dreyfuss G, Moukrim A, Rondelaud D et al (1994) Field observations concerning infection of *Lymnaea palustris* by *Fasciola hepatica*. J Helminthol 68(2):115–118. https://doi.org/10.1017/s0022149x00013626
Dreyfuss G, Abrous M, Rondelaud D (2000) The Susceptibility of *Lymnaea fuscus* to Experimental Infection with *Fasciola hepatica*. J Parasitol 86(1):158–160. https://doi.org/10.1645/0022-3395(2000)086[0158,tsolft]2.0.co;2
Dreyfuss G, Vignoles P, Abrous M et al (2002) Unusual snail species involved in the transmission of *Fasciola hepatica* in watercress beds in central France. Parasite 9(2):113–120. https://doi.org/10.1051/parasite/2002092113
Dreyfuss G, Vignoles P, Rondelaud D (2003) Natural infections of *Omphiscola glabra* (Lymnaeidae) with *Fasciola hepatica* in central France. Parasitol Res 91:458–491. https://doi.org/10.1007/s00436-003-0892-8
Dreyfuss G, Vignoles P, Rondelaud D (2005) *Fasciola hepatica*: epidemiological surveillance of natural watercress beds in central France. Parasitol Res 95:278–282. https://doi.org/10.1007/s00436-004-1269-3
Dreyfuss G, Vignoles P, Rondelaud D (2007) *Fasciola hepatica*: the infectivity of cattle-origin miracidia had increased over the past years in central France. Parasitol Res 101:1157–1160. https://doi.org/10.1007/s00436-007-0580-1
Dreyfuss G, Correa A, Djuikwo-Teukeng F et al (2015) Differences in the compatibility of infection between the liver flukes *Fascioloides magna* and *Fasciola hepatica* in a Colombian population of the snail *Galba* sp. J Helminthol 89:720–726. https://doi.org/10.1017/S0022149X14000509
Dung B, Doanh P, The D et al (2013) Morphological and molecular characterization of lymnaeid snails and their potential role in transmission of *Fasciola* spp. in Vietnam. Kor J Parasitol 51(6):657–662. https://doi.org/10.3347/kjp.2013.51.6.657
Esteban J, Flores A, Angles R et al (1997) A population-based coprological study of human fascioliasis in a hyperendemic area of the Bolivian Altiplano. Tropical Med Int Health 2(7):695–699. https://doi.org/10.1046/j.1365-3156.1997.d01-356.x
Estuningsih S, Copeman D (1996) Trematode larvae in *Lymnaea rubiginosa* and their definitive hosts in irrigated rice fields in West Java. Jurnal Ilmu Ternak dan Veteriner 1(3):200–205
Foreyt W, Todd A (1978) Experimental Infection of Lymnaeid Snails in Wisconsin with Miracidia of *Fascioloides magna* and *Fasciola hepatica*. J Parasitol 64(6):1132–1134
Fromsa A, Meharent B, Mekibib B (2011) Major trematode infections of cattle slaughtered at Jimma municipality Abattoir and the occurrence of the intermediate hosts in selected water bodies of the zone. J Anim Vet Adv 10(12):1592–1597
Gasnier N, Rondelaud D, Abrous M et al (2000) Allopatric combination of *Fasciola hepatica* and *Lymnaea truncatula* is more efficient than sympatric ones. Int J Parasitol 30:573–578. https://doi.org/10.1016/s0020-7519(00)00038-2

Glöer P (2019) The freshwater gastropods of the West-Palaearctis. Volume 1. Fresh- and brakish waters except spring and subterranean snails. Identification key, anatomy, ecology, distribution. The author, Hetlingen

Goater T, Cameron P, Esch G (2014) Parasitism: the diversity and ecology of animal parasites. Cambridge University Press, Cambridge, UK

Goll P, Scott J (1978) The interrelationship of *Lymnaea truncatula* and ovine fascioliasis in the Ethiopian central highlands. Br Vet J 134(6):551–555. https://doi.org/10.1016/S0007-1935(17)33336-5

Goumghar M, Dreyfuss G, Rondelaud D et al (2001) More efficient allopatric combinations of *Fasciola hepatica* and *Lymnaea truncatula* due to modification of redial development? Parasitol Res 87:1016–1019. https://doi.org/10.1007/s004360100493

Grabner D, Mohamed F, Nachev M et al (2014) Invasion biology meets parasitology: a case study of parasite spill-back with Egyptian *Fasciola gigantica* in the invasive snail *Pseudosuccinea columella*. PLoS One 9(2):e88537. https://doi.org/10.1371/journal.pone.0088537

Gupta S, Yadav S (1994) Emergence of *Fasciola gigantica* cercariae from naturally infected *Lymnaea auricularia*. Indian J Parasitol 18(1):53–56

Gutiérrez A, Pointier J, Fraga J et al (2003a) *Fasciola hepatica*: identification of molecular markers for resistant and susceptible *Pseudosuccinea columella* snail hosts. Exp Parasitol 105:211–218. https://doi.org/10.1016/j.exppara.2003.12.006

Gutiérrez A, Pointier J, Yong M et al (2003b) Evidence of phenotypic differences between resistant and susceptible isolates of *Pseudosuccinea columella* (Gastropoda: Lymnaeidae) to *Fasciola hepatica* (Trematoda: Digenea) in Cuba. Parasitol Res 90:129–134. https://doi.org/10.1007/s00436-002-0824-z

Gutiérrez A, Hernández D, Sánchez J (2005) Variation of snail's abundance in two water bodies harboring strains of *Pseudosuccinea columella* resistant and susceptible to *Fasciola hepatica* miracidial infection, in Pinar del Río Province, Cuba. Mém Inst Oswaldo Cruz 100(7):725–727. https://doi.org/10.1590/s0074-02762005000700007

Gutiérrez A, Vázquez A, Hevia Y et al (2011) First report of larval stages of *Fasciola hepatica* in a wild population of *Pseudosuccinea columella* from Cuba and the Caribbean. J Helminthol 85:109–111. https://doi.org/10.1017/S0022149X10000350

Hammami H, Hamed N, Ayadi A (2007) Epidemiological studies on *Fasciola hepatica* in Gafsa Oases (south west of Tunisia). Parasite 14(3):261–264. https://doi.org/10.1051/parasite/2007143261

Heneberg P (2013) Phylogenetic data suggest the reclassification of *Fasciola jacksoni* (Digenea: Fasciolidae) as *Fascioloides jacksoni* comb. nov. Parasitol Res 112:1679–1689. https://doi.org/10.1007/s00436-013-3326-2

Hodasi J (1972) The output of cercariae of *Fasciola hepatica* by *Lymnaea truncatula* and the distribution of metacercariae on grass. Parasitology 64:53–60. https://doi.org/10.1017/s0031182000044644

Howell A, Mugisha L, Davies J et al (2012) Bovine fasciolosis at increasing altitudes: Parasitological and malacological sampling on the slopes of Mount Elgon, Uganda. Parasites Vectors 5(1):196. https://doi.org/10.1186/1756-3305-5-196

Huang SY, Gong JZ, Yang B et al (2019) Development of a nest-PCR for detection of Fasciola hepatica DNA in the intermediate snail host, Radix cucunorica, and the prevalence in northwestern China. Infect Genet Evol 75:103984. https://doi.org/10.1016/j.meegid.2019.103984

Imani-Baran A, Yakhchali M, Malekzadeh Viayeh R et al (2012) Molecular study for detecting the prevalence of Fasciola gigantica in field-collected snails of Radix gedrosiana (Pulmonata: Lymnaeidae) in northwestern Iran. Vet Parasitol 189(2–4):374–377. https://doi.org/10.1016/j.vetpar.2012.04.027

Itagaki T, Fujiwara S, Mashima K et al (1988) Experimental infection of *Lymnaea* snails with Australian *Fasciola hepatica*. Jpn J Vet Sci 50(5):1085–1091. https://doi.org/10.1292/jvms1939.50.1085

Jackiewicz M (1998) European species of the family Lymnaeidae (Gastropoda: Pulmonata: Basommatophora). Genus 9(1):1–93

Jones R, Williams J, Dalesman S et al (2017) The prevalence and development of digenean parasites within their intermediate snail host, *Galba truncatula*, in a geographic area where the presence of *Calicophoron daubneyi* has recently been confirmed. Vet Parasitol 240:68–74. https://doi.org/10.1016/j.vetpar.2017.03.021

Kadhim JK, Altaif KI (1970) The experimental demonstration of *Lymnaea lagotis euphratica* as an intermediate host of *Fasciola gigantica* in Iraq. Ann Trop Med Parasitol 64(3):335–337. https://doi.org/10.1080/00034983.1970.11686702

Kaplan R, Dame J, Reddy G et al (1997) The prevalence of *Fasciola hepatica* in its snail intermediate host determined by DNA probe assay. Int J Parasitol 27(12):1585–1593. https://doi.org/10.1016/s0020-7519(97)00139-2

Kaset C, Eursitthichai V, Vichasri-Grams S et al (2010) Rapid identification of lymnaeid snails and their infection with *Fasciola gigantica* in Thailand. Exp Parasitol 126:482–488. https://doi.org/10.1016/j.exppara.2010.05.021

Kendall SB (1949) *Lymnaea stagnalis* as an intermediate host of *Fasciola hepatica*. Nature 163(4153):880. https://doi.org/10.1038/163880a0

Khallaayoune K, Hari E (1991) Variations saisonnières de l'infestation par *Fasciola hepatica* chez la chèvre dans la région du Haouz (Maroc). Ann Rech Vet 22(2):219–226

Khallaayoune K, Stromberg BE, Dakkak A et al (1991) Seasonal dynamics of Fasciola hepatica burdens in grazing Timahdit sheep in Morocco. Int J Parasitol 21(3):307–314. https://doi.org/10.1016/0020-7519(91)90032-3

Kleiman F, Pietrokovsky S, Paraense W et al (2004) Southernmost Finding of *Lymnaea viatrix* Orbigny, 1835 (Pulmonata: Lymnaeidae), Intermediate Host of *Fasciola hepatica* (Linnaeus, 1758) (Trematoda: Digenea), in Urban and Rural Areas of Patagonia, Argentina. Mem Inst Oswaldo Cruz 99(1):23–24. https://doi.org/10.1590/s0074-02762004000100004

Kleiman F, Pietrokovsky S, Prepelitchi L et al (2007) Dynamics of *Fasciola hepatica* transmission in the Andean Patagonian valleys, Argentina. Vet Parasitol 145:274–286. https://doi.org/10.1016/j.vetpar.2006.12.020

Knapp S, Dunkel A, Han K et al (1992) Epizootiology of fascioliasis in Montana. Vet Parasitol 42: 241–246. https://doi.org/10.1016/0304-4017(92)90065-h

Kock K, Joubert P, Pretorious S (1989) Geographical distribution and habitat preferences of the invader freshwater snail species *Lymnaea columella* (Mollusca: Gastropoda) in South Africa. Onderstepoort J Vet Res 56:271–275

Kozac M, Wędrychowicz H (2009) The performance of a PCR assay for field studies on the prevalence of *Fasciola hepatica* infection in *Galba truncatula* intermediate host snails. Vet Parasitol 168(1–2):25–30. https://doi.org/10.1016/j.vetpar.2009.10.014

Kruglov ND (1986) Snail (Pulmonata, Lymnaeidae) specificity to infestation by trematode parthenites and the analysis of this specificity for taxonomy and parasitology. Zool Zh 55(12): 1799–1807. [in Russian]

Lacourarie C (1996) Etude eco-éthologique des populations de *Lymnaea truncatula* Müller vivant le long des rivières: leur infestation par *Fasciola hepatica* Linné. PhD Thesis, University of Limoges, France

Lang B (1977) Snail and mammalian hosts for *Fasciola hepatica* in Eastern Washington. J Parasitol 63:938–939

Le T, De N, Agatsuma T et al (2008) Human fascioliasis and the presence of hybrid/introgressed forms of *Fasciola hepatica* and *Fasciola gigantica* in Vietnam. Int J Parasitol 38(6):725–730. https://doi.org/10.1016/j.ijpara.2007.10.003

Lee CG, Cho SH, Lee CY (1995) Metacercarial production of *Lymnaea viridis* experimentally infected with *Fasciola hepatica*. Vet Parasitol 58(4):313–318. https://doi.org/10.1016/0304-4017(94)00725-r

Leimbacher F (1973) Étude de l'effet de quelques facteurs climatiques sur l'évolution épizootique de Fasciola hepatica. Leur utilisation pour la mise au point d'une méthode de la prévision de l'incidence de la fasciolose en France. Haliotis 3:43–55

Leuckart R (1882) Zur entwicklungsgeschichte des leberegels (*Distomum hepaticum*). Arch Naturgesch 1(48):80–119

Londoñe P, Chávez A, Li O et al (2009) Presencia de caracoles Lymnaeidae con formas larvarias de *Fasciola hepatica* en altitudes sobre los 4000 msnm en la Sierra Sur del Perú. Revista de Investigaciones Veterinarias del Perú 20(1):58–65

Lotfy W, Brant S, DeJong R et al (2008) Evolutionary origins, diversification, and biogeography of liver flukes (Digenea, Fasciolidae). Am J Trop Med Hyg 79(2):248–255

Lounnas M, Correa AC, Vázquez AA et al (2017) Self-fertilization, long-distance flash invasion and biogeography shape the population structure of *Pseudosuccinea columella* at the worldwide scale. Mol Ecol 26:887–903. https://doi.org/10.1111/mec.13984

Lounnas M, Correa AC, Alda P et al (2018) Population structure and genetic diversity in the invasive freshwater snail *Galba schirazensis*. Can J Zool 96(5):425–435. https://doi.org/10.1139/cjz-2016-0319

Mage C, Bourgne H, Toullieu J et al (2002) *Fasciola hepatica* and *Paramphistomum daubneyi*: changes in prevalences of natural infections in cattle and in *Lymnaea truncatula* from central France over the past 12 years. Vet Res 33:439–447. https://doi.org/10.1051/vetres:2002030

Mahulu A, Clewing C, Stelbrink B et al (2019) Cryptic intermediate snail host of the liver fluke *Fasciola hepatica* in Africa. Parasites Vectors 12:573. https://doi.org/10.1186/s13071-019-3825-9

Mailles A, Capek I, Ajana F et al (2006) Commercial watercress as an emerging source of fascioliasis in Northern France in 2002: results from an outbreak investigation. Epidemiol Infect 134(5):942–945. https://doi.org/10.1017/S095026880600611X

Malatji MP, Mukaratirwa S (2019) Molecular detection of natural infection of *Lymnaea* (*Pseudosuccinea*) *columella* (Gastropoda: Lymnaeidae) with *Fasciola gigantica* (Digenea: Fasciolidae) from two provinces of South Africa. J Helminthol 94:e38. https://doi.org/10.1017/S0022149X19000129

Marques P, Caldeira R, Bernadete M et al (2006) Genetic variability of Brazilian populations of *Lymnaea columella* (Gastropoda: Lymnaeidae), an intermediate host of *Fasciola hepatica* (Trematoda: Digenea). Acta Trop 97:339–345. https://doi.org/10.1016/j.actatropica.2006.01.003

Martín PR, Ovando XMC, Seuffert ME (2016) First record of the freshwater snail *Pseudosuccinea columella* (Gastropoda: Lymnaeidae) in southern Pampas (Argentina) and assessment of future spread. Mollusc Res 36(3):213–221. https://doi.org/10.1080/13235818.2015.1128602

Martínez M, Robles D, Manuel J et al (2013) Prevalence of gastrointestinal nematodes and *Fasciola hepatica* in sheep in the northwest of Spain: relation to climatic conditions and/or man-made environmental modifications. Parasites Vectors 6(1):282. https://doi.org/10.1186/1756-3305-6-282

Martínez-Orti A (2013) Nuevo hallazgo del limnéido exótico *Pseudosuccinea columella* (Ƅlay, 1817) (Gastropoda: Pulmonata) en la Península Ibérica. Noticiario de la Sociedad Española de Malacología 60:41–42

Mas-Coma S (2005) Epidemiology of fascioliasis in human endemic areas. J Helminthol 79:207–216. https://doi.org/10.1079/JOH2005296

Mas-Coma S, Valero MA, Bargues MD (2009) *Fasciola*, lymnaeids and human fascioliasis, with a global overview on disease transmission, epidemiology, evolutionary genetics, molecular epidemiology and control. Adv Parasitol 69:41–146. https://doi.org/10.1016/S0065-308X(09)69002-3

Massoud J, Sadjadi S (1980) Susceptibility of different species of *Lymnaea* snails to miracidia of *Fasciola gigantica* and *F. hepatica* in Iran. J Helminthol 54(3):201–202. https://doi.org/10.1017/s0022149x00006593

McCann C, Baylis M, Williams D (2010) The development of linear regression models using environmental variables to explain the spatial distribution of *Fasciola hepatica* infection in dairy herds in England and Wales. Int J Parasitol 40:1021–1028. https://doi.org/10.1016/j.ijpara.2010.02.009

Medeiros C, Scholte R, D'Avila S et al (2014) Spatial distribution of Lymnaeidae (Mollusca, Bassomatophora), intermediate host of *Fasciola hepatica* Linnaeus, 1758 (Trematoda, Digenea) in Brazil. Revista do Instituto de Medicina Tropical de Sao Paulo 56(3):235–252. https://doi.org/10.1590/S0036-46652014000300010

Mekroud A, Benakhla A, Vignoles P et al (2004) Preliminary studies on the prevalences of natural fasciolosis in cattle, sheep, and the host snail (*Galba truncatula*) in north-eastern Algeria. Parasitol Res 92:502–505. https://doi.org/10.1007/s00436-004-1072-1

Mekroud A, Titi A, Benakhla A et al (2006) *Fasciola hepatica*: sensibilité des *Galba truncatula* du nord-est algérien à l'infestation expérimentale avec des miracidiums sympatriques. Revue de Médicine Vétérinaire 157(10):494–501

Meunier C (2002) Fonctionnement des populations hôtes et adaptation au parasite dans l'interaction limnée-douve (Mollusque-Trématode). PhD Thesis, Université de Montpellier 2

Meunier C, Tirard C, Hurtrez-Boussès S et al (2001) Lack of molluscan host diversity and the transmission of an emerging parasitic disease in Bolivia. Mol Ecol 10:1333–1340. https://doi.org/10.1046/j.1365-294x.2001.01284.x

Meunier C, Hurtrez-Boussès S, Durand P et al (2004) Small effective population sizes in a widespread selfing species, *Lymnaea truncatula* (Gastropoda: Pulmonata). Mol Ecol 13:2535–2543. https://doi.org/10.1111/j.1365-294X.2004.02242.x

Molloy J, Anderson G (2006) The distribution of *Fasciola hepatica* in Queensland, Australia, and the potential impact of introduced snail intermediate hosts. Vet Parasitol 137:62–66. https://doi.org/10.1016/j.vetpar.2005.12.024

Monny C (1995) Etude de quelques relations entre le mollusque *Myxas glutinosa* Müller (Lymnaeidae) et le trématode *Fasciola hepatica* Linné. Infestations éxperimentales et observations histologiques. PhD Thesis, Université de Limoges, Limoges, France

Morales G, Pino I (1982) Infection of *Lymnaea cubensis* by *Fasciola hepatica* in a high altitude region in Venezuela. Ann Parasitol Hum Comp 58:27–30

Moriena R, Alvarez J, Pietrokovsky S et al (2008) Presencia de *Pseudosuccinea columella* naturalmente infestada con *Fasciola hepatica* en Santo Tomé (Corrientes, Argentina). Rev Vet 19(2):147–149

Nguyen T, Le T, Dao T et al (2011) Bovine fasciolosis in the human fasciolosis hyperendemic Binh Dinh province in central Vietnam. Acta Trop 117:19–22. https://doi.org/10.1016/j.actatropica.2010.09.003

Novobilský A, Kašný M, Beran L et al (2013) *Lymnaea palustris* and *Lymnaea fuscus* are potential but uncommon intermediate hosts of *Fasciola hepatica* in Sweden. Parasites Vectors 6(1):251. https://doi.org/10.1186/1756-3305-6-251

Nyindo M, Lukambagire AH (2015) Fascioliasis: an ongoing zoonotic trematode infection. Biomed Res Int 2015:786195. https://doi.org/10.1155/2015/786195

Oliveira S, Fujii T, Filha E et al (2002) Ocorrencia de *Lymnaea columella* Say, 1817 infectada naturalmente por *Fasciola hepatica* (Linnaeus, 1758), no vale do Ribeira, São Paulo, Brasil. Aquivos do Instituto de Biologia, Saõ Paulo 1:29–37

Olsen W (1944) Bionomics of the lymnaeid snail *Stagnicola bulimoides techella*, the intermediate host of the liver fluke in Southern Texas. J Agric Res 69:389–403

Paraense W (1984) *Lymnaea diaphana*: a study of topotypic specimens (Pulmonata: Lymnaeidae). Mem Inst Oswaldo Cruz 79(1):75–81. https://doi.org/10.1590/S0074-02761984000100009

Paraense WL (1994) *Limnaea peregrina* Clessin, 1882, synonym of *Lymnaea columella* Say, 1817 (Gastropoda: Lymnaeidae). Mem Inst Oswaldo Cruz 89(4):561–566. https://doi.org/10.1590/s0074-02761994000400010

Pereira AE, Uribe N, Pointier JP (2020) Lymnaeidae from Santander and bordering departments of Colombia: morphological characterization, molecular identification and natural infection with

Fasciola hepatica. Vet Parasitol Reg Stud Rep 20:100408. https://doi.org/10.1016/j.vprsr.2020.100408

Perera G (1996) Ecologie des mollusques d'eau douce d'intérêt médical et vétérinaire à Cuba. PhD Thesis, Université de Perpignan, France

Perera G, Sánchez R, Yong M et al (1986) Ecology of some freshwater pulmonates from Cuba. Malacol Rev 19:99–104

Perera G, Ferrer J, Gutiérrez A et al (1995) Ecological structure and factors regulating the population dynamics of the freshwater snail populations in Hanabanilla Lake, Cuba. Malacol Rev 28:63–69

Pointier J (ed) (2015) Freshwater Molluscs of Venezuela and Their Medical and Veterinary Importance. ConchBooks, Harxheim

Pointier J, Marquet G (1990) Taxonomy and distribution of freshwater mollusks of French Polynesia. Venus 49(3):147–160. https://doi.org/10.18941/venusjjm.49.3_215

Pointier J, Yong M, Gutiérrez A (2005) Guide to the freshwater mollusks of Cuba. Conchbooks, Hackenheim

Pointier J, Coustau C, Rondelaud D, Théron A (2007) *Pseudosuccinea columella* (Say 1817) (Gastropoda, Lymnaeidae), snail host of *Fasciola hepatica*: first record for France in the wild. Parasitol Res 101:1389–1392. https://doi.org/10.1007/s00436-007-0656-y

Prepelitchi L, Kleiman F, Pietrokovsky S et al (2003) First report of *Lymnaea columella* Say, 1817 (Pulmonata: Lymnaeidae) naturally infected with *Fasciola hepatica* (Linnaeus, 1758) (Trematoda: Digenea) in Argentina. Mem Inst Oswaldo Cruz 98(7):889–891. https://doi.org/10.1590/s0074-02762003000700005

Pullan NB, Climo FM, Mansfield CB (1972) Studies on the distribution and ecology of the family lymnaeidae (Mollusca: Gastropoda) in New Zealand. J R Soc N Z 2(3):393–405. https://doi.org/10.1080/03036758.1972.10421824

Rakotondravao MA, Hourdin P et al (1992) Redial generations of *Fasciola gigantica* in the pulmonate snail *Lymnaea truncatula*. J Helminthol 66(3):159–166. https://doi.org/10.1017/s0022149x00014504

Rao MP (1966) On the comparative susceptibility of Lymnaea natalensis (Kraus) and L. rufescens (Gray) to infection with Fasciola gigantica (West African strain) and the tissue responses in the snails. J Helminthol 40(1):131–140. https://doi.org/10.1017/s0022149x00034143

Relf V, Good B, McCarthy E et al (2009) Evidence of *Fasciola hepatica* infection in *Radix peregra* and a mollusc of the family Succineidae in Ireland. Vet Parasitol 163:152–155. https://doi.org/10.1016/j.vetpar.2009.04.003

Relf V, Good B, Hanrahan J, McCarthy E, Forbes A, deWaal T (2011) Temporal studies on *Fasciola hepatica* in *Galba truncatula* in the west of Ireland. Vet Parasitol 175:287–292. https://doi.org/10.1016/j.vetpar.2010.10.010

Rognlie M, Dimke K, Potts R et al (1996) Seasonal transmission of *Fasciola hepatica* in Montana, USA, with detection of infected intermediate hosts using a DNA-based assay. Vet Parasitol 65: 297–305. https://doi.org/10.1016/s0304-4017(96)00957-0

Roldan C, Begovoeva M, Lopez-Olvera JR et al (2020) Endemic occurrence of Fasciola hepatica in an alpine ecosystem, Pyrenees, Northeastern Spain. Transbound Emerg Dis. https://doi.org/10.1111/tbed.13865

Rondelaud D (1993) Variabilité interpopulationelle de l'infestation fasciolienne chez le mollusque *Lymnaea truncatula* Müller. Influence du contact préalable de la population avec le parasite. Bull Soc Zool Fr 118:185–193

Rondelaud D, Morel-Veraille C (1975) Distribution estivale et survie des Limnées tronquées *Lymnaea* (*Galba*) *truncatula* Müller saines ou infestées par *Fasciola hepatica* L. Ann Parasitol 50(5):603–616

Rondelaud D, Vignoles P, Dreyfuss G (2000) First field observations on the aestivation of *Omphiscola glabra* (Gastropoda, Lymnaeida) uninfected or infected with *Fasciola hepatica* in central France. Int J Limnol 39(2):129–133. https://doi.org/10.1051/limn/2003010

Rondelaud D, Vignoles P, Abrous M et al (2001) The definitive and intermediate hosts of *Fasciola hepatica* in the natural watercress beds in central France. Parasitol Res 87:475–478. https://doi.org/10.1007/s004360100385
Rondelaud D, Hourdin P, Vignoles P et al (2005) The contamination of wild watercress with *Fasciola hepatica* in central France depends on the ability of several lymnaeid snails to migrate upstream towards the beds. Parasitol Res 95:305–309. https://doi.org/10.1007/s00436-004-1283-5
Rondelaud D, Vignoles P, Dreyfuss G (2009) La Limnée tronquée, un mollusque d'intérêt médical et vétérinaire. PULIM, Limoges
Rondelaud D, Sanabria R, Vignoles P (2013) *Fasciola hepatica*: variations in redial development and cercarial production in relation to the geographic origin of the parasite. Parasite 20:33. https://doi.org/10.1051/parasite/2013034
Sabourin E, Alda P, Vázquez A et al (2018) Impact of human activities on fasciolosis transmission. Trends Parasitol 34(10):891–903. https://doi.org/10.1016/j.pt.2018.08.004
Salazar L, Estrada V, Velázquez L (2006) Effect of the exposure to *Fasciola hepatica* (Trematoda: Digenea) on life history traits of *Lymnaea cousini* and *Lymnaea columella* (Gastropoda: Lymnaeidae). Exp Parasitol 114:77–83. https://doi.org/10.1016/j.exppara.2006.02.013
Sanabria R, Mouzet R, Courtioux B et al (2012) Intermediate snail hosts of French *Fasciola hepatica*: *Lymnaea neotropica* and *Lymnaea viatrix* are better hosts than local *Galba truncatula*. Parasitol Res 111:2011–2016. https://doi.org/10.1007/s00436-012-3049-9
Sazanov A (1979) On applicability of trematod specificity criterion in the taxonomy of the Lymnaeidae. In: Likharev I (ed) Molluscs: main results of their study. Sixth All-Union meeting on the investigation of molluscs. Nauka, Leningrad, pp 234–236. [in Russian]
Schillhorn van Veen TW (1980) Dynamics of *Lymnaea natalensis* populations in the Zaria area (Nigeria) and the relation to *Fasciola gigantica* infections. Acta Trop 37(2):183–194
Schniebs K, Glöer P, Vinarski MV et al (2017) A new alien species in Europe: First record of *Austropeplea viridis* (Quoy & Gaimard, 1833) in Spain. J Conchol 42(5):357–370
Schniebs K, Glöer P, Quiñonero-Salgado S et al (2018) The first record of *Galba cubensis* (L. Pfeiffer, 1839) (Mollusca: Gastropoda: Lymnaeidae) from open fields of Europe. Folia Malacol 26(1):3–15. https://doi.org/10.12657/folmal.026.002
Schweizer G, Meli M, Torgerson P et al (2007) Prevalence of *Fasciola hepatica* in the intermediate host *Lymnaea truncatula* detected by real time TaqMan PCR in populations from 70 Swiss farms with cattle husbandry. Vet Parasitol 150:164–169. https://doi.org/10.1016/j.vetpar.2007.08.006
Sindou P, Cabaret J, Rondelaud D (1991) Survival of snails and characteristic lesions of Fasciola hepatica infection in four European species of Lymnaea. Vet Parasitol 40(1–2):47–58. https://doi.org/10.1016/0304-4017(91)90082-7
Stadnichenko A (2006) Lymnaeidae and Acroloxidae of Ukraine: Methods of their sampling and study, biology, ecology, useful and harmful meaning: monograph. Zhytomyr, Ruta. [in Russian]
Takeuchi-Storm N, Denwood M, Petersen H et al (2018) Patterns of *Fasciola hepatica* infection in Danish dairy cattle: implications for on-farm control of the parasite based on different diagnostic methods. Parasites Vectors 11:674. https://doi.org/10.1186/s13071-018-3248-z
Thomas AP (1882) The rot in sheep, or the life-history of the liver-fluke. Nature 26(677):606–608. https://doi.org/10.1038/026606a0
Toet H, Piedrafita D, Spithill T (2014) Liver fluke vaccines in ruminants: strategies, progress and future opportunities. Int J Parasitol 44:915–927. https://doi.org/10.1016/j.ijpara.2014.07.011
Torgerson P, Claxton J (1999) Epidemiology and control. In: Dalton J (ed) Fasciolosis. CAB International, Dublin, pp 113–149
Vareille-Morel C, Rondelaud D, Dreyfuss G (2002) Experimental colonization of new habitats by *Galba truncatula* O.F. Müller (Gastropoda: Lymnaeidae) in central France and their susceptibility to experimental infection with *Fasciola hepatica*. Int J Limnol 38:35–40. https://doi.org/10.1051/limn/2002003

Vassiliades G (1978) Capacité de résistance à la sécheresse de la limnée (*Limnaea natalensis*), Mollusque hôte intermédiaire de *Fasciola gigantica*, au Sénégal. Rev Elev Méd Vét Pays Trop 31(1):57–62

Vázquez A, Gutiérrez A (2007) Ecología de moluscos fluviales de importancia médica y veterinaria en 3 localidades de La Habana. Rev Cubana Med Trop 59(2):149–153

Vázquez A, Sánchez J, Pointier J et al (2014) *Fasciola hepatica* in Cuba: compatibility of different isolates with two intermediate snail hosts, *Galba cubensis* and *Pseudosuccinea columella*. J Helminthol 88(4):434–440. https://doi.org/10.1017/S0022149X13000382

Vázquez A, Sánchez J, Alba A et al (2015) Natural prevalence in Cuban populations of the lymnaeid snail *Galba cubensis* infected with the liver fluke *Fasciola hepatica*: small values do matter. Parasitol Res 114:4205–4210. https://doi.org/10.1007/s00436-015-4653-2

Vázquez AA, Pilar A, Lounnas M et al (2018) Lymnaeid snails hosts of *Fasciola hepatica* and *Fasciola gigantica* (Trematoda: Digenea): a worldwide review. CAB Rev 13:1–15. https://doi.org/10.1079/PAVSNNR201813062

Vázquez A, Vargas M, Alba A et al (2019) Reviewing *Fasciola hepatica* transmission in the West Indies and novel perceptions from experimental infections of sympatric vs. allopatric snail/fluke combinations. Vet Parasitol 275:108955. https://doi.org/10.1016/j.vetpar.2019.108955

Vázquez A, Sabourin E, Alda P et al (2020) Genetic diversity and relationships of the liver fluke *Fasciola hepatica* (Trematoda) with native and introduced definitive and intermediate hosts. Transbound Emerg Dis 68(4):2274–2286. https://doi.org/10.1111/tbed.13882

Vignoles P, Ménard A, Rondelaud D et al (2001) *Fasciola hepatica*: the characteristics of experimental infections in *Lymnaea truncatula* subjected to miracidia differing in their mammalian origin. Parasitol Res 87:945–949. https://doi.org/10.1007/s004360100481

Vignoles P, Dreyfuss G, Rondelaud D (2015) *Fasciola hepatica*: comparative metacercarial productions in experimentally-infected *Galba truncatula* and *Pseudosuccinea columella*. Parasite 22(15):1–7. https://doi.org/10.1051/parasite/2015015

Vignoles P, Dreyfuss G, Rondelaud D (2017a) Detection of habitats colonized by Omphiscola glabra (Gastropoda: Lymnaeidae) on acid soils using indicator plants. Ann Limnol Int J Limnol 53:261–269. https://doi.org/10.1051/limn/2017010

Vignoles P, Rondelaud D, Dreyfuss G (2017b) Determination of zones at risk for fasciolosis in the department of Haute-Vienne, central France: a retrospective study on natural infections detected in 108,481 Galba truncatula for 37 years. Parasite 24:55. https://doi.org/10.1051/parasite/2017055

Villavicencio A, de Vasconcellos M (2005) First report of *Lymnaea cousini* Jousseaume, 1887 naturally infected with *Fasciola hepatica* (Linnaeus, 1758) (Trematoda: Digenea) in Machachi, Ecuador. Mem Inst Oswaldo Cruz 100(7):735–737. https://doi.org/10.1590/s0074-02762005000700010

Villavicencio A, Gorochov V, Carvalho M (2006) *Lymnaea truncatula* Müller, 1774 (Pulmonata: Lymnaeidae) infected with *Fasciola hepatica* (Linnaeus, 1758) (Trematoda: Digenea), in Moscow districts, Russian Federation. Rev Patol Trop 35(1):59–64

Vinarski MV, Grebennikov ME, Shishkoedova OS (2013) Past and present distribution of *Myxas glutinosa* (O.F. Müller, 1774) in the waterbodies of the Urals and Siberia. J Limnol 72(2): 336–342. https://doi.org/10.4081/jlimnol.2013.e27

Vinarski MV, Aksenova OV, Bolotov IN (2020) Taxonomic assessment of genetically-delineated species of radicine snails (Mollusca, Gastropoda, Lymnaeidae). Zoosyst Evol 96(2):577–608. https://doi.org/10.3897/zse.96.52860

Vázquez AA, Alba A, Alda P, Vittecoq M, Hurtrez-Bousses S (2022) On the arrival of fasciolosis in the Americas. Trends Parasitol 38(3):195–204. https://doi.org/10.1016/j.pt.2021.12.001

Yakhchali M, Baran I, Malekzadeh-Viayeh R (2015) Molecular detection of the infection with *Fasciola hepatica* in field-collected snails of *Galba truncatula* and *Lymnaea stagnalis* from West Azarbaijan, Iran. Arch Razi Inst 70(3):195–202

Yasuraoka K (1954) Ecology of the miracidium. II. On the behavior to light of the miracidium of *Fasciola hepatica*. Jpn J Med Sci Biol 7(2):181–192. https://doi.org/10.7883/yoken1952.7.181

Zumaquero J, Sarracent J, Rojas R et al (2013) Fascioliasis and intestinal parasitoses affecting schoolchildren in Atlixco, Puebla State, Mexico: epidemiology and treatment with nitazoxanide. PLoS Negl Trop Dis 7(11):e2553. https://doi.org/10.1371/journal.pntd.0002553

Chapter 14
Laboratory Cultures of Lymnaeidae for Parasitological Experiments

Gilles Dreyfuss, Philippe Vignoles, Daniel Rondelaud, Jorge Sánchez, and Antonio A. Vázquez

Abstract Many species of Digenea require a host snail for the development of their larval forms. Among the species of molluscs involved in the transmission of these parasitoses, Lymnaeidae serves as intermediate hosts for at least 71 species of parasites, distributed in 13 families. The breeding of lymnaeids under laboratory conditions allows specifying the dynamics of this larval development often complex or to obtain infective larvae (cercariae) capable of developing in their definitive host. As several breeding techniques have been published by the authors, the objective of this chapter is to review the different systems used, with a particular emphasis on the species of the genus *Galba* because their breeding is more difficult due to their amphibiosis. The different types of natural food that can be proposed to snails as well as food supplements used to have optimal growth of these snails are examined in a second step. The impact of these food supplements on the parasite production of two snail species capable of ensuring the larval development of *Fasciola hepatica* (liver fluke) is specified in a third step. Finally, the general principles of the infection of these snails by the larval forms (miracidia) of a trematode are briefly recalled at the end of this review.

G. Dreyfuss (✉) · P. Vignoles · D. Rondelaud
Laboratory of Parasitology, University of Limoges, Limoges, France
e-mail: gilles.dreyfuss@unilim.fr; philippe.jarne@cefe.cnrs.fr; daniel.rondelaud@unilim.fr

J. Sánchez
Laboratory of Malacology, Institute of Tropical Medicine "Pedro Kourí", Havana, Cuba
e-mail: jsnoda@ipk.sld.cu

A. A. Vázquez
Laboratory of Malacology, Institute of Tropical Medicine "Pedro Kourí", Havana, Cuba

MIVEGEC, University of Montpellier, CNRS, IRD, Montpellier, France

M. V. Vinarski, A. A. Vázquez (eds.), *The Lymnaeidae*, Zoological Monographs 7,
https://doi.org/10.1007/978-3-031-30292-3_14

14.1 Introduction

The Lymnaeidae family is known for its medical and veterinary importance. These freshwater snails act as intermediate hosts in the life cycle of various parasites by ensuring the development of their larval forms. More than 71 species of Digenea at least, distributed in 13 families, use these lymnaeids as intermediate hosts (Dawes 1968; Bargues et al. 2001; Correa et al. 2010). The best known of these parasites is *Fasciola hepatica*, which causes liver fluke disease. This distomatosis is recognized as a major veterinary problem because this parasite is responsible for losses in production capacity for meat, milk, wool, etc. (Hillyer and Apt 1997; Kaplan 2001). It is also a significant human disease with around 20 million cases worldwide (Mas-Coma et al. 2009).

At least 20 species of Lymnaeidae have been reported as host snails of *F. hepatica* (Torgerson and Claxton 1999; Hurtrez-Boussès et al. 2001; Vázquez et al. 2018). These species do not all have the same capacity to ensure the larval development of this digenean and four groups of lymnaeids ranging from species capable of being infected at any age to species resistant to infection have been described by Boray (1978). The suitability of these species as host snails can only be determined in the laboratory by carrying out experimental infections to specify whether these lymnaeids can provide cercarial shedding, ensure incomplete larval development limited to immature rediae or rediae with intraredial cercariae, or on the contrary be refractory to the infection with the degeneration of miracidia (Dreyfuss et al. 2015b). Carrying out such experiments requires raising these molluscs under laboratory conditions with methods that take into account their mode of life (strictly aquatic, semiaquatic, or amphibious species).

The breeding of aquatic lymnaeids is quite easy in the laboratory provided to respect the rules of aquaristics (reasonable number of snails according to the volume of water, chemical quality of it, food in abundance, regular cleaning of breeding containers). On the other hand, the rearing of amphibious and semiaquatic species is much more difficult due to their ability to get out of water and move across the wet emerging area. Another difficulty is the type of food for these molluscs when they are infected with a parasite. Indeed, high-quality food is required to ensure optimal shell growth and high parasitic production. Due to the difficulties in raising these amphibious lymnaeids, the paragraphs below are devoted to the reminder of some principles for raising these snails, the breeding systems used by the authors, the different types of food and their impact on the parasite production that the snail can provide.

As the nomenclature of Lymnaeidae is still subject to controversy, the names of the species cited in this chapter have been updated following the names of the taxa listed in MolluscaBase (2022). Species, whose names have changed, are cited by first indicating the old name, followed by the current name within brackets.

14.2 Principles of Snail Breeding for *Galba* Species

The method to be adopted must take into account the following three rules:

1. Amphibiosis of these species must be respected. These snails naturally come out of the water table to move on the walls of their breeding container and rest there at night. In addition, the presence of an unfavourable element in their environment (absence of food, presence of copepods, etc.) forces them to leave the water to travel through the emerged area in search of food or take refuge there. It is therefore useful that the breeding container has emerged areas on which the animals can move if they feel the need to. However, these areas must remain humid at all times, hence the need for an enclosed space (aquarium closed by a glass, for example). Indeed, the rapid disappearance of humidity under laboratory conditions, if the breeding container is not covered, limits the movement of snails in the emerged area and can even lead them to settle in a drier area, with retraction of the body into their shell and rapid death of the animals if the drying out persists.
2. By its diet, the snail consumes the plant food at its disposal, hence the production of excreta that accumulate rapidly in the environment if the animal lives in groups. It is therefore necessary to regularly renew the water polluted by the faeces of the snail by replacing it every day with clean spring water. In addition, weekly cleaning of the breeding container is required to remove unicellular algae and stains that develop on the walls.
3. If many lymnaeids are placed in the same living space, there is competition for the use of food at their disposal, which usually results in the rapid growth of one or more individuals in the group while the others have limited development. The number of snails to be placed in a breeding system must therefore take into account its "useful volume" (quantity of water available to the animals + wet emerged walls) and the size of the individuals (the number of snails which can be raised in a breeding container under good conditions is inversely proportional to their shell height).

Two points must also be considered. The first concerns the quality of spring water used for snails. According to the geographical origin of populations (on sedimentary or acidic soils), it is preferable to provide them with water which has the same content of dissolved calcium ions as that which runs in their habitat. Snails living on acidic soils tolerate eucalcic water (from 60 to 100 mg/L of Ca^{++}) without too much problem. In contrast, the snails from sedimentary soils quickly leak the water layer to take refuge on the emerged zones when subjected to oligocalcic water (<10 mg/L of Ca^{++}). If the process continues, these snails die quickly. Second, it is necessary to eliminate predators and snails naturally infected with a digenean or an ectosymbiont (*Chaetogaster limnaei*) when field-collected snails are used for experiments. The search for aquatic predators (larvae of *Dytiscus* spp., larvae of Diptera Sciomyzidae, etc.) is carried out by placing these snails in spring water for 2 or 3 days and ensuring regular surveillance every 6 h to detect intruders. The detection of parasites can only be performed by crushing or dissecting 20 to 50 adult snails (depending on the size of the population) to look for and identify larval forms of digeneans.

14.3 Breeding Systems Used for Amphibious Lymnaeids

14.3.1 Classical Methods

The oldest successful method was that of Schumacher (1938). This author raised *Galba truncatula* in terracotta dishes with a sloping surface and low-flow water circulation so that snails are kept on the surface of the mud while being immersed. The use of spring water was recommended, but accumulation of minerals through evaporation makes dishes unsuitable for snails and the mud must be changed every 8 days (Lämmler 1955). Another system widely used by later authors is that which Kendall (1953) described by placing snails on sloping mud, with a little water at the bottom in a crystallizer. The latter is covered with a glass in order to maintain permanent humidity. Unicellular algae belonging to Chlorophyceae and Chromophyceae (with a dominance of *Cosmarium* spp.) are inoculated on the muddy slope and are then subjected to constant conditions of temperature and light to allow optimal growth. When algal development has taken place, 50 to 60 snails are placed in each crystallizer to feed for 2 or 3 days. Beyond that, they are transferred to another crystallizer. This technique has the advantage of taking into account the amphibiosis of snails, thanks to the inclined plane formed by the sloping mud. According to Kendall (1953), snail eggs are deposited at the water-mud border each month of the year. On the other hand, the growth of the snail is maximum and the larval development of the parasite (*F. hepatica*, for example) takes place under good conditions. This technique has been widely used in laboratories which produced *F. hepatica* metacercariae using *G. truncatula* as a host snail (Ollerenshaw 1971; Osborn et al. 1982, for example). However, its implementation requires a significant technical time (handling crystallizers and animals) and difficulties to obtain a homogeneous mud layer with equal consistency.

An ingenious device was made by Pécheur (1974) by adapting a technique used in aquaristics. In an aquarium, a series of partitions simulates a "miniature river". This system therefore has an inlet and an outlet for water to circulate, ensuring the cleanliness of the breeding. The water circulates on a bed of large gravel and is then decanted, filtered, and re-oxygenated. This aquarium is located under a closed hood which ensures 100% humidity. Fluorescent tubes ensure the growth of unicellular algae for 12 h/day, and the entire device is maintained between 18 °C and 24 °C. With this device, Pécheur (1974) tried to recreate the natural habitat of snails under laboratory conditions. Despite this advance, the system proposed by this author also requires significant technical time.

Boray (1963, 1966, 1969) gained an interesting experience in rearing amphibious snails like *Austropeplea tomentosa*. Initially, this author used trays (1 m × 60 cm, 10 cm high) whose walls were covered with a plastic sheet. These containers are filled with mud collected from snail habitats and sterilized before its introduction into each tray. The mud is immersed in deionized water and furrows 3 to 4 cm deep are dug. A moderate slope is created at a corner of each tray and deionized water is poured into it at a flow rate of 100–200 mL/h to replace water lost by evaporation and remove snail excreta. A siphon located in the lowest area of each tray controls the water level and allows it to be removed if necessary. The containers are exposed to a natural or artificial photoperiod to facilitate the growth of algae. As Boray (1969) wished to standardize the breeding technique, he adopted a method derived from the previous one, with the use of smaller Plexiglas™ boxes, a quantity of mud reduced to a corner of each box, intense aeration, and the use of powdered food. The breeding boxes are placed at 22° C–24 °C in air-conditioned rooms and are subjected to artificial lighting.

The use of soil, coming from the habitat where the snails live, is still carried out today to constitute the bottom of aquariums in the laboratory. Moazeni et al. (2018) reported the constitution of an open-top aquarium with this type of soil. The *G. truncatula* used by these authors was subjected to artificial lighting for 12 h/day, with a temperature varying from 18° C to 32 °C and a relative humidity of 22% to 37% depending on the season.

14.3.2 Development of New Methods

The use of systems used by the above authors for raising amphibious or semiaquatic lymnaeids has several drawbacks. First, the soil and water in the tanks or aquariums must be changed periodically, which requires significant technical time due to the volume of the two components to be changed if there are several tanks or several aquariums in the breeding room. Second, not all soil samples taken from snail habitats were equally easy to form mud when immersed in spring water. This depends on their geological origin (crystallophyllian or sedimentary soils), which poses problems when the sediment used comes from acidic soils. Finally, the use of these tanks or aquariums sometimes raises a problem of space and therefore limits the size of the experiments to be carried out when experimental infections of snails are performed. Several lines of research have therefore been followed by authors to find solutions to solve one or the other of these problems. Three lines were studied.

As a small amount of mud in each breeding system had already been practiced by Boray (1969) in his Plexiglass™ boxes to raise *A. tomentosa*, Sánchez et al. (1995) used the method that Liang and van der Schalie (1975), Bruce and Liang (1992) adopted for the intermediate hosts of *Schistosoma* spp. Pre-sterilized wet mud is introduced into the centre of a Petri dish (diameter, 12 cm) to form a central mound measuring 1.5 cm in height at the centre and 6 cm in diameter at the base. Sterilized water is then added to each container and blue-green algae are inoculated on the central mound. The snails (15 per dish) are then kept at 28 °C under artificial light. Water and food are changed every 4 days. The use of a central mound to establish a food source was also carried out by Apostoloff (2001) to raise parasitized

G. truncatula in tanks (Fig. 14.1a). Marl is placed in each box (area, 0.66 m^2) to form a mound measuring 10 cm in diameter and 4 cm in height. Several stems of *Callitriche* spp. are introduced in this mound because they easily take root in the marl. The presence of this plant allows the collection of numerous young snails if necessary. In fact, the juveniles take refuge in the network formed by stems and floating roots. Three or four stems of grasses (*Glyceria* spp. in general) with dead leaves and dried lettuce leaves are placed between the marl mound and the pipe which provides aeration of the water. The presence of these grass leaves greatly limits the emersion of snails on the tank walls. The tanks are placed in an air-conditioned room at 24 °C, with a photoperiod of 12 h/day for a light intensity of 3000–4000 lux above the water surface. The water level is checked twice a week

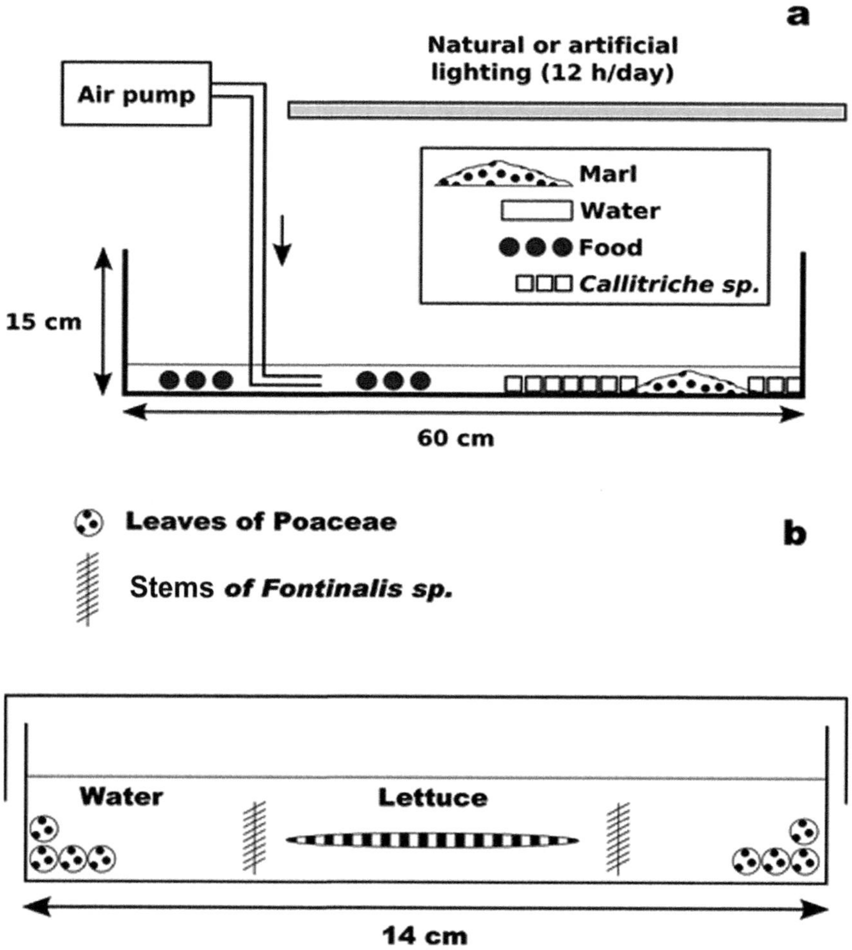

Fig. 14.1 Two systems used to raise *Galba truncatula* in the laboratory: (**a**) from Apostoloff (2001) and (**b**) from Rondelaud et al. (2007)

by adding rain water or deionized water. The cleaning of each tank is done on the 14th day, then every week until the end of each experiment.

Other authors have abandoned the mud and replaced it with another substrate. Whitlock et al. (1976, 1977) replaced the mud by a soil-bentonite agar substrate to provide habitat and food for *A. tomentosa*. Two-litre glass jars, containing this substrate and filled with artificial water receive compressed air, preheated to 25 °C to maintain thermoregulation, aeration, filtration, and circulation of the water. Small amounts of washed lettuce are given as a supplement.

Finally, the affinity of snails of the genus *Galba* for dead grass leaves led Rondelaud et al. (2007) to abandon mud in breeding vessels. These authors used single-use Petri dishes (diameter, 14 cm) to raise *G. truncatula* (Fig. 14.1b). Dead grass leaves (usually *Molinia caerulea*) are placed along the periphery of each dish. The centre is occupied by a fragment of dried lettuce, surrounded by four live stems of *Fontinalis* spp. (spring moss) arranged in a rectangle. Spring moss ensures oxygenation of the water, while grass leaves limit the emersion of snails. Spring water (60 mL) and 6 to 10 lymnaeids (the number depends on the initial height of snails) are added in each dish. The containers are placed in an air-conditioned room as above and are subjected to maintenance conditions (24 °C, 12-h daylight) with a daily change of water and food if necessary. This method has several advantages: (i) an obvious space-saving because the Petri dishes can be stacked on top of each other (up to five per column), (ii) an ease in the transport of these dishes containing the infected snails (after removing the water) if necessary, (iii) limitation of mortality because the snails are in small number in each dish, (iv) transmission of iridovirosis, if present, can be easily stopped by the sterilization of the soft forceps which are used to handle the animals, and (v) a better monitoring of the larval development of the parasite (when snails come from acidic soils, the presence of rediae, secondary sporocysts, or cercariae can be easily seen in the snail thanks to the transparency of the shell). This method has been used to breed several species of the genus *Galba*: *G. truncatula* (Rondelaud et al. 2007, 2014), *G. neotropica* and *G. viator*, var. *ventricosa* (Sanabria et al. 2012, 2013), *G. cubensis* (Vignoles et al. 2014), and *G. schirazensis* (Dreyfuss et al. 2015a). Successful results were also obtained with juveniles and preadults of *Omphiscola glabra* (Rondelaud et al. 2015).

14.4 Snail Food

14.4.1 Natural Foods

According to Hynes (in Kendall 1953), the snail is able of eating any vascular plant present in its environment. The most remarkable example is the dead leaves that *G. truncatula* rasps with its radula, as the snail only leaves a network of nervures. The dissection of numerous snails and the presence of vascular plant fragments in their intestine confirmed this fact in the laboratory (Rondelaud et al.

Table 14.1 Algae used by several authors for the food of *Galba* spp. under laboratory conditions

Taxa	Algal species	References
Chlorophyceae	Unidentified species	Taylor and Mozley (1948)
Chlorophyceae and Chromophyceae	Several species with dominance of *Cosmarium* sp.	Kendall (1953)
Cyanophyceae	*Oscillatoria formosa*	Sampaio Xavier et al. (1968)
Chlorophyceae	*Palmella* sp.	Sevo (1973)
Chlorophyceae	*Chlorella* sp.	Pécheur (1974)
Cyanophyceae	*Oscillatoria* sp.	Osborn et al. (1982)
Cyanophyceae	Genera *Lynghya, Leotolynghya, Phormidium,* and *Schimidleinema*	Sánchez et al. (1995)
Cyanophyceae	*Oscillatoria formosa*	Pereira Filipa et al. (2000)
Chlorophyceae	*Draparnaldia* sp., *Microspora* sp., *Monostroma* sp., *Oscillatoria* sp., *Spirogyra* sp., *Ulothrix* sp.	Valladon (2000), Rondelaud et al. (2002)

2009; Dreyfuss et al. 2015b). However, the animal is known to mainly feed on cyanobacteria (blue-green algae) or green algae that it ingests by raping substrates on which these cryptogams develop (Kendall 1953). Table 14.1 lists the species that several authors have used to raise *G. cubensis* or *G. truncatula* under laboratory conditions. Sampaio Xavier et al. (1968) and Pereira Filipa et al. (2000) used *Oscillatoria formosa* for their studies. In contrast, other algae used to feed snails have not been identified or have only been identified at the genus level. Valladon (2000) and Rondelaud et al. (2002) were successful in raising *G. truncatula* on a layer of filamentous and unicellular algae consisting of six species, with *Monostroma sp.* and *Oscillatoria sp.* as dominant algae. According to these authors, this lymnaeid use different species of algae for its food without worrying about the composition and structure of the algal cover. The geographic origin of snails studied by the authors may partly explain the different species of algae that have been proposed as food for amphibious lymnaeids. Indeed, Comerton and Houghton (1975) noted that green algae were preferentially consumed by *G. truncatula* living in Ireland. These authors frequently found diatoms in the intestine and faeces of the snail, while cyanobacterial filaments were more rarely observed. According to Kendall (1953), algae are necessary for the snail to have optimal growth. Ingestion of *Oscillatoria* sp. by *G. truncatula* in the laboratory allows an optimum growth of the shells because the maximum height of snails is reached in 6 to 8 weeks instead of 6 to 9 months in natural conditions (Osborn et al. 1982). Lee et al. (1993, 1994, 1995) also noted rapid growth of *Lymnaea viridis* (= *Orientogalba viridis*) when this species fed on cyanobacteria in the laboratory. According to these authors, these snails reached a shell height of 12 mm in 37 days at 20 °C and began to lay eggs 18 to 24 days after hatching.

The algae of the genus *Oscillatoria* appear to be the most used currently to feed amphibious snails under laboratory conditions (Iturbe-Espinoza and Muñiz-Pareja 2012; Hodgkinson et al. 2018, for example). More recently, another cyanobacterial species: *Arthrospira platensis* (spirulina) has prompted authors to use it to feed snails due to its high nutritional value (Belay et al. 1996). This species was currently used as a food supplement for raising amphibious species like *G. truncatula* (Lombardo et al. 2010; Moazeni et al. 2018) as for more aquatic lymnaeids (Prepelitchi et al. 2015).

All varieties of *Lactuca sativa* (leaf lettuce, romaine, iceberg, oak leaf, etc.) may be suitable for feeding lymnaeids. But it is essential that the salad grows without chemical treatment, which excludes the use of specimens from supermarkets. In addition, plants that have grown near major roads should not be used as food because of the risk of pollution (Dreyfuss et al. 2015b). Whether lettuce leaves are fresh does not have the same nutritional value for snails. According to Thompson (1984), fresh lettuce leaves are a very satisfying food source for maintaining populations of planorbids in breeding, while they induce poor nutrition when dried out. Fresh lettuce leaves are widely appreciated by aquatic lymnaeids for their food. In contrast, amphibious species of the genus *Galba* do not directly consume these fresh leaves. These must have evolved in the breeding container for several days for these snails to be interested. However, if young individuals belonging to a more aquatic species (*O. glabra* for example) are present with *G. truncatula* in the same container, their presence leads to a faster degradation of these green leaves so that *G. truncatula* uses them for personal consumption. Despite this interest, the presence of another species with *G. truncatula* can be fatal because juveniles of *O. glabra* grow faster and tend to remove the other species from its environment (Dreyfuss et al. 2006, 2015b). Lettuce macerated after a 5-day stay in spring water has been widely used in experiments that several authors have carried out with *G. truncatula* from central France between 1970 and 1998. Dried lettuce leaves as basic food are also consumed by snails from these same populations, and their use has become widespread with that of Petri dishes to breed five species of *Galba* from the 2000s (Dreyfuss et al. 2015b).

Rondelaud et al. (2009) and Dreyfuss et al. (2015b) reported three other types of natural food that *G. truncatula* from central France can consume in its environment. Indeed, these populations are able to feed on grass leaves or tree leaves when they are dead. One population living in the French department of Creuse had a marked preference for peach leaves which fall in its habitat in autumn, while abandoning other types of tree leaves. Dead grass leaves were therefore harvested in the snail habitat and placed into the Petri dishes used by our group after freezing them at −18 °C for at least 3 days. Similarly, peach leaves were harvested in autumn, frozen at −18 °C, and used as food when snails from the population of Creuse were used for experimental infections. In April and May, the algal cover commonly found in lymnaeid habitats on calcareous soils provides refuge and food for local snails. Samples of these algae can therefore be collected and frozen at −18 °C to serve as supplementary food during the winter months. However, these samples should be avoided at sites without lymnaeids because some species of cyanobacteria or green algae are toxic to snails.

14.4.2 Food Supplements

Under this term, we consider a high-quality food which results in optimal growth of the snail under laboratory conditions and maximum cercarial production if this snail is infected with a digenean. Several authors proposed different diets, and Table 14.2 shows their composition. These products contain powdered nutrients (alfalfa leaves, wheat germ, etc.) and mineral salts, particularly calcium that the snail uses for the growth of its shell. They are presented under a solid form (Malek 1962; Boray 1969) or a liquid (Pécheur 1974; Bruce and Liang 1992). They can contain different vitamins (Pécheur 1974) or agar (Bruce and Liang 1992).

Two other product categories can be added to the list detailed in Table 14.2. First, commercial fish foods have been widely used since the 1970s by the authors as a supplement to feed various freshwater snails under laboratory conditions. On the different products on the market, the two most commonly used seem to be Tetramin™ (Madsen 1992; Brendelberger 1997; Florin et al. 2000; Sandland and Minchella 2003; Guasch et al. 2016, for example) and Tetraphyll™ (El-Assal et al. 2005; Okoye et al. 2009; Tufi et al. 2016; Swart et al. 2020). In order to determine which product is most palatable to *G. truncatula*, Rondelaud et al. (2009) and

Table 14.2 Food supplements used by several authors to raise *Galba truncatula* and other freshwater molluscs in the laboratory

References	Composition	Quantity
Kendall (1949)	Lime powder	50%
	Oat flakes	50%
Malek (1962)	Cerophyl[a]	10 g
	Whole milk powder	2.5 g
	Wheat germ powder	5 g
	Sodium alginate	5 g
Boray (1969)	Lettuce powder	25%
	Alfalfa leaf powder	25%
	Wheat germ powder	50%
	Calcium sulphate	3%
Pécheur (1974)	Bemax (complex of vitamins)	1%
	Condensed milk	5%
	Lettuce powder	1%
	Sodium alginate	1%
	Calcium chloride	2%
	Distilled water	100 mL
Bruce and Liang (1992)	Soy powder	10 g
	Dehydrated mud	40 g
	Agar	3 g
	Cerophyl[a]	3 g
	Fish food	2 g
	Yeasts	2 g
	Distilled water	60 mL

[a]Leaves of dehydrated cereals

Dreyfuss et al. (2015b) selected 17 products that they proposed to several French populations of this snail during their breeding in the laboratory. Of these products, Tetraphyll™ was chosen for this species. It contains 47% crude protein, 8.5% crude fat, sugars, and several vitamins according to manufacturer's indications. Rondelaud et al. (2009) and Dreyfuss et al. (2015b) used this food as follows: 5 or 6 flakes per breeding box (about 4 cm^2 of food), 2 or 3 flakes per 14-cm diameter Petri dish (1.4 to 2 cm^2 of food), or 1 flake per 35-mm dish (0.2 to 0.3 cm^2). Twice a week, Tetraphyll™ is presented to snails for 12 h by placing food either in a corner of the breeding box or in Petri dishes separated from those in which *G. truncatula* lives. At the end of the 12 h, the food remains are removed, either by means of an electric pump (breeding box) or by washing the dishes with dechlorinated tap water. The palatability of Tetraphyll™ leads a number of *G. truncatula* to form small balls of food which they hold with their foot while consuming it. Despite the interest of this food in the cercarial production of digeneans (see below), this food was abandoned because of too much technical time (2 h per week for 100 infected snails) and therefore too high cost for the commercial production of metacercariae (Rondelaud et al. 2009; Dreyfuss et al. 2015b).

Since the 2000s, *Arthrospira platensis* (spirulina) has been used on several occasions as a complementary food in the breeding of aquatic snails. The interest of this cyanobacterium lies in its high nutritional value. Its chemical composition includes proteins (55%–70%), carbohydrates (15%–25%), essential fatty acids (18%), vitamins, minerals, and pigments like carotenes, chlorophyll a, and phycocyanin (Sánchez et al. 2003). The presence of this food in aquariums of *Pseudosuccinea columella*, in addition to lettuce, has had beneficial effects on the growth and reproduction of this snail. According to Prepelitchi et al. (2015), snails fed with lettuce and spirulina (33.3 mg/1 L of water) had a higher survival rate, faster growth, longer reproduction period, more eggs, and a higher hatching rate than their congeners fed with lettuce only. But these results are not true in other species of aquatic pulmonates. Spirulina thus reduces the reproductive period and lifespan of *Physella acuta* (Auld and Henkel 2014; Auld 2018). Its products also have a lethal effect on adults of *Biomphalaria alexandrina* by reducing or stopping their egg-laying period and decreasing the hatching rate of eggs (Mostafa and Gawish 2009; Saad et al. 2019). These differences can be partly explained by the amount of spirulina offered to snails because Lombardo et al. (2010) use commercial fish food flakes, enriched with 5% spirulina, to raise their *G. truncatula*. The effects of this cyanobacterium on the parasitic production of digeneans remain to be investigated because no study to our knowledge has yet been carried out in species of the genus *Galba* when they are feed on lettuce and spirulina.

14.5 High-Quality Diets and Parasite Production

The quality of the diet, provided as food to raise lymnaeid snails under laboratory conditions has had a significant effect on their growth, fecundity, and fertility, because well-fed snails became larger and laid more eggs (Skoog 1979; Islam et al. 2001; Dillon 2004; Eveland and Haseeb 2011). If these snails are infected with a digenean (*F. hepatica*, for example), the presence of high-quality food leads to greater production of parasites, especially for cercariae (Kendall 1949; Kendall and Ollerenshaw 1963; Boray 1969; Rondelaud et al. 2002, 2004). The efficacy of high-quality food on parasitic production of a snail can only be studied by comparing the effects of this diet with those of a basic food such as lettuce. In addition, this type of experiment must be performed with the same population of snails, the same type of infection, and the same breading conditions for the snails.

As this type of study was carried out mainly with amphibious or semiaquatic lymnaeids in the case of *F. hepatica*, the information detailed below comes from experiments performed by several authors on *A. tomentosa* and *G. truncatula*.

14.5.1 Impact of Algae and Food Supplement

Table 14.3 shows the cercarial production of *F. hepatica* that Boray (1969) noted in relation to the quality of the food given to *A. tomentosa*. The use of unicellular algae resulted in the production of 150 cercariae per snail. The presence of a food supplement improved the figures: 250 cercariae per snail if the food is mixed (algae + supplement), 460 to 510 if the supplement is used alone. When environmental conditions are good (reasonable density of snails, abundant food, permanent oxygenation of water), the mean amount of shed cercariae reached 622 per infected snail. The use of a food supplement allows a high cercarial production, and it can be done without algae. But the best factor is still the conditions in which the snail lives. Among these, the presence of a sufficient oxygenation of water seems necessary.

A similar study was also carried out by several authors in three French populations of *G. truncatula* with algae, wheat germ, or modified Boray's diet as food (Table 14.4). Compared to snails fed with lettuce, the number of *F. hepatica* cercariae was higher in snails that feed on unicellular algae (experiment 1). As this result was noted in the two populations used for these infections, the geographic origin of the snails had no influence on this parameter. If the nature of the food is changed in another experiment, the modified Boray's diet resulted in a higher cercarial production than those noted for lettuce or wheat germ, for which there was no significant difference between the mean values. In the third experiment, the results were greater with the modified Boray's diet than with lettuce, whatever the digenean species used. Compared to lettuce, the mean production of cercariae is increased by 164% or 178% when the algae are used as food, while wheat germ and

Table 14.3 Cercarial production of *Fasciola hepatica* by the snail *Austropeplea tomentosa* in relation to the quality of food proposed to snails. From Boray (1969)

Type of food	Number of infected snails	Mean cercarial production
Algae only	360	150
Algae + food supplement[a]	1348	250
Food supplement[a] only (two experiments)	895 2003	460 510
Food in abundance, permanent oxygenation	495	622

[a]Lettuce, alfalfa leaves, wheat germs, calcium salt

Table 14.4 Cercarial production of three digeneans in *Galba truncatula* in relation to food quality. From Maire (2000), Valladon (2000) and Rondelaud et al. (2002)

Experiment no.	Snail population	Parasite	Nature of food	Mean cercarial production
1	A	*Fasciola hepatica*	Lettuce	98.3
			Algae	175.1
	B	*F. hepatica*	Lettuce	91.3
			Algae	150.3
2	C	*F. hepatica*	Lettuce	104.7
	C	*F. hepatica*	Wheat germs	145.8
	C	*F. hepatica*	Modified Boray's diet[a]	213.7
3	C	*Fasciola gigantica*	Lettuce	56.7
			Modified Boray's diet[a]	95.3
	C	*F. hepatica*	Lettuce	112.6
			Modified Boray's diet[a]	188.7
	C	*Calicophoron daubneyi*	Lettuce	95.4
			Modified Boray's diet[a]	156.2

[a]Modified Boray's diet: 5 g dried lettuce leaves in powder, 5 g dried wheat germs in powder, 0.3 g calcium phosphate, and 60 mL spring water of oligocalcic origin (9 mg/L Ca^{++})

modified Boray's food allow an increase of 139% and 204.3%, respectively (Maire 2000).

14.5.2 Impact of Tetraphyll™

Three experiments were carried out by Guichard (2003), Belfaiza et al. (2004) and Rondelaud et al. (2004) by proposing lettuce alone or in association with a food

Table 14.5 Cercarial production of three digeneans in *Galba truncatula* in relation to the quality of food. From Guichard (2003), Belfaiza et al. (2004), and Rondelaud et al. (2004)

Experiment no.	Parasite	Nature of food	Mean cercarial production
1	*Fasciola hepatica*	Lettuce Lettuce + Tetraphyll™	112.4 358.9
	Fasciola gigantica	Lettuce Lettuce + Tetraphyll™	93.7 187.2
	Calicophoron daubneyi	Lettuce Lettuce + Tetraphyll™	77.3 246.8
2	*F. hepatica*	Lettuce Lettuce + Tetraphyll™	98.7 348.5
3	*F. hepatica*	Lettuce + Tetraphyll™	327.3
		Lettuce + modified Boray's diet	247.5
		Lettuce + algae	269.7

supplement (Tetraphyll™, modified Boray's diet, or unicellular algae) to infected *G. truncatula* (Table 14.5). In the first experiment, the number of cercariae shed by these snails is significantly higher in groups fed with lettuce and Tetraphyll™ than in those fed only with lettuce. The increase in production is thus 3.19 times for *F. hepatica*, 1.99 times for *Fasciola gigantica*, and 3.19 times for *Calicophoron daubneyi*. This mixed diet therefore induces the same effects on cercarial production, whatever the digenean used for snail infections (Guichard 2003). But the snail population also plays a role because there is interpopulation variability when snails are fed with lettuce and Tetraphyll™ (Rondelaud et al. 2004). The repetition of the experiment by Belfaiza et al. (2004) led to the same results as those of the first test and demonstrates that the association of lettuce and Tetraphyll™ causes the greatest increase in cercarial production in *G. truncatula*, at least in the case of *F. hepatica*.

More surprising was the increase in cercarial production that Belfaiza et al. (2004) have noted with the association of lettuce and Tetraphyll™. In infected controls that used only lettuce as food, the mean number of *F. hepatica* cercariae shed every day progressively decreased during the patent period. In contrast, the results were different when lettuce + Tetraphyll™ was used. As this food supplement was given to snails only twice a week, the number of cercariae was significantly higher on the first day of each use of Tetraphyll™. The following day, the number of shed cercariae decreased and was very low on the third day. To verify whether this finding was only due to Tetraphyll™, the above authors repeated this experiment and proposed lettuce leaves associated to the modified Boray's diet or unicellular algae to snails. The same results were obtained with the two types of food, metacercariae counted on the first day of the Tetraphyll™ distribution were clearly less numerous. According to Belfaiza et al. (2004), a first contact of an infected snail with Tetraphyll™ would result in a faster differentiation of intraredial cercariae and the accumulation of numerous free cercariae in the snail's body so that a second contact with this food 3 days later would allow free cercariae to accumulate glycogen and fatty acids before their shedding in mass into the water. However, another assumption, i.e. a temporary immobilization of the infected snail by Tetraphyll™ or rather its degradation products, followed by a massive release of cercariae cannot be totally ruled out.

14.6 General Principles of Cercarial Production

It seemed useful to us to recall the technique which is used to obtain cercariae or metacercariae of digeneans using host snails. This technique is not specific to lymnaeids and concerns all species of molluscs used in the laboratory to obtain larvae capable of infecting definitive hosts. Most of the information in the paragraphs below comes from the experience acquired by our team in the production of metacercariae for experimental or commercial purposes.

Several field-collected populations of snails (*G. truncatula*, for example) are necessary. The first three can come from lowlands, while the fourth population must come from a habitat in altitude in order to have 4-mm high snails over most months of the year (overwintering snails from highland populations lay their eggs later in the season: June or July, than their congeners living in the plains). The absence of digenean larval forms in these populations must be verified twice a year by dissection of adult snails. Parasite eggs must incubate (the length depends on the temperature) for miracidia to differentiate. In our experience, the results obtained with the eggs of *C. daubneyi*, *F. hepatica*, *F. gigantica*, *Fascioloides magna*, and *Haplometra cylindracea* are consistent with the incubation lengths that Ollerenshaw (1971) reported for *F. hepatica*, i.e. 20 days at constant 20 °C, 15 days at 22.5 °C, or 11 days at 25 °C. At the end of the incubation period, simple exposure of the eggs to sun or artificial light for 10 to 20 minutes is enough for the hatching of miracidia. With the use of a Pasteur pipette whose tip is stretched under the flame (the internal diameter must be twice the width of the miracidium), one or two miracidia (usually) are taken in order to put them in each dish, tube, or plate well containing the snail and spring water or dechlorinated tap water for 4 h. During this exposure, a regular surveillance is necessary to dip emerging snails into water.

Snails are then put into Petri dishes (diameter, 14 cm in our case) taking into account their shell height (10 snails per dish if their size at exposure is 4 mm). These dishes are stacked (up to 5 per column) before being placed in an air-conditioned room, with a constant temperature of 24 °C, a photoperiod of 12 diurnal hours, and an artificial lighting (fluorescent tubes) of 3000–4000 lux. Every day, water is changed and lettuce, live spring moss, or dead leaves of grass are added if necessary. Petri dishes are cleaned on day 14 post-exposure and weekly thereafter. Among the problems which may occur during this period, the development of iridovirosis in the case of *G. truncatula* is the most frequent: it is then necessary to individually isolate the survivors in small Petri dishes (diameter, 35 mm) and use the habitual rules of hygiene (handwashing, sterilization of soft forceps). On day 30, the survivors are put in 35 mm dishes with a single snail, 4.5 mL of spring water, and pieces of lettuce, dead grass, and spring moss per recipient. The dishes are followed every day to change the water and the recipient if cercarial shedding occurred. In this case, the dishes containing the metacercariae are kept in spring water at 4°-6 °C until these larvae are counted. The collection of these cysts is performed under a stereomicroscope and each larva is aspirated at the end of a flame-stretched Pasteur pipette whose internal diameter is lightly smaller than that of the cyst. Metacercariae are placed in groups of 100 each in 2.5 mL Eppendorf™ tubes and placed at 4° C–6 °C. These larvae must not remain in the cold for more than 12 weeks, even if Pécheur (1967, 1974) indicates that metacercariae of *F. hepatica* can survive more than 400 days at 4 °C or 257 days at 20 °C when submerged.

In the case of *F. hepatica*, cercariae exit from the host snail during shedding waves of variable length, separated by time intervals during which nothing occurs. In addition, the number of these larvae decreases throughout the patent period, with only a few cysts per day during the third week following the exit of the first cercariae (Dreyfuss et al. 2015b). Under these conditions, one can wonder if the dissection of infected snails on a given day of the patent period would not reduce the time required to obtain the same larvae during shedding. An experimental study with *C. daubneyi* was carried out by Vignoles et al. (2008) because 7 or 8 weeks at 20 °C are necessary to obtain metacercariae of this paramphistome. Four groups of *G. truncatula*, measuring 4, 5, 6, or 7 mm in height at miracidial exposure, were used for this study and snail samples were dissected twice a week up to day 70 of infection. The first cercariae encysted from day 35 in groups 6 and 7 mm, and from day 40 in the two others. In all cases, the number of these metacercariae increased gradually over time up to day 70. The dissection of snails at regular intervals allows thus to obtain metacercariae earlier than with cercarial shedding (a mean difference of 15 days).

The infectivity of such metacercariae originating from dissected snails was verified in another experiment through infection of a definitive host (lambs in this case). If these ruminants were infected with metacercariae coming from snails dissected on day 45 or day 60, the number of adult paramphistomes found in the paunch of the two sheep was 27 and 38. In contrast, the number of adult worms in sheep infected with metacercariae coming from snails dissected on day 35 was lower: 7 and 13 adults per animal (Vignoles et al. 2008). These last results indicate (i) that metacercariae collected from snails before day 45 were not sufficiently differentiated to produce adult flukes in the definitive host, and (ii) snails must be raised for at least 45 days (at 24 °C) to give infective larvae of paramphistomes. Dissection of infected *G. truncatula* is therefore an interesting technique for the production of metacercariae because it reduces the time required to obtain these larvae. The method can be applied to all digeneans with a metacercarial stage in their life cycle. According to Boray (1969), the viability of *F. hepatica* metacercariae and their infectivity in the definitive host were the same, whatever the method (cercarial shedding or snail dissection) used for their formation.

References

Apostoloff C (2001) Influence a posteriori de l'hôte définitif sur le développement rédien de *Fasciola hepatica* Linné et la productivité cercarienne chez le mollusque *Lymnaea truncatula* Müller. PharmD Thesis, University of Limoges

Auld JR (2018) The effects of diet and mating system on reproductive (and post-reproductive) life span in a freshwater snail. Ecol Evol 8:12260–12270. https://doi.org/10.1002/ece3.4689

Auld JR, Henkel JF (2014) Diet alters delayed selfing, inbreeding depression, and reproductive senescence in a freshwater snail. Ecol Evol 4:2968–2977. https://doi.org/10.1002/ece3.1146

Bargues MD, Vigo M, Horák P et al (2001) European Lymnaeidae (Mollusca: Gastropoda), intermediate hosts of trematodiases, based on nuclear ribosomal DNA ITS-2 sequences. Infect Genet Evol 1:87–107. https://doi.org/10.1016/S1567-1348(01)00019-3

Belay A, Kato T, Ota Y (1996) Spirulina (Arthrospira): potential application as an animal feed supplement. J Appl Phycol 8:303–311

Belfaiza M, Abrous M, Rondelaud D et al (2004) The use of Tetraphyll® as food for snails increases the intensity of cercarial shedding in *Galba truncatula* infected with *Fasciola hepatica*. Parasitol Res 94:86–90

Boray JC (1963) The ecology of *Fasciola hepatica* with particular reference to its intermediate host in Australia. In: Proceedings of the seventeenth world veterinary congress, Hannover, 14–21 August 1963 6:709–715

Boray JC (1966) Studies on the relative susceptibility of some lymnaeids to infection with *Fasciola hepatica* and *F. gigantica* and on the adaptation of *Fasciola spp.* Ann Trop Med Parasitol 60: 114–124

Boray JC (1969) Experimental fascioliasis in Australia. Adv Parasitol 7:95–210

Boray JC (1978) The potential impact of exotic *Lymnaea* spp. on fascioliasis in Australasia. Vet Parasitol 4:127–141

Brendelberger H (1997) Contrasting feeding strategies of two freshwater gastropods, *Radix peregra* (Lymnaeidae) and *Bithynia tentaculata* (Bithyniidae). Arch Hydrobiol 140:1–21

Bruce JJ, Liang YS (1992) Cultivation of schistosomes and snails for researchers in The United States of America and other countries. J Med Appl Malacol 4:13–30

Comerton M, Houghton JA (1975) A description of the algal constituents of the diet of the snail *Lymnaea truncatula*. Irish J Agric Res 14:55–70

Correa AC, Escobar JC, Durand P et al (2010) Bridging gaps in the molecular phylogeny of the Lymnaeidae (Gastropoda: Pulmonata), vectors of fascioliasis. BMC Evol Biol 10:381. https://doi.org/10.1186/1471-2148-10-381

Dawes B (1968) The Trematoda with special reference to British and other European forms. Cambridge University Press, London

Dillon RT (2004) The ecology of freshwater molluscs. Cambridge University Press, Cambridge

Dreyfuss G, Vignoles P, Mekroud A et al (2006) The presence of uninfected *Omphiscola glabra* in a breeding of infected *Galba truncatula* enhanced the characteristics of snail infections with *Fasciola hepatica*. Parasitol Res 99:197–199

Dreyfuss G, Correa AC, Djuikwo-Teukeng FF et al (2015a) Differences in the compatibility of infection between the liver flukes between *Fascioloides magna* and *Fasciola hepatica* for a Colombian population of *Galba* sp. J Helminthol 89:720–726

Dreyfuss G, Vignoles P, Rondelaud D et al (2015b) The mud snail (*Galba truncatula*). Ecology, parasitism and control. Lambert Academic Publishing, Saarbrücken

El-Assal FM, Shanab SMM, Abou-El-Hassan AA, Mahmoud KMA (2005) Effect of some algal species on the snail intermediate hosts of schistosomiasis in Egypt. II. Growth, infection and mortality rates. Egypt J Phycol 6:93–110

Eveland LK, Haseeb MA (2011) Laboratory rearing of *Biomphalaria glabrata* snails and maintenance of larval schistosomes *in vivo* and *in vitro*. In: Toledo R, Fried B (eds) *Biomphalaria* snails and larval trematodes. Springer, New York, pp 33–55

Florin L, Fried B, Reddy A (2000) Growth and fecundity of *Lymnaea elodes* (Gastropoda: Lymnaeidae) under laboratory conditions. Veliger 43:78–81

Guasch H, Ricart M, Lopez-Doval J et al (2016) Influence of grazing on triclosan toxicity to stream periphyton. Freshw Biol 61:2002–2012

Guichard T (2003) L'impact d'une alimentation mixte: salade + Tetraphyll®, sur le mollusque hôte: *Galba truncatula* O.F. Müller et la production cercarienne de plusieurs Digènes. PharmD Thesis, University of Limoges

Hillyer GV, Apt W (1997) Food-borne trematode infections in the Americas. Parasitol Tod 13:87–88

Hodgkinson JE, Cwiklinski K, Beesley N et al (2018) Clonal amplification of *Fasciola hepatica* in *Galba truncatula*: within and between isolate variation of triclabendazole-susceptible and -resistant clones. Parasites Vectors 11:363. https://doi.org/10.1186/s13071-018-2952-z

Hurtrez-Boussès S, Meunier C, Durand P et al (2001) Dynamics of host–parasite interactions: the example of population biology of the liver fluke (*Fasciola hepatica*). Microbes Infect 3:841–849

Islam MN, Port GR, McLachlan AJ (2001) The biology of *Lymnaea peregra* (Müller) (Gastropoda: Pulmonata: Basommatophora) with special reference to the effects of herbicides on its reproduction. J Biol Sci 1:532–540. https://doi.org/10.3923/jbs.2001.532.540

Iturbe-Espinoza P, Muñiz-Pareja F (2012) *Galba truncatula* induced to infection with miracidia of *Fasciola hepatica*, collected in Huayllapampa-San Jerónimo-Cusco, Peru. Neotrop Helminthol 6:211–217

Kaplan RM (2001) *Fasciola hepatica*: a review of the economic impact in cattle and considerations for control. Vet Ther 2:40–50

Kendall SB (1949) Nutritional factors affecting the rate of development of *Fasciola hepatica* in *Limnaea truncatula*. J Helminthol 23:179–190

Kendall SB (1953) The life-history of *Limnaea truncatula* under laboratory conditions. J Helminthol 27:17–28

Kendall SB, Ollerenshaw CB (1963) The effect of nutrition on the growth of *Fasciola hepatica* in its snail host. Proc Nutr Soc 22:41–46

Lämmler G (1955) Die Chemotherapie der Fasciolose, zugleich ein Beitrag zur experimentell-chemotherapeutischen Untersuchungsmethodik. Arzneim Forsch 5:497–502

Lee CG, Kim SK, Lee CY (1993) Laboratory cultivation of blue-green algae for use as a food for lymnaeids, the intermediate hosts of *Fasciola hepatica*. Korean J Vet Res 32:239–243

Lee CG, Kim SK, Lee CY (1994) Rapid growth of *Lymnaea viridis*, the intermediate host of *Fasciola hepatica*, under laboratory conditions. Vet Parasitol 51:327–331

Lee CG, Kim SK, Lee CY (1995) Metacercarial production of *Lymnaea viridis* experimentally infected with *Fasciola hepatica*. Vet Parasitol 58:313–318

Liang YS, van der Schalie H (1975) Cultivating *Lithoglyphopsis aperta* Temcharoen, a new snail host for *Schistosoma japonicum*, Mekong strain. J Parasitol 61:915–919

Lombardo P, Miccoli FP, Giustini M et al (2010) Diel activity cycles of freshwater gastropods under natural light: patterns and ecological implications. Ann Limnol Int J Lim 46:29–40

Madsen H (1992) Food selection by freshwater snails in the Gezira irrigation canals, Sudan. Hydrobiologia 228:203–217

Maire S (2000) Impact de trois types d'alimentation sur la croissance de *Lymnaea truncatula* Müller et la production cercarienne de trois Digènes. PharmD Thesis, University of Limoges

Malek EA (1962) Laboratory guide and notes for medical malacology. Burgess Publishing Company, Minneapolis, MN

Mas-Coma S, Valero MA, Bargues MD (2009) *Fasciola*, lymnaeids and human fascioliasis, with a global overview on disease transmission, epidemiology, evolutionary genetics, molecular epidemiology and control. Adv Parasitol 69:41–146

Moazeni M, Ahmadi A, Alavi AM (2018) A new method for laboratory rearing of *Galba truncatula*, the intermediate host of *Fasciola hepatica*. Vet Parasitol 253:12–15

MolluscaBase (2022) MolluscaBase. http://www.molluscabase.org. Accessed 07 January 2022. https://doi.org/10.14284/448

Mostafa SS, Gawish FA (2009) Towards to control *Biomphalaria alexandrina* snails and the free living larval stages of *Schistosoma mansoni* using the microalga *Spirulina platensis*. Aust J Basic Appl Sci 3:4112–4119

Okoye IC, Obiezue NR, Mgbemena US (2009) Formulated feed preference for survival and optimal growth of *Bulinus* species reared in the laboratory. Anim Res Int 6:1102–1106

Ollerenshaw CB (1971) Some observations on the epidemiology of fascioliasis in relation to the timing of molluscicide applications in the control of the disease. Vet Rec 88:152–164

Osborn GD, Gron N, Simmons D (1982) Maintenance and infection of the mud snail *Lymnaea truncatula* for *Fasciola hepatica* studies. J Inst Animal Technol 33:1–5

Pécheur M (1967) La cercaire de *Fasciola hepatica*. Le rôle de la couleur, de la lumière et des plantes sur le choix de l'endroit de fixation. La cercaire est-elle infestante? Ann Med Vet 6:349–355

Pécheur M (1974) Lutte stratégique contre la distomatose. Comptes-Rendus de Recherches, Travaux du Centre de Recherches sur les Maladies Parasitaires des Animaux Domestiques 38: 85–150

Pereira Filipa M, Delgado ML, Seixa Lopes AM et al (2000) Optimum conditions for growth in liquid medium of *Oscillatoria formosa* Bory used as the principal food in laboratory culture of intermediate hosts for schistosomosis and fasciolosis. Parasite 7:227–231

Prepelitchi L, Pujadas JM, Wisnivesky-Colli C (2015) The effect of spirulina (*Arthrospira platensis*) (Oscillatoriales: cyanobacteria) on the experimental breeding of *Pseudosuccinea columella* (Basommatophora: Lymnaeidae). Rev Biol Trop 63:479–489

Rondelaud D, Abrous M, Dreyfuss G (2002) The influence of different food sources on cercarial production in *Lymnaea truncatula* experimentally infected with Digenea. Vet Res 33:95–100

Rondelaud D, Denève C, Belfaiza M et al (2004) Variability in the prevalences of infections and cercarial production in *Galba truncatula* raised on a high quality diet. Parasitol Res 92:242–245

Rondelaud D, Fousi M, Vignoles P et al (2007) Optimization of metacercarial production for three digenean species by the use of Petri dishes for raising lettuce-fed *Galba truncatula*. Parasitol Res 100:861–865

Rondelaud D, Vignoles P, Dreyfuss G (2009) La Limnée tronquée, un mollusque d'intérêt médical et vétérinaire. PULIM, Limoges, France

Rondelaud D, Mouzet R, Vignoles P et al (2014) The production of mammalian trematode infective stages by the snail *Galba truncatula*. J Helminthol 88:105–111

Rondelaud D, Djuikwo Teukeng FF, Vignoles P et al (2015) *Lymnaea glabra*: progressive increase in susceptibility to *Fasciola hepatica* through successive generations of experimentally infected snails. J Helminthol 89:398–403

Saad AEHA, Hassannain MA, Darwish EM (2019) Molluscicidal effect of some red and green marine algae on *Lymnaea natalensis*, Egypt. J Aquat Biol Fish 23:483–490

Sampaio Xavier ML, Fraga de Azevedo J, Avelino I (1968) Importance d'*Oscillatoria formosa* Bory dans la culture au laboratoire des mollusques vecteurs de *Schistosoma haematobium*. Bull Soc Pathol Exot 61:52–66

Sanabria R, Mouzet R, Courtioux B et al (2012) Intermediate snail hosts of French *Fasciola hepatica*: *Lymnaea neotropica* and *Lymnaea viatrix* are better hosts than local *Galba truncatula*. Parasitol Res 111:2011–2016

Sanabria R, Mouzet R, Courtioux B et al (2013) Comparative strategies and success of sympatric and allopatric *Fasciola hepatica* infecting *Galba truncatula* of different susceptibilities. Parasitol Res 112:2255–2259

Sánchez R, Perera G, Sánchez J (1995) Cultivo de *Fossaria cubensis* (Pfeiffer) (Pulmonata: Lymnaeidae) hospedero intermediario de *Fasciola hepatica* (Linnaeus) en Cuba. Rev Cubana Med Trop 47(1):71–73

Sánchez M, Bernal-Castillo J, Rozo C et al (2003) Spirulina (Arthrospira): an edible microorganism: a review. Revista Universitas Scientiarum 8:7–24

Sandland GJ, Minchella DJ (2003) Effects of diet and *Echinostoma revolutum* infection on energy allocation patterns in juvenile *Lymnaea elodes* snails. Oecologia 134:479–486

Schumacher W (1938) Untersuchungen über den Wanderungsweg und die Entwicklung von *Fasciola hepatica* L. im endwirt. Z Parasitenkd 10:608–643

Sevo S (1973) Technique de production des métacercaires. Haliotis 3:185–189

Skoog G (1979) Influence of natural food items on growth and egg production in brackish water populations of *Lymnaea peregra* and *Theodoxus fluviatilis* (Mollusca). Oikos 31:340–348

Swart NC, Starkloff S, Ypenburg J et al (2020) The effect of mating on female reproduction across hermaphroditic freshwater snails. Invertebr Biol 139:e1227. https://doi.org/10.1111/ivb.12275

Taylor EL, Mozley A (1948) A culture method for *Lymnaea truncatula*. Nature 161:894

Thompson SN (1984) *Spirulina* as a nutrient source in experimental media for maintaining the schistosome vector *Biomphalaria glabrata*. Ann Trop Med Parasitol 78:547–548

Torgerson P, Claxton J (1999) Epidemiology and control. In: Dalton JP (ed) Fasciolosis. CABI Publishing, Oxon, pp 113–149

Tufi S, Wassenaar PNH, Osorio V et al (2016) Pesticide mixture toxicity in surface water extracts in snails (*Lymnaea stagnalis*) by an *in vitro* acetylcholinesterase inhibition assay and metabolomics. Environ Sci Technol 50:3937–3944. https://doi.org/10.1021/acs.est.5b04577

Valladon C (2000) Impact d'une alimentation à base d'algues sur la croissance du mollusque *Lymnaea truncatula* et la production cercarienne de *Fasciola hepatica*. PharmD Thesis, University of Limoges

Vázquez AA, Pilar A, Lounnas M et al (2018) Lymnaeid snails hosts of *Fasciola hepatica* and *Fasciola gigantica* (Trematoda: Digenea): a worldwide review. CAB Rev 13:1–15. https://doi.org/10.1079/PAVSNNR201813062

Vignoles P, Rondelaud D, Dreyfuss G (2008) *Paramphistomum daubneyi*: production dynamics and infectivity of metacercariae originating from snails dissected at regular intervals. J Helminthol 82:175–180

Vignoles P, Novobilský A, Höglund J et al (2014) *Lymnaea cubensis* (Pfeiffer, 1839), an experimental intermediate host for *Fascioloides magna*. Folia Parasitol 61:185–188

Whitlock HV, Chow DCM, Kelly JD (1976) The laboratory maintenance of field-collected *Lymnaea tomentosa* for the production of *Fasciola hepatica* metacercariae. Vet Parasitol 1: 317–325

Whitlock HV, Campbell NJ, Chow DCM et al (1977) A comparison of two laboratory methods of maintaining field-collected *Lymnaea tomentosa* for the production of *Fasciola hepatica* metacercariae. Vet Parasitol 3:75–83

Chapter 15
Control of Fasciolosis-Transmitting Lymnaeids in the Field

Daniel Rondelaud, Philippe Vignoles, Gilles Dreyfuss, Jean-Pierre Pointier, and Antonio A. Vázquez

Abstract The objective of this chapter is to review the different methods for field control of host snails involved in the transmission of fasciolosis. Environmental measures such as drainage of swampy soils or cutting of vegetation in watercourses can reduce the number of snails. Synthetic molluscicides are being used less and less because of their cost, toxicity, and contamination they cause in the environment. Conversely, research on plant extracts is being developed because many of them are natural molluscicides that are less toxic and more environmentally friendly. In the nature, several groups of vertebrates and invertebrates predate lymnaeids and a control technique of *Galba truncatula* has been developed in central France using predation by the terrestrial snail *Zonitoides nitidus*. Biological control can be performed using animal species which compete with lymnaeids for food. Pathogens such as parasites and other infectious agents can also be used. Finally, an integrated liver fluke control strategy for the control of liver fluke, associating deworming of the definitive hosts and control of host snails, is analysed with a review of results provided by this type of control.

D. Rondelaud · P. Vignoles · G. Dreyfuss (✉)
Laboratory of Parasitology, University of Limoges, Limoges, France
e-mail: daniel.rondelaud@unilim.fr; philippe.vignoles@unilim.fr; gilles.dreyfuss@unilim.fr

J.-P. Pointier
PSL Research University, USR 3278 CNRS–EPHE, CRIOBE Université de Perpignan, Perpignan, France
e-mail: pointier@univ-perp.fr

A. A. Vázquez
Laboratory of Malacology, Institute of Tropical Medicine "Pedro Kourí", Havana, Cuba

MIVEGEC, University of Montpellier, CNRS, IRD, Montpellier, France

M. V. Vinarski, A. A. Vázquez (eds.), *The Lymnaeidae*, Zoological Monographs 7,
https://doi.org/10.1007/978-3-031-30292-3_15

15.1 Introduction

Fasciolosis is a parasitosis caused by fluke species of the genus *Fasciola*. If *F. hepatica* is the main cause of this disease because of its very wide distribution over the five continents, *F. gigantica* seems to be of secondary importance because it is limited to the Old World (Mas-Coma et al. 2009). When present, these parasites are common in cattle and sheep, resulting in significant economic losses in the livestock industry due to their pathogenicity (Torgerson and Claxton 1999). This parasitosis also affects people. Human fasciolosis was considered a secondary disease until the late 1990s (Chen and Mott 1990). But the situation has changed considerably in recent years due to the increasing importance of this disease in Europe, Asia, Africa, as well as the North, Central, and South Americas (Mas-Coma et al. 2009).

Both species of *Fasciola* use freshwater snails of the Lymnaeidae family as intermediate hosts. The latter ensure the larval development of these parasites (Kendall 1965). At least twenty species of this family have been reported as host molluscs for *F. hepatica* (Boray 1978; Torgerson and Claxton 1999; Hurtrez-Boussès et al. 2001; Vázquez et al. 2018) and most species of the genus *Galba* (amphibious molluscs) are involved in the transmission of this fasciolosis (Alda et al. 2021). In the case of *F. gigantica*, transmission is mainly ensured by molluscs belonging to the genus *Radix* (Spithill et al. 1999; Mas-Coma et al. 2009; Vázquez et al. 2018). Among the means that have been proposed to fight the disease, the control of host snails has long been suggested (Mehl 1932; Taylor 1964).

The control application faces three types of difficulties. The first relates to the snail's lifestyle. Most species are strictly aquatic like *Lymnaea stagnalis*, while others are semi-aquatic like *Omphiscola glabra* or amphibious with the possibility of living in water or on wetlands. The treatment of these amphibious snails is more complex because the survivors do not hesitate to take refuge on emerged areas when a molluscicide or a biological control agent is introduced into their habitat (Dreyfuss et al. 2015). The second difficulty is the type of habitat, depending on whether it remains in water during the summer or dries out. In permanent watercourses, environmental measures or biological control agents will be preferred at the expense of chemical treatment as the latter is known to have a lethal effect on associated flora and fauna as well as animal and human health if this water is used for drink (Rozendaal 1997). In drying habitats, several lymnaeid species bury themselves during the summer months (aestivation) and this process should therefore be taken into account when deciding whether to apply environmental measures or to introduce biological control agents (Dreyfuss et al. 2015). The third difficulty is the number of snail generations per year. Most aquatic species have only an egg-laying period extending from spring to mid-summer depending on the climatic and environmental conditions of their habitat. Semi-aquatic and amphibious species often have two egg-laying periods, one in the spring and the other in the autumn at the end of aestivation. But in very wet years in temperate countries, a third generation of snails (summer generation) occurs as a result of the eggs laid by the spring generation (Taylor 1964; Dreyfuss et al. 2015).

Among the techniques proposed by the authors, environmental measures are the oldest and relate to modifications of the habitat in which the host snail lives. The most well-known example is drainage which allows faster flow of excess water into lymnaeid habitats (Taylor 1964). Synthetic or natural molluscicides, which humans can introduce into snail habitats, were mainly used up to the 1970s in Western Europe (Euzeby 1971). The disadvantages caused by the use of these products have led to a gradual decrease in their use over time and their abandonment in some countries (Rozendaal 1997; Bustinduy and King 2014). Finally, three types of biological agents, i.e. predators, competitors, and pathogens, have been proposed by the authors (Taylor 1964). The development of this type of control has been much slower than the application of environmental measures or molluscicides. The main difficulty is the often incomplete elimination of snails and these agents must be used for several successive years to have complete eradication of the snail (Dreyfuss et al. 2015). Whatever treatment chosen for the snails, it is necessary to couple this control with the treatment of definitive hosts in an integrated management concept in order to eliminate the reservoir of the disease in the definitive host (Mage et al. 1989; Dreyfuss et al. 2015).

The methods for controlling host snails, reported in the paragraphs below, are not exclusive to lymnaeids. They also concern the intermediate hosts of other diseases. Many articles related to the control of host snails of *Schistosoma* spp. have been published in the literature. As many aquatic lymnaeids are part of a larger malacological population (case of *Radix natalensis* with *Bulinus* spp. and/or *Biomphalaria* spp. in Africa), many references concerning intermediate hosts of *Schistosoma* spp. have been incorporated into this chapter.

As the nomenclature of Lymnaeidae is still controversial, the species names cited in this chapter have been updated following the names of the taxa listed in MolluscaBase (2022). Species, whose names have changed, are cited by first indicating the old name, followed by the current name in brackets.

15.2 Environmental Measures

15.2.1 Drainage of Lands

Soil drainage has a long history and apparently dates back to the earliest civilizations of Mesopotamia and Iran in the 4000s BC (Valipour et al. 2020). Nowadays, it is the most widely used method for drying the environment in a wet meadow or a swamp (Euzeby 1971). According to this author, drainage is a difficult, delicate, and costly operation which is rarely performed for the sole prophylaxis of parasitoses. However, if properly implemented, it is a radical means to eliminate *Galba truncatula* because the cleaned soil does not allow the snails to live there. Three techniques are used to do this drainage (Brouwer et al. 1985; Ritzema 1994).

The most common is the digging of a network of surface swales and ditches (surface drainage) using a trench plough (Taylor 1964). To prevent the walls of these swales and ditches from being destroyed by trampling of domestic ruminants, a fence must be put in place around the network (Taylor 1964) but this is rarely applied in central France for reasons of cost (Dreyfuss et al. 2015). To overcome this difficulty, it is useful to restore the drainage network every year in October or November, i.e. during the months when the passage of tractors is still possible on the ground to do this operation (Taylor 1964). In central France, the annual cleaning of this network is carried out only by a small number of farmers, while the others clean up swales and ditches only every 2 or 3 years when the network has been completely levelled by the passage of cattle or sheep (Dreyfuss et al. 2015). This first method only limits the development of snail populations and the results obtained depend on the date where the drainage is performed in the field.

The second technique is the digging of the subsoil using a mole plough (subsurface drainage) to create simple underground conduits, reinforced or not by pottery or PVC pipes. This technique has several advantages, namely the complete disappearance of snail populations on the surface which no longer have enough moisture to live and the easy passage of agricultural machines on the ground especially when the pipes are surrounded by gravel (Euzeby 1971; Blann et al. 2009). However, the purchase of pipes, whatever their nature, is expensive, which limits the diffusion of this technique. If the farmer does not use pipes, this technique can only be used in soils with sufficient structural stability to prevent the collapse of underground conduits during the passage of ruminants (Euzeby 1971). Fencing of these networks, at least in central France, was little used (Dreyfuss et al. 2015).

The third method is to do a system of shovel-dug ditches to remove most of the runoff in a meadow. In fairly flat areas, the application of this technique creates regular areas of pasture separated from each other by parallel ditches, oriented in the direction of the steepest land slope. Even if these ditches are deeper than simple drainage ditches, fencing is necessary to prevent them from collapsing under the passage of domestic animals. All this represents a financial cost that many farmers hesitate to assume in order to clean up their pastures (Dreyfuss et al. 2015).

15.2.2 Other Environmental Measures

In watercourses, several techniques have been proposed to control freshwater snails. Removal of aquatic vegetation in the canals reduces the number of snails. This cleaning can be very effective as it can be durable and selective against specific snail species. The most relevant requirement is to target all the habitats covered by macrophytes and algae where the snails live because a single snail can rapidly recolonize the whole irrigation system (Rozendaal 1997; Sabourin et al. 2018). Similarly, rising and falling water levels and increasing flows can disturb snail habitats and food sources (Brown 1994; Rozendaal 1997; Lardans and Dissous 1998; Secor 2014). However, results vary according to the species of snails. In

Morocco, Laamrani et al. (2001) found that the three methods used for irrigation siphons (covering, regular cleaning, and increase in water speed) were effective in reducing the number of *Bulinus truncatus* while other species such as *Radix peregra* (= *Peregriana peregra*) and *Physella acuta* were little affected. In the irrigation system of Akka (southern Morocco), the repeated cleaning of the network and the elimination of aquatic vegetation resulted in a significant reduction in the number of bulinids and their egg masses, indicating that these low-cost measures are effective and sustainable (Boelee and Laamrani 2004).

Agronomic measures have also been applied in the meadows where cattle and/or sheep are grazing. On acidic soils in central France, Dreyfuss et al. (2015) noted several series of measures that 141 breeders have applied over the past 40 years to prevent the risk of fasciolosis in their farms. The first series aims to isolate or destroy the habitats of *G. truncatula*. In pastures, the habitats of the snail, located at the peripheral end of the surface drainage swales, are isolated from the rest of each meadow by barbed wire so that cattle or sheep are not in contact with this vegetation. As these habitats on acidic soils are small (less than 3 m^2 per habitat: Dreyfuss et al. 2015), their isolation represents less than 1% of the total area of the pasture. In the case of sources, the method used depends on their status. If they are temporary, the vegetation is destroyed by rotary crushing so that there was currently a drastic decrease in their number in central France. Permanent sources are fenced off and running water sometimes circulates in underground pipes to the nearest drainage ditch or stream. The second series of measures aims to prohibit the access of cattle or sheep to permanent water collections that are present on pastures (pools, small ponds). The latter are increasingly fenced off so that animals cannot access them and are often watered using mobile or fixed water tanks. Trampling of animals around these water tanks can sometimes lead to the formation of snail habitats when there is a permanent flow of water towards a ditch, but this case is rather rare in central France. Finally, many meadows that are difficult to drain and known locally to be "fluke hotbeds" have simply been transformed into ponds (Dreyfuss et al. 2015).

In order to show that these techniques have changed since the 1970s, Table 15.1 gives the number of agronomic transformations that breeders in the 141 above farms have applied in their meadows. Surface drainage is still widely applied on these pastures while subsurface drainage, usually coupled with the laying of PVC pipes, has shown a marked increase since the 1980s. Digging deeper ditches using a shovel was more recent and did not really develop until 2000. The transformation of meadows into ponds is an incidental process that remains rare in central France. The other agronomic measures have been applied in various ways from the 1980s. The use of water reservoirs was widespread in the region (73.1% of meadows). In contrast, this is not the case for the fencing of *G. truncatula* habitats (38.1% of meadows owned by 49 breeders only). The destruction of hillside sources did not really develop until the 2000s. The other two measures, i.e. the laying of underground pipes for the water flow coming from permanent sources and the fencing of water collections, only interested a small number of pastures (Dreyfuss et al. 2015).

Table 15.1 Changes in agronomical measures applied to 776 meadows by 141 breeders in central France. From Dreyfuss et al. (2015)

Agronomical measure	Frequency (%) of meadows concerned by an agronomical measure between 1970 and 1979	1980 and 1989	1990 and 1999	2000 and 2008
Surface drainage	99.7	96.7	82.9	59.1
Subsurface drainage	0.001	2.8	13.0	30.5
Ditches (mechanical shovel)	0	0.002	3.7	9.6
Fencing of snail habitats	0	1.4	7.6	38.2
Gyro-crushing of rushes (temporary springs)	0.001	0.001	5.6	22.5
Buried pipes (permanent springs)	0	0	1.4	4.2
Installation of watering tanks	0.003	8.6	41.1	73.1
Fencing of water collections	0	0.001	3.0	9.1
Creation of a pond	0.001	0.001	0.002	0.006

15.2.3 Consequences of Environmental Measures

The effects of these agronomic measures are really well known only for surface drainage in swampy meadows and marshes. The results depend on the date on which this operation is carried out. If drainage is done in spring (March), the density of lymnaeids (*G. truncatula* in this case) decreases hyperbolically over time. Snails of the overwintering generation suffer considerable losses due to lack of water and are unable to reproduce, which increases the decline in population during the season (Worden et al. 1963). If this measure is applied later in summer, the effects of this operation are then different on the two generations of *G. truncatula*. Spring generation snails, resulting from the reproduction of molluscs that have spent the winter, are little affected by the realization of drainage. In contrast, those of the overwintering, older generation are more sensitive to drought and die gradually (Worden et al. 1963). Finally, if the drainage is done in October or November, the remaining *G. truncatula* have time to recolonize the habitat and lay eggs so that regular maintenance can keep the snails at high and substantially constant values. On the other hand, if the cleaning is irregular (every two, 3, or 4 years), the numbers of snails decrease, sometimes up to the disappearance of the species (Rondelaud 1977b). In any case, as Euzeby (1971) underlines it, there are always a few individuals which are able to repopulate the habitat. For this reason, surface drainage applied alone cannot be sufficient to completely eradicate the population of *G. truncatula*.

Unlike surface drainage, the data on the consequences of other agronomic measures are scarce, so it is difficult at present to do a synthesis on this point. The only information available relates to the isolation of snail habitats. In this case, the prevalence of natural infection with *F. hepatica* in cattle shows a drastic drop in the seven farms where this technique was used (Dreyfuss et al. 2015). But these results come from only one region and it is difficult to generalize them without a larger study to verify their validity.

15.3 Molluscicides

Two types of products, i.e. synthetic molluscicides to which copper salts can be added, and natural substances from various plant species have been used to control aquatic snails which intervene as intermediate hosts in the life cycle of digeneans. Copper salts have been little applied in the field due to their toxicity (Taylor 1964). Synthetic molluscicides have been widely used in programmes that many countries, especially in Africa (Brown 1994), have implemented to control snail hosts of *Schistosoma* spp. Conversely, these products have been little used to control the intermediate hosts of *Fasciola* spp. Research on natural molluscicides has developed considerably since the 1990s for the host snails of *Schistosoma* spp. and from the 2000s for those of *Fasciola* spp. But most of these studies have been carried out under laboratory conditions to determine the toxicity of these products on snails, and few of them have yet been carried out in the field.

15.3.1 General Principle of Use

Among these substances, there is no perfect product to eliminate snails. According to Levêque (1990), three criteria must be taken into account for the selection of a product: (i) the molluscicide must be fully effective against the target snail, (ii) it must not be toxic to humans and the upper vertebrates, and (iii) it must be of an attractive price and easy to use under operational conditions. To these criteria are added three other requirements relating to the aquatic environment. The first is not to use pesticides which produce toxic and persistent degradation products that accumulate in trophic chains. Second, the molluscicide must not affect fish at different stages of their life. Third, it must not cause a long-term imbalance of ecosystems under normal conditions of application.

Spreading programmes must also take into account two elements which are specific to amphibious snails, i.e. the density of the population and the nature of its habitat. First, the molluscicide should preferably be applied in spring or late summer when the population density is at its highest and snails are active. Second, these substances are easier to use in stagnant water because it is possible to maintain a constant concentration of molluscicide at any point and to obtain a longer persistence, resulting in an increased efficiency of the molecule. The spreading must also involve a large perimeter around the main site. On the other hand, these requirements are not very reproducible in running water where dilution problems arise. In addition, prior destruction of aquatic and/or hygrophilous vegetation on the application site is necessary because it interferes with the dispersion of the product.

15.3.2 Chemically Defined Substances

Many products have been used to control different species of freshwater snails in their natural habitats (see review by Euzeby 1971, for example). From this list, only a few have been studied to determine their efficacy against lymnaeids in the field. Table 15.2 lists the main products which were widely used during the twentieth century, their mode of application and their toxicity to snails and other zoological groups. This synthesis comes from several sources: Euzeby (1971), Pécheur (1974), Gayral and Cavier (1977), McCullough et al. (1980), Rondelaud (1988), Haseeb and Fried (1997), Torgerson and Claxton (1999), and World Health Organization (2002, 2017).

Copper sulphate is the oldest. However, it should be avoided because of its toxicity when cattle and/or sheep are present on pastures. The product is very active on snails but its penetrating power is low in muddy environments. It is toxic to fish and corrosive to the material used. Another copper salt, cupric chloride, was used at a sublethal dose (0.1 mg/L) by Rondelaud (1986, 1988) in wild watercress beds on acidic soils with a spray in March. According to this author, this toxic at a sublethal

Table 15.2 Main molluscicides used during the twentieth century to eliminate the snail, their mode of application and their toxicity. LC_{90}, lethal concentration at 90%

		LC_{90} (g/L) at 24 h		Toxicity	
Substances	Mode of application	Adult snails	Young snails	Fish	Mammals
Copper sulphate $CuSO_4$	Powder (3.5 g/m^2) or aqueous solution (3%)	0.8–4.2	0.3	Yes	Yes
Cupric chloride $CuCl_2$	Aqueous solution (0.1 mg/L)	0.5–3.1	0.2	Yes	Yes
Sodium pentachlorophenate	Aqueous solution (10% or 15%)	1–4	0.2	Yes	Yes
Calcium cyanamide	Liquid solution (125 or 250 L/ha), powder (300 kg/ha) or pearled form (500 kg/ha)	2–4	0.5	?	?
N-trityl-morpholine (Frescon™)[a]	Concentrated at 16.5% (3 L/ha, diluted in 400 L of water)	0.02–0.06	0.03	Yes	Yes
Niclosamide (Bayluscide™)	Powder (2.5–10 kg/ha) or aqueous solution (0.2–1%)	0.1–0.3	0.05	Yes	Yes

[a]This product is not manufactured at the present time

dose disturbs the reproductive activity of *G. truncatula* so that egg-laying occurs later in the spring and that juveniles born from these eggs can be eaten by a predatory snail, *Zonitoides nitidus*, in June. Despite its low concentration, this salt is also toxic to fish, at least to fry, and mammals.

A comparative study with the four other products listed in Table 15.2 was carried out by Pécheur (1974) on the Belgian habitats of *G. truncatula*. According to this author, sodium pentachlorophenate caused the complete elimination of snails after a single application when water was stagnant. But this product is highly toxic to vertebrates and especially to fish. Calcium cyanamide has excellent molluscicidal properties at a rate of 300 kg/ha when the treatment is applied in March–April on stagnant water habitats. The pearly form of this product is preferable to powder because the pearls are not retained by the plant cover and fall to the bottom of the habitat where the snails are located. Frescon® (N-trityl-morpholine) was a good molluscicide while its ovicidal action was low. Only the eggs located at the periphery of snail egg masses were destroyed. According to Pécheur (1974), the toxicity of this product at the doses used was null for fish and other vertebrates, while other authors reported relative toxicity. Finally, niclosamide gives good results in muddy conditions and stagnant waters, but it is toxic to fish and other vertebrates. Although these products were widely used between 1970 and 1990, this is no longer the case today, at least for snail habitats located in temperate countries. All these products are toxic to the macrofauna found in the snail habitats as well as to fish and other vertebrates.

The World Health Organization (2017) currently recommends that niclosamide be used as a 25% emulsifiable concentrate or 70% wettable powder. Other formulations have been used in China such as 50% wettable powder of niclosamide ethanolamine salt (Yang et al. 2010), 4% niclosamide ethanolamine salt powder (He et al. 2007), or a niclosamide derivative, called 2′,5-dichloro-4′--nitrosalicylanilide-quinoid salt (Xia et al. 2014). The toxicity of niclosamide to non-target plant and animal species was assessed by Andrews et al. (1983). The product is capable of killing freshwater molluscs and free larval forms of parasites (miracidia, cercariae). But it is toxic to fish and amphibians (frogs and toads). According to Dawson (2003), aquatic plants and agricultural crops do not appear to be negatively affected by the concentrations of niclosamide used to control lamprey or freshwater snails. Mayflies (*Hexagenia* sp.) are relatively resistant to the effects of niclosamide exposure. In addition, this author concludes that there is a minimal risk to humans and the environment provided that the product is properly applied and supervised. As niclosamide can be used safely in the presence of livestock or poultry when applied at the indicated doses (World Health Organization 2017), application of this product has been carried out in several countries such as Brazil (Coelho and Caldeira 2016), China (Chen 2003; He et al. 2007, 2017; Dai et al. 2008, 2015; Xia et al. 2014), Egypt (Ismail et al. 2019), Kenya (Kariuki et al. 2013), and Morocco (Khallaayoune et al. 1998; Belkacemi and Jana 2006; Barkia et al. 2011). Despite this, the use of niclosamide has gradually decreased in several countries such as Brazil where application of the product was stopped in 2002, mainly due to increasing global pressure to preserve the environment. In addition, the presence of snails (*Oncomelania hupensis*) probably resistant to niclosamide (Cao et al. 2012) poses a new problem in countries where this molluscicide has been used for several decades.

15.3.3 Plant Substances

Much of the research on natural molluscicides has been done to find products of plant origin capable of exerting toxic activity on the host snails of *Schistosoma* spp. As these products are potentially biodegradable in the environment, their possible use in the field can prevent long-term pollution (Kashyap et al. 2019). The idea of using plants to control disease-carrying molluscs dates back to the 1930s when Archibald (1933) and then Wagner (1936) used the fruits of *Balanites* in Sudan and South Africa. Numerous trials were subsequently carried out over the world which led Kloos and McCullough (1982, 1987) to review them to identify the most effective plant species against snails, particularly in tropical countries. Contrary to the host snails of schistosomes, research on plants with a molluscicidal effect on Lymnaeidae was rare until 1990 and only became important from the 2000s. Most of these studies focused on the screening of numerous plant species to determine their toxicity to snails. Lethal concentrations (LC) of these products and identification of active substances were then studied on several species of lymnaeids (Thakur et al. 2019; Chaturvedi et al. 2021). However, information on the application of these products in the field is still scarce today.

Medina and Woodbury (1979) carried out tests on 200 plant species from Puerto Rico to identify those which had a molluscicidal effect on *Galba cubensis* and *Pseudosuccinea columella*. Of the 30 species selected by these authors, the aqueous extracts from 16 of them were highly toxic, killing all snails in a range of 25 to 200 ppm. *Solanum nodiflorum* was then selected for their further studies because all parts of the plant are toxic (Medina and Ritchie 1980). The effects of other plants have also been studied on these snails. Tests were performed by Cruz-Reyes et al. (1989) with three concentrations (5, 25, and 50 ppm) of an aqueous extract of *Piqueria trinervia* on five species of lymnaeids, including *G. cubensis* and *P. columella*. A 100% mortality was obtained for all species at 50 ppm after 6 h of exposure. The same percentage was obtained at 25 ppm after 24 h, and mortality of 60% to 100% was observed at 5 ppm after 24 h of exposure. A hydroethanolic extract of *Sapindus saponaria* fruit with concentrations ranging from 20 to 100 mg/L was effective against *G. cubensis* with a LC_{50} of 39.8 mg/L and a LC_{90} of 67.9 mg/L for 72 h of exposure (Abreu Guirado et al. 2019). In Brazil, an aqueous extract of the latex of *Euphorbia splendens* var. *hislopii* at 5 mg/L killed 97.4% of *P. columella* in an irrigation ditch after 24 h of exposure (Vasconcelos and Amorim 2003a, b).

Table 15.3 lists several plant species that the authors have cited since the 2000s for their high molluscicidal power against *G. truncatula* and *Radix acuminata* (= *R. rufescens*). This activity depends on the plant species studied, the segment of the plant, and the degree of ripeness in the case of fruits. But it also depends on the

Table 15.3 Molluscicidal activity of some plant-derived substances against *Galba truncatula* and *Radix acuminata* from the 2000s, with indication of the solvent used to obtain the extract and the length of contact between the substance and snails. Only the highest activities of these plants are listed in this table. Ac., acetate; LC_{50}: lethal concentration at 50%; LC_{90}: lethal concentration at 90%

Snail and plant species	Segment of the plant (solvent; exposure time)	Values (mg/L)		References
		LC_{50}	LC_{90}	
Galba truncatula				
Hammada scoparia	Leaf (methanol; 48 h)	28.9	69.9	Mezghani-Jarraya et al. (2009)
Solanum nigrum	Immature fruit (methanol-H_2O; 48 h)	3.9	7.4	Hammami et al. (2011)
Atriplex inflata	Leaf (Ethyl Ac.; 48 h)	5.9	8.8	Hamed et al. (2015)
Capparis spinosa	Leaf (methanol; 48 h)	3.5	29.9	Njeh et al. (2015a)
Solanum elaeagnifolium	Seed (methanol; 48 h)	1.1	2.2	Njeh et al. (2015b)
Citrullus colocynthis	Leaf (Ethyl Ac.; 48 h)	11.7	15.9	Chawech et al. (2017)
Clematis flammula	Flower (Ethyl Ac.; 48 h)	11.6	46.4	Khanous et al. (2017)
Radix acuminata[a]				
Alstonia scholaris	Stem bark (H_20; 96 h) Bark (purified; 96 h)	138.3 70.6	– –	Singh and Singh (2003)
Bauhinia variegata	Leaf (ethanol; 96 h) Leaf (purified; 96 h)	14.4 5.9	– –	Singh et al. (2012)
Euphorbia hirta	Latex (–; 24 h)	7.3	–	Yadav and Singh (2011)
Lantana indica	Leaf (acetone; 96 h)	1.3	–	Chauhan and Singh (2009)
Mimusops elengi	Bark (ethanol; 96 h) Bark (purified; 96 h)	15.0 7.2	– –	Singh et al. (2012)
Myristica fragrans	Mace (ethanol; 96 h) Nutmeg (ethanol; 96 h)	28.6 36.9	– –	Jaiswal and Singh (2009)
Nerium indicum	Bark (ethanol; 96 h)	0.94	–	Singh and Singh (1998)
Piper nigrum	Fruit (ethanol; 96 h) Fruit (purified; 96 h)	6.2 3.7	– –	Srivastava et al. (2009)

[a](= *Radix rufescens*)

solvent used to obtain the extract, as Njeh et al. (2015a) or Khanous et al. (2017) have shown this by producing ethyl acetate, methanol, and hydro-methanol extracts from the same plant. The association of rutin, ellagic acid, and/or taraxerol with the latex of *Jatropha gossypifolia*, its bark or its leaves at sublethal doses results in a significant reduction in the fecundity of *R. acuminata*, hatchability of eggs, and survival of young individuals (Yadav and Singh 2014). The incorporation of various concentrations of natural molluscicides in baits consisting of starch and agar resulted in higher toxicity to snails than that of the same raw molluscicides. Storage of these baits for up to 4 weeks resulted in a marked reduction in their toxicity (Tiwari and Singh 2007). In addition, seasonal variations in toxicity of their active ingredients have been reported (Agrahari and Singh 2012).

Few studies have been carried out on natural molluscicides and their effect against *Radix natalensis* and *R. peregra* (= *Peregriana peregra*). Vassiliades (1984) showed that *Euphorbia tirucalli* (the whole plant) and *Jatropha curcas* (the seed almond) killed 100% of *R. natalensis* at concentrations of 0.3 g/L and 0.1–0.2 g/L, respectively. In Kenya, 100 or 50 mg/L of powdered sun-dried or freeze-dried berries of *Solanum aculeatum* killed more than 60% of *R. natalensis* in the laboratory (Mkoji et al. 1989). Exposure of this snail for 24 and 48 h to dried fruit and leaf powders or raw water extracts of *Solanum nigrum* powders revealed high molluscicidal activity with an average LC_{50} of 17.7 mg/L (Ahmed and Ramzy 1997). Essential oils from leaves of three Azorean species: *Hedychium gardnerianum*, *Juniperus brevifolia*, and *Laurus azorica*, showed molluscicidal activity on both juveniles and adults of *R. peregra*, with a LC_{50} varying between 15.4 and 44.6 mg/L for juveniles and 45.3 to 54.6 mg/L for adult snails (Teixeira et al. 2012). The essential oils of *Cuminum cyminum*, *Foeniculum vulgare*, and *Petroselinum crispum* (Apiaceae) at 50 mg/L were also very active against the eggs and adults of *R. peregra*. The estimated LC_{50}s ranged from 13.7 to 46.5 mg/L for a 48-h exposure and the old fruits of *P. crispum* had the most significant molluscicidal activity (Sousa et al. 2017).

The main problems with these natural molluscicides are the lack of field experiments and the lack of studies on non-target flora and fauna. In addition, the use of these products in the field would require a lot of labour and relatively skilled workers in many cases. Treatments should be applied regularly and arable land may be required for growing these plants (Hammond et al. 1994).

15.3.4 Long-Term Effects

They are known only for chemically defined molluscicides. Their effect on lymnaeids varies with the date of treatment (Worden et al. 1963). If applied in the spring, prior to the period of snail reproduction, the number of molluscs remains low and no growth in numbers like that seen in the control population is observed. If spreading is made in September, there is a drop in numbers, followed by their maintenance at low values. The density of snails can be modified depending on the nature of the habitat. Van den Bruel (1968) and Pécheur (1974) noted that the results obtained with the same molluscicide are not always overlapping from one *G. truncatula* habitat to another.

To be effective, chemical control must allow the complete elimination of the snail population. If a few individuals remain alive, this is sufficient to repopulate the habitat due to the high multiplying power of the species. It is therefore easy to imagine that the presence of shelters will allow snails to escape the action of treatment. Similarly, spreading over a small area often results in re-colonization of the habitat by snails coming from neighbouring untreated sites (Ximenes 1991).

15.4 Biological Control by Predation

15.4.1 Species of Predators

Several taxa of vertebrates and invertebrates are natural predators of freshwater molluscs, particularly Lymnaeidae. According to Dillon (2004), mammals, birds, turtles, salamanders, fish, crustaceans, insects, predatory molluscs, and leeches can feed on these snails. Several authors such as Pelseneer (1928), Mehl (1932), Michelson (1957), Taylor (1964), and Brown (1994) have reviewed the various groups of vertebrates and invertebrates involved in the predation of the snail hosts of *Schistosoma* spp. or *Fasciola* spp.

The predatory action of several groups of species resulted in a strong reduction in the number of lymnaeids. The malacophagous action of domestic ducks has been reported in several articles (Samson and Wilson 1973; Hull 2017, for example). If these birds are present in a meadow, the number of snails strongly decreased. According to SuSin (2001), a density of 5 to 10 ducks per ha in continuous grazing for a period of 1 to 2 months in irrigated rice significantly reduced the number of snails from 5/m^2 to less than 1 snail/m^2. However, not all palmipeds have the same ability to eat snails because it depends on their type of diet (Euzeby, 1971). Small waders, starlings, thrushes, and blackbirds incorporate *G. truncatula* in their diet but no figures have yet been reported to determine the amount of snails they can eat. According to Taylor (1964), the action of these birds also depends on the type of habitat in which *G. truncatula* lives because the predation of snails would be low or null in rush meadows.

The introduction of snail-eating fish like *Astatoreochromis alluaudi* (Cichlidae) into small dams in western Kenya and ponds in Cameroon has raised much hope. But the results were not satisfactory. According to Slootweg et al. (1993, 1994), no significant reduction in the number of bulinids was seen after the introduction of *A. alluaudi* in the water bodies of Cameroon. Other authors reported positive results in the elimination of snails when *Mylopharyngodon piceus* or *Trematocranus placodon* were used (Chiotha et al. 1991; Ben-Ami and Heller 2001), while the effects of native fish on snail reduction were limited in water bodies in the lower basin of the Senegal River (Arostegui et al. 2019). This variability in results may be explained by the fact that native or introduced fish species are only partially molluscivorous (Brown 1994).

The introduction of the crayfish *Procambarus clarkii* into irrigation canals in Egypt has resulted in a strong reduction and sometimes complete disappearance of snails (Ibrahim et al. 1995; Sleem and El-Hommossany 2008). Similar results were also obtained in Kenya (Mkoji et al. 1999). However, the introduction is not without danger because this crayfish also consumes aquatic plants (hence a partially molluscivorous diet) and undermines the earthen banks by digging burrows (Brown 1994). Another side effect could be observed due to the activity of many crayfish species as second intermediary hosts in the life cycle of *Paragonimus* spp.

The prawn *Macrobrachium vollenhovenii* is also capable of reducing the abundance and lifespan of snails in the Senegal River basin (Sokolow et al. 2015; Swartz et al. 2015) but this prawn is also partially molluscivorous (Jimoh et al. 2011). Predation by the water bug, *Sphaerodema urinator*, has been used to control host snails that transmit schistosomosis (Younes et al. 2017). According to Brown (1994), the only completely molluscivorous invertebrates are leeches of the Glossiphoniidae family and marsh flies of the Sciomyzidae family. Predation of *Glossiphonia complanata* was higher when leeches fed on juvenile *Stagnicola emarginata* (= *Ladislavella emarginata*).than on juveniles of *Physa gyrina* (Brönmark 1992). The Sciomyzidae had raised great hope to eliminate the intermediate hosts of *Fasciola gigantica* in the Hawaiian Islands in the 1950s. These flies had been released in these islands to eliminate the local host snail, but the results were not successful (Berg 1964). According to Knutson and Vala (2011), first, second, or third stage larvae often eat a limited part of the snail's body, i.e. the foot in most cases.

There are few aquatic predators of *G. truncatula* in central France. According to Rondelaud (1979), predation of adult dytiscids and their larvae was significantly greater than that of hirudinids and hemipterids. Sciomyzidae larvae were more common in small stagnant water collections such as pools and were not specific in the choice of their prey (Dreyfuss et al. 2002). On emerged areas, numerous terrestrial predators are present in the habitats of *G. truncatula* at the onset of summer drying. The larvae of *Lampyris noctiluca* (glowworm) are already known to be molluscivorous (Taylor 1964). The carabids *Anisodactylus binotatus*, *Platysma nigrita*, and *Poecilus cupreus* leave many empty and broken shells of *G. truncatula* at the end of their predation, while the shells remain intact in the case of other carabid species. This predation of prey occurred in the 2–3 weeks preceding the summer drying out of *G. truncatula* habitats (Rondelaud 1976). A land mollusc, *Zonitoides nitidus*, also consumes *G. truncatula* at the onset of summer drying (Rondelaud 1975). This ability of *Z. nitidus* was at the origin of biological control by predation that Rondelaud et al. (2006) developed for populations of *G. truncatula* living on acidic soils in central France.

15.4.2 Control of *Galba truncatula*

In central France, *Z. nitidus* is common in swampy meadows and along rivers on acidic soils. On sedimentary soils, it is only found on the banks of streams and rivers (Didier 1986). Its peculiarity comes from its diet. From September to May, it feeds on plant fragments as shown by the analysis of its intestinal content under a stereomicroscope (Rondelaud et al. 2006). From the month of June, the adult predates at the expense of other pulmonates and bivalves which live in the same habitat as it. Predation first affects *G. truncatula* and *Omphiscola glabra*, then *Succinea* sp. When the first cracks form in the drying soil, the *Pisidium* population pays its tribute. Finally, towards the end of July, *Z. nitidus* does not hesitate to eat its own congeners, either alive or dead (adults die during this period). According to Didier and Rondelaud (1989a, b), *Z. nitidus* needs this supply of animal proteins in order to ensure its reproduction (it lays in July). In addition, this snail has chitinase, which allows it to partially destroy areas in the shell of its victims.

Three techniques of snail control were used. The first consists of mowing the hygrophilous vegetation at the onset of summer drying and depositing it on the habitats of *G. truncatula*, which maintains humidity and attracts local predators. If necessary, *Z. nitidus* is introduced at a rate of 10 per m^2 of watercress bed and 20 per m^2 in other types of habitats. The second technique is not to mow the natural environment but a higher number of predators are introduced ($60/m^2$). The third technique consists of two steps: (i) a solution of cupric chloride (8 L to 0.1 mg/L) is poured into each habitat in April during the spring mating of *G. truncatula* and (ii) 20 *Z. nitidus*/m^2 are introduced or not at the beginning of summer drying (Ximenes 1991). These three techniques have been applied in 122 valley-bottom meadows, 59 hillside rush beds around temporary or permanent sources, 11 trampled areas, 37 river or pond banks, and 134 wild watercress beds (Rondelaud et al. 2006). Table 15.4 gives the survival rates of *G. truncatula* in September in several habitat types after 1, 2, 3, and 4 year(s) of control with *Z. nitidus*. Mowing of the hygrophilous vegetation in June and its deposition on the lymnaeid habitats resulted in the removal of *G. truncatula* after 2 years of control in the habitats located in swampy meadows and around temporary sources. In the case of wild watercress beds, 3 years

Table 15.4 Percentage of alive *Galba truncatula* counted in September in several types of habitats on acidic soils after a control of 1, 2, 3, and 4 years with *Zonitoides nitidus*. From Rondelaud et al. (2006)

Habitat type	Number of habitats	Percentage of alive *Galba truncatula* counted in September after a control of 1 year	2 years	3 years	4 years
Swampy meadows					
Controls	9	94.3	90.2	92.7	96.5
Treated with mowing	78	11.5	0.1	0	0
Treated without mowing	14	91.2	87.3	95.3	88.7
Rush beds with temporary source					
Controls	3	78.5	69.6	81.3	73.6
Treated with mowing	20	5.1	0.6	0	0
Treated without mowing	4	81.7	63.2	56.5	43.7
Rush beds with permanent source					
Treated with mowing	1	29.1	22.5	17.5	9.3
Treated without mowing	2	61.3	47.6	32.1	15.4
Wild watercress beds					
Treated with mowing	15	39.6	7.2	2.3	0

were necessary. In contrast, around the permanent sources, there was only a gradual decrease in the number of *G. truncatula* during the 4 years of control. In treated habitats without mowing, the populations of *G. truncatula* remained stable as controls (swampy meadows) or decreased slightly over time (both types of rush beds). A combination of predatory molluscs: *Z. nitidus* + *Oxychilus draparnaudi*, allowed the elimination of *G. truncatula* in a single year (rushes with temporary source) or 2 years (swampy meadows, wild watercress beds). Finally, in the 117 watercress beds treated with cupric chloride and *Z. nitidus*, no lymnaeid was observed in September after a single year of control, while in the 17 others, 2 years were required to eliminate *G. truncatula* (Rondelaud et al. 2006).

With the exception of a few owners who have introduced *Z. nitidus* into their watercress beds each year, the use of this zonitid to control *G. truncatula* has not become widespread in cattle and sheep farms in the French department of Haute Vienne. The reasons for this situation are probably the complexity of the techniques used to apply this control in the field by non-specialists and the difficulty for selecting the date of this control at the end of June due to the frequent rains (Rondelaud et al. 2006).

15.4.3 Long-Term Effects of this Control

Table 15.5 indicates, for each habitat type, the number of sites that were recolonized by *G. truncatula* in the years following the last application of biological control. In habitats located in swampy meadows and on river banks, most of them were re-invaded in the third or fourth post-control year. For areas around temporary sources and watercress beds, the re-colonization of most habitats occurred 6 or 7 years after the last application of control. In both cases, several habitats have not been recolonized by *G. truncatula* at the end of investigations (2005) and this fact can be easily explained by the location of these habitats at the peripheral end of the surface drainage networks.

The predator *Z. nitidus* showed numerical variations after its introduction into the habitats of *G. truncatula*. In swampy meadows, predators decreased in numbers for 2 years and then showed constant values. In the other three types of habitats, the predator populations disappeared during the first year (areas around sources), second year (river banks), or third year (watercress beds) after the last application of control. More surprising are the numerical variations that *O. glabra* presents in the treated habitats (*G. truncatula* and *O. glabra* often live in the same swampy meadows but each species occupies its own habitat: Vareille-Morel et al. 1999). Indeed, the number of snails per population and the area of habitat increased from the last application of biological control until the third year after. Subsequently, there was a gradual decrease in numbers while the habitat area did not change over time. As the extension of *O. glabra* occurs after the disappearance of *G. truncatula*, this finding suggests that there is probably competition for food between the two species. An

Table 15.5 Number of habitats recolonized by *Galba truncatula* in the years following the last application of biological control. n, total number of habitats studied. From Rondelaud et al. (2006)

Number of years after the last year of biological control (1978)	Number of *Galba truncatula* habitats recolonized			
	Swampy meadows ($n = 83$)	Rush beds with temporary source ($n = 25$)	Pond and river banks ($n = 21$)	Watercress beds ($n = 26$)
3	21	0	3	1
4	33	2	10	2
5	11	5	2	3
6	4	8	1	2
7	3	4	2	4
8	0	1	2	7
9	0	1	1	3
10	5	1	0	0
12	2	0	0	0
15	2	0	0	0
No re-colonization in 2005	2	3	0	4

argument supporting this interpretation is provided by Dreyfuss et al. (2006). According to these authors, when adults of both species are raised together in the laboratory, there is competition between the two lymnaeids, which causes the death of many *G. truncatula*.

As most field experiments were carried out on 12 farms over a 25-year period, it was interesting to determine the impact of this control on the prevalence of natural *F. hepatica* infection in cattle. Blood samples were therefore taken during the 4 years of control (1975–1978) on the ruminants from these herds and analysed serologically for antibodies against *F. hepatica*. In the first year, the frequency of infected animals was 11.7% (out of a total of 1514 cattle). In subsequent years, the prevalence fell to 0.26%, 0.15%, and 0.11%, respectively (Rondelaud et al. 2006).

15.5 Biological Control by Competition

Biological control of the host snails of schistosomes and other digeneans has been considered in recent decades as an alternative to molluscicides. Several groups of organisms have been proposed to control the host snails, but very few have proven their efficacy in the field. Competitive snails can be considered the most effective biological control agents and numerous promising laboratory studies and field experiments have been carried out, mainly in the Caribbean. Two species of Caenogastropoda snails belonging to the families of Ampullariidae (*Marisa cornuarietis*) and Thiaridae (*Melanoides tuberculata*) succeeded in eliminating or reducing the populations of schistosome-transmitting snails, especially *Biomphalaria glabrata* in several different habitats in Saint Lucia, Martinique, and Guadeloupe (Pointier et al. 2011). Three other species belonging to Physidae

(*Aplexa hypnorum*, *Physella acuta*) and Tateidae (*Potamopyrgus antipodarum*) can also be cited due to their competition with *G. truncatula*.

Unlike the intermediate hosts of schistosomes, there is much less work on the effects of these competitors on Lymnaeidae. Under laboratory conditions, *M. cornuarietis* can compete with *Lymnaea caillaudi* (= *Radix natalensis*), a planorbid and a bulinid by feeding on aquatic plants in the environment. The competitor is also capable of predatory activity by consuming egg masses, newborns, juveniles, and adults of the three snail species (Demian and Lutfy 1965, 1966). The introduction of *M. cornuarietis* in 1972 into an ornamental pond resulted in the disappearance of *P. columella* and a planorbid (Peebles et al. 1972). This result was confirmed by Nguma et al. (1982) who reported the disappearance of *R. natalensis* and other pulmonates in a small man-made dam 12 months after the introduction of *M. cornuarietis*. No pulmonate was found in subsequent snail samples collected by the authors over the next 2.5 years (Nguma et al. 1982). The omnivorous diet of *M. cornuarietis*, together with its large body mass, high reproductive efficiency and often high density, means that these snails can rapidly alter the structure of the macrophytic community, with consequent disturbances in nutritional balance, turbidity, and trophic structure of water bodies. The use of *M. cornuarietis* for such purposes is no longer encouraged at this time due to the adverse environmental effects of the species (Barker 2016). The role of *M. tuberculata* as a competitor of pulmonate gastropods is still controversial. In several Caribbean islands, this species has significantly reduced the populations of the planorbid *Biomphalaria glabrata* (Perera et al. 1990; Pointier and Augustin 2000; Pointier 2001) while the results are negative in irrigation canals in Sudan (Madsen et al. 1988). In Kenya, *M. tuberculata* lives in equilibrium with *R. natalensis* and other freshwater pulmonates (Mkoji et al. 1992). However, the presence of *M. tuberculata* in Tunisia causes a decrease in the number of annual generations in *G. truncatula* (2 instead of 4 in the control habitat) so that the authors conclude that there is competition between the thiarid and the lymnaeid (Ghouaidia and Hammami 2013).

The competition of *P. acuta* has already been reported by Michelson (1957) and Graber and Euzéby (1975). This species is able to exclude *G. truncatula* from its habitat when both species live in the same environment (Rondelaud 1978). Another species, *A. hypnorum*, is also able to compete with lymnaeids. If it is introduced into a ditch colonized by *G. truncatula*, its presence ultimately results in a displacement of the *G. truncatula* population upstream of the water collection and a progressive decrease in the size of the lymnaeid population over the 7 years of the study (Rondelaud et al. 2016). The competition of these two physids was only observed in small habitats of *G. truncatula* with stagnant or low-flowing waters and it is questionable whether this competition between physids and native species exists in other types of habitats. Although *P. acuta* was considered a competitor in Egypt (El-Hassan 1974), populations of this species often co-existed with planorbids in other African countries (Brown 1994). In addition, this competition with *G. truncatula* is not exclusive and has been observed with other species of freshwater snails. Several authors have reported that the presence of *P. acuta* in several countries resulted in competition with native snail species such as *Physastra variabilis* (syn. *Glyptophysa variabilis*) in New Zealand (Winterbourn 1980), *Bulinus tropicus* in South Africa (Brackenbury and Appleton 1993), or *Glyptophysa gibbosa* in South Australia (Zukowski and Walker 2009). In the USA, *A. hypnorum* also competes for food and habitat type with *Stagnicola elodes* (= *Ladislavella elodes*) and *Physa gyrina* (Brown 1982).

The invasive snail *Potamopyrgus antipodarum* is known to be a potential competitor with native snails such as two hydrobiid species in a freshwater spring in the USA (Richards et al. 2001) or another hydrobiid in Australia (Sardiña et al. 2015). In the presence of another competitor such as *P. acuta*, the growth and reproductive performance of both snail species in New Zealand are influenced more by the density of conspecifics than the presence and density of the other species (Cope and Winterbourn 2004). However, this competitor also acts on Lymnaeidae as reported by Rondelaud (1977a) for *G. truncatula* in several temporary habitats on French acidic and sedimentary soils. When there is a high concentration of *P. antipodarum* in habitats, *G. truncatula* cannot reproduce and there is a gradual decrease in the initial number of lymnaeid snails. If *G. truncatula* lays its eggs, the number of young individuals issuing from these masses is low. In addition, the presence of the competitor in the submerged zone forced the lymnaeid to emerge earlier in the season and daily migrations showed a short immersion time of the snails. The use of large groups of *Potamopyrgus* allows therefore to obtain an effective competition against *G. truncatula* during the immersion of habitats.

15.6 Biological Control with Pathogens

Like other animal groups, Lymnaeidae are sometimes infected by parasites, bacteria, and/or viruses. They can also host commensal species. As the list of these pathogens includes many species, the information presented in the paragraphs below mainly concerns the amphibious snails of the genus *Galba*.

Within Oligochaeta, *Chaetogaster limnaei* has an unusual parasitic relationship with freshwater pulmonate snails. Two subspecies are known: *C. limnaei limnaei* is an ectosymbiont and is found inside the mantle cavity of the snail, while *C. limnaei vaghini* is a parasite and lives in the snail's kidney. This oligochaete infects most species belonging to the Lymnaeidae family such as *G. truncatula* (Muñiz-Pareja and Iturbe-Espinoza 2018), *Lymnaea stagnalis* (Buse 1971), *P. columella* (Martins and Alves 2010), or *R. natalensis* (Fashuyi and Williams 1977; Ibrahim 2007). Most authors consider *C. l. limnaei* as an ectosymbiont because the animal consumes various small organisms (algae, rotifers, miracidia, trematode cercariae) that it manages to suck on its host (Michelson 1964; Fried et al. 2008). This efficacy of *C. l. limnaei* led Muñiz-Pareja and Iturbe-Espinoza (2018) to consider this oligochaete as a controller of *Fasciola hepatica* during experimental infections of snails. In France, *C. l. limnaei* was found in six populations of *G. truncatula* on acidic soils (out of a total of 317 populations studied). The infection of three *C. l. limnaei*-carrying populations with *F. hepatica* resulted in the death of snails between day 14 and day 21[t] post-exposure at 20 °C (Dreyfuss et al. 2015). Under these conditions, one can wonder whether this oligochaete would not be an ectoparasite which weakens the snail by its presence and causes the death of *G. truncatula* when the snail is subjected to an experimental infection with *F. hepatica.*

Terrestrial nematodes are capable of infecting molluscs of the Lymnaeidae family and can be used to control intermediate hosts of trematodes. Morley (2010), in his review, identified 18 species of nematodes which use lymnaeids as auxiliary intermediate hosts. One of these species, *Phasmarhabditis hermaphrodita*, is commercially available as a biological control agent for slugs and land snails. Among nematodes listed by Morley (2010), *Angiostrongylus cantonensis*, *Muellerius capillaris*, *Cystocaulus ocreatus*, *Elaphostrongylus rangiferi*, *Syngamus trachea*, and *Angiostrongylus dujardini* have the widest range of lymnaeids as intermediate hosts: 11, 8, 4, 4, 4, and 3 species, respectively. *Galba truncatula* may be infected with *C. ocreatus*, *E. rangiferi*, *M. capillaris*, *Neostrongylus linearis*, *S. trachea*, and *Varestrongylus sagittatus* (Morley 2010). The first stage (L1) larvae of these nematodes enter the snail, transform into second stage larvae (L2) and then into third stage larvae (L3) which exit from the snail before being ingested by the definitive host. But all these species of lymnaeids do not support the larval development of these nematodes from L1 to L3. Several *N. linearis* reached the L3 stage in *G. truncatula* subjected to monospecific infections while *M. capillaris* did not exceed the L2 stage (Hourdin et al. 1991). On the other hand, if *G. truncatula* is subjected to co-infections with *F. hepatica* and *M. capillaris*, or vice versa, L3 larvae of *M. capillaris* have been observed in snails with an abortive *F. hepatica* infection and the formation of these L3 in this case would be due to a phenomenon of facilitation (Hourdin et al. 1990).

The control of *Fasciola*-infected snails by another digenean has been the subject of few investigations. *Echinostoma audyi* has been proposed by Hoa et al. (1970) to control *F. gigantica* because the rediae of the former digenean predate on the sporocysts of *F. gigantica* within the snail. This control method was applied in the field by Estuningsih (1991, 1998) and Suhardono et al. (2006). According to the latter authors, the introduction of duck faeces containing *Echinostoma revolutum* eggs and cattle faeces with *F. gigantica* eggs into rice fields resulted in a decrease in the number of *Radix rubiginosa* infected with *F. gigantica* compared to control rice fields. According to Lim and Heyneman (1972), locally available parasites could be used to control the number of host snails in the field. For these methods to work, these authors indicated that the definitive host of the parasite intended for control must be present in sufficient numbers and be easily maintained and infected in order to produce enough parasites to overwhelm the snails. For such a control zone, a specific antagonist system must be put in place because a single system is not applicable in all cases (Lim and Heyneman 1972; Torgerson and Claxton 1999).

No bacteria capable of killing molluscs have yet been reported in Lymnaeidae. On the other hand, several bacteria have been tested against the vector snails of schistosomes. *Bacillus thuringiensis kurstaki* has negative effects on snail populations through molluscicidal activity and by preventing the hatching of *Biomphalaria alexandrina* eggs (Gamalat et al. 2011). *Brevibacillus laterosporus* has also been reported as a pathogen against juveniles of *Biomphalaria glabrata* (de Oliveira et al. 2004). A preliminary study suggested the potential pathogenicity of *Bacillus brevis* towards *Biomphalaria pfeifferi* and *Bulinus truncatus* (Singer et al. 1994), but this has never been tested in the field. Finally, a new bacterium, *Paenibacillus glabratella*, affected both adult snails and embryonic stages of *B. glabrata*, and has caused significant mortality. This bacterium invades most snail tissues and proliferates, causing high lethality, and can be transmitted both vertically and horizontally to other snails causing their death within 30 days. This latter bacterium may be promising as a biological control agent to limit the transmission of schistosomosis and other parasitoses in the field (Duval et al. 2015). To these pathogens, we must add the discovery of an iridovirus in a French population of *G. truncatula* after the sudden occurrence of numerous deaths of snails and the examination of their corpses by transmission electron microscopy (Barthe et al. 1984). Field investigations were then carried out by Rondelaud and Barthe (1992) in 11 French populations of *G. truncatula* to determine the prevalence of this viral infection. Prevalence ranged from 3% to 87% and was rather constant in eight snail populations. The three others showed variations in frequency over time (from 23% to 69%). *Iridovirus sp.* is a common infection in snails living in central France, suggesting an equilibrium between the virus and its host (Rondelaud and Barthe 1992). Further studies are still needed to determine whether this virus affects other lymnaeid species such as *O. glabra* that live in the same grasslands as *G. truncatula* and whether it can be proposed as a biological control agent.

Several species of fungi have been used to study their activity on the intermediate hosts of *Schistosoma* spp. Contact of *B. alexandrina* with sublethal concentrations of an extract of *Aspergillus fumigatus* for 24 h before, during, or after exposure of snails to *Schistosoma mansoni* resulted in a significant reduction in the rate of infection and a decrease in the total number of cercariae shed by surviving snails (Gamalat et al. 2013). The same type of contact between *B. alexandrina* and sublethal concentrations of a filtrate of *Aspergillus terreus* or *Penicillium janthinellum* caused adverse effects on the gonad of snails with degeneration or deformation of the ova and spermatozoa (Saad et al. 2014). Development of *B. glabrata* eggs was significantly slowed down when hyphae or conidia of *Beauveria bassiana* and *Metarhizium anisopliae* were present (Duarte et al. 2015). Conversely, no report on the effects of these fungi on Lymnaeidae is yet available in the literature.

15.7 Integrated Liver Fluke Control Strategy

No single control strategy such as drainage or application of a molluscicide can effectively reduce the transmission of fasciolosis. The objective of this strategy is therefore to use the various prophylactic means available to eradicate fasciolosis in some farms and limit the frequency of the disease in most other farms (Mage and Rondelaud 1983). An integrated parasite control programme is therefore needed to control this parasitosis. Several studies were conducted to test a strategic integrated control programme to reduce the prevalence of fasciolosis in central France (Mage et al. 1989, 1995; Vignoles et al. 2016) and in Peru (Claxton et al. 1998; Raunelli and Gonzalez 2009).

The first step in this strategy is to determine the presence of the parasite in cattle by coproscopy or serology. The topography of the land, bearing capacity of the soil, runoff circulation, and pasture management must also be analysed. The snail habitats are isolated in a second step with a fence and deworming of the definitive host is carried out using an anthelmintic. The third step is the choice of the technique to eliminate the populations of snails. The isolation of habitats must be carried out in the absence of any other possibility. Very swampy areas must be transformed into ponds. Surface drainage must be carried out when the areas to be treated are large. Small areas such as rush beds should be treated with molluscicide or predatory snails.

A first trial took place in a cattle farm in the department of Corrèze (central France). In 1986, the herd consisted of 30 bovines and the coproscopies for *F. hepatica* were all positive. *G. truncatula* was only found in one area where cattle waited before entering stalls. The various measures in the concept were applied to the herd and the waiting area. The *G. truncatula* disappeared at the end of 1988 and the coproscopies performed after the cleaning of the waiting area were all negative (Mage et al. 1989). Further trials were carried out in the summer meadows in mountains. Three dairy farms in the French department of Cantal have simultaneously applied fasciolicide to 164 cattle and eliminated the sources of infection present on these pastures. At the end of the second year, the percentage of cattle naturally infected with *F. hepatica* was only 11.1% at the time of grazing, whereas it was 57.4% in the 10 farms which used only fasciolicide alone as therapeutic prevention (Mage et al. 1995). Another trial was performed on two farms between 2011 and 2014 by isolating the habitats of *O. glabra* from the rest of the pastures, as the area of these habitats on acidic soils is generally small. Isolation of *O. glabra* in both farms resulted in a decrease in the prevalence of snail infections in 2012 and the disappearance of *F. hepatica* larval forms in snails in 2013 and 2014. In cattle, the prevalence decreased gradually to its negativation in 2014. These results show that it is possible to safeguard the populations of the host snail during an integrated control in livestock while interrupting the development cycle of the parasite (Vignoles et al. 2016).

In Peru, a first control programme involving two doses of triclabendazole together with the use of niclosamide was evaluated for a single year. This double treatment did not significantly reduce the overall parasite burden, although there was a significant reduction in the number of intermediate host snails after the application of molluscicide (Claxton et al. 1998). Another study carried out by Raunelli and Gonzalez (2009) in the same country over a 2-year period involved three fasciolicide treatments per year, strategically timed according to the epidemiological cycle of the disease, and environmental control activities aimed at irrigation ditches and paddock drainage. The application of this integrated control led to a decrease in the prevalence of *F. hepatica* in cattle (from 63% to 14%), a 38% increase in the live weight of cattle, and a 75% increase in milk production over the 2 years of the study.

References

Abreu Guirado OA, Pérez MP, Diéguez Fernández L et al (2019) Effecto molusquicida del fruto de *Sapindus saponaria* sobre *Galba cubensis*, hospedero intermediario de fasciolosis en Cuba. Rev Prod Anim 31:47–54

Agrahari P, Singh DK (2012) Seasonal variation in abiotic factors and ferulic acid toxicity in snail-attractant pellets against the intermediate host snail *Lymnaea acuminata*. Zoonoses Public Health 60:478–486

Ahmed AH, Ramzy RM (1997) Laboratory assessment of the molluscicidal and cercaricidal activities of the Egyptian weed, *Solanum nigrum* L. Ann Trop Med Parasitol 91:931–937

Alda P, Lounnas M, Vázquez AA et al (2021) Systematics and geographical distribution of *Galba* species, a group of cryptic and worldwide freshwater snails. Mol Phylogenet Evol 157:107035. https://doi.org/10.1016/j.ympev.2020.107035

Andrews P, Thyssen J, Lorke D (1983) The biology and toxicology of molluscicides, Bayluscide. Pharmacol Ther 19:245–295

Archibald RG (1933) The use of the fruit of the tree *Balanites aegyptiaca* in the control of schistosomiasis in The Sudan. Trans R Soc Trop Med Hyg 27:207–210

Arostegui MC, Wood CL, Jones IJ et al (2019) Potential biological control of schistosomiasis by fishes in the lower Senegal River basin. Am J Trop Med Hyg 100:117–126

Barker GM (2016) *Marisa cornuarietis* (giant ramshorn). https://www.cabi.org/isc/datasheet/32526e. Accessed 22 October 2020

Barkia H, Barkia A, Nhammi H et al (2011) La schistosomiase au Maroc: de sa découverte à l'après-élimination. East Med Health J 17:250–256

Barthe D, Rondelaud D, Faucher Y et al (1984) Infection virale chez le Mollusque Pulmoné *Lymnaea truncatula* Müller. C R Acad Sci Ser III Sci Vie 298:513–515

Belkacemi M, Jana M (2006) Curage et traitement molluscicide pour la lutte contre la schistosomiase. East Med Health J 12:129–136

Ben-Ami F, Heller J (2001) Biological control of aquatic pest snails by the black carp *Mylopharyngodon piceus*. Biol Control 22:131–136

Berg CO (1964) Snail control in trematode diseases: the possible value of sciomyzid larvae, snail-killing Diptera. Adv Parasitol 2:259–309

Blann KL, Anderson JL, Sands GR et al (2009) Effects of agricultural drainage on aquatic ecosystems: a review. Crit Rev Environ Sci Technol 39:909–1001

Boelee E, Laamrani H (2004) Environmental control of schistosomiasis through community participation in a Moroccan oasis. Trop Med Int Health 9:997–1004

Boray JC (1978) The potential impact of exotic *Lymnaea* spp. on fascioliasis in Australasia. Vet Parasitol 4:127–141

Brackenbury TD, Appleton CC (1993) Recolonization of the Umsindusi River, Natal, South Africa, by the invasive gastropod, *Physa acuta* (Basommatophora, Physidae). J Med Appl Malacol 5: 39–44

Brönmark C (1992) Leech predation on juvenile freshwater snails: effects of size, species and substrate. Oecologia 91:526–529

Brouwer C, Goffeau A, Heibloem M (1985) Irrigation water management: training manual no. 1 - introduction to irrigation. FAO, Roma, Italy

Brown KM (1982) Resource overlap and competition in pond snails: an experimental analysis. Ecology 63:412–422

Brown DS (1994) Freshwater snails of Africa and their medical importance. Taylor & Francis, London. https://doi.org/10.1201/9781482295184

Buse AB (1971) Population dynamics of *Chaetogaster limnaei vaghini* Gruffydd (Oligochaeta) in a field population of *Lymnaea stagnalis* L. Oikos 22:50–55

Bustinduy AL, King CH (2014) Schistosomiasis. In: Farrar J, Hotex P, Junghanss T et al (eds) Manson's tropical diseases. Elsevier, New York, pp 698–725

Cao ZG, Wang TP, Zhang SQ (2012) Experimental study on the resistance of *Oncomelania* snails to niclosamide. J Pathog Biol 7:352–353
Chaturvedi D, Soni N, Singh VK (2021) Ethnobotanical molluscicides. Eur J Biol Res 11:417–433
Chauhan S, Singh A (2009) Molluscicidal potential of *Lantana indica* and *Alstonia scholaris* plants against freshwater snail *Lymnaea acuminata*. Internet J Toxicol 7:2. https://doi.org/10.5580/18cb
Chawech R, Njeh F, Hamed N et al (2017) A study of the molluscicidal and larvicidal activities of *Citrullus colocynthis* (L.) leaf extract and its main cucurbitacins against the mollusc *Galba truncatula*, intermediate host of *Fasciola hepatica*. Pest Manag Sci 73:1473–1477
Chen C (2003) Molluscicides and their application in China (in Chinese). Chin J Schistosomiasis Control 2003:321–322
Chen MG, Mott KE (1990) Progress in assessment of morbidity due to *Fasciola hepatica* infection: a review of recent literature. Trop Dis Bull 87:R1–R38
Chiotha SS, McKaye HR, Stauffer Jr (1991) Use of indigenous fishes to control schistosome snail vectors in Malaŵi, Africa. Biol Control 1:316–319
Claxton JR, Zambrano H, Ortiz P (1998) Strategic control of fasciolosis in the inter-Andean valley of Cajamarca, Peru. Vet Rec 143:42–45
Coelho PMZ, Caldeira RL (2016) Critical analysis of molluscicide application in schistosomiasis control programs in Brazil. Infect Dis Poverty 5:57. https://doi.org/10.1186/s40249-016-0153-6
Cope NJ, Winterbourn MJ (2004) Competitive interactions between two successful molluscan invaders of freshwaters: an experimental study. Aquat Ecol 38:83–91
Cruz-Reyes A, Chavarin C, Campos Arias MP et al (1989) Actividad molusquicida del Piquerol A aislado de Piqueria trinervia (compositae) sobre ocho especies de caracoles pulmonados. Mem Inst Oswaldo Cruz 84:35–40
Dai JR, Wang W, Liang YS et al (2008) A novel molluscicidal formulation of niclosamide. Parasitol Res 103:405–412
Dai JR, Li YZ, Wang W et al (2015) Resistance to niclosamide in *Oncomelania hupensis*, the intermediate host of *Schistosoma japonicum*: should we be worried? Parasitology 142:332–340
Dawson VK (2003) Environmental fate and effects of the lampricide Bayluscide: a review. J Great Lakes Res 29:475–492
de Oliveira EJ, Rabinovitch L, Monnerat R et al (2004) Molecular characterization of *Brevibacillus laterosporus* and its potential use in biological control. Appl Environ Microbiol 70:6657–6664
Demian ES, Lutfy RG (1965) Predatory activity of *Marisa cornuarietis* against *Biomphalaria alexandrina* under laboratory conditions. Ann Trop Med Parasitol 59:337–339
Demian ES, Lutfy RG (1966) Factors affecting the predation of *Marisa cornuarietis* on *Bulinus* (*B.*) *truncatus*, *Biomphalaria alexandrina* and *Lymnaea caillaudi*. Oikos 17:212–230
Didier B (1986) Contribution à l'étude écologique et écophysiologique d'un mollusque prédateur, *Zonitoides nitidus* Müller. PhD Thesis, University of Limoges
Didier B, Rondelaud D (1989a) Premières données sur le régime alimentaire de *Zonitoides nitidus* Müller (Mollusque, Gastéropode, Pulmoné). Bull Soc Hist Nat Toulouse 125:55–60
Didier B, Rondelaud D (1989b) Les caractéristiques des proies consommées par le mollusque *Zonitoides nitidus* Müller et leur dynamique en juin, juillet et août. Bull Soc Hist Nat Toulouse 125:111–117
Dillon RT Jr (2004) The ecology of freshwater molluscs (Cambridge studies on ecology). Cambridge University Press, Cambridge, pp. 509
Dreyfuss G, Vignoles P, Rondelaud D (2002) Les larves de Diptères Sciomyzidae dans le département de la Haute-Vienne. Leur impact sur le système *Galba truncatula-Fasciola hepatica*. Ann Sci Limousin 13:39–46
Dreyfuss G, Vignoles P, Mekroud A et al (2006) The presence of uninfected *Omphiscola glabra* in a breeding of infected *Galba truncatula* enhanced the characteristics of snail infections with *Fasciola hepatica*. Parasitol Res 99:197–199
Dreyfuss G, Vignoles P, Rondelaud D et al (2015) The mud snail (*Galba truncatula*). Ecology, parasitism and control. Lambert Academic Publishing, Saarbrücken

Duarte GF, Rodrigues J, Fernandes EKK et al (2015) New insights into the amphibious life of *Biomphalaria glabrata* and susceptibility of its egg masses to fungal infection. J Invertebr Pathol 125:31–36

Duval D, Galinier R, Mouahid G et al (2015) A novel bacterial pathogen of *Biomphalaria glabrata*: a potential weapon for schistosomiasis control? PLoS Negl Trop Dis 9:e0003489. https://doi.org/10.1371/journal.pntd.0003489

El-Hassan AA (1974) *Helisoma tenuis* and *Physa acuta* snails as biological means of control against *Bulinus truncatus* and *Biomphalaria alexandrina*. In: Proceedings of the Third International Congress of Parasitology, München, 25–31 August 1974 3:1597–1598

Estuningsih SE (1991) Studies on trematodes infecting *Lymnaea rubiginosa* in West Java. MSc Thesis, James Cook University of North Queensland

Estuningsih SE (1998) Studi tentang penggunaan larva cacing *Echinostoma revolutum* sebagai agen kontrol biologis cacing *Fasciola gigantica*. Jurnal Ilmu Ternak dan Veteriner 3:129–134

Euzeby J (1971) Les maladies vermineuses et leurs incidences sur la pathologie humaine. Volume II: Maladies dues aux Plathelminthes. Part 2: Trématodes. Book 1: Généralités. Distomatoses hépato-biliaires. Vigot frères, Paris

Fashuyi SA, Williams MO (1977) The role of *Chaetogaster limnaei* in the dynamics of trematode transmission in natural populations of freshwater snails. Z Parasitenkd 54:55–60

Fried B, Peoples RC, Saxton TM et al (2008) The association of *Zygocotyle lunata* and *Echinostoma trivolvis* with *Chaetogaster limnaei*, an ectosymbiont of *Helisoma trivolvis*. J Parasitol 94:553–554

Gamalat YO, Ahmed MM, Ahmed AK et al (2011) Biological studies on *Biomphalaria alexandrina* snails treated with *Furcraea selloa marginata* plant (family: Agavaceae) and *Bacillus thuringiensis kurstaki* (Dipel-2x). J Appl Pharm Sci 1:47–55

Gamalat YO, Ahmed MM, Ahmed AK et al (2013) Biological and biochemical impacts of the fungal extract of *Aspergillus fumigatus* on *Biomphalaria alexandrina* snails infected with *Schistosoma mansoni*. Res Rev Biosci 7:473–484

Gayral P, Cavier R (1977) Actualité et perspectives d'avenir des molluscicides. Actual Chim Thér 5:179–209

Ghouaidia N, Hammami H (2013) Interaction between the intermediate host of fascioliasis in Tunisia, *Galba truncatula* and a possible competitor, *Melanoides tuberculata* (Muller, 1774): a field study. Afr J Ecol 52:328–333

Graber M, Euzéby J (1975) Lutte biologique contre les mollusques vecteurs de trématodoses humaines et animales. Etude de l'action compétitive de *Physa acuta* Draparnaud à l'égard de *Biomphalaria glabrata* Say. Bull Soc Sci Vet Med Comp 5:321–324

Hamed N, Njeh F, Damak M et al (2015) Molluscicidal and larvicidal activities of *Atriplex inflata* aerial parts against the mollusk *Galba truncatula*, intermediate host of *Fasciola hepatica*. Rev Inst Med Trop São Paulo 57:473–479

Hammami H, Mezghani-Jarraya R, Damak M et al (2011) Molluscicidal activity of various solvent extracts from *Solanum nigrum* var. *villosum* L. aerial parts against *Galba truncatula*. Parasite 18:63–70

Hammond JA, Fielding D, Nuru H (1994) *Eucalyptus*: a sustainable self-delivery molluscicide? Vet Res Commun 18:359–365

Haseeb MA, Fried B (1997) Modes of transmission of trematode infections and their control. In: Fried B, Graczyk TK (eds) Advances in trematode biology. CRC Press, Boca Raton, FL, pp 31–56

He JX, Wang H, Wu MY et al (2007) Observation on the molluscicidal effect of niclosamide ethanolamine salt dustable powder by dusting in fields. J Trop Dis Parasitol 5:153–154. (In Chinese)

He P, Wang W, Sanogo B et al (2017) Molluscicidal activity and mechanism of toxicity of a novel salicylanilide ester derivative against *Biomphalaria* species. Parasit Vectors 10:383. https://doi.org/10.1186/s13071-017-2313-3

Hoa KE, Lie KJ, Yong OYC (1970) Predation of sporocysts of *Fasciola gigantica* by rediae of *Echinostoma audyi*. Southeast Asian J Trop Med Public Health 1:429

Hourdin P, Rondelaud D, Cabaret J (1990) Interactions entre *Fasciola hepatica* L. et *Muellerius capillaris* Müller chez le mollusque *Lymnaea truncatula* Müller. Bull Soc Fr Parasitol 8:111–118

Hourdin P, Rondelaud D, Cabaret J (1991) Initial studies on infection of *Lymnaea truncatula* by *Muellerius capillaris* and by *Neostrongylus linearis* (Nematoda: Protostrongylidae). Parasitol Res 77:273–274

Hull J (2017) Ducks and the control of liver fluke in sheep. Vet Rec 181:270–271

Hurtrez-Boussès S, Meunier C, Durand P et al (2001) Dynamics of host–parasite interactions: the example of population biology of the liver fluke (*Fasciola hepatica*). Microbes Infect 3:841–849

Ibrahim MM (2007) Population dynamics of *Chaetogaster limnaei* (Oligochaeta: Naididae) in the field populations of freshwater snails and its implications as a potential regulator of trematode larvae community. Parasitol Res 101:25–33

Ibrahim AM, Khalil MT, Mubarak FM (1995) On the feeding behavior of the exotic crayfish *Procambarus clarkii* in Egypt and its prospects in the biocontrol of local vector snails. J Union Arab Biol 4:321–340

Ismail M, Sedek MN, El-Said KM et al (2019) Effect of water bodies on the efficacy of molluscicides against *Biomphalaria alexandrina* snails with emphasis to their integrated control measures. Egypt J Aquat Biol Fish 23:267–283

Jaiswal P, Singh DK (2009) Molluscicidal activity of nutmeg and mace (*Myristica fragrans* Houtt.) against the vector snail *Lymnaea acuminata*. Int J Geogr Inf Syst 15:177–186

Jimoh AA, Clarke EO, Whenu OO et al (2011) Food and feeding habits of the African river prawn (*Macrobrachium vollenhovenii*, Herklots, 1857) in Epe lagoon, Southwest Nigeria. Int J Fish Aquac 3:10–15

Kariuki HC, Madsen H, Ouma JH et al (2013) Long term study on the effect of mollusciciding with niclosamide in stream habitats on the transmission of schistosomiasis mansoni after community-based chemotherapy in Makueni District, Kenya. Parasites Vectors 6:107. https://doi.org/10.1186/1756-3305-6-107

Kashyap S, Khagta S, Guleria K et al (2019) Plants as molluscicides: a recent update. Int J Bot Stud 4:50–56

Kendall SB (1965) Relationships between the species of *Fasciola* and their molluscan hosts. Adv Parasitol 3:59–98

Khallaayoune K, Madsen H, Laamrani H (1998) Evaluation of three methods to control *Bulinus truncatus*, the intermediate host of *Schistosoma haematobium* in an irrigation scheme, Tessaout-Amont, Morocco. Acta Trop 69:61–63

Khanous L, Allah SK, Hamdi B et al (2017) Antifungal, molluscicidal and larvicidal assessment of anemonin and *Clematis flammula* L. extracts against mollusc *Galba truncatula*, intermediate host of *Fasciola hepatica* in Tunisia. Asian Pac J Trop Med 10:967–973

Kloos H, McCullough FS (1982) Plant molluscicides. J Med Plant Res 46:195–209

Kloos H, McCullough FS (1987) Plants with recognized molluscicidal activity. In: Mott KE (ed) Plant molluscicides. Wiley, New York, pp 45–108

Knutson LV, Vala JC (2011) Biology of snail-killing Sciomyzidae flies. Cambridge University Press, Cambridge

Laamrani H, Khallaayoune K, Boelee E et al (2001) Evaluation of environmental methods to control snails in an irrigation system in Central Morocco. Tropical Med Int Health 5:545–552

Lardans V, Dissous C (1998) Snail control strategies for reduction of schistosomiasis transmission. Parasitol Tod 14:413–417

Levêque C (1990) Impact de la lutte antivectorielle sur l'environnement aquatique. Ann Parasitol Hum Comp 65(Suppl. 1):119–124

Lim HK, Heyneman D (1972) Intramolluscan inter-trematode antagonism: a review of factors influencing the host-parasite system and its possible role in biological control. Adv Parasitol 10: 191–268

Madsen H, Daffalla AA, Karoum KO et al (1988) Distribution of freshwater snails in irrigation schemes in The Sudan. J Appl Ecol 25:854–866

Mage C, Rondelaud D (1983) Réflexions sur la prévention de la fasciolose en France. Dossiers de l'Elevage 5:25–28

Mage C, Reynal P, Rondelaud D et al (1989) Mise en pratique du contrôle de l'infestation par *Fasciola hepatica* chez des bovins limousins. Bulletin des Groupements Techniques Vétérinaires 347:5–10

Mage C, Mauran A, Moisset M (1995) Contrôle de l'infestation de la Grande Douve des bovins: une organisation et une application en fermes. Bulletin des Groupements Techniques Vétérinaires 506:11–17

Martins R, Alves R (2010) Occurrence of *Chaetogaster limnaei* K. von Baer, 1927 (Oligochaeta, Naididae) associated with Gastropoda mollusks in horticultural channels in southeastern Brazil. Braz J Biol 70:1055–1057

Mas-Coma S, Valero MA, Bargues MD (2009) *Fasciola*, lymnaeids and human fascioliasis, with a global overview on disease transmission, epidemiology, evolutionary genetics, molecular epidemiology and control. Adv Parasitol 69:41–146

McCullough FS, Gayral P, Duncan J et al (1980) Les molluscicides dans la lutte contre la schistosomiase. Bull World Health Org 59:17–26

Medina FM, Ritchie LS (1980) Molluscicidal activity of the Puerto Rican weed, *Solanum nodiflorum*, against snail hosts of *Fasciola hepatica*. Econ Bot 34:368–375

Medina FM, Woodbury R (1979) Terrestrial plants molluscicidal to lymnaeid hosts of fascioliasis hepatica in Puerto Rico. J Agric Univ Puerto Rico 63:366–376

Mehl S (1932) Die Lebensbedingungen der Leberegelschnecke (*Galba truncatula* Müller). Untersuchungen über Schale, Verbreitung, Lebensgeschichte, natürliche Feinde und Bekämpfungsmöglichkeiten. Arbeiten aus der Bayerischen Landesanstalt für Pflanzenbau und Pflanzenschutz 2:1–177

Mezghani-Jarraya R, Hammami H, Ayadi A et al (2009) Molluscicidal activity of *Hammada scoparia* (Pomel) Iljin leaf extracts and the principal alkaloids isolated from them against *Galba truncatula*. Mem Inst Oswaldo Cruz 104:1035–1038

Michelson E (1957) Study on the biological control of schistosome bearing snails. Predators and parasites of freshwater Mollusca: a review of the literature. Parasitology 47:413–426

Michelson E (1964) The protective action of *Chaetogaster limnaei* on snails exposed to *Schistosoma mansoni*. J Parasitol 50:441–444

Mkoji GM, Njunge K, Kimani G et al (1989) Molluscicidal activity of *Solanum aculeatum* (family: Solanaceae) berries against *Biomphalaria pfeifferi*, *Bulinus globosus* and *Lymnaea natalensis*. Trop Med Parasitol 40:119–120

Mkoji GM, Mungai BN, Hofkin BV et al (1992) Does the snail *Melanoides tuberculata* have a role in biological control of *Biomphalaria pfeifferi* and other medically important African pulmonates? Ann Trop Med Parasitol 86:201–204

Mkoji GM, Hofkin BV, Kuris AM et al (1999) Impact of the crayfish *Procambarus clarkii* on *Schistosoma haematobium* transmission in Kenya. Am J Trop Med Hyg 61:751–759

MolluscaBase (2022) MolluscaBase. http://www.molluscabase.org. Accessed 07 January 2022. https://doi.org/10.14284/448

Morley J (2010) Aquatic molluscs as auxiliary hosts for terrestrial nematode parasites: implications for pathogen transmission in a changing climate. Parasitology 137:1041–1056

Muñiz-Pareja FS, Iturbe-Espinoza PA (2018) Effectiveness of *Chaetogaster limnaei* as a controller of *Fasciola hepatica* in experimental infections of *Galba truncatula*. Trop Parasitol 8:88–93

Nguma JFM, McCullough FS, Masha E (1982) Elimination of *Biomphalaria pfeifferi*, *Bulinus tropicus* and *Lymnaea natalensis* by the ampullarid snail, *Marisa cornuarietis*, in a man-made dam in northern Tanzania. Acta Trop 39:85–90

Njeh F, Hamed N, Ayadi A et al (2015a) Molluscicidal and larvicidal activities of *Capparis spinosa* aerial parts against *Galba truncatula* intermediate host of *Fasciola hepatica*. Indian J Pharm Educ Res 49:242–249

Njeh F, Feki H, Koubaa I et al (2015b) Molluscicidal activity of *Solanum elaeagnifolium* seeds against *Galba truncatula* intermediate host of *Fasciola hepatica*: identification of β–solamarine. Pharm Biol 54:726–731

Pécheur M (1974) Lutte stratégique contre la distomatose. Comptes-Rendus de Recherches, Travaux du Centre de Recherches sur les Maladies Parasitaires des Animaux Domestiques. IRSIA, Bruxelles, 38:85–150

Peebles CR, Oliver-Gonzalez J, Ferguson FF (1972) Apparent adverse effect of *Marisa cornuarietis* upon *Lymnaea columella* and *Biomphalaria glabrata* in an ornamental pond in Puerto Rico. Hawaiian Entomol Soc 21(2):247–256

Pelseneer P (1928) Les parasites des mollusques et les mollusques parasites. Bull Soc Zool Fr 53: 158–189

Perera G, Yong M, Ferret JR et al (1990) Effectiveness of three biological control agents against intermediate hosts of snail-mediated parasites in Cuba. Malacol Rev 23:47–52

Pointier JP (2001) Invading freshwater snails and biological control in Martinique Island, French West Indies. Mem Inst Oswaldo Cruz 96(Suppl):63–74

Pointier JP, Augustin D (2000) Biological control and invading freshwater snails. A case study. C R Acad Sci Ser III Sci Vie 322:1093–1098

Pointier JP, David P, Jarne P (2011) The biological control of the snail hosts of schistosomes: the role of competitor snails and biological invasions. In: Toledo R, Fried B (eds) *Biomphalaria* snails and larval trematodes. Springer, New York, pp 215–238

Raunelli F, Gonzalez S (2009) Strategic control and prevalence of *Fasciola hepatica* in Cajamarca, Peru. A pilot study. Int J Appl Res Vet Med 7:145–152

Richards DC, Cazier LD, Lester GT (2001) Spatial distribution of three snail species, including the invader *Potamopyrgus antipodarum*, in a freshwater spring. West North Am Nat 61:375–380

Ritzema H (1994) Drainage principles and applications. International Institute for Land reclamation and Improvement, Wageningen, The Netherlands

Rondelaud D (1975) Contribution à l'étude expérimentale de la prédation de *Lymnaea* (*Galba*) *truncatula* Müller par *Zonitoides nitidus* Müller (Mollusques Gastéropodes Pulmonés). Ann Parasitol Hum Comp 50:275–286

Rondelaud D (1976) Analyse biocénotique et activité prédatrice des Carabiques dans les habitats de *Lymnaea* (*Galba*) *truncatula* Müller dans le nord-ouest du Limousin. Bull Mus Natl Hist Nat 385:57–66

Rondelaud D (1977a) Données expérimentales sur les possibilités compétitives de *Potamopyrgus jenkinsi* Smith vis-à-vis de *Lymnaea* (*Galba*) *truncatula* Müller en Vienne et Haute-Vienne. Ann Parasitol Hum Comp 52:131–139

Rondelaud D (1977b) L'évolution démographique de *Lymnaea* (*Galba*) *truncatula* Müller en Haute-Vienne. A propos de quatre années d'observations. Ann Parasitol Hum Comp 52:511–520

Rondelaud D (1978) Les effets d'un groupement de Mollusques prédateurs: Zonitidae-Physidae, dans le contrôle biologique de *Lymnaea* (*Galba*) *truncatula* Müller. Ann Parasitol Hum Comp 53:511–517

Rondelaud D (1979) Les espèces animales prédatrices de *Lymnaea truncatula* Müller dans ses habitats immergés en Haute-Vienne. Etude expérimentale de leur prédation. Bull Soc Hist Nat Toulouse 115:99–105

Rondelaud D (1986) Le contrôle mixte et alterné de *Lymnaea truncatula* Müller par voie chimique et biologique. Premiers essais expérimentaux sur le terrain. Ann Rech Vet 17:15–20

Rondelaud D (1988) Les effets d'une concentration sublétale de molluscicide ($CuCl_2$) sur l'activité reproductrice et les déplacements du mollusque hôte, *Lymnaea truncatula* Müller. Ann Rech Vet 19:273–278

Rondelaud D, Barthe D (1992) Observations épidémiologiques sur l'iridovirose de *Lymnaea truncatula*, mollusque vecteur de *Fasciola hepatica*. C R Acad Sci Ser III Sci Vie 314:609–612

Rondelaud D, Vignoles P, Dreyfuss G et al (2006) The control of *Galba truncatula* (Gastropoda: Lymnaeidae) by the terrestrial snail *Zonitoides nitidus* on acid soils. Biol Control 39:290–299

Rondelaud D, Vignoles P, Dreyfuss G (2016) *Aplexa hypnorum* (Gastropoda: Physidae) exerts competition on two lymnaeid species in periodically-dried ditches. Ann Limnol Int J Lim 52: 379–386

Rozendaal JA (1997) Vector control: methods for use by individuals and communities. World Health Organization, Geneva, Switzerland

Saad AEHA, Khalil MT, Ragab FMA et al (2014) Efficacy of the fungi *Aspergillus terreus* and *Penicillium janthinellum* as biological control agents against *Biomphalaria alexandrina* snails. Int J Environ Sci Eng 5:25–37

Sabourin E, Alda P, Vázquez A et al (2018) Impact of human activities on fasciolosis transmission. Trends Parasitol 34:891–903

Samson KS, Wilson GL (1973) The use of ducks as biological control agents of *Fasciola hepatica*. Proc Helminthol Soc Wash 40:292–293

Sardiña P, Beringer J, Roche D et al (2015) Temperature influences species interactions between a native and a globally invasive freshwater snail. Freshw Sci 34:933–941

Secor WE (2014) Water-based interventions for schistosomiasis control. Pathog Glob Health 108: 246–254

Singer S, Bair T, Hammill TB et al (1994) Fermentation and toxin studies of the molluscicidal strains of *Bacillus brevis*. J Ind Microbiol 13:112–119

Singh S, Singh DK (1998) Molluscicidal activity of *Nerium indicum* bark. Braz J Med Biol Res 31: 951–954

Singh SK, Singh A (2003) Molluscicidal and anti-cholinesterase activity of *Alstonia scholaris* plant against freshwater snail *Lymnaea acuminata*. Pak J Biol Sci 6:1442–1446

Singh KL, Singh DK, Singh VK (2012) Characterization of the molluscicidal activity of *Bauhinia variegata* and *Mimusops elengi* plant extracts against the *Fasciola* vector *Lymnaea acuminata*. Rev Inst Med Trop São Paulo 54:135–140

Sleem SH, El-Hommossany K (2008) Can the freshwater crayfish (*Procambarus clarkii*) help control schistosomiasis and fascioliasis in Egypt? Egypt J Aquat Biol Fish 12:175–183

Slootweg R, Vroeg PA, Wiersma SJ (1993) Effects of molluscivorous fish, water quality and pond management on the development of schistosomiasis vector snails in aquaculture ponds. Aquac Fish Manag 24:123–128

Slootweg R, Malek EA, McCullough FS (1994) The biological control of snail intermediate hosts of schistosomiasis by fish. Rev Fish Biol Fish 4:67–90

Sokolow SH, Huttinger E, Jouanard N et al (2015) Infection with schistosome parasites in snails leads to increased predation by prawns: implications for human schistosomiasis control. J Exp Biol 112:9650–9655

Sousa RMOF, Rosa JS, Cunha AC et al (2017) Molluscicidal activity of four Apiaceae essential oils against the freshwater snail *Radix peregra*. J Pest Sci 90:971–984

Spithill TW, Smooker PM, Copeman DB (1999) *Fasciola gigantica*: epidemiology, control, immunology and molecular biology. In: Dalton JP (ed) Fasciolosis. CABI Publishing, Oxon, pp 465–525

Srivastava P, Kumar P, Singh VK et al (2009) Molluscicidal activity of *Piper nigrum* (black) against the snail *Lymnaea acuminata* and *Indoplanorbis exustus* in the control of fascioliasis. J Herb Med Toxicol 3:81–86

Suhardono M, Roberts JA, Copeman DB (2006) Biological control of *Fasciola gigantica* with *Echinostoma revolutum*. Vet Parasitol 140:166–170

SuSin T (2001) Evaluation of different duck varieties for the control of the golden apple snail (*Pomacea canaliculata*) in transplanted and direct seeded rice. Crop Prot 20:599–604

Swartz SJ, De Leo GA, Wood CL et al (2015) Infection with schistosome parasites in snails leads to increased predation by prawns: implications for human schistosomiasis control. J Exp Biol 218: 3962–3967

Taylor EL (1964) Fascioliasis and the liver-fluke. FAO Agric Stud 64:1–235

Teixeira T, Rosa JS, Rainha N et al (2012) Assessment of molluscicidal activity of essential oils from five Azorean plants against *Radix peregra* (Müller, 1774). Chemosphere 87:1–6

Thakur S, Kumar Y, Sharma V et al (2019) Plants as molluscicides. Res J Pharmacogn Phytochem 11:8–13

Tiwari F, Singh DK (2007) Toxicity test of plant derived molluscicides with attractant food pellets against snail, *Lymnaea acuminata*. Iran J Pharmacol Ther 6:103–107

Torgerson P, Claxton J (1999) Epidemiology and control. In: Dalton JP (ed) Fasciolosis. CABI Publishing, Oxon, pp 113–149

Valipour M, Krasilnikof J, Yannopoulos S et al (2020) The evolution of agricultural drainage from the earliest times to the present. Sustainability 12:416. https://doi.org/10.3390/su12010416

Van den Bruel WH (1968) Rapport d'activité 1968 (Groupe de travail pour l'étude de la lutte contre les limnées). Centre Agronomique de Gembloux, Belgium

Vareille-Morel C, Dreyfuss G, Rondelaud D (1999) The characteristics of habitats colonized by three species of *Lymnaea* in swampy meadows on acid soil: their interest for fasciolosis control. Ann Limnol Int J Lim 35:173–178

Vasconcelos MC, Amorim A (2003a) Activity of *Euphorbia splendens* var. *hislopii* N.E.B. (Euphorbiaceae) latex against *Lymnaea columella* (Say, 1817) (Pulmonata: Lymnaeidae), intermediate host of *Fasciola hepatica* Linnaeus, 1758 (Trematoda: Fasciolidae). 1: test in laboratory. Mem Inst Oswaldo Cruz 98:544–563

Vasconcelos MC, Amorim A (2003b) Activity of *Euphorbia splendens* var. *hislopii* N.E.B. (Euphorbiaceae) latex against *Lymnaea columella* (Say, 1817) (Pulmonata: Lymnaeidae), intermediate host of *Fasciola hepatica* Linnaeus, 1758 (Trematoda: Fasciolidae). 2: limited field-testing. Mem Inst Oswaldo Cruz 98:981–985

Vassiliades G (1984) Note sur les propriétés molluscicides de deux Euphorbiacées: *Euphorbia tirucalli* et *Jatropha curcas*. Essais en laboratoire au Sénégal. Rev Elev Méd Vét Pays Trop 37: 32–34

Vázquez AA, Pilar A, Lounnas M et al (2018) Lymnaeid snails hosts of *Fasciola hepatica* and *Fasciola gigantica* (Trematoda: Digenea): a worldwide review. CAB Rev 13:1–15. https://doi.org/10.1079/PAVSNNR201813062

Vignoles P, Rondelaud D, Dreyfuss G (2016) Contrôle intégré de deux parasitoses basé sur l'isolement des habitats colonisés par *Omphiscola glabra* dans deux fermes de la Haute-Vienne. Ann Sci Limousin 27:31–37

Wagner VA (1936) The possibility of eradicating bilharzia by extensive planting of the tree *Balanites*. South Afr Med J 10:10–11

Winterbourn MJ (1980) The distribution and biology of the freshwater gastropods *Physa* and *Physastra* in New Zealand. J Malacol Soc Austral 4:233–234

Worden AA, Sellers KC, Tribe DF (1963) Animal health, Production and pasture. Longmans, Green and Co, London, UK

World Health Organization (2002) WHO specifications and evaluations for public health pesticides: niclosamide. World Health Organization, Geneva, Switzerland

World Health Organization (2017) Field use of molluscicides in schistosomiasis control programmes: an operational manual for programme managers. World Health Organization, Geneva, Switzerland

Xia J, Yuan Y, Xu X et al (2014) Evaluating the effect of a novel molluscicide in the endemic schistosomiasis japonica area of China. Int J Environ Res Public Health 11(10):10406–10418. https://doi.org/10.3390/ijerph111010406

Ximenes TF (1991) Le contrôle biologique de *Lymnaea truncatula* Müller, hôte intermédiaire de *Fasciola hepatica* L. Possibilités et perspectives. VMD Thesis, Alfort, University of Créteil

Yadav RP, Singh A (2011) Efficacy of *Euphorbia hirta* latex as plant derived molluscicides against freshwater snails. Rev Inst Med Trop São Paulo 53:101–106

Yadav RP, Singh A (2014) Effects of single, binary and tertiary combinations with *Jatropha gossypifolia* and other plant-derived molluscicides on reproduction and survival of the snail *Lymnaea acuminata*. Rev Inst Med Trop São Paulo 56:421–426

Yang GJ, Li W, Sun LP (2010) Molluscicidal efficacies of different formulations of niclosamide: result of meta-analysis of Chinese literature. Parasit Vectors 3:84. https://doi.org/10.1186/1756-3305-3-84

Younes A, El-Sherief H, Gawish F et al (2017) Biological control of snail hosts transmitting schistosomiasis by the water bug, *Sphaerodema urinator*. Parasitol Res 116:1257–1264

Zukowski S, Walker KF (2009) Freshwater snails in competition: alien *Physa acuta* (Physidae) and native *Glyptophysa gibbosa* (Planorbidae) in the river Murray, South Australia. Mar Freshw Res 60:999–1005

Chapter 16
Conservation of the Lymnaeidae

Maxim V. Vinarski

Abstract Like many other groups of freshwater Mollusca, the lymnaeid snails are prone to extinction, the main cause of which is currently habitat degradation. Unfortunately, only a modest portion of species of this species have been assessed from the conservationist's point of view, and the available information on distribution, ecology, and population trends of many of the assessed species is not enough to determine their current conservation status. The chapter deals with the state of the art of lymnaeid conservation and reviews the existing advances and challenges in the field. A quantitative analysis of the patterns of a rarity among the Lymnaeidae, based on the available IUCN assessments of their conservation status, is presented. A single case of a presumed rarity at the global scale (*Myxas glutinosa*) is reviewed and discussed. The article provides a brief overview of the eight shortfalls "impairing knowledge and conservation of freshwater molluscs" (Lopes-Lima et al., Hydrobiologia 848(11–12):2831–2867, 2021), with discussion of their application to protection and conservation of the lymnaeid snails.

It is a very well-known fact that the conservationists' efforts have historically been biased toward the protection of the so-called charismatic species of animals, i.e. a group of taxa, which easily get public attention (McKinney 1999; Clark and May 2002; Colléony et al. 2017; Delso et al. 2021). This group includes, according to Skibins et al. (2017, p. 157), primarily those creatures that are "cute and cuddly, or exotic and alluring" (or, following an alternate rating, "Rare, Endangered, Beautiful, Cute, Impressive, and Dangerous"; see Albert et al. 2018). In other words, these charismatic taxa belong to relatively large vertebrates (especially mammals, birds, and reptiles) and are characterized by either increased attractivity to laymen or high media coverage, or both (Albert et al. 2018; Berti et al. 2020; Mammola et al. 2020).

M. V. Vinarski (✉)
Laboratory of Macroecology & Biogeography of Invertebrates, Saint-Petersburg State University, Saint-Petersburg, Russia

S.I. Vavilov Institute for the History of Science and Technology, St. Petersburg Branch, Russian Academy of Sciences, Saint-Petersburg, Russia

M. V. Vinarski, A. A. Vázquez (eds.), *The Lymnaeidae*, Zoological Monographs 7,
https://doi.org/10.1007/978-3-031-30292-3_16

The issue of how to increase the attention of decision-makers and the public to the lesser-known and less charismatic groups of animals, including most taxa of insects and many groups of relatively unseen and "non-attractive" invertebrates, forms a hot challenge for conservation biology. These relatively neglected animal groups are especially prone to what was called "quiet extinction" (Eisenhauer et al. 2019), and the disappearance of their members often is almost unnoticed by both the scientific community and the public (Régnier et al. 2009, 2015; Cowie et al. 2017). Freshwater mollusks, with the probable exception of some species popular among aquarists, also belong to animals relatively unknown to the public that results in insufficient effort to protect them, including the lack of legislative measures.

Nevertheless, the scientific and public awareness of the problem is continuously increasing, and, during several last decades, a special field of malacology, *conservation malacology*, has formed. Special publications stressing the need for the protection of freshwater snails and bivalves started to appear around a century ago (for example, in Poland, see Poliński 1927; Urbański 1932). Already in these pioneering works, a significant idea was developed: the protection of molluscan communities as a whole (and their habitats) is a much more effective conservation measure than the protection of an individual molluscan species. A body of publications concerning various aspects of freshwater mollusks conservation is today growing increasingly, and, during the last 25 years, a series of valuable review studies on this subject have appeared (see Bouchet et al. 1999; Lydeard et al. 2004; Perez and Minton 2008; Régnier et al. 2009, 2015; Johnson et al. 2013; Cowie et al. 2017, 2022; Lopes-Lima et al. 2017, 2018, 2021; Ferreira-Rodríguez et al. 2019; Böhm et al. 2021; Neubauer and Georgopoupou 2021; Neubauer et al. 2021; and references therein). These studies demonstrate very high extinction rates for freshwater Mollusca as well as an urgent need for the sound taxonomy of this group, the lack of which hampers the efforts made by conservationists. The deficiency of reliable knowledge of other aspects of freshwater malacology (distribution, abundance, ecology and competition, and current population trends) is also evident (Lopes-Lima et al. 2021).

The Lymnaeidae have very rarely (if any) been an object of conservation efforts and research related to it. Most representatives of the family remain virtually neutral in relation to protective measures, whereas some of the pond snails (e.g., the dwarf pond snail, *Galba truncatula*) were subjected to eradicative rather than protective actions (see Zhadin and Pankratova 1931; Rondelaud 1978; Rondelaud and Vareille-Morel 1994; Tunholi et al. 2017).

In this review article, I try to outline the current state of the Lymnaeidae conservation at a global scale and discuss some perspectives for further work in this direction.

The world's most authoritative data source in animal conservation is the IUCN Red List of Threatened Species, which is accessible through the website (https://www.iucnredlist.org/) and is updated steadily. Despite its title, it contains not only truly threatened taxa but also widespread and abundant species, whose survival raises no concern (the so-called least concern species, or LC). Each animal species, included in this list, is classified as belonging to one of the following basic categories

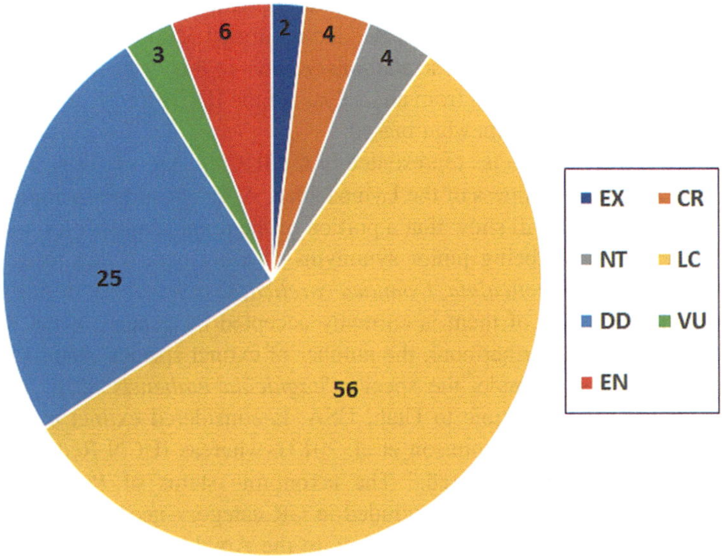

Fig. 16.1 Extinction risk in the Lymnaeidae (based on 100 Red List conservation assessments). See text for the abbreviations. The numbers correspond to number of lymnaeid species belonging to each conservation category

(in addition to LC): EX—extinct; EW—extinct in the wild; CR—critically endangered; EN—endangered; VU—vulnerable; NT—near threatened; DD—data deficient. The latter category contains species the information about which is too scarce to make a conservation assessment. The assessment of the conservation status of species and their placement in a certain Red List category are being made using a set of criteria, including the number of known locations, the current population trend, the known threats, existing protective efforts, etc. The data that are currently available for the taxon across its entire global range form the basis for the assessment (see https://www.iucnredlist.org/assessment/process for more detail).

As of 30 September 2021, the IUCN Red List database includes 100 lymnaeid species, which constitutes, roughly, 2/3 of the global species richness of this family. Most assessed species have been classified as belonging to LC ($n = 56$) and DD ($n = 25$) categories (Fig. 16.1). According to these data, only 19% of recent lymnaeid species are threatened by extinction or extinct. The high proportion of DD species among the Lymnaeidae is by no means surprising. Many species of pond snails have attracted too little attention from conservationists and ecologists, and little is known about their current distribution, abundance, and population trends. The high proportion of DD species is characteristic for the freshwater Mollusca as a whole (Lopes-Lima et al. 2021): for example, the percentage of these species in freshwater Gastropoda is as high as 40.8 (Böhm et al. 2021). The two lymnaeid species considered extinct by the IUCN Red List are *Galba vancouverensis* F.C. Baker, 1939 and *Hinkleyia pilsbryi* (Hemphill, 1890), both endemic to North America.

The numbers discussed above may give an impression that the Lymnaeidae are not seriously threatened at a global scale, and no urgent actions to protect them are needed. However, I must add some cautionary notes to this discussion to show that the overall picture, as it is seen from an analysis of the IUCN Red List, may be quite misleading, or at least is somewhat biased.

First of all, the 100 species represented in the IUCN database form, in fact, less than 2/3 of the global richness of the Lymnaeidae. A survey of the included species made by a systematist will show that a portion of the pond snail species assessed to the day, are not valid, being junior synonyms of other taxa. I can mention such species as *Omphiscola reticulata*, *Lymnaea carelica*, *L. maroccana*, or *Radix lilli* as examples of this. None of them is currently accepted by experts in the field (see MolluscaBase 2021). Furthermore, the number of extinct species seems to be a bit underestimated. For example, the species *Stagnicola utahensis* (Call, 1884), the thickshell pond snail, endemic to Utah, USA, is considered extinct by the North American conservationists (Johnson et al. 2013), whereas IUCN Red List categorizes it as "critically endangered." The taxonomic status of *Radix arachleica* (Kruglov & Starobogatov, 1989) included in CR category needs to be elucidated. Probably, this snail, thought to be endemic to the Arakhlei Lake in Transbaikalia (south Siberia, Russia), is a synonym of another widespread species—the genetic information required to check its validity is still unavailable (Vinarski, unpublished data).

Given that a substantial portion of the actual lymnaeid diversity remains not evaluated, there are all reasons to consider the situation with conservation of the pond snails family far from being clear, and substantial efforts are required to fulfill the gap between the actual knowledge and the information currently provided by the IUCN website. I should like to make it clear that the critical remarks toward the current state of the IUCN Red List presented above do not mean I underrate the efforts which the experts involved in the process of assessment of species' conservation status are being made. Their work is extremely important in the "Sixth Extinction" epoch and should be praised as very urgent. I must admit that the practitioners of the IUCN Red List assessments fully realize that the data, available today from the IUCN Red List, need a serious update, and some actions are undertaken now to actualize this information (David Allen, pers. comm.). Hopefully, the quality of taxonomic and conservation information concerning lymnaeid snails provided by the IUCN website will increase in the nearest future.

Let us proceed to a survey of the lymnaeid conservation status at the level of the world's continents and ecoregions. The primary data for this survey was taken from a series of IUCN regional assessment projects covering territories of Europe (Cuttelod et al. 2011), Pan-Africa (Darwall et al. 2009), Western Ghats (Molur et al. 2011), and Indo-Burma (Köhler et al. 2012). These data were checked and updated using the IUCN website. For North America (the USA and Canada excluding Mexico) I used the data represented in Johnson et al. (2013) though the conservation categories used in the latter publication do not fully coincide with those of IUCN (see below).

The IUCN assessments data show that the percentage of lymnaeid species threatened with extinction varies drastically from region to region (Table 16.1).

Table 16.1 Red List status of lymnaeid species by region[a]

Region	No. of species	IUCN category							% threatened
		EX/RE[b]	CR	EN	VU	NT	LC	DD	
Europe	20	0	0	2	0	2	13	3	10.0
Mediterranean	10	0	0	3	0	1	4	2	40.0
Pan-Africa	7	1	0	3	0	0	3	0	66.6
North Asia	27	0	1	0	0	1	20	5	7.4
West and Central Asia	19	0	0	1	0	0	16	2	5.3
Indo-Burma	19	0	0	0	1	0	11	4	5.2
Western Ghats	4	0	0	0	0	0	2	2	0.0
Oceania	7	0	0	0	1	0	6	0	14.3

[a]The primary data for this table were taken from a series of IUCN publications (Darwall et al. 2009; Molur et al. 2011; Cuttelod et al. 2011; Köhler et al. 2012) as well as from the IUCN Red List website

[b]RE (regionally extinct) denotes species disappeared from a particular region but survived in the other parts of its range

Table 16.2 Conservation status of the North American lymnaeid species assessed by AFS and IUCN

Authority	No. of species	Conservation status (category)							% threatened
AFS	60	*X*	*Xp*	*E*	*T*	*V*	*CS*	*U*	61.7
		2	8	13	8	6	23	0	
IUCN	42	*EX/RE*	*CR*	*EN*	*VU*	*NT*	*LC*	*DD*	19.0
		2	2	2	1	1	27	7	

Different parts of the same continent (e.g., Europe, Asia) may demonstrate different portions of threatened and not threatened (LC, DD) species.

The system of conservation categories developed by the AFS Endangered Species Committee (AFS = the American Fisheries Society) and used by Johnson et al. (2013) resembles that of the IUCN; below I list these seven categories and propose their probable correspondences to the IUCN categories: X (EX)—extinct; Xp (EW/CR)—probably extinct; E (EN)—endangered; T (VU)—threatened; V (NT)—vulnerable; CS (LC)—currently stable; U (DD)—unknown. Table 16.2 contains a comparison between two series of assessments of the conservation status of the North American lymnaeids made independently by the AFS and IUCN experts. As the table shows, the percentages of the threatened pond snail species provided by the two organizations are dramatically different, with the AFS estimate being much more "pessimistic" than that of IUCN. Most probably, such a disparity reflects some substantial differences between the assessment procedures realized by IUCN and AFS. This numerical discrepancy, in my opinion, gives a good illustration of the current uncertainty of the conservation status of the Lymnaeidae at the global scale (see above).

Fig. 16.2 Shells of some lymnaeid species with conservation status other than LC or DD. (**a**) *"Stagnicola" idahoensis* (CR). USA, Idaho, Bear Lake (NHMW). (**b**) *Radix pinteri* (EN), the holotype. North Macedonia, Prespa Lake (SMF). (**c**) *Stagnicola montenegrinus* (NT). Montenegro, Skadar lake, Karuč (LMBI). (**d**) *Erinna newcombi* (VU). USA, Hawaii, Henda River (NHMUK). (**e**) *Racesina ovalior* (VU). India, Calcutta (NHMUK). (**f**) *Omphiscola glabra* (NT). UK, Dewsbury (NHMUK). The global extinction risk is given according to IUCN Red List of Threatened Species. Scale bars 2 mm (**d, f**), 5 mm (**a–c**, **e**). Museum acronyms: LMBI—Laboratory of Macroecology and Biogeography of Invertebrates, Saint-Petersburg State University, Russia; NHMUK—Natural History Museum of the United Kingdom, London; NHMW—Natural History Museum of Vienna, Austria; SMF—The Naturmuseum Senckenberg, Frankfurt am Main, Germany

What is common in the lymnaeid species considered threatened (EX, EN, VU, and NT categories) by the IUCN (Fig. 16.2)? What properties do they share?

Foremost, the vast majority of the threatened pond snails are very narrowly distributed, being sometimes found in a very restricted area only. For example, *Lantzia carinata* (CR) is known from a single locality, i.e. from a waterfall at or near the type locality on Réunion (Van Damme 2016). The ranges of *Stagnicola utahensis* (CR), *S. idahoensis* (EN) are confined to a single state of the USA (Utah and Idaho, respectively). Two lymnaeids classified as endangered (*Radix pinteri* and *R. skutaris*) are endemics to the great ancient lakes of Balkans (south Europe). *Kutikina hispida* and *Erinna newcombi* (both VU) are endemics to islands (Tasmania and Hawaii, respectively). Not surprisingly, the high extinction risk in this family is related mainly to the narrow range and a low number of localities. Such properties of the species make them extremely prone to extinction, since the chances of the total or partial degradation of their habitats are especially high. Among the threatened

Fig. 16.3 Living specimens of *Myxas glutinosa* (left) and the general view of their habitat (right)—Russia, Chelyabinsk City, the Miass River. The photo was taken on 16.09.2009. The shell of the largest snail in this picture is 11.5 mm height. Photo: Olga S. Shishkoedova. After Vinarski et al. (2013), with modifications

lymnaeid species, only two, *Acella haldemani* in North America and *Omphiscola glabra* in Europe, are characterized by relatively vast ranges; both species are classified as NT by IUCN Red List, and no of the two is in imminent danger of extinction.

A substantial portion of the threatened lymnaeids belongs to monotypic or very small genera and subgenera (*Acella*, *Erinna*, *Fisherola*, *Kutikina*, *Lantzia*, *Omphiscola*) which may indicate, though indirectly, that the ecological specialization and phylogenetic distinctness may enhance the extinction risks within the Lymnaeidae. In other words, members of more speciose, widespread, and (presumably) ecologically plastic lymnaeid genera are less susceptible to extinction.

However, even those pond snail species that are not considered threatened by the IUCN Red List may, in fact, be of great conservation interest being endangered at either country or region levels. Some of such species are still qualified as LC or DD by the IUCN Red List, which requires urgent attention. Below, a single example is discussed in detail to illustrate this.

Myxas glutinosa (O.F. Müller, 1774), the glutinous pond snail (Fig. 16.3), is widespread in Europe, except for its northern and southern parts (Welter-Schultes 2012). This species' range also includes Western and some parts of Central Siberia (Prozorova and Sharyi-ool 1999; Kruglov 2005; Andreeva et al. 2010; Vinarski et al. 2013). *M. glutinosa* has been assessed as a data deficient species by the IUCN at the global scale, whereas the European Red List of Non-marine Molluscs (Cuttelod et al. 2011) lists it as LC, which was considered a surprising decision by some authors (Welter-Schultes 2012; Vinarski et al. 2013; Mouthon and Vimpère 2014).

Historically, the glutinous pond snail has been recorded from 25 European countries and Kazakhstan (Falkner et al. 2001; Hubendick 1951; Lazareva 1968; Glöer 2002, 2019; Vinarski et al. 2013). The current situation of this species in each country of occurrence is summarized below (in alphabetical order).

Austria *M. glutinosa* was listed in the malacofauna of this country by Klemm (1960), who recorded it for Lower Austria and East Tirol. Now this species is considered extinct in Austria (Falkner et al. 2001; Welter-Schultes 2012).

Belarus Layenko (2012) is the most recent monograph on the Belarus aquatic malacofauna. The author cites three localities in Belarus where *M. glutinosa* was found between 2005 and 2008. No data on the abundance of this mollusk in Belarus are provided in this book, and the current status of the glutinous snail there remains unclear. San'ko (2007) recorded shells of *M. glutinosa* from the Holocene deposits of Belarus.

Belgium Adam (1960) mentions this species for several parts of the country; no information of the current state of *M. glutinosa* in Belgium is accessible.

Bulgaria Neither Hubendick (1951) nor Welter-Schultes (2012) included Bulgaria in the range of *M. glutinosa*, however, in 1994 this snail was discovered in the Skomlya River situated in the northwestern part of this country (Hubenov 2007; Georgiev 2014). This location is, possibly, the southernmost finding of *M. glutinosa* in Europe. There is no other data on the distribution and abundance of the glutinous snail in Bulgaria, but it seems highly likely that this species is extremely rare there.

The Czech Republic Ložek (1955) mentioned several regions of this country where the species had been living in the middle of the past century. However, *M. glutinosa* was rare in the Czech Republic already 150 years ago (Beran 2002), and today it is recognized as extinct (Welter-Schultes 2012; Horsák et al. 2013). In the second half of the twentieth century, there were no reliable findings of *M. glutinosa* in the Czech Republic (Flasar 1998; Beran 2002; Horsák et al. 2013).

Denmark This country is the type locality of *M. glutinosa*, which was described from the environs of Copenhagen (see Vinarski and Kantor 2016 for details). Mandahl-Barth (1949) characterizes it as a rare species for Denmark. According to Welter-Schultes (2012), since 1949 there were no reliable findings of *M. glutinosa* in Denmark.

Estonia Except for some old data (Schlesch 1942; Zhadin 1952), almost nothing is known about the abundance and distribution of *M. glutinosa* in Estonia. Since this snail is relatively common in the Pskov Region of Russia, which is adjacent to the territory of Estonia, it may be assumed that the snail is still present in the malacofauna of the latter.

Finland The historical recordings of this species in the country have been made from different regions, including a single locality situated at 69° N (Luther 1901; Carlsson 2000). Luther (1901) classified *M. glutinosa* as a common species in Finland. According to Welter-Schultes (2012), *M. glutinosa* occurs only in the southern part of Finland, where around 20 lake populations of this snail are known. On the other hand, Carlsson (2000, p. 105) noted that the glutinous snail in Finland "seems to have a stronghold" and has not been listed as an endangered species. Carlsson (2000) found this species in 4 lakes of 51 visited on the Åland Islands.

France Mouthon and Vimpère (2014) have recently reviewed the current state of *M. glutinosa* in France. Their data shows the "dramatic regression" of this snail in France during the last century. Since 1950, *M. glutinosa* inhabits only 6 departments, instead of 30 departments where its occurrence was recorded in old literary sources.

Germany German malacologists of the nineteenth—first half of the twentieth centuries recorded *M. glutinosa* from various areas of their country and, as a rule, did not consider it rare (Clessin 1884; Geyer 1927; Ehrmann 1933). Although Goldfuss (1900) mentioned that the glutinous snail is rare, he nevertheless listed about a dozen habitats of this species from Central Germany. Currently, *M. glutinosa* has become extremely rare throughout the country and is included in the Red Data Books of almost all federal states of Germany (Glöer 2015) where it is ranked exclusively as an extinct (category 0) or endangered (category 1) species.

Ireland In this country, the glutinous snail has been recorded mainly in the central part of the island, but is rare everywhere and occurs only sporadically. In the second half of the twentieth century, there was a sharp decline in the number of populations of the species. So, if in the 1970s, it was quite common in a number of places (Kerney 1999), then from 1985 to 2002 there was not a single reliable finding of living individuals of *M. glutinosa* in Ireland (Beckmann 2006). On the other hand, judging from the Kerney (1999), Holyoak (2005), and Beckmann (2006) information, the glutinous snails are still to be found in Ireland, and it is estimated that Ireland maintains up to 50% of the global population of *M. glutinosa* (Byrne et al. 2009).

Kazakhstan A few localities of the glutinous pond snail were identified in some regions of western (Smirnova 1967), northern, and central Kazakhstan (Lazareva 1968; Frolova 1984; Vinarski et al. 2013). The author of this work collected *M. glutinosa* in October 2002, from the Ulkendamdy stream, located in the central part of the country (Kostanay District). In 2012, a new finding of this snail from central Kazakhstan (Akmola Region, Kulanutpes stream in the Nura river drainage basin) was reported, which represents the southernmost locality of *M. glutinosa* in Asia (Krainyuk 2012). In total, around 10 localities of this species have been discovered in Kazakhstan during the last 60 years (Smirnova 1967; Lazareva 1968; Krainyuk 2012; Vinarski et al. 2013). In 1968, Lazareva (1968) reported the abundance of *M. glutinosa* in the Tobol River floodplain equal to 36 ind/m^2; unfortunately, both the current state of these populations and the population trends are unknown.

Latvia *Myxas glutinosa* is listed as a rare and protected species in the recent checklists of the Latvia malacofauna (Rudzīte et al. 2010, 2018). However, the current abundance estimates and population trends of the glutinous snail in Latvia are unknown.

Lithuania Zettler et al. (2005) discussed some recent findings of *M. glutinosa* in this Baltic country and concluded that "it appears that this species finds relatively good life conditions in Lithuania" (Zettler et al. 2005, p. 38). In 2004, the snail was found here in three locations, in one of which it reached high abundance (Zettler et al. 2005).

Moldova *Myxas glutinosa* has been recorded in Moldova in the Dniester River (Balashov et al. 2020); no data on its current abundance and population trends are available.

The Netherlands Like in many other European countries, the population of *M. glutinosa* in the Netherlands has experienced a pronounced decline in the last half of the twentieth century (Gittenberger et al. 2004). Welter-Schultes (2012) reports (without a reference) a 90% decline in the Netherlands since 1960. Today, the species is patchily distributed in this country, being found mainly in its central part.

Norway According to Økland's (1990) data, the species is very rare in Norway. It was first discovered in the waterbodies of this country in the middle of the twentieth century, and at the end of the century lived in three lakes on the southeastern coast.

Poland Before World War II, this snail was not rare in Poland (Feliksiak 1939), but the situation has quickly become much worsened, and some prominent Polish workers on freshwater snails, including Maria Jackiewicz (1920–2018), had never found living specimens of *M. glutinosa* (Szarowska and Falniowski 2006). Nevertheless, the most recent monograph on the Polish freshwater Mollusca states that the glutinous snail in Poland is "fairly common in the Pomeranian and Mazurian lakelands, in the Wielkopolsko-Kujawska and Mazovian Lowlands"; it is absent from the mountains (Piechocki and Wawrzyniak-Wydrowska 2016, p. 115). Unfortunately, no quantitative estimates of the species abundance and population trends have been provided by the authors. Arguably, *M. glutinosa* is now among the rarest freshwater Gastropoda species of the Polish fauna (Szarowska and Falniowski 2006).

Romania Grossu (1955) mentioned this species from a few localities of this country. However, later on, the author (Grossu 1987) excluded *M. glutinosa* from the Romanian malacofauna explaining that the previous record was a result of misidentification. Nonetheless, the presence of this snail in adjacent Bulgaria indicates that *M. glutinosa* can be living in Romania but is so extremely rare that it is overlooked both by professional malacologists and amateur naturalists.

Russian Federation In Russia, numerous localities of this snail have been discovered during the last half of the past century, mainly from the northern part of European Russia, including Karelia and the Pechora River basin (Sokolova 1965; Leshko 1998). In particular, Sokolova (1965) lists as many as 26 large lakes and rivers of the Republic of Karelia where the glutinous snails occurred in the mid-twentieth century. Though I am not aware of the recent studies which deal with the abundance and distribution of *M. glutinosa* in European Russia, it seems likely that the snail is still more or less common in some parts of Karelia and adjacent areas (Ivan Nekhaev, pers. communication), in the Pskov Region (Dmitry Palatov, pers. communication) as well as in the Pechora River basin (Leshko 1998). The southernmost recently discovered locality of *M. glutinosa* in European Russia lies, possibly, in the eastern part of the Moscow Region (Dmitry Palatov, pers. communication), whereas the northernmost locality is situated in the Kola Peninsula, at 68° 51′N, i.e. north of the Polar circle (Nekhaev 2021). The southern boundary of the species' range in European Russia needs to be clarified. The presence of this species in the Urals and Siberia has been documented by various authors (Mozley 1936; Ioganzen 1951; Gundrizer 1979; Prozorova and Sharyi-ool 1999; Kruglov 2005; Khokhutkin et al. 2009; Vinarski et al. 2013). It is very rare and sporadically distributed in Western Siberia (Vinarski et al. 2013), whereas in central Siberia, its presence in the Tuva Republic and the Lower Yenisei basin is known (Gundrizer 1979; Prozorova and Sharyi-ool 1999). The single record of the glutinous snail from the Lena River basin (Yakutia) made in the 1960s (Belimov 1969) remains enigmatic and, most likely, was based on a misidentification. *Myxas glutinosa* is not included in the last edition of the Red Data Book of Russia and thus is not protected at the federal level. However, recently this species was listed in the Red Data books of the Omsk and Chelyabinsk regions.

Slovakia Hubendick (1951) included this country into the range of *M. glutinosa*. Nevertheless, no data on its occurrence in Slovakia are given in faunistic monographs published during the last 70 years (Ložek 1955; Lisický 1991; Horsák et al. 2013). Both the past and present status of the glutinous snail in Slovakia remains, thus, totally unclear.

Spain *Myxas glutinosa* was once recorded for the northeastern part of this country (the Pyrenees) [Jeffreys 1862]. It is unclear if the glutinous snail still occurs in Spain; at least Welter-Schultes (2012) does not include Spain in the range of *M. glutinosa*.

Sweden Only seven localities of the glutinous snail are known in Sweden, all situated in the southern part of the country (Nilsson et al. 1998; Welter-Schultes 2012). *M. glutinosa* is included in the national Red List of threatened invertebrates (von Proschwitz 1997).

Switzerland Included in the range of *M. glutinosa* by Hubendick (1951). No recent data on the presence of the glutinous snail here is available.

Ukraine The most recent data on the occurrence and ecology of the glutinous snail in this country have been summarized by Stadnichenko (2004, 2006). These data are based on observations made by the author during the 1960–early 2000s. According to Stadnichenko, *M. glutinosa* is distributed throughout the entire Ukraine territory, but the specific data on the abundance of *M. glutinosa* are absent from the monographs quoted above. It can be assumed that in Ukraine the species has not yet become as rare as it happened in neighbouring Poland, however, some more recent research has revealed the apparent decline of the glutinous snail in different regions of the country. For instance, a survey of the malacofauna of the Zhytomir Region did not reveal *M. glutinosa* there (Zhitova et al. 2006) although, according to A.P. Stadnichenko (2004), about 40 years ago the species occurred in this area.

The United Kingdom Jeffreys (1862, p. 102) characterized this snail as "a local species [in Britain] although abundant where it occurs" and considered it neither especially rare nor declining. Today's naturalists qualify *M. glutinosa* as the "Britain's rarest freshwater snail" (Willing et al. 2014, p. 673). Notably enough, already 85 years ago, Boycott (1936) was, essentially, of the same opinion. This species was feared to be completely extinct of all its British location before it was rediscovered in 1998 in a single lake in North Wales (Willing et al. 2014). For more details on the past and present distribution of the glutinous snails in the UK, see Boycott (1936), Whitfield et al. (1998), and Kerney (1999). The latest record in North Ireland is dated 1900 (Welter-Schultes 2012).

There are no historical data on the occurrence of the glutinous snail in Croatia, Greece, Hungary, Iceland, Italy, Portugal, Serbia, Slovenia, and some other European countries, and, most probably, their territories should not be included in the native range of this snail. An anecdotal report of this species from Syria (Turton 1857) can, evidently, be rejected as based on a misidentification.

This detailed survey shows that in most European states lying within its historical range, *M. glutinosa* has either became extinct or is experiencing strong population decline accompanied by the decrease in the number of known localities. Regrettably, the current information on the status of this snail in some countries is unavailable, which means that no national monitoring programs are working in these areas. The case studies on ecology, distribution, and abundance of *M. glutinosa* are very scarce [see Carlsson (2000) and Willing et al. (2014) for an example of a much-needed study of this kind].

The general conclusion from the above survey of the available data is that since the 1900s, *M. glutinosa* has been experiencing a steady decline throughout its entire range and the future survival of this species cannot be secured unless its habitats are protected. One of the most urgent measures is to change its IUCN category from DD/LC to VU or EN.

The causes of the global decline of the species are rather obscure. There is no common agreement among researchers on some important points of the species ecology, and sometimes contradictory statements are issued. Many researchers have reported data on the low tolerance of *M. glutinosa* to different environmental factors. For example, it is registered to be susceptible to biodegradable pollution (Mouthon and Charvet 1999) and sensitive to eutrophication (Whitfield et al. 1998; Donohue et al. 2009) as well as to a high degree of the water's hardness (Beriozkina et al. 1980). Zhadin (1952), on the contrary, believes *M. glutinosa* inhabits dystrophic lakes and oxygen-deficient waterbodies. Carlsson (2001) and Briers (2003) mentioned the species among calcephile mollusks that do not inhabit calcium-deficient waterbodies. Moreover, *M. glutinosa* is not tolerant to low values of ambient pH (Salazkin 1969; Berezina 2001).

Some of these alleged causes of decline can however be ruled out. Namely, the lymnaeid species *Lymnaea ovata* (Drapanaud, 1805) [= *Ampullaceana balthica* in the current nomenclature] that is not considered to be rare has equal with *M. glutinosa* range of pH tolerance, 6.0–9.0 (Berezina 2001). According to the data of Vinarski et al. (2013), collected in June 2010, the water salinity also does not limit the glutinous snail distribution. In the South Urals, it inhabits waterbodies where salinity varies from 90 to 593 ppm. Furthermore, Carlsson (2001) found *M. glutinosa* in eutrophic habitats in the Åland Islands; therefore, its alleged intolerance to eutrophication (Whitfield et al. 1998) may be ruled out. A recent finding of *M. glutinosa* in a polluted area of a large industrial city of Chelyabinsk, South Urals, Russia (see Vinarski et al. 2013; Fig. 16.3) affirms that this species is not critically dependent on the purity of water.

At last, the rarity of *M. glutinosa* may be partially explained by some peculiarities of its life cycle. Feliksiak (1939) and Willing et al. (2014) reported that this species lives only one year (until spring) and in the summer season only juvenile individuals occur. Due to their small sizes, the juveniles can be overlooked by collectors to give the impression of the absence of this species in a waterbody. Special efforts are needed to learn if this assumption is true.

Some authors believe habitat destruction must be considered the most important cause of the global decline of this species (e.g., Welter-Schultes 2012).

Lymnaeidae belong to Hygrophila, a group of freshwater molluscs that, in general, are less prone to extinction than gill-breathing snails classified within Caenogastropoda (Neubauer and Georgopoupou 2021). Nonetheless, the list of actual and potential factors threatening the survival of the lymnaeid snails (and other aquatic pulmonates) at a global scale is rather long. It includes a number of factors such as extensive farming, water pollution, acidification of snail environment, destruction or degradation of inland waterbodies, invasion of alien species of mollusks and other animals (non-indigenous species of fish, aquatic macroinvertebrates).

Recently, Lopes-Lima et al. (2021) have summarized the eight major shortfalls "impairing knowledge and conservation of freshwater molluscs," each named in honor of a prominent scientist in the field of ecology, biodiversity, or biological conservation (Box 16.1). The statement made by the authors (Lopes-Lima et al. 2021, p. 2832), that "our basic knowledge of [freshwater Mollusca] is still highly incomplete, which hampers the development and implementation of effective and timely conservation strategies for these rapidly disappearing animals" is surely be applied to the family Lymnaeidae. Though the lymnaeid snails form one of the most well-studied families of freshwater molluscs, much scientific work is left to be done in order to improve the efficacy of conservation efforts on this taxon.

Box 16.1 The Eight Major Shortfalls Impairing Knowledge and Conservation of Freshwater Molluscs*

Title	Brief description	Eponym
Linnaean	Knowledge gaps in taxonomy of a studied group	Carl Linnaeus (1707–1778), the founder of modern biological systematics
Wallacean	Deficiency of geographical distribution data for many species	Alfred R. Wallace (1823–1913), co-discoverer of the natural selection principle, an outstanding biogeographer
Prestonian	Lack of knowledge on the abundance of species and its population dynamics in space and time	Frank W. Preston (1896–1989), author of pioneering works on species commonness and rarity
Darwinian	Lack of knowledge about the tree of life and evolution of lineages, species, and traits	Charles R. Darwin (1809–1882), the founder of modern evolutionary theory
Raunkiaeran	Lack of knowledge about ecologically relevant species traits	Christen Raunkiaer (1860–1938), one of the founders of modern plant ecology, the creator of an influential plant life-form classification
Hutchinsonian	Knowledge gaps in abiotic tolerances of particular species of a studied group, including their life histories, functional roles and responses to habitat changes	George E. Hutchinson (1903–1991), who established the modern concept of the ecological niche
Eltonian	Lack of knowledge about interactions among species or among groups of species	Charles S. Elton (1900–1991), the pioneer of the concept of food chains and food webs; also known as an early student of biological invasions
Ostromian	Lack of knowledge about the application and effectiveness of conservation assessments, methods, funding, and policies	Elinor Ostrom (1933–2012), author of ground-breaking works on common resources governance, and its impacts on biodiversity management and policies

*After Hortal et al. (2015) and Lopes-Lima et al. (2021)

Let us discuss, one by one, the eight major shortfalls delineated by Lopes-Lima et al. (2021) as applied to the Lymnaeidae conservation.

1. *Linnaean Shortfall*. Despite the substantial advancements of the lymnaeid taxonomy made during the last 10–20 years, a large number of questions remain unresolved. The validity of a large fraction of nominal species and genera of recent lymnaeid snails has yet to be reassessed using molecular techniques, and,

until this is done, a significant portion of hitherto described taxa will be classified as "taxa inquirenda" (see MolluscaBase 2021). Among the objective causes of such a situation, one can mention the inadequacy of original descriptions published 150–200 years ago, the loss of the type specimens, and the destruction of the type localities which prevents the sampling of topotypic specimens. The synonymy rate in the Lymnaeidae is one of the highest among all families of freshwater Mollusca (Lopes-Lima et al. 2021). On the other hand, several papers aiming to lower the degree of taxonomic uncertainty in Lymnaeidae have recently appeared. Some of these papers focused on discussion of the type materials of previously described species, usually with high-quality illustrations (i.e., Sitnikova et al. 2014; Vinarski 2016), whereas others provide the "integrative" reassessment of nominal species introduced by taxonomists in the past, which may result either in their synonymization (e.g., Vinarski et al. 2016, 2021; Aksenova et al. 2017) or, sometimes, in the re-establishment of old names long considered be synonyms (e.g., Mahulu et al. 2019).

2. *Wallacean Shortfall.* This shortfall, as applied to the lymnaeid snails, seems to be relatively relaxed. On my personal estimate, the state of our knowledge on the range and peculiarities of distribution of many species of pond snails varies from excellent to satisfying. Distribution maps are available for many species and many continents (see, for example, Hubendick 1951; Clarke 1973; Welter-Schultes 2012; Glöer 2019). Of course, we still are dealing with only a rough picture illustrating the wide-scale distribution patterns of the lymnaeid snails. What is urgently needed is small-scale research that would map localities of different lymnaeid species within the relatively restricted areas, e.g. provinces, river basins, state regions. Examples of such maps exist in some West- and Central European countries (e.g., Lisický 1991; Flasar 1998; Gittenberger et al. 2004) and they may be indispensable for the conservation measures applied within a particular country. The further progress of GIS technologies and the development of online occurrences databases (like GBIF—Global Biodiversity Information Facility) will, probably, help to ameliorate the Wallacean Shortfall (see Lopes-Lima et al. 2021 for more details on this subject).
3. *Prestonian Shortfall.* Comparing with the previous one, this shortfall is much more demanding. For most lymnaeid species, we lack any reliable and updated data on their abundance and current population dynamics. Exceptions are species of particular practical interest (such as the dwarf pond snail, *Galba truncatula*; see Relf et al. 2011; Charlier et al. 2014; Jones et al. 2021), some invasive taxa (e.g., *Pseudosuccinea columella*), or species/populations attracting a specific interest from conservationists (e.g., *Myxas glutinosa* in Llyn Tegid, North Wales; see Willing et al. 2014). The deficiency of such data explains a high fraction of lymnaeid species categorized as DD by the IUCN experts, which, in itself, constitutes a serious hamper for effective conservation planning and activities. Such a situation is by no means unique for the Lymnaeidae, and it characterizes the state of our knowledge of almost every genus or family of freshwater Mollusca (Lopes-Lima et al. 2021). Some newly developed methodologies, such as environmental DNA (eDNA) analysis, are thought to be useful in making

the study of the abundance of freshwater snails easier and more efficient. There are some examples of eDNA surveys applied to the study of economically important lymnaeid species (Davis et al. 2020; Jones et al. 2021; Rathinasamy et al. 2021). The use of this promising methodology for the surveys of rare and threatened species of the family will be quite desirable.

4. *Darwinian Shortfall.* The evolution and phylogeny of the Lymnaeidae have been extensively studied during the last 100–120 years, by both paleontologists and neontologists. It was possible because lymnaeid shells are vastly represented in many paleontological collections, and their evolutionary history can be traced back to the Jurassic. By the start of the molecular revolution in the lymnaeid taxonomy which took place around 1997, an impressive body of phylogenetic facts and hypotheses had been accumulated, and the extensive use of molecular information is currently helping to integrate all these data in a comprehensive integrated picture. Though many questions are still left open, the last 15–20 years have witnessed a great advance in this field, and there is little doubt that, within the next several years, the phylogenetic relationships between the majority of recent genera, subgenera and, to a lesser extent, species will be elucidated more or less satisfyingly. In my opinion, there are no serious obstacles for the tree of life for the recent Lymnaeidae to be built to the end of the current decade based on the multi-omics approach; the evolutionary relationships between many extinct lymnaeid taxa will, apparently, remain problematic much longer.
5. *Hutchinsonian Shortfall.* The lymnaeid snails, many of which belong to the most widely distributed, abundant, and conspicuous species of freshwater snails, have attracted naturalists since long ago. Some of the lymnaeids (e.g., *Galba truncatula* and *Lymnaea stagnalis*) were studied ecologically in many regions of the world and during many decades. The abiotic tolerances of the European Lymnaeidae have been discussed in many publications (Boycott 1936; Fromming 1956; Russell-Hunter 1978; Beriozkina and Starobogatov 1988; Økland 1990, etc.). Analogous information is available for some of the representatives of the family inhabiting other continents (e.g., Lynch 1965; Hunter 1975; Monzon et al. 1993; Abdul Aziz and Raut 1996) but, in general, these data are prone to geographic bias (i.e., species of "exotic" faunas are understudied as compared with species of West and Central Europe and North America). As for the lymnaeid species of conservationists' interest, the data on their abiotic interactions are extremely scarce, and the lack of relevant information constrains the measures toward the protection of these molluscs. One of the possible ways to ameliorate the Hutchinsonian Shortfall is the realization of expansive laboratory trials, which may provide useful information on the abiotic tolerances of particular lymnaeid species.
6. *Raunkiaeran Shortfall.* Almost all that was said about the previous shortfall can be applied to this one. As compared with taxonomy and distribution, the ecological traits of the Lymnaeidae, including the parameters of their life cycles, are relatively poorly studied. Though some species of this family have long served as model objects for both field observations and laboratory trials and these data can be cautiously extrapolated to the other lymnaeid species, the state of our

knowledge on the biological and ecological traits of the endangered pond snails remains dissatisfying. Once again, it is typical for other groups of freshwater Mollusca, and the Lymnaeidae only exemplify this widespread situation.

7. *Eltonian Shortfall.* Not surprisingly, most of the available data on the biotic interactions of the Lymnaeidae was obtained from studies on *Galba truncatula*, *Austropeplea tomentosa*, and a handful of other lymnaeid species acting as the intermediate hosts for parasitic Trematoda. The details of the host–parasite relationships in these biotic systems have been profoundly studied, whereas the other sorts of biotic interactions such as predation, competition, and facilitation have been even less researched. Numerous pond snail species are almost unknown in this respect.
8. *Ostromian Shortfall* was originally defined as "a lack of knowledge about the application and effectiveness of conservation assessments, methods, funding, and policies" (Lopes-Lima et al. 2021, p. 2848). The existing gap between the taxonomy of the pond snails and their conservation assessment within the IUCN Red List framework is discussed above. In many countries, for example in Russia, freshwater Mollusca, including the lymnaeids, are underrepresented in the regional Red Data books (Grebennikov and Vinarski 2009), whereas in other (e.g., Germany) the members of this family are present in all regional lists of endangered animals (see Glöer 2015 for review). In general, the global lack of conservation knowledge about the Lymnaeidae, which constitutes the Ostromian Shortfall, can be considered substantial and worrying. The deficiency of taxonomic, phylogenetic, and especially ecological information about the recent pond snails is the main cause of this. The Ostromian Shortfall cannot be ameliorated until the other seven shortfalls are overcome.

The general conclusions which can be made from the facts and discussion presented above are as follows.

- The Lymnaeidae suffer globally from the same conservation shortfalls as most other families of freshwater gastropods and bivalves. The high rate of uncertainty in the available taxonomic and ecological information has resulted in a high fraction of species that are categorized as "data deficient" and thus belong to the "dark matter" of biological conservation.
- The lack of ecological information on the rare and endangered species of the pond snails is today more essential for conservation planning than the existing gaps in our knowledge on the systematics, phylogeny, and distribution of the family. The reliable data on current abundance and population dynamics are unavailable for most lymnaeid species in most continents and regions. The global shortage of taxonomists, field ecologists, and conservationists working with freshwater Mollusca (including Lymnaeidae) is the main cause of this situation.
- A prominent geographic bias characterizes the available data on the rare and endangered lymnaeid species, when taxa of European and North American distribution being, in general, much more studied than those of Africa, Asia, South America, and other continents (with probable exception to Australia).

- Considering that it may take a long time to fill gaps in available knowledge about concrete species of pond snails in need of protection, it would be more rational to put more attention on the conservation of specific habitats of freshwater fauna, rather than particular species of molluscs. This will allow better use of available limited resources, including financial ones, and optimize efforts to conserve freshwater communities as a whole. At the same time, it is necessary to develop special programs for the monitoring and protection of those species of pond snails that were classified in the categories with the highest risk of extinction (CR, EN), and to provide measures for the reintroduction of the endangered species into those parts of their ranges where they have become extinct by now.

Acknowledgments The partial funding for preparing this article was obtained from the Russian Scientific Fund (grant No. 19-14-00066-P). I wish to thank Peter Glöer (Hetlingen, Germany), Maxim Grebennikov (Yekaterinburg, Russia), Ivan Nekhaev (St.-Petersburg, Russia), Dmitry Palatov (Moscow, Russia), and Olga Shishkoedova (Chelyabinsk, Russia), who have sent me their observations on the distribution of *M. glutinosa* and/or copies of some rare publications which helped me to elucidate the current distribution of this species.

References

Abdul Aziz MD, Raut SK (1996) Thermal effect on the life-cycle parameters of the medically important freshwater snail species *Lymnaea* (*Radix*) *luteola* (Lamarck). Mem Inst Oswaldo Cruz 91(1):119–128. https://doi.org/10.1590/S0074-02761996000100022

Adam W (1960) Mollusques. I. Mollusques terrestres et dulcicoles. Faune de Belgique, Bruxelles

Aksenova O, Vinarski M, Bolotov I et al (2017) Two *Radix* spp. (Gastropoda: Lymnaeidae) endemic to thermal springs around Lake Baikal represent ecotypes of the widespread *Radix auricularia*. J Zool Syst Evol Res 55(4):298–309. https://doi.org/10.1111/jzs.12174

Albert C, Luque GM, Courchamp F (2018) The twenty most charismatic species. PLoS One 13(7): e0199149. https://doi.org/10.1371/journal.pone.0199149

Andreeva SI, Andreev NI, Vinarski MV (2010) Key to freshwater gastropods of Western Siberia (Mollusca: Gastropoda). V. 1. Gastropoda: Pulmonata. Fasc. 1. Families Acroloxidae and Lymnaeidae. The authors, Omsk [in Russian]

Balashov IA, Son MO, Coadă V et al (2020) An updated annotated checklist of the mollusks of the Republic of Moldova. Folia Malacologica 21(3):175–181. https://doi.org/10.12657/folmal.021.021

Beckmann KH (2006) Die Mantelschnecke *Myxas glutinosa* (O.F. Müller 1774) in Irland (Gastropoda: Basommatophora: Lymnaeidae). Heldia 6:221–223

Belimov GT (1969) Malacofauna of the floodplain waterbodies of the Middle Lena. In: Ioganzen BG (ed) Voprosy malakologii Sibiri. Tomsk State University Press, Tomsk, pp 72–73. [in Russian]

Beran L (2002) Vodní měkkýši České republiky – rozšíření a jeho změny, stanoviště, šíření, ohrožení a ochrana, červený seznam. Sborník přírodovědného klubu v Uh. Hradišti 10:1–258. https://ia802807.us.archive.org/31/items/VodniMekkysi/VodniMekkysi.pdf

Berezina NA (2001) Influence of ambient pH on freshwater invertebrates under experimental conditions. Russ J Ecol 32:343–351. https://doi.org/10.1023/A:1011978311733

Beriozkina GV, Starobogatov YI (1988) Reproductive ecology and egg clusters of freshwater Pulmonata. Trudy Zoologicheskogo Instituta AN SSSR 174:1–306. [in Russian]

Beriozkina GV, Izakeinaite AP, Kiseleva LN (1980) Some peculiarities of calcium content in shells of lymnaeid snails. In: Andrievskiy AV (ed) Ekologiya zhivotnykh Smolenskoi i sopredelnykh oblastei. Smolensk State Pedagogical Institute Press, Smolensk, pp 45–49. [in Russian]
Berti E, Monsarrat S, Munk M et al (2020) Body size is a good proxy for vertebrate charisma. Biol Conserv 251:108790. https://doi.org/10.1016/j.biocon.2020.108790
Böhm M, Dewhurst-Richman NI, Seddon M et al (2021) The conservation status of the world's freshwater molluscs. Hydrobiologia 848(11–12):3231–3254. https://doi.org/10.1007/s10750-020-04385-w
Bouchet P, Falkner G, Seddon M (1999) Lists of protected land and freshwater molluscs in the Bern convention and European habitats directive: are they relevant to conservation? Biol Conserv 90: 21–31. https://doi.org/10.1016/S0006-3207(99)00009-9
Boycott AE (1936) The habitats of fresh-water Mollusca in Britain. J Anim Ecol 5(1):116–186. https://doi.org/10.2307/1096
Briers RA (2003) Range size and environmental calcium requirements of British freshwater gastropods. Glob Ecol Biogeogr 12:47–51. https://www.jstor.org/stable/3697469
Byrne A, Moorkens EA, Anderson R (2009) Ireland red list no. 2 – non-marine Molluscs. National Parks and Wildlife Service, Department of Environment, Heritage and Local Government, Dublin https://researchrepository.ucd.ie/bitstream/10197/5318/1/Byrne_et_al._2009_Mollusc_red_list.pdf
Carlsson R (2000) Ecology and lifecycle of *Myxas glutinosa* (Müller) in lakes of the Åland Islands, southwestern Finland. J Conchol 37:105–118
Carlsson R (2001) Freshwater snail communities and lake classification. An example from the Åland Islands, southwestern Finland. Limnologica 31:129–138. https://doi.org/10.1016/S0075-9511(01)80007-4
Charlier J, Soenen K, De Roeck E et al (2014) Longitudinal study on the temporal and micro-spatial distribution of *Galba truncatula* in four farms in Belgium as a base for small-scale risk mapping of *Fasciola hepatica*. Parasites Vectors 7:528. https://doi.org/10.1186/s13071-014-0528-0
Clark JA, May RM (2002) Taxonomic bias in conservation research. Science 297(5579):191–192. https://doi.org/10.1111/acv.12586
Clarke AH (1973) The freshwater molluscs of the Canadian Interior Basin. Malacologia 13(1–2): 1–509
Clessin S (1884) Deutsche Excursions-Mollusken–Fauna. Bauer et Raspe, Nürnberg
Colléony A, Clayton S, Couvet D (2017) Human preferences for species conservation: animal charisma trumps endangered status. Biol Conserv 206:263–269. https://doi.org/10.1016/j.biocon.2016.11.035
Cowie RH, Régnier C, Fontaine R et al (2017) Measuring the sixth extinction: what do mollusks tell us? Nautilus 131(1):3–41
Cowie RH, Bouchet P, Fontaine B (2022) The sixth mass extinction: fact, fiction or speculation? Biol Rev 97:640–663. https://doi.org/10.1111/brv.12816
Cuttelod A, Seddon M, Neubert E (2011) European red list of non-marine Molluscs. Publications Office of the European Union, Luxembourg. https://portals.iucn.org/library/sites/library/files/documents/RL-4-014.pdf
Darwall W, Smith K, Allen D et al (2009) Freshwater biodiversity: a hidden resource under threat. In: Vie JC, Hilton-Taylor C, Stuart SN (eds) Wildlife in a changing world: an analysis of the 2008 IUCN red list of threatened species. IUCN, Gland, pp 43–53
Davis CN, Tyson F, Cutress D et al (2020) Rapid detection of *Galba truncatula* in water sources on pasture-land using loop-mediated isothermal amplification for control of trematode infections. Parasites Vectors 13:1–11. https://doi.org/10.1186/s13071-020-04371-0
Delso A, Fajardo J, Muñoz J (2021) Protected area networks do not represent unseen biodiversity. Sci Rep 11:12275. https://doi.org/10.1038/s41598-021-91651-z
Donohue I, Donohue LA, Ainin BN et al (2009) Assessment of eutrophication pressure on lakes using littoral invertebrates. Hydrobiology 633:105–122. https://doi.org/10.1007/s10750-009-9868-8

Ehrmann P (1933) Mollusca Die Tierwelt Mitteleuropas 2(1):1–264
Eisenhauer N, Bonn A, Guerra CA (2019) Recognizing the quiet extinction of invertebrates. Nat Commun 10:50. https://doi.org/10.1038/s41467-018-07916-1
Falkner G, Bank RA, von Proschwitz T (2001) Check-list of the non-marine molluscan species-group taxa of the states of Northern, Atlantic and Central Europe (CLECOM I). Heldia 4:1–76
Feliksiak S (1939) Über Biologie und Morphologie der Mantelschnecke, *Radix glutinosa* (O.F. Müller). Zool Jahrb Syst Ökol Geogr Tiere 72:17–70
Ferreira-Rodríguez N, Akiyama YB, Aksenova OV et al (2019) Research priorities for freshwater mussel conservation assessment. Biol Conserv 231:77–87. https://doi.org/10.1016/j.biocon.2019.01.002
Flasar I (1998) Die Gastropoden Nordwestböhmens und ihre Verbreitung. Heldia 3(4):1–210
Frolova ES (1984) Freshwater molluscs of northern Kazakhstan and their role in the benthos biomass of the natural complexes. In: Petlina AP (ed) Zametki po faune i flore Sibiri. Tomsk State University Press, Tomsk, pp 42–50. [in Russian]
Frömming E (1956) Biologie der mitteleuropäischen Süßwasserschnecken. Dunker-Humblot, Berlin
Georgiev D (2014) The freshwater Mollusca of Bulgaria. Paisii Lhilendaski Press, Plovdiv. [in Bulgarian]
Geyer D (1927) Unsere Land– und Süsswasser-mollusken. Lutz, Stuttgart
Gittenberger E, Janssen AW, Kuiper JGJ et al (2004) De Nederlandse Zoetwatermolluscen. Recente en fossiele Weekdieren uit Zoet en Brak Water. Nederlandse Fauna, vol 2. Nationaal Natuurhistorisch Museum Naturalis, KNNV Uitgeverij & EIS-Nederland, Leiden, pp 1–288
Glöer P (2002) Die Sußwassergastropoden Nord- und Mitteleuropas: Bestimmungschlussel, Lebenweise, Verbreitung. Die Tierwelt Deutschlands, vol 73. Conchbooks, Hackenheim, pp 1–327
Glöer P (2015) Süsswassermollusken. Ein Bestimmungschlüssel für die Bundesrepublik Deutschland. 14. überarbeitene Auflage. Deutscher Jugendbund für Naturbeobachtung, Hamburg
Glöer P (2019) The freshwater gastropods of the West-Palaearctis. Volume 1. Fresh- and brackish waters except spring and subterranean snails. Identification key, anatomy, ecology, distribution. The author, Hetlingen
Goldfuss O (1900) Die Binnenmollusken Mittel-Deutschlands mit besonderer Berücksichtigung der Thüringer Lande, der Provinz Sachsen, des Harzes, Braunschweigs und der angrenzenden Landestheile. W. Engelmann, Leipzig
Grebennikov ME, Vinarski MV (2009) Non-marine molluscs species in the regional Red data books of the Urals and Siberia (Russian Federation). Tentacle. The Newsletter of the IUCN/SSC Mollusc Specialist Group 18:17–20
Grossu AV (1955) Gastropoda. Pulmonata. In: Fauna Republicii Populare Romîne. Mollusca. Editura Acad. RPR, Bucureşti 3(1):1–520.
Grossu AV (1987) Gastropoda Romaniae. 2. Subclasa Pulmonata. I Ordo Basommatophora. II Ordo Stylommatophora. Superfamiliile: Succinacea, Cochlicopacea, Pupillacea. Editura litera, Bucureşti
Gundrizer VA (1979) The European elements in the lower Yenisei malacofauna. In: Likharev IM (ed) Mollyuski, osnovnyye rezul'taty ikh izucheniya. Vsesoyuznoye soveshchaniye po izucheniyu mollyuskov. Avtoreferaty dokladov. Nauka Publishers, Leningrad, pp 201–202. [In Russian]
Holyoak GA (2005) New records of *Myxas glutinosa* and *Catinella arenaria* (Mollusca: Gastropoda) in Ireland. Irish Nat J 28:175
Horsák M, Juricková L, Picka J (2013) Měkkýši České a Slovénské republiky. Kabourek, Zlín
Hortal J, de Bello F, Diniz-Filho AF et al (2015) Seven shortfalls that beset large-scale knowledge of biodiversity. Annu Rev Ecol Evol Syst 46:523–549. https://doi.org/10.1146/annurev-ecolsys-112414-054400

Hubendick B (1951) Recent Lymnaeidae. Their variation, morphology, taxonomy, nomenclature and distribution. Kungliga Svenska Vetenskapsakademiens Handlingar. Fjärde Serien 3(1): 1–223

Hubenov Z (2007) Fauna and zoogeography of marine, freshwater, and terrestrial mollusks (Mollusca) in Bulgaria. In: Fet V, Popov A (eds) Biogeography and ecology of Bulgaria. Springer, Heidelberg, pp 141–198. https://doi.org/10.1007/978-1-4020-5781-6_6

Hunter RD (1975) Growth, fecundity, and bioenergetics in three populations of *Lymnaea palustris* in upstate New York. Ecology 56:50–63. https://doi.org/10.2307/1935299

Ioganzen BG (1951) Freshwater Mollusca of the environs of Tomsk City. Trudy Tomskogo Gosudarstvennogo Universiteta 115:291–302. [in Russian]

Jeffreys GW (1862) British conchology. Vol. 1. Van Voorst, London

Johnson PD, Bogan AE, Brown KM et al (2013) Conservation status of freshwater gastropods of Canada and the United States. Fisheries 38(6):247–282. https://doi.org/10.1080/03632415.2013.785396

Jones RA, Davis CN, Jones DL et al (2021) Temporal dynamics of trematode intermediate snail host environmental DNA in small water body habitats. Parasitology 148(12):1490–1496. https://doi.org/10.1017/S0031182021001104

Kerney M (1999) Atlas of the land and freshwater Molluscs of Britain and Ireland. Great Horkesley, Colchester

Khokhutkin IM, Vinarski MV, Grebennikov ME (2009) Molluscs of the Urals and the adjacent areas. The family Lymnaeidae (Gastropoda, Pulmonata, Lymnaeiformes). Fasc. 1. Goshchitskiy Publishers, Yekaterinburg. [in Russian]

Klemm W (1960) Kreis: Mollusca. In: Catalogus Faunae Austriae, vol VIIa. Springer, Wien, pp 1–60

Köhler F, Seddon M, Bogan AE et al (2012) The status and distribution of freshwater molluscs of the Indo-Burma region. In: Allen DJ, Smith KG, Darwall WRT (eds) The status and distribution of freshwater biodiversity in Indo-Burma. IUCN, Gland, pp 66–88. https://www.iucn.org/sites/dev/files/import/downloads/chapter_4_molluscs.pdf

Krainyuk VN (2012) A new finding of the glutinous snail *Lymnaea glutinosa* (L., 1758) (Gastropoda: Lymnaeidae) in Central Kazakhstan. In: Materialy Mezhdunarodnoi nauchnoi konferentsii "Zhivotnyi mir Kazakhstana i sopredelnykh territoriy". Almaty, pp 125–126 [in Russian]

Kruglov ND (2005) Lymnaeid snails of Europe and Northern Asia. Smolensk State Pedagogical University Press, Smolensk. [in Russian]

Layenko TM (2012) The fauna of aquatic molluscs of Belarus. Belarusskaya navuka, Minsk. [in Russian]

Lazareva AI (1968) The pond snails of Kazakhstan. Unpublished PhD Thesis. Leningrad [in Russian]

Leshko YV (1998) Mollusks. In: Fauna Evropeiskogo severo-vostoka Rossii. Tom V, chast' 1. Nauka, St. Petersburg. [in Russian]

Lisický MJ (1991) Mollusca slovenska. VEDA, Bratislava

Lopes-Lima M, Sousa R, Geist J et al (2017) Conservation status of freshwater mussels in Europe: state of the art and future challenges. Biol Rev 92:572–607. https://doi.org/10.1111/brv.12244

Lopes-Lima M, Burlakova LE, Karatayev AY et al (2018) Conservation of freshwater bivalves at the global scale: diversity, threats and research needs. Hydrobiologia 810:1–14. https://doi.org/10.1007/s10750-017-3486-7

Lopes-Lima M, Riccardi N, Urbanska M et al (2021) Major shortfalls impairing knowledge and conservation of freshwater molluscs. Hydrobiologia 848(11–12):2831–2867. https://doi.org/10.1007/s10750-021-04622-w

Ložek V (1955) Klíč československých měkkýšů. Vydavatel'stvo Slovenskej Akadémie vied, Bratislava

Luther A (1901) Bidrag till kännedomen om land- och sötvattengastropodernas utbredning i Finland. Acta societatis pro Fauna et Flora Fennica 20(3):1–125

Lydeard C, Cowie RH, Ponder WF et al (2004) The global decline of nonmarine mollusks. Bioscience 54(4):321–330. https://doi.org/10.1641/0006-3568(2004)054[0321:TGDONM]2.0.CO;2
Lynch JJ (1965) The ecology of *Lymnaea tomentosa* (Pfeiffer, 1855) in South Australia. Aust J Zool 13:461–473. https://doi.org/10.1071/ZO9650461
Mahulu A, Clewing C, Stelbrink B et al (2019) Cryptic intermediate snail host of the liver fluke *Fasciola hepatica* in Africa. Parasite Vectors 12:573. https://doi.org/10.1186/s13071-019-3825-9
Mammola S, Riccardi N, Prié V et al (2020) Towards a taxonomically unbiased European Union biodiversity strategy for 2030. Proc R Soc B 287:20202166. https://doi.org/10.1098/rspb.2020.2166
Mandahl-Barth C (1949) Bløddyr. III. Ferskvands-bløddyr. G.E.C. Gads Verlag, Copenhagen
McKinney ML (1999) High rates of extinction and threat in poorly studied taxa. Conserv Biol 13(6):1273–1281. https://doi.org/10.1046/j.1523-1739.1999.97393.x
MolluscaBase (2021) MolluscaBase. http://www.molluscabase.org. Accessed 21 September 2021. https://doi.org/10.14284/448
Molur S, Smith KG, Daniel BA et al (2011) The status and distribution of freshwater biodiversity in the Western Ghats. IUCN and Zoo Outreach Organisation, India. http://data.iucn.org/dbtw-wpd/edocs/RL-540-001.pdf
Monzon RB, Kitikoon V, Thammapalerd N (1993) Ecological observations on *Lymnaea (Bullastra) cumingiana*. Southeast Asian J Trop Med Public Health 24(3):563–569
Mouthon J, Charvet S (1999) Compared sensitivity of species, genera and families of Molluscs to biodegradable pollution. Ann Limnol 35:31–39. https://doi.org/10.1051/limn/1999009
Mouthon J, Vimpère J (2014) *Myxas glutinosa* (Mollusca: Gastropoda), espèce mal connue menacée: état des connaissances sur sa repartition passé et actuelle en France. Folia conchyliologica 27:14–20
Mozley A (1936) The freshwater and terrestrial Mollusca of Northern Asia. Trans R Soc Edinburgh 58:605–695
Nekhaev IO (2021) Freshwater gastropods of the western part of the Kola peninsula and northern Karelia (northern Europe). Ruthenica Russ Malacol J 31(4):147–175. https://doi.org/10.35885/ruthenica.2021.31(4).1
Neubauer TA, Georgopoupou E (2021) Extinction risk is linked to lifestyle in freshwater gastropods. Divers Distrib 27(12):2357–2368. https://doi.org/10.1111/ddi.13404
Neubauer TA, Hauffe T, Silvestro D et al (2021) Current extinction rate in European freshwater gastropods greatly exceeds that of the late Cretaceous mass extinction. Commun Earth Environ 2:97. https://doi.org/10.1038/s43247-021-00167-x
Nilsson C, Ericsson U, Medin M et al (1998) Sötvattensnäckor i södra Sverige – en jämförelse med 1940-talet. Naturvårdsverket, Rapport 4903:1–77
Økland J (1990) Lakes and snails. Environment and Gastropoda in 1,500 Norwegian lakes, ponds and rivers. Universal Book Services/Dr. W. Backhuys Publ, Oegstgeest
Perez KE, Minton RL (2008) Practical applications for systematics and taxonomy in North American freshwater gastropod conservation. J N Am Benthol Soc 27(2):471–483. https://doi.org/10.1899/07-059.1
Piechocki A, Wawrzyniak-Wydrowska B (2016) Guide to freshwater and marine Mollusca of Poland. Bogucki Wydawnictwo Naukowe, Poznań
Poliński W (1927) Znaczenie zoogeograficzne mieczaków Polski i konieczność ochrony ich zespołów. Ochrona Przyrody 7:45–53
Prozorova LA, Sharyi-Ool MO (1999) Aquatic pulmonate molluscs (Gastropoda: Pulmonata) of Tuva. Bull Russ Far East Malac Soc 3:11–25. [in Russian]
Rathinasamy V, Tran L, Swan J et al (2021) Towards understanding the liver fluke transmission dynamics on farms: detection of liver fluke transmitting snail and liver fluke-specific environmental DNA in water samples from an irrigated dairy farm in Southeast Australia. Vet Parasitol 291:109373. https://doi.org/10.1016/j.vetpar.2021.109373

Régnier C, Fontaine B, Bouchet P (2009) Not knowing, not recording, not listing: numerous unnoticed mollusk extinctions. Conserv Biol 23(5):1214–1221. https://doi.org/10.1111/j.1523-1739.2009.01245.x

Régnier C, Achaz G, Lambert A et al (2015) Mass extinction in poorly known taxa. Proc Natl Acad Sci USA 112(25):7761–7766. https://doi.org/10.1073/pnas.1502350112

Relf V, Good B, Hanrahan JP (2011) Temporal studies on *Fasciola hepatica* in *Galba truncatula* in the west of Ireland. Vet Parasitol 175:287–292. https://doi.org/10.1016/j.vetpar.2010.10.010

Rondelaud D (1978) The effects of an association of predatory snails (Zonitidea-Physidea) in biological control of *Lymnaea* (*Galba*) *truncatula* Müller. Ann Parasit Hum Comp 53:511–517

Rondelaud D, Vareille-Morel C (1994) The chemical and biological control of *Lymnaea truncatula* in natural watercress beds in the Limousin region (France). Parasite 1(1):89–92. https://doi.org/10.1051/parasite/1994011089

Rudzīte M, Drejers E, Ozoliņa-Moll L et al (2010) Latvijas gliemji: Sugu noteicējs. A Guide to the Molluscs of Latvia. LU Akadēmiskais apgāds, Rīga

Rudzīte M, Boikova E, Drejers E et al (2018) Distribution and protection of the molluscs of Latvia. Schr Malakozool 30:19–28

Russell-Hunter WD (1978) Ecology of freshwater pulmonates. In: Fretter V, Peake J (eds) Pulmonates, vol 2A. Academic Press, London, pp 335–384

Salazkin AA (1969) On certain peculiarities of gastropod distribution in different types of waterbodies of the humid zone of Western Siberia. In: Ioganzen BG (ed) Voprosy malakologii Sibiri. Tomsk State University Press, Tomsk, pp 57–60. [in Russian]

San'ko AF (2007) The quaternary freshwater Mollusca of Belarus and adjacent regions of Russia, Lithuania, and Poland. Institute of Geology and Geophysics of the National Academy of Sciences of Belarus, Minsk. [in Russian]

Schlesch H (1942) Die Land- und Süsswassermollusken Lettlands mit Berücksichtigung der in den Nachbargebieten vorkommenden Arten. Korrespondenzblatt des Naturforscher-Vereins zu Riga 64:245–360

Sitnikova TY, Sysoev AV, Prozorova LA (2014) Types of freshwater gastropods described by Ya.I. Starobogatov, with additional data on the species: family Lymnaeidae. Zoologicheskie Issledovania 16:7–37

Skibins JC, Dunstan E, Pahlow K (2017) Exploring the influence of charismatic characteristics on flagship outcomes in zoo visitors. Hum Dimens Wildl 22(2):157–171. https://doi.org/10.1080/10871209.2016.1276233

Smirnova VA (1967) Freshwater Mollusca of western Kazakhstan as intermediate hosts for trematodes. Unpublished PhD Thesis. Alma-Ata [in Russian]

Sokolova VA (1965) Gastropods of the lakes of Karelia. In: Polyansky YI (ed) Fauna ozior Karelii: Bespozvonochnye. Nauka Publishers, Moscow-Leningrad, pp 85–95. [in Russian]

Stadnichenko AP (2004) Pond snails and limpet snails (Lymnaeidae and Acroloxidae) of Ukraine. Tsentr uchebnoi literatury, Kiev. [in Russian]

Stadnichenko AP (2006) Lymnaeidae and Acroloxidae of Ukraine: sampling and studying methods, biology, ecology, their practical importance. Ruta Press, Zhytomyr. [in Russian]

Szarowska M, Falniowski A (2006) Disappearance of freshwater gastropods in Niepołomice forest (South Poland). Tentacle. The Newsletter of the IUCN/SSC Mollusc Specialist Group 14:16–17

Tunholi VM, Lorenzoni PO, da Silva YH (2017) Molluscicidal potential of *Heterorhabditis baujardi* (Rhabditida: Heterorhabditidae), strain LPP7, on *Lymnaea columella* (Gastropoda: Pulmonata): an alternative for biological control of fasciolosis. Acta Trop 173:23–29. https://doi.org/10.1016/j.actatropica.2017.05.024

Turton W (1857) Manual of the land- and fresh-water shells of the British Islands. Longman et al, London

Urbański J (1932) Godne ochrony gatunki i zespoly mięczaków województwa poznańskiego. Ochrona Przyrody 12:37–40

Van Damme D (2016) *Lantzia carinata*. The IUCN Red List of Threatened Species, https://doi.org/10.2305/IUCN.UK.2016-3.RLTS.T11302A91286059.en

Vinarski MV (2016) Annotated type catalogue of lymnaeid snails (Mollusca, Gastropoda) in the collection of the Natural History Museum, Berlin. Zoosyst Evol 92(1):131–152. https://doi.org/10.3897/zse.92.8107

Vinarski MV, Kantor YI (2016) Analytical catalogue of fresh and brackish water molluscs of Russia and adjacent countries. A.N. Severtsov Institute of Ecology and Evolution of RAS, Moscow

Vinarski MV, Grebennikov ME, Shishkoedova OS (2013) Past and present distribution of *Myxas glutinosa* (O.F. Müller, 1774) in the waterbodies of the Urals and Siberia. J Limnol 72(2): 336–342. https://doi.org/10.4081/jlimnol.2013.e27

Vinarski MV, Aksenova OV, Bespalaya YV et al (2016) *Ladislavella tumrokensis*: the first molecular evidence of a Nearctic clade of lymnaeid snails inhabiting Eurasia. Syst Biodivers 14(3):276–287. https://doi.org/10.1080/14772000.2016.1140244

Vinarski MV, Aksenova OV, Bespalaya YV et al (2021) One Beringian genus less: a re-assessment of *Pacifimyxas* Kruglov & Starobogatov, 1985 (Mollusca: Gastropoda: Lymnaeidae) questions the current estimates of Beringian biodiversity. J Zool Syst Evol Res 59:44–59. https://doi.org/10.1111/jzs.12411

von Proschwitz T (1997) Manteldammsnäckan – en för Dalsland ny, sällsynt sötvattenssnäcka. Natur på Dal 1:16–18

Welter-Schultes F (2012) European non-marine molluscs: a guide for species identification. Planet Poster Editions, Göttingen

Whitfield M, Carlsson R, Biggs J et al (1998) The ecology and conservation of the glutinous snail *Myxas glutinosa* (Müller) in Great Britain: a review. J Conch 2:209–221

Willing MJ, Holyoak DT, Holyoak GA (2014) Ecology and annual cycle of *Myxas glutinosa* (O.F. Müller, 1774) (Gastropoda: Lymnaeidae) in Llyn Tegid, North Wales. J Conch 41(6): 673–683

Zettler M, Zettler A, Daunys D (2005) Bemerkenswerte Süßwassermollusken aus Litauen. Aufsammlungen vom September 2004. Malak Abh 23:27–40

Zhadin VI (1952) Fresh- and brakishwater mollusks of the USSR. Sovetskaya Nauka, Moscow. [in Russian]

Zhadin VI, Pankratova VA (1931) Studies on biology of mollusks – vectors of fasciolosis, and the development of measures against them. Raboty Okskoi Biologicheskoi Stantsii 6:79–157. [in Russian]

Zhitova EP, Astakhova LE, Burlak LV et al (2006) Biodiversity of the pond snails of the Zhytomir Polisse. Ekologo-funktsionalni ta faunistichni aspekti doslidzhenn'a moluskiv, ikh rol' u bioindikatsii stanu navkolishnyogo seredovischcha: Zbirnik naukovikh prats. I. Franko Zhytomir State University, Zhytomir 2:91–93. [in Ukrainian]

Chapter 17
Perspectives: An Integrated Approach on Future Studies of the Lymnaeidae

Antonio A. Vázquez and Maxim V. Vinarski

Abstract In this last chapter, we have briefly summarized several topics treated within the main chapters of this book on the Lymnaeidae. A call of attention is placed upon the integrativeness of research towards this group of snails, from classical to the state-of-the-art approaches. While certain areas of the Lymnaeidae study have been thoroughly covered, some have only been moderately or slightly explored and would need comprehensive research. Hence, we expose some challenges lying ahead for scientists and students that may find themselves passionate with the study of such an interesting family of freshwater snails.

17.1 On the Acknowledgment of Lymnaeid Species: From Identification to Phylogeny

"Integrativeness" is a keyword of current research in many fields of life sciences, from taxonomy to ecology and genetics. Sometimes, it is referred to as the "Age of Integration", when the multidisciplinary and multi-faceted approaches to research are most welcome (Clamp and Lynn 2017). For research, to be "integrative" means to use data of various origins, taken from different sources and obtained following various methods and techniques. The trend towards integrativeness is markedly expressed in biological systematics, which is thought to rescue this discipline from long-lasting disregard and neglect as well as to overcome the so-called taxonomic impediment (Gomes et al. 2015; Cao et al. 2016; Vinarski 2020).

A. A. Vázquez (✉)
Laboratory of Malacology, Institute of Tropical Medicine "Pedro Kourí", Havana, Cuba

MIVEGEC, University of Montpellier, CNRS, IRD, Montpellier, France

M. V. Vinarski
Laboratory of Macroecology & Biogeography of Invertebrates, Saint-Petersburg State University, Saint-Petersburg, Russia

S.I. Vavilov Institute for the History of Science and Technology, St. Petersburg Branch, Russian Academy of Sciences, Saint-Petersburg, Russia

M. V. Vinarski, A. A. Vázquez (eds.), *The Lymnaeidae*, Zoological Monographs 7,
https://doi.org/10.1007/978-3-031-30292-3_17

This concluding chapter aims to review the existing challenges in the lymnaeid research and outline the benefits that different aspects of research can gain from the integrative approach.

The explicit application of the integrative taxonomic approach for classification of the pond snails has started relatively recently (Vinarski et al. 2016; Aksenova et al. 2018), however, the most prominent workers in the field always tried to utilize as many independent sources of data as possible, not using, of course, the very term "integrative taxonomy" (Hubendick 1951; Kruglov 2005). In other words, "taxonomy has been integrative for most of its history" (Valdecasas et al. 2008, p. 211).

Until now, only a very limited number of lymnaeid genera and species complexes has been studied on the "integrative" basis. We still lack molecular information on some small genera of the Lymnaeidae, whereas the phylogenetic relationships and taxonomic position of some other genera cannot be satisfactorily resolved because the available information is too scarce. The new classification scheme for the living Lymnaeidae published in this book (Aksenova et al. 2023, this volume) remains not finalized as the position of some taxa is not clear, and a large deal of further work is required to fulfil the task of classification.

Another challenge for the current lymnaeid systematics is the overload of synonyms and nominal species names in this family, which is among the highest among freshwater Mollusca (Lopes-Lima et al. 2021). Hubendick (1951) compiled a list of about 1150 nominal taxa (including *nomina nuda*) of the species and below-species rank which were established in this family between 1758 and 1950. Today, such a list is remarkably lengthier and approaches 1500 items (Vinarski, unpublished data). Apparently, only a minority of these names correspond to valid species, however, most of the nomenclaturally available names have not been reassessed since their description, and at least part of these can be suitable for the naming of cryptic species and other entities, which molecular taxonomy can reveal. One of the most urgent tasks of the integrative lymnaeid taxonomy is to link the genetically defined clades of the species rank to already existing species names. The use of old available names is much more preferable than introducing new ones, but such an approach requires a qualification of a "classical" taxonomist, who can work with museum collections and has experience with resolving issues related to zoological nomenclature. Unfortunately, it is not an uncommon situation today, when a team of "integrative" researchers does not include a professional taxonomist, which results in the inability to determine the proper taxonomic position of the revealed species-level clades or describe new species when it is needed. Hence, many current studies (not only in the taxonomy of freshwater Mollusca) suffer from evident incompleteness: their authors are unable to place their findings in a proper taxonomic context, i.e. name and describe biological species revealed by sophisticated analysis. The opposite examples are, of course, available (Aksenova et al. 2018; Mahulu et al. 2019; Vinarski et al. 2020, 2022).

The authors of such publications prefer to deal with "putative" species or "candidate species" rather than to undertake their formal description in accordance with the rules of zoological nomenclature (to give them binomial Latin names). Sometimes it is due to lack of information needed for the proper description, but not rarely does it indicate the absence of an experienced taxonomist in the authors' team. This practice, however, should not be welcomed because the absence of taxonomic formalization of species hampers future progress in both the studies of biological diversity and conservation efforts (Pante et al. 2015; Vinarski 2020).

Thus, the most desirable kind of taxonomic work with lymnaeids is what combines molecular and morphological study with a thorough exploration of existing literature and available museum collections. Needless to say the truly integrative system of the family Lymnaeidae, both at the genus and below-genus levels, when it is reached will constitute a firm basis for biogeographic and ecological studies of the pond snails as well as for conservation purposes. The use of fossil-calibrated molecular trees with estimates of the absolute ages of certain evolutionary events will allow researchers to put the evolution of the family in a broad paleogeographic context (see, for example, Aksenova et al. 2018). The integrative approach is crucial for the studies of host–parasite systems which are formed by the pond snails and various trematode taxa, for example, for a deepened understanding of coevolution of these groups and the patterns of host specificity of helminths parasitizing the Lymnaeidae (Vázquez et al. 2022).

Up to now, virtually all integrative and molecular research on the pond snail systematics was based on either single-gene (like DNA barcoding) or multi-locus approaches, which has some apparent limitations. The mighty possibilities of the use of complete genomes for phylogeny and taxonomy of the family are still weakly used, and the complete mitochondrial genomes are available for only a few pond snail species (Feldmeyer et al. 2010, 2015; Liu et al. 2012; Qin et al. 2019). The time is ripe for extensive usage of the "omics-approach" for the purpose of lymnaeid classification.

The future of the Lymnaeidae taxonomy will probably benefit from the further advancements of the most innovative approaches, such as the use of machine learning and artificial intelligence (see Quenu et al. 2020; Klinkenbuß et al. 2020 for examples of the application of these techniques to systematic malacology). Today, such traditional tasks of classification as species delimitation are increasingly relegated to computers, which even creates a perspective of the predictable "dehumanization" of biological systematics, the extinction of the scientific community of taxonomists (Vinarski 2022). Are they doomed to repeat the fate of mammoths and the recent victims of the Sixth Extinction? Perhaps, the solution is that, after its "integrative" renewal, systematics will become a constituent part of a synthetic science of biodiversity (or *metataxonomy*), where taxonomic revisions and descriptions of new taxa, based on an integrated approach, are inseparable from reconstructions of phylogenies, historical biogeographical models, and other products of "non-descriptive" science (Vinarski 2020, 2022).

17.2 Lymnaeid Snails in Applied Medical Malacology

From an applied perspective, in this book we have reviewed several aspects of the lymnaeid studies including experimental models, biological invasions, or disease transmission by the lymnaeid snails. The processes leading to the introduction of lymnaeid snails, whether of natural or anthropic nature, are of particular importance and need further research (Lounnas et al. 2017; Schniebs et al. 2017). However, introduction not necessarily means invasion, and some alien lymnaeid species are unable to expand in the new site (López-Soriano and Quiñonero-Salgado 2020). Invasions are critical events bringing perils to local ecosystem health, threatening not only biodiversity but public health as well when the species is of medical importance. Examples exist within the Lymnaeidae in which introduced species end by resulting in main or relevant hosts of several parasites by either boosting an ongoing transmission (parasite spill-back) (Molloy and Anderson 2006; Carolus et al. 2019) or by bringing new pathogens into the introduction site (parasite spill-over) (Schols et al. 2021).

It is a fact that several lymnaeid species are important hosts of relevant parasites of wildlife, some of great zoonotic relevance (see Hurtrez-Boussès et al. 2023; Vázquez et al. 2023). However, we are still far from the complete understanding of the plethora of species interacting with lymnaeids, particularly in the microbial world (e.g. bacteria and protozoa). We have, to some extent, advanced in elucidating several host–parasite interactions of medical and/or veterinary interest, but mainly in helminth parasites. Within the latter, in the case of lymnaeid snails, the liver flukes (*Fasciola* spp.) have been thoroughly studied because of their relevance to both medical/veterinary health and meat/milk production (Dreyfuss et al. 2006; Rondelaud et al. 2014; Vázquez et al. 2018). However, we still lack several details on transmission dynamics (mainly related to environmental/climatic interaction), species–species (and genotypic–genotypic) compatibility, ecological risks, etc., that would ultimately determine the presence, susceptibility, and transmissibility by a given lymnaeid population. Following this line, if we dig deeper into the biology of lymnaeid–parasite interactions, research should be aimed at disentangling the effects of the environmental gradients over the population and transmission dynamics. How do differential levels of pollution (chemical, organic, etc.), human perturbations of heterogeneous sorts (habitat transformation, water diversion, etc.), climatic settings (tropical, temperate, alpine, etc.) affect lymnaeids? At the individual level, attention should be steered to the host's microbiota interactions with pathogens, as we know it might play a major role in driving host–parasite interactions (Le Clec'h et al. 2022). Funding is, however, a big issue for this kind of investigation, but by succeeding in this, we will not only add to the general knowledge on lymnaeid biology, but also better tackle disease transmission.

In an attempt to support such research activities in the laboratory, we have devoted two chapters to explicitly applied works related to lymnaeid snails. Laboratory culture of snails is a fundamental activity to undertake much of the research we have discussed in several chapters. However, such a task has proved to be challenging to researchers around the globe. Conditions needed to rear lymnaeids in the laboratory are quite heterogeneous, as we understand that across the phylogeny of Lymnaeidae we stumble upon species with contrasted patterns of ecological requirements. Thus, the study of lymnaeid ecology is fundamental and more studies should be heading this way. We can share some experiences in some model species (e.g. *Galba truncatula*, *Lymnaea stagnalis*, *Pseudosuccinea columella*, etc.) that range from aquatic to amphibious behaviour, but only solid ecological knowledge would enhance our understanding of more precise conditions that would ultimately translate in better results of our experiments. In this sense, performing experimental infections are critical steps for the general understanding of lymnaeid–parasite interactions, and many studies are routinely carried out in this regard (see Dreyfuss et al. 2023). Detailed conditions for lymnaeid infection, particularly with *F. hepatica*, in order to measure several parasitological variables are presented in this work.

All in all, we are confident this book will serve as a reference for scientists and students related to the study of the Lymnaeidae, but a broader range of readers such as teachers, parasitologists, malacologists, epidemiologists, paleobiologists, and surveillance and vector control authorities would benefit from the works detailed in their chapters.

References

Aksenova OV, Bolotov IN, Gofarov MY et al (2018) Species richness, molecular taxonomy and biogeography of the radicine pond snails (Gastropoda: Lymnaeidae) in the Old World. Sci Rep 8:11199. https://doi.org/10.1038/s41598-018-29451-1

Aksenova MV, Bolotov IN, Kondakov AV, Vinarski MV (2023, this volume) Phylogeny and taxonomy of the family Lymnaeidae. In: Vinarski MV, Vázquez A (eds) The Lymnaeidae. A handbook on their natural history and parasitological significance. Springer, Cham, pp 67–101

Cao X, Liu J, Chen J et al (2016) Rapid dissemination of taxonomic discoveries based on DNA barcoding and morphology. Sci Rep 6:37066. https://doi.org/10.1038/srep37066

Carolus H, Muzarabani KC, Hammoud C et al (2019) A cascade of biological invasions and parasite spillback in man-made Lake Kariba. Sci Total Environ 659:1283–1292. https://doi.org/10.1016/j.scitotenv.2018.12.307

Clamp JC, Lynn DH (2017) Investigating the biodiversity of ciliates in the 'age of integration'. Eur J Protistol 61:314–322. https://doi.org/10.1016/j.ejop.2017.01.004

Dreyfuss G, Vignoles P, Mekroud A et al (2006) The presence of uninfected *Omphiscola glabra* in a breeding of infected *Galba truncatula* enhanced the characteristics of snail infections with *Fasciola hepatica*. Parasit Res 99:197–199. https://doi.org/10.1007/s00436-006-0147-6

Dreyfuss G, Vignoles P, Rondelaud D et al (2023, this volume) Rearing and maintenance of lymnaeid snails. In: Vinarski MV, Vázquez A (eds) The Lymnaeidae. A handbook on their natural history and parasitological significance. Springer, Cham, pp 395–414

Feldmeyer B, Hoffmeier K, Pfenninger M (2010) The complete mitochondrial genome of *Radix balthica* (Pulmonata, Basommatophora), obtained by low coverage shot gun next generation sequencing. Mol Phylogenet Evol 57:1329–1333. https://doi.org/10.1016/j.ympev.2010.09.012

Feldmeyer B, Greshake B, Funke E et al (2015) Positive selection in development and growth rate regulation genes involved in species divergence of the genus *radix*. BMC Evol Biol 15(1):164. https://doi.org/10.1186/s12862-015-0434-x

Gomes LC, Pessali TC, Sales NG et al (2015) Integrative taxonomy detects cryptic and overlooked fish species in a neotropical river basin. Genetica 143:581–588. https://doi.org/10.1007/s10709-015-9856-z

Hubendick B (1951) Recent Lymnaeidae. Their variation, morphology, taxonomy, nomenclature and distribution. Kungliga Svenska Vetenskapsakademiens Handlingar. Fjärde Serien 3(1): 1–223

Hurtrez-Boussès, S, Vittecoq, M, Alba A et al (2023, this volume) Overview of interactions between parasitic Trematoda and lymnaeid snails. In: Vinarski MV, Vázquez A (eds) The Lymnaeidae. A handbook on their natural history and parasitological significance. Springer, Cham, pp 331–357

Klinkenbuß D, Metz O, Reichert J et al (2020) Performance of 3D morphological methods in the machine learning assisted classification of closely related fossil bivalve species of the genus *Dreissena*. Malacologia 63(1):95–105. https://doi.org/10.4002/040.063.0109

Kruglov ND (2005) Lymnaeid snails of Europe and northern Asia. Smolensk State Pedagogical University Press, Smolensk. [in Russian]

Le Clec'h W, Nordmeyer S, Anderson TJC et al (2022) Snails, microbiomes, and schistosomes: a three-way interaction? Trends Parasitol 38:353. https://doi.org/10.1016/j.pt.2022.01.012

Liu GH, Wang SY, Huang WY et al (2012) The complete mitochondrial genome of *Galba pervia* (Gastropoda: Mollusca), an intermediate host snail of *Fasciola* spp. PLoS One 7:e42172. https://doi.org/10.1371/journal.pone.0042172

Lopes-Lima M, Riccardi N, Urbanska M et al (2021) Major shortfalls impairing knowledge and conservation of freshwater molluscs. Hydrobiologia 848(11–12):2831–2867. https://doi.org/10.1007/s10750-021-04622-w

López-Soriano J, Quiñonero-Salgado S (2020) Status of the populations of two alien species of Lymnaeidae in the Ebro Delta (NE Iberian Peninsula). Elona, Revista de Malacología Ibérica 2: 17–23

Lounnas M, Correa AC, Vázquez AA et al (2017) Self-fertilization, long-distance flash invasion and biogeography shape the population structure of *Pseudosuccinea columella* at the worldwide scale. Mol Ecol 26:887–903. https://doi.org/10.1111/mec.13984

Mahulu A, Clewing C, Stelbrink B et al (2019) Cryptic intermediate snail host of the liver fluke *Fasciola hepatica* in Africa. Parasites Vectors 12:573. https://doi.org/10.1186/s13071-019-3825-9

Molloy J, Anderson G (2006) The distribution of *Fasciola hepatica* in Queensland, Australia, and the potential impact of introduced snail intermediate hosts. Vet Parasitol 137:62–66. https://doi.org/10.1016/j.vetpar.2005.12.024

Pante E, Schoelinck C, Puillandre N (2015) From integrative taxonomy to species description: one step beyond. Syst Biol 64:152–160. https://doi.org/10.1093/sysbio/syu083

Qin DM, Huang XC, Yang LM et al (2019) Complete mitochondrial genome of the radicine pond snail *Radix plicatula* (Gastropoda: Lymnaeidae). Mitochondr DNA B 4(2):2861–2862. https://doi.org/10.1080/23802359.2019.1661300

Quenu M, Trewick SA, Brescia F et al (2020) Geometric morphometrics and machine learning challenge currently accepted species limits of the land snail *Placostylus* (Pulmonata: Bothriembryontidae) on the isle of pines, New Caledonia. J Mollus Stud 86:35–41. https://doi.org/10.1093/mollus/eyz031

Rondelaud D, Titi A, Vignoles P et al (2014) Adaptation of *Lymnaea fuscus* and *Radix balthica* to *Fasciola hepatica* through the experimental infection of several successive snail generations. Parasite Vector 7:296. https://doi.org/10.1186/1756-3305-7-296

Schniebs K, Glöer P, Vinarski M et al (2017) A new alien species in Europe: first record of *Austropeplea viridis* (Quoy & Gaimard, 1833) (Mollusca, Gastropoda, Lymnaeidae) in Spain. J Conchol 42(5):357–370

Schols R, Carolus H, Hammoud C et al (2021) Invasive snails, parasite spillback, and potential parasite spillover drive parasitic diseases of *Hippopotamus amphibius* in artificial lakes of Zimbabwe. BMC Biol 19(1):160. https://doi.org/10.1186/s12915-021-01093-2

Valdecasas AG, Williams D, Wheeler QD (2008) 'Integrative taxonomy' then and now: a response to Dayrat (2005). Biol J Linn Soc 93:211–216. https://doi.org/10.1111/j.1095-8312.2007.00919.x

Vázquez AA, Alda P, Lounnas M et al (2018) Lymnaeid snails hosts of *Fasciola hepatica* and *Fasciola gigantica* (Trematoda: Digenea): a worldwide review. CAB Rev 13:1–15. https://doi.org/10.1079/PAVSNNR201813062

Vázquez AA, Alba A, Alda P et al (2023, this volume) Lymnaeid snails and the transmission of fasciolosis: understanding the differential risks from local to global scale. In: Vinarski MV, Vázquez A (eds) The Lymnaeidae. A handbook on their natural history and parasitological significance. Springer, Cham, pp 359–394

Vázquez AA, Alba A, Alda P et al (2022) On the arrival of fasciolosis in the Americas. Trends Parasitol 38(3):195–204. https://doi.org/10.1016/j.pt.2021.12.001

Vinarski MV (2020) Roots of the taxonomic impediment: is the "integrativeness" a remedy? Integr Zool 15:2–15. https://doi.org/10.1111/1749-4877.12393

Vinarski MV (2022) Today's taxonomy: a way to a new synthesis or an endless deadlock? Biol Bull Rev 12(6):561–569. https://doi.org/10.1134/S2079086422060093

Vinarski MV, Aksenova OV, Bespalaya YV et al (2016) *Radix dolgini*: the integrative taxonomic approach supports the species status of a Siberian endemic snail (Mollusca, Gastropoda, Lymnaeidae). C R Biol 339(1):24–36. https://doi.org/10.1016/j.crvi.2015.11.002

Vinarski MV, Aksenova OV, Bolotov IN (2020) Taxonomic assessment of genetically-delineated species of radicine snails (Mollusca, Gastropoda, Lymnaeidae). Zoosyst Evol 96(2):577–608. https://doi.org/10.3897/zse.96.52860

Vinarski MV, von Oheimb PV, Aksenova OV et al (2022) Trapped on the roof of the world: taxonomic diversity and evolutionary patterns of Tibetan plateau endemic freshwater snails (Gastropoda: Lymnaeidae: *Tibetoradix*). Integr Zool 17:825. https://doi.org/10.1111/1749-4877.12600

GPSR Compliance

The European Union's (EU) General Product Safety Regulation (GPSR) is a set of rules that requires consumer products to be safe and our obligations to ensure this.

If you have any concerns about our products, you can contact us on ProductSafety@springernature.com

In case Publisher is established outside the EU, the EU authorized representative is:

Springer Nature Customer Service Center GmbH
Europaplatz 3
69115 Heidelberg, Germany

Batch number: 10755214

Printed by Printforce, the Netherlands